Advances in Experimental Medicine and Biology

For further volumes:
http://www.springer.com/series/5584

William V. Holt • Janine L. Brown
Pierre Comizzoli
Editors

Reproductive Sciences in Animal Conservation

Progress and Prospects

Springer

Editors
William V. Holt
Academic Department of Reproductive and Developmental Medicine
The University of Sheffield
Sheffield, UK

Janine L. Brown
Center for Species Survival
Smithsonian Conservation Biology Institute
Front Royal, VA, USA

Pierre Comizzoli
Center for Species Survival
Smithsonian Conservation Biology Institute
Front Royal, VA, USA

ISSN 0065-2598 ISSN 2214-8019 (electronic)
ISBN 978-1-4939-5350-9 ISBN 978-1-4939-0820-2 (eBook)
DOI 10.1007/978-1-4939-0820-2
Springer New York Heidelberg Dordrecht London

Printed on acid-free paper

Springer is part of Springer Science+Business Media (www.springer.com)

Foreword

During my time as an environment correspondent with the BBC, people often asked me for advice on how they should prepare for a career attempting to conserve nature or "save the environment" generally. Increasingly, the first thing I said was "become a specialist". The environment/conservation movement still needs people who will construct placards, cold-call potential funders and make the tea; but increasingly it depends on the expertise of economists, lawyers, scientists and the politically connected to make progress.

An unspoken question running richly through the pages of this book is whether reproductive biology fits into the conservation movement's pantheon of useful specialities; whether knowing your frontalin from your farnesene enables you to make a meaningful contribution to conserving species, ecosystems and biodiversity.

I suspect that many people outside the community for which this book is intended would be mildly surprised to find that this is a significant question. In normal usage, the term "reproductive biology" conjures up a vision centred on people—of in vitro fertilisation and the overcoming of human childlessness. In general, this is a happy image. For practitioners, there is the comforting thought that they are in a growth industry, in terms of both demand and tools. Driven by growing demand, science advances; ICSI, for example, has gone from laboratory proof-of-concept to everyday clinical reality in little more than 20 years. On the "client" side, the image would surely feature happy parents, a bouncing baby and, in the background, satisfied white-coated scientists; not so much the white heat of technology as the warm glow.

The picture of the conservation context is somewhat darker, in that the thrust is not so much to enrich peoples' lives as to retard the seemingly inexorable progress of environmental decline. However, like chocolate, what is darker is also, on closer inspection, far richer and more profoundly satisfying. Part of its richness is that the field is so varied, as this book makes clear, ranging from traditional captive breeding, through investigation of hormonal cascades and imaginative species-specific interventions, to cloning, and even—in a fascinating final chapter by Pasqualino Loi and colleagues—to the prospects for "de-extinction" of long-forgotten species.

A couple of years back, I had the chance to observe at first-hand what you might call "front-line" reproductive biology in action. This was one of those rare times as a journalist when you forget tight deadlines, narrow-minded editors, incessant 18-hour days and other unwelcome aspects of the daily grind. Instead you remember the fascination that drew you into the field, and wonder: "They're paying me to be here? I'd do it for free." (Not that you communicate that last bit to the boss, of course!)

In Panama, to report on the annual meeting of the International Whaling Commission meeting for the BBC, I took a half-day out with Adrian Benedetti of the Smithsonian Tropical Research Institute to visit some newly rescued amphibians. Some of the species had not been formally described, being freshly arrived from the eastern end of the country; the principle is to gather up individuals from the wild before the arrival of *Batrachochytrium dendrobatidis* (*Bd*), the often-lethal fungus that is inexorably progressing west-to-east across the isthmus. They were housed in a converted shipping container; and is this not the great irony of the amphibian extinction crisis, that while global trade simultaneously symbolises humanity's increasing demand for habitat and resources and spreads *Bd*, the shipping container, the main vehicle of global trade, also provides sanctuary for so many imperilled species?

Once rescued, the priorities were to find out how to keep the frogs alive, and then how to facilitate their breeding. There was no manual; there was, however, a well of cumulative experience and goodwill within the amphibian breeding community on which to draw. The froglets appeared to be doing quite well, and this is fortunate: failure in the shipping container would probably mean the end of the species. No pressure, then.

Apart from simply keeping species alive, the dire situation facing amphibians also illustrates the importance of reproductive biology as a research tool. On another BBC trip, this time to Japan in conjunction with Conservation International, I saw how techniques for captive breeding developed decades ago at institutions such as Asa Zoological Park in Hiroshima are being deployed in the wild to aid in situ conservation of the spectacular giant salamander. In the zoo, researchers had constructed a number of artificial nests in which to attempt captive breeding. The most successful designs were then deployed in the wild, along rivers where the natural breeding habitat has been concreted over.

In addition, studying species' reproduction can help to forecast what lies ahead for them in a rapidly changing world. Certainly the natural world is changing, in ways that are both rapid and profound. And because the changes are vastly more rapid and profound than funding for science acknowledges, their implications are also poorly understood.

The broadest and most disturbing glimpse of the future in the book before you lies in chapter three, where Cynthia Carey shows us that today's biodiversity challenges are but a foretaste of what is to come in a climatically changing world. About 18–34 % of species are forecast to be extinct by 2050 (and that figure is based on projections for emission growth that have since been exceeded), she relates; rapid warming will lead to novel climates on about a third of landmasses by the end of the century; climate change and ocean acidification are adding to established drivers of decline and extinction, such as habitat loss. Turtles with temperature-dependent sex ratios may come to be dominated by a single sex; birds may find their migratory patterns disrupted, challenging reproduction; Arctic mammals may lose the entire basis of their life cycle on a scale of decades. And everywhere, phenology is increasingly uncertain. As Dr Carey concludes, "The indications are alarming that impending climate change, possibly beyond the 'tipping point' and therefore irreversible, is likely to cause widespread extinctions of animals and plants, reorganisation of interactions among species in existing communities, and disappearances of existing ecosystems".

So whatever the roles that reproductive biology plays in conservation now, the demands on it are only likely to increase in coming decades. And these are decades in which other stressors bearing down on biodiversity, be they invasive species, habitat destruction, chemical contamination or disease, are likely to increase, especially in the rapidly developing regions of Asia and Latin America that are home to myriad biodiversity hotspots.

There may be laboratory-bound scientists in the reproduction field who give climate change, habitat destruction and the other well-documented drivers of natural decline hardly a thought. Conservation is not the only valid reason for becoming a reproductive biologist, or for adding to the specialism's store of expertise; simply increasing the sum of human knowledge is motivation enough for many a scientist, and our society is the richer for it. But evidence that threatened species are a little less threatened because of your work would surely be a powerful spur to many researchers.

Where resources are available, the science and the technology (the "plumbing", perhaps) can be described in extraordinary detail. One of the editors' aims in compiling this book was to provide a state-of-the-art manual for reproductive biology in species other than our own; and for some, such as the elephant and the koala, contributors have provided "recipe books" that range from deciphering the cascade of hormones that indicate pregnancy to designing an artificial vagina. And I was struck (as I was inside the Panamanian shipping container) by how pragmatic this brand of science has to be. To most people, being ejaculated over by a randy koala would make for a very bad day at the office; but for the reproductive biologist, the fact that it happens regularly is (apparently) just another phenomenon to be documented, written into the literature, and exploited for the betterment of the species.

However, the very depth of expertise on these abundant species serves to point up our paucity of knowledge and options regarding those that are most threatened. The chapters on cats and whales, for example, illustrate the mismatch between need and knowledge; species such as the Amur leopard (about 25 individuals believed still in existence), the vaquita (a few hundred at most) and the north Atlantic right whale (again, a few hundred) are apparently just a disease outbreak away from extinction, but there is no toolkit in the reproductive biologist's book capable of rescuing them in the same way that the black-footed ferret or the Wyoming toad have been rescued. In fact, the chances that anyone will ever be able to write a manual for these species as Janine L. Brown (Chap. 8) has for the elephant, or Steve Johnston and Bill Holt (Chap. 9) for the koala, are virtually zero—partly, ironically, because of their dire conservation status. There is also the question of what utility such a manual would have, even if it did exist, for a species such as the north Atlantic right whale.

One reality of the field, undoubtedly, is that money has been invested where there is a commercial imperative. That is why the porcine, the equine and the bovine have munched most of the collective research budget for non-human reproductive biology. That is also why Gabriela Mastromonaco (Chap. 18) can note the agency of aquaculture as a driver for research on cell culture in fish. However, social and cultural factors can also drive funding, enabling Katarina Jewgenow and Nucharin Songsasen (Chap. 10) to detail the huge strides made with artificial insemination in giant pandas, in contrast to other bears.

Does the future hold much prospect of change? Presumably the manuals for already well-funded species will continue to be updated and improved. Perhaps other species will be added to that list; might the polar bear, for example, be a target for intensive research given its iconic status in culture and precarious prospects in nature? In Chap. 15, we glimpse the very beginnings of the manual on coral, a group severely threatened by climate change, as Mary Hagedorn and Rebecca Spindler reveal the state of the cryopreservation art (ticks for sperm and embryonic cells, crosses for oocytes and larvae). And simple captive breeding will presumably be the sole means of survival for an increasing number of species.

But it is hard to conceive (apologies for the phrase!) that the situation facing biodiversity will change markedly, short of a major transformation of societal priorities such that elections are fought on the basis of parties' policies for "averting the sixth great mass extinction". (I know, I know—and pigs may fly—especially cloned ones.) Does the future hold a world in which science is able to keep many species alive indefinitely behind closed doors that will not, because of the broader sweep of political and economic development, be able to thrive ever again in the wild?

London, UK Richard Black

Contents

Contributors

Manal Alsaadi, PhD Department of Industrial Pharmacy, Faculty of Pharmacy, University of Tripoli, Tripoli, Libya

A. Bielanski, DVM, PhD Animal Diseases Research Institute, Ottawa, ON, Canada

Richard Black Former Environment Correspondent, BBC News, London, UK

Janine L. Brown, PhD Center for Species Survival, Smithsonian Conservation Biology Institute, Front Royal, VA, USA

Heather Branvold-Faber, MS, DVM Southside Animal Hospital, Anchorage, AK, USA

Elizabeth A. Burgess, BSc (Hons), MSc, PhD School of Biological Sciences, The University of Queensland, Brisbane, QLD, Australia

Cynthia Carey, AB, MA PhD Department of Integrative Physiology, University of Colorado, Boulder, CO, USA

John Clulow, PhD School of Environmental and Life Sciences, University of Newcastle, Newcastle, Australia

Pierre Comizzoli, DVM, PhD Smithsonian Conservation Biology Institute, National Zoological Park, Washington, DC, USA

Agustín F. Fernández, PhD Cancer Epigenetics Laboratory, Instituto Universitario de Oncología del Principado de Asturias (IUOPA), HUCA, Universidad de Oviedo, Oviedo, Asturias, Spain

Ignacio Arnott Fernández, MD Gynecology, FIV4-Instituto de Reproducción Humana, Oviedo, Asturias, Spain

Valerie A. Ferro, BSc, PhD Strathclyde Institute of Pharmacy and Biomedical Sciences, University of Strathclyde, Glasgow, Scotland, UK

Melissa Filice, MSc Reproductive Physiology, Toronto Zoo, Toronto, ON, Canada

Department of Biomedical Sciences, University of Guelph, Guelph, ON, Canada

Mario F. Fraga, PhD Cancer Epigenetics Laboratory, Instituto Universitario de Oncología del Principado de Asturias (IUOPA), HUCA, Universidad de Oviedo, Oviedo, Asturias, Spain

Department of Immunology and Oncology, Centro Nacional de Biotecnología/ CNB-CSIC, Cantoblanco, Madrid, Spain

Deborah Garside, BSc, PhD Department of Medicine, Imperial College London, South Kensington, London, UK

Ayman Gebril, BVSc, MSc Strathclyde Institute of Pharmacy and Biomedical Sciences (SIPBS), University of Strathclyde, Glasgow, Scotland, UK

Jaime Gosálvez, BSc, PhD Faculty of Sciences, Department of Biology, University Autónoma de Madrid, Madrid, Spain

L. Antonio González-Grajales, DVM Reproductive Physiology, Toronto Zoo, Toronto, ON, Canada

Department of Biomedical Sciences, University of Guelph, Guelph, ON, Canada

Mary Hagedorn, PhD Smithsonian Conservation Biology Institute, Smithsonian Institution, Washington, DC, USA

Hawaii Institute of Marine Biology, University of Hawaii, Kaneohe, HI, USA

William V. Holt, PhD Academic Department of Reproductive and Developmental Medicine, University of Sheffield, Sheffield, UK

Katarina Jewgenow, PhD Department for Reproduction Biology, Leibniz-Institute for Zoo and Wildlife Research, Berlin, Germany

Warren E. Johnson, PhD Smithsonian Conservation Biology Institute, National Zoological Park, Front Royal, VA, USA

Stephen D. Johnston, BSc (Hons), PhD Wildlife Biology Unit, The School of Agriculture and Food Science, The University of Queensland, Gatton, QLD, Australia

Klaus Koepfli, PhD Center for Conservation and Evolutionary Genetics, Smithsonian Conservation Biology Institute, National Zoological Park, Washington, DC, USA

Andrew J. Kouba, PhD Conservation and Research Department, Memphis Zoo, Memphis, TN, USA

Emmelianna Kumar, BSc, MSc Institute for the Environment, Brunel University, Uxbridge, Middlesex, UK

Abel Gayo Lana, PhD Embryology, FIV4-Instituto de Reproducción Humana, Oviedo, Asturias, Spain

Janet M. Lanyon, BSc (Hons), PhD School of Biological Sciences, The University of Queensland, Brisbane, QLD, Australia

Travis M. Livieri, MS Prairie Wildlife Research, Wellington, CO, USA

Pasqualino Loi, PhD Department Comparative Biomedical Sciences, University of Teramo, Teramo, Italy

Paul E. Marinari, MS Center for Species Survival, Smithsonian Conservation Biology Institute, National Zoological Park, Front Royal, VA, USA

Gabriela F. Mastromonaco, MSc, PhD Reproductive Physiology, Toronto Zoo, Toronto, ON, Canada

Department of Biomedical Sciences, University of Guelph, Guelph, ON, Canada

Steven L. Monfort, DVM, PhD Smithsonian Conservation Biology Institute, National Zoological Park, Front Royal, VA, USA

Grazyna Ptak, PhD Department of Comparative Biomedical Sciences, University of Teramo, Teramo, Italy

Rachel M. Santymire, MD, PhD Conservation and Science Department, Davee Center for Epidemiology and Endocrinology, Lincoln Park Zoo, Chicago, IL, USA

Joseph Saragusty, DVM, PhD Department of Reproduction Management, Leibniz Institute for Zoo and Wildlife Research, Berlin, Germany

Nucharin Songsasen, DVM, PhD Center for Species Survival, Smithsonian Conservation Biology Institute, National Zoological Park, Front Royal, VA, USA

Rebecca Spindler, PhD Taronga Conservation Society Australia, Mosman, Australia

Estela García Toraño, PhD Cancer Epigenetics Laboratory, Instituto Universitario de Oncología del Principado de Asturias (IUOPA), HUCA, Universidad de Oviedo, Oviedo, Asturias, Spain

Vance L. Trudeau, MSc, PhD Department of Biology, Centre for Advance Research in Environmental Genomics, University of Ottawa, Ottawa, ON, Canada

Rocío González Urdinguio, PhD Cancer Epigenetics Laboratory, Instituto Universitario de Oncología del Principado de Asturias (IUOPA), HUCA, Universidad de Oviedo, Oviedo, Asturias, Spain

Part I
Introduction

Chapter 1
Reproductive Science as an Essential Component of Conservation Biology

William V. Holt, Janine L. Brown, and Pierre Comizzoli

Abstract In this chapter we argue that reproductive science in its broadest sense has never been more important in terms of its value to conservation biology, which itself is a synthetic and multidisciplinary topic. Over recent years the place of reproductive science in wildlife conservation has developed massively across a wide and integrated range of cutting edge topics. We now have unprecedented insight into the way that environmental change affects basic reproductive functions such as ovulation, sperm production, pregnancy and embryo development through previously unsuspected influences such as epigenetic modulation of the genome. Environmental change in its broadest sense alters the quality of foodstuffs that all animals need for reproductive success, changes the synchrony between breeding seasons and reproductive events, perturbs gonadal and embryo development through the presence of pollutants in the environment and drives species to adapt their behaviour and phenotype. In this book we explore many aspects of reproductive science and present wide ranging and up to date accounts of the scientific and technological advances that are currently enabling reproductive science to support conservation biology.

Keywords Biobanking • Biodiversity • Endocrinology • Environmental change • Epigenetics • Inbreeding • Nutrition • Pollution

---

W.V. Holt, Ph.D. (✉)
Academic Department of Reproductive and Developmental Medicine, University of Sheffield, Jessop Wing, Tree Root Walk, Sheffield S10 2SF, UK
e-mail: Bill.holt@sheffield.ac.uk

J.L. Brown
Center for Species Survival, Smithsonian Conservation Biology Institute, 1500 Remount Road, Front Royal, VA, USA

P. Comizzoli
Smithsonian Conservation Biology Institute, National Zoological Park, Washington, DC, USA

W.V. Holt et al. (eds.), *Reproductive Sciences in Animal Conservation*, Advances in Experimental Medicine and Biology 753, DOI 10.1007/978-1-4939-0820-2_1,

1 Introduction

Wildlife conservation is an incredibly broad topic that encompasses a myriad of activities, ranging from the protection of whole ecosystems to the conservation of a few plants and butterflies in a local park or garden. However, these activities all have an overarching objective, namely to prevent or mitigate the loss of species caused by human activities. This objective is often seen as a moral obligation or duty of "Stewardship of natural resources" (Worrell and Appleby 2000). Global examples of why these efforts are important are well known: major oil spills that devastate wildlife across large areas of coastline, increased atmospheric greenhouse gases, which are widely believed to cause global warming and acidification of the oceans, and the extreme weather events that result from such changes can wreak havoc on whole nations. At a more local scale, urbanization, industrialization, agriculture, forestry, mining, etc. all are involved with the destruction of ecosystems, habitats, and the consequent loss of species, and conservation biologists are often tasked with finding ways to ameliorate such problems. Actions that mitigate the worst effects of environmental damage have a major economic benefit that has been valued at around 33 trillion dollars per annum (Costanza et al. 1997; Gomez-Baggethun and Ruiz-Perez 2011).

As a discipline, conservation biology is one of the most difficult scientific endeavours of our time, because it encompasses so many specialties, both inside and outside of what are normally considered in the context of "biology". It also includes economics, social sciences, ethics, geography and politics, because global problems require global solutions, with the inevitable involvement of disparate countries, governments and cultures, and their various vested interests. There is even a sub-discipline of conservation biology that tries to work out quantitatively whether mitigation measures are actually effective; this is "evidence-based" conservation. Sutherland and colleagues, who championed this approach (Sutherland et al. 2004), published a paper in 2004 stating that about 77 % of conservation interventions are based solely on anecdotal evidence rather than on scientific data. In this sense, it is easier to measure the success of small and focused conservation projects, where the objectives can be easily identified and the appropriate methodologies tested, whether the projects are related to restoring a habitat, a waterway, or even providing road crossings that help prevent vehicle–wildlife collisions (Litvaitis and Tash 2008).

Programmes that help with conservation problems where they exist in the wild are widely regarded as taking place "*in situ*". They occur all over the world and can have diverse objectives. In some cases, a specific project may be the only practical way to help the survival of individual species, especially if habitat protection is going to solve the problem. However, in other cases there may be an argument for trying to maintain a species in captivity, possibly in the expectation that one day in the future it will be returned to its original habitat. These "*ex situ*" programmes usually take place within zoos, aquariums and wildlife parks, where the projects are supervised by managers whose primary aims include the prevention of inbreeding, the avoidance of genetically related diseases and the maintenance of healthy populations with the capacity to thrive in the long-term. These objectives are discussed in detail in this book by Steve Monfort (Chap. 2).

Reproductive sciences feature at all levels of conservation biology, whether it is to help understand the consequences of pollutants on the development and survival of animals in the marine environment, or to predict how global warming might change (or is changing!) the availability of nutrients and thus affect normal reproductive processes. The purpose of this book is to provide readers with a broadly based perspective of how this discipline is interwoven with nearly all aspects of conservation biology and we, the editors, want to stress that it should not simply be regarded as a stand-alone set of techniques aimed at breeding a few endangered species. This is unfortunately the way in which reproductive sciences tend to be viewed by many conservation biologists whose only window on reproductive biology may be via the sensational news headlines that accompany announcements that some kind of endangered species has been produced using a hi-tech method, or even worse, that someone is merely "planning" to breed an endangered species using a hi-tech method. The sensational headlines tend to miss the point that the conservation of dwindling populations is best served when technologies focus upon supporting the preservation of genetic diversity, thus enabling the recovering populations to continue breeding and thriving into the future when the technological support is no longer needed. Producing the occasional offspring rather randomly will not achieve this goal, while using reliable reproductive technologies in well planned breeding programmes can only be beneficial.

2 Environmental Change and Its Consequences

Reproduction is undeniably key to the survival of all species on earth. Technology aside, the study of reproductive processes in animal species remains dauntingly broad, ranging though the details of gametogenesis, fertilisation and the subsequent processes of embryonic development, growth and sexual differentiation, endocrinology and aspects of behaviour and brain function. As if this list were not broad enough, modern scientific advances have enabled us to drill down into the intricate details of gene expression, protein synthesis and the immune system as it affects each of the processes mentioned above. Importantly, animals evolve and adapt to their environment to optimize fitness, and the science of understanding these interactions has led to the realisation that phenomena such as temperature, photoperiod and seasonality have massive impacts on reproductive function. It is increasingly realised that environmental changes, both global and local, can affect the health and wellbeing of animals and humans alike during their entire life and even beyond (Gluckman et al. 2007; Jablonka and Raz 2009). One outstanding example in this category includes the realisation that epigenetics represent a profound, but hitherto rather unsuspected, influence in responses to environmental change. Grandparent's smoking behaviour and also the quality of their diet is now known to influence the body mass index of grandchildren through sex-specific germ-line inheritance mechanisms (Pembrey et al. 2006), and transgenerational changes in the behaviour and reproductive success of laboratory animals are induced by the action of endocrine disrupting chemicals (Anway et al. 2006a, b). Given these recent findings, what are the implications for reproductive success, long-term health and evolutionary adaptations in the face of climate change? Relatively few researchers have considered the relevance of these recent developments to wildlife, especially as they are now thought to involve not only direct modifications of the genome through DNA and chromatin methylation, regarded as "epigenetic" modifications (Turner 2009, 2011), but also non-genomic "soma-to-soma" inheritance mechanisms that do not require direct modification of the genome. In fact, it is worth quoting a relevant sentence from Jablonka's review (Jablonka 2012) where she goes so far as stating that:

> …it is safe to maintain that, as far as our idea of heredity is concerned, the view that inherited differences must involve differences in DNA base sequence is now recognized to be wrong.

Realisation that there is more to inheritance than the strict confines of a DNA sequence has meant there has been an explosion of relevant studies over the last decade. Incredibly, a literature search for papers in PubMed using the terms "epigenetics" and "environment", resulted in 648 references, of which only a single one was published prior to the year 2000. Adding the word "evolution" retrieved 51 references, and it is interesting that a few of these articles explicitly suggested that environmental changes, including climate change, might induce adaptations and evolutionary changes through epigenetic effects (Silvestre et al. 2012; Crews and Gore 2012). As the combination of climate change, epigenetics and adaptation provides an important and overarching context with links to most other aspects of reproductive sciences, we were keen to include an authoritative overview of climate change and reproduction here in this book (Chap. 3; Cynthia Carey) to understand some of its consequences.

The relevance of research in epigenetics, and its close relative "genomic imprinting", to reproduction, especially in terms of foetal–maternal interactions and their pre- and post-conception influences on phenotypic development, means that epigenetic effects are increasingly regarded as significant modulators of reproductive success. Given that the uterine environment in which a mammalian embryo develops can influence the onset of diabetes, heart disease and arteriosclerosis in later adult life (Henry et al. 2012; Turner 2012), what might be the long-term effects of producing and growing embryos in culture dishes? In view of changing environmental conditions, we wanted to explore and explain the ways that factors such as food availability and quality and the presence in the environment of certain chemicals with hormone-like activities might be affecting species (Chap. 6; Agustin Fernández et al. and Chap. 4; Emmelianna Kumar and William Holt). These are major subjects in their own right but the extensive degree of linkage among the topics is becoming increasingly clear. Genomics is another closely related and rapidly advancing field that is yielding insights into the way in which the genome functions. What were previously regarded as sequences of non-functional DNA, often regarded as "junk" DNA, are now known to contain unsuspected but functional sequences (e.g. enhancers; Zhang et al. 2013) with specific roles in controlling gene expression. Thus, the integration of advanced genomic insights into conservation programmes is becoming more important (Chap. 5; Warren Johnson and Klaus Koefli).

As part of the "big picture" we also wanted to review progress with one of the most successful technologically-managed wildlife conservation actions of the last few decades, namely the case of the black-footed ferret (*Mustela nigripes*). Once considered extinct in the USA until a small remnant population was discovered in 1981, the black-footed ferret has been the focus of an intensive captive breeding programme for reintroduction into its original habitats. Rachel Santymire and her colleagues (Chap. 7) have reviewed this programme for us, and declare that the outcome is rather mixed. Without the initial technological inputs by the late JoGayle Howard and her colleagues (Howard et al. 2003), the species would undoubtedly not have survived. Since 2001, however, continued monitoring of the black footed ferret population has revealed problems caused by inbreeding depression and environmental effects. This is rather disappointing, but this case provides some general lessons about the interface between conservation and reality.

3 How Has "Conservation-Based" Reproductive Science Progressed Over the Last Decade?

From the outset, one of our major objectives in producing this book has been to present a comprehensive progress report about various wildlife research programmes that involve aspects of reproductive biology. Here we present a set of six chapters that represent a huge variety of species, ranging from corals to elephants. These chapters also show how progress is dependent on well-focused, sustained research programmes. These attributes are well illustrated by studies in the elephant (Janine L. Brown; Chap. 8) and the koala (Steve Johnston and William Holt, Chap. 9), carnivores (Katarina Jewgenow and Nucharin Songsasen; Chap. 10) and the corals (Mary Hagedorn; Chap. 13). Studies of this type are interesting because of the way they eventually extend their scope into areas not originally foreseen. Endocrinology in the elephants is now linking reproduction with body condition, the control of appetite and even shows parallels of relevance to human clinical medicine (i.e., obesity research). Similarly, the problems of semen cryopreservation in marsupials led to the initiation of novel research directions on DNA fragmentation and semen quality in humans and domestic livestock. As a research topic this was almost non-existent before 2000, although it had been explored to a certain extent 10–15 years earlier (Ballachey et al. 1987; Royère et al. 1988, 1991). Studies on carnivores have been substantial and helped to describe new mechanisms (such as the persistence of corpora lutea in Lynx species) over the past 10 years (Katarina Jewgenow; Chap. 10). The importance of integrating laboratory and field studies are well exemplified by the chapters on marine mammals (Janet Lanyon; Chap. 11), amphibians (John Clulow et al.; Chap. 12) and corals (Mary Hagedorn; Chap. 13). These species are under threat from environmental change, the marine mammals mainly from pollution, shipping and even the use of sonar in naval activities (Piantadosi and Thalmann 2004), the corals from ocean acidification and bleaching, and the amphibians from the global spread of the destructive fungal disease, chytridiomycosis. Reproductive monitoring in wild sea mammals had hardly been thought possible a decade ago, but ingenious ways of collecting faecal samples and identifying individual animals have now been developed, using combinations of reproductive technologies and genetic methods that allow samples to be collected and hormones measured. Research by Mary Hagedorn has led to a suitable method for cryopreserving coral cells, so that they can be kept as a genuine genetic resource bank and used to repopulate threatened corals in their marine habitats. Similarly, the amphibian research is multifaceted and is also aimed at being able to maintain cryopreserved gametes, so that live biosecure, and therefore isolated, populations of endangered amphibians can at least receive as much genetic support as possible while treatments to mitigate the chytrid infections and habitat contaminations are being sought.

The application of reproductive science to conservation biology can be roughly envisaged as a pyramid of endeavours, with a solid set of scientific knowledge and research at its base, and various upper levels that represent interfaces between that knowledge and the practical application of science in the real world. As with any building, if the foundation is shaky the whole structure is likely to collapse. It therefore follows that a great deal of background, basic knowledge is required for any practical application to be successful, and we felt it worthwhile to bring readers up to date with developments in the basic sciences that underpin our interests in conservation. We therefore present a series of chapters that describe some of this advanced research, much of which has been developed in the last decade. Pierre Comizzoli's chapter (Chap. 14) about advanced cryopreservation methods describes an array of approaches to the preservation of gametes and reproductive tissues that are, even now, finding their uses in human medicine as well as wildlife conservation. Banking ovarian and testicular tissues has been something of a pipedream for many years, but such preservation methods are fast becoming a reality. Just as human oncology patients can now benefit from post-radiation therapy recovery of functional testicular and ovarian tissues, threatened populations will benefit from the same research progress because their vital genetic contribution to the next generation will not inevitably be lost forever. Even seemingly intransigent problems such as freezing fish and amphibian oocytes, whose large size spells disastrous survival during cryopreservation, are being tackled by techniques that seek to bypass the problem by freezing primordial germ cells and reviving them later in other individuals of the same or even closely related species. The importance of maintaining banks of frozen cells from species and individuals cannot be underestimated and is highlighted in the chapter by Gabriela Mastromonaco and Pierre Comizzoli (Chap. 16), who discuss the role and realities of attempting to integrate cloning technologies into conservation breeding programmes. These techniques, although at present they often result in offspring with poor survival, will develop and become more successful as time goes by. However, unless the raw materials, namely germ cells and somatic cells from genetically important individuals, are preserved now, much of the diverse genetics within small and threatened populations will disappear.

It is gratifying to see that long-term banking projects focused on diverse biomaterials have dramatically increased over the last decade and even that professional societies focused on topics of shared interest across different specialisms have been established. The International Society for Biological and Environmental Repositories (ISBER) is a US-based society, established in 1999, with a wide remit that includes human, animal, plant and microbiological specimens. A sister society has also been established more recently (2010), catering to users in Europe, the Middle East and Africa [European, Middle Eastern and African Society for Biopreservation and Biobanking (ESBB)]. Much of the impetus for this upsurge of interest has arisen out of biomedical and industrial research needs, but it is gratifying to see that various biological and environmental purposes are also included. These societies are now engaged in considering various formalities associated with the acquisition and management of samples, rather than being concerned only with technical procedures involved in cryopreservation. A recent statement by the President of ISBER (Fay Betsou) underlined these sentiments and emphasised the growing importance of Biobanking as follows:

> We are no longer the housemaid of pathology and taxonomy, and we are not to be swallowed by the diffuse area of –omics science… our strength comes from our ability to develop evidence-based procedures to provide biospecimens for effective research applications. Biobanking is a science on its own.

Growing international awareness that samples can be regarded as a form of national asset or wealth, especially if they or the genes within can be exploited for commercial purposes, has produced a culture where national governments are keen to prevent the export of their materials, even if only for purposes of research. This has produced some paradoxical difficulties for biobanks and museums, which are unable at present to accept samples that are not accompanied by evidence of their legal provenance. Even worse, they may be required to discard valuable historical samples that lack the requisite paperwork. Sample acquisition policy is therefore now a serious business, in contrast to the situation in past years when samples were often removed from one country to another without much thought for such niceties.

Related to the topic of sample preservation and repositories are the chapters by Jaime Gosalvez et al. (Chap. 15) and Andrzeij Bielanski (Chap. 17) who discuss different, but important, aspects of quality control in frozen cells and tissues. Jaime Gosalvez has long been a pioneer of understanding how to assess DNA damage in many types of cells, including spermatozoa. DNA is such a centrally important part of the reproduction process, that any damage has the capacity not only to prevent the process from happening, but perhaps more subtly to let reproduction proceed and result in the production of poor quality offspring. DNA repair mechanisms are known to operate as check points during fertilization and early development, but influences from the environment, including the presence of chemical pollutants, can interfere with them. Using laboratory methods to screen sperm DNA after cryopreservation is now recognized as important in clinical infertility research and veterinary science, and the methodology is replacing sperm motility as the principal sperm parameter to be assessed in research projects. Andrzeij Bielanski's chapter focuses, by contrast, on the problem of making sure that any frozen materials are not contaminated by microorganisms that could be passed on to other samples or to the recipients of frozen gametes and embryos. Until about 20 years ago, bacterial and viral contamination of samples was recognized, but the dangers associated with contaminated liquid nitrogen containers and even liquid nitrogen itself, were largely dismissed as unimportant.

A greater understanding of these issues has been invaluable, not only for conservation research, but also for agriculture, where embryos and sperm samples are regularly shipped around the world. Regulatory authorities recognise the importance of this work in preventing the transfer of diseases between countries, and our purpose in including this chapter was to ensure that practitioners in conservation also know about the potential risks of storing and transporting wild animal gametes and tissues. Many of the risks associated with wild species are actually unknown, and hence a general policy of caution is recommended until more information becomes available.

While much of conservation biology is concerned with ensuring the survival of species, it is increasingly apparent that if species are breeding successfully in the wrong places they can cause major damage to their environment. During the eighteenth and nineteenth centuries, when countries such as Australia and New Zealand were being colonised by European settlers, many non-native species were transported around the world and introduced for various reasons such as agriculture and sports. Australia now has populations of foxes, rabbits, hares, feral pigs and others, that cause havoc among the local wildlife, while feral cats are among the world's most effective bird killers, especially if the birds are located on tropical islands. Similar problems exist in other countries, even with local wildlife, especially where the wild species are in conflict with local human communities. Thus, there is considerable interest in developing methods to control such nuisance populations. Culling methods such as shooting are not always successful, and often are regarded as inhumane, and there is a worldwide need for fertility control methods that can be applied widely, but targeted specifically against a single species. Advances in contraception technology for wildlife were discussed at some length in a previous incarnation of this book (Rodger 2003), and at the time immunocontraception was a topic of considerable interest and potential. Efforts to develop contraceptive technologies for wildlife have moved on to a certain extent, and some technologies, such as zona pellucida proteins, are actively being used for the control of fertility in some species (for review, see Kirkpatrick et al. 2011) However, we know that this field is still developing and therefore we asked Debbie Garside and Valerie Ferro to bring us up to date with the latest developments (Chap. 18).

In the 1980s and 1990s there had been a great deal of interest in embryo transfer techniques (for review, see Loskutoff 2003), and cross-species surrogacy had been suggested as a method of increasing the number of offspring that could potentially be obtained from genetically valuable animals. Today only a few embryo technology programmes continue to be actively pursued by researchers focusing either on particular groups of species, especially felidae and equidae (Swanson 2012; Smits et al. 2012; Pope et al. 2012), or on wild species, such as Sika deer (Wang et al. 2012) that have some potential for commercial exploitation. However, the successful use of nuclear transfer and cloning technologies for threatened mammalian species has, by default, to involve species- optimised embryo transfer. The pitfalls associated with this technology seem to have been ignored in the numerous cloning projects that are now appearing in the literature, but trans-species embryo transfer is sometimes used successfully for these projects (Hajian et al. 2011).

Many of the problems with nuclear transfer are manifested as poor embryo survival and perinatal mortality, and it can be argued that the source of these problems lie with the "unknowns" of embryo transfer in species that have not previously been characterised sufficiently. Some biotechnologists are nevertheless actively promoting the use of cloning technology, not only for reproducing threatened species, but for resurrecting extinct species. As editors we felt a need to obtain an authoritative and realistic view of such intentions, and we therefore asked Pasqualino Loi and his colleagues (Chap. 19) to present a hypothetical strategy for resurrecting the mammoth, with due consideration given to the source of oocytes and the practicalities of undertaking such a project with any hope of success. This chapter highlights the difference between theory and practice, and also demonstrates the likelihood that attempting to produce a single mammoth might require thousands of female elephants and egg donors, thereby decimating the extant and highly endangered Asian elephant. This updates a previous review article with much the same perspective (Critser et al. 2003), published shortly after the initial demonstration of successful nuclear transfer in sheep.

4 Concluding Remarks

When compiling this book we were keen to emphasise and present the many remarkable dimensions of reproductive science and to show how they can and should inform the practice of wildlife conservation. The last decade has seen a change in the dynamics of that interaction, which has moved on from a position where reproductive science is used largely to help with ex-situ animal breeding and management, to a situation where it is able both to provide invaluable insights into the globally important consequences of environmental change and to mitigate some of the problems caused by human activities. We therefore chose to focus attention on topics that are timely, have shown exceptional progress, or that otherwise justify an update. This approach inevitably means that some fields of research were omitted. The cryopreservation of fish and amphibian oocytes and embryos, despite having been studied extensively, has remained stubbornly difficult to achieve, largely because they are so large and encased in impermeable vestments that prevent entry of cryoprotectants. The status of this particular technology has remained largely static, although as explained in Comizzoli's chapter, some novel approaches involving the vitrification and transfer of primordial germ cells are showing promise; such sophisticated developments are driven mainly by the important potential value of fish models for biomedical research and the applications in commercial aquaculture rather than the applications in conservation biology. We were unable to include a chapter on advanced reproductive technologies in birds, but we should nevertheless mention the demonstration that inter-species ovarian transplantation between Muscovy (*Cairina moschata*) and Pekin (*Anas platyrhynchos*) ducks resulted in donor derived offspring (Song et al. 2012) and similar outcomes with transplantation of vitrified testicular tissues from the Japanese Quail (*Coturnix japonica*). The authors proposed that these could be strategies for the conservation of endangered avian species and interested readers should consult a recent review of this topic (Liu et al. 2013).

Specific research into the application of reproductive technologies in reptiles is not included either, but this is not because little is known about their reproductive biology. Some snakes, alligators and turtles are known to store spermatozoa in the female reproductive tract for periods that exceed 2–3 years (Birkhead and Møller 1993; Holt and Lloyd 2010). If we had greater understanding of the mechanisms of long-term sperm storage in vivo it may be possible to develop exciting new technologies for use in multiple species, Furthermore, as described in the ecotoxicology chapter (Chap. 4; Emmelianna Kumar and William Holt), sexual differentiation in reptiles is profoundly affected by endocrine disrupting compounds in the environment as well as, in some cases, by environmental temperature (Lance 2003, 2009).

As implied by the title of this chapter, the editors firmly believe that it is extremely difficult to engage seriously in wildlife conservation without understanding how everything in the environment exerts an impact on reproduction. Moreover, we hope to make the case that reproductive technologies can only be developed and improved by taking multidisciplinary approaches that benefit from expertise derived from many different fields. This is why we favoured the term "reproductive science" when considering a title for this book. In keeping with this approach we are delighted with the breadth of topics we were able to include and would like to express our gratitude not only to the chapter authors, but also to the numerous reviewers for their willingness to help, their insights and their constructive comments.

References

Anway MD, Leathers C, Skinner MK. Endocrine disruptor vinclozolin induced epigenetic transgenerational adult-onset disease. Endocrinology. 2006a;147(12):5515–23. doi:10.1210/en.2006-0640.

Anway MD, Memon MA, Uzumcu M, Skinner MK. Transgenerational effect of the endocrine disruptor vinclozolin on male spermatogenesis. J Androl. 2006b;27(6):868–79. doi:10.2164/jandrol.106.000349.

Ballachey BE, Hohenboken WD, Evenson DP. Heterogeneity of sperm nuclear chromatin structure and its relationship to bull fertility. Biol Reprod. 1987;36:915–25. doi:10.1095/biolreprod36.4.915.

Birkhead TR, Møller AP. Sexual selection and the temporal separation of reproductive events: sperm storage data from reptiles, birds and mammals. Biol J Linn Soc. 1993;50:295–311. doi:10.1111/j.1095-8312.1993.tb00933.x.

Costanza R, D'Arge R, De Groot R, Farber S, Grasso M, Hannon B, Limburg K, Naeem S, O' Neill RV, Paruelo J, Raskin RG, Sutton P, Van Den Belt M. The value of the world's ecosystem services and natural capital. Nature. 1997;387(6630):253–60. doi:10.1038/387253a0.

Crews D, Gore AC. Epigenetic synthesis: a need for a new paradigm for evolution in a contaminated world. F1000 Biol Rep. 2012;4:18. doi:10.3410/B4-18.

Critser JK, Riley LK, Prather RS. Application of nuclear transfer technology to wildlife species. In: Holt WV, Pickard AR, Wildt DE, editors. Reproductive science and integrated conservation, vol. 8. Cambridge: Cambridge University Press; 2003. p. 195–208.

Gluckman PD, Hanson MA, Beedle AS. Early life events and their consequences for later disease: a life history and evolutionary perspective. Am J Hum Biol. 2007;19(1):1–19. doi:10.1002/ajhb.20590.

Gomez-Baggethun E, Ruiz-Perez M. Economic valuation and the commodification of ecosystem services. Prog Phys Geogr. 2011;35(5):613–28. doi:10.1177/0309133311421708.

Hajian M, Hosseini SM, Forouzanfar M, Abedi P, Ostadhosseini S, Hosseini L, Moulavi F, Gourabi H, Shahverdi AH, Dizaj AVT, Kalantari SA, Fotouhi Z, Iranpour R, Mahyar H, Amiri-Yekta A, Nasr-Esfahani MH. "Conservation cloning" of vulnerable Esfahan mouflon (*Ovis orientalis isphahanica*): *in vitro* and *in vivo* studies. Eur J Wildl Res. 2011;57(4):959–69. doi:10.1007/s10344-011-0510-5.

Henry SL, Barzel B, Wood-Bradley RJ, Burke SL, Head GA, Armitage JA. Developmental origins of obesity-related hypertension. Clin Exp Pharmacol Physiol. 2012;39(9):799–806. doi:10.1111/j.1440-1681.2011.05579.x.

Holt WV, Lloyd RE. Sperm storage in the vertebrate female reproductive tract: how does it work so well? Theriogenology. 2010;73(6):713–22. doi:S0093-691X(09)00286-6 [pii]10.1016/j.theriogenology.2009.07.002.

Howard JG, Marinari PE, Wildt DE. Black-footed ferret: model for assisted reproductive technologies contributing to *in situ* conservation. In: Holt WV, Pickard AR, Rodger JC, Wildt DE, editors. Reproductive science and integrated conservation, vol. 8. Cambridge: Cambridge University Press; 2003. p. 249–66.

Jablonka E. Epigenetic variations in heredity and evolution. Clin Pharmacol Ther. 2012;92(6):683–8. doi:10.1038/clpt.2012.158.

Jablonka E, Raz G. Transgenerational Epigenetic inheritance: prevalence, mechanisms, and implications for the study of heredity and evolution. Q Rev Biol. 2009;84(2):131–76. doi:10.1086/598822.

Kirkpatrick JF, Lyda RO, Frank KM. Contraceptive vaccines for wildlife: a review. Am J Reprod Immunol. 2011;66(1):40–50. doi:10.1111/j.1600-0897.2011.01003.x.

Lance VA. Reptile reproduction and endocrinology. In: Holt WV, Pickard AR, Wildt DE, editors. Reproductive science and integrated conservation, Conservation Biology Series, vol. 8. Cambridge, UK: Cambridge University Press; 2003. p. 338–58.

Lance VA. Is regulation of aromatase expression in reptiles the key to understanding temperature-dependent sex determination? J Exp Zool A Ecol Genet Physiol. 2009;311A(5):314–22. doi:10.1002/jez.465.

Litvaitis JA, Tash JP. An approach toward understanding wildlife-vehicle collisions. Environ Manage. 2008;42(4):688–97. doi:10.1007/s00267-008-9108-4.

Liu JN, Cheng KM, Silversides FG. Fundamental principles of cryobiology and application to ex situ conservation of avian species. Avian Biol Res. 2013;6(3):187–97. doi:10.3184/175815513x13740778695007.

Loskutoff NM. Role of embryo technologies in genetic management and conservation of wildlife. In: Holt WV, Pickard AR, Wildt DE, editors. Reproductive science and integrated conservation, Conservation biology, vol. 8. Cambridge: Cambridge University Press; 2003. p. 183–94.

Pembrey ME, Bygren LO, Kaati G, Edvinsson S, Northstone K, Sjöström M, Golding J. Sex-specific, male-line transgenerational responses in humans. Eur J Hum Genet. 2006;14(2):159–66. doi:10.1038/sj.ejhg.5204538.

Piantadosi CA, Thalmann ED. Pathology: whales, sonar and decompression sickness. Nature, 2004. 428: p. 1 p following 716; discussion 2 p following 716.

Pope CE, Gomez MC, Galiguis J, Dresser B. Applying embryo cryopreservation technologies to the production of domestic and black-footed cats. Reprod Domest Anim. 2012;47 Suppl 6:125–9. doi:10.1111/rda.12053.

Rodger JC. Fertility control for wildlife. In: Holt WV, Pickard AR, Wildt DE, editors. Reproductive science and integrated conservation, Conservation biology, vol. 8. Cambridge: Cambridge University Press; 2003. p. 281–90.

Royère D, Hamamah S, Nicolle JC, Barthelemy C, Lansac J. Freezing and thawing alter chromatin stability of ejaculated human spermatozoa: fluorescent acridine orange staining and Feulgen-DNA cytophotometric studies. Gamete Res. 1988;21:51–7. doi:10.1002/mrd.1120210107.

Royère D, Hamamah S, Nicolle JC, Lansac J. Chromatin alterations induced by freeze-thawing influence the fertilizing ability of human sperm. Int J Androl. 1991;14:328–32. doi:10.1111/j.1365-2605.1991.tb01100.x.

Silvestre F, Gillardin V, Dorts J. Proteomics to assess the role of phenotypic plasticity in aquatic organisms exposed to pollution and global warming. Integr Comp Biol. 2012;52(5):681–94. doi:10.1093/icb/ics087.

Smits K, Hoogewijs M, Woelders H, Daels P, Van Soom A. Breeding or assisted reproduction? Relevance of the horse model applied to the conservation of endangered equids. Reprod Domest Anim. 2012;47 Suppl 4:239–48. doi:10.1111/j.1439-0531.2012.02082.x.

Song Y, Cheng KM, Robertson MC, Silversides FG. Production of donor-derived offspring after ovarian transplantation between Muscovy (Cairina moschata) and Pekin (Anas platyrhynchos) ducks. Poult Sci. 2012;91(1):197–200. doi:10.3382/ps.2011-01672.

Sutherland WJ, Pullin AS, Dolman PM, Knight TM. The need for evidence-based conservation. Trends Ecol Evol. 2004;19(6):305–8. doi:10.1016/j.tree.2004.03.018.

Swanson WF. Laparoscopic oviductal embryo transfer and artificial insemination in felids–challenges, strategies and successes. Reprod Domest Anim. 2012;47 Suppl 6:136–40. doi:10.1111/rda.12069.

Turner BM. Epigenetic responses to environmental change and their evolutionary implications. Philos Trans R Soc Lond B Biol Sci. 2009;364(1534):3403–18. doi:10.1098/rstb.2009.0125.

Turner BM. Environmental sensing by chromatin: an epigenetic contribution to evolutionary change. FEBS Lett. 2011;585(13):2032–40. doi:10.1016/j.febslet.2010.11.041.

Turner S. Perinatal programming of childhood asthma: early fetal size, growth trajectory during infancy, and childhood asthma outcomes. Clin Dev Immunol. 2012;2012:962923. doi:10.1155/2012/962923.

Wang L, Zhou GB, Shi WQ, Shi JM, Tian XZ, Gao C, Zhang L, Zhu SE, Zhang TT, Zeng SM, Liu GS. First live offspring born in superovulated sika deer (Cervus nippon) after embryo vitrification. Theriogenology. 2012;78(7):1627–32. doi:10.1016/j.theriogenology.2012.07.008.

Worrell R, Appleby MC. Stewardship of natural resources: definition, ethical and practical aspects. J Agri Environ Ethics. 2000;12(3):263–77. doi:10.1023/a:1009534214698.

Zhang Y, Wong CH, Birnbaum RY, Li G, Favaro R, Ngan CY, Lim J, Tai E, Poh HM, Wong E, Mulawadi FH, Sung WK, Nicolis S, Ahituv N, Ruan Y, Wei CL. Chromatin connectivity maps reveal dynamic promoter-enhancer long-range associations. Nature. 2013. doi:10.1038/nature12716.

Chapter 2
"Mayday Mayday Mayday", the Millennium Ark Is Sinking!

Steven L. Monfort

Abstract Despite exceptional advances in ensuring the health and well-being of animals in human care, zoos of the twenty-first century are ill-prepared and overwhelmed by the sheer number of species requiring conservation support. Furthermore, small population management paradigms have failed to achieve the demographic and genetic targets required to sustain most endangered species in human care. Predictions made in the 1980s regarding the potential of a "millennium ark"—aided by the use of assisted reproductive technologies (ARTs)—for saving species have proven to be wildly over-optimistic. ARTs continue to be touted as a panacea for saving endangered species and even for resurrecting extinct ones. And yet, while the first successful interspecies embryo transfer in a wildlife species occurred 30 years ago, there still is not a single example of embryo-based technologies being used to consistently manage a conservation-reliant species. The limited contribution of ARTs to species conservation to date principally stems from the lack of knowledge of species biology, as well as inadequate facilities, space, expertise, and funding needed for their successful application. ARTs could and should be an important tool in our conservation toolbox, but we cannot fall into the trap of believing that we can "assist" or clone our way out of the present biodiversity crisis. Reproductive technologists overstate the potential of ARTs for saving endangered species, zoos overestimate their ability to sustain genetically and demographically viable captive populations with existing resources, and conservationists underestimate their need for zoos in the face of failing efforts to sustain species in nature. Unless all parties concerned—reproductive technologists, zoo biologists and conservationists—adopt parallel efforts to sustain wild populations and places, zoos risk becoming living museums exhibiting relic species that no longer exist in nature.

S.L. Monfort, D.V.M., Ph.D. (✉)
Smithsonian Conservation Biology Institute, National Zoological Park,
1500 Remount Road, Front Royal, VA 22630, USA
e-mail: monforts@si.edu

W.V. Holt et al. (eds.), *Reproductive Sciences in Animal Conservation*,
Advances in Experimental Medicine and Biology 753, DOI 10.1007/978-1-4939-0820-2_2,

Keywords Zoos • Assisted reproductive technologies • Endangered species • Millennium ark • Conservation

1 Introduction

Zoos and aquariums have evolved over the past century from primitive menageries to modern zoological parks, with naturalistic exhibits and habitats designed to ensure the health and well-being of animals in human care (Wemmer 1995a; Hoage and Deiss 1996). London Zoo, founded in 1826, was the first zoo established to support scientific study (Wemmer and Thompson 1995), but it was not until the 1960s that stand-alone zoo-based research departments were established (e.g., London Zoo, Smithsonian's National Zoo, San Diego Zoo) with expanded research portfolios in disciplines we now know as "Zoo Biology"—reproductive biology, genetics, behavior and animal health and husbandry sciences (Benirschke 1984). Today, roughly 20 % of accredited European and American zoos have research departments, but a much smaller number employ full-time Ph.D.-level scientists and conservationists (Reid et al. 2008).

There is no question that modern zoos, working with diverse partners and stakeholders, have become champions for conservation. Zoos have pioneered the concepts of conservation breeding linked to species reintroduction and restoration of species like the golden-lion tamarin (Kleiman et al. 1986), California condor (Toone and Wallace 1994; Walters et al. 2010), black-footed ferret (Lockhart et al. 2005), and Wyoming toad (Dreitz 2006), among others. And the global zoo community currently invests more than $350 million per year in field conservation programs (WAZA 2005; Penning et al. 2009; Gusset and Dick 2010). But the challenge facing zoos in conserving species is daunting. The IUCN estimates that 25 % of mammals, 12 % of birds, 20 % of reptiles, 30 % of amphibians, 20 % of fishes, and 30 % and 68 % of invertebrate and plant species evaluated to date, respectively, are threatened with extinction (IUCN 2012). Faced with this accelerating global loss of species (Collen et al. 2008), zoos are forced to perform triage in determining which species to save, and which will be left to fend for themselves in nature (Conway 2003; Nijhuis 2012).

Despite exceptional advances, zoos of the twenty-first century have yet to achieve their potential as "conservation centers" or "environmental resource centers" focused on holistic conservation that emphasizes both species and the habitats they require for survival (Rabb 1994; Conway 1996, 2003; IUDZG/CBSG 1993). Furthermore, well-intentioned cooperative population management efforts designed to slow the inevitable loss of genetic diversity that occurs in small, closed populations, have largely failed due to insufficient numbers of founders, inadequate space, poor reproductive management, and inadequate knowledge of species biology (Lacy 2013). Numerous analyses have revealed that "most zoo populations are not being managed at adequate population sizes, reproductive rates, genetic diversity levels, and projected long-term viability that would allow them to contribute positively to species conservation." (CBSG 2011).

2 The Millennium Ark

In a landmark paper, Soulé et al. (1986) predicted that if environmental destruction rates continued unabated, virtually all primates, large carnivores, antelopes, rhinoceros, wild equids, and hundreds of species of birds, reptiles and amphibians would effectively disappear from the wild within 100 years. As a response to this impending biodiversity apocalypse, a group of scientists proposed the creation of "millennium arks" to buy time for the more than 2,000 wildlife species that would likely survive only under human care. The concept called for establishing zoo-maintained populations consisting of at least 20 founders per target species with a goal of sustaining 90 % of the genetic variation of the original founder population for a period of at least 200 years. While the authors recognized the limitations of the ark model, their great faith in unforeseen breakthroughs clearly was evident: "The captive breeding of so many species will saturate the available space and resources, but, hopefully, advances in cryogenics and similar technologies will obviate the need to maintain all of these at one time as living organisms." These authors went on to predict that, "…based on the recent successes in bovids, equids, and primates, we consider it likely that traditional captive breeding programs for many species in these groups will be obsolete in a few decades (given reliable refrigeration)." It is especially noteworthy that none of the authors of this paper were reproductive biologists.

Now, only 30 years later, some have concluded that the millennium ark is sinking (Lees and Wilcken 2009). Even relaxing the goal of sustaining 90 % genetic variation from 200 to 100 years does not alter the grim facts that: (1) less than 50 % of all of the worlds' zoo-managed animal populations are breeding to replacement levels; and (2) only 55 % are sustaining more than 90 % gene diversity. Presently, about 75 % of zoo-based programs for birds and 66 % of those for mammals are not achieving specified demographic and genetic targets (de Man 2011). Overall, 30 % of all zoo-maintained populations are declining and 30 % have fewer than 20 founders (Long 2011). Simply stated, zoos are overwhelmed and ill-equipped to manage more than 500 high-priority species programs due to lack of space, specialized facilities, technical expertise, and funding.

We possess an appalling lack of fundamental scientific knowledge of species biology. In a review of the roughly 250 wildlife species referenced in the reproductive sciences literature, 75 % of these species were represented by three or fewer references (Wildt et al. 2003). Additionally, only three species classified by IUCN as endangered (African elephants, Asian elephants, and cheetah) were considered relatively "well-studied", having more than ten peer-reviewed publications (Wildt et al. 2003). The routine application of research tools like noninvasive endocrine and genetic methods has increased the number of species studied over the past decade, but efforts are heavily skewed towards mammals and birds (Monfort 2003; Schwartz and Monfort 2008), and our knowledge of the reproductive biology of the vast majority of species in the animal kingdom remains rudimentary or non-existent.

3 The Application of ARTs for Conserving Endangered Species

Early breakthroughs in the application of ARTs to endangered species were stunning, including successful interspecies embryo transfers from eland (Dresser et al. 1982) and gaur (Stover and Evans 1984) to domestic cow, and bongo to eland (Dresser et al. 1985)—successes that soon raised great expectations that ARTs would revolutionize the management of endangered species. During these heady times, the concept of "Frozen Zoos"—biorepositories of frozen tissues—was introduced (Benirschke 1984; Clarke 2009), long before the benefits of such collections were fully understood, and before the name "Genome Resource Bank (GRB)" entered our lexicon (Wildt et al. 1997).

ARTs, including artificial insemination (AI), the use of sex-sorted sperm, embryo collection and transfer (ET), in vitro oocyte maturation (IVM) and fertilization (IVF), and cloning have been widely promoted as having tremendous potential for enhancing breeding management and the genetic and demographic sustainability of small populations of rare species (Pukazhenthi and Wildt 2004; Holt and Lloyd 2009). A wide range of ancillary methods and tools have been developed and applied, including hormonal and behavioral assessments for developing fundamental knowledge in diverse species (e.g., ovulatory mechanisms, seasonality, pregnancy, infertility), manipulating (e.g., superovulation, estrous synchronization), augmenting or overcoming blocks to reproduction (e.g., AI, ET, IVF), suppressing fertility (e.g., contraception, aggression control), and establishing biorepositories for capturing extant genomic diversity (e.g., cryoprotectant evaluations and cryopreservation methods).

Despite an early emphasis on embryo technologies in the 1980s, and recent interest in cloning and other genomic approaches for "rescuing" or even resurrecting extinct species (Zimmer 2013), major technical and ecological challenges remain for their application in conservation. This is reflected in the fact that 30 years after the first successful interspecies embryo transfer in a wildlife species, there is not a single example of embryo-based technologies having been used to consistently produce or manage a conservation-reliant species. The simple explanation for this is that reproductive mechanisms are incredibly diverse, and what works in one species likely will not be directly applicable to another species—even among closely related species in the same taxonomic group (Wildt et al. 2009). The problem has been summed up succinctly as follows: "Cow AI technology does not work in a cheetah or a gorilla. But, why should it? Each species is evolutionarily distinct, having developed highly specialized reproductive adaptations. It is the job of reproductive biologists to understand the diverse ways that animals reproduce, because reproduction is the essence of species survival." (Wildt and Wemmer 1999).

The time has come to stop and take stock in why we have generally underperformed in applying even the most basic ARTs such as AI for routinely producing offspring and managing the genetics and demography of wildlife species. We are in an age when genomes can be wholly reconstructed, and biodiversity genomics will soon be yet another tool to add to the ART toolbox. But what good are new or better tools to a mechanic when he or she has absolutely no idea of how the engine was designed to operate in the first place? The trap for the reproductive technologist—especially those with zero experience or knowledge of wildlife biology—is ignorance in believing that any ART can be successfully applied to any species. While history demonstrates that this is a specious notion, the latest technological applications continue to attract attention disproportionate to their potential for sustainably managing reproduction in endangered species, much less resurrecting extinct species (The Long Now Foundation 2013). Whether it is the successful application of AI or the use of cloning to sustain an endangered living species or resurrect an extinct one, success is dependent upon knowledge of a species' biology, ecology, social structure, reproductive cycle, seasonality, implantation, placentation, gestation, parturition, maternal behavior, neonatal care, nutrition, disease susceptibilities, and causes of endangerment. Failure to appreciate the need for this fundamental information is an epic miscalculation that dooms the application of ARTs to certain failure, at least in a practical sense.

4 Case Studies

While this chapter is not intended to provide a comprehensive overview of ART applications in endangered species, a few examples follow that demonstrate both the promise of these approaches, as well as the very real challenges to their practical application.

4.1 ARTs in Endangered Fish

One of the oldest applications of ARTs was invented in the mid-nineteenth century when Joseph Remy and Antoine Géhin harvested eggs and milt from trout and then artificially propagated them by the thousands in vitro (Halverson 2010). This is essentially the method that remains in use today for cultivating diverse species such as carp, salmon, trout, catfish, and tilapia, among others. For example, more than five billion hatchery-reared juvenile salmonids are released annually into the Pacific Ocean from North American hatcheries, alone (Flagg and Nash 1999). In addition, hormone-induced spawning at commercial levels has been practiced for decades (Mylonas 2010), and while fish embryo cryopreservation remains challenging (Hagedorn et al. 2002), sperm has been cryopreserved in more than 200 freshwater and 40 marine fish species worldwide, with routine offspring production using frozen-thawed sperm (Chew and Zulkafli 2012). As the numbers of threatened or endangered fish species increases, "conservation aquaculture," including the use of ARTs, has emerged as a strategy for conserving the genotypes, phenotypes and behaviors of locally-adapted fish populations in support of comprehensive recovery strategies (Anders 1998). However, new research suggests that this approach is not without risks, as the impacts of large-scale mixing of hatchery-produced fish with wild stocks have been shown to reduce overall fitness in species like salmon (Reisenbichler and Rubin 1999) and trout (Araki et al. 2007). Nonetheless, conservation hatcheries, augmented by ARTs, are likely to become increasingly important for recovering critically endangered fish populations—especially those of commercial value—to avoid reductions in population size and the loss of genetic diversity that could increase the risk of extinction (Drauch Schreier et al. 2012).

Zoos and aquariums are increasingly being called upon to help conserve endangered fish species using both ex situ and in situ approaches (Reid et al. 2013). After more than a century of management practice, it now appears that simply producing and releasing large numbers of hatchery-reared fish is not sufficient to sustain and/or recover fish populations. Conservation aquaculture is in its infancy, and its clear that more research is required to understand the impacts of diverse factors such as genetics (inbreeding, outbreeding), broodstock sourcing, maturation and development, growth rate modulation, environmental enrichment, anti-predator conditioning, as well as an improved understanding of anthropogenic impacts on aquatic environments, such as habitat loss/fragmentation, pollution, and climate change (Flagg and Nash 1999, Reid et al. 2013). To maximize their conservation impact, zoos and aquariums will need to make new capital investments in space, infrastructure and scientific expertise, as well as to leverage extant resources to create new and novel partnerships with governments, universities, fish hatcheries, aquaculturists and other technical experts, as required to achieve success.

4.2 ARTs in Endangered Birds

Intravaginal AI has been used in the domestic poultry industry for more than a half-century (Quinn and Burrows 1936), and today nearly 300 million turkeys are produced annually in the United States, alone (USDA Statistical Service 2012). AI has now been used to produce chicks in numerous species of raptors, cranes, waterfowl, psittacines, and passerines (Gee 1995), and this technology has played a key role in successful species recovery programs for the Peregrine falcon (Hoffman 1998), houbara bustard (Saint Jalme et al. 1994), and whooping crane (Ellis et al. 1996). The success of these excellent programs was underpinned by systematic research in diverse disciplines, including behavior, genetics, animal husbandry, veterinary medicine, and chick rearing (Ellis et al. 1996). While AI in wild or rare birds can be incredibly challenging, this approach has tremendous potential for augmenting reproduction in endangered birds for maintaining gene diversity in small populations, and especially when natural breeding is not possible due to behavioral incompatibility, reproductive asynchrony, physiological stress, poor libido, physical abnormalities, among other causes. For all bird species, successful application of AI requires pre-emptive research in semen collection and processing, access to sufficient numbers of birds for basic and applied research, baseline knowledge of species' biology, and appropriate facilities and expertise (Blanco et al. 2009).

An incredibly successful example of the application of ARTs to the conservation of an endangered bird species can be found with houbara bustards. Since the mid-1980s, scientists in Saudi Arabia (Saint Jalme et al. 1994; Seddon et al. 1995) and the United Arab Emirates (International Fund for Houbara Conservation 2012) have conducted extensive research on houbara bustards in the areas of behavior, genetics, reproductive biology, veterinary medicine, as well as the ecology, status, distribution and wild population trends. Since 1996, the Emirates-led program has released a total of more than 111,000 houbara in North Africa, with 20,310 released in 2013, alone; the long-term goal is to release 50,000 birds per year (International Fund for Houbara Conservation 2012). Success of this magnitude has required massive long-term financial investments in facilities infrastructure, scientific and husbandry expertise and logistical support motivated, in large part, by the desire to restore sustainable wild houbara bustard populations to support a strongly ingrained cultural interest in falconry. While conservation breeding programs of this magnitude are clearly out of reach of the zoological community, there are many valuable lessons to be learned from such programs that could be scaled appropriately to conserve zoo-maintained endangered bird species.

4.3 ARTs in Endangered Ungulates

It is not surprising that initial successes were achieved in the Bovidae, as many of the ARTs were developed and applied in domestic cattle in efforts to refine their reproductive management for economic benefit. The simplest of these techniques, AI, has now been successfully applied to produce live offspring in 14 species of non-domestic bovids and seven cervid species (Morrow et al. 2009). Yet, despite tremendous strides in developing this technology, AI is used to routinely manage the genetics of only a single zoo-maintained endangered ungulate, the Eld's deer (*Rucervus eldi*), and only in a very small number of individuals (Monfort et al. 1993).

The case studies of two endangered species—Eld's deer (critically endangered with fewer than 1,500 animals in the wild) and the scimitar-horned oryx (*Oryx dammah*, extinct in the wild)—illustrate some of the challenges in applying ARTs to the genetic management of small populations held in zoos. Both species were the subject of comprehensive research programs that successfully characterized ovarian cycles, developed estrous synchronization methods, semen collection and sperm cryopreservation protocols, and were found useful for routine offspring production (~50 % conception rate) following a single insemination with frozen-thawed sperm (Monfort et al. 1993; Morrow et al. 2000). Despite the clear potential for these methods for enhancing the genetics and demographics of ex situ populations, few zoos possessed the facilities or expertise to permit animals to be safely handled twice to permit insertion and removal of intra-vaginal progesterone-releasing devices during the prescribed 12- to 14-day estrous synchronization interval; followed by anesthesia and laparoscopic AI. In the early 1990s the author contacted a veterinarian at another zoo, which held the second largest Eld's deer population (of six AZA zoos managing this species), to inquire about the possibility of conducting AI to manage the genetics of their inbred Eld's deer population. The veterinarian conveyed that the risk of injury and/or mortality associated with simply darting (anesthetizing) the deer was too great, making this approach impractical. Thus, despite years of systematic research and proven success, ARTs could not be applied due to the limitations imposed by existing facilities and management schemes typical of most zoos. A decade later, AI is still only used to manage Eld's deer reproduction at the Smithsonian's National Zoological Park, which maintains a GRB for Eld's deer sperm, and has facilities that permit safe handling and manipulation of this species.

4.4 ARTs in Endangered Carnivores

The cheetah is a highly charismatic endangered species that has been maintained in human care for literally thousands of years, and yet cheetah populations are not sustainable in zoo-maintained collections worldwide. More than half of all captive cheetahs fail to ever reproduce, and despite more than 30 years of intensive research, the reasons for this remain elusive, although husbandry, management, behavior, health, and age-related infertility likely all contribute to poor reproductive success in zoos (Wielebnowski et al. 2002). While notable reproductive milestones have been achieved in the cheetah, including the birth of offspring following AI using both fresh (Howard et al. 1992) and frozen-thawed sperm imported from South Africa (Howard et al. 1997), these methods are not reliable for routinely producing offspring. A major insight into the reproductive biology of cheetah occurred when noninvasive fecal hormone assessments and behavioral observations revealed that females housed together often experience suppressed ovarian activity linked to agonistic behaviors (Wielebnowski et al. 2002). This fortuitous discovery led to a major shift in ex situ management practices to better mimic the social structure observed for this species in the wild, i.e., females are maintained alone or with their offspring, males are housed in small groups or coalitions, and social introductions are managed to permit natural mate choice and reproduction. The results have been impressive since implementing these changes. For example, at the Smithsonian's National Zoo, seven litters have been born during the last three years compared with only two litters being born over the Zoo's previous 125-year history. This is a case where fundamental reproductive knowledge (i.e., noninvasive endocrinology, animal husbandry, mate choice) has been far more significant in moving towards the goal of cheetah population sustainability than has heretofore been possible using ARTs.

A highly successful example of the practical use of ARTs for augmenting the conservation of an endangered carnivore species is the black-footed ferret. The species, which had declined to only 18 living individuals in the 1980s, has since been brought back from the brink of extinction as a result of cooperative management and breeding programs amongst zoos, state and federal government agencies (Howard et al. 2003; Lockhart et al. 2005). Basic research conducted at the Smithsonian's

National Zoo focused on understanding ferret reproduction and seasonality, semen collection and sperm cryopreservation methods, and laparoscopic AI of females that have not produced offspring via natural breeding (Howard and Wildt 2009). To date, more than 150 kits (60 % success with fresh sperm) have been produced by AI, including multiple litters of kits that have been produced from frozen founder sperm stored for as long as 20 years. Many of the individuals produced by AI have subsequently reproduced and some of their offspring have been reintroduced into the wild, representing a direct example of how ARTs have tangibly contributed to a successful species recovery program. Since 1987, more than 8,000 black-footed ferrets have been produced and more than 3,000 of these have been released into prairie dog colonies across North America.

5 Why Aren’t There More Success Stories?

One thing is clear: we have grossly underestimated the complexity and diversity of reproduction in the animal kingdom, and we have certainly overestimated our ability to develop and apply ARTs that can be used to aid reproductive management and contribute to biodiversity conservation (Wildt et al. 2009; Holt and Lloyd 2009). In fact, the barrier to the successful application of ARTs is not a shortage of new techniques, but rather a fundamental lack of “conservation capital”—trained scientists, sufficient numbers of research subjects, funding, and appropriate facilities designed specifically to study and manage nondomestic species.

Scientists who work with rodent, primate, and dog or cat models appreciate the requirements for appropriate facilities, handling devices, trained staff, appropriate nutrition, adequate veterinary care, and standards of humane care. Farmers and ranchers similarly understand that excellent production and profits require appropriate investments in husbandry, care and management. And all animal scientists appreciate the decades of research and hundreds of millions of dollars invested in research, and the armies of scientists, lab managers, farmers and ranch hands that moved the state of the art to where it is today. With this solid history and understanding of the importance of methodically and systematically studying species’ biology and management, we remain surprisingly ignorant about the biology of the hundreds of species of wildlife whose very survival is inextricably dependent upon human care.

The reproductive technologists are not the only ones underestimating the challenges they face in being relevant to ensuring species survival. The zoo community has been too slow to recognize that current management paradigms are insufficient for sustaining hundreds of species across diverse taxa. Zoos also lack sufficient knowledge of the biology of the majority of species under their care, and in many cases, maintain animals in facilities that suffer from limited space, an absence of handling/manipulation facilities, and insufficient flexibility to mimic and/or manipulate social groupings or to deal with multiple male aggression. Likewise, conservationists have often minimized the role of zoos and resisted biotechnology at a time when their own efforts to stem the loss of biodiversity and wild places have fallen short. Reproductive technologists, zoo professionals, and conservation biologists all want the same thing—to save species and the ecosystems they require for survival. Success will require collective efforts to identify extant limitations and fundamental gaps in knowledge, both intellectual and practical, and joint efforts to secure long-overdue improvements.

Conservation biologists are beginning to recognize the value of ex situ species management programs arguing that "minimal management" of wild species in their natural habitats is no longer realistic (Conde et al. 2011; Redford et al. 2012). Although the genetic, phenotypic and behavioral consequences of captivity support the notion that captive management should not be the first option for species recovery, the time to master a species' biology is when they are not rare (Snyder et al. 1996). Having extremely small founder populations (e.g., black-footed ferrets, 18; Przewalski's horses, 14; California condors, 14) severely restricts access to animals for most forms of research, hinders the design of experiments likely to yield statistically-valid results, and saddles the species with depauperate genetic diversity in perpetuity (Holt and Lloyd 2009). Fortunately, this scenario can be avoided because zoos have access to multiple planning (e.g., Population and Habitat Viability Assessment, Lacy 1993/1994), and database tools (e.g., Red Data List, IUCN 2012; computer modeling, ISIS 2013) that can be used to identify and prioritize species in need of basic and applied conservation science to aid in their survival and recovery. New concepts articulate the need to manage species along a conservation continuum with differing levels of intervention, from controlled captive breeding to metapopulation strategies that employ large spaces to managing extractive reserves to protected areas that require minimal intervention (CBSG 2011; Lacy 2013). Most conservationists agree that the list of conservation-reliant species will continue to grow unabated. These trends present a great challenge to the conservation community, but also a wealth of opportunities for reproductive technologies to contribute to species conservation.

Conservation Centers for Species Survival (C2S2 2013) is a new model that provides space, specialized facilities and expertise for the sustainable management of select endangered species (Sawyer et al. 2010). Established in 2005, C2S2 is a group of five Association of Zoos and Aquariums [AZA]-accredited institutions that collectively manage more than 25,000 acres of land that constitutes more than 60 percent of all land holdings of the entire AZA membership, which includes roughly 225 accredited zoos in North America. C2S2's mission is to "conduct science to understand biology and conservation complexities of species. Breeding to ensure availability of sustainable source populations—for recovery, reintroduction and managed populations." This is an innovative and welcome approach to addressing the need for increased knowledge and new models for sustaining species. This model recognizes that reproduction is fundamental to species survival, and that there are no shortcuts to developing a comprehensive understanding of the diverse factors that impact reproductive fitness, including endocrinology, genetics, developmental biology, animal behavior, health, nutrition, and the social factors needed to maximize natural breeding or to develop or apply ARTs. Additionally, the success of this model is dependent on providing appropriate environmental and social milieu, supported by highly trained, competent professionals. There is no doubt that the zoo and conservation communities need more such facilities, and not just in North America, but also in lesser-developed countries that are being challenged to respond to increasing numbers of endangered species emergencies. New and nimble alliances are needed to facilitate effective peer-based species survival programs deployed across a conservation continuum (Conway 2010). The success of the C2S2 program and other similar programs and alliances (e.g., Amphibian Ark, National Elephant Center, Turtle Survival Alliance) benefit from novel business models and cost sharing within and among the zoo and conservation community. It is worth noting that for species like amphibians, where space and facilities are less of a barrier to implementing effective conservation breeding and research programs, organizations like the Amphibian Ark (Amphibian Ark 2013) and the Panama Amphibian Rescue and Conservation Project (PARC 2013) have made tremendous progress in demonstrating zoo-based conservation leadership, including utilizing ARTs (Browne et al. 2006; Kouba and Vance 2009). Likewise, basic research and the application of ARTs in a zoo context shows great promise for breeding coral and collecting and raising coral larvae that may one day be used to out-plant sexually-derived coral for restoring reefs (Hagedorn et al. 2006). In summary, because the task facing zoos is immense, solutions must realistically rely on (1) forming new alliances among conservationists, scientists, and animal managers; (2) securing the space and specialized facilities needed to facilitate and manage reproduction; and (3) conducting the scientific research required to achieve sustainability targets across the continuum of extensively managed populations for conservation.

With some notable exceptions (e.g., amphibians), traditional zoos are currently not designed or equipped to utilize ARTs for routine animal management, nor even to support hypothesis-driven research that utilizes appropriate numbers of research subjects. New facilities and programs, perhaps supported through consortia and cost-sharing agreements, should be developed to specifically meet the strategic conservation needs of the zoo and conservation communities. And because of the unique and vital role that zoos can play, more effort should focus on engaging governments, bilateral agencies and civil society organizations to join with zoos to make the investments in infrastructure and human capital needed for zoos to affect greater global leadership—including outside of North America and Europe—in sustaining the biodiversity that benefits current and future human societies.

6 Good Science and Effective Conservation Practice Are Good for Zoo Business

Managing and sustaining species in human care is the mandate of the modern zoo. And the zoo community often speaks of educating and inspiring the public to care—to develop empathy for species and their conservation, and to inspire people to take actions in their own personal lives that will lead to tangible conservation outcomes (Rabb and Saunders 2005). There is no doubt that this is a noble and worthwhile goal, but there is another "social contract" that is implicit between modern zoos and their publics—that zoos will be champions in taking direct actions designed to save species from extinction. In essence, the general public may or may not take direct conservation actions to save species or ecosystems themselves, but increasingly, they will not excuse zoos for failing to do so. As former Wildlife Conservation Society Director, Bill Conway, wrote more than a decade ago, "If zoos do not act to help save nature now, much wildlife will be lost that might have been saved. The zoo's moment will have passed. It's relevance will disappear." (Conway 1996).

Zoos have made much progress in recognizing the importance of pursuing a conservation mission, but strategies employed to date have failed to achieve the goal of sustaining genetically diverse, demographically stable assurance populations. It seems clear that zoos, working in partnership with donors, governments and the wider conservation community, must vastly increase their investments in space, facilities, technical and scientific expertise, as well as their investments in supporting field conservation.

In fundamental ways, the zoo business is no different than any other business in that it relies on a "product pipeline" (animals) to generate the revenues required to sustain capital (e.g., exhibits, infrastructure) and operational (e.g., staff, maintenance) expenditures. In the zoo business, losing control of the supply and quality of the "animal pipeline" that zoos depend upon for the success of their business models would be catastrophic, potentially leading to an industry-wide contraction driven by the law of supply and demand. Many of the most sought after zoo exhibit animals, including okapi, elephants, cheetah, to name a few, are declining in nature, and zoo-based breeding programs cannot keep up with extant demand (Lees and Wilcken 2009; Lacy 2013). Combined with moral, ethical and legal restrictions associated with harvesting animals from nature, or even importing them from other zoos, sustaining animal populations has become an existential challenge for the zoo community that must be addressed urgently.

The choice for zoos is really quite clear: increase the supply of animals by alternate, sustainable means, or watch animal availability plummet and the price of doing business skyrocket. With this in mind, investments in conservation, just like investments in new exhibits and infrastructure, would appear to make good business sense for zoos. And while this process will undoubtedly increase costs in the near term, these actions will likely stabilize the cost of doing business in the future, and secure long-term institutional viability. While new animal and conservation costs are unwelcome, arguments against creating "pay to play" systems for acquiring animals are rather unsophisticated given that there already are real costs of producing and providing animals, the burden of animal importation and production is disproportionately borne by a relatively small number of large zoological institutions, and some species already come with great costs of acquisition (e.g., giant pandas, okapi, Asian elephants, golden monkeys). The inconvenient truth facing zoos today is that they must make a choice between paying now or paying more later, and ceasing to be relevant or even ceasing to exist at all. In summary, investments to ensure sustainability of wildlife populations provide at least two essential long-term benefits for zoos: (1) providing a steady supply of diverse animal species to fill exhibits so that zoos can continue to provide their customers with up-close-and-personal encounters with inspiring wildlife, while generating the revenues needed to meet expenditures; and (2) ensuring that genetically-diverse, demographically-stable, and behaviorally-competent populations of animals are available to support conservation-oriented goals including restoring, exchanging or bolstering wild populations of species of critical conservation importance. These goals are fully within the grasp of zoos, but success will first require an unflinching recognition that the problem exists, matched by outstanding leadership, a tenacious commitment to developing long-term solutions, and increased financial investments in conservation capital—animal management facilities, science and husbandry, including the factors that influence the reproductive fitness of conservation-reliant species.

And so what does "zoo business" have to do with the application of ARTs in wildlife species? Quite simply, improvements in zoo management schemes, and an increased emphasis on gaining new fundamental knowledge of species biology will make it increasingly possible to successfully utilize ARTs—some of which have been available for nearly a century. To a very large extent ARTs have outstripped the capacity of zoos to implement them. Aligning technological capability with good animal management and sound conservation principles will make it increasingly possible to apply ARTs to increase reproductive efficiency; to readily transport gametes (sperm, eggs, embryos), raw DNA or genomes to overcome increasingly onerous international animal importation restrictions; to facilitate zoo-to-zoo animal exchanges (e.g., elephant AI already serves this purpose, Brown et al. 2004); and eventually to permit the routine exchange of genetic material between zoo and wild populations (Holt and Lloyd 2009). The justification for a return to building basic knowledge boils down to this: what is the ultimate value of using ARTs to produce endangered animals, or even resurrect extinct species, if we lack the capacity to manage and sustain these species in the first place? If we cannot now sustainably manage an oryx, Eld's deer or cheetah with or without ARTs, then what chance do we have of sustaining resurrected woolly mammoth, guagga or dodo in the future? Our strategy and focus must change or the true potential of ARTs for managing endangered species will never be fully realized.

7 Can We Rescue the Millennium Ark from Sinking?

Many zoo biologists and managers adopted an unquestioned belief in the philosophies so eloquently articulated by zoo directors Bill Conway, George Rabb, and others in the 1980s when they spoke of the on-going evolution of zoos into conservation organizations. Their clarion calls to action are as relevant today as they were nearly three decades ago. It may be too late for some species, but by pursuing progressive animal management strategies, and investing in new conservation capital and human resources, as well as embracing zoo and conservation biology more broadly, the zoo community still has the potential to match the rhetoric of conservation with measurable conservation outcomes. The millennium ark may yet be salvageable, although it may be smaller than originally envisioned. But ignorance and arrogance remain our worst enemies, and it is not especially visionary to predict that ignoring this advice for another 30 years will jeopardize the very survival of zoos themselves—or at least those that fail to evolve—and severely diminish their value as relevant cultural, scientific and conservation organizations.

There can be no disputing that wildlife will ultimately be managed across a conservation continuum whereby animals—in zoos and in nature—will increasingly be managed under human care. Nor should there be any doubt that our publics expect zoos to demonstrate a direct link between the animals under their care and the role they play in sustaining their counterparts in nature. The role and relevance of ARTs for contributing to species conservation is inextricably linked to whether or not zoos invest in developing an improved understanding of overall species' biology, and reproduction, in particular. But while reproductive biology is a vital piece of the conservation puzzle, we should not fool ourselves with the ignorant notion that we can "assist" or clone our way out of the biodiversity crisis. Technology combined with sound husbandry and management, appropriate facilities, and parallel efforts to sustain wild populations and places, offers the best chance for conservation success. Zoos must adopt such holistic conservation strategies or they risk becoming living museums exhibiting relic species that no longer exist in nature, and the resurrection biologists will have more work than they ever bargained for.

References

Amphibian Ark. 2013. http://www.amphibianark.org. Accessed 22 Mar 2013.

Anders PJ. Conservation aquaculture and endangered species. Fisheries. 1998;23:28–31.

Araki H, Cooper B, Blouin MS. Genetic effects of captive breeding cause a rapid, cumulative fitness decline in the wild. Science. 2007;318:100–3.

Benirschke K. The frozen zoo concept. Zoo Biol. 1984;3:325–8.

Brown JL, Göritz F, Pratt-Hawkes N, et al. Successful artificial on an Asian elephant at the National Zoological Park. Zoo Biol. 2004;23:45–63.

Blanco JM, Wildt DE, Hofle U, et al. Implementing artificial insemination as an effective tool for ex situ conservation of endangered avian species. Theriogenology. 2009;71:200–13.

Browne RK, Seratt J, Vance C, et al. Hormonal priming, induction of ovulation and in-vitro fertilization of the endangered Wyoming toad *(Bufo baxteri)*. Reprod Biol Endocrinol. 2006;4:1–11.

CBSG. Intensively managed populations for conservation workshop report. Apple Valley, MN: IUCN/SSC Conservation Breeding Specialist Group; 2011.

Chew PC, Zulkafli A. Sperm cryopreservation of some freshwater fish species in Malaysia. In: Katkov I editor. Current frontiers in cryopreservation. InTech; 2012. ISBN: 978-953-51-0302-8. http://www.intechopen.com/books/current-frontiers-in-cryopreservation/spermcryopreservation-of-some-freshwater-fish-species-in-malaysia.

Clarke AG. The frozen ark project: the role of zoos and aquariums in preserving the genetic material of threatened species. Int Zoo Yb. 2009;43:222–30.

Collen B, Loh N, Whitmee S, et al. Monitoring change in vertebrate abundance: the living planet index. Conserv Biol. 2008;23:317–27.

Conde DA, Flesness N, Colchero F, et al. An emerging role of zoos to conserve biodiversity. Science. 2011;331:1390–1.

Conservation Centers for Species Survival. 2013. http://www.conservationcenters.org. Accessed 2 Jan 2013.

Conway W. The conservation park: a new zoo synthesis for a changed world. In: Wemmer C editor. The ark evolving: zoos and aquariums in transition. Front Royal: Smithsonian Institution Conservation and Research Center, National Zoological Park; 1996. pp. 259–76.

Conway W. The role of zoos in the 21st century. Int Zoo Yb. 2003;38:7–13.

Conway WG. Buying time for wild animals with zoos. Zoo Biol. 2010;30:1–8.

de Man D. Evaluation of self-sustainability of EAZA bird and mammal programs. Intensively managed populations for conservation workshop report. Apple Valley: IUCN/SSC Conservation Breeding Specialist Group; 2011.

Drauch Schreier A, Rodzen J, Ireland S, et al. Genetic techniques inform conservation aquaculture of the endangered Kootenai river white sturgeon Acipenser transmontanus. Endang Species Res. 2012;16:65–75.

Dreitz VJ. Issues in species recovery: an example based on the Wyoming toad. Bio Sci. 2006; 56:765–71.

Dresser BL, Kramer L, Pope E, et al. Superovulation of African eland *(Taurotragus oryx)* and interspecies embryo transfer to Holstein cattle. Theriogenology. 1982;17:86.

Dresser BL, Pope CE, Kramer L, et al. Birth of bongo antelope *(Tragelaphus euryceros)* to eland antelope *(Taurotragus oryx)* and cryo-preservation of bongo embryos. Theriogenology. 1985;23:190.

Ellis DH, Gee GF, Mirande CM. In: Cranes: their biology, husbandry, and conservation 1996. US Washington DC: Department of Interior, National Biological Service, Baraboo, WI: International Crane Foundation, in cooperation with the Fish and Wildlife Service; 1996.

Flagg T, Nash C. A conceptual framework for conservation hatchery strategies for Pacific salmonids. US Department of Commerce, NOAA Tech Memo NMFS-NWSC-38. 1999. http://www.nwfsc.noaa.gov/assets/25/5377_06172004_101351_tm38.pdf

Gee GF. Artificial insemination and cryopreservation of semen from non-domestic birds. In: Bakst MR, Wishart GJ, editors. First international symposium on the artificial insemination of poultry. Poultry Science Association; 1995. pp. 262–79.

Gusset M, Dick G. The global reach of zoos and aquariums in visitor numbers and conservation expenditures. Zoo Biol. 2010;29:1–10.

Hagedorn M, Lance SL, Fonseca DM, et al. Altering fish embryos with aquaporin-3: an essential step toward successful cryopreservation. Biol Reprod. 2002;67:961–6.

Hagedorn M, Pan R, Cox EF, et al. Coral larvae conservation: physiology and reproduction. Cryobiology. 2006;52:33–47.

Halverson A. An entirely synthetic fish: how rainbow trout beguiled America and overran the world. New Haven: Yale University Press; 2010.

Hoage RJ, Deiss WA. New worlds, new animals: from menagerie to zoological park in the nineteenth century. Baltimore: Johns Hopkins University Press; 1996.

Hoffman C. Peregrine to soar off endangered species list. Endang Species Bull. 1998;23:20–1.

Holt WV, Lloyd RE. Artificial insemination for the propagation of CANDES: the reality! Theriogenology. 2009;71:228–35.

Howard JG, Donoghue AM, Goodrowe KL, et al. Successful induction of ovarian activity and laparoscopic intrauterine artificial insemination in the cheetah *(Acinonyx jubatus)*. J Zoo Wildl Med. 1992;23:288–300.

Howard JG, Roth TL, Swanson WF et al. Successful intercontinental genome resource banking and artificial insemination with cryopreserved sperm in cheetahs. J Androl Suppl. 1997;P-55 [abstract 123]

Howard JG, Marinari PE, Wildt DE. Black-footed ferret: model for assisted reproductive technologies contributing to *in situ* conservation. In: Holt WV, Pickard AR, Rodger JC, Wildt DE, editors. Reproductive science and integrated conservation. Cambridge: Cambridge University Press; 2003. p. 249–66.

Howard JG, Wildt DE. Approaches and efficacy of artificial insemination in felids and mustelids. Theriogenology. 2009;71:130–48.

International Fund for Houbara Conservation Annual Report. 2012. http://houbarafund.org/flipbook/en/. Accessed 2 Jan 2014.
International Species Information System. 2013. http://www2.isis.org. Accessed 22 Mar 2013.
IUDZB/CBSG (IUCN/SSC). The world zoo conservation strategy: the role of zoos and aquaria of the world in global conservation. Chicago: Chicago Zoological Society; 1993.
IUCN. The IUCN red list of threatened species. Version 2012.2. 2012. http://www.iucnredlist.org. Accessed 22 Mar 2013.
Kleiman DG, Beck BB, Dietz JM, et al. Conservation program for the golden lion tamarin: captive research and management, ecological studies, educational strategies, and reintroduction. In: Primates. New York: Springer; 1986. p. 959–79.
Kouba AJ, Vance CK. Applied reproductive technologies and genetic resource banking for amphibian conservation. Reprod Fertil Dev. 2009;21:719–37.
Lacy RC. What is population (and habitat) viability analysis? Primate Cons. 1993/1994; 14/15:27–33.
Lacy RC. Achieving true sustainability of zoo populations. Zoo Biol. 2013;32:19–26.
Lees CM, Wilcken J. Sustaining the ark: the challenges faced zoos in maintaining viable populations. Int Zoo Yb. 2009;43:6–18.
Lockhart JM, Thorne ET, Gober DR, et al. A historical perspective on recovery of the black-footed ferret and the biological and political challenges affecting its future. In: Roelle JE, Miller BJ, Godbey JL, Biggins DE, editors. Recovery of the black-footed ferret: progress and continuing challenges. United States geological survey scientific investigations report; 2005. p. 6–19.
Long S. Status of association of zoos and aquarium's cooperatively managed populations. Intensively managed populations for conservation workshop report. Apple Valley: IUCN/SSC Conservation Breeding Specialist Group; 2011.
Monfort SL, Asher GW, Wildt DE, et al. Successful intrauterine insemination of Eld's deer *(Cervus eldi thamin)* with frozen-thawed spermatozoa. J Reprod Fertil. 1993;99:459–65.
Monfort SL. Non-invasive endocrine measures of reproduction and stress in wild populations. In: Holt WV, Pickard AR, Rodger JC, Wildt DE, editors. Reproductive science and integrated conservation. Cambridge: Cambridge University Press; 2003. p. 147–65.
Morrow CJ, Wolfe BA, Roth TL, Wildt DE, Bush M, Blumer ES, Atkinson MW, Monfort SL. Comparing ovulation synchronisation protocols for artificial insemination in the scimitar-horned oryx (*Oryx dammah*). Anim Reprod Sci. 2000;59:71–86.
Morrow CJ, Penfold LM, Wolfe BA. Artificial insemination in deer and non-domestic bovids. Theriogenology. 2009;71:149–65.
Mylonas CC. Broodstock management and hormonal manipulations of fish reproduction. Gen Comp Endocrinol. 2010;165:516–34.
Nijhuis M. Conservationists use triage to determine which species to save or not. Sci Am. 2012;307:74–9.
Penning M, Reid G, McG KH, et al. Turning the tide: a global aquarium strategy for conservation and sustainability. Bern: World Association of Zoos and Aquariums Executive Office; 2009.
PARC: Panama amphibian rescue and conservation project. 2013. http://amphibianrescue.org. Accessed 2 Jan 2013.
Pukazhenthi B, Wildt DE. Which reproductive technologies are most relevant to studying, managing and conserving wildlife. Reprod Fertil Dev. 2004;16:33–46.
Quinn JP, Burrows WH. Artificial insemination in fowls. J Heredity. 1936;27:31–7.
Rabb GB. The changing roles of zoological parks in conserving biological diversity. Am Zool. 1994;34:159–64.
Rabb GB, Saunders CD. The future of zoos and aquariums: conservation and caring. Int Zoo Yb. 2005;39:1–26.
Redford KJH, Jensen DB, Breheny JJ. Integrating the captive and the wild. Science. 2012;228:1157–8.
Reid GMG, Macdonald AA, Fidgett AL, et al. Developing the research potential of zoos and aquaria: the EAZA research strategy. Amsterdam: European Association of Zoos and Aquariums Executive Office; 2008.

Reid GMG, Contreras MacBeath T, Csatadi K. Global challenges in freshwater-fish conservation related to public aquariums and the aquarium industry. Int Zoo Yb. 2013;47:6–45.

Reisenbichler RR, Rubin SP. Genetic changes from artificial propagation of Pacific salmon affect the productivity and viability of supplemented populations. ICES J Mar Sci. 1999;56:459–66.

Saint Jalme M, Gaucher P, Paillat P. Artificial inseminations in houbara bustards *(Chlamydotis undulata):* influence of the number of spermatozoa and insemination frequency on fertility and ability to hatch. J Reprod Fertil. 1994;100:93–103.

Sawyer R, Cavender N, Shurter S. Conservation centers for species survival. AZA Connect Magazine, November; 2010. pp. 10–3.

Schwartz MK, Monfort SL. Genetic and endocrine tools for carnivore surveys. In: Long RA, MacKay P, Ray JC, Zielinski WJ, editors. Noninvasive survey methods for North American carnivores. Washington, DC: Island; 2008. p. 228–50.

Seddon PJ, Saint Jalme M, van Heezik Y, et al. Restoration of houbara bustard populations in Saudi Arabia: developments and future directions. Oryx. 1995;29:136–42.

Snyder NFR, Derrickson SR, Beissinger SR, et al. Limitations of captive breeding in endangered species recovery. Conserv Biol. 1996;10:338–48.

Soulé M, Gilpin M, Conway W, et al. The millennium ark: how long a voyage, how many staterooms, how many passengers? Zoo Biol. 1986;5:101–13.

Stover J, Evans J. Interspecies embryo transfer from gaur *(Bos gaurus)* to domestic Holstein cattle *(Bos taurus)* at the New York Zoological Park. Tenth international congress on animal reproduction and artificial insemination. 1984;2:243.

The Long Now Foundation. 2013. http://longnow.org/revive/. Accessed 4 Jan 2014.

Toone WD, Wallace MP. The extinction in the wild and reintroduction of the California condor *(Gymnogyps californianus).* In: Olney PJS, Mace GM, Feistner ATC, editors. Creative Conservation: Interactive management of wild and captive animals. Netherlands: Springer; 1994. p. 411–9.

USDA National Agricultural Statistics Service. Hatchery production summary. 2012. Summary. http://usda01.library.cornell.edu/usda/current/HatcProdSu/HatcProdSu-04-15-2013.pdf.

Walters JR, Derrickson SR, Fry DM. Status of the California condor *(Gymnogyps californianus)* and efforts to achieve its recovery. Auk. 2010;127:969–1001.

WAZA. Building a future for wildlife – the world zoo and aquarium conservation strategy. Berne, Switzerland: WAZA Executive Office; 2005.

Wemmer CM, editor. The ark evolving: zoos and aquariums in transition. Front Royal: Smithsonian Institution Conservation and Research Center, National Zoological Park; 1995a.

Wemmer C, Thompson S. A short history of scientific research in zoological gardens. In: Wemmer C, editor. The ark evolving: zoos and aquariums in transition. Front Royal: Smithsonian Institution Conservation and Research Center, National Zoological Park; 1995.

Wielebnowski NC, Ziegler K, Wildt DE, et al. Impact of social management on reproductive, adrenal and behavioural activity in the cheetah *(Acinonyx jubatus).* Anim Conserv. 2002;5:291–301.

Wildt DE, Wemmer C. Sex and wildlife: the role of reproductive science in conservation. Biodiv Conserv. 1999;8:965–76.

Wildt DE, Rall WF, Critser JK, et al. Genome resource banks. Bioscience. 1997;47:689–98.

Wildt DE, Ellis SE, Janssen D, et al. Toward more effective reproductive science in conservation. In: Holt WV, Pickard AR, Rodger JC, Wildt DE, editors. Reproductive science and integrated conservation. Cambridge: Cambridge University Press; 2003. p. 2–20.

Wildt DE, Comizzoli P, Pukazhenthi B, et al. Lessons from biodiversity—the value of nontraditional species to advance reproductive science, conservation, and human health. Mol Reprod Dev. 2009;77:397–409.

Zimmer C. Bringing them back to life. The revival of an extinct species is no longer a fantasy. But is it a good idea? 2013. http://ngm.nationalgeographic.com/2013/04/125-species-revival/zimmer-text. Accessed 2 Jan 2014.

Part II
The Big Picture: Can Species Survive and Adapt in a Changing World?

Chapter 3
Climate Change, Extinction Risks, and Reproduction of Terrestrial Vertebrates

Cynthia Carey

Abstract This review includes a broad, but superficial, summary of our understanding about current and future climate changes, the predictions about how these changes will likely affect the risks of extinction of organisms, and how current climate changes are already affecting reproduction in terrestrial vertebrates. Many organisms have become extinct in the last century, but habitat destruction, disease and man-made factors other than climate change have been implicated as the causal factor in almost all of these. Reproduction is certain to be negatively impacted in all vertebrate groups for a variety of reasons, such as direct thermal and hydric effects on mortality of embryos, mismatches between optimal availability of food supplies, frequently determined by temperature, and reproductive capacities, sometimes determined by rigid factors such as photoperiod, and disappearance of appropriate foraging opportunities, such as melting sea ice. The numbers of studies documenting correlations between climate changes and biological phenomena are rapidly increasing, but more direct information about the consequences of these changes for species survival and ecosystem health is needed than is currently available.

Keywords Climate change • Vertebrate reproduction • Mammals • Birds • Reptiles • Amphibians • Extinction

C. Carey, A.B., M.A., Ph.D. (✉)
Department of Integrative Physiology, University of Colorado,
Boulder, CO 80309-0354, USA
e-mail: careyc@colorado.edu

W.V. Holt et al. (eds.), *Reproductive Sciences in Animal Conservation*,
Advances in Experimental Medicine and Biology 753, DOI 10.1007/978-1-4939-0820-2_3,

1 Global Climate Change

Increasing concerns are being expressed within the scientific scholarly world, and more recently even by the non-scientists and world governments, about global climate change. However, climate change, per se, is not new. Global climates have been changing since the evolution of early life. In fact, it appears that climate change has been the rule, rather than the exception. Dramatic, and sometimes abrupt, changes in climate have occurred frequently throughout the evolution of living organisms (Alley et al. 1997; Overpeck and Webb 2000; Jones et al. 2001; Zachos et al. 2001). Worldwide temperatures in the last Millenium fluctuated between cool periods in the seventeenth and nineteenth centuries, to warmer ones in the eleventh, twelfth, and twentieth centuries (Jones et al. 1997). The twentieth century was the warmest period of the last 1,000 years, averaging 0.2 °C above the mean temperature of the past 500 years. The most rapid warming in the last Millenium occurred in the last 30 years of the twentieth century (Jones et al. 2001). However, the global changes within the last 40 years or so are differentiated from prehistorical and historical changes because they appear to be caused by humans rather than by natural fluctuations and they are changing much more rapidly than the world has experienced in the past.

A draft report, recently issued by the National Climate Assessment and Development Advisory Committee (NCADAC) of the United States (2013) details the newest data and modeling results regarding changes in air temperature over land and oceans, water vapor pressure in air, sea surface temperatures, increases in mean sea level, and decreases in both glacier size and depth, and winter snow cover. The conclusions are that (1) much more certainty exists than previously thought (IPCC, 2001) that recent global climate change are due to human activities, (2) these changes will continue to occur into the next century, (3) the average temperature in the United States has increased about 1.5 °F since 1985, (4) the length of the growing season has increased considerably since 1980, (5) some western and southwestern areas of the United States have recently experienced severe droughts, while other areas in the US have received more precipitation than normal, (6) extreme weather events such as heavy downpours and flooding, strong hurricanes, and severe heat waves have become more common and intense, (7) global sea level has risen about 8° since about the 1880s and is expected to rise 1-4 ft by the next century, and (8) winter snow cover, the extent and thickness of sea ice, and the thickness and size of glaciers are all decreasing. Finally, oceans are becoming more acidic as they absorb about one fourth of the carbon dioxide produced annually throughout the world. This review will not address further the biological ramifications of increasing ocean temperatures and acidity, but interested readers are directed to the following papers as representative of this developing field (Maier et al. 2011; Carstillo et al. 2012; Milazzo et al. 2012; Bednarsek et al. 2012).

Current and near future alterations in air temperatures and precipitation patterns are expected to induce major shifts in the climates of the world. Some existing climates are likely to disappear, while novel climates (not existing now nor at any time in the past) are likely to appear. By AD 2100, about a third of the world's existing landmasses are predicted to experience novel climates (Williams et al. 2007).

1.1 Biological Effects of Climate Change

The models that are currently being used to predict temperature and precipitation changes on each continent and in the oceans are becoming increasingly more sophisticated and accurate. In contrast, predicting the ecological impacts of climate change is, at this point, difficult at best (Mustin et al. 2007; Sheldon et al. 2011). The accuracy with which the direct and indirect effects of climate change on living organisms can be anticipated is complicated many factors, some of which include: (1) each species within a community is likely to respond differently to local climatic changes, (2) natural variation and fluctuation in biological phenomena are being exacerbated by man-made factors other than climate change, such as habitat destruction, exposure to xenobiotics, and introduced or invasive species, and (3) the genetic plasticity with which to adapt to new situations and alter annual phenological events, such as reproduction, varies among species. Many studies suggesting a link between climate change and biological phenomena are merely correlations (McCarty 2001). Although the number of new papers suggesting links between current climate change and biological effects has dramatically increased in the last 5 years, many of them are based on only a few years of observations and are limited in their ability to predict the long-term survival of various species. The best studies are those that use long-term data sets (>20+ years), and additional evidence besides correlative observations (Carey and Alexander 2003). Added together, however, the preponderance of evidence from ever increasing number of studies suggesting links between climate change and biological phenomena cannot be ignored. Biological shifts that have already been noted in species distributions and phenology of annual events appear to be generally consistent with what would be predicted in a warming world (Root et al. 2003; Parmesan and Yohe 2003; Parmesan 2006).

As climates change, species must adapt, change their geographical distribution, or go extinct. Community structures will be disrupted because those species moving to other regions to follow their climate space will move at varying speeds, extinctions of species within the community will alter species associations, such as food availability, predator-prey relationships or competitive interactions, while other species will adapt to the new conditions.

2 Climate Change and Risk of Extinction

Extinctions of animals, plants and microbiotic species have occurred since the dawn of life, about 4 billion years ago. According to the fossil record, the number of terrestrial and marine organisms has substantially increased in the last 600 million years. Several episodes (sometimes called "mass extinctions"), in which the rates of organismal extinction were impressively higher than the rates in the preceding and following periods, have occurred (Mayhew 2012). Eighty-one and 53 % of marine genera went extinct in the end-Permian and end-Cretaceous episodes, respectively. However, not all of these "mass extinctions" have been correlated with dramatic shifts in climate. Global climate, as judged by several different kinds of proxy data (i.e. ice cores, pollen records, etc.), appears to have fluctuated between warm, "greenhouse" modes and "ice house modes" with a cycle length of approximately 140 million years (Mayhew 2012). Although the relationships between climate change and pre-historical biological extinctions are more complex than can be addressed here, those periods associated with the highest temperatures were generally associated with high rates of extinctions, although one "ice house" (end Ordovician) event also had high extinction rates (Mayhew 2012).

It is estimated that the rate of extinction of living organisms is greater now than at any time in the last 100,000 years. (Wilson 1992) and the current rate probably falls within the high range of extinction events in the past (Mayhew 2012). Barnosky et al. (2011) and Wake and Vernberg (2008) have proposed that we are now seeing the sixth mass extinction event. Recent extinctions have certainly stemmed from many factors, most of which are anthropogenic, but they are not exclusively, or even necessarily, linked to climate change. Clearly, anthropomorphic environmental degradation does not affect all species similarly in a given geographical area. Some species exhibit rapid declines, others none at all. Understanding how and why certain species respond differently than others to the challenges of the future may inform conservation efforts and assist in understanding the limits of various species in coping with stress.

A provocative paper published by Williams et al. (2004) made the first attempt to model how predicted climatic changes might affect the risk of extinction for living organisms. Using climatic models existing up to the time of publication and data on a number of species, the study estimated that climate change will drive 18–34 % of the approximately 5 M terrestrial species on earth extinct or nearly extinct by 2050. The model focused on the direct effects of climate change on the species in question, rather than indirect variables such as vegetational changes. A number of factors, such as genetic variation, birth and death rates, immigration and emigration, the appearance of new invasive species, changes in habitat caused by human land uses, the evolution of novel pathogens, etc. were not considered. Because emissions of CO_2 have now exceeded the values used in the model of Williams et al., the predictions of the paper are even more alarming than when it was originally published (Cameron 2012). The idea that climate change could cause extinctions of upwards of over 1 M species has sparked considerable controversy and additional research. Interested readers are referred to a new publication that not only analyzes in depth the original paper by Williams et al. (2004), but provides much new information about the risk of extinctions of living organisms (Hannah 2012).

3 Reproduction of Terrestrial Vertebrates in Changing Climates

While many questions exist concerning how climate change will affect organisms, this review will provide a broad but superficial account focused on two questions: Will climatic changes in temperature and/or water availability exceed lethal limits of a given species and directly cause mortality and lead to species extinctions? Or, will climate changes cause reproductive failure, leading to species extinctions?

The ability of a species to survive over time requires successful breeding. Breeding activities of terrestrial vertebrates in temperate and high latitude areas have been selected to occur when temperatures, food supplies and other necessities are optimal for survival of young. In tropical areas, breeding of many such animals is likewise synchronized with dry/wet cycles in a manner that coordinates maximum food availability with development of young.

Variations in temperature and/or precipitation within habitats may indeed ultimately exceed lethal levels of some species and cause extinctions. But, currently, most focus has been placed on the problems associated with reproduction in changing conditions. Because reproduction of most organisms occurs at times of the year in which temperatures, food supplies, and precipitation are optimal, climate change is expected to pose severe challenges for reproduction, in part because some aspects of the phenology of annual events may be timed by rigid factors, such as photoperiod, whereas other aspects may be temperature dependent. Increasing global temperatures might enhance the rate of thermally- dependent phenomena, such as seed germination, flowering, etc., while photoperiodically controlled phenomena, such as onset of mating behavior, development of gonads, etc. would remain more or less constant. Mismatches in thermally determined and genetically fixed aspects of reproduction are now becoming evident in a number of systems and the consequences of such mismatches in reducing reproductive success are now becoming apparent. See for instance, Visser and Holleman (2001), McLaughlin et al. 2002; Dixon (2003), Both and Visser (2005), and Anderson et al. (2012).

3.1 Amphibians

Over the last 30 years, more population declines and species extinctions have been recorded for amphibians than for any other terrestrial vertebrate Class (Wake and Vernberg 2008). While the potential causes of declines are numerous, and possibly interrelated (Carey et al. 2003a, b; Blaustein et al. 2010), disease and habitat destruction are currently ranked among the most important causes of population declines, and even species extinctions (Stuart et al. 2004; Skerratt et al. 2007)

Future variations in temperature and moisture associated with climate change are expected to pose significant challenges for amphibians because these factors are the two most important environmental variables for amphibian biology. As ectothermic animals, their body temperature is directly determined by air/substrate temperature. Biochemical, cellular, and physiological rate processes (such as resting and maximal metabolic rate, digestive rate, heart rate, respiratory rate, etc.) generally increase or decrease two to threefold for each 10 °C change in body temperature (Rome et al. 1992). Most amphibians are active at night and consequently have relatively little behavioral ability to vary their body temperature. Those species active during the day may have some opportunity to seek microclimates and maintain preferred temperatures for some portion of the day (Carey 1978). Many amphibian adults have some ability to adjust both lethal body temperatures and/or metabolic or other physiological rate processes through physiological acclimation (Carey 1979; Rome et al. 1992). Lethal temperatures of larvae are generally correlated with their thermal environment: Larvae developing in aquatic situations in hot climates have a significantly higher temperature tolerance than those in more moderate climates (Salthe and Mecham 1974). Whether thermal tolerances of amphibian larvae can be modified by acclimation to higher temperatures is unclear. It is speculative at this point whether acclimation or changes in behavioral microclimate selection could foster survival of a species subjected to a rise of 3–4 °C in average air temperature, if all other factors like water and food supply, remained constant. However, because amphibians have survived as a Class for over 250 million years, extant amphibians have descended from ancestors that persisted through many global climatic changes. Therefore, most living amphibians presumably possess at least some abilities to survive rapid climate change.

An example of how thermal change can impact amphibians is reflected in a study evaluating the effects of body temperature change on the ability of several amphibian species to jump, a very important factor in their ability to capture prey and/or escape predators. The length of a jump varies with body temperature, from short leaps at low body temperature to longer jumps at higher body temperatures. The thermal range at which the longest jumps occurred varied among species: some groups (*Rana pipiens* and *R. calamitans)* exhibited broad thermal independence of the peak distance, while maximal leap distances of other species *(Limnodynastes tasmanienseis,* and *Xenopus laevis)* occurred over a very narrow range of body temperatures. However, in all cases studied, jump length decreased precipitously at body temperatures higher than those at which peak performances were measured. Therefore, if amphibians were forced by a warming climate to be active at body temperatures above those at which they are active now, escape from predators or the ability to catch food might be compromised (Whitehead et al. 1989).

Many amphibian populations probably died during the droughts in the United States in the 1930s and 1950s, but records are lacking. Droughts of this magnitude have occurred about 1–2 times per century in the last millennium (deMenocal et al. 2000). However, prolonged and very severe droughts, such as a 22-year drought in the late 1500s and a 26-year drought in the 1200s have recurred about every 500 years in the United States (deMenocal et al. 2000). Models used by Diffenbaugh and Ashfaq (2010) predict a substantial increase in the number and duration of very hot air temperature episodes, coupled with prolonged droughts and decreases in soil moisture over next three decades. If these predictions are realized, the results could be catastrophic for many amphibian populations, and possibly species, Amphibians possess such limited ability to restrict evaporation of water through their highly permeable skins that moisture availability, rather than temperature, is the major determinant of distributions of most amphibian species (Duellman 1999). Additionally, because most species require standing water for breeding, eggs and larvae are particularly at risk for desiccation (Duellman and Trueb 1985). Eggs and larvae of species laid in water appear to have no resistance to desiccation. Adults of only a few species have some ability to curtail water loss and endure periods of dryness. These particular species not only can curtail cutaneous water loss to levels comparable to reptiles, they also can conserve water by excreting uric acid rather than urea. (Shoemaker and McClanahan 1975; Drewes et al. 1977; Shoemaker et al. 1992). A few species, such as *Scaphiopus couchii,* have the ability to burrow in desert sand and survive for several years between periods of rainfall (McClanahan 1967). However, most amphibians do not possess the behavioral and physiological attributes that would promote survival in even a mild (1–3 year) drought.

To my knowledge, no studies have yet shown that recent climatic factors have been the direct cause of extinction or declines in population sizes in an amphibian species. A number of studies have shown a correlation between a climatic event and amphibian declines, but few studies of this kind have made an attempt to determine to what extent the decline was directly or indirectly caused by the climatic event (Carey and Alexander 2003). Pounds et al. (2005) have suggested that climate change has indirectly caused amphibian population extinctions and extinction of one species by fostering outbreaks of the pathogenic chytrid fungus *Batrachochytrium dendrobatidis.* This suggestion has generated considerable controversy, however, and confirming support has been lacking (Lips et al. 2008; Bustamante et al. 2010).

3.1.1 Amphibian Reproduction in Changing Climates

Most amphibians have a complex life cycle, with aquatic egg and larval stages, followed by metamorphosis into an adult terrestrial or quasi-terrestrial form. The complexity of both an aquatic and terrestrial stage of many species exposes amphibians to a greater number of risks than direct developing vertebrates, such as reptiles, mammals, and birds (Wilbur 1980). Amphibians employ the largest variety of reproductive modes compared to other terrestrial vertebrates. Most amphibians lay eggs in aquatic situations (streams, ponds or lakes) but a few species lay eggs in cavities of trees or bromeliads. Other species lay eggs in terrestrial foam nests, or on dirt, rocks or other substrates from which larvae wiggle their way to water to complete development. Still yet other species employ ovoviparity, in which eggs are retained in the oviduct and nourished by an egg yolk, while a few other species employ true viviparity, in which secretions from the oviduct nurture the young. Additionally, a few species have adopted very novel means of reproduction, such as by incubating embryos in the stomach of the male or in pits on the backs or legs of males (Duellman and Trueb 1986).

Amphibian "breeding seasons" are typically defined to include the period of time in which male calling and egg laying occurs (Salthe and Mecham 1974). Tropical species may mate more than once a year, commonly in association with periods of rainfall. Temperate species typically breed once in the late spring or early summer. In the latter case, rising temperatures in the spring and/or heavy spring rains appear to be the cue that initiates breeding. Fixed environmental factors, such as photoperiod, appear to have limited control over spermatogenesis, perhaps because most amphibians are nocturnal (Salthe and Meecham 1974). Spawning of a few species may be triggered odors of particular species of algae, which grow in the few months following rainfall and signal food availability for larval growth (Savage 1961).

Because the frequency, volume and timing of annual precipitation can have profound effects of reproduction in amphibians, climate change resulting in prolonged droughts, flooding or severe thunderstorms are expected to disrupt amphibian breeding. Flooding at critical times in egg and larval development can cause egg and larval mortality (Carey et al. 2003a, b). Insufficient snow melt prior to breeding or lack of adequate precipitation during the aquatic larval stages can cause substantial, if not complete reproductive failure. Small temporary ponds can dry before the larvae can metamorphose (Rowe and Dunson 1995) and eggs that are normally laid in deep water can be exposed to harmful UVB radiation if laid in atypically shallow ponds (Palen and Schindler 2010). Reduction in pond size affects food supply, density of tadpoles, competition among tadpoles, size at metamorphosis and the efficiency with which predators can capture tadpoles (Morin 1983). Concentration of xenotiobics by evaporation of pond water can increase toxic effects on larvae that normally might develop normally in more dilute solutions of toxins (Carey and Bryant 1995).

Since temperature and rainfall appear to be the major cues for the onset of amphibian breeding seasons, it is not surprising that some species in areas experiencing warmer springs track the earlier arrival of spring in temperate environments by breeding earlier (see review by Carey and Alexander 2003). For example, breeding of the common toad *(Bufo bufo)* in England began up to 7 weeks earlier in warmer springs than in colder ones (Reading 1998). Early breeding may not necessarily be advantageous. In years in which common toads laid earlier in the spring, the larval period, in which mortality can be higher that after metamorphosis, lasted up to 30 days longer (Reading and Clarke 1999). The ultimate consequences of earlier breeding or longer larval periods for reproductive success need further research.

Some studies correlating spring temperatures and amphibian breeding have found contradictory results. A 30-year data set on amphibian breeding in a wetland community in South Carolina, USA, was used to examine correlations between the initiation of breeding in 10 species of amphibians. Mean arrival dates at breeding sites proved to be a better predictor of timing of reproduction than the first appearance of males at breeding sites, the onset of the first calling by males or the first date of egg laying. Three salamander and one anuran species exhibited significant changes in mean arrival date since 1979. Both species that breed in the fall arrived significantly later in the fall, whereas two species that mate in winter arrived significantly earlier in the spring. The biological consequences, if any, of such changes were not evaluated. Six other species exhibited no significant change in the timing of their arrival in breeding ponds.

3.2 *Reptiles*

Amphibians and reptiles, separated by about 300 million years of evolution (Pough et al. 1998) share some morphological, behavioral and physiological attributes, such as a relatively low metabolism (compared to mammals and birds), body temperatures that are generally set by the thermal environment, and temperature-sensitive biochemical, cellular and physiological rate processes. More mobile than typical amphibians, some reptilian species have home ranges that range upwards of hundreds of square kilometers (Brown 1993), and some sea turtles may traverse more than half the world's oceans in a year (Ernst and Barbour 1989). Additionally, the relatively impermeable scaly skin of reptiles fosters more independence of hydric conditions than amphibians have. These factors may allow reptiles greater flexibility in adjusting to climate change than most amphibians.

Population sizes of certain reptilian species have been observed to suffer declines in recent decades (Gibbons et al. 2000). Direct effects of climate change on reptile populations are difficult to assess, due to confounding factors of habitat destruction, exposure to man-made chemicals, overharvesting for pets, zoos, and human food, introduced invasive species that serve as predators or competitors, disease or parasitism (Gibbons et al. 2000).

The thermal ecology of reptiles has received intensive study over the last 40 years (Huey 1982). Diurnally active, terrestrial reptiles can behaviorally thermoregulate to a much greater extent than amphibians by shuttling back and forth between appropriate microclimates, with the result that body temperatures can be maintained at a relatively constant level as long as environmental conditions provide suitable conditions. Regulation of body temperature is not an end in itself, because this process co-evolved with physiological rate processes that generally reach optimal levels at or near the preferred body temperature (Huey 1982). As a result, the ability to seek and maintain body temperatures that foster optimal performance aids in feeding, reproduction, and evasion of predators. As noted by Huey (1982), understanding the role of body temperature in one simple ecological problem, catching prey, is difficult because the rate of processes, like prey detection, catching the prey, and digestion do not speed up or slow down at the same rates during changes in body temperatures change.

Like amphibians, reptiles have upper and lower thermal lethal limits. Few reptiles, even those living in deserts, prefer to be active at temperatures near their upper lethal limits. But, in many species, the optimal performance of a particular function is closer to the upper lethal limit than the lower one (Huey 1982). These findings suggest that an increase in average temperatures in a species' habitat may not directly cause death, but could impact the number of hours that a species could be active at or near its optimal temperatures for feeding and digestion. Extinction of 12 % of lizard populations at 200 locations in Mexico has been linked to thermal changes that prevented foraging at their thermal optima by Sinervo et al. 2010. However, alternative explanations for local population extinctions were not examined in this paper.

Droughts or other variations in the hydric environments would be expected to have less direct impact on terrestrial reptiles than amphibians, due not only to the features of the skin that allow reptiles to conserve water, but also the ability of many species to conserve water by excreting uric acid, rather than urea. Even so, reptiles, of course, are dependent on food resources that could be so impacted by drought or high temperatures that particular reptilian species could go extinct.

3.2.1 Reptilian Reproduction in Changing Climates

Reptiles employ fewer reproductive modes than amphibians, but oviparity, ovoviparity and viviparity occur in various groups of reptiles. Oviparity, in which eggs contain all the nutrients needed to produce viable hatchlings, is the most common form of reproduction. Eggs are formed with either a flexible or rigid, calcareous shells (Packard and Packard 1988).

Reproduction in reptiles may be severely impacted by changes in temperature and moisture patterns. Generalizations about the types of nests in which reptiles lay their eggs are impossible, because of the vast variety of substrates and locations (Packard and Packard 1988). However, the ability of each species to select a nest site that provides the best thermal and hydric environments for development of the young has undoubtedly been under intense selection. Metabolic rates of embryos are temperature-dependent, as are those of adults, and therefore, higher nest temperatures should generally result in faster rate of growth. Embryos of most species undoubtedly experience some, if not considerable, daily fluctuation in temperature. Climate changes leading to higher soil temperatures may cause embryonic mortality if they exceed tolerance levels.

The sex of offspring of most species of vertebrates is determined genotypically by sex chromosomes, a mechanism that usually yields roughly equal sex rations. However, the sex of some species of reptiles is determined not by inheritance of sex chromosomes, as in other reptiles, birds and mammals, but by temperatures of the embryos prior to hatching (Ewert and Nelson 1991; Janzen 1994). Therefore, thermal conditions in the nest can yield uneven sex ratios that can have negative consequences for future reproduction of the population (Mitchell et al. 2009). In some species of turtles, cooler incubation temperatures produce males and warmer ones produce females. In other turtles, females are produced at cooler and warmer temperatures, while males are produced at intermediate temperatures (Ewert and Nelson 1991). In some alligators and lizards, mostly males are produced at warmer temperatures, whereas females at cooler ones (Bull 1983). The temperature of peak production of males varies among species. The range of temperatures at which males are produced in one turtle species, *Stenotherhus odoratus* occurred between 24 and 29 °C, with a peak production of males occurred near 25 °C. Incubation at about 32 °C of eggs of another turtle, *Pelomedisa subrufa,* produced 100 % males (Bull 1991). These data suggest that even a modest change in temperature of soil surrounding the nest could result in the production of young of all one sex. If this pattern were to continue over several breeding seasons, successful reproduction of that population would be prevented because individuals of only one sex would remain in the population. However, a recent study (Warner and Shine 2010) that incubated eggs of the lizard *Amphibolurus muricatus* under conditions of either fluctuating or constant mean temperatures indicated that interacting thermal effects may cancel each other out, with the result that sex ratios of the hatchlings mirrored the same approximate sex ratio of 50:50 as in genetically determined species. These data suggest that much more research is needed before conclusions about future risks to those species in which sex ratios are thermally determined.

Some reptilian species also require specific hydric conditions during egg incubation in order to produce viable hatchlings. For instance, eggs of some species must absorb significant amounts of water from the soil in order for the embryo to develop and hatch. Embryos of painted turtles *(Chrysemys picta)* held in dry conditions in the middle of incubation had significantly reduced hatching success than embryos in eggs in wet conditions (Guttzke and Packard 1986). Eggs incubated initially in dry conditions were able to take up enough water if incubated in wet conditions later in incubation so that hatching was minimally impacted. Therefore, hatching success of painted turtle eggs depends importantly on the soil moisture content during incubation. Drought conditions could cause nest failure and population declines.

Not surprisingly, temperature and hydric conditions in the nest can interact to affect hatching success and hatchling mass, the latter of which has a significant effect on the future fitness of the hatchling (Gutzke and Packard 1987). Embryos of bull snakes *(Pituophis melanoceucus)* incubated at moderate temperatures were significantly larger than those incubated at colder or hotter temperatures. Although incubation in varying hydric environments had no effect on hatching success, eggs in moist environments produced larger hatchlings than those in drier ones.

3.3 Birds

The biology of birds, numbered at about 10,000 species, has been studied in much greater detail than that of other terrestrial vertebrates, due in large part to their largely diurnal behavior and great visibility. Extinctions of a number of species, such as the passenger pigeon *(Ectopistes migratorius)* and the great auk (*Pinguinus impennis*), have been caused by human activities in the nineteenth and twentieth centuries. Approximately 800 species are globally threatened with extinction (Bird Life International 2013). Habitat loss is considered to be the major cause of bird population declines.

Standard metabolic rates of birds and mammals average seven to tenfold the rate of equivalently-sized reptiles or amphibians, when measured at the same body temperature. By obtaining a balance between endogenous heat production and the rate of heat loss, birds maintain a high and relatively constant body temperature around 40-43 °C, a range that averages slightly higher than body temperature of most mammalian species (Marsh and Dawson 1989). These temperatures are within a few degrees of lethal temperatures, but behavior, evaporative cooling, and the ability to fly to less extreme areas are likely to prevent lethal overheating. Because of their mobility, birds are also able to fly to water and/or other habitats to avoid droughts. Therefore, birds may be less impacted, at least over the short term, by variations in temperature and precipitation than other terrestrial vertebrates.

Because food intake must support the costs of maintaining high rates of heat production, cold temperatures, short photoperiods for foraging, and reduced food supplies in temperate and polar winters pose severe challenges for resident birds (Marsha and Dawson 1989). However, as temperate winters have become thermally more moderate over the last 50 years, wintering ranges of some birds in mid-temperate areas have moved north and other species that commonly migrated south in the winter have stopped migrating and remain resident throughout the year (Root and Weckstein 1994; Böhning-Gaese and Lemoine 2004).

While a few birds may utilize daily torpor or even hibernate, as some mammals can (Wang 1989), when food supplies are diminished in seasonal environments, the mobility of most birds allows them another option: migration. A number of patterns of migratory behavior found among avian species, from local movements to migrations involving thousands of kilometers, sometimes over oceans (Berthold 2001). The abilities to fly long distances without refueling, to navigate successfully over land and water between wintering and breeding grounds, and the morphological, physiological and behavioral attributes fostering migration are primarily genetically controlled (Berthold 2001). Many birds wintering at latitudes in which sufficient differences in photoperiod occur seasonally use photoperiod as the cue to prepare for migration (molt, pre-migratory fattening) and departure for the breeding ground. Birds wintering close to the equator, where photoperiod varies only to a limited degree throughout the year, depend upon internal circannual rhythms to time their departure on spring migrations (Gwinner 1977).

3.3.1 Avian Reproduction in Changing Climates

Birds breed in some of the most inhospitable environments on earth, including the Arctic and Antarctica, high altitudes up to 6,500 m in the Himalayas, and the hottest deserts on earth (see review by Carey 2002). Avian annual cycles, including molt, fattening, and breeding, have been under intense selection so that breeding occurs at the time of year in which moderate temperatures and optimal food supplies maximize the opportunity for reproductive success (Berthold 2001). Mistakes in the timing of arrival on breeding grounds and breeding could have severe consequences on reproductive success. Photoperiod plays a key role in determining the phenology of migration, molt and fattening for a number of species because it is a highly accurate predictor of the time of year.

However, increasing numbers of studies now document that, while arrivals on breeding grounds are staying relatively constant for many birds (which probably are using photoperiod as their cue to begin migration), higher air temperatures in the springs result in earlier food availability, resulting in mismatches between production of young and food (Visser et al. 2004). How climate change affects food availability for breeding will be species-specific. For instance, neither the arrival time of migrating Broad-tailed hummingbirds *(Selasphorus platycercus)* nor the dates of flowering of plants they visit for nectar has changed significantly over the past few decades in the southern part of their breeding range. But, in the northern end of the range of their breeding range, the dates of the first flowering and peak flowering have advanced in spring, yet the hummingbirds are continuing to arrive at the same time. If this trend continues, the arrival and breeding of hummingbirds at the northern end of their distribution will be out of synchrony with their food supply and reduced reproductive success is likely to occur (Anderson et al. 2012).

It is becoming clear that the control of migration and breeding may be much more flexible than previously imagined and may result more from phenotypic plasticity rather than genetic changes (Gienapp et al. 2007). Formerly sedentary species are becoming semi-migratory, whereas some migratory populations are now becoming much more sedentary (Berthold 2001). Advancement of the arrival date on breeding grounds is occurring in a number of species but there is considerable interspecific and intraspecific variation in these changes (Leihonen et al. 2004). For instance, in an analysis of spring arrival times of 22 species, three species were arriving significantly earlier, and four were arriving significantly later. The averages differ by only a few days from their previous mean (Ellwood et al. 2010). Since the arrival dates of a population can vary substantially over a period of weeks, the biological significance, if any, of an advancement or delay in arrival date of a few days needs further examination. A recent paper (Knudsen et al. 2011) examines in depth many different issues regarding the relation between climate change and avian migration, and interested readers are directed to this paper.

Many species have a time gap of some duration between the arrival on breeding grounds and laying of the first egg. Therefore, some species may have flexibility in adjusting to local food supplies once they arrive. However, once the first egg is laid, the sequences of events between clutch completion, hatching and the requirements of nestlings for food are fairly invariate. It is at the latter point that mismatches between food availability and requirements of the young for food become most critical (Both and Visser 2005). The impact on climate change on avian breeding is proving to be just as variable interspecifically and intraspecifically as in migration. The date at which the first egg is laid has varied with temperature in spring for some species, but not for others (see review by Carey 2009). In most cases, the biological importance of such variation has not been demonstrated. New information indicates that air temperature may not only affect the development of food resources, it may also affect the timing of egg production, at least in one non-migratory species. Great tits *(Parus major),* which breed earlier in warmer springs than colder ones, varied the timing of egg laying when experimentally exposed to colder or warmer temperatures (differing by only 4 °C). These data suggest that, at least for this species, both the timing of the development of food resources and egg laying may be coordinated by temperature in the spring in a manner than can avoid mismatches between breeding and food availability (Visser et al. 2009).

3.4 Mammals

Over 4,000 species of mammals live and breed in oceans and on all continents. Some live in the harshest environments on earth, including the Arctic, Antarctic, and deserts. Nearly 173 terrestrial mammalian species are known to be declining in size; extinctions of many populations appear to have been caused by man-made habitat destruction (Ceballos and Ehrlich 2002). The effects of habitat destruction and other anthropogenic activities on mammalian populations and species make it difficult to separate out any potential direct effects of climate change on mammals.

As endothermic organisms, mammals share with birds the necessity to fuel high metabolic rates. In temperate and polar climates, hibernation is employed by a variety of species for surviving periods of food restriction and/or cold temperatures. Some mammals aestivate, or become inactive with a slightly reduced body temperature, during periods low food and water supplies in extremely hot and dry conditions in deserts. The ability to hibernate has evolved in a number of different mammalian taxa (Gieser 1998), and has required a suite of biochemical, cellular, physiological, and neurological adaptations to foster not only the ability to reduce metabolism and body temperatures to near ambient temperatures, but also to initiate the procedures of rewarming to normothermia at appropriate intervals during hibernation and at the end of the hibernation season (Carey et al. 2003a, b). Interestingly, about 94 % of 61 mammalian species that have recently become extinct were non-hibernators, but by comparison, only about 7 % of species that have become extinct in the same time period were hibernators (Geiser and Turbill 2009). While the causes of extinctions of these species remain to be understood, these data suggest that those species that have evolved the ability to become torpid, either daily or seasonally, may be able to survive the challenges of climate change far more successfully than non-hibernators.

Several studies have evaluated factors that are correlated with population and species extinction in mammals. Species with small adult body masses appear to be more resistant than larger ones to extinction (Cardillo 2003). Small litter sizes, large home ranges, small geographical distribution and exposure to habitat loss or invasive species are other factors that are correlated with extinctions (Russell et al. 1998; Gonzalez-Suarez and Revilla 2013).

3.4.1 Mammalian Reproduction in Changing Climates

Most mammalian species breed when thermoregulatory costs of the adults are minimal and food supplies are optimal for females during gestation and growth of weaned young (Bronson 2009). Costs of reproduction are usually comparatively minor for males, but the energy expenditures of females are very high. For instance, female coyotes *(Canis latrans)* must consume about 18 % more prey during one breeding cycle over annual costs (Laundre and Hernandez (2003). In most mammalian species, these energy requirements have selected for seasonal breeding, as opposed to breeding throughout the year.

Birth of temperate mammalian offspring usually occurs in the spring or early summer when food supplies are optimal for supporting costs of maternal lactation and growth of the young. In long-lived, high latitude mammals, reproduction in many short-lived mammals is less dependent on photoperiod, perhaps because many live in burrows and/or be nocturnal. Reproduction in many temperate desert mammals occurs after rainfall and appears to be cued, at least in some species, by organic chemicals produced by forage plants soon after rain (Bronson 2009). The largest number of mammalian species, however, live in the tropical regions. There, seasonal temperatures and moisture vary less than in temperate and polar regions.

Unfortunately, what environmental cues signal the advent of reproduction in tropical species are generally unknown.

Temporal mismatches between births of young and the peaks of food supplies have been recorded thus far for two mammals. Yellow-bellied marmots *(Marmota flaviventris)* are emerging from their hibernacula up to 38 days earlier in spring than they did nearly three decades ago, but snow cover at the time of emergence is still deep enough to prevent growth of emergent vegetation. Since lactating females are unable to find adequate food, litter size has declined (Inouye et al. 2000).

Climate warming in Greenland has reduced the spatial variability in forage for migrating caribou *(Rangifer tarandus)* (Post et al. 2008). Migration normally fosters the ability of caribou to follow the spatial front of the emergence of vegetation, providing lactating females and young with the most nutritious and highly digestible plant matter. However, the advance of spring warmth is causing the earlier onset of plant emergence. Because the timing of the calving season has not advanced commensurately, caribou reproductive success has been declining (Post and Forchhammer 2008).

Many species living at high latitudes are adapted to or dependent on snow and ice cover for foraging, reproduction and survival. In the Arctic, caribou *(Rangifer tarandus)*, arctic fox (*Vulpes lagopus),* and most species of smaller mammals, such as voles (*Microtus)* and lemmings *(Lemmus),* reproduce in the warmer parts of the year, but polar bears (*Ursus maritimus)* hibernate during the summer and give birth in winter while in their hibernacula (Bronson 2009). Downward trends in population sizes have been noted in a number of arctic species, leading to predictions that the most likely species to go extinct are polar bears, walruses, narwals, and ivory gulls (Post and Forchhammer 2008). Because polar bears require stable sea ice for reproduction, most foraging and most of developmental period of young occur at sea during the winter. Successful hunting of their primary prey, ringed seals *(Pusa hispida),* has been become progressively difficult as arctic warming thins the sea ice and increases the number of breathing holes used by the seals. Female body condition, birth rates and proportion of yearling bears in populations have declined significantly since 1980s and extinction of polar bears is likely if all sea ice disappears (Stirling et al. 1999; DeRocher et al. 2004). Extinction of polar bears and consequent increases in ringed seal populations will have consequences on cod populations and marine food webs (Post and Brodie. 2012). Estimates of the approximate year in which arctic sea ice will completely disappear range from 2020–2040, depending on the type of model and data used (Overland and Wang 2013).

4 Conclusions

The indications are alarming that impending climate change, possibly beyond the "tipping point" and therefore irreversible, is likely to cause widespread extinctions of animals and plants, reorganization of interactions among species in existing communities, and disappearances of existing ecosystems. Conservation efforts to reduce the rate of extinctions and preservation of key biodiversity sites are underway and financial support for such efforts has been pledged by world governments. However, the costs of preserving 211 threatened bird species alone has been estimated at around 1 billion US dollars yearly (McCarthy et al. 2012). Preservation of biodiversity sites important for conservation of both birds and other taxa would raise the estimated costs to roughly US $76 billion annually. It is unlikely, if not impossible, for these goals to be met by even high-income countries. The impact to humans of global change, such as shortages in food and water, destruction of property by increasingly severe hurricanes, floods, droughts, rising sea levels, and evolution of new pathogens, etc. are more likely to determine the formulation of governmental policies and determination of financial commitments than protection of biodiversity. Yet, one has to wonder how many ecosystems have to fail before the survival of humans is in question.

References

Alley RB, Mayewski PA, Sowers T. Holocene climate instability: a prominent, widespread event 8200 yr ago. Geology. 1997;25:483–6. doi:10.1130/0091-7613(1997)025<0483:HCIAPW>2.3CO;2.

Anderson JT, Inouye DW, McKinney AM et al. Phenotypic plasticity and adaptive evolution contribute to advancing flowering phenology in response to climate change. 2012.

Barnosky AD, Matzke N, Tomiya S, et al. Has the Earth's six mass extinction already arrived? Nature. 2011;471:51–7.

Bednarsek N, Tarling GA, Bakker DCE, et al. Extensive dissolution of live pteropods in the southern ocean. Nat Geosci. 2012. doi:10.1038/NGEO1635.

Berthold P. Bird migration: a general survey. Oxford: Oxford University Press; 2001.

Bennett AF, Dawson WR. Metabolism. In: Gans C, Dawson WR, editors. Biology of the reptilia, Physiology A, vol. 5. New York: Academic; 1976.

Bird Life International. IUCN red list for birds. 2013. www.birdlife.org .Accessed 12 Feb, 2013.

Blaustein AR, Walls SC, Bancroft BA. Direct and indirect effects of climate change on amphibian populations. Diversity. 2010;2:281–313.

Bronson FH. Climate change and seasonal reproduction in mammals. Philos Trans R Soc Lond B Biol Sci. 2009;364:3331–40.

Böhning-Gaese K, Lemoine N. Importance of climate change for ranges, communities and conservation of birds. Adv Ecol Res. 2004;35:211–36. doi:10.1016/S0065.2504(04)35010-5.

Both C, Visser ME. The effect of climate change on the correlation between avian life-history traits. Global Change Biol. 2005;11:1606–13. doi:10.1111/j.1365-2846.01038x.

Bull JJ. Evolution of sex determining mechanisms. Menlo Park, CA: Benjamin Cummings; 1983.

Bustamante HM, Livo LJ, Carey C. Effects of temperature and hydric environment on survival of the Panamanian golden frog with a pathogenic chytrid fungus. Integr Zool. 2010;1:142–52.

Cameron A. Refining risk estimates using models. In: Hannah L, editor. Saving a million species. Extinction risk from climate change. Washington, DC: Island; 2012.

Cardillo M, Mace GM, Gittleman JM. Latent extinction risk and future battlegrounds of mammalian conservation. Proc Natl Acad Sci U S A. 2006;103:4157–61.

Carey C. Factors affecting body temperatures of toads. Oecologia. 1978;35:197–219.

Carey C. Effect of constant and fluctuating temperatures on resting and active oxygen consumption of toads, *Bufo boreas*. Oecologia. 1979;39:201–12.

Carey C. Incubation in extreme environments. In: Deeming DC, editor. Avian incubation: behavior, environment and evolution. Oxford: Oxford University Press; 2002.

Carey C. The impacts of climate change on the annual cycles of birds. Philos Trans R Soc Lond B Biol Sci. 2009;364:3321–30. doi: 10.1098/rstb.2009.0182.

Carey C, Alexander MA. Climate change and amphibian declines: is there a link? Divers Distrib. 2003;9:111–21.
Carey C, Bradford DF, Brunner JF, et al. Biotic factors in amphibian declines. In: Linder G, Sparling DW, Krest SK, editors. Multiple stressors and declining amphibian populations: evaluating cause and effect. Boca Raton, FL: Society of Environmental Toxicology and Chemistry; 2003a.
Carey C, Bryant CJ. Possible interrelations among environmental toxicants, amphibian development, and decline of amphibian populations. Environ Health Perspect. 1995;103 Suppl 4:13–7.
Carey HV, Andrews MT, Martin SL. Mammalian hibernation: cellular and molecular responses to depressed metabolism and low temperature. Physiol Rev. 2003b;83:1153–81.
Carstillo KD, Ries JB, Weiss JM, et al. Decline of forereef corals in response to recent warming linked to history of thermal exposure. Nat Clim Chang. 2012. doi:10.1038/nclimate1577.
Ceballos G, Ehrlich PR. Mammal population losses and extinction crisis. Science. 2002;296: 904–7.
deMenocal P, Ortiz P, Guilderson T, et al. Coherent high- and low latitude variability during the Holocene warm period. Science. 2000;288:2198–202.
Derocher AE, Lunn NJ, Stirling I. Polar bears in a warming climate. Integr Comp Biol. 2004;44:163–76.
Diffenbaugh AS, Ashfaq M. Intensification of hot extremes in the United States. Geophys Res Lett. 2010;37, L15701.
Dixon A. Climate change and phenological asynchrony. Ecol Entomol. 2003;28:380–91. doi:10.1046/j1365-2311.203.00509.x.
Drewes RC, Hillman SS, Putnam RW. Water, nitrogen and ion balance in the African treefrog *Chiromantis petersi* Boulenger (Anura: Rahcophoridae) with comments on the structure of the integument. J Comp Physiol. 1977;116:257–67.
Duellman WE, editor. Patterns of distributions of amphibians. Baltimore, MD: Johns Hopkins University Press; 1999.
Duellman WE, Trueb L. Biology of the amphibians. New York: McGraw-Hill; 1985.
Ellwood ER, Primack RB, Talmadge ML. Effects of climate change on spring arrival times of birds in Thoreau's Concord from 1851 to 2007. Condor. 2010;112:754–62.
Ernst CH, Barbour RW. Turtles of the world. Washington, DC: Smithsonian Institute Press; 1989.
Ewert MA, Nelson CE. Sex determination in turtles: diverse patterns and some possible adaptive values. Copeia. 1991;1991:50–69.
Gibbons JW, Scott DE, Ryan TJ, et al. The global decline of reptiles, déjà vu amphibians. BioScience. 2000;50:652–66.
Gieser F. Evolution of daily torpor and hibernation in birds and mammals: importance of body size. Clin Exp Pharmacol Physiol. 1998;25:736–40.
Gieser F, Turbill C. Hibernation and daily torpor minimize mammalian extinction. Naturwissenschaften. 2009;96:1235–40. doi:10.1007/s00114-009-0583-0.
Glenapp P, Leimu R, Merila J. Responses to climate change in avian migration time—microevoluton versus phenotypic plasticity. Clim Res. 2007;35:25–35. doi:10.3354/cr00712.
Gonzalez-Suarez M, Revilla E. Variability in life-history and ecological traits is a buffer against extinction in mammals. Ecol Lett. 2013;16:242–51.
Gutzke WHN, Packard GC. Sensitive periods for the influence of the hydric environment on eggs and hatchlings of painter turtles *(Chrysemys picta)*. Physiol Zool. 1986;59:337–43.
Gwinner E. Circannual rhythms in bird migration. Annu Rev Ecol Systemat. 1977;8:381–405.
Gutzke WHN, Packard GC. Influence of the hydric and thermal environments on eggs and hatchlings of bull snakes *Pituophis melanoleucus*. Physiol Zool. 1987;60:9–17.
Hannah L. Saving a million species: extinction risk from climate change. Washington, DC: Island; 2012.
Huey RB. Temperature, physiology and the ecology of reptiles. In: Gans C, Pough FH, editors. Biology of the reptilia, Physiology C, vol. 12. New York: Academic; 1982.
Inouye DS, Barr B, Armitage KB, et al. Climate change is affecting altitudinal migrants and hibernating species. Proc Natl Acad Sci U S A. 2000;97:1630–3. doi:10.1073/pnas.97.4.1630.

Intergovernmental Panel on Climate Change (IPCC). Climate Change 2001: the scientific basis, contribution of working group I to the third assessment report of the intergovernmental panel on climate change. Cambridge, UK: Cambridge University Press; 2001.

Intergovernmental Panel on Climate Change (IPCC). Fourth assessment report of the intergovernmental on climate change, vol. 4. Cambridge, UK: Cambridge University Press; 2007.

Janzen FJ. Climate change and temperature-dependent sex determination in reptiles. Proc Natl Acad Sci U S A. 1997;91:7487–90.

Jonzén N, Ergon T, Lindén A, et al. Bird migration and climate: the general picture and beyond. Clim Res. 2007;35:177–80. doi: 10:3354/cr00724.

Jones PE, Osborn TJ, Briffa KR. The evolution of climate over the last Millennium. Science. 2001;292:662–6.

Knudsen E, Lindén A, Both C, et al. Challenging claims in the study of migratory birds and climate change. Biol Rev. 2011;86:928–46. doi:10.1111/j.l469-185X.2011.00179x.

Leihonen E, Sparks TH, Zalakavicis M. Arrival and departure dates. Adv Ecol Res. 2004;35:1–31. doi:10.1016/S0065-2504(04)35001-4.

Lips KR, Diffendorfer J, Mendelson III JR, et al. Riding the wave: reconciling the roles of disease and climate change in amphibian declines. PLoS Biol. 2008;6:441–54. doi:10.1371/journal.pbio.0060072.

Mayhew PJ. Extinctions in deep time. In: Hannah L, editor. Saving a million species: extinction risk from climate change. Washington, DC: Island; 2012.

McCarthy DP, Donald PF, Scharlemann PW, et al. Financial costs of meeting global biodiversity conservation targets: current spending and unmet needs. Science. 2012;338:946–494.

McCarty JP. Ecological consequences of recent climate change. Conserv Biol. 2001;15:320–31.

McClanahan LL. Adaptations of the spadefoot toad, *Scaphiopus couchii,* to desert environments. Comp Biochem Physiol. 1967;20:73–99.

McKinney AM, CaraDonna PJ, Inouye DW, et al. Asynchronous changes in phenology of migrating broad-tailed hummingbirds and their early-season nectar resources. Ecology. 2012; 93:1987–193.

McLaughlin JF, Hellman JJ, Boggs CL, et al. Climate change hastens population extinctions. Proc Natl Acad Sci U S A. 2002;99:6070–4.

Marsh RL, Dawson WR. Avian adjustments to cold. In: Wang LCH, editor. Animal adaptation to cold. Advances in comparative environmental physiology, vol 4. Berlin: Springer; 1989.

Milazzo M, Mirto S, Dominici P et al. Climate change exacerbates interspecific interactions in sympatric coastal fishes. J Anim Ecol. 2012; doi: 10/1111/j.1365-2656.2012.02034.x.

Mitchell NJ, Allendorf FA, Keall S, et al. Demographic effects of temperature-dependent sex determination: will tuatara survive global warming? Global Change Biol. 2009;16:60–72.

Morin PJ. Predation, competition and the composition of larval anuran guilds. Ecol Monogr. 1983;53:119–38.

Mustin K, Sutherland WJ, Gill JA. The complexity of predicting climate-induced ecological impacts. Clim Res. 2007;35:165–75.

National Climate Assessment and Development Advisory Committee (NCADAC). Draft report. http://ncadac.globalchange.gov. 2013. Accessed 20 Jan, 2013.

Laundre JW, Hernandez L. Total energy budget and prey requirements of free-ranging coyotes in the Great Basin desert of the western United States. J Arid Environ. 2003;55:675–89. doi:10.1016/S0140-1963(02)00316-6.

Maier C, Watremez P, Taviani M, et al. Calcification rates and the effect of ocean acidification on Mediterranean cold-water corals. Proc R Soc B. 2011. doi:10.1098/rspb.2011.1763.

Overland JE, Wang M. When will the summer arctic be nearly sea ice free? Geophys Res Lett. 2013. doi:10.1002/grl.50316.

Overpeck J, Webb R. Nonglacial rapid climate events: past and future. Proc Natl Acad Sci U S A. 2000;97:1335–8.

Packard GC, Packard MJ. The physiological ecology of reptilian eggs and embryos. In: Gans C, Huey RB, editors. Biology of the reptilia, Ecology B, defense and life history, vol. 16. New York: Alan R Liss; 1988.

Palen WJ, Schindler DE. Water clarity, maternal behavior, and physiology combine to eliminate UV radiation risk to amphibians in a montane landscape. Proc Natl Acad Sci. 2010;107:9701–6.

Parmesan C. Ecological and evolutionary responses to recent climate change. Annu Rev Ecol Evol Syst. 2006;37:637–69. doi:10.1146/aanurev.ecol.syst.37.091305.110100.

Parmesan C, Yohe G. A globally coherent fingerprint of climate change impacts across natural systems. Nature. 2003;421:37–42. doi:10.1038/nature 01286.

Post E, Brodie J. Extinction risk at high latitudes. In: Hannah L, editor. Saving a million species. Washington, DC: Island; 2012. doi:10.5822/978-1-61901-182-5_6.

Post E, Forchhammer MC. Climate change reduces reproductive success of an Arctic herbivore through tropic mismatch. Philos Trans R Soc Lond B Biol Sci. 2008;363:2369–75. doi: 10.1098/rstb2007.2207.

Post E, Pedersen C, Wilmers CC, Forchhammer MC. Warming, plant phenology, and the spatial dimension of tropic mismatch for large herbivores. Proc. Roy Soc B Biol Sci. 2008; 275:2005–13.

Pough FH, Andrews RM, Cadle JE, et al. Herpetology. New Jersey: Prentice-Hall; 1998.

Pounds JA, Bustamante MR, Coloma LA, et al. Widespread amphibian extinctions from epidemic disease driven by global warming. Nature. 2005;439:161–7. doi:10.1038/nature04246.

Reading CJ. The effect of winter temperature on the timing of breeding activity in the common toad *Bufo bufo*. Oecologia. 1998;117:469–75.

Reading CJ, Clarke RT. The effects of density, rainfall and environmental temperature on body condition and fecundity in the common toad, *Bufo bufo*. Oecologia. 1995;102:453–9.

Rome LC, Stevens ED, John-Alder HB. The influence of temperature and thermal acclimation on physiological function. In: Feder ME, Burggren WW, editors. Environmental physiology of the amphibian. Chicago: University of Chicago Press; 1992.

Rowe CL, Dunson WA. Impacts of hydroperiod on growth and survival of larval amphibians in temporary ponds of central Pennsylvania, USA. Oecologia. 1995;102:397–403.

Root TL, Weckstein JD. Changes in distribution patterns of select wintering North American birds from 1901 to 1989. Stud Avian Biol. 1994;15:191–201.

Root TL, Price JT, Hall KR, et al. Fingerprints of global warming on wild animals and plants. Nature. 2003;421:57–60. doi:10.1038/nature01333.

Russell GJ, Brooks TM, McKinney MM, et al. Present and future taxonomic selectivity in bird and mammal extinctions. Conserv Biol. 1998;12:1365–76.

Saino N, Ambrosini R, Rubolini D, et al. Climate warming, ecological mismatch at arrival and population decline in migratory birds. Proc R Soc B. 2010;278:835–42. 1098/rspb.2010.1778.

Salthe SN, Mecham JS. Reproductive and courtship patterns. In: Lofts BL, editor. Physiology of the amphibia, vol. 2. New York: Academic; 1974.

Savage RM. The ecology and life history of the common frog *(Rana temporaria temporaria)*. London: Pittman; 1961.

Sheldon KS, Yang S, Tewksbury JJ. Climate change and community dissembly: impacts of warming on tropical and temperate montane community structure. Ecol Lett. 2011;14:1191–200. doi:10.1111/j.1461-0248.2011.01689.x.

Shoemaker VH, Hillyard SD, Jackson DC, et al. Exchange of water, ions, and respiratory gases in terrestrial amphibians. In: Feder ME, Burggren WW, editors. Environmental physiology of the amphibians. Chicago, Ill: University of Chicago Press; 1992.

Shoemaker VH, McClanahan L. Evaporative water loss, nitrogen excretion and osmoregulation in phyllomedusine frogs. J Comp Physiol. 1975;100:331–45.

Sinervo B, Mendez-de-la-Cruz F, Miles DB. Erosion of lizard diversity by climate change and altered thermal niches. Science. 2010;328:894–9.

Skerratt LF, Berger L, Speare R, et al. Spread of chytridiomycosis has caused the rapid global decline and extinction of frogs. EcoHealth. 2007;4:125–34.

Stirling I, Lunn NJ, Iacozza J. Long term trends in the population ecology of polar bears in western Hudson Bay in relation to climate change. Arctic. 1999;52:294–306.

Stuart SN, Chanson JS, Cox NA, et al. Status and trends of amphibian declines and extinctions worldwide. Science. 2004;2004(306):1783–876.

Thomas CD, Cameron A, Green RE, et al. Extinction risk from climate change. Nature. 2004;427: 145–8.

Todd BD, Scott DE, Pechmann JHK, et al. Climate change correlates with rapid delays and advancements in reproductive timing in an amphibian community. Proc R Soc B. 2011; 278:2191–7. doi:10.1098/rspb.2010.1768.

Visser ME, Holleman LJM. Warmer springs disrupt the synchrony of oak and winter moth phenology. Proc R Soc B. 2001;268:289–94. doi: 10:1098/rspb.2000.1363.

Visser ME, Holleman LJM, Caro SP. Temperature has a causal effect on avian timing of reproduction. Proc R Soc B. 2009;276:2323–31. doi:10.198/rspb.2009.0213.

Visser ME, Both C, Lambrechts MM. Global climate change leads to mistimed avian reproduction. Adv Ecol Res. 2004;35:89–110. doi:10.1016/S0065-2504(4)35005-1.

Wake DB, Vredenburg VT. Are we in the midst of the sixth mass extinction? A review from the world of amphibians. Proc Natl Acad Sci U S A. 2008;105:11466–73.

Wang LCH. Ecological, physiological, and biochemical aspects of torpor in mammals and birds. In: Wang LCH, editor. Animal adaptation to cold. Advances in comparative environmental physiology, vol. 4. Berlin: Springer; 1989.

Warner DA, Shine R. Interactions among thermal parameters determine offspring sex under temperature-dependent sex determination. Proc R Soc B. 2011;278:256–65. doi:10.1098/rspb.2010.1040.

Whitehead PJ, Puckridge JT, Leigh CM, et al. Effect of body temperature on jump performance of the frog *Limnodynastes tasmaniensis*. Physiol Zool. 1989;62:937–49.

Wilbur HM. Complex life cycles. Annu Rev Ecol Syst. 1980;11:67–93.

Williams JW, Jackson ST, Kutzback JE. Projected distributions of novel and disappearing climates by 2100 A.D. Proc Natl Acad Sci U S A. 2007;104:5738–42.

Wilson EO. The diversity of life. Cambridge, MA, USA: Harvard University Press; 1992.

Zachos J, Pagani M, Sloan L, et al. Trends, rhythms and aberrations in global climate 65 Ma to present. Science. 2001;292:686–7.

Chapter 4
Impacts of Endocrine Disrupting Chemicals on Reproduction in Wildlife

Emmelianna Kumar and William V. Holt

Abstract The European Environment Agency (The Weybridge+15 (1996–2011) report. EEA Technical report, vol 2. Copenhagen, 2012) and the United Nations Environment programme together with the World Health Organisation (State of the science of endocrine disrupting chemicals-2012. Geneva, Switzerland) both recently published major and highly authoritative reviews of endocrine disrupting chemicals in the natural environment and their effects on reproduction and health in both humans and wildlife. One surprising conclusion to emerge from these reviews was that there are relatively few well documented reports of endocrine disruption (ED) in wild mammals, mainly because much of the available evidence is correlative and does not conclusively demonstrate that the chemicals in question cause the physiological and phenotypic problems attributed to them. However, based on strong evidence from studies of wild birds, reptiles, invertebrates, and laboratory animals, it is difficult to imagine that wild mammals would be the exception. This chapter is therefore included to emphasize the point that the role of reproductive science within wildlife conservation is much broader than a narrow focus on artificial breeding technologies.

Keywords PCB • Flame retardants • Marine mammals • Birds • Fish • Invertebrates

---

E. Kumar, B.Sc., M.Sc. (✉)
Institute for the Environment, Brunel University, Kingston Lane, Uxbridge, Middlesex UB8 3PH, UK
e-mail: emmelianna.kumar@brunel.ac.uk

W.V. Holt, Ph.D.
Academic Department of Reproductive and Developmental Medicine, University of Sheffield, Jessop Wing, Tree Root Walk, Sheffield S10 2SF, UK

W.V. Holt et al. (eds.), *Reproductive Sciences in Animal Conservation*,
Advances in Experimental Medicine and Biology 753, DOI 10.1007/978-1-4939-0820-2_4,

1 Introduction

The last decade has seen an increase in evidence for relationships between exposure to certain man-made chemicals and endocrine disruption in wildlife UNEP/WHO (2012); of particular concern are flame retardants, organochlorine pesticides and polychlorinated biphenyls (PCBs), which are a class of compounds now banned due to their harmful effects but historically used in various industrial applications. A range of toxic persistent organic pollutants (POPs), such as PCBs, polychlorinated dibenzop-dioxins (PCDDs), polychlorinated dibenzofurans (PCDFs), dichlorodiphenyldichloroethylene (p,p′-DDE) and polybrominated diphenyl ethers (PBDEs), continue to contaminate the environment due to past and present human activities. Many of the endocrine disrupting effects reported in wildlife can be reproduced in laboratory studies where animals are experimentally exposed to EDCs, adding support to the hypothesis that exposure to these chemicals is associated with adverse health effects. Here we briefly describe some of the recent comparative research findings, which are of interest to reproductive biologists in the context of a changing environment. These chemicals represent another form of environmental stress, adding to the stresses that wildlife already experience because of human population growth, habitat degradation and climate change.

In addition to the chemicals mentioned above, attention is being increasingly paid to the effects on reproduction of the ever growing number of antimicrobial compounds used in both human and veterinary medicine. Between 1992 and 1999, over 700 antibacterial products, the majority of which contained triclosan (TCS) entered the consumer market. A major review of TCS, which described its structure, occurrence, degradation products and both toxicity and endocrine disrupting effects (Dann and Hontela 2011) illustrated the biological background of the compound and provided figures to show that, in Sweden alone, about 2 t are used in personal care products and toothpaste and that global production of TCS has now exceeded 1,500 t per year. A considerable amount of this material eventually finds its way into sewage, where it exerts endocrine disrupting effects on algae and other aquatic invertebrates. Experimentally it also exerts endocrine disrupting effects, both estrogenic and androgenic, on fish (Pinto et al. 2013), although fortunately at doses that are usually higher than those typically found in the environment. Although TCS is noted for its toxicity, endocrine disrupting effects mediated via its resemblance to thyroid hormones have recently been described in a Pacific tree frog, *Pseudacris regilla*, (Marlatt et al. 2013) and endocrine disrupting effects in mice have been attributed to its estrogenic action (Crawford and de Catanzaro 2012). These examples highlight the growing realisation that pharmaceutical compounds, as well as industrial chemicals, are becoming more prevalent in the environment. Fortunately regulatory authorities are becoming more aware of the attendant problems and need for vigilance, and a new field of "ecopharmacovigilance" has recently emerged (Holm et al. 2013). Authorities concerned with reproductive aspects of human clinical medicine have also recently voiced their concern. For example, the American College of Obstetricians and Gynaecologists and the American Society for

Reproductive Medicine have issued a joint Committee Opinion (ACOG/ASRM 2013) advocating for government policy changes to identify and reduce exposure to toxic environmental agents. It included the following statement: "*Every pregnant woman in America is exposed to many different chemicals in the environment; prenatal exposure to certain chemicals is linked to miscarriages, stillbirths, and birth defects. Many chemicals that pregnant women absorb or ingest from the environment can cross the placenta to the fetus. Exposure to mercury during pregnancy, for instance, is known to harm cognitive development in children. Toxic chemicals in the environment harm our ability to reproduce, negatively affect pregnancies, and are associated with numerous other long-term health problems.*"

Understanding of the effects of EDCs on animal development and reproduction can only be achieved by increasing our knowledge of basic reproductive biology. Determining how compounds exert their influence across the whole spectrum of life on the planet requires comparative knowledge of developmental and reproductive mechanisms in everything from marine invertebrates around coastal waters to large terrestrial mammals. Indeed, as these species are intimately linked to each other within complex food webs, there is a myriad of ways in which these man-made chemicals and their products can influence reproductive processes.

2 Overview of Comparative Effects

Research into the effects of environmental chemical exposure in both laboratory and wild species has been ongoing for decades, but was brought into sharp focus by observations in the 1960s that organochlorine pesticides, mainly dichlorodiphenyl-trichloroethane (DDT), caused eggshell thinning and reduced reproductive success in birds (for review, see Hellou et al. 2013). For example, organochlorines endangered the populations of a top predator, the peregrine falcon (*Falco peregrinus*) in the northern hemisphere and led to its extinction in the most heavily impacted areas of North America and Europe between the 1950s and 1970s. Similar observations (Hernandez et al. 2008) have been made in other top predators, including the Spanish Imperial Eagle, an iconic but threatened Iberian species. The reasons behind these effects, especially pronounced in top predators through bioaccumulation from dietary exposure, included disruption to calcium metabolism, neurotoxicity and behavioural changes. This type of research ultimately prompted a ban on the use of DDT in North America and Europe, leading to a reduction in body burdens in birds, an improvement in eggshell thickness, and the subsequent recovery of many of the affected populations. Recent studies have shown encouraging evidence of long-term declines in organochlorine concentrations within the eggs of these birds and correlative increases in shell thickness (Vorkamp et al. 2009; Falk et al. 2006).

Nevertheless, high concentrations of pesticides have occasionally been released by accident into the environment, with disastrous consequences. When industrial pesticides (DDT and its metabolites DDD and DDE, plus the acaricide dicofol; *p*, *p*′-dichlorodiphenyl-2,2,2,-trichloroethanol) were accidentally spilled into a tributary of Lake Apopka in Florida, USA in 1980, dramatic declines were observed in juvenile recruitment in the resident American alligator (*Alligator mississippiensis*) population. Abnormal ovarian morphology, large numbers of polyovular follicles and polynuclear oocytes were reported in female alligators. Reduced phallus size and altered plasma testosterone concentrations also were seen in males. Plasma estradiol concentrations were almost double those in female alligators from a reference lake (Guillette and Moore 2006; Guillette et al. 1996). Contaminant exposure was regarded as the most likely explanation for the abnormalities observed. Although unfortunate for the American alligators affected by the chemical spillage, the event stimulated considerable interest in the response of reptiles to endocrine disruption because of the plasticity and variety of their sex determination mechanisms (Sarre et al. 2004, 2011). The turtle, *Trachemys scripta elegans*, has subsequently been proposed as a laboratory model for assaying the estrogenic effects of exogenous chemicals (Gale et al. 2002) because the species depends on both environmental temperature and the hormonal environment for sex determination. Although this is an interesting idea, regulatory authorities may dispute its effectiveness because endocrine disruption might be confounded by the effects of temperature.

It is, however, important to recognise the conflict that occurs when chemicals are banned and the original purposes of these chemicals are still required. For example, DDT has had a major impact on the control of insect borne diseases, such as malaria, across the world, but unless suitable alternatives are available, the diseases will continue to ravage human populations. In fact, the apparent absence of suitable alternatives has resulted in the reintroduction of DDT in South Africa for use in controlling mosquito populations near Johannesburg. This step, which involves enforced spraying within the houses of the local human populations had had devastating reproductive effects on the humans themselves, i.e. reduced sperm quality and external urogenital birth defects, and the local wildlife (Bornman and Bouwman 2012; de Jager et al. 2002), with the reappearance of disorders related to abnormal sexual differentiation.

In contrast to DDT and PCBs, some other EDCs are increasing, or at least not showing signs of decrease, in the environment (e.g. perfluorinated alkyl compounds and replacements for banned brominated flame retardants). Perfluoroalkyl acids (PFAAs), a sub-class of fluorochemicals with fully fluorinated carbon chains, have gained increasing attention as an emerging category of pollutants (Jensen and Leffers 2008; Joensen et al. 2009). As such, they have become the target for risk evaluation and reduced production. These synthetic fluorinated organic chemicals have extensive industrial applications, including as surfactants and emulsifiers often in the production of other fluorinated chemicals, as well as grease and stain-repellents, friction reducers (wiring, computers), water-proofing and insulating agents, and in fire extinguishing foam (Benskin et al. 2012). Recent studies (White et al. 2011) have shown that these compounds are found globally in human tissues, including human milk and human cord blood from individuals in North America, China, and various European countries. White et al. (2011) also pointed out the probable interactions between the presence of chemicals such as PFAAs during embryonic and postnatal development in mammals, and the influence of the maternal intrauterine environment on the health of adults later in life. Galatius et al. (2011) studied temporal trends in harbour porpoises from the Danish North Sea collected between 1980 and 2005, and found no evidence of a population decline. This is encouraging, but it is nevertheless likely that more sensitive studies might find evidence of correlations between adult fitness in porpoises and early life exposure to endocrine disrupters.

There is growing evidence for long-term interactions between the foetal environment and adult health, i.e. that the original sources of adult disease can often be attributed to what happened "*in utero*" or even earlier during gamete development (Thornburg et al. 2010). The relationships between the intrauterine environment and the embryo during the early life of mammals and the onset of adult diseases, such as cardiovascular disease, hypertension and diabetes were initially identified from epidemiological observations on human populations, but have since been confirmed in experimental mammals. These observations, often collectively known as the Barker hypothesis (Barker 1995), repeatedly show that if embryos undergo different forms of "stress" during early development, children are likely to be underweight at birth and will then show phenotypic symptoms of disease as they develop into adults. Most attention has been paid to dietary stress, where the embryo initially seems to adapt its metabolic functions to make the most of the limited resources available. If conditions improve later in life, this individual cannot cope with the better lifestyle and tends to become obese and a develop a suite of late onset diseases (Barker et al. 2010; Barker 1995). Logically the presence of EDCs within the foetal environment, embryo, newborn or juveniles and the female reproductive tract could exert additional stress or have important influences on embryonic growth and development in wild species. In support of this hypothesis we can point to recent evidence from a Swedish human population showing that prenatal exposure to PCBs was associated with higher birth weight, and PBDE exposure with lowered birth weights (Lignell et al. 2013). Although these effects are complex and difficult to clarify, such human population studies suggest that wild species, whose body burden of such chemicals is often higher, will also be affected; extensive and detailed data on contaminant concentrations in arctic wild species, including mammals, birds and fishes, reported by Letcher et al. (2010) lend support to this idea. In fact, these authors commented that evidence of defective neurological development in some polar bears might be attributed to such long-term effects, but they could not be sure because of the difficulties involved in obtaining relevant data.

This information becomes more pertinent when it is considered that chemical analyses of several large mammals (seals, porpoises, whales and polar bears) have demonstrated high body burdens of hydrophobic contaminants, such as PCBs and brominated flame retardants (for review, see Sonne 2010). Typically, these species acquire pollutants via their diet and bioaccumulate EDCs, which tend to be lipophilic compounds, within their body fat so that the concentrations increase together with increasing age. When females begin to suckle their offspring, their milk is enriched with EDCs, and the EDCs are transferred to their newborns, with probable effects on survival and reproductive development (Hall et al. 2009). Surprisingly, however, it has proven difficult to demonstrate that PCBs and other lipophilic compounds actually cause impaired reproductive development (Letcher et al. 2010), although population and pharmacokinetic modelling studies of East Greenland polar bears, backed up by experimental studies of Greenland sledge dogs, predict the imminent occurrence of negative population impacts (Sonne 2010). Evidence for reduced female reproductive efficacy has also been found in studies of the European harbour porpoise and short-beaked common dolphin (Munson et al. 1998). In harbour porpoises, high persistent organic pollutant burdens tended to be associated with few ovarian scars, suggesting that high contaminant levels may be inhibiting ovulation; however, the significance of ovarian scars has recently been re-evaluated (Dabin et al. 2008). Because there was no evidence of an age-related increase, the authors cast some doubt on the usefulness of this parameter. Several species of pinnipeds have experienced recent population declines; Alaskan populations of northern fur seal, the Galápagos sea lion, *Zalophus wollebaeki* (Alava et al. 2009) and the Steller sea lion (*Eumetopias jubatus*) (Trites and Donnelly 2003) have all suffered from reduced pupping rates. In the fur seal these reproductive failures are suspected to be associated with bioaccumulation of environmental contaminants in maternal body tissues (Fillman et al. 2007).

Many studies have been published on the potential impacts of contaminants on thyroid function in various large marine mammals. Schnitzler et al. (2008) studied thyroid histology in relation to trace metals (Cd, Fe, Zn, Cu, Se, and Hg) and showed that there were largely negative relationships between concentrations of cadmium, selenium and copper and thyroid fibrosis. They concluded that there was insufficient evidence from their study to support the hypothesis that these elements have adverse effects on thyroid function. Nevertheless, the European Environment Agency report drew attention to several reports that individual contaminants (including PCBs, dieldrin and chlordane) negatively affect thyroid function in seals (Routti et al. 2008), sea lions (Debier et al. 2005), beluga whales in the St Lawrence estuary (Deguise et al. 1995) and polar bears (Braathen et al. 2004). The relationships between thyroid function and reproduction are complex; they interact with other components of the endocrine system, are involved in growth and bone formation, and contaminants with ED activity will undoubtedly exert a broad range of physiological effects.

Wild amphibians are known to be sensitive to water-borne endocrine disruption because of their highly permeable integument and the possibility of exposure during critical periods of development (embryonic and larval). Owing to the importance of thyroid function during amphibian metamorphosis there is a growing body of work relating to thyroid disrupting chemicals, including Organization for Economic Co-operation and Development (OECD) methodology for detecting EDCs (Pickford 2010). Like similarly affected fish species (see section below), amphibians showing gonadal intersex, feminisation of secondary sexual characteristics and altered sex hormone concentrations have been observed at sites contaminated by agricultural pesticides across Italy, South Africa, parts of Florida, Ontario and Michigan (Carr and Patino 2011; Norris 2011; Papoulias et al. 2013). The exact causes are still regarded as uncertain, but extensive research efforts have been invested in the use of amphibian species as laboratory models for studying and detecting endocrine disruption in these species (Miyata and Ose 2012; Olmstead et al. 2012).

2.1 Endocrine Disruption in Fish

Searching Web of Science (in August 2013) with the terms "fish+endocrine+disruption" resulted in the retrieval of 1,147 scientific papers, and refining the search using the additional term "reproduction" found 301 papers. The oldest papers related to reproduction dated back to 1995, thus demonstrating how much this particular field has advanced in less than two decades. Within this short review it is impossible to cite all of the available work in this taxa, so we will summarise by focusing on major reviews; 70 in total, including recent ones by McNair et al. (2012), Soffker and Tyler (2012) and Waye and Trudeau (2011). The European Environment Agency report (2012) also provides a useful and succinct summary of this area.

Endocrine disruption in fish is clearly widespread; the best studied example is that of feminised male roach (*Rutilus rutilus*), a cyprinid (carp) fish in United Kingdom rivers (Rodgers-Gray et al. 2001), and a second example, the gudgeon (*Gobio gobio*) was later also identified (van Aerle et al. 2001). These males exhibited abnormal reproductive characteristics associated with exposure to effluents from sewage treatment works, including elevated plasma vitellogenin (a female-specific egg laying protein) concentrations, and many had eggs developing in their testes (intersex) or feminised sperm ducts. Following a nationwide survey, these effects were attributed to natural and synthetic estrogens in the sewage effluent (Jobling et al. 2006). Nonylphenol was also identified as an important EDC in some locations, and eventually an effect map that related the incidence of sexually disrupted fish to estrogenic activities in more than 2,000 sewage effluent outlet locations was constructed (Williams et al. 2009). The map showed that 39 % of the modelled river reaches in the UK were predicted not to be at risk from ED, and most of the remainder were predicted to be at risk (with 1–3 % were predicted to be at high risk).

Feminised fish have also been found in other European countries, such as Denmark, France, Italy, Germany and the Netherlands. Again, these were associated with sewage effluent outlets. Studies in North America support the European findings to a certain extent. A study of 16 species in nine river basins found that only 3 % of the fish, from four species examined at 111 sites, exhibited intersex (Hinck et al. 2009).

Although feminisation has now been widely reported in aquatic systems, the converse effect of female masculinisation has also been described both experimentally and in field studies. A study of Eastern mosquitofish (*Gambusia holbrooki*) in the St John river, Florida (Bortone and Cody 1999) found that females showed significant elongation of the anal fin and the gonopodium, an anal fin that is modified into an intromittent organ in males of the *poecilidae*, such as this. A similar study in China that focused on another mosquitofish (*Gambusia affinis*) (Hou et al. 2011) found evidence of masculinising effects on the anal fin but also detected increased testis mass. Interestingly, these study sites both received effluents from a local paper mill. Paper mill effluents have since been implicated as causing female masculinization at other sites (Deaton and Cureton 2011), such as changes in female mating behaviour, altered offspring sex ratios, diminished body size in masculinized females and lower fecundity (for review, see Soffker and Tyler 2012). A laboratory study in which both of these species were exposed to the androgen agonist 17-β trenbolone (TB) (Brockmeier et al. 2013) detected masculinising effects after exposure to 0.1, 1 or 10 μg TB/L, including a series of gene activation effects in the tip of the anal fin. Although Kovacs et al. (2013) concluded that chemical mixtures in paper mill effluents were too complex to understand in terms of physiological effects, most evidence strongly suggests that they do contain steroid analogues that directly interfere with sexual differentiation.

Endocrine disruption has also been reported in marine fish: Kirby et al. (2004) detected intersex and elevated vitellogenin concentrations in flounder (*Platichthys flesus*) in many coastal regions, especially estuaries receiving effluents from industrial and domestic sources. Similar effects have been noted in various marine species, for example bigeye tuna (*Thunnus obesus*) around the coast of Japan (Hashimoto et al. 2003), killifish (*Fundulus heteroclitus*) (Bugel et al. 2010) in Newark bay, New Jersey in the USA and marine top predators in the Mediterranean sea (Fossi et al. 2007).

Understanding precisely which causative agents are responsible for feminisation in fish presents something of a puzzle. 17β-Ethinyl estradiol (EE2), which is a component of the human contraceptive pill, has been studied extensively and is known to possess powerful estrogenic action. Although efficient water purification systems are used widely to extract it before water is recycled back into rivers, detailed studies have shown that even some advanced water purification systems do not eliminate all oestrogenic activities (Baynes et al. 2012). Many other chemicals possess endocrine disrupting actions. Apart from the pollutants mentioned above, these include nonylphenol, octylphenol ethoxylate surfactants, bisphenol A, phthalates, phyto-oestrogens and endogenous estrogens excreted from women. Moreover, work by Jobling et al. (2009) has suggested that estrogenic compounds (such as steroid hormones) in association with anti-androgenic activity (measured by in vitro techniques, and as yet of undetermined source) are statistically correlated to intersex induction in the UK. This illustrates the complexity of the problem and highlights the difficulties involved in attempting mitigation strategies.

3 Endocrine Disruption in Invertebrates

Aquatic invertebrates are key parts of food webs that underpin the life of all other species and are highly abundant in the world's oceans. For example, the density of a polychaete (*Nereis diversicolor*) living around coastal and estuarine habitats has been recorded as 3,700 m^3 (Scaps 2002). Their importance in marine ecology has been outlined by Lawrence and Soame (2009), who pointed out that as suspension feeders in fjords, populations filter the whole water mass up to three times per day and reduce the phytoplankton biomass by 50 % in less than 5 h. This species and others are therefore in a prime position for exposure to pollutants with varying effects. Their evolutionary histories are, however, very different and because they

Table 4.1 Endocrine disruption in invertebrates

	Main effects and comments
Annelida	These species produce and respond to estrogens. Numerous EDCs activate or antagonise the estrogen receptors (ER) and modulate vitellogenin production (Keay and Thornton 2009; Matozzo et al. 2008)
Mollusca	Molluscs appear to have estrogen-like receptors, but they apparently are not activated by vertebrate estrogen, estradiol, or by other known vertebrate EDCs. Nevertheless, mud snails responded to 12.5 and 25 % sewage by increased embryo production, while higher sewage concentration (50 %) reduced it (Jobling et al. 2004). Similar studies on sewage exposure have detected increased vitellogenin-like proteins in males, feminised sex ratios and low gonadosomic indices. Varying degrees of intersex also reported in over 20 % of individual bivalves (*Scrobicularia plana*) sampled from 17 out of 23 British estuaries (Gomes et al. 2009; Chesman and Langston 2006)
	Bisphenol A (BPA) was reported to act as an estrogen receptor agonist in ramshorn snails because effects were antagonized by co-exposure to ER-antagonists (Oehlmann et al. 2006). These results could not be replicated by Forbes et al. (2008) and remain controversial
	Potent androgen receptor agonists and aromatase inhibitors, as well as marine anti-fouling paint component tributyl tin (TBT), induce "imposex" in female gastropod molluscs at concentrations as low as parts per billion (Horiguchi 2006). This is where the penis "imposes" on the normal female anatomy, blocking the oviduct and inducing sterility (Pascoal et al. 2013)
Crustacea	Control of development requires neuropeptides, ecdysone and methyl farnesoate, but little is known about the identities of chemicals in the environment that may disrupt the signalling processes at relevant concentrations (European Environment Agency 2012)
Cnidarians	The phylum *Cnidaria* contains four extant classes, the *Hydrozoa* (e.g., hydras), *Scyphozoa* ("true" jellyfishes), *Cubozoa* (box jellies) and *Anthozoa* (e.g., corals and anemones). Endocrine disruption has not been documented in cnidarians (Tarrant 2007; Armoza-Zvuloni et al. 2012). Hormonal signalling pathways are poorly characterized and few appropriate endpoints have been established
Terrestrial invertebrates	Little attention has been paid to EDC effects in terrestrial invertebrates, although several isolated studies have been published. A cell line from *Drosophila melanogaster* was developed for use in a rapid screening assay for ecdysteroid receptor agonists and antagonists (Dinan et al. 2001a, b). The only pharmaceutical showing detectable EDC activity activity was 17alpha-ethynylestradiol. Many compounds were inactive over a wide concentration range or cytotoxic at high concentrations. However, antagonistic activity was associated with several classes of compounds: cucurbitacins and withanolides, phenylalkanoids and certain alkaloids described for the first time

have developed different endocrinological signalling systems, their responses to pollutants are diverse and not necessarily predictable. Endocrine disruption is therefore a distinct possibility in some species, while others seem more tolerant to known vertebrate EDCs. The whole field is so diverse that it is impossible to provide detailed information about all classes of invertebrates; however, some of the main effects that have been described are summarised in Table 4.1.

4 What Should Be Done?

The title of this subsection is unashamedly copied from a review by the late Dr Stuart Rhind (Rhind 2009) who presented it by invitation at a symposium of the Zoological Society of London in 2009 organized by one of the present authors (WVH). Apart from providing an excellent overview of the entire field, Dr Rhind memorably discussed an experiment in which sheep grazed on land that had been fertilized twice yearly using sewage sludge were compared with sheep that were grazed on untreated grass. Analyses showed that soil levels of contaminants such as phthalate and alkyl phenol PCB and PBDE were initially very low and were increased only minimally by the sewage treatment (Rhind et al. 2002). Nevertheless, when the reproductive performance of the experimentally exposed sheep was investigated, it was found that the testes of their fetuses contained fewer Leydig and Sertoli cells than the controls, coupled with lower blood concentrations of the hormones testosterone and inhibin (Paul et al. 2005). There were also fewer oocytes in the fetal ovaries (Fowler et al. 2008) and an altered balance of pro- and anti-apoptotic proteins towards apoptosis. This remarkable outcome can be regarded as a "real world" effect that probably applies not only to grazing domestic sheep but also many other terrestrial species, especially those whose habitats are likely to have suffered any form of airborne or waterborne pollution. Subtleties such as the reduction of oocyte and Sertoli cell production (which would both result in lowered gamete production) by mammalian fetuses are likely to be undetectable in wild and threatened species, because, by definition, these species are not intensely studied. Nevertheless, the outcome of such effects will ultimately be reflected in lowered fertility, an undesirable outcome under the circumstances.

As discussed elsewhere in this book, however, the way in which different species are affected cannot necessarily be predicted, given the huge diversity of reproductive mechanisms that have evolved to cope with different, and often very adverse, conditions. Improving our understanding of comparative reproductive mechanisms is therefore as essential in this, as it is in related fields. The outcomes of many field observations, especially those involving complex mixtures of chemicals, underline the crudeness of our understanding of mixtures, and the way in which they affect reproductive mechanisms. This is understandable because experimental laboratory scientists typically prefer to make sure they understand the variables in their treatments. Although regulatory initiatives such as that introduced in 2006 by the European Union, namely Registration, Evaluation, Authorization and Restriction of chemical substances (REACH), will provide basic toxicity data on the all chemicals produced in Europe or imported into Europe in amounts that exceed 100 t per annum, the enormous number of chemicals that REACH is expected to evaluate (143,000 were pre-registered with REACH in 2008) will preclude all but the most limited of testing regimes. In fact, under the REACH protocols all substances are only tested once. This is a massive undertaking and it is interesting to see that the policy itself has been criticised because of the extensive need for animal testing (Hartung and Rovida 2009); these authors suggested that 54 million vertebrate animals would be used under REACH and that the costs would be around €9.5 billion.

One conclusion to be drawn is that there is a pressing need for the further development of reliable tests that can be used in vitro to assess the toxicity of chemicals, thereby avoiding animal use. Some authors such as Schrattenholz et al. (2012) have considered that multifactorial systems biology may be useful for this purpose because of the possibility of integrating data across transcriptomics, proteomics, epigenomics and metabolomics. Others such as Lee et al. (2012) have proposed the use of whole embryo culture and mouse embryonic stem cells as alternative models for the study of developmental toxicology. Focusing on species of most ecological relevance has led some authors, such as Scholz et al. (2013), to concentrate on fish and amphibian cells for toxicity testing, while others have applied the same principle to the evaluation of chemicals that would be particularly relevant in terms of marine species such as corals (Shafir et al. 2003: Howe et al. 2012).

The studies cited in this short chapter underline and emphasise the vast amount of work that has been carried out over the past few decades, and it is apparent that although international regulatory authorities are now taking note of the need to prevent some of the worst chemicals from reaching the environment, the problems are global, multifactorial and difficult.

Acknowledgements The authors are grateful to Dr Alice Baynes (Brunel University, Uxbridge, UK) for her constructive comments during preparation of this article.

References

ACOG, ASRM. Exposure to toxic environmental agents. Fertil Steril. 2013. doi: 10.1016/j.fertnstert.2013.08.043.

Alava JJ, Ikonomou MG, Ross PS, Costa D, Salazar S, Aurioles-Gamboa D, Gobas F. Polychlorinated biphenyls and polybrominated diphenyl ethers in Galapagos sea lions (*Zalophus wollebaeki*). Environ Toxicol Chem. 2009;28(11):2271–82. doi:10.1897/08-331.s1.

Armoza-Zvuloni R, Kramarsky-Winter E, Rosenfeld H, Shore LS, Segal R, Sharon D, Loya Y. Reproductive characteristics and steroid levels in the scleractinian coral *Oculina patagonica* inhabiting contaminated sites along the Israeli Mediterranean coast. Mar Pollut Bull. 2012;64(8):1556–63. doi:10.1016/j.marpolbul.2012.05.020.

Barker DJ, Gelow J, Thornburg K, Osmond C, Kajantie E, Eriksson JG. The early origins of chronic heart failure: impaired placental growth and initiation of insulin resistance in childhood. Eur J Heart Fail. 2010;12(8):819–25. hfq069 [pii] 10.1093/eurjhf/hfq069.

Barker DJP. Fetal origins of coronary heart disease. BMJ. 1995;311(6998):171–4.

Baynes A, Green C, Nicol E, Beresford N, Kanda R, Henshaw A, Churchley J, Jobling S. Additional treatment of wastewater reduces endocrine disruption in wild fish-a comparative study of tertiary and advanced treatments. Environ Sci Technol. 2012;46(10):5565–73. doi:10.1021/es204590d.

Benskin JP, Li B, Ikonomou MG, Grace JR, Li LY. Per- and polyfluoroalkyl substances in landfill leachate: patterns, time trends, and sources. Environ Sci Technol. 2012;46(21):11532–40. doi:10.1021/es302471n.

Bornman MS, Bouwman H. Environmental pollutants and diseases of sexual development in humans and wildlife in South Africa: harbingers of impact on overall health? Reprod Domest Anim. 2012;47:327–32. doi:10.1111/j.1439-0531.2012.02094.x.

Bortone SA, Cody RP. Morphological masculinization in poeciliid females from a paper mill effluent receiving tributary of the St. Johns River, Florida, USA. B Environ Contam Tox. 1999; 63(2):150–6.

Braathen M, Derocher AE, Wiig O, Sormo EG, Lie E, Skaare JU, Jenssen BM. Relationships between PCBs and thyroid hormones and retinol in female and male polar bears. Environ Health Perspect. 2004;112(8):826–33. doi:10.1289/ehp.6809.

Brockmeier EK, Ogino Y, Iguchi T, Barber DS, Denslow ND. Effects of 17 beta-trenbolone on Eastern and Western mosquitofish (*Gambusia holbrooki* and *G affinis*) anal fin growth and gene expression patterns. Aquat Toxicol. 2013;128:163–70. doi:10.1016/j.aquatox.2012.12.007.

Bugel SM, White LA, Cooper KR. Impaired reproductive health of killifish (*Fundulus heteroclitus*) inhabiting Newark Bay, NJ, a chronically contaminated estuary. Aquat Toxicol. 2010;96(3):182–93. doi:10.1016/j.aquatox.2009.10.016.

Carr JA, Patino R. The hypothalamus-pituitary-thyroid axis in teleosts and amphibians: endocrine disruption and its consequences to natural populations. Gen Comp Endocrinol. 2011;170(2):299–312. doi:10.1016/j.ygcen.2010.06.001.

Chesman BS, Langston WJ. Intersex in the clam *Scrobicularia plana*: a sign of endocrine disruption in estuaries? Biol Lett. 2006;2(3):420–2. doi:10.1098/rsbl.2006.0482.

Crawford BR, de Catanzaro D. Disruption of blastocyst implantation by triclosan in mice: impacts of repeated and acute doses and combination with bisphenol-A. Reprod Toxicol. 2012;34(4):607–13. doi:10.1016/j.reprotox.2012.09.008.

Dabin W, Cossais F, Pierce G, Ridoux V. Do ovarian scars persist with age in all Cetaceans: new insight from the short-beaked common dolphin (*Delphinus delphis* Linnaeus, 1758). Mar Biol. 2008;156(2):127–39. doi:10.1007/s00227-008-1070-4.

Dann AB, Hontela A. Triclosan: environmental exposure, toxicity and mechanisms of action. J Appl Toxicol. 2011;31(4):285–311. doi:10.1002/jat.1660.

de Jager C, Bornman MS, Bailey JL. Male reproductive health and DDT: sufficient evidence to discontinue its use? Proceeding of 9th international symposium on spermatology. Capetown, South Africa. Monduzzi Editore, Bologna. 2002;191–6.

Deaton R, Cureton JC. Female masculinization and reproductive life history in the western mosquitofish (*Gambusia affinis*). Environ Biol Fish. 2011;92(4):551–8. doi:10.1007/s10641-011-9878-z.

Debier C, Ylitalo GM, Weise M, Gulland F, Costa DP, Le Boeuf BJ, de Tillesse T, Larondelle Y. PCBs and DDT in the serum of juvenile California sea lions: associations with vitamins A and E and thyroid hormones. Environ Pollut. 2005;134(2):323–32. doi:10.1016/j.envpol.2004.07.012.

Deguise S, Martineau D, Beland P, Fournier M. Possible mechanisms of action of environmental contaminants on St Lawrence Beluga whales (*Delphinapterus leucas*). Environ Health Perspect. 1995;103:73–7.

Dinan L, Bourne P, Whiting P, Dhadialla TS, Hutchinson TH. Screening of environmental contaminants for ecdysteroid agonist and antagonist activity using the *Drosophila melanogaster* B(II) cell *in vitro* assay. Environ Toxicol Chem. 2001a;20(9):2038–46.

Dinan L, Bourne PC, Meng Y, Sarker SD, Tolentino RB, Whiting P. Assessment of natural products in the *Drosophila melanogaster* B(II) cell bioassay for ecdysteroid agonist and antagonist activities. Cell Mol Life Sci. 2001b;58(2):321–42.

European Environment Agency. The impact of endocrine disrupters on wildlife, people and their environments. The Weybridge +15 (1996–2011) report. EEA technical report, vol 2. Copenhagen; 2012.

Falk K, Moller S, Mattox WG. A long-term increase in eggshell thickness of Greenlandic Peregrine Falcons *Falco peregrinus tundrius*. Sci Total Environ. 2006;355(1–3):127–34. doi:10.1016/j.scitotenv.2005.02.024.

Fillmann G, Hermanns L, Fileman TW, Readman JW. Accumulation patterns of organochlorines in juveniles of *Arctocephalus australis* found stranded along the coast of Southern Brazil. Environ Pollut. 2007;146(1):262–7. doi:10.1016/j.envpol.2006.02.034.

Forbes VE, Warbritton R, Aufderheide J, Van Der Hoeven N, Caspers N. Effects of bisphenol a on fecundity, egg hatchability, and juvenile growth of *Marisa cornuarietis*. Environ Toxicol Chem. 2008;27(11):2332–40. doi:10.1897/07-622.1.

Fossi MC, Casini S, Marsill L. Potential toxicological hazard due to endocrine-disrupting chemicals on Mediterranean top predators: state of art, gender differences and methodological tools. Environ Res. 2007;104(1):174–82. doi:10.1016/j.envres.2006.06.014.

Fowler PA, Dora NJ, McFerran H, Amezaga MR, Miller DW, Lea RG, Cash P, McNeilly AS, Evans NP, Cotinot C, Sharpe RM, Rhind SM. *In utero* exposure to low doses of environmental pollutants disrupts fetal ovarian development in sheep. Mol Hum Reprod. 2008;14(5):269–80. doi:10.1093/molehr/gan020.

Galatius A, Dietz R, Riget FF, Sonne C, Kinze CC, Lockyer C, Bossi R. Temporal and life history related trends of perfluorochemicals in harbor porpoises from the Danish North Sea. Mar Poll Bull. 2011;62(7):1476–83. doi:10.1016/j.marpolbul.2011.04.011.

Gale RW, Bergeron JM, Willingham EJ, Crews D. Turtle sex determination assay: mass balance and responses to 2,3,7,8-tetrachlorodibenzo-p-dioxin and 3,3',4,4',5-pentachlorobiphenyl. Environ Toxicol Chem. 2002;21(11):2477–82.

Gomes T, Gonzalez-Rey M, Bebianno MJ. Incidence of intersex in male clams *Scrobicularia plana* in the Guadiana Estuary (Portugal). Ecotoxicology. 2009;18(8):1104–9. doi:10.1007/s10646-009-0359-5.

Guillette Jr LJ, Moore BC. Environmental contaminants, fertility, and multioocytic follicles: a lesson from wildlife? Semin Reprod Med. 2006;24(3):134–41. doi:10.1055/s-2006-944419.

Guillette Jr LJ, Pickford DB, Crain DA, Rooney AA, Percival HF. Reduction in penis size and plasma testosterone concentrations in juvenile alligators living in a contaminated environment. Gen Comp Endocrinol. 1996;101(1):32–42.

Hall AJ, Thomas GO, McConnell BJ. Exposure to persistent organic pollutants and first-year survival probability in gray seal pups. Environ Sci Technol. 2009;43(16):6364–9. doi:10.1021/es9004398.

Hartung T, Rovida C. Chemical regulators have overreached. Nature. 2009;460(7259):1080–1. doi:10.1038/4601080a.

Hashimoto S, Kurihara R, Strussmann CA, Yamasaki T, Soyano K, Hara A, Shiraishi H, Morita M. Gonadal histology and serum vitellogenin levels of bigeye tuna *Thunnus obesus* from the Northern Pacific Ocean–absence of endocrine disruption bio-indicators. Mar Poll Bull. 2003;46(4):459–65. doi:10.1016/S0025-326X(02)00460-5.

Hellou J, Lebeuf M, Rudi M. Review on DDT and metabolites in birds and mammals of aquatic ecosystems. Environ Rev. 2013;21(1):53–69. doi:10.1139/er-2012-0054.

Hernandez M, Gonzalez LM, Oria J, Sanchez R, Arroyo B. Influence of contamination by organochlorine pesticides and polychlorinated biphenyls on the breeding of the Spanish Imperial Eagle (*Aquila adalberti*). Environ Toxicol Chem. 2008;27(2):433–41. doi:10.1897/07-308r.1.

Hinck JE, Blazer VS, Schmitt CJ, Papoulias DM, Tillitt DE. Widespread occurrence of intersex in black basses (*Micropterus spp.*) from U.S. rivers, 1995–2004. Aquat Toxicol. 2009;95(1):60–70. doi:10.1016/j.aquatox.2009.08.001.

Holm G, Snape JR, Murray-Smith R, Talbot J, Taylor D, Sorme P. Implementing ecopharmacovigilance in practice: challenges and potential opportunities. Drug Saf. 2013;36(7):533–46. doi:10.1007/s40264-013-0049-3.

Horiguchi T. Masculinization of female gastropod mollusks induced by organotin compounds, focusing on mechanism of actions of tributyltin and triphenyltin for development of imposex. Environ Sci. 2006;13(2):77–87.

Hou LP, Xie YP, Ying GG, Fang ZQ. Developmental and reproductive characteristics of western mosquitofish (*Gambusia affinis*) exposed to paper mill effluent in the Dengcun River, Sihui, South China. Aquat Toxicol. 2011;103(3–4):140–9. doi:10.1016/j.aquatox.2011.02.018.

Howe PL, Reichelt-Brushett AJ, Clark MW. *Aiptasia pulchella*: a tropical cnidarian representative for laboratory ecotoxicological research. Environ Toxicol Chem. 2012;31(11):2653–62. doi:10.1002/etc.1993.

Jensen AA, Leffers H. Emerging endocrine disrupters: perfluoroalkylated substances. Int J Androl. 2008;31(2):161–9. doi:10.1111/j.1365-2605.2008.00870.x.

Jobling S, Casey D, Rogers-Gray T, Oehlmann J, Schulte-Oehlmann U, Pawlowski S, Baunbeck T, Turner AP, Tyler CR. Comparative responses of molluscs and fish to environmental estrogens and an estrogenic effluent. Aquat Toxicol. 2004;66(2):207–22.

Jobling S, Williams R, Johnson A, Taylor A, Gross-Sorokin M, Nolan M, Tyler CR, van Aerle R, Santos E, Brighty G. Predicted exposures to steroid estrogens in U.K. rivers correlate with widespread sexual disruption in wild fish populations. Environ Health Perspect. 2006;114 Suppl 1:32–9.

Jobling S, Burn RW, Thorpe K, Williams R, Tyler C. Statistical modeling suggests that antiandrogens in effluents from wastewater treatment works contribute to widespread sexual disruption in fish living in english rivers. Environ Health Perspect. 2009;117(5):797–802. doi:10.1289/ehp.0800197.

Joensen UN, Bossi R, Leffers H, Jensen AA, Skakkebaek NE, Jorgensen N. Do perfluoroalkyl compounds impair human semen quality? Environ Health Perspect. 2009;117(6):923–7. doi:10.1289/ehp.0800517.

Keay J, Thornton JW. Hormone-activated estrogen receptors in annelid invertebrates: implications for evolution and endocrine disruption. Endocrinology. 2009;150(4):1731–8. doi:10.1210/en.2008-1338.

Kirby MF, Allen YT, Dyer RA, Feist SW, Katsiadaki I, Matthiessen P, Scott AP, Smith A, Stentiford GD, Thain JE, Thomas KV, Tolhurst L, Waldock MJ. Surveys of plasma vitellogenin and intersex in male flounder (*Platichthys flesus*) as measures of endocrine disruption by estrogenic contamination in United Kingdom estuaries: temporal trends, 1996 to 2001. Environ Toxicol Chem. 2004;23(3):748–58. doi:10.1897/03-166.

Kovacs TG, Martel PH, O'Connor BI, Hewitt LM, Parrott JL, McMaster ME, MacLatchy DL, Van der Kraak GJ, Van den Heuvel MR. A survey of Canadian mechanical pulp and paper mill effluents: insights concerning the potential to affect fish reproduction. J Environ Sci Health A Tox Hazard Subst Environ Eng. 2013;48(10):1178–89. doi:10.1080/10934529.2013.776440.

Lawrence AJ, Soame JM. The endocrine control of reproduction in Nereidae: a new multihormonal model with implications for their functional role in a changing environment. Phil Trans Roy Soc Lond B Biol Sci. 2009;364(1534):3363–76. doi:10.1098/rstb.2009.0127.

Lee HY, Inselman AL, Kanungo J, Hansen DK. Alternative models in developmental toxicology. Syst Biol Reprod Med. 2012;58(1):10–22. doi:10.3109/19396368.2011.648302.

Letcher RJ, Bustnes JO, Dietz R, Jenssen BM, Jorgensen EH, Sonne C, Verreault J, Vijayan MM, Gabrielsen GW. Exposure and effects assessment of persistent organohalogen contaminants in arctic wildlife and fish. Sci Total Environ. 2010;408(15):2995–3043. doi:10.1016/j.scitotenv.2009.10.038.

Lignell S, Aune M, Darnerud PO, Hanberg A, Larsson SC, Glynn A. Prenatal exposure to polychlorinated biphenyls (PCBs) and polybrominated diphenyl ethers (PBDEs) may influence birth weight among infants in a Swedish cohort with background exposure: a cross-sectional study. Environ Health. 2013;12. doi:44 10.1186/1476-069x-12-44.

Marlatt VL, Veldhoen N, Lo BP, Bakker D, Rehaume V, Vallee K, Haberl M, Shang DY, van Aggelen GC, Skirrow RC, Elphick JR, Helbing CC. Triclosan exposure alters postembryonic development in a Pacific tree frog (*Pseudacris regilla*) Amphibian Metamorphosis Assay (TREEMA). Aquat Toxicol. 2013;126:85–94. doi:10.1016/j.aquatox.2012.10.010.

Matozzo V, Gagne F, Marin MG, Ricciardi F, Blaise C. Vitellogenin as a biomarker of exposure to estrogenic compounds in aquatic invertebrates: a review. Environ Int. 2008;34(4):531–45. doi:10.1016/j.envint.2007.09.008.

McNair A, Lim JN, Nakagawa S. The fitness consequences of environmental sex reversal in fish: a quantitative review. Biol Rev. 2012;87(4):900–11. doi:10.1111/j.1469-185X.2012.00230.x.

Miyata K, Ose K. Thyroid hormone-disrupting effects and the amphibian metamorphosis assay. J Toxicol Pathol. 2012;25(1):1–9. doi:10.1293/tox.25.1.

Munson L, Calzada N, Kennedy S, Sorensen TB. Luteinized ovarian cysts in Mediterranean striped dolphins. J Wildl Dis. 1998;34(3):656–60.

Norris DO. Endocrine disruption of reproduction in amphibians. Hormones and reproduction of vertebrates, Vol 2: Amphibians. In: Norris DO, Lopez KH, editors. London: Academic Press. 2011;2:203–211.

Oehlmann J, Schulte-Oehlmann U, Bachmann J, Oetken M, Lutz I, Kloas W, Ternes TA. Bisphenol A induces superfeminization in the ramshorn snail *Marisa cornuarietis*(Gastropoda: Prosobranchia) at environmentally relevant concentrations. Environ Health Perspect. 2006;114 Suppl 1:127–33.

Olmstead AW, Kosian PA, Johnson RN, Blackshear PE, Haselman J, Blanksma C, Korte JJ, Holcombe GW, Burgess E, Lindberg-Livingston A, Bennett BA, Woodis KK, Degitz SJ.

Trenbolone causes mortality and altered sexual differentiation in *Xenopus tropicalis* during larval development. Environ Toxicol Chem. 2012;31(10):2391–8. doi:10.1002/etc.1965.

Papoulias DM, Schwarz MS, Mena L. Gonadal abnormalities in frogs (Lithobates spp.) collected from managed wetlands in an agricultural region of Nebraska, USA. Environ Pollut. 2013;172:1–8. doi:10.1016/j.envpol.2012.07.042.

Pascoal S, Carvalho G, Vasieva O, Hughes R, Cossins A, Fang Y, Ashelford K, Olohan L, Barroso C, Mendo S, Creer S. Transcriptomics and in vivo tests reveal novel mechanisms underlying endocrine disruption in an ecological sentinel, *Nucella lapillus*. Mol Ecol. 2013;22(6):1589–608. doi:10.1111/mec.12137.

Paul C, Rhind SM, Kyle CE, Scott H, McKinnell C, Sharpe RM. Cellular and hormonal disruption of fetal testis development in sheep reared on pasture treated with sewage sludge. Environ Health Perspect. 2005;113(11):1580–7.

Pickford DB. Screening chemicals for thyroid-disrupting activity: a critical comparison of mammalian and amphibian models. Crit Rev Toxicol. 2010;40(10):845–92. doi:10.3109/10408444.2010.494250.

Pinto PIS, Guerreiro EM, Power DM. Triclosan interferes with the thyroid axis in the zebrafish (*Danio rerio*). Toxicol Res. 2013;2(1):60–9. doi:10.1039/c2tx20005h.

Rhind SM. Anthropogenic pollutants: a threat to ecosystem sustainability? Phil Trans Roy Soc Lond B Biol Sci. 2009;364(1534):3391–401. doi:10.1098/rstb.2009.0122.

Rhind SM, Smith A, Kyle CE, Telfer G, Martin G, Duff E, Mayes RW. Phthalate and alkyl phenol concentrations in soil following applications of inorganic fertiliser or sewage sludge to pasture and potential rates of ingestion by grazing ruminants. J Environ Monit. 2002;4(1):142–8.

Rodgers-Gray TP, Jobling S, Kelly C, Morris S, Brighty G, Waldock MJ, Sumpter JP, Tyler CR. Exposure of juvenile roach (*Rutilus rutilus*) to treated sewage effluent induces dose-dependent and persistent disruption in gonadal duct development. Environ Sci Technol. 2001;35(3):462–70.

Routti H, Nyman M, Jenssen BM, Backman C, Koistinen J, Gabrielsen GW. Bone-related effects of contaminants in seals may be associated with vitamin D and thyroid hormones. Environ Toxicol Chem. 2008;27(4):873–80. doi:10.1897/07-139.1.

Sarre SD, Ezaz T, Georges A. Transitions between sex-determining systems in reptiles and amphibians. Annu Rev Genomics Hum Genet. 2011;12:391–406. doi:10.1146/annurev-genom-082410-101518.

Sarre SD, Georges A, Quinn A. The ends of a continuum: genetic and temperature-dependent sex determination in reptiles. BioEssays. 2004;26(6):639–45. doi:10.1002/bies.20050.

Scaps P. A review of the biology, ecology and potential use of the common ragworm *Hediste diversicolor* (O.F. Muller) (Annelida: Polychaeta). Hydrobiologia. 2002;470(1–3):203–18. doi:10.1023/a:1015681605656.

Schnitzler JG, Siebert U, Jepson PD, Beineke A, Jauniaux T, Bouquegneau JM, Das K. Harbor porpoise thyroids: histologic investigations and potential interactions with environmental factors. J Wildl Dis. 2008;44(4):888–901.

Scholz S, Renner P, Belanger SE, Busquet F, Davi R, Demeneix BA, Denny JS, Leonard M, McMaster ME, Villeneuve DL, Embry MR. Alternatives to in vivo tests to detect endocrine disrupting chemicals (EDCs) in fish and amphibians–screening for estrogen, androgen and thyroid hormone disruption. Crit Rev Toxicol. 2013;43(1):45–72. doi:10.3109/10408444.2012.737762.

Schrattenholz A, Soskic V, Schopf R, Poznanovic S, Klemm-Manns M, Groebe K. Protein biomarkers for in vitro testing of toxicology. Mutat Res. 2012;746(2):113–23. doi:10.1016/j.mrgentox.2012.02.008.

Shafir S, Van Rijn J, Rinkevich B. The use of coral nubbins in coral reef ecotoxicology testing. Biomol Eng. 2003;20(4–6):401–6.

Soffker M, Tyler CR. Endocrine disrupting chemicals and sexual behaviors in fish – a critical review on effects and possible consequences. Crit Rev Toxicol. 2012;42(8):653–68. doi:10.3109/10408444.2012.692114.

Sonne C. Health effects from long-range transported contaminants in Arctic top predators: an integrated review based on studies of polar bears and relevant model species. Environ Int. 2010;36(5):461–91. doi:10.1016/j.envint.2010.03.002.

Tarrant AM. Hormonal signaling in cnidarians: do we understand the pathways well enough to know whether they are being disrupted? Ecotoxicology. 2007;16(1):5–13. doi:10.1007/s10646-006-0121-1.

Thornburg KL, Shannon J, Thuillier P, Turker MS. In utero life and epigenetic predisposition for disease. Adv Genet. 2010;71:57–78. doi:10.1016/B978-0-12-380864-6.00003-1.

Trites AW, Donnelly CP. The decline of Steller sea lions *Eumetopias jubatus* in Alaska: a review of the nutritional stress hypothesis. Mammal Rev. 2003;33(1):3–28. doi:10.1046/j.1365-2907.2003.00009.x.

UNEP/WHO. State of the science of endocrine disrupting chemicals — 2012. Geneva, Switzerland: United Nations Environment Programme/World Health Organization;2012.

van Aerle R, Nolan TM, Jobling S, Christiansen LB, Sumpter JP, Tyler CR. Sexual disruption in a second species of wild cyprinid fish (the gudgeon, *Gobio gobio*) in United Kingdom freshwaters. Environ Toxicol Chem. 2001;20(12):2841–7.

Vorkamp K, Thomsen M, Moller S, Falk K, Sorensen PB. Persistent organochlorine compounds in peregrine falcon (*Falco peregrinus*) eggs from South Greenland: levels and temporal changes between 1986 and 2003. Environ Int. 2009;35(2):336–41. doi:10.1016/j.envint.2008.08.008.

Waye A, Trudeau VL. Neuroendocrine disruption: more than hormones are upset. Toxicol Environ Health B Crit Rev. 2011;14(5–7):270–91. doi:10.1080/10937404.2011.578273.

White SS, Fenton SE, Hines EP. Endocrine disrupting properties of perfluorooctanoic acid. J Steroid Biochem Mol Biol. 2011;127(1–2):16–26. doi:10.1016/j.jsbmb.2011.03.011.

Williams RJ, Keller VD, Johnson AC, Young AR, Holmes MG, Wells C, Gross-Sorokin M, Benstead R. A national risk assessment for intersex in fish arising from steroid estrogens. Environ Toxicol Chem. 2009;28(1):220–30. doi:10.1897/08-047.1.

Chapter 5
The Role of Genomics in Conservation and Reproductive Sciences

Warren E. Johnson and Klaus Koepfli

Abstract Genomics, the study of an organism's genome through DNA analyses, is a central part of the biological sciences and is rapidly changing approaches to animal conservation. The genomes of thousands of organisms, including vertebrates, invertebrates, and plants have been sequenced and the results annotated, augmented and refined through the application of new approaches in transcriptomics, proteomics, and metabolomics that enhance the characterization of messenger RNA, proteins, and metabolites. The same computational advances that are catalyzing "-omic" technologies and novel approaches to address fundamental research questions are facilitating bioinformatic analysis and enabling access of primary and derivative data and results in public and private databases (Zhao and Grant. Curr Pharm Biotechnol 12:293–305, 2011). These tools will be used to provide fundamental advances in our understanding of reproductive biology across vertebrate species and promise to revolutionize our approach to conservation biology.

The vulnerability of animal populations and their genetic diversity is well documented, as are the myriad of causes and threats to their persistence, including habitat degradation and loss, overexploitation, pollution, invasive alien species, and climate change. Of the 64,283 vertebrates assessed by the International Union for Conservation of Nature in their 2012 Red List of Threatened Species, 7,250 or ~11 % are threatened with extinction, a percentage that has been increasing steadily for at least the last decade (www.iucnredlist.org). Among many of these species, important genetic diversity has been lost, thereby increasing their vulnerability as

W.E. Johnson, Ph.D. (✉)
Smithsonian Conservation Biology Institute, National Zoological Park,
1500 Remount Rd., Front Royal, VA 22630, USA
e-mail: johnsonwe@si.edu

K. Koepfli, Ph.D.
Center for Conservation and Evolutionary Genetics, Smithsonian Conservation Biology Institute, National Zoological Park, Washington, DC, USA

W.V. Holt et al. (eds.), *Reproductive Sciences in Animal Conservation*,
Advances in Experimental Medicine and Biology 753, DOI 10.1007/978-1-4939-0820-2_5,

genetically diverse populations have higher fitness, generally are more resilient to environmental challenges, and have more adaptive potential (Reed and Frankham Conserv Biol 17:230–237, 2003; Luikart et al. Nat Rev Genet 4:981–994, 2003). In turn, genetic variation within and among populations may be essential to maintaining functional ecosystems, evolutionary process and will impact future food supplies, human health, biomaterial development and geopolitics (Myers and Knoll Proc Natl Acad Sci U S A 98:5389–5392, 2001; Templeton et al. Proc Natl Acad Sci U S A 98:5426–5432, 2001). Therefore, conservation of genetic diversity is a social, cultural, scientific, and economic prerogative and is the key to adaptation in the uncertain future of a human-dominated environment. Once lost, genetic resources are nearly impossible to regain, increasing the urgency of fundamental global approaches (e.g. www.cbd.int/sp/targets).

In this chapter we provide a review of current research and recent advances in biotechnology and genomic approaches for animal conservation and the management of genetic resources, with an emphasis on reproductive sciences. It is intended to provide information and insights for research and to provoke thoughts on how to take advantage of these opportunities.

Keywords Conservation genetics • Adaptation • Selection • Genomics

1 Commonly Used Modern Molecular Genetic Approaches

The essence of conservation genetics is understanding patterns of genetic variation, generally through the use of molecular markers, or heritable DNA sequences that are located at specific locations within a genome (DeSalle and Amato 2004). Several types of molecular markers have been commonly used for assessing and understanding genetic variation (reviewed in Duran et al. 2009; Teneva 2009). These are often categorized as being either neutral or non-neutral (under selection), generally depending upon whether they are located in the coding or non-coding regions of the genome. Most genetic markers are fixed in terms of their allelic composition at fertilization and can thus be assessed at any stage in an organism's life. However, no single molecular genetic marker is suitable for all studies and their utility depends on technical and practical considerations including cost, time constraints, availability of known markers, number and type of samples to be screened, available labor, and repeatability. Most importantly, marker choice depends on the research question and the genetic characteristics and evolutionary history of the individuals, populations, and species being studied and the resulting data should be interpreted within this context.

One of the most widely used markers to estimate genetic diversity in animals are microsatellites or short tandem repeats (STRs). These generally are defined as stretches of DNA of 2–6 bp of DNA that are repeated in tandem from about 5 to 40 times. STRs have a much higher rate of mutation than most neutral markers, which is reflected in the number of repeats and thus the length of the PCR fragment (Luikart and England 1999; Selkoe and Toonen 2006). STRs are ubiquitous throughout animal genomes, have co-dominant inheritance (from both parents), are relatively easy to interpret and analyze, can be utilized in a wide range of genetic diversity applications (see below), and produce repeatable results that can be compared across populations and labs.

Sex-linked markers are another important subclass. These are transmitted through maternal (e.g. mitochondrial DNA or mtDNA) and paternal lineages (e.g. markers located on the Y chromosome in mammals and the Z chromosome in birds). Analyses of these markers provide more nuanced understanding of evolutionary and population history and differences between patterns of males and females.

Research focus is increasingly shifting to non-neutral markers because these sometimes retain evidence of positive and negative selection that can then be associated with traits or functions of interest. This change in emphasis has been catalyzed by rapidly decreasing costs of next generation sequencing (NGS) and accompanying advances in targeted sequencing of specific genomic regions and in bioinformatic analytical approaches. Together, these tools are facilitating the direct analysis of sequence variation across whole genomes, including single-nucleotide polymorphisms (SNPs).

SNPs have many attractive characteristics as genetic markers. For example, they are common throughout the genome and generally are well annotated, often with a known location in the genomic landscape. They are also easy to genotype with low error rates with either SNP panels or direct sequencing and easy to replicate among samples and labs (Brumfield et al. 2003; Morin et al. 2004). Analytically, SNP variation generally follows simple mutation models that permit analyses of neutral and non-neutral selection across the entire genome and with very large sample sizes. SNPs are sufficiently variable to provide useful population and evolutionary markers, and on average 3–8 biallelic SNPs are as informative as one microsatellite marker (Rosenberg et al. 2003; Schopen et al. 2008). In model or agricultural species, for which large numbers of SNPs have been described, it is clear that SNP variation is associated with numerous evolutionary and biological processes, which facilitates their simultaneous use in a wide-variety of analyses and applications. SNP panels and genotyping systems allowing the simultaneous analysis of tens-of-thousands of SNPs are commercially available for many model or agriculturally important species (e.g. humans, mouse, cattle, domestic dog, chicken). These panels generally are designed from SNPs obtained from whole genome sequencing data, and thus provide genome-wide measures of genetic variation. In addition, many SNPs will be related to traits of interest in genome-wide association studies (GWAS). However, as the costs of whole-genome sequencing decrease and systems of data analyses become easier, indepth sequence analysis will become the norm, and is already the most common approach used for small genomes such as those of many viruses or bacteria (Maclean et al. 2009; Didelot et al. 2012). As such, sequence-based methodologies and SNP analyses are rapidly complementing and/or replacing other marker systems in most species (Vignal et al. 2002; Schlotterer 2004; Soller et al. 2006). This trend will continue to accelerate as whole genome assemblies of species of conservation interest (or their close relatives) are completed (see e.g. Table 5.1).

Table 5.1 List of vertebrate species classified as vulnerable, endangered, or critically endangered according to the IUCN Red List that have been the subject of whole genome sequencing or population genomic studies

Common name	Species	Reference
Atlantic cod	*Gadus morhua*	Star et al. (2011)
Chinese alligator	*Alligator sinensis*	Star et al. (2011)
Galapagos tortoise	*Chelonoidis nigra*	Loire et al. (2013)
Saker falcon	*Falco cherrug*	Zhan et al. (2013)
Puerto Rican parrot	*Amazonia vittata*	Oleksyk et al. (2012)
Tasmanian devil	*Sarcophilus harrisi*	Miller et al. (2011), Murchison et al. (2012)
Chimpanzee	*Pan troglodytes*	Mikkelsen et al. (2005)
Bonobo	*Pan paniscus*	Prüfer et al. (2012)
Gorilla	*Gorilla gorilla*	Scally et al. (2012)
Sumatran orang-utan	*Pongo abelii*	Locke et al. (2011)
Bornean orang-utan	*Pongo pygmaeus*	Locke et al. (2011)
Giant panda	*Ailuropoda melanoleuc*	Li et al. (2009)
Polar bear	*Ursus maritimus*	Miller et al. (2012), Cahill et al. (2013)
Tiger	*Panthera tigris*	Cho et al. (2013)
Tibetan antelope	*Pantholops hodgsonii*	Ge et al. (2013)
Yangtze River dolphin	*Lipotes vexillifer*	Zhou et al. (2013)

2 Common Applications of Molecular Markers in Conservation Genetics

The first and arguably most-important step in genetic management is the clear defining of objectives, including what to manage and how. This multi-faceted process ideally includes a thorough understanding of the ecology, evolutionary and population history of the population or species and a clear consensus of the management goals. The determination of "what to manage" can be based on criteria such as morphological and behavioral traits, geographic distribution, molecular genetic variables, uniqueness, ecology, economic concerns, cultural importance, population structure and size, and probability of extinction. Here we will focus on direct measures of genetic variation and briefly summarize how genetic markers and next-generation technologies can provide insights that help inform and assist in the conservation management of genetic resources.

Conservation genetics has historically focused on observed differences (or levels of genetic divergence) among individuals and populations, subspecies, or breeds of domestic animals as might be quantified by (1) estimating mean observed and expected heterozygosity averaged over the typed loci, (2) the average number of alleles, or (3) allelic richness (Luikart and Cornuet 1998). When groups of individuals do not have a large number of fixed differences (for example populations that are recently isolated or relatively inbred), molecular differentiation is estimated by differences in allele frequencies among populations as specific differences will be rare and most common alleles will be shared across groups (e.g. MacHugh et al. 1998; Balloux and Lugon-Moulin 2002; Laval et al. 2002). However, sometimes demographic events will be as or more important than time, as genetic drift and shifts in allele frequencies will become more significant through incidences of inbreeding, genetic bottlenecks, and/or increasing amounts of admixture (Lande 1988).

Fixed genetic differences are used as the basis to classify and identify species, especially those that are difficult to distinguish from only morphological features (e.g. in cryptic species). For example, DNA barcoding, which uses sequence variation from standardized regions of the mitochondrial DNA, has become a widely used system for cataloging animal biodiversity and has been instrumental in the discovery of several new species. Barcoding is being organized globally through several international initiatives, including the International Barcode of Life project and the Consortium for the Barcode of Life (CBOL) and DNA barcode databases such as the Barcode of Life Data Systems and the International Nucleotide Sequence Database Collaboration.

When subspecies, populations, or breeds are difficult to define based on geographic location, morphological, ecological or evolutionary criteria (Waples and Gaggiotti 2006), reliance on molecular genetic criteria becomes even more crucial. It is also often advantageous to identify groups without prior information of their genetic structure (e.g. without preassigned population or subspecific assignment), and to identify individuals with genetic heritage from more than one of these groups. Multi-locus clustering analyses, such as employed in the program STRUCTURE (Pritchard et al. 2000) use multi-locus genotypes and specific ancestry models to estimate the fraction the genome of each individual that belongs to each cluster. They can also be used to assign 'unknown' individuals to populations (Manel et al. 2003; Paetkau et al. 2004; Stella et al. 2008; Toro et al. 2009) and are especially useful when natural barriers to gene flow are not obvious, such as with marine species (Primmer 2009). For example, management of commercial fisheries such as Atlantic salmon (Griffiths et al. 2010) and lake sturgeon (Bott et al. 2009) have utilized genetic analyses to facilitate the identification of source populations and thus help avoid overexploitation. Similar approaches have been used for monitoring the source of animal products being sold in markets (Baker 2008; Chapman et al. 2009; Kochzius et al. 2010) or distinguishing among similar looking species, either as adults or only during specific stages of development (Kon et al. 2007; Ogden 2008).

These methods are also very effective for identifying escaped animals from captivity into the wild, confirm the origin of these escapes, and establish the extent of introgression into the wild population. For example, Kidd et al. (2009) documented hybridization between domestic mink (*Neovison vison*) that had escaped from farms and wild mink using a panel of microsatellites and admixture analyses, thereby altering the evolutionary integrity of the wild populations. As another example, genetic markers were used to document genetic introgression in the Florida panther (*Puma concolor coryi*) from individuals released of Central American origin, from a captive group of pumas from a small animal exhibit, in addition to the intentional release of pumas from Texas (Johnson et al. 2010).

Once genetic groups have been defined, the classic Wright's F-statistic is commonly used to partition genetic variation into a within-subpopulation (average subpopulation inbreeding coefficient F_{IS}) and a between-subpopulation component (fixation index F_{ST}), depending upon the genetic markers used, their mutations rates, and sampling scheme (see e.g. Cockerham and Weir 1984; Holsinger and Weir 2009). Although these estimates should not be strictly compared with each other or among studies, FST values from 0.05 and 0.3 are typical among populations or breeds, with values above 0.15 often interpreted as evidence of significant differentiation (Frankham et al. 2002). For comparisons among populations with sequence data, analysis of molecular variance (AMOVA) (Excoffier et al. 1992) is often the most-appropriate method.

3 Inbreeding, Relatedness, Effective Population Size, and Gene Flow

An assessment of inbreeding is one of the primary concerns of all efforts to conserve genetic diversity. Depending upon the type of available data, inbreeding is estimated from pedigrees or from molecular data with the Wright F_{IS} inbreeding coefficient (Frankham et al. 2002). With high-density genome-wide SNP or sequence data, comparatively-long stretches of homozygosity are a sign of inbreeding (McQuillan et al. 2008; Kirin et al. 2010). Inbreeding is inversely correlated with effective population size (N_e), or the number of individuals for which random breeding in an ideal population would generate the same dispersion of allele frequencies or amount of inbreeding as that observed in the real population (Wang and Whitlock 2003; Charlesworth and Willis 2009). N_e is extensively used as a criterion for determining the risk status of populations and is invariably much less than the census population size. N_e is also correlated with the degree of relatedness among individuals and to some extent, populations, and is best estimated using a large number of genetic markers (Oliehoek et al. 2006; Toro et al. 2009). For example, Tapio et al. (2010) used this approach to estimate relatedness among non-pedigreed cryo-banked Yakutian cattle bulls, a breed of cattle native to Siberia with only ~1,200 purebred individuals remaining, and showed that these cryo-banked samples harbored unique allelic variation of potential use to enhancing the genetic diversity of the remaining purebred population.

Past population dynamics, such as population expansions and bottlenecks can also be inferred from patterns of genetic variation. Most recently, these approaches have included the analysis of individual whole-genome sequences with a pairwise sequentially Markovian coalescent model (PSMC, Li and Durbin 2011) or with the program BEAST using a Bayesian coestimation of time to most recent common ancestor, evolutionary rates, and past population dynamics (Drummond et al. 2005). This approach demonstrated, for example, that the domesticated water buffalo, yak, gayal, and bovine recently experienced a rapid population increase that was not observed in the wild African buffalo (Finlay et al. 2007).

Following the identification of populations, conservation units, species or other subdivisions, managers are often most interested in estimated levels of gene flow. Gene flow may be advantageous, for example in efforts to increase N_e and reduce the effects of inbreeding, but can be problematic if it leads to hybridization and undesired admixture or introgression. Gene flow or introgression can be detected by discordant results between autosomal and sex-linked markers, or from clustering analyses described above, as implemented in STRUCTURE (Pritchard et al. 2000).

Because introgression or hybridization can be a major threat to conservation in the form of outbreeding depression (e.g., Templeton 1986), early detection is fundamental for effective conservation strategies. For example, statistical analyses of genetic data from reintroduced Arabian oryx (*Oryx leucoryx*) demonstrated that outbreeding depression was affecting juvenile survival (Marshall and Spalton 2000). Molecular marker data have also demonstrated gene flow from wild populations to domesticated animals, for instance, from jungle fowl to domesticated populations of Vietnamese chicken (Berthouly et al. 2009). There are also an increasing number of examples of natural introgression and hybridization, for example among California tiger salamanders (Fitzpatrick et al. 2009), between American bison and domestic cattle (Halbert and Derr 2007), among Darwin's finches (Grant and Grant 2010), and between African and Asian elephants (Roca et al. 2005) and brown and polar bears (Miller et al. 2012) See Box 5.1.

Box 5.1 Genomics and Population Genomics of the Giant Panda

The giant panda (*Ailuropoda melanoleuca*) is an endangered ursid found in mountain habitats across several provinces of western China. Unique among the bear family, giant pandas are almost entirely herbivorous, feeding almost exclusively on bamboo. The population size of the species is estimated to be between 2,500 and 3,000 individuals based on molecular genetic analyses (Zhan et al. 2006) and with a low rate of fecundity combined with loss of habitat, the giant panda faces a precarious future. As a result, the conservation management of the giant panda in both the wild and in captivity has received much attention.

In 2010, the genome of the giant panda became the first mammalian genome to be sequenced and assembled *de novo* (Li et al. 2010). Genome size was estimated to be 2.4 gigabases and in comparison with the genomes of the dog and human, it was found that the giant panda had a relatively low rate of divergence. Remarkably, however, the giant panda genome was found to have a high number (2.7 million) of heterozygous SNPs, with a rate of heterozygosity almost twice that found in the human genome. This finding confirmed earlier molecular genetic studies based on many fewer markers that suggested that giant pandas still retain a relatively high level of genetic diversity and little evidence of inbreeding (Zhang et al. 2007). Given the low rate of fecundity, the genome was also used to identify many genes involved in reproduction and gonad development.

(continued)

Box 5.1 (continued)

Once a reference genome of a species has been generated, additional individuals from different populations can be sequenced at lower coverage and mapped against the reference, thereby providing a population genomic perspective of genetic diversity and historical demography. This was done for the giant panda by Zhou et al. (2013), who sequenced 34 pandas at ~4.7-fold coverage from the three main areas where this species is found in western China. Although only two subspecies of giant panda have been recognized, analyses of genetic structure resolved giant pandas into three genetic clusters, with the isolated population in the Qinling Mountains, being the most distinct and estimated to have diverged about 300,000 years ago. The other subspecies was resolved into two genetic populations that diverged more recently, about 2,700 years ago. Analysis of historical demography suggested that giant pandas underwent two rounds of population bottlenecks and expansions each. Most interestingly, however, analyses of genome-wide SNP diversity made it possible to distinguish signatures of local adaptation from neutral diversity by locating genes that showed evidence of directional selection among the three populations. Population genomic studies such as this are of great interest to conservationists because the ability to identify genes of adaptive significance within species can be quantified and used to prioritize populations of conservation importance, as the neutral and adaptive components of genomic diversity may not always be correlated (Bonin et al. 2006, 2007).

4 Adaptation and Selection

Conservation strategies that focus on genetic variation are increasingly also interested in identifying genotypes associated with advantageous traits or phenotypes and the preservation of adaptive variation (or with maladaptive deleterious traits associated with dis-advantageous phenotypes). Fundamentally, this involves distinguishing positive or negative selection from neutral variation that is the product of genetic drift (Joost et al. 2007; Novembre and Di Rienzo 2009), or distinguishing between events that affect only a specific region of the genome (selection) versus the entire genome (drift).

A traditional method to identify positive selection is to compare allele frequencies of different populations with markers near genes of interest. With the increased availability of variable genetic markers from larger numbers of unrelated individuals (e.g. 30–50) from contrasting groups and available software packages such as Bayescan (Foll and Gaggiotti 2008), Lositan (Antao et al. 2008) and Mcheza (Antao and Beaumont 2011) Fst values that differ significantly from the rest of the genome are used to identify selection (high values suggest positive or negative selection and low values suggest balancing selection (Slatkin 2008)). These approaches were used to detect candidate loci for adaptation along a gradient of altitude in the common frog (*Rana temporaria*). Other methods taking advantage of genomic sequence data and analyses of mutational patterns in extended regions of linkage disequilibrium (LD) are also going to be increasingly accessible (Prasad et al. 2008; Joost et al. 2007; Rubin et al. 2010).

5 Molecular Markers in Conservation and Reproductive Sciences

Molecular genetic techniques are steadily becoming integrated into the reproductive sciences. Traditionally, reproductive management has relied upon phenotypic variation or measurable inherited characteristics for assisted selection or breeding for desired traits, starting with the earliest efforts of animal domestication. The use of genetic markers has largely been pioneered in agricultural species and model organisms such as mice, chicken, cattle, pig, horse, and dog. However, with the increased accessibility of genomic resources for non-model organisms and through comparative research, the opportunities and feasibility for integrating and enhancing reproductive technologies with genomic approaches are increasing rapidly.

5.1 Molecular Marker-Assisted Selection

Marker-assisted selection (MAS) is one of the most promising tools for linking reproductive techniques and genetics. MAS requires a genetic marker linked to a gene or genomic region containing quantitative trait loci. These markers can also be used to identify quantitative traits and also be used for the selection of mating pairs. Marker-assisted mate selection are especially useful to increase the efficiency of genetic improvements, especially when phenotype screening is difficult (e.g. with resistance to an infectious disease), is expressed late in life (e.g. late-onset diseases), or is expressed only in one sex.

To date, MAS has mostly been implemented in large-scale agriculture systems in developed countries (e.g. Dekkers 2004), but it may become an effective technique for group or population management to monitor wild and semi-wild populations, especially in gregarious species such as birds, fish, and ungulates to estimate the genomic breeding values or for species where following pedigrees is impractical. One of the best-described early cases of MAS involved genetic disease resistance in domestic animals to transmissible spongiform encephalopathy in sheep, where polymorphisms in the PrP gene were linked with susceptibility to scrapies, and which led to breeding programs for disease resistant sheep in the European Union (Hunter et al. 1996). In fish, molecular markers have been used to study and identify individuals with resistance to infectious pancreatic necrosis and infectious Salmon anemia in Atlantic salmon (Moen et al. 2009; Jieying Li et al. 2011). In addition,

Box 5.2 The Tasmanian Devil

The Tasmanian devil (*Sarcophilus harrisii*) is the largest carnivorous marsupial in the world and an island endemic. Populations of the species are declining due to various threats such disease epidemics and loss of habitat. The greatest threat facing Tasmanian devils, however, is a highly infectious and transmissible cancer known as Devil Facial Tumor Disease (DFTD). This clonal cancer is transmitted through the natural physical contact among Tasmanian devils (biting), and individuals that contract the cancer develop infections within months and suffer a 100 % mortality rate. Without intervention to stop the spread of DFTD, Tasmanian devils may go extinct (McCallum et al. 2007).

Two independent research groups sequenced the genome of the Tasmanian devil and its DFTD cancer (Miller et al. 2011; Murchison et al. 2012). Among the major findings was that genome-wide genetic diversity is quite low in devils, but that population genetic substructure exists across the range of the species, providing useful information that can be applied in captive breeding of healthy individuals not yet exposed to DFTD. The genome of the cancer revealed a large set of unique SNPs and copy-number variants, along with chromosomal rearrangements, that together suggest a distinct mutational process shaping the DFTD genome. Moreover, examination of protein-coding genes revealed 138 amino acid variants found only in the tumor genome compared to the normal genome of the host. The studies of the Tasmanian devil and DFTD genomes, which also include assessing genetic diversity within the species, provide a particularly strong example of the multifaceted applications of genomic data (Ryder 2005).

markers have been developed in fish that trace influence of growth, spawning time, sex determination, abiotic stress tolerance, and disease resistance (Loukovitis et al. 2011). Improved methods to isolate, sequence and interpret differences among pathogens will increase the power of diagnostics, efficacy of treatments, and the ability to monitor wildlife diseases such as rabies, distemper, blue tongue disease, avian influenza, foot-and-mouth disease, and CSV and to link these to variation and outcomes in affected individuals (Hoffmann et al. 2009) See Box 5.2.

The implementation of genetic improvement using molecular tools in the conservation context has largely focused on highly managed captive populations, including the avoidance of inbreeding, selection against maladaptive traits, increasing genetic variation in highly inbred populations, maintaining "pure" individuals and populations that represent recognized subspecies and species. For example, more tigers live in captivity than in the wild, and the captive population hold genetic variation in pure and hybrid subspecies that has not been documented in the wild (Luo et al. 2008). Marker-assisted introgression might also be an efficient method of introducing desirable traits, such as disease resistance into wild populations. For example, over 200 amphibian species worldwide are declining from a fungal skin disease caused by

Batrachochytrium dendrobatidis (Berger et al. 1998; Lips et al. 2006). However, since intraspecific and interspecific response to the disease varies within (Tobler and Schmidt 2010; Kriger and Hero 2006) and among (Stuart et al 2004; Woodhams et al 2007) species, the identification of resistance genes or gene markers would provide management options for improving amphibian conservation strategies.

However, because we know only a very few of the many possible loci that are critical to individual fitness and population viability and are not likely to be able to predict what genetic variation will be important for adaptation, survival and overall fitness in the future, it is prudent to act with caution. Variation that is advantageous in one environment (e.g. a captive setting) might well be linked with adaptations that are deleterious in some wild conditions, and vice versa. Therefore, if we were to model conservation programs on the basis of only a few loci about which we some knowledge, it is quite likely to affect genetic variability in unknown ways at other important loci (Hedrick 2001; Lacy 2000).

6 Genomics and Advancing Reproductive Sciences

6.1 Genetic Management and Reproductive Technologies

Reproductive technologies are an important and increasingly relied-upon management tool of both wild and captive populations. These include assisted reproduction techniques which have been used to enhance gene flow between isolated wild populations, between captive and wild populations, between different institutions, and to ensure the genetic representation of individuals that otherwise would not breed naturally through artificial insemination, in vitro methods, etc. (Pukazhenthi and Wildt 2004; Comizzoli et al. 2009; Wildt et al. 2010). Among the growing number of examples of captive populations that are have had important roles in augmenting or establishing wild populations are Puerto Rican parrots (Brock and White 1992), California condors (Geyer et al. 1993), Micronesian kingfishers (Haig et al. 1995), whooping cranes (Jones et al. 2002) and primate species [e.g., lion-tailed macaques (Morin and Ryder 1991), bonobos (Reinartz and Boese 1977)], black-footed ferrets (Cain et al. 2011), and Iberian lynx (Vargas et al. 2008; Gañán et al. 2010).

In addition to avoiding the loss of genetic variation, the major concerns of highly managed populations is the risk of genetic drift, which can result in loss of alleles, and of adaptation to non-natural conditions. With captive animals, this includes attempting to prevent or mitigate adaptation to conditions of captivity (i.e. they must retain a certain degree of wildness) in addition to preventing the loss of overall genetic diversity. However, we have almost no understanding of specific levels of genetic variation or specific genotypes that are associated with survival in the wild, especially in changing environments. Therefore, the preservation of the most genetic variation possible or of equal representation of founder stock has been the default goal of genetic captive management plans in most cases.

Increasingly, genetic potential for future generations is being promoted through the establishment of germ-plasm banks and viably frozen cell lines and through testing of advanced assisted reproductive techniques. For example, cryobanking of germplasm is being used in almost all livestock species (Mazur et al. 2008) and increasingly in wildlife species as well (Comizzoli et al. 2009; Swanson et al. 2007). However, the implementation of these tools has been slow, in part because of the lack of fundamental knowledge of the complex reproductive biologies of these species (Andrabi and Maxwell 2007).

One of the most fertile areas for genomics and conservation is the development of tools and approaches that will facilitate the discovery of mechanisms of important life history and adaptive traits in populations and species. This is occurring most rapidly in model organisms and closely related species. For example, genetic markers and genes of complex traits associated with growth rate, milk production, and disease are being identified in many domestic animals (Fan et al. 2010). Similarly, comparative genomic techniques are being used to identify candidate genes involved in life history traits, development, and behavior in fish species such as the Atlantic salmon (Li et al. 2011; Miller et al. 2011; Sarropoulou and Fernandes 2011) and Bluefin tuna (Nakamura et al. 2013). Increased efforts are needed to develop assemblies and sequences from non-model organisms (Ekblom and Galindo 2011) specifically with the goal of elucidating the genomic underpinning of important evolutionary traits.

6.2 Genomics and Insights on Functional and Adaptive Variation of Reproductive Traits

At the cellular level, animal reproduction, and thus fitness and survival are fundamentally tied to the sperm, egg and to producing offspring, that in turn successfully propagate. Therefor, genomic techniques will probably have their most significant and fundamental influence on conservation by contributing to our understanding of reproductive biology across the wide diversity of plants and animals. These genome-level approaches will include proteomic and transcriptomic methods to enhance our understanding of reproductive physiology and the evolutionary mechanisms involved in reproductive isolation, gamete incompatibility, and associated pathologies.

One of the most powerful approaches will be to leverage the power of comparative genomics, or the study of patterns of variation across a range of individuals and/or organisms. These comparative methods allow insights into large-scale genomic re-arrangements, the conservation of functional elements and the tracking of evolutionary phylogenies through the examination of both closely and distantly related species. As an example, the characterization of marsupial genomes is providing insights on the shared and unique evolutionary history of reproductive genes in marsupials and eutherians, including the identification of highly modified reproductive genes, mammary gland-specific genes, and genes likely associated with other unique reproductive traits including long embryonic diapause (Frankenberg et al. 2011; Renfree et al. 2011; Pharo et al. 2012). Across diverse groups, especially groups like the carnivores with well-described model organisms (e.g. the domestic cat and dog; Table 5.1), comparing and contrasting reproductive patterns will be especially informative (Amstislavsky et al. 2012).

The process of comparative genomics is iterative, because once candidate genes are identified in one species they can be tested in others. For example variable markers from 14 candidate genes, some shared among diverse species from fly to human, have helped lead to the identification of genes associated with female and male fertility rates (Li et al 2012). Correlating conserved and divergent phenotypes with their corresponding genetic patterns, including differences among rapidly and slowly evolving genes and loss, the number of gene copies, and the number of intact functional genes in gene families will then provide hypotheses for formal testing. When combined or followed up with analyses of proteomic data this approach will also provide hypotheses for interactions among proteins, such as those involved in sperm-egg interactions.

Comparative genomic methods take advantage of the multiple mechanisms by which species maximize adaptive potential under diverse evolutionary scenarios. For example, among vertebrates there are a wide range of patterns of varying degrees of reproductive isolation, with some species diverging rapidly and developing strong methods of reproductive isolation (e.g. hybrid infertility) compared with other groups, such as parthenogenetic lizards, where hybridization may be a common recurrent mechanism for maintaining evolutionary potential and mitigating the effects of inbreeding (Fujita and Moritz 2009). Other areas of comparative genomic research, such as among normal and diseased tissues will also provide a synergistic approach of study that will assist in the management of inherited diseases through improved diagnosis and therapies (e.g. in horses as in Rosnaha et al. 2010). Finally, comparative genomic techniques and the application of metagenomic technologies and approaches to the study of whole "ecosystems" or biomes, such as the NIH Human Microbiome project, will also provide insights on the range of functional and abnormal systems and the role of microbiota in diverse settings such as in reproductive systems (Aagaard et al. 2012).

6.3 Sex Determination

Among vertebrates, gonadal development at the cellular level is conserved. However, the embryonic gonad is the only organ that is capable of producing two unique and complex adult organs as it can produce either the testis or the ovary through two distinct pathways. In mammals and birds, chromosomal sex determination is virtually universal, but in other groups sex is determined or strongly influenced by environmental factors, such as temperature, hormones, and a variety of chemicals (Parma and Radi 2012; Ungewitter and Yao 2013). These differences have large evolutionary repucusions. For example, in the fish species where population sex ratios are controlled by inherited, environmental, and biochemical elements, population dynamics and selection patterns can vary greatly both temporally and geographically (Piferrer et al. 2012).

In most species sex determination is closely tied with the equally sophisticated processes of sperm and egg production, whether occurring in the fetus or adult. Genomic methods are beginning to elucidate many of the steps involved in these processes, largely through a process of documenting the genes that are expressed in reproductive tissues and linking these patterns with genetic variation. These approaches have helped determine that at the molecular level, vertebrate gonad-specific genes generally evolve more rapidly, and thus are more diverged, than ovary genes. In turn, reproductive genes appear to evolve significantly faster than non-reproductive genes. However, functional orthologs of reproductive genes have thus far shown similar rates of evolutionary divergence across all vertebrate orders (Grassa and Kulathinal 2011).

6.4 Spermatogenesis, Oogenesis, and Fertilization

Spermatogenesis varies among species, but occurs in a series of complex steps involving hundreds of genes that are functionally active at specific times in specific tissues during development (Chocu et al. 2012). In mammals, sperm cells start forming during embryonic development and the pool of sperm stem cells are established shortly after birth (Govindaraju et al. 2012). Although many of these processes occur within the testis, they also include post-gonadal modifications controlled by genetic variation that influence sperm motility, interuterine interactions with the female, sperm capacitation, egg binding, and sperm penetration that in aggregate will determine levels of male fertility. Because a successful sperm also interacts with a wide variety of environments and must match a specific female genotype, individual male success also depends on maintaining a certain level of genetic and phenotypic variation while preserving many conserved functional motifs.

In aggregate, this complexity ensures that male fertility (and infertility, a common concern of conservation genetics) is multigenic, and that normal function can be altered in numerous ways. Comparative genomic techniques to elucidate differences among normal and abnormal spermatozoa and the associated metabolic and signaling pathways promise to improve our understanding of these fundamental processes and to provide biomarkers to assist managers and scientists in predicting the probability of successful fertilization. Our understanding of male reproductive biology is being empowered and is increasing at a more-rapid pace through new methods such as single cell (single sperm) sequencing and by an increased number of Y-chromosomes that are being sequenced. Traditionally, Y-chromosomes have not been completely sequenced because their highly repetitive genomic architecture can be difficult to interpret (Hughes and Rozen 2012).

In contrast with sperm, the structure and contents of the egg have been relatively conserved across vertebrates for millions of years, and these features are the main factors impacting successful zygotic growth. However, there are specific details, especially those related with sperm-egg interactions, that tend to be very species-specific and more rapidly evolving (i.e. less conserved) (Claw and Swanson 2012). For example, the rapid evolution of the egg's extracellular barriers suggests that this is an important evolutionary feature and mechanism for ensuring species-specificity and the establishment of pre-zygotic barriers (Swanson et al. 2001; Swanson and Vacquier 2002).

The intricate steps involved in the binding of sperm with the egg probably evolved a very long time ago, as is evidenced by similarities (highly conserved features) in the three-dimensional protein structure and sequence conservation in key gene families among vertebrates, invertebrates, and unicellular eukaryotes. However, quite strikingly, many of the sperm and egg proteins involved in sperm-egg interactions have patterns of rapid evolution, which offers the opportunity to use comparative genomic approaches and functional studies to better understand gene function and constraints and to gain insights on interspecific and intraspecific reproductive strategies at both the cellular and organismal level (Swanson et al. 2011).

Reproductive success may also be dependent on interactions between the immunological and reproductive systems, since sperm, the developing fetus, and the parents often must successfully distinguish between specific cell types in a very complex environment. There are two main immune systems that are probably involved. The innate immune system, which employs various cells and molecules such as phagocytes, natural killer cells, and defensins to identify pathogenic targets, probably predates the divergence of plants from animals and includes non-specific mechanisms to protect hosts from infection. In contrast, the adaptive immune system arose much later in animal evolution and targets specific pathogens through the major histocompatibility complex (MHC) and diverse immunoglobins (Igs). More recently, the complement immune system has also been identified through proteomic studies as being actively involved in the female reproductive tract. This third major element of vertebrate immunity expanded notably in vertebrates and has been linked with sperm survival and fertilization (Nonaka and Kimura 2006) and to the complex molecular dialogue between the maternal tract and the embryo as the mother must have complex interactions with the blastocyst during implantation while simultaneously continuing to fight foreign infection (Almiñana and Fazeli 2012; Dorus et al. 2012).

6.5 Epigenetics

Genomic tools are contributing to the emerging understanding of the increasingly important field of epigenetics, or the study of how gene function can be changed, activated, or inactivated without altering DNA sequence, but through chemical reactions that can be cell specific and turn parts of the genome on or off at specific times and locations. The mechanisms involved, including DNA methylation, RNA interference, and post-translational modification of histones, can have subtle or major effects on the inheritance and development of innumerable traits of interest (Hong et al. 2011). In the field of reproductive genetics, epigenetic factors are being studied and used to determine the steps involved in reprogramming cells, such as for the development of pluripotent embryonic stem cells or induced pluripotent stem (iPS) cells that can regain their capacity to become respecialize into other cell types.

7 Applied Reproductive Biotechnologies and Genomics

Management of genetic resources has been part of our human heritage since the at least the beginning of agriculture and plant and animal domestication through the selection of phenotypic markers of interest or utility, or that were associated with a desired trait. With the development of the earliest genetic markers, it became possible to link traits with specific genotypes that most often did not segregate perfectly with the desired phenotype.

Comparative genomic methods have greatly enhanced the efficiency with which gene markers for candidate genes can be identified and tested. For example, gene markers from known candidate genes and gene pathways were tested and shown to influence fertility characteristics such as litter size, time between litters, and age of first litter in several domestic pig breeds (Sironen et al. 2010). Similar approaches have been used to link markers associated with the transforming growth factor beta (TGFB) family with ovulation rates and follicular development (Juengel et al. 2011) in sheep. And in pigs, this approach helped identify genes expressed in reproductive tissues that are involved in fat regulation and which are linked with reproduction traits such as total lifetime number of offspring born (Onteru et al. 2011). Similarly, the most efficient or most readily available biomarkers may not be based directly on genetic differences, but instead be indirectly based on an expressed protein. For example, candidate genes and their associated proteins found in sperm and seminal plasma have been associated with semen quality and fertility in stallions (Novak et al. 2010). Once identified, genetic markers such as these can be used to directly select breeding regimes without further understanding of the mechanisms involved.

The role of genetic markers in monitoring and managing the health of individuals will increase as we learn more about fundamental biological processes, in large part because this will allow the design of diagnostic tools that are cheaper, more sensitive, and which provide more direct and predictive information. For example, as comparative genomic techniques provide fundamental insights on the genetic pathways involved in reproduction (e.g. Huang et al. 2010), there will be a wider range of sophisticated tools developed for increasing efficiency of assisted reproduction protocols, including methods to more rapidly and precisely identify normal and abnormal sperm and embryos, to distinguish between male and females to preferential produce one sex, and to monitor normal development of ovarian follicles and embryos in vitro (e.g. Aydiner et al. 2010, Scott and Treff 2010, Grado-Ahuir et al., 2011). These will complement and improve other modern reproductive biotechnologies and assisted reproductive techniques including artificial insemination, in vitro fertilization, embryo transfer/sexing, semen sexing, gamete/embryo micromanipulation, and somatic cell nuclear transfer (cloning) in conservation programs for endangered mammalian species (Andrabi and Maxwell 2007).

Genomics research will change our approach to many aspects of managing and monitoring the reproductive process. A more precise understanding of the genetic pathways involved in basic reproductive processes will lead to better tools and alternative approaches and targets for contraception or treating for infertility. For example, this will include methods to intervene or promote the complex process of blastocyst implantation with the endometrium, by alteration of the chemoattractants that are secreted by the egg, or by alteration of the proteins acquired on the sperm surface.

Most promising, genomic approaches are increasing our understanding of the mechanisms of establishing induced pleuropotent cells (iPC), along with improved techniques of viably freezing immortal cells lines. This will increase the available options for the management of genetic variation in current and future populations and will lead to more reliable methods of cloning and genetic engineering. Perhaps more importantly for conservation purposes, pleuropotent cells will provide many more relatively inexpensive and efficient options for the short and long term management of genetic variation across a wide range of species.

8 Population Management and Genomics

For most species, for the foreseeable future, intensive management will only be desirable and possible not at the individual level, but only at the population or even species level. However, even this scale of genetic management will mostly consist of periodic genetic monitoring and rare direct management of overall genetic variation patterns. It will simply be a long time before we fully understand gene function and the interactions of genes and gene pathways, the role of genetic variation, and the interplay with diverse environment to employ these approaches in wholesale management. Therefore, genomic tools will be used most commonly for monitoring of genetic variation to minimize divergence of populations through random drift or undesired selective processes for unintended traits, for identification of significant population-scales, and as diagnostic tools to monitor pathogen exposure (Schwartz et al. 2007; Luikart et al. 2003; Myers et al. 2001; Reed and Frankham 2003; Templeton et al. 2001). This will most often be of concern in small or isolated populations such as those in captive settings or in fragmented habitats. Perhaps most importantly, these populations will provide opportunities for studying processes of selection in relatively controlled environments. When combined with epidemiological studies, these will enable the study of the genetic basis of disease risk and the chance to improve diagnostic and predictive modeling tools. It is in scenarios such as these as well, whether in situ or ex situ, when it will be appropriate to apply the labor-intensive and costly interventions requiring advanced reproductive techniques. In is in cases such as these when cryopreservation of embryos and gametes will be the most effective at slowing evolution by utilizing germ plasm of under represented or long dead donors as parents in future generations.

9 Conclusions

Genomic applications have become an integral part of all biological sciences through increased accessibility of the techniques and lower costs. However, the introduction of third generation of sequencing tools that promise longer, more accurate reads of DNA at even lower costs will inexorably increase the potential of integrating these techniques into basic research and wildlife conservation efforts (Kohn et al. 2006; Allendorf et al. 2010; Ouborg et al. 2010; Govindaraju et al. 2012; Zhao and Grant 2011). Other techniques, such as real-time single molecule sequencing will enable rapid and precise assessment of genomic interactions, such as between the sperm and egg or between the embryo and mother. New approaches will become available, enabling for example, the study of post-transcriptional modifications that influence the links between genotype and phenotype, the interplay of pathogens with adaptive variation, and the links between behavior, social interactions, stress, the environment and reproduction.

However, because conservation in general, and conservation genomics in particular are multidisciplinary in nature, the biotechnological applications described here will only be successful if they are part of broader conservation efforts and programs (e.g. Lacy 2012; Steiner et al. 2013). These genomic tools can assist in identifying and determining what resources to conserve and through what methods. Although conservation genomics will include novel assisted-reproductive techniques and improved diagnostic tools, these will not replace more traditional methods. Emphasis will still have to be on maintaining functional ecosystems and populations with sufficient adaptive capacity to adjust to future environmental changes and demographic threats such as novel pathogens. This will still require an increased understanding of the basic biology of the species being conserved and of their unique evolutionary history and biological features.

The potential and pace of developing genomic technologies will depend on several factors, but most certainly will require training programs to efficiently make the tools readily available among specialists and non-specialist citizen scientists and the development of accessible computational approaches and complementary storage capacity and connectivity that will allow the analysis of the large amount of data that will be produced. This is especially true since our ability to produce genomic data is outpacing technological innovation to store and analyze the genetic data (Kahn 2011). But most importantly, we must continue to broaden our scope of biological enquiry by focusing more on non-traditional model organisms and systems, and by continuing to train scientists and empower citizen scientists to explore, document, and preserve the vast diversity and mysteries of life that remain to be discovered.

References

Aagaard K, Riehle E, Ma J, Segata N, Mistretta T-A, Coarfa C, Raza S, Rosenbaum S, VandenVeyver I, Milosavljevic A, Gevers D, Huttenhower C, Petrosino J, Versalovic J. A metagenomic approach for characterization of the vagial microbiome signature in pregnancy. PLoS One. 2012;7(6):e36466.

Allendorf FW, Hohenlohe PA, Luikart G. Genomics and the future of conservation genetics. Nat Rev Genet. 2010;11:697–709.

Almiñana C, Fazeli A. Exploring the application of high-throughput genomics technologies in the field of maternal-embryo communication. Theriogenology. 2012;77:717–37.

Amstislavsky S, Lindeberg H, Luvoni GC. Reproductive technologies relevant to the genome resource bank in Carnivora. Reprod Domest Anim. 2012;47:164–75.

Andrabi SMH, Maxwell WMC. A review on reproductive biotechnologies for conservation of endangered mammalian species. Anim Reprod Sci. 2007;99:223–43.

Antao T, Beaumont MA. Mcheza: a workbench to detect selection using dominant markers. Bioinformatics. 2011;27:1717–8.

Antao T, Lopes A, Lopes RJ, Beja-Pereira A, Luikart G. Lositan: a workbench to detect molecular adaptation based on a Fst-outlier method. Bioinformatics. 2008;9:323.

Aydiner F, Yetkin CE, Seli E. Perspectives on emerging biomarkers for non-invasive assessment of embryo viability in assisted reproduction. Curr Mol Med. 2010;2010:206–15.

Balloux F, Lugon-Moulin N. The estimation of population differentiation with microsatellite markers. Mol Ecol. 2002;11:155–65.

Baker CS. A truer measure of the market: the molecular ecology of fisheries and wildlife trade. Mol Ecol. 2008;17:3985–98.

Berger L, Speare R, Daszak P, Green DE, Cunningham AA, Gonnin CL, Slocombe R, Ragan MA, Hyat AD, McDonald KR, Nines HB, Lips KR, Marantelli GM, Parkes H. Chytridiomycosis causes amphibian mortality associated with population declines in the rain forests of Australia and Central America. Proc Natl Acad Sci U S A. 1998;95:9031–6.

Berthouly C, Leroy G, Van Nhu T, Hoang Thanh H, Bed'Hom B, Nguyen BT, Vu Chi C, Monicat F, Tixier-Boichard M, Verrier E, Maillard JC, Rognon X. Genetic analysis of local Vietnamese chickens provides evidence of gene flow from wild to domestic populations. BMC Genet. 2009;10:1.

Bonin A, Taberlet P, Miaud C, Pompanon F. Explorative genome scan to detect candidate loci for adaptation along a gradient of altitude in the common frog (Rana temporaria). Mol Biol Evol. 2006;23:773–83.

Bonin A, Nicole F, Pompanon F, Miaud C, Taberlet P. Population adaptive index: a new method to help measure intraspecific genetic diversity and prioritize populations for conservation. Conserv Biol. 2007;2:697–708.

Bott K, Kornely GW, Donofrio MC, Elliott RF, Scribner KT. Mixed-stock analysis of lake sturgeon in the Menominee River sport harvest and adjoining waters of Lake Michigan. N Am J Fish Manag. 2009;29:1636–43.

Brock MK, White BN. Application of DNA fingerprinting to the recovery program of the endangered Puerto Rican parrot. Proc Natl Acad Sci U S A. 1992;89:11121–5.

Brumfield RT, Beerli P, Nickerson DA, Edwards SV. The utility of single nucleotide polymorphisms in inferences of population history. Trends Ecol Evol. 2003;18:249–55.

Cahill JA, Green RE, Fulton TL, Stiller M, Jay F, et al. Genomic evidence for island population conversion resolves conflicting theories of polar bear evolution. PLoS Genet. 2013;9:e1003345.

Cho YS, Hu L, Hou H, Lee H, Xu J, et al. The tiger genome and comparative analysis with lion and snow leopard genomes. Nat Commun. 2013;4:1–7.

Cain CM, Livieri TM, Bradley J, Swanson J. Genetic evaluation of a reintroduced population of black-footed ferrets (Mustela nigripes). J Mammal. 2011;92:751–9.

Chapman DD, Pinhal D, Shivji MS. Tracking the fin trade: genetic stock identification in western Atlantic scalloped hammerhead sharks Sphyrna lewini. Endanger Species Res. 2009;9:221–8.

Charlesworth D, Willis JH. The genetics of inbreeding depression. Nat Rev Genet. 2009;10:783–96.
Chocu S, Calvel P, Rolland AD, Pineau C. Spermatogenesis in mammals: proteomic insights. Syst Biol Reprod Med. 2012;58:179–90.
Claw KG, Swanson WJ. Evolution of the egg: new findings and challenges. Annu Rev Genom Human Genet. 2012;13:109–25.
Cockerham CC and Weir BS. Covariances of Relatives Stemming from a Population Undergoing Mixed Self and Random Mating Biometrics pp. 1984;40:157–164.
Comizzoli P, Wildt DE, Pukazhenthi BS. In Vitro Compaction of Germinal Vesicle Chromatin is Beneficial to Survival of Vitrified Cat Oocytes. 2009;44 Suppl 2:47–52. doi:10.1111/j.1439-0531.2009.01372.x.
Dekkers JCM. Commercial application of marker- and gene- assisted selection in livestock: strategies and lessons. J Anim Sci. 2004;82:E313–28.
DeSalle R, Amato G. The expansion of conservation genetics. Nat Rev Genet. 2004;75:04–712.
Didelot X, Bowden R, Wilson DJ, Peto TE, Crook DW. Transforming clinical microbiology with bacterial genome sequencing. Nat Rev Genet. 2012;13(9):601–12. doi:10.1038/nrg3226. Epub 2012 Aug 7.
Dorus S, Skerget S, Karr TL. Proteomic discovery of diverse immunity molecules in mammalian spermatozoa. Syst Biol Reprod Med. 2012;58:218–28.
Drummond AJ, Rambaut A, Shapiro B, Pybus OG. Bayesian coalescent inference of past population dynamics from molecular sequences. Mol Biol Evol. 2005;22:1185–92.
Duran C, Appleby N, Edwards D, Batley J. Molecular genetic markers: discovery, applications, data storage and visualisation. Curr Bioinform. 2009;4:1–27.
Ekblom R, Galindo J. Applications of next generation sequencing in molecular ecology of non-model organisms. Heredity. 2011;107:15.
Excoffier L, Smouse PE, Quattro JM. Analysis of molecular variance inferred from metric distances among DNA haplotypes: application to human mitochondrial DNA restriction data. Genetics. 1992;131:479–91.
Fan B, Du Z-H, Gorbach DM, Rothschild MF. Development and application of high-density SNP arrays in genomic studies of domestic animals. Asian-Aust J Anim Sci. 2010;23:833–47.
Finlay EK, Gaillard C, Vahidi SM, Mirhoseini SZ, Jianlin H, Qi XB, El-Barody MA, Baird JF, Healy BC, Bradley DG. Bayesian inference of population expansions in domestic bovines. Biol Lett. 2007;3:449–52.
Fitzpatrick BM, Johnson JR, Kump DK, Shaffer HB, Smith JJ, Voss SR. Rapid fixation of non-native alleles revealed by genome-wide SNP analysis of hybrid tiger salamanders. BMC Evol Biol. 2009;9:176.
Foll M, Gaggiotti O. A genome-scan method to identify selected loci appropriate for both dominant and codominant markers: a bayesian perspective. Genetics. 2008;180:977–93.
Frankenberg S, Fenelon J, Dopheide B, Shaw G, Renfree MB. A novel MSMB-related microprotein in the postovulatory egg coats of marsupials. BMC Evol Biol. 2011;11:373.
Frankham R, Ballou JD, Briscou DA. Introduction to conservation genetics. Cambridge, UK: Cambridge University Press; 2002.
Fujita MK, Moritz C. Origin and evolution of parthenogenetic genomes in lizards: current state and future directions. Cytogenet Genome Res. 2009;127:261–72.
Gañán N, Sestelo A, Garde JJ, Martínez F, Vargas A, Sánchez I, Pérez-Aspa MJ, López-Bao JV, Palomares F, Gomendio M, Roldan ERS. Reproductive traits in captive and free-ranging males of the critically endangered Iberian lynx (Lynx pardinus). Reproduction. 2010;139:275–85.
Ge R-L, Cai Q, Shen Y-Y, San A, Ma L, et al. Draft genome sequence of the Tibetan antelope. Nat Commun. 2013;4:1858.
Geyer JC, Ryder OA, Chemnick LG, Thompson EA. Analysis of relatedness in the California Condors from DNA fingerprints. Mol Biol Evol. 1993;10:571–89.
Govindaraju A, Dogan S, Rodriguez-Osorio N, Grant K, Kaya A, Memili E. Delivering value from sperm proteomics for fertility. Cell Tissue Res. 2012;349:783–93.
Grado-Ahuir JA, Aad PY, Spicer LJ. New insights into the pathogenesis of cystic follicles in cattle: microarray analysis of gene expression in granulosa cells. J Anim Sci. 2011;89:1769–86.

Grant BR, Grant PR. Conspecific versus heterospecific gene exchange between populations of Darwin's finches. Philos Trans R Soc B Biol Sci. 2010;365:1065–76.

Grassa CJ, Kulathinal RJ. Elevated evolutionary rates among functionally diverged reproductive genes across deep vertebrate lineages. Int J Evol Biol. 2011;2011:9.

Griffiths AM, Machado-Schiaffino G, Dillane E, Coughlan J, Horreo JL, Bowkett AE, Minting P, Toms S, Roche W, Gargan P, et al. Genetic stock identification of Atlantic salmon (Salmo salar) populations in the southern part of the European range. BMC Genet. 2010;11:31.

Haig SM, Ballou JD, Casna NJ. Genetic identification of kin in Micronesian kingfishers. J Hered. 1995;86:423–1.

Halbert ND, Derr JN. A comprehensive evaluation of cattle introgression into US federal bison herds. J Hered. 2007;98:1–12.

Hedrick PW. Conservation genetics: where are we now? Trends Ecol Evol. 2001;16:629–36.

Hoffmann B, Beer M, Reid SM, Mertens P, Oura CAL, van Rijn A, Slomka MJ, Banks J, Brown IH, Aleander DJ, King DP. Disease diagnostics A review of RT-PCR technologies used in veterinary virology and disease control: sensitive and specific diagnosis of five livestock diseases notifiable to the World Organisation for Animal Health. Vet Microbiol. 2009;139:1–23.

Holsinger KE, Weir BS. Fundamental concepts in genetics Genetics in geographically structured populations: defining, estimating and interpreting FST. Nature Reviews Genetics. 2009;10: 639–50, doi:10.1038/nrg2611.

Hong CP, Park J, Roh T-Y. Epigenetic regulation in cell reprogramming revealed by genome-wide analysis. Epigenomics. 2011;3:73–81.

Huang W, Yandell BS, Khatib H. Transcriptomic profiling of bovine IVF embryos revealed candidate genes and pathways involved in early embryonic development. BMC Genomics. 2010;11:23.

Hughes JF, Rozen S. Genomics and genetics of human and primate Y chromosomes. Annu Rev Genomics. 2012;13:83–108.

Hunter N, Foster J, Goldmann W, Stear M, Hope J, Bostock C. Natural scrapie in a closed flock of Cheviot sheep occurs only in specific PrP genotypes. Arch Virol. 1996;141:809–24.

Jieying Li, Keith A, Boroevich, Koop BF, Davidson WS. Comparative Genomics Identifies Candidate Genes for Infectious Salmon Anemia (ISA) Resistance in Atlantic Salmon (Salmo salar) Mar Biotechnol. 2011;13:232–241. DOI 10.1007/s10126-010-9284-0.

Johnson WE, Onorato DP, Roelke ME, Land ED, Cunningham M, Belden C, McBride R, Jansen D, Lotz M, Shindle D, Howard J, Wildt DE, Penfold LM, Hostetler JA, Oli MK, O'Brien SJ. Genetic restoration of the Florida panther. Science. 2010;329:1641–5.

Joost S, Bonin A, Bruford M, Despres L, Conord C, Erhardt G, Taberlet P. A spatial analysis method (SAM) to detect candidate loci for selection: towards a landscape genomics approach to adaptation. Mol Ecol. 2007;16:3955–69.

Jones KL, Glenn TC, Lacy RC, Pierce JR, Unruh M, Mirande CM, Chavez-Ramirez F. Refining the whooping crane studbook by incorporating microsatellite DNA and leg-banding analyses. Conserv Biol. 2002;16:789–99.

Juengel JL, O'Connell AR, French MC, Proctor LE, Wheeler R, Farquhar PA, Dodds KG, Galloway SM, Johnstone PD, Davis GH. Identification of a line of sheep carrying a putative autosomal gene increasing ovulation rate in sheep that does not appear to interact with mutations in the transforming growth factor beta superfamily. Biol Reprod. 2011;85:113–20.

Kahn SD. On the future of genomic data. Science. 2011;331:728–9.

Kidd AG, Bowman J, Lesbarreres DS. Mink introgression hybridization between escaped domestic and wild American mink (*Neovison vison*). Mol Ecol. 2009;18:1175–86.

Kirin M, McQuillan R, Franklin CS, Campbell H, McKeigue PM, Wilson JF. Genomic runs of homozygosity record population history and consanguinity. PLoS One. 2010;5:e13996.

Kochzius M, Seidel C, Antoniou A, Botla SK, Campo D, Cariani A, Vazquez EG, Hauschild J, Hervet C, Hjörleifsdottir S, et al. Identifying fishes through DNA barcodes and microarrays. PLoS One. 2010;5:e12620.

Kohn MH, Murphy WJ, Ostrander EA, Wayne RK. Genomics and conservation genetics. Trends Ecol Evol. 2006;21:629–37.

Kon T, Yoshino T, Mukai T, Nishida M. DNA sequences identify numerous cryptic species of the vertebrate: a lesson from the gobioid fish Schindleria. Mol Phylogenet Evol. 2007;44:53–62.
Kriger KM, Hero JM. Survivorship in wild frogs infected with Chytridiomycosis. Ecohealth. 2006;3:171.
Lacy RC. Structure of the VORTEX simulation model for population viability analysis. Ecol Bull. 2000;48:191–203.
Lacy RC. Achieving true sustainability of zoo populations. Zoo Biol. 2012;32:19–26.
Lande, R. Genetics and demography in biological conservation. Science. 1988;241:1455–9.
Laval G, SanCristobal M, Chevalet C. Measuring genetic distances between breeds: use of some distances in various short term evolution models. Genet Sel Evol. 2002;34:481–507.
Loire E, Chiari Y, Bernard A, Cahais V, Romiguier J, et al. Population genomics of the endangered giant Galapagos tortoise. Genome Biol. 2013;14:R136.
Li G, Peñagaricano F, Weigel KA, Zhang Y, Rosa G, Khatib H. Comparative genomics between fly, mouse, and cattle identifies genes associated with sire conception rate. J Dairy Sci. 2012;95:6122–9.
Li H, Durbin R. Inference of human population history from individual whole-genome sequences. Nature. 2011;475:493–6.
Li J, Boroevich KA, Koop BF, Davidson WS. Comparative genomics identifies candidate genes for Infectious Salmon Anemia (ISA) resistance in Atlantic salmon (Salmo salar). Mar Biotechnol. 2011;13:232–41.
Li R, Fan W, Tian G, Zhu H, He L, et al. The sequence and de novo assembly of the giant panda genome. Nature. 2010;463:311–7.
Lips KR, Brem F, Brenes R, Reeve JD, Alford RA, Voyles J, Carey C, Livo L, Pessier AP, Collins JP. Emerging infectious disease and the loss of biodiversity in a Neotropical amphibian community. Proc Natl Acad Sci U S A. 2006;103:3165–70.
Locke DP, Hillier LW, Warren WC, Worley KC, Nazareth LV, et al. Comparative and demographic analysis of orang-utan genomes. Nature. 2011;469:529–33.
Loukovitis D, Sarropoulou E, Tsigenopoulos CS, Batargias C, Magoulas A, Apostolidis AP, Chatziplis D, Kotoulas G. Quantitative trait loci involved in sex determination and body growth in the gilthead sea bream (Sparus aurata L.) through targeted genome scan. PLoS One. 2011;6:16599.
Luikart G, Cornuet JM. Empirical evaluation of a test for identifying recently bottlenecked populations from allele frequency data. Conserv Biol. 1998;12:228–37.
Luikart G, England PR. Statistical analysis of microsatellite DNA data. Tree. 1999;14:253–6.
Luikart G, England PR, Tallmon D, Jordan S, Taberlet P. The power and promise of population genomics: from genotyping to genome typing. Nat Rev Genet. 2003;4:981–94.
Luo S, Johnson WE, Martenson J, Agostinho A, Martelli P, Uphyrkina O, Traylor-Holzer K, Smith JLD, O'Brien SJ. Subspecies genetic assignments of worldwide captive tigers increase conservation value of captive populations. Curr Biol. 2008;18:592–6.
MacHugh DE, Loftus RT, Cunningham P, Bradley DG. Genetic structure of seven European cattle breeds assessed using 20 microsatellite markers. Anim Genet. 1998;29:333–40.
MacLean D, Jones JD, Studholme DJ. Application of 'next-generation' sequencing technologies to microbial genetics. Nat Rev Microbiol7, 2009;287–96.
Manel S, Schwartz MK, Luikart G, Taberlet P. Landscape genetics: combining landscape ecology and population genetics. Trends Ecol Evol. 2003;18:189–97.
Marshall TC, Spalton JA. Simultaneous inbreeding and outbreeding depression in reintroduced Arabian oryx. Anim Conserv. 2000;3:241–8.
Mazur P, Leibo SP, Seidel Jr GE. Cryopreservation of the germplasm of animals used in biological and medical research: importance, impact, status, and future directions. Biol Reprod. 2008; 78:2–12.
McCallum H, Tompkins DM, Jones M, Lachish S, Marvanek S, et al. Distribution and impacts of Tasmanian devil facial tumor disease. Ecohealth. 2007;4:318–25.
McQuillan R, Leutenegger AL, Abdel-Rahman R, et al. Runs of homozygosity in European populations. Am J Hum Genet. 2008;83:359–72.

Mikkelsen TS, Hillier LW, Eichler EE, The Chimpanzee Sequencing and Analysis Consortium. Initial sequence of the chimpanzee genome and comparison with the human genome. Nature. 2005;437:69–87.
Miller KM, Li S, Kaukinen KH, Ginther N, Hammill E, Curtis JMR, Patterson DA, Sierocinski T, Donnison L, Pavlidis P, et al. Genomic signatures predict migration and spawning failure in wild Canadian salmon. Science. 2011a;331:214–7.
Miller W, Hayes VM, Ratan A, Petersen DC, Wittekindt NE, et al. Genetic diversity and population structure of the endangered marsupial Sarcophilus harrisii (Tasmanian devil). Proc Natl Acad Sci U S A. 2011b;108:12348–53.
Miller W, Schuster SC, Welch AJ, Ratan A, Bedoya-Reina OC, et al. Polar and brown bear genomes reveal ancient admixture and demographic footprints of past climate change. Proc Natl Acad Sci U S A. 2012a;109:14295–6.
Miller W, Schuster SC, Welch AJ, Ratan A, Bedoya-Reina OC, et al. Polar and brown bear genomes reveal ancient admixture and demographic footprints of past climate change. Proc Natl Acad Sci U S A. 2012b;109:E2382–90.
Murchison EP, Schulz-Trieglaff OB, Ning Z, Alexandrov LB, Bauer MJ, et al. Genome sequencing and analysis of the Tasmanian devil and its transmissible cancer. Cell. 2012;148:780–91.
Moen T, Baranski A, Sonesson AK, Kjoeglum S. Confirmation and fine mapping of a major QTL for resistance to infectious pancreatic necrosis in Atlantic salmon (Salmo salar): population level association between markers and trait. BMC Genomics. 2009;10:368.
Morin PA, Ryder OA. Founder contribution and pedigree inference in a captive breeding colony of lion-tailed macaques using mitochondrial-DNA and DNA fingerprint analyses. Zoo Biol. 1991;10:341–52.
Morin PA, Luikart G, Wayne RK, Grp WSNP. SNPs in ecology, evolution and conservation. Trends Ecol Evol. 2004;19:208–16.
Myers N, Knoll AH. The biotic crisis and the future of evolution. Proc Natl Acad Sci U S A. 2001;98:5389–92.
Nakamura Y, Morib K, Saitoha K, Oshimac K, Mekuchia M et al. Evolutionary changes of multiple visual pigment genes in the complete genome of Pacific bluefin tuna. 2013;11061–11066, doi: 10.1073/pnas.1302051110.
Nonaka M, Kimura A. Genomic view of the evolution of the complement system. Immunogenetics. 2006;58:701–13.
Novak S, Smith TA, Paradis F, Burwash L, Dyck MK, Foxcroft GR, Dixon WT. Biomarkers of in vivo fertility in sperm and seminal plasma of fertile stallions. Theriogenology. 2010;74: 956–67.
Novembre J, Di Rienzo A. Spatial patterns of variation due to natural selection in humans. Nat Rev Genet. 2009;10:745–55.
Ogden R. Fisheries forensics: the use of DNA tools for improving compliance, traceability and enforcement in the fishing industry. Fish Fish. 2008;9:462–72.
Oliehoek PA, Windig JJ, van Arendonk JA, Bijma P. Estimating relatedness between individuals in general populations with a focus on their use in conservation programs. Genetics. 2006;173:483–96.
Oleksyk TK, Pombert J-F, Siu D, Mazo-Vargas A, Ramos B, et al. A locally funded Puerto Rican parrot (Amazona vittata) genome sequencing project increases avian data and advances young researcher education. Gigascience. 2012;1:14.
Onteru SK, Fan B, Nikkil MT, Garrick DJ, Stalder KJ, Rothschild MF. Whole-genome association analyses for lifetime reproductive traits in the pig. J Anim Sci. 2011;89:988–95.
Ouborg NJ, Pertoldi C, Loeschcke V, Bijlsma R, Hedrick PW. Conservation genetics in transition to conservation genomics. Trends Genet. 2010;26:177–87.
Paetkau D, Slade R, Burden M, Estoup A. Genetic assignment methods for the direct, real-time estimation of migration rate: a simulation-based exploration of accuracy and power. Mol Ecol. 2004;13:55–65.
Parma P, Radi O. Molecular mechanisms of sexual development. Sex Dev. 2012;6:7–17.

Pharo EA, De Leo AA, Renfreee MB, Thomson PC, Lefevre CM, Nicholas KR. The mammary gland-specific marsupial ELP and eutherian CTI share a common ancestral gene. BMC Evol Biol. 2012;12:80.

Piferrer F, Ribas L, Diaz N. Genomic approaches to study genetic and environmental influences on fish sex determination and differentiation. Mar Biotechnol. 2012;14:591–604.

Pukazhenthi BS, Wildt DE. Which reproductive technologies are most relevant to studying, managing and conserving wildlife? Reprod Fertil Dev. 2004;16:33–46.

Prasad A, Schnabel RD, McKay SD, Murdoch B, Stothard P, Kolbehdari D, Wang Z, Taylor JF, Moore SS. Linkage disequilibrium and signatures of selection on chromosomes 19 and 29 in beef and dairy cattle. Anim Genet. 2008;39:597–605.

Primmer CR. From conservation genetics to conservation genomics. Ann N Y Acad Sci. 2009;1162:357–68.

Pritchard JK, Stephens M, Donnelly P. Inference of population structure using multilocus genotype data. Genetics. 2000;155:945–59.

Prüfer K, Munch K, Hellmann I, Akagi K, Miller JR, et al. The bonobo genome compared with the chimpanzee and human genomes. Nature. 2012;486:527–31.

Reed DH, Frankham R. Correlation between fitness and genetic diversity. Conserv Biol. 2003;17:230–7.

Renfree MB, Papenfuss AT, Deakin JE, Lindsay J, Heider T, et al. Genome sequence of an Australian kangaroo, Macropus eugenii, provides insight into the evolution of mammalian reproduction and development. Genome Biol 2011;12:R81.

Reinartz GE, Boese G. Bonobo conservation: the evolution of a Zoological Society Program. In: Wallis J, editor. Primate conservation: the role of zoological parks. USA: American Society of Primatologists; 1997.

Roca AL, Georgiadis N, O'Brien SJ. Cytonuclear genomic dissociation in African elephant species. Nat Genet. 2005;37:96–100.

Rosnaha MM, Brooks SA, Antczak DF. Equine clinical genomics: a clinician's primer. Equine Vet J. 2010;42:658–70.

Rosenberg NA, Li LM, Ward R, Pritchard JK. Informativeness of genetic markers for inference of ancestry. Am J Hum Genet. 2003;73:1402–22.

Rubin CJ, Zody MC, Eriksson J, et al. Whole-genome resequencing reveals loci under selection during chicken domestication. Nature. 2010;464:587–91.

Ryder OA. Conservation genomics: applying whole genome studies to species conservation efforts. Cytogenet Genome Res. 2005;108:6–15.

Sarropoulou E, Fernandes JMO. Comparative genomics in teleost species: knowledge transfer by linking the genomes of model and non-model fish species. Comp Biochem Physiol D Genomics Proteomics. 2011;6:92–102.

Scally A, Dutheil JY, Hillier LW, Jordan GE, Goodhead I, et al. Insights into hominid evolution from the gorilla genome sequence. Nature. 2013;483:169–75.

Schlotterer C. The evolution of molecular markers – just a matter of fashion? Nat Rev Genet. 2004;5:63–9.

Schopen GC, Bovenhuis H, Visker MH, van Arendonk JA. Comparison of information content for microsatellites and SNPs in poultry and cattle. Anim Genet. 2008;39:451–3.

Scott Jr RT, Treff NR. Assessing the reproductive competence of individual embryos: a proposal for the validation of new "-omics" technologies. Fertil Steril. 2010;94(3):791–4.

Schwartz MK, Luikart G, Waples RS. Genetic monitoring as a promising tool for conservation and management. Trends Ecol Evol. 2007;22:25–33.

Selkoe KA, Toonen RJ. Microsatellites for ecologists: a practical guide to using and evaluating microsatellite markers. Ecol Lett. 2006;9:615–29.

Sironen AI, Uimari P, Serenius T, Mote B, Rothschild M, Vilkki J. Effect of polymorphisms in candidate genes on reproduction traits in Finnish pig populations. J Anim Sci. 2010;88:821–7.

Slatkin M. Linkage disequilibrium—understanding the evolutionary past and mapping the medical future. Nat Rev Genet. 2008;9:477–85.

Soller M, Weigend S, Romanov MN, Dekkers JCM, Lamont SJ. Strategies to assess structural variation in the chicken genome and its associations with biodiversity and biological performance. Poult Sci. 2006;85:2061–78.

Star B, Nederbragt AJ, Jentoft S, Grimholt U, Malmstrøm M, et al. The genome sequence of Atlantic cod reveals a unique immune system. Nature. 2011;477:207–10.

Steiner CC, Putnam AS, Hoeck PEA, Ryder OA. Conservation genomics of threatened animal species. Annu Rev Anim Biosci. 2013;1:261–81.

Stella A, Panzitta F, Gandini G, Boettcher PJ. Use of linked loci as individuals or haplotypes for marker-assisted breed assignment. Anim Genet. 2008;39:8–14.

Stuart S, Chanson JS, Cox NA, Young BE, Rodrigues AS, Fischman DL, Waller R. Status and trends of amphibian declines and extinctions worldwide. Science. 2004;306:1783–6.

Swanson WF, Magarey GM, Herrick JR. Sperm cryopreservation in endangered felids: developing linkage of in situ-ex situ populations. Soc Reprod Fertil Suppl. 2007;65:417–32.

Swanson WJ, Yang Z, Wolfner MF, Aquadro CF. Positive Darwinian selection drives the evolution of several female reproductive proteins in mammals. Proc Natl Acad Sci U S A. 2001;98: 2509–14.

Swanson WJ, Vacquier VC. The rapid evolution of reproductive proteins. Nat Rev Genet. 2002;3:137–44.

Swanson WJ, Aagaard JE, Vacquier VD, Monne M, Sadat Al Hosseini H, Jovine L. The molecular basis of sex: linking yeast to human. Mol Biol Evol. 2011;28:1963–66.

Tapio I, Tapio M, Li MH, Popov R, Ivanova Z, Kantanen J. Estimation of relatedness among non-pedigreed Yakutian cryo-bank bulls using molecular data: implications for conservation and breed management. Genet Sel Evol. 2010;42–28.

Templeton AR. Coadaptation and outbreeding depression. In: Conservation Biology: The Science of Scarcity and Diversity (ed. Soulé ME), Sunderland, Massachusetts. 1986;105–116.

Templeton AR, Robertson RJ, Brisson J, Strasburg J. Disrupting evolutionary processes: the effect of habitat fragmentation on collared lizards in the Missouri Ozarks. Proc Natl Acad Sci U S A. 2001;98:5426–32.

Teneva A. Molecular markers in animal genome analysis. Biotechnol Anim Husbandry. 2009;25: 1267–84.

Tobler U, Schmidt BR. Within- and among-population variation in chytridiomycosis-induced mortality in the toad Alytes obstetricians. PLoS One. 2010;5:e10927.

Toro MA, Fernandez J, Caballero A. Molecular characterization of breeds and its use in conservation. Livest Sci. 2009;120:174–95.

Ungewitter EK, Yao H-C. How to make a Gonad: cellular mechanisms governing formation of the testes and ovaries. Sex Dev. 2013;7:7–20.

Vargas A, Sanchez I, Martinez F, Rivas A, Godoy JA, Roldan E, Simon MA, Serra R, Perez MJ, Enseñat C, Delibes M, Aymerich M, Sliva A, Breitenmoser U. The Iberian lynx *Lynx pardinus* conservation breeding program. Int Zoo Yearbook. 2008;42:190–8.

Vignal A, Milan D, SanCristobal M, Eggen A. A review on SNP and other types of molecular markers and their use in animal genetics. Genet Sel Evol. 2002;34:275–305.

Wang J, Whitlock M. Estimating effective population size and migration rates from genetic samples over space and time. Genetics. 2003;163:429–46.

Waples RS, Gaggiotti O. What is a population? An empirical evaluation of some genetic methods for identifying the number of gene pools and their degree of connectivity. Mol Ecol. 2006;15:1419–39.

Wildt DE, Comizzoli P, Pukazhenthi B, Songsasen N. Lessons from biodiversity-the value of non-traditional species to advance reproductive science, conservation, and human health. Mol Reprod Dev. 2010;77:397–409.

Woodhams DC, Ardipradja K, Alford RA, Marantelli G, Reinert LK, Rollins-Smith A. Resistance to chytridiomycosis varies among amphibian species and is correlated with skin peptide defenses. Anim Conserv. 2007;10:409–17.

Zhan X, Li M, Zhang Z, Goossens B, Chen Y, et al. Molecular censusing doubles giant panda population estimate in a key nature reserve. Curr Biol. 2006;16:R451–2.

Zhan X, Pan S, Wang J, Dixon A, He J, et al. Peregrine and saker falcon genome sequences provide insights into evolution of a predatory lifestyle. Nat Genet. 2013;45:563–6.
Zhang B, Li M, Zhang Z, Goossens B, Zhu L, et al. Genetic viability and population history of the giant panda, putting an end to the "evolutionary dead end"? Mol Biol Evol. 2007;24:1801–10.
Zhao J, Grant SFA. Advances in whole genome sequencing technology. Curr Pharm Biotechnol. 2011;12:293–305.
Zhou X, Sun F, Xu S, Fan G, Zhu K, et al. Baiji genomes reveal low genetic variability and new insights into secondary aquatic adaptations. Nat Commun 4, 2013. doi:10.1038/ncomms3708

Chapter 6
The Epigenetic Basis of Adaptation and Responses to Environmental Change: Perspective on Human Reproduction

Agustín F. Fernández, Estela García Toraño, Rocío González Urdinguio, Abel Gayo Lana, Ignacio Arnott Fernández, and Mario F. Fraga

Abstract Not only genetic but also epigenetic mechanisms regulate gene expression, cellular differentiation and development processes. Additionally, "environmental epigenetics" studies the interaction between the environment and the epigenome, and its potential role in the regulation of gene activity. Several studies have shown that the impact of environmental exposures on the epigenome takes on more importance during early fertilization and embryonic development, given that during these periods epigenetic reprogramming occurs and the new epigenetic profile of the offspring is established. Epigenetic alterations in the germline are especially relevant since they can be transmitted trans-generationally and could be associated with a

A.F. Fernández, Ph.D. • E.G. Toraño, Ph.D. • R.G. Urdinguio, Ph.D.
Cancer Epigenetics Laboratory, Instituto Universitario de Oncología del Principado de Asturias (IUOPA), HUCA, Universidad de Oviedo, C/Dr. Emilio Rodríguez Vigil, s/n., Bloque polivalente A, 4° planta, Oviedo 33006, Asturias, Spain
e-mail: affernandez@hca.es

A.G. Lana, Ph.D.
Embryology, FIV4-Instituto de Reproducción Humana, Oviedo, Asturias, Spain

I.A. Fernández, M.D.
Gynecology, FIV4-Instituto de Reproducción Humana, Oviedo, Asturias, Spain

M.F. Fraga, Ph.D. (✉)
Cancer Epigenetics Laboratory, Instituto Universitario de Oncología del Principado de Asturias (IUOPA), HUCA, Universidad de Oviedo, C/Dr. Emilio Rodríguez Vigil, s/n., Bloque polivalente A, 4° planta, Oviedo 33006, Asturias, Spain

Department of Immunology and Oncology, Centro Nacional de Biotecnología/CNB-CSIC, Cantoblanco, Madrid 28049, Spain
e-mail: mffraga@cnb.csic.es

W.V. Holt et al. (eds.), *Reproductive Sciences in Animal Conservation*, Advances in Experimental Medicine and Biology 753, DOI 10.1007/978-1-4939-0820-2_6,

wide range of diseases including several reproductive disorders. In this chapter we review some epigenetic mechanisms, focusing mainly on DNA methylation and histone modifications, which are related to reproductive aspects, and we discuss the controversies in the literature surrounding how environmental conditions, such as exposure to toxic substances or treatment with assisted reproductive techniques (ART), may be involved in epigenetic alterations that affect reproductive success.

Keywords Environmental epigenetics • DNA methylation • Reproduction • Infertility • Imprinting

1 Introduction

Epigenetics involves the study of heritable changes affecting gene expression produced without any change in DNA sequence (Holliday 1987). This area of knowledge became an important player in cancer research 20 years ago when its central role in tumor development was revealed (Feinberg and Tycko 2004). Today, epigenetic mechanisms are considered one of the main molecular mediators in most differentiation and development processes. We now know that epigenetics, which includes the methylation status of DNA, posttranslational modifications (PTM) of histones, and non-coding RNAs (ncRNAs), among other things, has an important function in normal cellular processes (Fig. 6.1). Alterations in the normal functioning of these epigenetic mechanisms have been found in different diseases in mammals (Jones and Baylin 2002; Esteller 2008; Melo et al. 2009; Fernandez et al. 2012).

The best-known epigenetic mark is DNA methylation (Esteller 2008). This is a dynamic process that takes place throughout the course of development in multicellular organisms and ensures the maintenance of normal expression patterns. DNA methylation is involved in many processes including genomic imprinting (Feinberg et al. 2002), the gene-dosage reduction involved in X-chromosome inactivation in females (Payer and Lee 2008), and silencing parasitic and foreign elements (Doerfler 1991), among other processes.

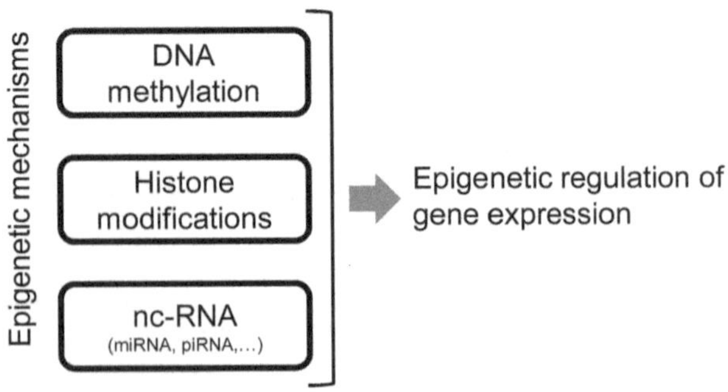

Fig. 6.1 Epigenetic mechanisms. DNA methylation, histone modifications and non-coding RNAs (nc-RNA) are often associated with changes in transcriptional activity

DNA methylation is carried out by a family of enzymes called DNA methyltransferases (DNMTs), which transfer a methyl group from the donor S-adenosylmethionine (SAM) to the DNA base. In mammals, this family of enzymes includes DNMT1, DNMT2, DNMT3A, DNMT3B, and DNMT3-like (DNMT3L) (Bestor 2000), each of which has a specific function although only DNMT1, DNMT3A, and DNMT3B have demonstrated DNA methyltransferase activity (Okano and Li 2002). DNMT1 is considered the principal enzyme responsible for the maintenance of DNA methylation patterns following cellular replication (Fuks 2005), while DNMT3A and DNMT3B have been linked to *de novo* methylation during embryonic development (Okano et al. 1999; Chen et al. 2003). DNA methylation occurs in cytosines that precede guanines, which are called CpG dinucleotides. CpGs are not randomly distributed in the human genome, rather there are CpG-rich regions, known as CpG islands, which span the 5′ end of the regulatory region of many genes. Whilst these islands are usually non-methylated in normal cells, DNA methylation alterations related to different human pathologies, developmental processes and aging have been found (Urdinguio et al. 2009; Fernandez et al. 2012). In particular, DNA methylation alterations have been widely studied in cancer, where it has been found that specific hypermethylation in the CpG islands of tumor suppressor genes causes their inactivation and, the consequent loss of their protective function in tumor development (Jones and Baylin 2002; Esteller 2008; Fernandez et al. 2012). DNA methylation has been studied extensively in relation to reproductive biology, where not only has it been characterized in germ cells and during the different stages of development, but it has also been related to various disorders, such as those linked to imprinting (Sandhu 2010).

Apart from DNA methylation, other important epigenetic mechanisms are involved in chromatin regulation, including gene expression control resulting from the reversible modification of the amino-terminal tail of histones (Fraga and Esteller 2005). There are in fact at least eight types of posttranslational modifications found on histones, of which lysine and arginine acetylation, lysine methylation, serine phosphorylation and lysine ubiquitination are the most thoroughly studied. The combination of the different histone modifications constitutes the histone code (Jenuwein and Allis 2001). This code determines chromatin function in the nucleus context regarding, for instance, the chromatin packaging state and, therefore, gene activity in specific chromatin regions. Thus, different histone modifications have been associated with activation or repression of transcription. The most extensively researched histone modification is, in fact, acetylation, which is generally associated with active gene transcription (Allfrey et al. 1964), whereas methylation is linked to either activation (i.e., H3K4, H3K36, and H3K79) or repression (i.e., H3K9, H3K27, and H4K20), depending on the conditions or residue modified (Kouzarides 2007). These modifications are mediated by histone acetyl transferases (HATs), histone methyl transferases (HMTs), histone deacetylases (HDACs), and histone demethylases (HDMs), among other enzymes. The characterization of histone modifications in relation to reproduction and the different stages of development is on-going although it has been associated with the processes of gametogenesis and embryogenesis (Sasaki and Matsui 2008).

In recent years the non-protein-coding portion of the genome has been shown to be crucial for normal development and function as well as for disease (Mercer et al. 2009; Esteller 2011). Its functional relevance is especially noticeable in the case of a class of small non-coding RNAs (ncRNAs) called microRNAs (miRNAs) (He and Hannon 2004; Mendell 2005). Both epigenetic and genetic defects in miRNAs, as well as their processing machinery, have been found in human diseases, particularly cancer (Esquela-Kerscher and Slack 2006; Hammond 2007; Croce 2009; Nicoloso et al. 2009). However, miRNAs are just a small element of a complex and as yet poorly understood picture, and other ncRNAs might also play a role in the development of many different disorders (Mercer et al. 2009; Esteller 2011). ncRNAs consist of various RNA species that are not translated and evolutionarily conserved among organisms. One single ncRNA may control hundreds of genes. In complex organisms, the developmental and tissue specific regulation of many genomic sequences (Carninci et al. 2005; Kapranov et al. 2007) has promoted the characterization of the different types of ncRNAs that are transcribed in human cells. Based on their length, ncRNAs can be divided into short ncRNAs including small interfering RNAs, miRNAs and PIWI-interacting RNAs (piRNAs); intermediate ncRNAs like small nucleolar RNAs (snoRNAs); and the heterogeneous group of lncRNAs (including large intergenic non-coding RNAs (lincRNAs) and transcribed ultraconserved regions (T-UCRs) among others). Thus far, most of the work has focused on short RNAs, such as miRNAs, however long non-coding RNAs (lncRNAs) (Kapranov et al. 2007), are recently attracting attention.

In terms of reproductive biology, ncRNAs are not as well-studied as DNA methylation or histone PTM. However, several short ncRNAs have been related to germ cell development (Banisch et al. 2012), specially miRNAs have been demonstrated to participate in the physiology and development of gonadal cells in mammalian reproduction (Hossain et al. 2012). Recently, the role of lncRNAs in this process has also begun to be revealed (Sendler et al. 2013).

In this chapter, it will be explained the epigenetic mechanisms that are related to reproductive aspects and how environmental factors, such as exposure to toxic substances or treatment with assisted reproductive techniques (ART), may be involved in epigenetic alterations that affect the reproduction process.

2 Environmental Epigenetics

Epigenomes change during ontogenic development and aging. Some of these changes have natural biological functions, such as defining specific developmental stages (Reik 2007; Feil and Fraga 2012), In addition to natural epigenetic changes associated with development, there are also apparently random variations with, seemingly, no biological function. These random epigenetic changes may be mediated by both intrinsic and extrinsic factors (environment), and are considered to be epigenetic alterations ("epimutations") (Feil and Fraga 2012).

The term "environmental epigenetics" refers to the interaction between the environment and the epigenome, which is susceptible to undergoing modifications. Although how environmental factors can cause negative epigenetic changes remains largely unknown, alterations in DNA methylation or histone modification patterns may induce changes in normal gene expression, which in turn could be associated with a wide range of diseases including various reproductive disorders (Cortessis et al. 2012). Given the absence of mechanisms to repair epimutations, at least none that are known, it is to be expected that the effect of disturbances caused by the environment are greater in the epigenome than in the genome (McCarrey 2012).

Modifications can occur in the epigenome throughout life, but to better understand the impact the environment has it is necessary to consider two different scenarios: embryonic development and adult life (Aguilera et al. 2010). The most vulnerable period is during embryogenesis due to high level of cell division, and the fact that it is when epigenetic marks are undergoing critical modifications (Dolinoy et al. 2006). Moreover epigenetic alterations in this period can be transmitted over consecutive mitotic divisions and affect more cells than those occurring in adults during postnatal development. The placenta is especially important during fetal development, and its functions can be altered or influenced by the environment which may result in pregnancy problems such as early pregnancy loss, preterm birth, intrauterine growth restriction (IUGR), congenital syndromes, and preeclampsia, which have all been linked to epigenetic alterations (Robins et al. 2011).

There is some evidence that epigenetic alterations underlie the associations found between adverse environmental conditions during early developmental stages and later adult disease. Developmental Origins of Health and Disease (DoHAD) is a hypothesis based on the concept of "developmental plasticity" (Hales and Barker 2001), which could explain how environmental exposures during developmental periods can cause alterations which may increase the risk of disease and dysfunction later in life (Barouki et al. 2012). As such, the mother's lifestyle—diet, obesity and alcohol—and tobacco consumption—during pregnancy or the characteristics of the placenta—size of the uterus, availability of nutrients could affect the epigenome of the offspring.

The DoHAD hypothesis is supported by evidence that dietary restrictions during early development in mammals have been associated with the onset of various diseases during lifetime, including cardiovascular or metabolic disorders, or even cancer (Perera and Herbstman 2011).

Different epigenetic mechanisms may be mediating the effect of such nutritional conditions on the appearance of altered phenotypes. A very good example of these associations comes from studies of the offspring of women pregnant during the Dutch Hunger Winter (1944–1945) (a famine that took place in the Netherlands during World War II). Famine exposure in the peri-conceptional period, which is particularly sensitive to changes in the diet of the mother, led to adverse metabolic phenotypes and mental phenotypes in the next generation (Heijmans et al. 2008). Individuals prenatally exposed to famine during this period presented less DNA methylation of the imprinted insulin-like growth factor 2 (*IGF2*) gene compared with their unexposed siblings (Heijmans et al. 2008). Interestingly this gene may play a role in the development of certain diseases, including coronary heart diseases, one of the diseases that has been found associated with food deprivation during the gestational period. Another classic example that illustrates the effect of diet on the phenotype during the period of gestation comes from the Agouti viable yellow (A^{vy}) mice model (Cropley et al. 2006). The Agouti gene is responsible for determining whether a mouse's coat is banded (agouti) or of a solid color (non-agouti), and is regulated by the DNA methylation status of an intra-cisteral A particle (IAP) inserted upstream of the canonical wild-type transcription start site. Methyl-donor supplementation during gestation has been demonstrated to affect the epigenetic status of fetal germ cells and, accordingly, the coat color of the offspring. These epigenetic changes induced by maternal diet were maintained in gametogenesis and embryogenesis of the progeny. Furthermore, the mentioned diet supplementation was shown to affect not only the F1 but also the F2 (Cropley et al. 2006).

Due to their adverse effects on reproduction it is important to emphasize the importance of exposure to endocrine disruptors during pregnancy. Endocrine disruptors are chemicals that at certain doses can interfere with the endocrine (or hormone) system in mammals. They can be classified by their chemical composition into: pesticides (DDT and methoxychlor), fungicides (vinclozolin), herbicides (atrazine), industrial chemicals (PCBs, dioxins), plastics (phthalates, bisphenol A (BPA), alkylphenols) and plant hormones (phytoestrogens) (Skinner et al. 2011). Apart from these chemical products, some pharmaceuticals, personal care products and nutriceuticals are also known to be endocrine disruptors (Daughton and Ternes 1999). The majority of endocrine disruptors are not actually able to alter DNA sequence, but their most significant long term action appears to be related with alterations in the epigenome where they can affect normal reproductive physiological development and functions by acting as weak estrogenic, antiestrogenic, or anti-androgenic compounds. According to the DoHAD hypothesis, abnormal actions of endocrine disruptors during pregnancy can have drastic effects with regards diseases in later life. Females exposed to an excess of androgens early in gestation exhibit increased susceptibility to diseases such as polycystic ovaries in adult life (Abbott et al. 2005). In adult males, perinatal or pubertal exposure to compounds such as estradiol and BPA alters the prostate epigenome and heightens susceptibility to carcinogenesis in adult males (Prins et al. 2008). The mechanisms ways in which such endocrine disruptor exposure in early life is able to promote an adult onset effect in an organ are assumed to involve, at least in part, epigenetic mechanisms.

Besides endocrine disruptors, other pollutants can also alter epigenetic patterns during the prenatal stages: For example, in humans, arsenic exposure in uterus has been associated with increased lung cancer in adulthood. Exposure to tobacco smoke has also been associated with several diseases such as respiratory and metabolic diseases or cancer in children exposed in the womb, which in turn showed abnormalities of methylation patterns (Perera and Herbstman 2011). In addition, exposure to atmospheric pollutants like polycyclic aromatic hydrocarbons (PAHs) can also have adverse effects on fetal growth and have been associated with a decrease in global DNA methylation in cord blood cells of newborns exposed in the womb (Perera and Herbstman 2011).

Although the epigenome of differentiated cells is relatively stable, it too can be altered by environmental conditions during the postnatal stages. It is important to take into account that epigenetic patterns are tissue or cell specific and that not all tissues are equally exposed, hence the effects of environmental factors on the epigenome of an organism may depend on tissue type. The effects of environmental factors during adulthood may also depend on lifestyle conditions including diet, living place and/or workplace, pharmacological treatments, and unhealthy habits (Aguilera et al. 2010). For example, many dietary components have been shown to be linked to DNA methylation changes, and others to have the capacity to influence the activity of HDACs (Feil and Fraga 2012). Useful examples of how the environment affects the epigenome come from monozygotic twin (same genotype) studies, showing that in addition to the increase in epigenetic differences between identical twins over time, different lifestyles may contribute to heighten these differences, and even explain the differential appearance of diseases (Fraga et al. 2005). These epigenetic changes are small and most probably cumulative and develop over the lifetime of the individual, making it difficult to establish the relationship between environmental factors and epigenetic changes (Baccarelli and Bollati 2009).

Apart from nutrition, there are other environmental exposures such as to air pollutants, metals, tobacco smoke, drugs, sun, or alcohol that can alter the epigenetic patterns in the adult stage (Belinsky et al. 2002; Bleich et al. 2006; Baccarelli and Bollati 2009; Christensen et al. 2009; Gronniger et al. 2010; Langevin et al. 2011). Epidemiological studies have determined that exposure to air pollutants may cause not only alterations in global DNA methylation in blood (Baccarelli and Bollati 2009; Baccarelli et al. 2009), but also specific DNA hypermethylation of tumor suppressor genes such as the tumor protein p53 (*p53*) and cyclin-dependent kinase inhibitor 2A (*p16*) in arsenic exposed subjects (Chanda et al. 2006). Exposure to the carcinogen benzene has also been associated with changes in DNA methylation, including a significant reduction in global methylation and hypermethylation of cyclin-dependent kinase inhibitor 2B (*p15*) and hypomethylation of the melanoma antigen family A, 1 (*MAGE*-1) (Bollati et al. 2007). Another environmental contaminant of note is the endocrine disruptor BPA, and as we have seen in relation to the prenatal stages, exposure to this compound may be associated with epigenetic alterations (Hanna et al. 2012). In addition, exposure to this product has been related to breast and prostate cancer, polycystic ovarian syndrome and male infertility, among other diseases (Markey et al. 2001; Ho et al. 2006; Kandaraki et al. 2011; Li et al. 2011).

One of the most important aspects of the effect of environment or environmental conditions on the epigenome is the impact that these can have on trans-generational inheritance, i.e. the transmission of epigenetic changes induced by the environment from one generation to another. At this point it is important to distinguish between trans-generational epigenetic effects and trans-generational epigenetic inheritance. The former include the effects of environmental exposures on adults, which is capable of altering the phenotype of their offspring via the placenta or breastmilk in mammals. The latter, which will be discussed in more detail in the following sections of this chapter, refers to transmission via the gametes. Although most epigenetic alterations that occur in germ cells are reversed during epigenetic reprogramming in gametogenesis, some are able to evade this control and are thus transmitted to the next generation.

3 Epigenetic Reprogramming in Germ Cells and Early Embryo Development

In mammals, genome-wide epigenetic reprogramming mainly occurs at two stages of development: during primordial germ cell (PGC) development and during the early stages of embryonic development following fertilization (Sasaki and Matsui 2008; Feng et al. 2010; Kota and Feil 2010) (Fig. 6.2). Environmental exposures during these two periods can consequently particularly affect the offspring. In both stages, DNA methylation patterns are erased and followed by re-methylation. Together with these DNA methylation changes, there are also rearrangements in the post-translational modifications (PTMs) of some histones. However, changes in DNA methylation and PTMs differ slightly depending on the gender of the germ cell that will be generated (Sasaki and Matsui 2008).

An increased knowledge of the epigenetic mechanisms involved in those processes will help to understand, for instance, infertility-associated pathologies or failures occurring during the application of assisted reproductive technology (ART) in humans.

In mammals, PGCs originate in the epiblast during embryonic development. In mice, it has been shown that germ cells begin to suffer epigenetic changes such as histone H3K9 dimethylation (H3K9me2) loss or DNA methylation decrease from embryonic day 7.5 (E7.5) until approx. E11.5; changes which coincide with germ cells migration to the developing gonads (Fig. 6.2). Concurrently, an increase in H3K27me2 has been detected until E13.5 (Sasaki and Matsui 2008). When PGCs

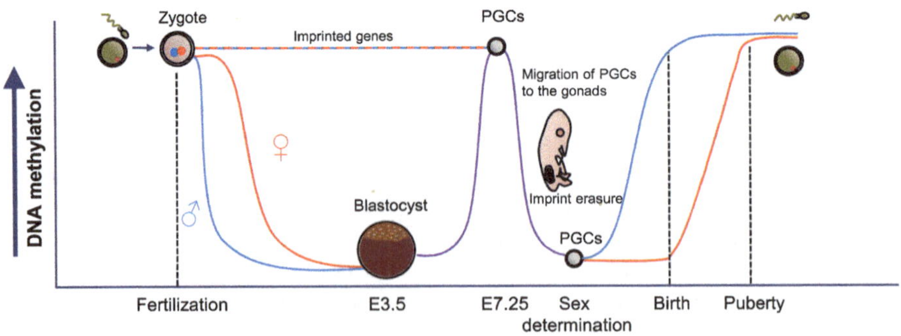

Fig. 6.2 DNA methylation changes during epigenetic reprogramming in mammalian germ cells and early embryo development. *PGCs* Primordial germ cells

have settled in the gonads (E11.5), the erasure of parental imprints occurs and the inactive female X chromosome is thought to be reactivated. Although at this stage almost all DNA sequences have suffered demethylation, some sequences such as intra-cisteral A particle (IAP) and a few long terminal repeats (LTR) sequences avoid the complete loss of DNA methylation (Sasaki and Matsui 2008; Smith and Meissner 2013) indicating that some epigenetic marks related to these regions, and their possible alteration in response to environmental factors, could be heritable and transmitted to the next generation. A classic example of this heritability is the Agouti viable yellow (A^{vy}) mice mentioned in previous sections.

It is still unclear how the DNA demethylation process is regulated during this phase, given that it could be related to either passive or active demethylation mechanisms. The latter would involve DNA deaminases or the 5-methylcytosine (5mC) dioxygenases TET1 and TET2, which trigger, in response, the repair of pathways, especially base excision repair (BER) (Feng et al. 2010).

After the erasure of parental imprints, there is a re-establishment of DNA methylation patterns (Fig. 6.2), which are dependent on the gender of the generated germ cell. While paternal imprinting is established during early stages of spermatogenesis in the fetus from E14.5 until birth, maternal imprinting only occurs after birth during oocyte growth and finishes before maturation, during puberty (Sasaki and Matsui 2008; Smith and Meissner 2013). During this process not all sequences from both types of gametes re-methylate in the same way. While some repeats, such as LINE1, are more methylated in male than in female gametes, the latter show higher methylation levels in IAP sequences than male gametes. Although the mechanisms involved in these re-methylation processes are still unknown, it seems that *de novo* DNMTs and small non-coding RNA molecules could be involved (Sasaki and Matsui 2008; Smith and Meissner 2013). Defects in the re-methylation process of male germ cells in genes, such as the imprinted maternally expressed transcript H19 and mesoderm specific transcript MEST, may be associated with infertility (Rajender et al. 2011) and abnormal methylation of several imprinted genes in human sperm, with oligozoospermia (Marques et al. 2008; Kota and Feil 2010), indicating the importance of epigenetic mechanisms in the regulation of reproduction.

Apart from differential DNA re-methylation, post-translational modifications of histones also differ between male and female germ cells during meiosis and gamete maturation processes. Male germ cells suffer several changes in histone methylation and acetylation patterns during the pre-meiotic stage and the initial stages of Prophase I in the first meiotic division (Sasaki and Matsui 2008). Later, during male gamete maturation (spermiogenesis), as well as changes in some histone mark patterns, there is also a process of histone-protamine exchange, which induces spermatozoa chromatin compactness and seems to function to give the genome of the germ cell protection against environmental damage (Sasaki and Matsui 2008; Kota and Feil 2010). This process whereby histones are replaced by protamines is not complete, and 5–15 % of the histones are retained in mature sperm and play an important role in development (Hammoud et al. 2009). In contrast, changes in histone methylation and acetylation patterns in female germ cells occur from final the stages of Prophase I of the first meiotic division until oocyte maturation following puberty (Sasaki and Matsui 2008; Kota and Feil 2010).

The other important round of epigenetic reprogramming is produced in early embryo development, immediately after fertilization and zygote formation (Fig. 6.2). At this point, global DNA methylation loss (with the exception of imprinted genes) begins and lasts until the blastocyst stage (E3.5). Afterwards, DNA methylation gain occurs until approximately E7.5, when a new cycle of reprogramming starts (Perera and Herbstman 2011; McCarrey 2012; Smallwood and Kelsey 2012; Smith and Meissner 2013) (Fig. 6.2). Similar to the demethylation processes in PGCs, it seems that base excision repair (BER) is associated with DNA demethylation in the zygote (Kota and Feil 2010). Furthermore, like the epigenetic reprogramming in PGCs, these demethylation processes are coupled with changes in histone mark patterns although in this case, however, they are greater in the paternal than in the maternal genome (Seisenberger et al. 2013; Smith and Meissner 2013).

Given all the above, it seems reasonable to assume that these fine epigenetic reprogramming processes may be particularly susceptible to changes that might even compromise reproductive success. Although the mechanisms by which environmental factors may influence the success of this "epigenetic reset" are not clear, the most important point to note is that changes occurring in germ cell precursors or gametes can be transmitted to the next generation.

4 Epigenetic Alterations in Germ Cells

Although environmental factors can modify the epigenome of somatic cells, only when these changes occur in the germ cell line can they be trans-generationally transmitted (Skinner et al. 2011). The transfer of heritable material from parents to progeny through the germ line is known as "transgenerational" change and to be defined as such changes must continue at least through to the F3 generation (Anway et al. 2005). When an F0 is exposed to environmental factors during gestation, the F1 embryo is also exposed and changes in its germ line can result in the transmission of the alteration to a subsequent F2. However, when these changes are present in F3, further transmission occurs in the absence of direct exposure, and the resulting transgenerational phenotypes may be maintained for generations (Fig. 6.3).

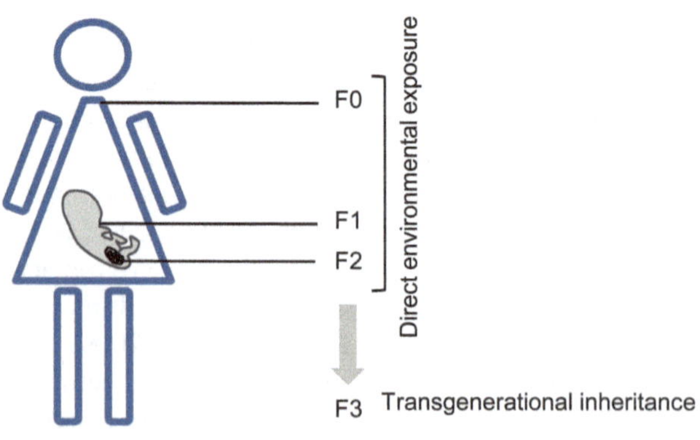

Fig. 6.3 Transgenerational epigenetic inheritance. Scheme showing how direct exposure to environmental factors during pregnancy can affect the epigenome of the F0 (mother), F1 (embryo), and F2 (germ-line). Trans-generational inheritance refers to the transfer of epigenetic changes to the F3 generation (without its direct exposure)

In germ line development, permanent alteration in epigenetic programming appears to be the mechanism involved in the transgenerational phenotype (Jirtle and Skinner 2007). Nonetheless further research in this area is required to determine how epigenetic alterations are transmitted down the generations.

Both male and female gametes may, in theory, be susceptible to epigenetic alterations, but in practice most of studies have been conducted on male gametes due to the low number of female gametes produced. Identified epigenetic alterations in male gametes come mainly from studies which have analyzed samples from individuals with low seminal quality, which may compromise their fertility and most studies have focused on imprinted regions or imprinted genes.

Genomic imprinting refers to the monoallelic expression of a subset of genes in a conserved parent-of-origin fashion, orchestrated by the timely placement of epigenetic signals including DNA methylation and histone modification (Tycko and Morison 2002). Genomic imprinting has been observed only in eutherian mammals and angiosperms. These apparently divergent groups share an interesting similarity in the development of extra-embryonic structures (i.e., the placenta and the endosperm), which connect the embryo to the maternal parent for the purpose of nourishment. Imprinted genes identified in plants are expressed exclusively in the endosperm and/or contribute to endosperm development (Berger et al. 2006). As explained earlier, in mammals female germline imprinting occurs postnatally, whereas male germline imprinting starts prenatally and continues into the postnatal period (Fig. 6.2) (Kerjean et al. 2000; Geuns et al. 2003; Sasaki and Matsui 2008). Gene clusters subjected to genomic imprinting are regulated by differentially methylated regions or domains (DMRs) and the majority of such genes have a role in placenta and embryonic development, as well as neurological functions. During gametogenesis, epigenetic modifications of alleles of imprinted genes are established, and then inherited (Paoloni-Giacobino et al. 2007). Misregulation of this phenomenon may result in developmental and neurological disorders when it occurs during early development. Specifically, imprinting disorders have been linked to disorders such as Angelman's (AS), Prader-Willi (SPW) and Beckwith-Wiederman (BWS) syndromes, cancer, autism and other neurological syndromes (Sandhu 2010). Examples of parental-specific imprinting are *IGF2*, *H19*, *SNRPN*, *KCNQ1OT1*, *LIT1*, *RASGRF1* and *GTL2* loci (Li et al. 2004; Market-Velker et al. 2010). *IGF2* and *H19* were the first imprinting genes to be characterized. They are linked and reciprocally imprinted; *IGF2* is paternally expressed and acts as mitogen implicated in embryonic growth, while *H19* is maternally expressed, and downregulates cell proliferation (Bartolomei et al. 1991; DeChiara et al. 1991). When *H19* is unmethylated in the maternal allele, the CCCTC-binding factor (CTCF) insulator protein may bind to the DMR, which prevents access of *IGF2* to enhancers, thus allowing *H19* expression and inhibiting *IGF2*. In contrast, when *H19* is methylated in the paternal allele, binding of CTCF is blocked, allowing *IGF2* expression and inhibiting *H19* (Arney 2003).

Early studies that analyzed male gametes in individuals with fertility problems identified aberrant DNA methylation in imprinted genes. Recent studies at genome-wide level have identified DNA methylation alterations in other genes involved in such important processes as spermatogenesis. In humans, male infertility is implicated in 40–50 % of cases of infertility, (Hamada et al. 2012) although no more than 15 % of these cases can be explained by genetic causes (Ferlin et al. 2006).

Most infertile men present disturbed spermatogenesis or sperm abnormalities and incorrect genomic imprinting could be associated with the former. Analysis of two imprinting genes—*MEST* (paternally expressed) and *H19* (maternally expressed)—in sperm DNA established a relationship between defective *H19* methylation and poor quality sperm in oligozoospermic patients, while there was no relation between *MEST* and sperm quality (Marques et al. 2004). Kobayashi and collaborators also found, in oligospermic individuals, abnormal DNA methylation of the paternal imprinting of *IGF2/H19* and *GTL2*, and abnormalities of maternal DMRs at *PEG1*, *LIT1*, *ZAC*, *PEG3* and *SNRPN* (Kobayashi et al. 2007). In addition, Poplinski and collaborators found a strong association between infertility and hypermethylation of *MEST* DMR and hypomethylation of IGF2/*H19* (Poplinski et al. 2010). Apart from imprinting abnormalities, spermatogenic failure has also been associated with the hypermethylation of several genes, such as *PAX8*, *NTF3*, *SFN* and *HRAS* (Houshdaran et al. 2007), or with both hypomethylation and hypermethylation of hundreds of genes as has recently been demonstrated by Pacheco and collaborators using genome-wide promoter methylation arrays (Pacheco et al. 2011). It is interesting to note that DNA methylation alterations have been found in methylene tetrahydrofolate reductase (*MTHFR*) in sperm DNA obtained from infertile patients (Wu et al. 2010) as this gene codifies for a enzyme with a significant role in folate metabolism, which is very important for spermatogenesis (Wu et al. 2010). Furthermore the role of DNMTs in sperm production has been demonstrated in mice (Yaman and Grandjean 2006): Defective DNMT3L leads to impaired spermatogenesis, and mutations in DNMT3B result in very low numbers of spermatocytes (Bourc'his and Bestor 2004).

It has also been shown that mammalian testes have a specific genome-wide DNA methylation pattern which is far more hypomethylated than that of somatic cells (Oakes et al. 2007). Testicular samples of human infertile male patients with spermatogenic disorders present alterations of DNA methylation profiles. A study by Heyn and collaborators found about 600 genes to be differentially methylated in testicular biopsies of men with several spermatogenic disorders as compared with controls (with conserved spermatogenesis). These included the hypermethylation of the germline-specific genes piwi-like RNA-mediated gene silencing 2 (*PIWIL2*) and tudor domain containing 1 (*TDRD1*), two genes involved in piRNA processing machinery that may have a role in the regulation of spermatogenesis (Heyn et al. 2012). Moreover, it has also been found in testicular biopsies that patients with non-obstructive azoospermia showed DNA hypermethylation of *MTHFR* (Khazamipour et al. 2009).

Not only alterations in DNA methylation but also histone modifications have been found to be associated with male infertility. In the postmeiotic stage of spermatogenesis (spermiogenesis), histones are replaced by protamines, basic proteins that facilitate the packaging of DNA in sperm. Epigenetic changes, such as an increase in histone acetylation, occur during this histone-protamine exchange in order to facilitate the transition, and mistakes in this process can lead to sperm aberrations and infertility (Nanassy et al. 2011; Rajender et al. 2011; Dada et al. 2012). It seems that around 15 % of nucleosomes are retained after this exchange, but more important is that these nucleosomes are enriched in genes which are important for development (including imprinted genes), and present specific histone marks (Hammoud et al. 2009). In mature sperm, the promoters of developmental genes are enriched in trimethylation of histone H3 at lysine 4 (H3k4me3) and trimethylation of histone H3 at lysine 27 (H3k27me3), a "bivalent mark" previously identified in embryonic stem (ES) cells, which, it has been proposed, is associated with the silencing of developmental genes in ES cells and keeping them poised for further activation (Bernstein et al. 2006). Changes in histone modifications of this "bivalent mark" in sperm could have clinical implications in embryo outcome and/or infertility, and, for example, in infertile men, a reduction of H3K27me or H3K4me in some imprinted genes and transcription factors associated with development has been found (Hammoud et al. 2011). Mono- and dimethylation of histone H3 at lysine 9 (H3k9me2/1) also plays an important role in spermatogenesis, and has been evidenced in studies in which mice deficient in specific demethylase JHDM2A (JmjCdomain-containing histone demethylase 2A) exhibit defects in sperm chromatin condensation that impair spermatogenesis (Okada et al. 2007).

It seems clear that alterations in epigenetic mechanisms in male germ cells can compromise fertility and therefore reproductive success. The next step will be to identify the extent to which environmental factors contribute to the occurrence of these alterations.

5 Epigenetic Alterations Mediated by Environmental Factors and Fertility

In previous sections of this chapter we have shown how epigenetic mechanisms may mediate the appearance of phenotypic alterations produced by different environmental exposures. This section will discuss how the environment can affect fertility through epigenetic alterations. There are several studies that associate lifestyle or exposure to different compounds with the incidence of male infertility in humans (Sharpe 2000; Miyamoto et al. 2012), but nevertheless, there are not many works that demonstrate that epigenetic mechanisms mediate these effects.

Aberrant DNA methylation patterns of male germ cells have been related with early developmental exposure to drugs/endocrine disruptors such as 5-azacytidine, alcohol, the fungicide vinclozolin, the pesticide methoxychlor, tamoxifen (selective estrogen receptor modulator, SERM) in rodents. This aberrant DNA methylation was found in promoter genes of early developmental genes, and also at the DMR of imprinted genes such as *IGF2*, and *H19* (Doerksen et al. 2000; Oakes et al. 2007; Pathak et al. 2009; Pathak et al. 2010; Stouder et al. 2011).

Several studies examining the neonatal exposure of rats to environmental factors such as estrogens, diethylstilbestrol or BPA suggest that this period is one of the most critical in terms of long lasting effects on sperm count, motility, spermatogenesis, and fertility later in life (Sharpe et al. 1998; Goyal et al. 2003; Salian et al. 2009). In the case of BPA, a recent study conducted by Doshi and collaborators demonstrated that neonatal exposure of male rats (F0) to BPA produced an alteration in *IGF2/H19* imprinting in sperm that, interestingly, is inherited by the embryo (F1) and ultimately leads to post-implantation loss (Doshi et al. 2013).

Several environmental factors that alter phenotypes have been shown to propagate through generations, but only a few have been shown to be transgenerational. Notable among these are several studies of the transgenerational effect of endocrine disruptors (vinclozolin and methoxychlor) on the male germ line of rats and mice. Both substances are able to produce transgenerational defects in spermatogenic and fertility capacity, due to their anti-androgenic endocrine disrupting action, which is transmitted through four generations (Anway et al. 2005; Guerrero-Bosagna et al. 2012).

In humans, chemotherapy treatment with temozolomide, an oral alkylating agent used for glioblastoma and melanoma (Neyns et al. 2010), showed aberrant DNA methylation of the *H19* locus in sperm, which was accompanied by the risk of impaired fertility (Berthaut et al. 2013).

After highlighting the importance of the effect of environmental conditions on the disturbances that can be caused in relation to fertility, and the mediation of epigenetic mechanisms in these processes, it is of great interest to identify the potential risks associated with the use of assisted reproductive technologies (ART) in the infertility treatments.

6 Assisted Reproductive Technology (ART) and Epigenetics

Assisted reproductive technology (ART) is a general term referring to methods used to achieve pregnancy by artificial or partially artificial means. It is used in infertility treatments, and is involved in many of the steps leading to conception; from the stimulation of gamete production to the ex vivo culture of embryos. Between 1 and 2 % of all children born in developed countries result from the use of ART for the treatment of human subfertility/infertility. The manipulations carried out for ARTs include, for instance, the use of hormones to stimulate the ovary for supernumerary oocyte production (superovulation), in vitro maturation of oocytes, intrauterine insemination, in vitro fertilization (IVF), intracytoplasmic sperm injection (ICSI), in vitro culture of pre-implantation embryos before transferring back to the uterus, or cryopreservation of either gametes or embryos. An alteration in any of these steps could produce epigenetic abnormalities, such as imprinting disorders.

Environmental factors seem to be crucial for epigenetic regulation in the first stages of mammalian development (Robins et al. 2011); Feil and Fraga 2012; Hales and Barker 2001; Perera and Herbstman 2011; Heijmans et al. 2008) (see Sect. 2). Additionally, as major epigenetic events occur during both germ-cell development and the pre-implantation stages when ART procedures are being performed, there is a high possibility that these manipulations may interfere with the proper establishment and maintenance of genomic imprints. This is why an increasing number of studies are investigating the possibility that imprinting diseases could be related to the use of ART, and exactly how ART influences development (Gosden et al. 2003).

Different studies suggest that some aspects of the ART procedure increase the frequency of epigenetic abnormalities leading to congenital malformation syndromes, and specially in relation to imprinting disorders. For example, an increased prevalence of ART in patients with the human overgrowth syndrome Beckwith-Wiedemann syndrome (BWS) has been found (DeBaun et al. 2003; Gicquel et al. 2003; Maher et al. 2003). Also ART has been reported to be associated with Silver-Russell syndrome (SRS) (Hiura et al. 2012). Furthermore, Angelman syndrome has been observed to occur after ICSI, with the suggestion that ICSI interferes with the establishment of the maternal imprint in the oocyte or pre-embryo (Cox et al. 2002). Other works have found different results depending on the ART procedure considered, such as the fact that ICSI is significantly related to increased risk of birth defects while IVF is not (Davies et al. 2012). Several of these studies have indicated that the cause of the imprinting disorders was to the result of epigenetic disruptions rather than genetic reasons.

On the other hand, there are studies where no association between imprinting disorders and ARTs could be ascertained. For instance, no abnormal methylation of the region related to Angelman and Prader-Willi syndromes at chromosome 15 could be found in a study of 92 ICSI cases (Manning et al. 2000). Additionally, the methylation analyses of multiple imprinted regions in 161 children born after ART showed no imprinting errors (Zheng et al. 2011). And more recently, another study showed no significant difference in DNA methylation levels of several imprinting control regions in children conceived spontaneously or those conceived after ART (Puumala et al. 2012). Furthermore, it should be taken into consideration that methylation levels of the imprinted genes related to BWS and Silver-Russell syndrome (SRS), *H19* and *IGF2*, have been found to vary among phenotypically healthy children (Rancourt et al. 2013).

All this controversy is currently under discussion (Savage et al. 2011). The authors of a recent review reached the conclusion that, although a higher prevalence of imprinting disorders can be observed after IVF or ICSI, studies should be corrected for fertility problems and in this manner the authors consider that there is a highly improbable causal association between ARTs and imprinted diseases in humans (Vermeiden and Bernardus 2013). An undeniable potential confounding factor in this regard is that the underlying condition of infertility/subfertility of the parents who conceive using ART procedures could be, at least partially, responsible for some of the defects observed in children born after ART. This is supported by a recent study of women who conceived once with the help of ART and once spontaneously, in which unfavorable pregnancy and birth outcomes were correlated to infertility, not to any ART procedures (Romundstad et al. 2009). Another study revealed the same risk for AS in children of subfertile couples who conceived spontaneously, via ICSI, or following superovulation only (Ludwig et al. 2005), suggesting that increased AS risk may be a function of the condition of infertility, instead of the exogenous hormone treatment. Apart from this, the low frequency of imprinting disorders dictates that many such studies are conducted with small sample sizes, thereby complicating the establishment of associations and reducing the ability of researchers to draw biologically meaningful conclusions. Additionally, although a three- to six-fold increase in the risk of BWS was observed in ART births (DeBaun et al. 2003; Gicquel et al. 2003; Murrell et al. 2004; Poole et al. 2012), the absolute risk is still quite low. Imprinting disorders are very rare and, even with a relative increase in incidence of these disorders, most children conceived through ART are healthy.

The specific contribution of ART to adverse outcomes for the resulting offspring is difficult to ascertain due to the many confounding variables, including maternal age, inherent parental infertility, underlying parental medical conditions such as diabetes or obesity, and maternal diet (Grace and Sinclair 2009). Clearly, ART stresses developing embryos during a period of epigenetic vulnerability. The complex interaction of genetic and environmental factors with epigenetics is not fully elucidated and future research will shed light on this controversial issue.

References

Abbott DH, Barnett DK, Bruns CM, Dumesic DA. Androgen excess fetal programming of female reproduction: a developmental aetiology for polycystic ovary syndrome? Hum Reprod Update. 2005;11:357–74.

Aguilera O, Fernandez AF, Munoz A, Fraga MF. Epigenetics and environment: a complex relationship. J Appl Physiol. 2010;109:243–51.

Allfrey VG, Faulkner R, Mirsky AE. Acetylation and methylation of histones and their possible role in the regulation of Rna synthesis. Proc Natl Acad Sci U S A. 1964;51:786–94.

Anway MD, Cupp AS, Uzumcu M, Skinner MK. Epigenetic transgenerational actions of endocrine disruptors and male fertility. Science. 2005;308:1466–9.

Arney KL. H19 and Igf2—enhancing the confusion? Trends Genet. 2003;19:17–23.

Baccarelli A, Bollati V. Epigenetics and environmental chemicals. Curr Opin Pediatr. 2009;21:243–51.

Baccarelli A, Wright RO, Bollati V, Tarantini L, Litonjua AA, Suh HH, et al. Rapid DNA methylation changes after exposure to traffic particles. Am J Respir Crit Care Med. 2009;179:572–8.

Banisch TU, Goudarzi M, Raz E. Small RNAs in germ cell development. Curr Top Dev Biol. 2012;99:79–113.

Barouki R, Gluckman PD, Grandjean P, Hanson M, Heindel JJ. Developmental origins of non-communicable disease: implications for research and public health. Environ Health. 2012;11:42.

Bartolomei MS, Zemel S, Tilghman SM. Parental imprinting of the mouse H19 gene. Nature. 1991;351:153–5.

Belinsky SA, Palmisano WA, Gilliland FD, Crooks LA, Divine KK, Winters SA, et al. Aberrant promoter methylation in bronchial epithelium and sputum from current and former smokers. Cancer Res. 2002;62:2370–7.

Berger F, Grini PE, Schnittger A. Endosperm: an integrator of seed growth and development. Curr Opin Plant Biol. 2006;9:664–70.

Bernstein BE, Mikkelsen TS, Xie X, Kamal M, Huebert DJ, Cuff J, et al. A bivalent chromatin structure marks key developmental genes in embryonic stem cells. Cell. 2006;125:315–26.

Berthaut I, Montjean D, Dessolle L, Morcel K, Deluen F, Poirot C, et al. Effect of temozolomide on male gametes: an epigenetic risk to the offspring? J Assist Reprod Genet. 2013;30:827–33.

Bestor TH. The DNA methyltransferases of mammals. Hum Mol Genet. 2000;9:2395–402.

Bleich S, Lenz B, Ziegenbein M, Beutler S, Frieling H, Kornhuber J, et al. Epigenetic DNA hypermethylation of the HERP gene promoter induces down-regulation of its mRNA expression in patients with alcohol dependence. Alcohol Clin Exp Res. 2006;30:587–91.

Bollati V, Baccarelli A, Hou L, Bonzini M, Fustinoni S, Cavallo D, et al. Changes in DNA methylation patterns in subjects exposed to low-dose benzene. Cancer Res. 2007;67:876–80.

Bourc'his D, Bestor TH. Meiotic catastrophe and retrotransposon reactivation in male germ cells lacking Dnmt3L. Nature. 2004;431:96–9.

Carninci P, Kasukawa T, Katayama S, Gough J, Frith MC, Maeda N, et al. The transcriptional landscape of the mammalian genome. Science. 2005;309:1559–63.

Cortessis VK, Thomas DC, Levine AJ, Breton CV, Mack TM, Siegmund KD, et al. Environmental epigenetics: prospects for studying epigenetic mediation of exposure-response relationships. Hum Genet. 2012;131:1565–89.

Cox GF, Burger J, Lip V, Mau UA, Sperling K, Wu BL, et al. Intracytoplasmic sperm injection may increase the risk of imprinting defects. Am J Hum Genet. 2002;71:162–4.

Croce CM. Causes and consequences of microRNA dysregulation in cancer. Nat Rev Genet. 2009;10:704–14.

Cropley JE, Suter CM, Beckman KB, Martin DI. Germ-line epigenetic modification of the murine A vy allele by nutritional supplementation. Proc Natl Acad Sci U S A. 2006;103: 17308–12.

Chanda S, Dasgupta UB, Guhamazumder D, Gupta M, Chaudhuri U, Lahiri S, et al. DNA hypermethylation of promoter of gene p53 and p16 in arsenic-exposed people with and without malignancy. Toxicol Sci. 2006;89:431–7.

Chen T, Ueda Y, Dodge JE, Wang Z, Li E. Establishment and maintenance of genomic methylation patterns in mouse embryonic stem cells by Dnmt3a and Dnmt3b. Mol Cell Biol. 2003;23: 5594–605.

Christensen BC, Houseman EA, Marsit CJ, Zheng S, Wrensch MR, Wiemels JL, et al. Aging and environmental exposures alter tissue-specific DNA methylation dependent upon CpG island context. PLoS Genet. 2009;5:e1000602.

Dada R, Kumar M, Jesudasan R, Fernandez JL, Gosalvez J, Agarwal A. Epigenetics and its role in male infertility. J Assist Reprod Genet. 2012;29:213–23.

Daughton CG, Ternes TA. Pharmaceuticals and personal care products in the environment: agents of subtle change? Environ Health Perspect. 1999;107 Suppl 6:907–38.

Davies MJ, Moore VM, Willson KJ, Van Essen P, Priest K, Scott H, et al. Reproductive technologies and the risk of birth defects. N Engl J Med. 2012;366:1803–13.

DeBaun MR, Niemitz EL, Feinberg AP. Association of in vitro fertilization with Beckwith-Wiedemann syndrome and epigenetic alterations of LIT1 and H19. Am J Hum Genet. 2003;72: 156–60.

DeChiara TM, Robertson EJ, Efstratiadis A. Parental imprinting of the mouse insulin-like growth factor II gene. Cell. 1991;64:849–59.

Doerfler W. Patterns of DNA methylation—evolutionary vestiges of foreign DNA inactivation as a host defense mechanism. A proposal. Biol Chem Hoppe Seyler. 1991;372:557–64.

Doerksen T, Benoit G, Trasler JM. Deoxyribonucleic acid hypomethylation of male germ cells by mitotic and meiotic exposure to 5-azacytidine is associated with altered testicular histology. Endocrinology. 2000;141:3235–44.

Dolinoy DC, Weidman JR, Waterland RA, Jirtle RL. Maternal genistein alters coat color and protects Avy mouse offspring from obesity by modifying the fetal epigenome. Environ Health Perspect. 2006;114:567–72.

Doshi T, D'Souza C, Vanage G. Aberrant DNA methylation at Igf2-H19 imprinting control region in spermatozoa upon neonatal exposure to bisphenol A and its association with post implantation loss. Mol Biol Rep. 2013;40:4747–57.

Esquela-Kerscher A, Slack FJ. Oncomirs—microRNAs with a role in cancer. Nat Rev Cancer. 2006;6:259–69.

Esteller M. Epigenetics in cancer. N Engl J Med. 2008;358:1148–59.

Esteller M. Non-coding RNAs in human disease. Nat Rev Genet. 2011;12:861–74.

Feil R, Fraga MF. Epigenetics and the environment: emerging patterns and implications. Nat Rev Genet. 2012;13:97–109.
Feinberg AP, Cui H, Ohlsson R. DNA methylation and genomic imprinting: insights from cancer into epigenetic mechanisms. Semin Cancer Biol. 2002;12:389–98.
Feinberg AP, Tycko B. The history of cancer epigenetics. Nat Rev Cancer. 2004;4:143–53.
Feng S, Jacobsen SE, Reik W. Epigenetic reprogramming in plant and animal development. Science. 2010;330:622–7.
Ferlin A, Arredi B, Foresta C. Genetic causes of male infertility. Reprod Toxicol. 2006;22:133–41.
Fernandez AF, Assenov Y, Martin-Subero JI, Balint B, Siebert R, Taniguchi H, et al. A DNA methylation fingerprint of 1628 human samples. Genome Res. 2012;22:407–19.
Fraga MF, Ballestar E, Paz MF, Ropero S, Setien F, Ballestar ML, et al. Epigenetic differences arise during the lifetime of monozygotic twins. Proc Natl Acad Sci U S A. 2005;102:10604–9.
Fraga MF, Esteller M. Towards the human cancer epigenome: a first draft of histone modifications. Cell Cycle. 2005;4:1377–81.
Fuks F. DNA methylation and histone modifications: teaming up to silence genes. Curr Opin Genet Dev. 2005;15:490–5.
Geuns E, De Rycke M, Van Steirteghem A, Liebaers I. Methylation imprints of the imprint control region of the SNRPN-gene in human gametes and preimplantation embryos. Hum Mol Genet. 2003;12:2873–9.
Gicquel C, Gaston V, Mandelbaum J, Siffroi JP, Flahault A, Le Bouc Y. In vitro fertilization may increase the risk of Beckwith-Wiedemann syndrome related to the abnormal imprinting of the KCN1OT gene. Am J Hum Genet. 2003;72:1338–41.
Gosden R, Trasler J, Lucifero D, Faddy M. Rare congenital disorders, imprinted genes, and assisted reproductive technology. Lancet. 2003;361:1975–7.
Goyal HO, Robateau A, Braden TD, Williams CS, Srivastava KK, Ali K. Neonatal estrogen exposure of male rats alters reproductive functions at adulthood. Biol Reprod. 2003;68:2081–91.
Grace KS, Sinclair KD. Assisted reproductive technology, epigenetics, and long-term health: a developmental time bomb still ticking. Semin Reprod Med. 2009;27:409–16.
Gronniger E, Weber B, Heil O, Peters N, Stab F, Wenck H, et al. Aging and chronic sun exposure cause distinct epigenetic changes in human skin. PLoS Genet. 2010;6:e1000971.
Guerrero-Bosagna C, Covert TR, Haque MM, Settles M, Nilsson EE, Anway MD, et al. Epigenetic transgenerational inheritance of vinclozolin induced mouse adult onset disease and associated sperm epigenome biomarkers. Reprod Toxicol. 2012;34:694–707.
Hales CN, Barker DJ. The thrifty phenotype hypothesis. Br Med Bull. 2001;60:5–20.
Hamada A, Esteves SC, Nizza M, Agarwal A. Unexplained male infertility: diagnosis and management. Int Braz J Urol. 2012;38:576–94.
Hammond SM. MicroRNAs as tumor suppressors. Nat Genet. 2007;39:582–3.
Hammoud SS, Nix DA, Hammoud AO, Gibson M, Cairns BR, Carrell DT. Genome-wide analysis identifies changes in histone retention and epigenetic modifications at developmental and imprinted gene loci in the sperm of infertile men. Hum Reprod. 2011;26:2558–69.
Hammoud SS, Nix DA, Zhang H, Purwar J, Carrell DT, Cairns BR. Distinctive chromatin in human sperm packages genes for embryo development. Nature. 2009;460:473–8.
Hanna CW, Bloom MS, Robinson WP, Kim D, Parsons PJ, vom Saal FS, et al. DNA methylation changes in whole blood is associated with exposure to the environmental contaminants, mercury, lead, cadmium and bisphenol A, in women undergoing ovarian stimulation for IVF. Hum Reprod. 2012;27:1401–10.
He L, Hannon GJ. MicroRNAs: small RNAs with a big role in gene regulation. Nat Rev Genet. 2004;5:522–31.
Heijmans BT, Tobi EW, Stein AD, Putter H, Blauw GJ, Susser ES, et al. Persistent epigenetic differences associated with prenatal exposure to famine in humans. Proc Natl Acad Sci U S A. 2008;105:17046–9.
Heyn H, Ferreira HJ, Bassas L, Bonache S, Sayols S, Sandoval J, et al. Epigenetic disruption of the PIWI pathway in human spermatogenic disorders. PLoS One. 2012;7:e47892.

Hiura H, Okae H, Miyauchi N, Sato F, Sato A, Van De Pette M, et al. Characterization of DNA methylation errors in patients with imprinting disorders conceived by assisted reproduction technologies. Hum Reprod. 2012;27:2541–8.
Ho SM, Tang WY, Belmonte de Frausto J, Prins GS. Developmental exposure to estradiol and bisphenol A increases susceptibility to prostate carcinogenesis and epigenetically regulates phosphodiesterase type 4 variant 4. Cancer Res. 2006;66:5624–32.
Holliday R. The inheritance of epigenetic defects. Science. 1987;238:163–70.
Hossain MM, Sohel MM, Schellander K, Tesfaye D. Characterization and importance of microRNAs in mammalian gonadal functions. Cell Tissue Res. 2012;349:679–90.
Houshdaran S, Cortessis VK, Siegmund K, Yang A, Laird PW, Sokol RZ. Widespread epigenetic abnormalities suggest a broad DNA methylation erasure defect in abnormal human sperm. PLoS One. 2007;2:e1289.
Jenuwein T, Allis CD. Translating the histone code. Science. 2001;293:1074–80.
Jirtle RL, Skinner MK. Environmental epigenomics and disease susceptibility. Nat Rev Genet. 2007;8:253–62.
Jones PA, Baylin SB. The fundamental role of epigenetic events in cancer. Nat Rev Genet. 2002;3:415–28.
Kandaraki E, Chatzigeorgiou A, Livadas S, Palioura E, Economou F, Koutsilieris M, et al. Endocrine disruptors and polycystic ovary syndrome (PCOS): elevated serum levels of bisphenol A in women with PCOS. J Clin Endocrinol Metab. 2011;96:E480–4.
Kapranov P, Cheng J, Dike S, Nix DA, Duttagupta R, Willingham AT, et al. RNA maps reveal new RNA classes and a possible function for pervasive transcription. Science. 2007;316:1484–8.
Kerjean A, Dupont JM, Vasseur C, Le Tessier D, Cuisset L, Paldi A, et al. Establishment of the paternal methylation imprint of the human H19 and MEST/PEG1 genes during spermatogenesis. Hum Mol Genet. 2000;9:2183–7.
Khazamipour N, Noruzinia M, Fatehmanesh P, Keyhanee M, Pujol P. MTHFR promoter hypermethylation in testicular biopsies of patients with non-obstructive azoospermia: the role of epigenetics in male infertility. Hum Reprod. 2009;24:2361–4.
Kobayashi H, Sato A, Otsu E, Hiura H, Tomatsu C, Utsunomiya T, et al. Aberrant DNA methylation of imprinted loci in sperm from oligospermic patients. Hum Mol Genet. 2007;16:2542–51.
Kota SK, Feil R. Epigenetic transitions in germ cell development and meiosis. Dev Cell. 2010;19:675–86.
Kouzarides T. Chromatin modifications and their function. Cell. 2007;128:693–705.
Langevin SM, Houseman EA, Christensen BC, Wiencke JK, Nelson HH, Karagas MR, et al. The influence of aging, environmental exposures and local sequence features on the variation of DNA methylation in blood. Epigenetics. 2011;6:908–19.
Li DK, Zhou Z, Miao M, He Y, Wang J, Ferber J, et al. Urine bisphenol-A (BPA) level in relation to semen quality. Fertil Steril. 2011;95(625–30):e1–4.
Li JY, Lees-Murdock DJ, Xu GL, Walsh CP. Timing of establishment of paternal methylation imprints in the mouse. Genomics. 2004;84:952–60.
Ludwig M, Katalinic A, Gross S, Sutcliffe A, Varon R, Horsthemke B. Increased prevalence of imprinting defects in patients with Angelman syndrome born to subfertile couples. J Med Genet. 2005;42:289–91.
Maher ER, Brueton LA, Bowdin SC, Luharia A, Cooper W, Cole TR, et al. Beckwith-Wiedemann syndrome and assisted reproduction technology (ART). J Med Genet. 2003;40:62–4.
Manning M, Lissens W, Bonduelle M, Camus M, De Rijcke M, Liebaers I, et al. Study of DNA-methylation patterns at chromosome 15q11-q13 in children born after ICSI reveals no imprinting defects. Mol Hum Reprod. 2000;6:1049–53.
Market-Velker BA, Zhang L, Magri LS, Bonvissuto AC, Mann MR. Dual effects of superovulation: loss of maternal and paternal imprinted methylation in a dose-dependent manner. Hum Mol Genet. 2010;19:36–51.
Markey CM, Luque EH, Munoz De Toro M, Sonnenschein C, Soto AM. In utero exposure to bisphenol A alters the development and tissue organization of the mouse mammary gland. Biol Reprod. 2001;65:1215–23.

Marques CJ, Carvalho F, Sousa M, Barros A. Genomic imprinting in disruptive spermatogenesis. Lancet. 2004;363:1700–2.

Marques CJ, Costa P, Vaz B, Carvalho F, Fernandes S, Barros A, et al. Abnormal methylation of imprinted genes in human sperm is associated with oligozoospermia. Mol Hum Reprod. 2008;14:67–74.

McCarrey JR. The epigenome as a target for heritable environmental disruptions of cellular function. Mol Cell Endocrinol. 2012;354:9–15.

Melo SA, Ropero S, Moutinho C, Aaltonen LA, Yamamoto H, Calin GA, et al. A TARBP2 mutation in human cancer impairs microRNA processing and DICER1 function. Nat Genet. 2009;41:365–70.

Mendell JT. MicroRNAs: critical regulators of development, cellular physiology and malignancy. Cell Cycle. 2005;4:1179–84.

Mercer TR, Dinger ME, Mattick JS. Long non-coding RNAs: insights into functions. Nat Rev Genet. 2009;10:155–9.

Miyamoto T, Tsujimura A, Miyagawa Y, Koh E, Namiki M, Sengoku K. Male infertility and its causes in human. Adv Urol. 2012;2012:384520.

Murrell A, Heeson S, Cooper WN, Douglas E, Apostolidou S, Moore GE, et al. An association between variants in the IGF2 gene and Beckwith-Wiedemann syndrome: interaction between genotype and epigenotype. Hum Mol Genet. 2004;13:247–55.

Nanassy L, Liu L, Griffin J, Carrell DT. The clinical utility of the protamine 1/protamine 2 ratio in sperm. Protein Pept Lett. 2011;18:772–7.

Neyns B, Tosoni A, Hwu WJ, Reardon DA. Dose-dense temozolomide regimens: antitumor activity, toxicity, and immunomodulatory effects. Cancer. 2010;116:2868–77.

Nicoloso MS, Spizzo R, Shimizu M, Rossi S, Calin GA. MicroRNAs—the micro steering wheel of tumour metastases. Nat Rev Cancer. 2009;9:293–302.

Oakes CC, La Salle S, Smiraglia DJ, Robaire B, Trasler JM. A unique configuration of genome-wide DNA methylation patterns in the testis. Proc Natl Acad Sci U S A. 2007;104:228–33.

Okada Y, Scott G, Ray MK, Mishina Y, Zhang Y. Histone demethylase JHDM2A is critical for Tnp1 and Prm1 transcription and spermatogenesis. Nature. 2007;450:119–23.

Okano M, Bell DW, Haber DA, Li E. DNA methyltransferases Dnmt3a and Dnmt3b are essential for de novo methylation and mammalian development. Cell. 1999;99:247–57.

Okano M, Li E. Genetic analyses of DNA methyltransferase genes in mouse model system. J Nutr. 2002;132:2462S–5.

Pacheco SE, Houseman EA, Christensen BC, Marsit CJ, Kelsey KT, Sigman M, et al. Integrative DNA methylation and gene expression analyses identify DNA packaging and epigenetic regulatory genes associated with low motility sperm. PLoS One. 2011;6:e20280.

Paoloni-Giacobino A, D'Aiuto L, Cirio MC, Reinhart B, Chaillet JR. Conserved features of imprinted differentially methylated domains. Gene. 2007;399:33–45.

Pathak S, Kedia-Mokashi N, Saxena M, D'Souza R, Maitra A, Parte P, et al. Effect of tamoxifen treatment on global and insulin-like growth factor 2-H19 locus-specific DNA methylation in rat spermatozoa and its association with embryo loss. Fertil Steril. 2009;91:2253–63.

Pathak S, Saxena M, D'Souza R, Balasinor NH. Disrupted imprinting status at the H19 differentially methylated region is associated with the resorbed embryo phenotype in rats. Reprod Fertil Dev. 2010;22:939–48.

Payer B, Lee JT. X chromosome dosage compensation: how mammals keep the balance. Annu Rev Genet. 2008;42:733–72.

Perera F, Herbstman J. Prenatal environmental exposures, epigenetics, and disease. Reprod Toxicol. 2011;31:363–73.

Poole RL, Leith DJ, Docherty LE, Shmela ME, Gicquel C, Splitt M, et al. Beckwith-Wiedemann syndrome caused by maternally inherited mutation of an OCT-binding motif in the IGF2/H19-imprinting control region, ICR1. Eur J Hum Genet. 2012;20:240–3.

Poplinski A, Tuttelmann F, Kanber D, Horsthemke B, Gromoll J. Idiopathic male infertility is strongly associated with aberrant methylation of MEST and IGF2/H19 ICR1. Int J Androl. 2010;33:642–9.

Prins GS, Tang WY, Belmonte J, Ho SM. Perinatal exposure to oestradiol and bisphenol A alters the prostate epigenome and increases susceptibility to carcinogenesis. Basic Clin Pharmacol Toxicol. 2008;102:134–8.

Puumala SE, Nelson HH, Ross JA, Nguyen RH, Damario MA, Spector LG. Similar DNA methylation levels in specific imprinting control regions in children conceived with and without assisted reproductive technology: a cross-sectional study. BMC Pediatr. 2012;12:33.

Rajender S, Avery K, Agarwal A. Epigenetics, spermatogenesis and male infertility. Mutat Res. 2011;727:62–71.

Rancourt RC, Harris HR, Barault L, Michels KB. The prevalence of loss of imprinting of H19 and IGF2 at birth. FASEB J. 2013;27:3335–43.

Reik W. Stability and flexibility of epigenetic gene regulation in mammalian development. Nature. 2007;447:425–32.

Robins JC, Marsit CJ, Padbury JF, Sharma SS. Endocrine disruptors, environmental oxygen, epigenetics and pregnancy. Front Biosci (Elite Ed). 2011;3:690–700.

Romundstad LB, Romundstad PR, Sunde A, von During V, Skjaerven R, Vatten LJ. Assisted fertilization and breech delivery: risks and obstetric management. Hum Reprod. 2009;24:3205–10.

Salian S, Doshi T, Vanage G. Neonatal exposure of male rats to Bisphenol A impairs fertility and expression of sertoli cell junctional proteins in the testis. Toxicology. 2009;265:56–67.

Sandhu KS. Systems properties of proteins encoded by imprinted genes. Epigenetics. 2010;5:627–36.

Sasaki H, Matsui Y. Epigenetic events in mammalian germ-cell development: reprogramming and beyond. Nat Rev Genet. 2008;9:129–40.

Savage T, Peek J, Hofman PL, Cutfield WS. Childhood outcomes of assisted reproductive technology. Hum Reprod. 2011;26:2392–400.

Seisenberger S, Peat JR, Reik W. Conceptual links between DNA methylation reprogramming in the early embryo and primordial germ cells. Curr Opin Cell Biol. 2013;25:281–8.

Sendler E, Johnson GD, Mao S, Goodrich RJ, Diamond MP, Hauser R, et al. Stability, delivery and functions of human sperm RNAs at fertilization. Nucleic Acids Res. 2013;41:4104–17.

Sharpe RM. Lifestyle and environmental contribution to male infertility. Br Med Bull. 2000;56:630–42.

Sharpe RM, Atanassova N, McKinnell C, Parte P, Turner KJ, Fisher JS, et al. Abnormalities in functional development of the Sertoli cells in rats treated neonatally with diethylstilbestrol: a possible role for estrogens in Sertoli cell development. Biol Reprod. 1998;59:1084–94.

Skinner MK, Manikkam M, Guerrero-Bosagna C. Epigenetic transgenerational actions of endocrine disruptors. Reprod Toxicol. 2011;31:337–43.

Smallwood SA, Kelsey G. De novo DNA methylation: a germ cell perspective. Trends Genet. 2012;28:33–42.

Smith ZD, Meissner A. DNA methylation: roles in mammalian development. Nat Rev Genet. 2013;14:204–20.

Stouder C, Somm E, Paoloni-Giacobino A. Prenatal exposure to ethanol: a specific effect on the H19 gene in sperm. Reprod Toxicol. 2011;31:507–12.

Tycko B, Morison IM. Physiological functions of imprinted genes. J Cell Physiol. 2002;192:245–58.

Urdinguio RG, Sanchez-Mut JV, Esteller M. Epigenetic mechanisms in neurological diseases: genes, syndromes, and therapies. Lancet Neurol. 2009;8:1056–72.

Vermeiden JP, Bernardus RE. Are imprinting disorders more prevalent after human in vitro fertilization or intracytoplasmic sperm injection? Fertil Steril. 2013;99:642–51.

Wu W, Shen O, Qin Y, Niu X, Lu C, Xia Y, et al. Idiopathic male infertility is strongly associated with aberrant promoter methylation of methylenetetrahydrofolate reductase (MTHFR). PLoS One. 2010;5:e13884.

Yaman R, Grandjean V. Timing of entry of meiosis depends on a mark generated by DNA methyltransferase 3a in testis. Mol Reprod Dev. 2006;73:390–7.

Zheng HY, Shi XY, Wang LL, Wu YQ, Chen SL, Zhang L. Study of DNA methylation patterns of imprinted genes in children born after assisted reproductive technologies reveals no imprinting errors: a pilot study. Exp Ther Med. 2011;2:751–5.

Chapter 7
The Black-Footed Ferret: On the Brink of Recovery?

Rachel M. Santymire, Travis M. Livieri, Heather Branvold-Faber, and Paul E. Marinari

Abstract In an attempt to save the species from extinction, the last remaining 18 black-footed ferrets (Mustela nigripes) were trapped up from the wild to initiate a captive breeding program. Nearly 30 years later more than 8,000 black-footed ferrets have been produced in captivity and approximately 4,100 animals have been reintroduced into 20 sites in eight US states (Arizona, New Mexico, Utah, Colorado, Kansas, Wyoming, South Dakota and Montana), Mexico and Canada. However, full recovery of the species has yet to be achieved, mainly due to limited viable habitat, disease and reduced fecundity. This chapter will highlight the advances in the black-footed ferret recovery program over the last 10 years including: (1) adaptive management techniques employed for the captive population; (2) development of new reintroduction sites and associated challenges facing wild black-footed ferrets; and (3) optimization of assisted reproductive techniques to secure the future of this rare species.

Keywords Endangered species • Captive breeding • Black-footed ferrets • Assisted reproductive techniques • Inbreeding depression • Fecundity • Recovery

---

R.M. Santymire, M.D., Ph.D. (✉)
Conservation and Science Department, Davee Center for Epidemiology and Endocrinology, Lincoln Park Zoo, 2001 N. Clark St., Chicago, IL 60614, USA
e-mail: rsantymire@lpzoo.org

T.M. Livieri, M.S.
Prairie Wildlife Research, P.O. Box 308 Wellington, CO 80549, USA
e-mail: tlivieri@prairiewildlife.org

H. Branvold-Faber, M.S., D.V.M.
Southside Animal Hospital, 12000 Industry Way, Anchorage, AK 99504, USA

P.E. Marinari, M.S.
Center for Species Survival, Smithsonian Conservation Biology Institute, Smithsonian National Zoological Park, 1500 Remount Rd., Front Royal, VA 22630, USA
e-mail: marinarip@si.edu

W.V. Holt et al. (eds.), *Reproductive Sciences in Animal Conservation*, Advances in Experimental Medicine and Biology 753, DOI 10.1007/978-1-4939-0820-2_7,

1 Introduction and Objectives

The black-footed ferret (*Mustela nigripes*) is the only ferret species endemic to North America. This endangered species' population was decimated during the first half of the twentieth century due to extensive poisoning of their main prey species, prairie dogs (*Cynomys* spp.), conversion of prairie grasslands to agriculture, and the arrival of sylvatic plague (*Yersinia pestis*) in North America. Black-footed ferrets were thought to be extinct until 1981, when a small population was discovered near Meeteetse, Wyoming (Miller et al. 1996). Between 1981 and 1984, the Meeteetse population grew to 129 animals. In 1985, disease including canine distemper virus (CDV), fatal to black-footed ferrets, and sylvatic plague, fatal to both black-footed ferrets and prairie dogs, was detected in Meeteetse population (Carr 1986). The subsequent rapid decline in ferret numbers drove biologists from the United States Fish and Wildlife Service (USFWS), Wyoming Game and Fish and partners to take action. With recommendations from the World Conservation Union's Conservation Breeding Specialist Group (CBSG), the last remaining 18 black-footed ferrets were removed from the wild (Carr 1986; Miller et al. 1996) to start a captive-breeding program.

In 1988, the USFWS developed a revised Black-Footed Ferret Recovery Plan which emphasized preservation of the species through natural breeding, a multi-institutional propagation program, and development of assisted reproductive techniques such as artificial insemination (AI), with fresh or thawed sperm, to promote retention of existing genetic diversity (Wildt and Goodrowe 1989; Howard et al. 1991), and help ensure that founders are equally represented in lineages (Howard et al. 2001). However, the ability to develop successful reproductive techniques in rare species is dependent on the amount of basic reproductive information known for that species. Consequently, a model species is often used to optimize development of assisted reproductive techniques. Because the domestic ferret (*Mustela putorius furo*) and the black-footed ferret are not only in the same genus but are genetically similar (O'Brien et al. 1989), the domestic ferret was used in development of reproductive techniques before application to its endangered counterpart (Howard et al. 1989, 1991; Wildt et al. 1989). Over the years, methods for semen collection (Wildt et al. 1989; Howard 1993), semen cryopreservation (Atherton et al. 1989; Howard et al. 1991) and AI using fresh (Howard et al. 1989; Wildt et al., 1989) and frozen-thawed semen (Howard et al. 1991), have been developed, as well as a black-footed ferret Genome Resource Bank (or GRB; defined as an organized repository of cryopreserved sperm; Wildt 1994).

To date more than 8,000 black-footed ferrets have been produced in captivity and approximately 4,100 animals have been reintroduced into 20 sites in eight US states (Arizona, New Mexico, Utah, Colorado, Kansas, Wyoming, South Dakota and Montana), Mexico and Canada (Table 7.1; Fig. 1). Additionally, there are currently about 300 animals managed at six captive-breeding facilities. Although the captive black-footed ferret breeding program has been successful at rearing sufficient numbers of kits for reintroduction projects and to sustain captive breeding efforts, full recovery of the species has yet to be achieved, mainly due to limited viable habitat, disease and

Table 7.1 Black-footed ferret reintroduction sites (adapted from Gober 2009)

Site (year initiated)	Prairie dog species[a]	Land ownership	Total # BFFs released	Estimated 2012 population
[1]Shirley Basin, WY (1991)	Wtpd	Private, federal, state	513	239 (2009)
[2]UL Bend Natl. Wild. Refuge, MT (1994)	Btpd	Federal	252	18
[3]Badlands Natl. Park, SD (1994)	Btpd	Federal	225	30
[4]Aubrey Valley, AZ (1996)	Gpd	Tribal, private	399	123[b]
[5]Conata Basin, SD (1996)	Btpd	Federal	161	72
[6]Fort Belknap Indian Reservation, MT (1997)	Btpd	Tribal	180	0
[7]Coyote Basin, UT (1999)	Wtpd	Federal	424	10[b]
[8]Cheyenne River Indian Reservation, SD (2000)	Btpd	Tribal	350	26[b]
[9]BLM 40-Complex, MT (2001)	Btpd	Federal	92	0
[10]Wolf Creek, CO (2001)	Wtpd	Federal	254	0
[11]Janos, Mexico (2001)	Btpd	Private, community	299	17 (2006)
[12]Rosebud Indian Reservation, SD (2003)	Btpd	Tribal	150	30 (2006)
[13]Lower Brule Indian Reservation, SD (2006)	Btpd	Tribal	107	25[b]
[14]Wind Cave Natl. Park, SD (2007)	Btpd	Federal	61	67
[15]Espee Ranch, AZ (2007)	Gpd	Private	70	0
[16]Butte Creek/Smoky Valley Ranch, KS (2007)	Btpd	Private	125	39[b]
[17]Northern Cheyenne Indian Reservation, MT (2008)	Btpd	Tribal	88	0
[18]Vermejo Park Ranch, NM	Btpd	Private	255	5[b]
[19]Grasslands Natl. Park, Canada (2009)	Btpd	Federal	75	12[b]
[20]Vermejo Park Ranch, NM (2012)	Gpd	Private	20	11
Total			4,100	

[a]Wtpd = white-tailed prairie dog (*Cynomys leucurus*), Btpd = black-tailed prairie dog (*Cynomys ludovicianus*), Gpd = Gunnison's prairie dog (*Cynomys gunnisoni*)
[b]Estimate made prior to release of additional animals included in previous column

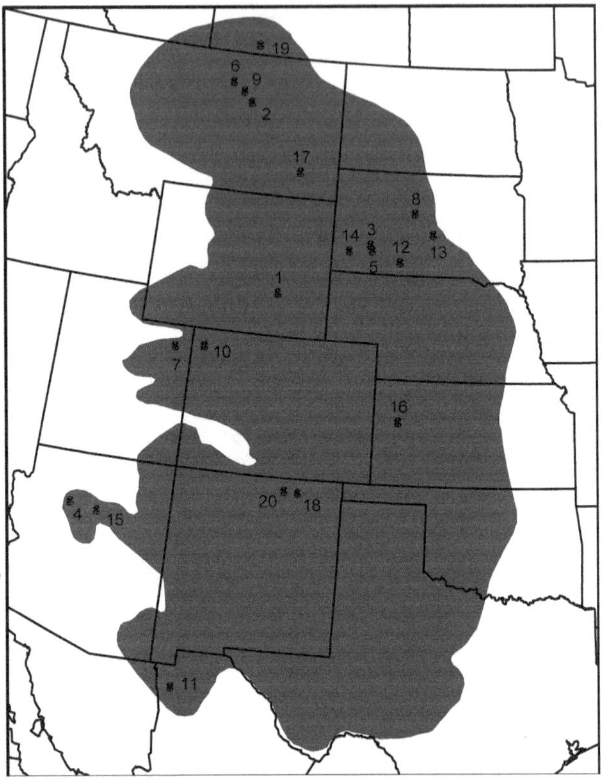

Fig. 7.1 Historical range (*shaded area*) of the black-footed ferret in North America and all reintroduction sites through 2012 (refer to Table 7.1).

reduced fecundity. This chapter will highlight the advances in the black-footed ferret recovery program over the last 10 years including: (1) adaptive management techniques employed for the captive population; (2) development of new reintroduction sites and associated challenges facing wild black-footed ferrets; and (3) optimization of assisted reproductive techniques to secure the future of this rare species.

2 Management of the Black-footed Ferret Recovery Program

The Black-footed Ferret Species Survival Plan® (SSP) collectively manage a core breeding population and house a small group of ferrets not suitable for reintroduction. The SSP's primary goal is to produce as many black-footed ferret kits as possible in order sustain future captive breeding and supply excess captive-born animals for ongoing reintroduction efforts (Garelle et al. 2012). Since 2000, six facilities across North America comprise the SSP, including: the USFWS's National Black-footed Ferret Conservation Center (Colorado), Smithsonian Conservation Biology Institute (Virginia, of the Smithsonian's National Zoological Park), Louisville

Zoological Garden (Kentucky), Cheyenne Mountain Zoo (Colorado), Phoenix Zoo (Arizona) and Toronto Zoo (Ontario, Canada). The number and demography of the SSP population has been maintained at relatively consistent levels since the late 1990s; however, new genetic software programs have altered decision making processes surrounding pairings and identifying reintroduction candidates. Prioritizing or determining which male should be paired with each female is accomplished using MateRx (Ballou et al. 2001). This analytical software program, developed jointly by the Smithsonian National Zoological Park (Washington, DC) and Lincoln Park Zoo (Chicago, IL) provides captive breeding facilities a numerical rating for every possible breeding pair. These ratings or Mate Suitability Indices (MSI) integrate several genetic factors, including the expected change in genetic diversity due to resultant offspring, the relative rareness or commonness of the parent's genetic make-up, inbreeding coefficient of offspring produced by a pair, and proportion, if any, of unknown pedigree (Garelle et al. 2012). The target number of ferrets allocated to reintroduction sites is determined by the USFWS prior to the annual SSP Master Planning meeting. Dynamic culling is utilized to designate individuals for release until the desired age and sex structure for the SSP is attained. A collaborative workshop sponsored by USFWS, the Black-footed Ferret Recovery Implementation Team (BFFRIT) and CBSG was held in 2003 and addressed recovery challenges facing captive and field populations.

Black-footed ferrets were extinct in the wild by 1987 after the remaining 18 were removed and placed into captivity. Successful captive breeding efforts have produced enough kits that wild reintroductions began in 1991 and have continued today. Partners and sites request black-footed ferrets from the USFWS through an annual allocation process (Jachowski and Lockhart 2009). Initial years of reintroduction focused on rearing and acclimation strategies (Biggins et al. 1998), habitat selection and use and techniques to monitor populations. Recent efforts have shifted towards disease management/mitigation in addition to finding and developing new release sites. Reintroduction sites occur on a variety of land ownership patterns including federal, state, tribal and private lands in eight US States, Mexico and Canada (Table 7.1; Fig. 1).

A site in Chihuahua, Mexico, at the southernmost portion of the black-footed ferret range, is within the Janos Biosphere Reserve (List et al. 2010) and includes private and community-owned lands. This site once contained the largest remaining complex of black-tailed prairie dogs (i.e. black-footed ferret habitat) in contemporary North America, but extreme drought has caused degradation and desertification of much of those grasslands. In recent years area drug violence has prevented biologists from assessing the status of reintroduced black-footed ferrets at the Mexico site.

Tribal lands are important for black-footed ferret recovery and account for seven of the 20 black-footed ferret reintroduction sites. Releases on private lands in Wyoming, Arizona, Kansas and New Mexico have also occurred and private land partnerships are regarded as vital to ultimate recovery success. Eco-capitalist Ted Turner and the Turner Endangered Species Fund have put considerable effort into prairie dog restoration and subsequent releases of black-footed ferrets onto his Vermejo Park Ranch in New Mexico. With the release of black-footed ferrets into

Grasslands National Park, Saskatchewan in 2009, black-footed ferrets were returned to the Canadian prairies, near the northern limit of their historical range.

A total of 4,100 black-footed ferrets were released 1991–2012 with 213 of those being of wild origin, captured, and translocated to other reintroduction sites. Wild kits used for direct translocation originated primarily from Conata Basin, South Dakota prior to the spread of plague into that population and subsequent site impacts. Translocation of wild black-footed ferrets has been a successful tool, and is regarded as one of the most effective means of initiating new reintroduction projects or supplementing existing sites (Biggins et al. 2001). Five of 20 reintroduction sites are currently considered devoid of black-footed ferrets, mostly because of plague, but may be re-considered if plague can be mitigated.

Within reintroduction sites, we observed both genotypic and phenotypic differences based on the length of time of population persistence since captive-born black-footed ferrets were first released, and if subsequent augmentations of captive-reared ferrets occurred. Based on nine microsatellite loci, the Wyoming reintroduction site, which had a slow establishment and population growth and had not been augmented with captive black-footed ferrets for >10 years, had reduced heterozygosity and fewer polymorphic loci compared with the Conata Basin, South Dakota reintroduced population (rapid population growth following initial establishment and augmentation within the last 5 years) and Aubrey Valley, Arizona (yearly augmentation). The Wyoming black-footed ferrets also had phenotypic changes, specifically shorter limbs and smaller overall body size, than other three populations (Wisely et al. 2007).

Previous morphological research demonstrated that captive black-footed ferrets were on average 4–10 % smaller and differently shaped based on skull morphology than historical museum specimens (Wisely et al. 2002). Because captive black-footed ferrets are the source population of all reintroduced black-footed ferrets, we expected that wild-born black-footed ferrets would be smaller than historical specimens; however, it was recently determined that wild-born individuals from reintroduction areas were 2–5 % larger than their captive-born counterparts and had returned to historical size, suggesting that reduced size was an environmental not a genetic effect (Wisely et al. 2005). Interestingly, based on these morphometric data, it was also determined that canine width can be used to age black-footed ferrets, which assists with demographic assessments of reintroduction sites (Santymire et al. 2012). These results demonstrate the importance of monitoring the genetic health and signs of inbreeding in the wild populations so that management strategies, such as translocation or augmentation with additional captive black-footed ferrets, can be employed.

3 Disease Management

Disease has limited the sustainability of both captive and wild black-footed ferret populations. Specifically, CDV has long been known to cause morbidity and mortality in this species, and contributed to the decline of the original Meeteetse population. Black-footed ferrets in captivity also succumbed to CDV (Williams et al. 1998), and others administered a modified-live CDV vaccine died from vaccine-induced infection (Carpenter et al. 1976). Even with supportive care, the mortality rate of CDV approaches 100 %. PureVax® Ferret Distemper Vaccine, a live, monovalent canarypox-vectored vaccine developed by Merial, Inc., Athens, GA, has proven safe and effective in the Siberian polecat (Wimsatt et al., 2003) and has been subsequently tested in the black-footed ferret. These trials have indicated that a minimum of two doses of vaccine produce protective neutralizing antibody titers (Marinari and Kreeger 2004). Currently, all captive born ferrets and many wild ferrets receive two doses of PureVax® Ferret Distemper Vaccine beginning as early as 60 days of age.

Plague is caused by the bacterium *Yersinia pestis*. It can be transmitted via flea vector, aerosol or ingestion of contaminated food. Plague entered North America's west coast in the early 1900s via ships carrying flea-infested rats. It has since been spreading eastward, infecting and killing many native species, including prairie dogs and black-footed ferrets. Interestingly, the domestic ferret appears resistant to plague (Williams et al. 1994), and it was initially believed that the black-footed ferret was as well. But in the 1990s, high susceptibility (with virtually 100 % mortality) to plague was discovered by Williams et al. (1994) and later affirmed by others (Rocke et al. 2004). This led to substantial research into developing a plague vaccine. The U.S. Geological Survey's National Wildlife Health Center (Madison, WI) in collaboration with United States Army Medical Research Institute for Infectious Diseases (Frederick, MD) was instrumental in developing a vaccine consisting of the plague antigens F1 and V (Rocke et al. 2004). Challenge studies were conducted using two doses of F1-V vaccine, with a 69 % survival rate (Rocke et al. 2004). Now, all captive born and many wild black-footed ferrets are given two doses of F1-V vaccine as part of the recovery management strategy. Unfortunately, vaccination of black-footed ferrets alone is insufficient to eliminate the threat of plague. A major obstacle to black-footed ferret recovery is the high susceptibility and wholesale loss of large prairie dog populations to plague, making control of this disease, both for predator and prey, a high priority. The BFFRIT is currently working with partners on the development of a safe and effective, oral bait plague vaccine for prairie dogs.

Direct manipulation of fleas has been an effective but labor intensive method to manage plague in the wild. Application of powdered insecticides directly into prairie dog burrows can be an effective short term solution but many of these chemicals lose their effectiveness quickly (Barnes et al. 1972; Karhu and Anderson 2000). More recently, the chemical deltamethrin, formulated into a waterproof dust as DeltaDust, has demonstrated protection of both prairie dog and black-footed ferrets for up to 10 months post-application (Biggins et al. 2010; Matchett et al. 2010). Dusting however is labor intensive, costly and has potential secondary effects on non-target species (Cully et al. 2006). Recent evidence from Conata Basin, South Dakota suggests that dusting alone may not be sufficient to maintain black-footed ferret populations during a plague epizootic and vaccination of black-footed ferrets with F1-V significantly increases survivorship in dusted prairie dog colonies (Livieri pers. comm.).

4 Advanced Assisted Reproductive Technology

With over 20 years of development, assisted reproductive technology has maintained the genetic diversity of the black-footed ferret captive population. Specifically, from 1996 through 2008, nearly 140 individuals have been produced through artificial insemination using fresh or frozen/thawed semen (Howard and Wildt 2009). These offspring are from sires that would not have breed naturally due to behavioral problems and/or were given multiple chances to breed on their own, but failed to produce a litter (Howard and Wildt 2009). Excitingly, frozen/thawed semen from the black-footed ferret GRB that had been stored for greater than 10 years was used successfully in two AI procedures resulting in two genetically valuable kits (Howard and Wildt 2009). These successful AIs clearly demonstrate that the black-footed ferret GRB is an integral part of the recovery and conservation program. In response to these successful breeding, efforts are being made to improve semen cryopreservation techniques. Specifically, it was determined that black-footed ferrets produce ejaculates with high osmolality (~500 mOsm) compared to serum levels (320 mOsm; Santymire et al. 2006), and compared to semen osmolality (~300 mOsm) from an array of other species (boar, bull, dog, cat, human, stallion, birds species; Gao et al. 1997; Blanco et al. 2000; Pukazhenthi et al. 2000). Black-footed ferret spermatozoa have also demonstrated sensitivity to osmotic stress with hyperosmotic conditions resulting in reduced sperm motility and acrosomal integrity (Santymire et al. 2006). Additionally, when designing a black-footed ferret specific semen cryopreservation protocol, it was determined that the spermatozoa were sensitive to cooling (from 37 to 4 °C) and required a slower rate (0.12 °C/min; Santymire et al. 2007) than what was previously used (0.20 °C/min; Atherton et al. 1989). Furthermore, it was demonstrated that an egg yolk based medium used with a pellet method of freezing achieved the highest post-thaw sperm viability in domestic ferrets (Howard et al. 1991) and black-footed ferrets (Santymire et al. 2007).

5 Current Captive Population Challenges

Loss of fecundity is one of the issues impeding the black-footed ferret captive breeding program (Howard et al. 2006). According to the black-footed ferret SSP (Garelle et al. 2012), there has been a decrease in the captive population's fecundity, indicated by a decrease in whelping rates (from 60 to 46 %) in females and normal sperm (from 50 to 16 %) in males (Wolf et al. 2000; Santymire et al. 2006, 2007). The causes of these physiological changes are unknown; both nutritional and genetic hypotheses are under examination. However, due to its limited gene pool (starting from 18 black-footed ferrets of unknown relatedness of which 15 successfully bred), issues with inbreeding depression are inevitable (Reading and Clark 1996). Negative implications of genetic loss on seminal quality have been observed in lions (*Panthera leo*; Wildt et al. 1987; Munson et al. 1996), cheetahs (*Acinonyx jubatus*;

Wildt et al. 1983; O'Brien et al. 1985) and Florida panthers (*P. concolor coryi*; Roelke et al. 1993; Barone et al. 1994).

Because nutrition can be a limiting factor for reproductive success in wild and domesticated mammals, it is important to investigate its attribution to declining fitness in the black-footed ferret captive population. The role of oxidative stress, a condition associated with increased reactive oxygen species (ROS), has been implicated as a cause of male infertility (Agarwal and Saleh 2002). Mammalian spermatozoa are rich in polyunsaturated fatty acids making them susceptible to ROS attack, which can result in decreased sperm motility and viability (Lamirande and Gagnon 1992; Sikka 1995, 2004). Vitamin E is an antioxidant which can promote sperm viability by eliminating ROS, improving sperm motility and increasing the number of spermatozoa/ejaculate {Brezezinska-Slebodzinska et al. 1995; Suleiman et al. 1996; Comhaire et al. 2000). An imbalance of vitamin A can also adversely affect reproduction causing low conception rates, stillbirths and abnormal sperm (NRC 1987) and interfering with vitamin E and selenium absorption (Combs 1976; Vahl and Van't Klooser 1987; Mazzaro et al. 1995; Surai et al. 1998).

Records reveal that a decrease in normal sperm (%) in captive black-footed ferrets occurred 1 year following the conversion of diet in 2001, from a manually-prepared rabbit meat-based diet to a commercial horsemeat-based diet (Toronto Small Carnivore, Milliken Meats, Canada). A preliminary study was conducted in 2007 on 23 male black-footed ferrets housed at the USFWS's National Black-footed Ferret Conservation Center (Carr, Colorado, USA) to determine whether serum levels were comparable to other closely related species, and to investigate the relationship between serum levels and vitamin E and A. Blood results demonstrated that mean serum levels of vitamin E (4.4 ± 0.8 μg/ml; range, 0.3–16.2 μg/ml) were lower in black-footed ferrets compared to mink and domestic ferrets (13–21 μg/ml); but the males had similar levels of vitamin A (0.36 ± 0.03 μg/ml; range, 0.09–0.61 μg/ml to other mustelids, 0.3–0.7 μg/ml; NRC 1987; Santymire et al. 2009). When investigating the relationship of male serum vitamin E level and ejaculate traits, it was determined that both sperm motility (%) and intact acrosomal membranes (%) were not related to serum vitamin E concentration. However, serum vitamin A had a negative impact on sperm motility (Fig. 7.2) and intact acrosomal membranes (Fig. 7.3). The percentage of normal spermatozoa was negatively affected by both vitamin E (Fig. 7.4) and A (Fig. 7.5). Due to these inconclusive results, further research is needed to determine etiology with declining fecundity in the captive black-footed ferret population, and this relationship to nutrition, husbandry and genetics.

Wildlife living in captive settings may experience stressors that can limit reproduction and health. Prolonged and/or chronic stress, as indicated by high levels of glucocorticoids, can negatively affect the body by suppressing the immune system resulting in increased susceptibility to disease and inhibit reproduction. Chronic exposure to glucocorticoids can cause severe protein loss (muscle wasting), disruption of normal behaviors, neuronal cell malfunction, suppression of growth and various other pathologic conditions in humans and other vertebrates (Wingfield and Romero 2001; Boonstra 2005; Wingfield 2005). Additionally, glucocorticoids can impact the reproductive system by inhibiting the gonadotrophin releasing hormone

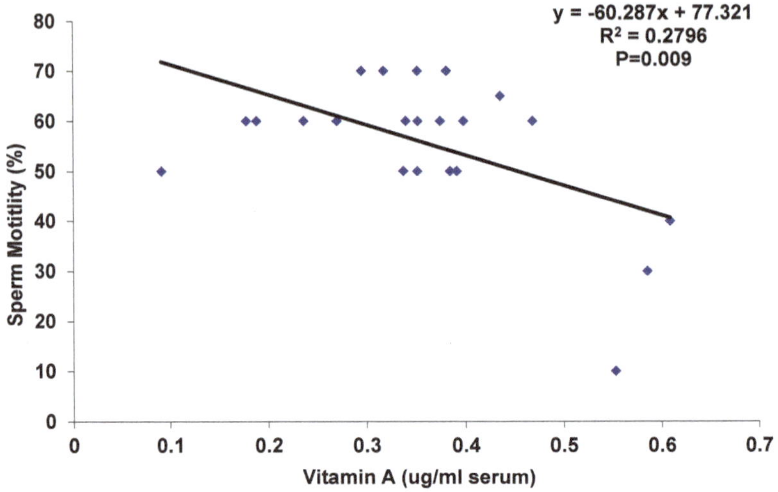

Fig. 7.2 Relationship between sperm motility (%) and serum vitamin A levels (μg/ml) in 23 male black-footed ferrets housed at the USFWS's National Black-footed Ferret Conservation Center (Carr, CO).

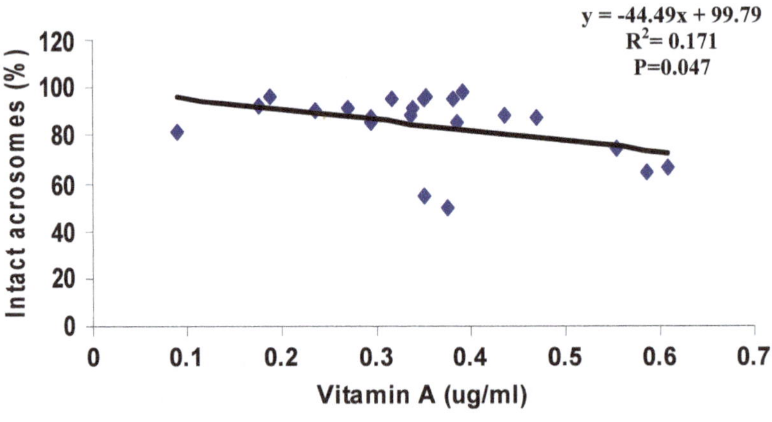

Fig. 7.3 Relationship between intact acrosomes (%) and serum vitamin A levels (μg/ml) in 23 male black-footed ferrets housed at the USFWS's National Black-footed Ferret Conservation Center (Carr, CO).

(GnRH), and suppressing the follicle-stimulating hormone (FSH) and luteinizing hormone (LH), which leads to depressed testosterone production in rats, bulls and men (Welsh et al. 1999). Chronically elevated cortisol can increase in immature sperm and abnormal sperm (Welsh et al. 1999). Interestingly, those individuals who already possess low sperm quality may be at greater risk of stressor-induced depression of spermatogenesis (Welsh et al. 1999).

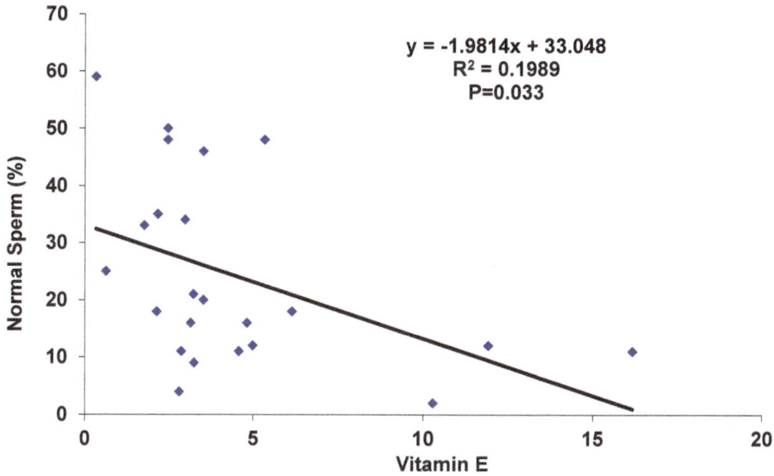

Fig. 7.4 Relationship between normal sperm (%) and serum vitamin E levels (μg/ml) in 23 male black-footed ferrets housed at the USFWS's National Black-footed Ferret Conservation Center (Carr, CO).

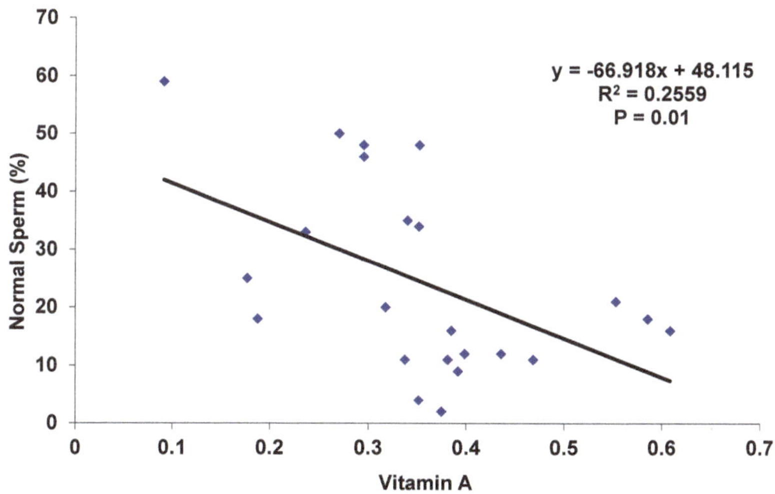

Fig. 7.5 Relationship between normal sperm (%) and serum vitamin A levels (μg/ml) in 23 male black-footed ferrets housed at the USFWS's National Black-footed Ferret Conservation Center (Carr, CO).

One successful method managers use to improve general well-being is to provide environmental enrichment, which is the practice of providing captive managed animals with environmental stimuli. Enrichment can reduce the impact of stress by encouraging social interaction, and lowering aggression and abnormal behavior (Shepherdson et al. 1993); by promoting natural foraging strategies (obtaining, manipulating and exploring acts; Lindburg 1988); and, preventing physiology, morphological changes (O'Regan and Kitchener 2005). A recent study conducted on captive black-footed ferrets demonstrated that environmental enrichment benefitted captive juvenile male ferrets by reducing adrenocortical activity, but increased fecal glucocorticoid metabolites in adult females and had no effect on juvenile females and adult males (Poessel et al. 2011). More research is needed to investigate the relationship among fecal glucocorticoid metabolites, environmental enrichment and reproductive traits in captive black-footed ferrets.

6 Priorities for the Future

The black-footed ferret is one of the most endangered mammals in North America and has required intensive management of both *in situ* and *ex situ* populations for over 20 years. Today, there are approximately 300 black-footed ferrets managed in captivity and the estimated number of wild animals ranges from 400 to 800. Unfortunately, full recovery of the species still hangs in the balance and significant challenges persist. With assisted reproductive technology we have monitored declining captive reproduction which could be attributed to inbreeding depression and/or environmental factors. Because this decline has occurred at a rapid rate since 2001 and we suspect that not just inbreeding depression is responsible, but could be caused by environmental effects, such as nutrition, genetic management, facility lighting and/or stress. Since captive produced ferrets are the main source of animals for ongoing and future reintroduction efforts, declining fecundity in the captive population is a significant program issue.

Another future program priority is to monitor and compare captive and wild black-footed ferret fecundity. From 2002 to 2006, we conducted a biomedical survey on wild, reintroduced black-footed ferret populations. Results demonstrated that wild black-footed ferret semen traits were significantly improved (35–45 % normal sperm) compared to captive (20 % at that time; Santymire et al. 2004). And over the last 6 years, captive black-footed ferret fecundity has continued to decline. Consequently, assessing the current status and reproductive health of both captive and wild black-footed ferret populations is becoming increasingly critical to species recovery.

Ultimately, the success of black-footed ferret recovery can be achieved only with the establishment of sustainable wild populations. This requires a consolidated effort of continued captive propagation and expanded support by public, private and tribal land managers to proactively develop and maintain adequate prairie dog habitats for ferrets across the species historical range. The BFFRIT must continue to confront issues of paramount importance to species persistence and population expansion; including disease management, increased public education, prairie dog vaccination trials and landowner incentive programs; and develop and ensure long range management of new reintroduction sites. Together, over 35 federal, state, tribal, and non-government organizations have joined with private landowners to ensure that wild populations of black-footed ferret continue to persist in a changing world.

Acknowledgements The authors acknowledge the USFWS and members of the BFFRIT especially the partner institutions comprising the Black-footed Ferret SSP for continued support. We are indebted to Dr. Julie Kreeger for providing veterinary care to animals housed at the National Black-footed Ferret Conservation Center from 1996 to 2012. The late Drs. JoGayle Howard, Elizabeth Williams and E. Tom Thorne will forever be an inspiration to all those working on the recovery of black-footed ferrets.

References

Agarwal A, Saleh RA. Role of oxidants in male infertility: rationale, significance, and treatment. Urol Clin N Am. 2002;29:1–12.

Atherton RW, Straley M, Curry P, Slaughter R, Burgess W, Kitchin RM. Electroejaculation and cryopreservation of domestic ferret sperm. In: Seal US, Thorne ET, Anderson SH, Bodan MA, editors. U.S. Conservation biology and the black-footed ferret. New Haven, CT: Yale University Press; 1989. p. 177–89.

Ballou JD, Earnhardt J, Thompson S. MATERX: Genetic Management Software. Smithsonian Institution, National Zoological Park, Washington, DC; 2001.

Barnes AM, Ogden LJ, Campos EG. Control of the plague vector, *Opisocrostis hirsutus*, by treatment of the prairie dog (*Cynomys ludovicianus*) burrows with 2% carbaryl dust. J Med Entomol. 1972;11:83–87.

Barone MA, Roelke ME, Howard JG, Brown JL, Anderson AE, Wildt DE. Reproductive characteristics of male Florida panthers: comparative studies from Florida, Texas, Colorado, Latin America, and North American Zoos. J Mammal. 1994;75(1):150–62.

Biggins DE, Godbey JL, Hanebury LR, Luce B, Marinari PE, Matchett MR, Vargas A. The effect of rearing methods on survival of reintroduced black-footed ferrets. J Wildlife Manage 1998;62: 643–653.

Biggins DE, Godbey JL, Horton BM, Livieri TM. Movements and survival of black-footed ferrets associated with an experimental translocation in South Dakota. J Mammal. 2001;92:742–750.

Biggins DE, Godbey JL, Gage KL, Carter LG, Montenieri JA. Vector control improves survival of three species of prairie dogs (*Cynomys*) in areas considered enzootic for plague. Vector-borne Zoonot. 2010;10:17–26.

Blanco JM, Gee G, Wildt DE, Donoghue AM. Species variation in osmotic, cryoprotectant, and cooling rate tolerance in poultry, eagle, and peregrine falcon spermatozoa. Biol Reprod. 2000;63:1164–71.

Boonstra R. Equipped for life: the adaptive role of stress axis in male mammals. J Mammal. 2005;86:236–47.

Brezezinska-Slebodzinska E, Slebodzinski AB, Pietras B, Wieczorek G. Antioxidant effect of vitamin E. and glutathione on lipid peroxidation in boar semen plasma. Biol Trace Elem Res. 1995;47:69–74.

Carpenter JW, Appel MJG, Erickson RC, Novilla MN. Fatal vaccine-induced canine distemper virus infection in black-footed ferrets. J American Veterin Medic Associ 1976:169:961–964.

Carr A. Introduction. Gt Basin Nat Mem. 1986;9:1–7.

Combs GFJ. Differential effects of high dietary levels of vitamin A on the vitamin E-selenium nutrition of young and adult chickens. J Nutr. 1976;106:967–75.

Comhaire FH, Christophe AB, Zalata AA, Dhooge WS, Mahmoud AM, Depuydt CE. The effects of combined conventional treatment, oral antioxidants and essential fatty acids on sperm biology in subfertile men. Prostaglandins Leukot? Essential Fatty Acids. 2000;63:159–65.

Cully JF, Biggins DE, Seery DB. Conservation of prairie dogs in areas with plague. In: Conservation of the black-tailed prairie dog, Hoogland, JL (editor), Washington DC: Island Press. 2006;157–168

Gao D, Mazur P, Critser JK. Fundamental cryobiology of mammalian spermatozoa. In: Karow AM, Critser JK, editors. Reproductive tissue banking: scientific principles. London: Academic Press; 1997. p. 263–328.

Garelle D, Marinari P, Lynch C. Breeding and transfer plan: Black-footed ferret (*Mustela nigripes*) Species Survival Plan. (ed Population Management Center). Association of Zoos and Aquariums, Chicago. 2012.

Gober P. A challenging future for the Black-footed Ferret. Endangered Species Bulletin 2009;34: 38–39.

Howard JG. Semen collection and analysis in nondomestic carnivores. In: Fowler ME, editor. Zoo and wild animal medicine: current therapy II. Philadelphia, PA: W.B. Saunders Co.; 1993. p. 390–9.

Howard JG, Bush M, Morton C, Morton F, Wildt DE. Comparative semen cryopreservation in ferrets (Mustela putorius furo) and pregnancies after laparoscopic intrauterine insemination with frozen-thawed spermatozoa. J Reprod Fertil. 1991;92:109–18.

Howard JG, Hurlbut SL, Morton C, Morton F, Bush M, Wildt DE. Pregnancies in the domestic ferret after laparoscopic artificial insemination with frozen-thawed spermatozoa. J Androl (Suppl.). Abstract 135:52, 1989.

Howard JG, Marinari PE, Wildt DE. Mustelid reproductive biology and integration of assisted reproductive technology in species recovery. In: The first international symposium on assisted reproductive technology for the conservation and genetic management of wildlife, p. 159–165. XY Inc., Omaha, Nebraska, 2001.

Howard JG, Santymire RM, Marinari PE, Kreeger JS, Williamson L, Wildt DE. Use of reproductive technology for black-footed ferret recovery. In: Proceedings of the black-footed ferret: progress and continuing challenges, p. 28–36. USGS. Fort Collins, CO, 2006.

Howard JG, Wildt DE. Approaches and efficacy of artificial insemination in felids and mustelids. Theriogenology. 2009;71:130–48.

Jachowski DS, Lockhart JM. Reintroducing the black-footed ferret *Mustela nigripes* to the Great Plains of North America. Small Carnivore Conserv. 2009;41:58–64.

Karhu R, Anderson S. Effects of pyriproxyfen spray, powder, and oral bait treatments on the relative abundance of fleas (Siphonaptera: Ceratophyllidae) in black-tailed prairie dog (Rodentia: Sciuridae) towns. J Med Entomol. 2000;37:864–871.

Lamirande ED, Gagnon C. Reactive oxygen species and human spermatozoa I. Effects on the motility of inact spermatozoa and on sperm. J Androl. 1992;13:368–78.

Lindburg DG. Improving the feeding of captive felines through application of field data. Zoo Biol. 1988;7:211–8.

List R, Pacheco J, Ponce E, Sierra-Corona R, Ceballos G. The Janos Biosphere Reserve, Northern Mexico. International Journal of Wilderness. 2010;16:35–41.

Marinari PE, Kreeger JS. An adaptive management approach for black-footed ferrets in captivity. In: Roelle JE, Miller BJ, Godbey JL, Biggins DE. Symposium on the Status of the Black-footed Ferret and Its Habitat Scientific Investigations Report 2005-5293. Fort Collins: U.S. Department of the Interior. 2004;23–27.

Matchett MR, Biggins DE, Carlson V, Powell B, Rocke T. Enzootic plague reduces black-footed ferret (*Mustela nigripes*) survival in Montana. Vector-borne Zoonot. 2010;10:27–35.

Mazzaro LM, Dunn L, Furr HC, Clark RM. Study of vitamin A supplementation in captive northern fur seals (*Callorhinus ursinus*) and its effect on serum vitamin E. Mar Mamm Sci. 1995; 11:545–53.

Miller BJ, Reading R, Forrest S. Prairie night: black-footed ferrets and the recovery of endangered species. Washington, DC: Smithsonian Institution Press; 1996.

Munson L, Brown JL, Bush M, Packer C, Janssen D, Reiziss SM, Wildt DE. Genetic diversity affects testicular morphology in free-ranging lions (*Panthera leo*) of the Serengeti Plains and Ngorongoro Crater. J Reprod Fertil. 1996;108:11–5.

National Research Council (NRC). Vitamin tolerance of animals. Washington, DC: National Academy Press; 1987.

O'Brien SJ, Martenson JS, Eichelberger MA, Thorne ET, Wright F. Genetic variation and molecular systematics of the black-footed ferret. In: Seal US, Anderson SH, Bogan MA, editors.

Conservation Biology and the Black-Footed Ferret. New Haven, CT: Yale University Press; 1989. p. 21–33.

O'Brien SJ, Roelke ME, Marker L, Newman A, Winkler CA, Meltzer D, Colly L, Evermann JF, Bush M, Wildt DE. Genetic basis for species vulnerability in the cheetah. Science. 1985; 227:1428–34.

O'Regan HJ, Kitchener AC. The effects of captivity on the morphology of captive, domesticated and feral mammals. Mamm Rev. 2005;35:215–30.

Poessel S, Biggins D, Santymire R, Livieri T, Cooks K, Angeloni L. Environmental enrichment affects adrenocortical stress responses in the endangered black-footed ferret. Gen Comp Endocrinol. 2011;172:526–33.

Pukazhenthi BS, Noiles E, Pelican K, Donoghue A, Wildt DE, Howard JG. Osmotic effects on feline spermatozoa from normospermic versus teratospermic donors. Cryobiology. 2000; 40:139–50.

Reading RP, Clark TW. Carnivore reintroductions: an interdisciplinary examination. In: Gittleman JL, editor. Carnivore behavior, ecology, and evolution, vol. 2. Ithaca, NY: Cornell University Press; 1996. p. 296–336.

Roelke ME, Martenson JS, O'Brien SJ. The consequences of demographic reduction and genetic depletion in the endangered Florida panther. Curr Biol. 1993;3(6):340–50.

Rocke TE, Mencher J, Smith SR, Friedlander AM, Andrews GP, Baeten LA. Recombinant F1-V fusion protein protects black-footed ferrets (*Mustela nigripes*) against virulent Yersinia pestis infection. J Zoo Wildlife Med. 2004;35:142–146.

Santymire RM, Howard JG, Wisely SM, Kreeger JS, Marinari PE, Wildt DE. Seminal characteristics of wild black-footed ferrets (*Mustela nigripes*). In: Defenders of wildlife carnivores-expanding partnerships in carnivore conservation, pp. 200. Santa Fe, NM; 2004.

Santymire RM, Marinari PE, Kreeger JS, Wildt DE, Howard JG. Sperm viability in the black-footed ferret (*Mustela nigripes*) is influenced by seminal and medium osmolality. Cryobiology. 2006;53:37–50.

Santymire RM, Marinari PE, Kreeger JS, Wildt DE, Howard JG. Slow cooling prevents cold-induced damage to sperm motility and acrosomal integrity in the black-footed ferret (*Mustela nigripes*). Reprod Fertil Dev. 2007;19:652–63.

Santymire R, Marinari P, Kreeger J. Determining the relationship among serum vitamin E and A levels and seminal quality in male black-footed ferrets. The Wildlife Society 16th Annual Conference, September 20–24, 2009 in Monterey, California.

Santymire RM, Wisely SM, Livieri TM, Howard JG. Using canine width to determine age in the black-footed ferret *Mustela nigripes*. Small Carnivore Conserv. 2012;46:17–21.

Santymire RM, Branvold-Faber H, Marinari PE. Recovery of the Black-footed Ferret. In: Fox JG editor. Biology and diseases of the ferret. 3rd ed. Philadelphia, PA: Lippincott Williams & Wilkins. 2014;219–231.

Shepherdson DJ, Carlstead K, Mellen JD, Seidensticker J. The influence of food presentation on the behavior of small cats in confined environments. Zoo Biol. 1993;12:203–16.

Sikka SC. Role of oxidative stress and antioxidants in andrology and assisted reproductive technology. J Androl. 2004;25:5–17.

Sikka SC, Rajasekaran M, Hellstrom WJG. Role of oxidative stress and antioxidants in male infertility. J Androl. 1995;16:464–8.

Suleiman SA, Ali ME, Zaki ZMS, El-Malik EMA, Nasr MA. Lipid peroxidation and human sperm motility: Protective role of viamin E. J Androl. 1996;17:530–7.

Surai PF, Ionov IA, Kuklenko TV, Kostjuk IA, Acpherson AM, Speake BK, Noble RC, Sparks NHC. Effect of supplementing the hen's diet with vitamin A on the accumulation of vitamins A and E, ascorbic acid and carotenoids in the egg yolk and in the embryonic liver. Br Poultry Sci. 1998;39:257–63.

Vahl HA, Van't Klooser AT. Effects of excessive vitamin A levels in broiler rations. J Anim Physiol Anim Nutr. 1987;57:204–18.

Welsh THJ, Kemper-Green CN, Livingston KN. Stress and reproduction. In: Neill JD, Knobil E, editors. Encyclopedia of reproduction. New York, NY: Academic Press; 1999. p. 662–74.

Wildt DE. Endangered species spermatozoa: diversity, research, and conservation. In: Bartke A, editor. Function of somatic cells in the testes. New York, NY: Springer-Verlag; 1994. p. 1–24.
Wildt DE, Bush M, Goodrowe KL, Packer C, Pusey AE, Brown JL, Joslin P, O'Brien SJ. Reproductive and genetic consequences of founding isolated lion populations. Nature. 1987;329:328–31.
Wildt DE, Bush M, Howard JG, O'Brien SJ, Meltzer D, Van Dyk A, Ebedes H, Brand DJ. Unique seminal quality in the South African cheetah and a comparative evaluation in the domestic cat. Biol Reprod. 1983;29:1019–25.
Wildt DE, Goodrowe K. The potential for embryo technology in the black-footed ferret. In: Seal US, Thorne ET, Anderson SH, Bogan MA, editors. Conservation biology and the black-footed ferret. New Haven, CT: Yale University Press; 1989. p. 160–76.
Wildt DE, Bush M, Morton C, Morton F, Howard JG. Semen characteristics and testosterone profiles in ferrets kept in long-day photoperiod, and the influence of hCG timing and sperm dilution on pregnancy rate after laparoscopic insemination. J Reprod Fertil. 1989;86:349–358.
Williams ES, Mills K, Kwiatkowski DR, Thorne ET, Boerger-Fields A. Plague in a black-footed ferret (*Mustela nigripes*). J Wildlife Dis. 1994;30:581–585.
Williams ES, Thorne ET, Appel MJG, Belitsky DW. Canine distemper in black-footed ferrets (*Mustela nigripes*) from Wyoming. J Wildlife Dis. 1998;24:385–398.
Wimsatt J, Biggins DE, Innes K, Taylor B, Garrell D. Evaluation of Oral and Subcutaneous Delivery of an Experimental Canarypox Recombinant Canine Distemper Vaccine in the Siberian Polecat (*Mustela eversmanni*). J Zoo Wildlife Med. 2003;34:25–35.
Wingfield JC. The concept of allostasis: coping with a capricious environment. J Mammal. 2005; 86:248–54.
Wingfield JC, Romero LM. Adrenocortical responses to stress and their modulation in free-living vertebrates. In: McEwen BS, Goodman HM, editors. Handbook of physiology: a critical, comprehensive presentation of physiological knowledge and concepts. Oxford: Oxford University Press; 2001. p. 211–34.
Wisely SM, Ososky JJ, Buskirk SW. Morphological changes to black-footed ferrets (*Mustela nigripes*) results from captivity. Can J Zool. 2002;80:1562–8.
Wisely SM, Sanytmire RM, Livieri TM, Marinari PE, Kreeger JS, Wildt DE, Howard JG. Environment influences morphology and development for *in situ* and *ex situ* populations of the black-footed ferret (*Mustela nigripes*). Anim Conserv. 2005;8:321–8.
Wisely SM, Santymire RM, Livieri TM, Mueting SA, Howard JG. Genotypic and phenotypic consequences of reintroduction history: case study of the black-footed ferret. Conserv Genet. 2007;89:87–96.
Wolf KN, Wildt DE, Vargas A, Marinari PE, Kreeger JS, Ottinger MA, Howard JG. Age-dependent changes in sperm production, semen quality, and testicular volume in the black-footed ferret (*Mustela nigripes*). Biol Reprod. 2000;63:179–87.

Chapter 8
Comparative Reproductive Biology of Elephants

Janine L. Brown

Abstract The ability to serially collect blood samples and conduct ultrasound examinations in Asian and African elephants has provided unique opportunities to study the biology of these endangered species. As a result, many unique aspects of elephant reproduction have been identified. For females, there are interesting differences in luteal steroidogenic activity, follicular maturation, pituitary gonadotropin secretion, fetal development and reproductive tract anatomy, while males exhibit the unique phenomenon of musth and an unusual reproductive anatomy (internal testes, ampullary semen storage). However, problems associated with uterine and ovarian pathologies hamper captive propagation efforts. Older, nulliparous cows are particularly susceptible, leading to speculation that continuous ovarian cyclicity of non-bred females in zoos is having a negative and cumulative effect on reproductive health. There are notable species differences in reproductive mechanisms as well (e.g., ovarian acyclicity, prolactin secretion, sperm cryosensitivity), implying that species-specific approaches to management and application of assisted reproductive techniques are needed for maximal reproductive efficiency and enhancement of genetic management.

Keywords African and Asian elephant • Reproduction • Endocrinology • Ultrasonography • Ovarian cycle • Bull physiology • Pregnancy • Semen cryopreservation

J.L. Brown (✉)
Center for Species Survival, Smithsonian Conservation Biology Institute,
1500 Remount Road, Front Royal, VA, USA
e-mail: brownjan@si.edu

W.V. Holt et al. (eds.), *Reproductive Sciences in Animal Conservation*,
Advances in Experimental Medicine and Biology 753, DOI 10.1007/978-1-4939-0820-2_8,

1 Introduction and Objectives

There are two genera of living elephants (Order *Proboscidea*, Family *Elephantidae*): Asian (*Elephas*) and African (*Loxodonta*). The African elephant (*Loxodonta africana*) is categorized as Vulnerable and listed as two distinct species: savanna (*L. a. africana*) and forest (*L. a. cyclotis*). There are approximately 500,000 elephants free-ranging in Africa, about 25 % of which are the forest subspecies. The Asian elephant (*Elephas maximus*) is more at risk and listed as Endangered, with <50,000 remaining in 13 range states; approximately 60 % are in India. The Asian elephant is divided into three subspecies: Indian (*E. maximus indicus*), Sri Lankan (*E. m. maximus*) and Sumatran (*E. m. sumatranus*), although mitochondrial and microsatellite analyses suggest the Borneo elephant is a fourth subspecies (*E. m. borneensis*) (Fernando et al. 2003).

Elephants are keystone species and modify habitat by converting forests to grassland, creating water holes in times of drought, and spreading the seeds of plants. They also are umbrella species, as the conservation of elephants preserves not only habitat, but other species therein. As iconic flagship species, elephants raise awareness for action and funding of broader conservation efforts. Despite their importance however, wild elephant populations are under siege. A major threat is the loss and fragmentation of habitat due to human expansion and agricultural land conversion, which leads to human-elephant conflict (Lee and Graham 2006; Fernando et al. 2008). Poaching for ivory also is a serious threat for both species, although more for African than Asian elephants (Stiles 2009; Wasser et al. 2009; Jackson 2013). In 2011, 25,000 African elephants were killed for ivory, and more than 30,000 in 2012. Over 60 % of forest elephants in Africa have been killed for the ivory trade in the last decade (Maisels et al. 2013).

With so many species on the endangered species list today, including elephants, captive breeding is increasingly viewed as a means of maintaining important populations as "insurance" against environmental or anthropomorphic catastrophe (Hoffman et al. 2010; Conde et al. 2011). Globally, there are about 1,000 African elephants in captivity, mostly in zoos, and upwards of 16,000 Asian elephants in zoos, circuses, sanctuaries, logging and tourist camps. Unfortunately, most captive elephant populations are not self-sustaining due to high mortality and low birth rates, and supplementation by wild capture and/or importation is widespread. As reviewed by Thitaram (2012), captive breeding programs throughout Asia in particular (i.e., Thailand, Sri Lanka, India, Indonesia. Lao and Myanmar) have had poor success and are not sustainable without wild offtake. In the U.S., only 3.5 births to five deaths have occurred annually over the past decade (Faust and Marti 2011a, b). Based on demographic modeling, six to nine offspring per species per year are needed to maintain current U.S. elephant population sizes, with increasing reproduction being more important than decreasing mortality for sustainability (Faust and Marti 2011a, b). Thus, efforts are centered on increasing reproductive output and breeding all reproductively viable elephants. This has not been without its challenges. Some problems are logistical, such as not housing fertile males and females together. Others are physiological, including ovarian cycle abnormalities, uterine pathologies, gestational difficulties and bull subfertility. And there are behavioral issues, such as mate incompatibility and poor libido that thwart breeding success. Through advances in endocrine monitoring, ultrasonographic imaging and semen collection techniques, we are beginning to understand some of the complex mechanisms involved in controlling reproductive function in elephants. As a result, through over two decades of study, an expansive database now exists for both Asian and African elephants. Several reproductive traits appear unique to *Elephantidae*, which have at times helped and at other times hampered breeding efforts. Adapting assisted reproductive techniques developed for domestic, laboratory and other wildlife to elephants has not always been straightforward; however, a successful artificial insemination approach has been developed that relies on luteinizing hormone (LH) analyses to time ovulation and transrectal ultrasound to guide semen deposition.

This chapter reviews current knowledge of Asian and African elephant reproductive biology and highlights species differences in reproductive function, fertility problems, and how use of assisted reproductive techniques are enhancing reproductive efficiency and genetic management. Future high priority research needs also are identified.

2 State of the Art

2.1 Female Reproductive Cycle

There is a clear age difference in the onset of puberty between captive and wild females, particularly for Asian elephants. Many Asian females begin cycling by the age of 5 years, and have conceived as young as 4 years of age, which is much younger than wild counterparts that reach sexual maturity between 10 and 12 years of age (Sukumar 2003; Glaeser et al. 2012). An age difference also exists for African elephants, although it is not as dramatic, with puberty occurring at ~8 years of age in zoos (Brown 2000) compared to 10–12 years in the wild (Sukumar 2003). The reason for this sexual shift is not clear, but could be related to higher levels of nutrition in zoos. In this regard, it may be similar to early puberty onset associated with increased body fat in girls (Kaplowitz 2008). It is not known if early puberty presents health concerns for elephant females in the long-term, but it can be a management problem for zoos housing bulls as it is vital that females are not bred before they are physically ready. There may also be trade-offs between reproduction and survival for elephants. Evaluation of an extensive longitudinal dataset of semi-captive timber elephants in Myanmar ($n=8{,}006$) found an association between reproduction and adult survival, being positive in early life, but negative in later life (Robinson et al. 2012). Reproduction and survival trade-offs were greater after peak reproduction was achieved, and investing in offspring after the age of 30 years decreased survival probabilities. Thus, long-lived females produced fewer offspring over their lifetimes.

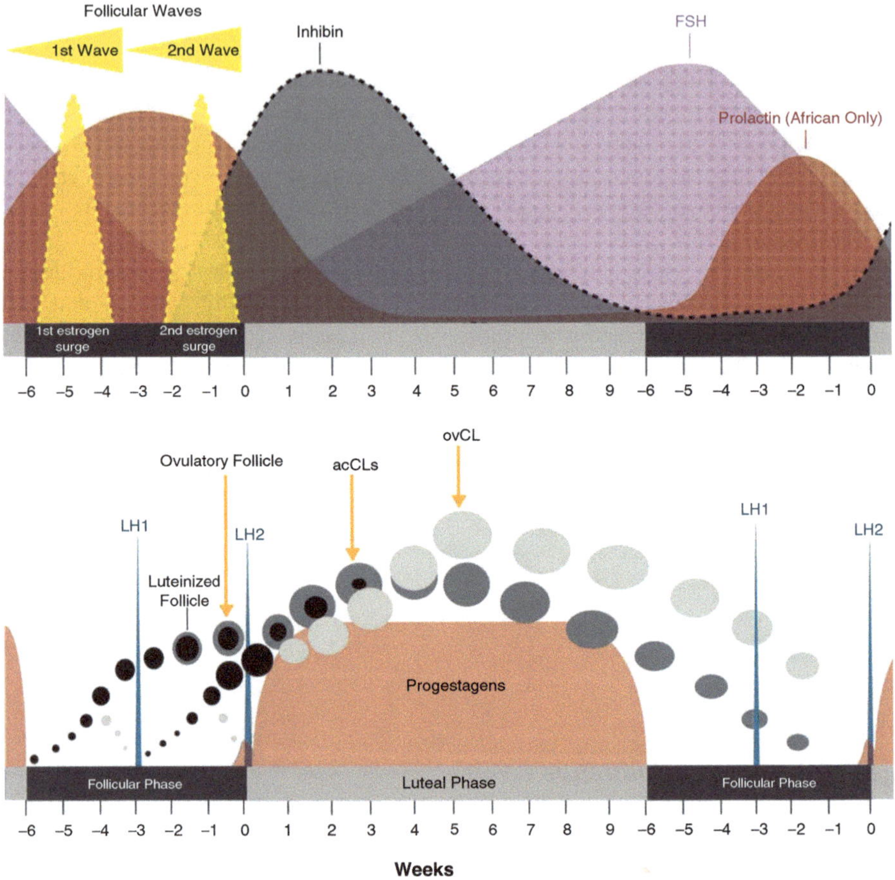

Fig. 8.1 Model of the elephant ovarian cycle, showing relationships among the secretion of estradiol, progestagens, LH, FSH, prolactin and inhibin throughout a follicular phase preceding and following a luteal phase. Schematic shows relationships among the two follicular waves that each culminate in an LH surge (first surge, LH1; second surge, LH2), and the development of follicles and corpora lutea (CL), both ovulatory (ovCL) and accessory (acCL). Prolactin is presented for African elephants only, as it is not influenced by stage of the cycle in Asian elephants. Adapted from Brown (2000), Hildebrandt et al. (2006) and Lueders et al. (2010, 2011)

Hormonal patterns of the elephant estrous cycle are now well established (see reviews, Brown 2006; Hildebrandt et al. 2011, 2012a) and summarized in Fig. 8.1. Females are polyestrous, and exhibit the longest spontaneous cycle of any mammal; 13–17 weeks in duration, with an 8–10 week luteal phase and 4–7 week follicular phase. An unusual characteristic is the production of 5α-reduced pregnanes by the corpus luteum rather than progesterone (e.g., 5α-pregnane-3,20-dione, 5α-pregnane-3-ol-20 one, 17α-hydroxyprogesterone), which will be referred to herein as 'progestagens'. Elephants are easily trained for non-stressful blood collection; however, ovarian activity also can be monitored noninvasively, based on analysis of progestagens metabolites excreted in feces and urine, and estrogen conjugates in urine (Wasser et al. 1996; Czekala et al. 2003). In general, the first pubertal cycle is similar to subsequent ones (Brown 2000; Glaeser et al. 2012; J. Brown, unpubl). A distinguishing characteristic is the "double LH surge"; two precisely timed LH surges associated with two follicular waves that occur 3 weeks apart during the follicular phase, referred to herein as LH1 (anovulatory) and LH2 (ovulatory). These two surges occur in both species (African: Kapustin et al. 1996; Asian: Brown et al. 1999), but generally are more distinct in Asian elephants, with peak concentrations being up to 3-fold higher than in Africans (Brown et al. 2004a). Female Asian elephants excrete a urinary signal, (*Z*)-7-dodecenyl acetate (*Z*7-12:Ac), which stimulates male breeding behavior (Rasmussen et al. 1997; Rasmussen 2001). Concentrations of *Z*7-12:Ac increase throughout the follicular phase, and are highest just before ovulation. African elephants do not produce *Z*7-12:Ac, but excrete frontalin, exo-brevicomin, endo-brevicomin, (E,E)-alpha-farnesene and (E)-beta-farnesene in urine leading up to and during estrus (Goodwin et al. 2006). Thus, both species appear to advertise impending fertility, but the chemical signals differ.

Elephants are not obligatory seasonal breeders, but conceptions can be influenced by rainfall and resource availability, as noted in African elephants (Sukumar 2003). Observations of elephants in zoos support a seasonal trend under some conditions; for example, a group of African elephants housed indoors because of extreme cold weather showed prolonged non-luteal phases before re-initiating normal ovarian cycles in the spring (Schulte et al. 2000). Seasonal effects on Asian elephant reproduction, if any, are more subtle. Glaeser et al. (2012) found no seasonality in ovarian cycle lengths in a study of nine females over a 20-year period, whereas a detailed study of elephants in Thailand showed females cycled year round, but exhibited slightly longer follicular phases during the rainy season (Thitaram et al. 2008). Specifically, it was the duration between the progestagens drop and LH1 that varied seasonally, averaging 33 days during the rainy season, compared to 22 days in the winter and 19 days during the summer seasons. By contrast, the interval between LH1 and LH2 was consistent at 19 days, similar to other studies. Thus, seasonal variation in estrous cycle length may be mediated by events during the period leading up to LH1 (Thitaram et al. 2008), perhaps due to more variation in the completion of the first follicular wave. Variability in duration of the follicular phase also tends to be greater than that of the luteal phase, at least in Asian elephants (Thitaram et al. 2008; Glaeser et al. 2012). In those two studies, luteal and follicular phase durations were negatively correlated, suggesting a possible regulatory role of the follicular phase in maintaining relatively consistent cycle duration within individuals. In the study of Glaeser et al. (2012), Asian elephants were highly resilient to numerous major life events (births, deaths, transfers in and out, changes in herd structure), which had a minimal effect on cycle dynamics. That Asian elephants maintain consistent cycles despite a variety of management changes agrees with about 20 years of unpublished data on over 60 Asian females evaluated at SCBI (J. Brown). By contrast, data on over 80 African females monitored at SCBI provide numerous examples of events related to temporary or permanent suppression of estrous cyclicity (e.g., translocations, changing blood collection frequency, altered herd dynamics, keeper changes). This suggests that Asian and African elephants may differ in responsiveness to changes in the captive environment, something that deserves further investigation in the context of optimizing conditions based on species-specific needs.

An elegant series of studies (Lueders et al. 2010, 2011) combined transrectal ovarian ultrasound and hormone measurements in Asian elephants to develop a novel theory for the double LH surge in relation to dominant follicle selection and luteal development. One enigma has been the observation of multiple corpora lutea (CL) during each cycle despite elephants being monovular. During the follicular phase, two distinct waves of follicles develop, each of which is terminated by an LH surge. During the first wave, multiple follicles develop, but none ovulate. Originally it was believed these all regress after LH1 (Hermes et al. 2000); however, serial transrectal ultrasound examinations revealed that some of these follicles do in fact luteinize, and are referred to as accessory CL (acCL) (Lueders et al. 2012). During the second wave, multiple follicles develop, but only one becomes dominant, ovulates and forms the ovulatory CL (ovCL). Thus, there appears to be two modalities for the development of acCL and ovCL. Follicle luteinization is apparent within 10 days of each LH surge (Lueders et al. 2010), but the acCLs remain dormant and do not secrete progestagens, possibly due to lack of 3β hydroxysteroid dehydrogenase activity in the lutein cells (African: Stansfield and Allen 2012). After ovulation, the acCLs reach a maximum diameter within 30 days, whereas the single ovCL attains a significantly larger size 10–15 days later (Lueders et al. 2010). All CLs are visible throughout the follicular phase, with some of the larger ones remaining in subsequent luteal periods. By repeatedly forming two distinct types of CLs during every reproductive cycle, elephants may have developed a mechanism to ensure there is sufficient luteal capacity for maintaining a 22-month gestation should conception occur, at least in Asians (Lueders et al. 2010).

Simultaneous with luteinized follicle formation after LH1, immunoreactive (ir) inhibin concentrations increase, preceding progestagens by about 2 weeks in both species (Brown et al. 1991; Kaewmanee et al. 2011a; Lueders et al. 2011; Yamamoto et al. 2012a) (Fig. 8.1). Immunohistochemistry has shown that inhibin α and β subunits are present in granulosa cells of antral follicles in Asian elephants, as are high concentrations of immunoreactive and bioactive inhibin in the follicular fluid (Kaewmanee et al. 2011a), similar to other species (Medan et al. 2007; see Kaewmanee et al. 2011a). However, in the study of Lueders et al. (2011), ir-inhibin never increased before LHI, even though large follicles were present and attained diameters within the range of ovulatory follicles. In other species, both small and large estrogenic follicles are a significant source of ir-inhibin (Campbell et al. 1991), so if inhibin is derived solely from follicles in elephants, there should be measurable levels prior to LH1 because estrogens are detectable before both LH peaks (Czekala et al. 2003). It is puzzling then, that ir-inhibin increases only after follicle luteinization post LH1, about 9 days before LH2 (Lueders et al. 2011; Kaewmanee et al. 2011a). The investigators theorize that rather than granulosa cells, it is the cells of luteinized follicles and acCLs that are the source of inhibin (Lueders et al. 2011). The absence of luteinized follicles before LH1 would explain why inhibin concentrations are low, and also why there is a significant correlation between luteinized follicle/acCL diameter and increasing ir-inhibin concentrations. Dominant follicle deviation begins about 5–6 days before ovulation when inhibin concentrations are high. From these observations, authors further propose that inhibin itself may be important for dominant follicle selection. Although inhibin was not measured, a study investigating the follicular response to GnRH would appear to support this concept (Thitaram et al. 2009). When administered at different times during the follicular phase, GnRH always stimulated LH release (i.e., LH1); however, a spontaneous secondary surge ~20 days later (i.e. LH2) only occurred if LH1 was induced between days 13–42 of the follicular phase. By contrast, no LH2 occurred when GnRH was administered before day 12. So, if LH1 is induced too early, follicles are not mature enough to luteinize and produce inhibin. As a result, there is no dominant follicle selection during the second follicular wave. Alternatively, rather than directly affecting follicle selection, increased inhibin concentrations may simply indicate that follicles have reached a level of maturity that permits deviation and subsequent ovulation.

In all likelihood, inhibin's role in follicle selection is through its control over FSH secretion. In both species, the FSH secretory pattern is protracted and inversely related to inhibin (Brown et al. 1991; Brown et al 1999; Kaewmanee et al. 2011a) (Fig. 8.1). FSH is highest towards the end of the luteal phase and throughout the first follicular wave when inhibin is low, then decreases to nadir concentrations before LH2 as inhibin is rising. Comparatively, the FSH pattern in elephants differs somewhat from other mammals, where concentrations typically are elevated coincident with the pre-ovulatory LH surge, with a secondary FSH surge sometimes occurring after ovulation (Downey 1980). Instead, FSH secretion in the elephant is more like that of the horse, where concentrations are highest at the end of the luteal phase and decrease progressively towards ovulation (Ginther 1992). However, the follicular phase in other mammals, including horses, is considerably shorter (<1 week) than that in the elephant, so these comparisons may not be relevant. Nevertheless, because FSH plays a key role in follicle recruitment and growth in other species, a prolonged stimulation may be necessary for two successive follicular waves to occur in the elephant. As in other species, the reduction in FSH likely facilitates dominant follicle deviation, with the transition of follicles from an FSH-dependent to an independent state being key to ovulatory selection (Baird 1983). FSH concentrations are low after ovulation and then rise during the latter part of the luteal phase, just behind the decline in inhibin (Brown et al. 1991; Kaewmanee et al. 2011a). This inhibin is of luteal origin, as no antral follicles are present post-ovulation (African: Hermes et al. 2000; Asian: Lueders et al. 2010), and immunohistochemical staining has localized inhibin α and β subunits to the lutein cells of CLs, similar to that observed in primates (Yamoto et al. 1991, 1992).

Prolactin is folliculogenic in several species (Freeman et al. 2000; Frasor and Gibori 2003), and in African elephants is elevated during the nonluteal phase of the cycle, inversely related to progestagens, and phase-shifted by about 4 weeks (Yamamoto et al. 2010; Dow and Brown 2012) (Fig. 8.1). However, in Asian elephants, prolactin concentrations are unvaried throughout the cycle and remain at baseline concentrations, representing a major species difference (Brown et al. 2004a).

As discussed below, it may be significant that ovarian cycle problems associated with abnormal prolactin secretion are common in African, but not Asian elephants (Dow and Brown 2012).

2.2 *Pregnancy and Parturition*

Once again, transrectal ultrasound and endocrine monitoring have been key to better understanding the physiology of pregnancy in elephants, which have the longest gestation period, lasting 20–23 months on average. The placenta is chorioallantoic, zonary endothelialchorial; implantation is central and superficial, with a mesometrical orientation of the yolk sac; the embryo is antimesometrial in location (see review, Allen 2006; Hildebrandt et al. 2006). Although the placenta itself is endocrinologically inert, the fetal gonads, which enlarge during the second half of gestation, synthesize 5α-dihydroprogesterone and other 5α-pregnane derivatives from cholesterol and pregnenolone (Allen et al. 2005; Allen 2006; Stansfield and Allen 2012). Placentation occurs during the second to third month of gestation (Drews et al. 2008). Based on longitudinal transrectal ultrasound monitoring, both species exhibit an initial period of comparatively slow embryonic development that has been compared to delayed implantation (Hildebrandt et al. 2006; Drews et al. 2008). The embryonic vesicle is visible at ~8 weeks post-conception, which is much smaller (~10 mm) than that of other species at that stage (cattle, 40 mm; sheep, 70 mm; horse, 40 mm). Likewise, time of implantation is estimated to be ≤20 days in human, dog and sheep, but ≥50 days in the elephant. In other delayed implanters (e.g., mustelids, bears, roe deer), a significant rise in serum progestagens occurs at implantation, which is observed in elephants at 6–8 weeks post-conception (Meyer et al. 2004). The embryo then doubles in size between the fourth and fifth month, increasing from 60 mm to 120 mm (Drews et al. 2008). Organogenesis is completed by about 110–120 days, when the end of the embryonic period is reached. It is possible to sex the fetus after about a year of gestation with near 100 % accuracy by measuring circulating maternal testosterone concentrations, at least in Asian elephants (Duer et al. 2002; Brown et al. 2004b). Presumably elevated testosterone is of fetal testicular origin, although the CL could also be a source (Castracane et al 1998). Interestingly, this technique has proven less accurate for African elephants (J. Brown, unpubl).

A model for the endocrinology of pregnancy is depicted in Fig. 8.2. Diagnosis and monitoring of pregnancy is easily done by longitudinal analysis of 5α-reduced pregnanes in the bloodstream or the relevant metabolites in urine or feces (see reviews, Brown 2000; Hildebrandt et al. 2006). Elephants and horses share some gestational traits, such as the presence of multiple large CL in the maternal ovaries (reviewed by Stansfield and Allen 2012). However, there are notable differences. In the elephant, acCLs are produced throughout the follicular phase of preceding cycles (Lueders et al. 2010, 2011), and there are no additional CLs produced during gestation (Lueders et al. 2012). By contrast, the mare produces one CL at ovulation, with additional CLs formed as the result of ovulations after conception (Squires and Ginther 1975). Gross examination and histology of African elephant

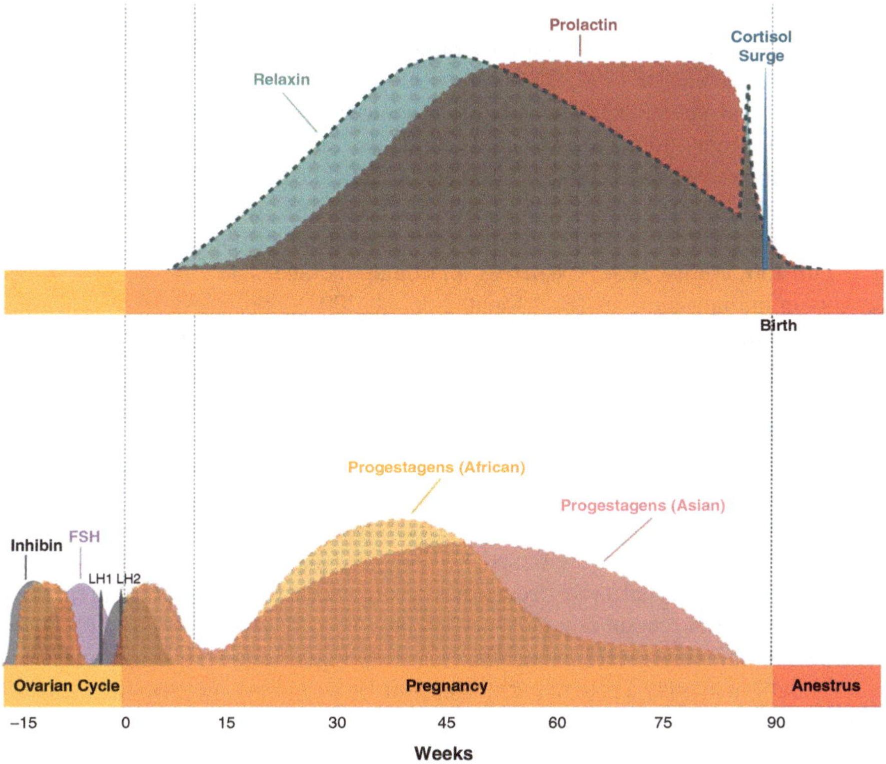

Fig. 8.2 Model of the relationships among the secretion of progestagens, LH, FSH, prolactin, relaxin, cortisol and inhibin during gestation and a prior estrous cycle in the elephant. Adapted from Meyer et al. (2004), Yamamoto et al. (2012a) and Lueders et al. (2012)

ovaries indicates acCL form by luteinization of follicles, with or without ovulation; stigmata are clearly visible on some pregnancy CLs (Stansfield and Allen 2012). By contrast, ultrasound examinations of pregnant Asian elephants suggest acCLs are the result of luteinization of unruptured follicles only (Lueders et al. 2010, 2011). It is not clear if this is a species difference. In both Asian and African elephants, acCLs and the ovCL begin to regress about 5–6 weeks after conception (the normal luteal phase lifespan), but then rebound and grow significantly larger than those in a non-conceptive luteal phase (Lueders et al. 2012), commensurate with the marked secondary rise in progestagen concentrations in the maternal circulation after the second gestational month (see Meyer et al. 2004; Lueders et al. 2012). In the pregnant mare, acCLs develop equally on both ovaries due to the LH-like activity of equine chorionic gonadotropin (eCG), which causes ovulation/luteinization of mature Graafian follicles (Urwin and Allen 1982). By contrast, mature follicles never develop in pregnant elephants (see Lueders et al. 2012), nor is there evidence of gestational gonadotropin-like activity in serum or placental extracts (Meyer et al. 2004; Allen 2006).

After 4–7 months of gestation, prolactin immunoactivity (ir-prolactin) increases up to 100-fold, peaks at 11–14 months and remains high until birth in both species (Brown and Lehnhardt 1995; Meyer et al. 2004; Yamamoto et al. 2011, 2012b). Prolactin and placental lactogens are luteotrophic in other species, and enhance CL progestagen production (Freeman et al. 2000; see Takahashi 2006). This would be important for elephants because the placenta is steroidogenically inactive (African: Allen et al. 2002). The source of high ir-prolactin during gestation is primarily placental (Yamamoto et al. 2011), similar to lactogenic hormones in other species (Forsyth and Wallis 2002). Whereas placental lactogens are derived from the desidua in humans and rats (Ben-Jonathan et al. 2008), ir-prolactin in elephants is immunolocalized in the trophoblast cells of both species (Yamamoto et al. 2011). Measurement of serum ir-prolactin past 7 months of gestation is a reliable pregnancy test, even on a single sample, unlike progestagens, which require longitudinal sampling. Unfortunately, ir-prolactin has not been detected in urine, so noninvasive pregnancy diagnosis by endocrine means has so far not been feasible (Brown et al. 2010). Serum relaxin also can be used diagnostically in both species (Meyer et al. 2004; Niemuller et al. 1998), as concentrations are elevated after 5 months of gestation. Levels peak at about 10 months and then gradually decline until a few weeks before birth, when a sharp rise occurs just before parturition. The CL of pregnancy is a main source of relaxin in many species, but in others the decidua also produces considerable amounts (MacLennan 1981). The source of relaxin in elephants is not known, but it may play a role in parturition similar to that in other species by facilitating a softening of the cervix and loosening of pelvic ligaments, and ensuring synchrony in uterine muscles after labor begins (MacLennan 1981).

In a large comparative study, there was a broad range of individual variation in gestation length (Asian, 623–729 days; African, 640–673 days) (Meyer et al. 2004). That study also identified several notable species differences in gestational hormone patterns. While overall mean progestagen concentrations were similar, temporal profiles differed. Concentrations were higher in African elephants during the first half of gestation, but then declined to levels below those observed in Asian elephants (Fig. 8.2). There was a fetal gender effect in Asian, but not African elephants, with progestagen concentrations being higher in Asian cows carrying male calves as compared to those carrying females. It is curious that significant fetal gender differences in maternal steroids (androgens and progestagens) are observed only in Asian elephants, suggesting a species difference in gonadal and/or placental function. Both species have zonary placentation; however, far more information is available on African elephant placental function, and detailed comparative studies at the level needed to identify species differences in steroidogenic activity have not been conducted. A fetal sex difference in progestagens may also be related to testicular steroid production. During sexual differentiation, progesterone produced by fetal Leydig cells is converted to testosterone to complete male duct system development (Lejeune et al. 1998), but why this would occur in Asian, but not African elephants is not known. Comparatively, overall prolactin concentrations were higher in Asian than in African elephants between 8 and 15 months of gestation, but there was no species difference in the secretory patterns of relaxin (Meyer et al. 2004). In both species, the observation of significant surges in serum cortisol between 8 and 11 days before parturition, and again on the day of parturition (Meyer et al. 2004), suggests an important role in the initiation of parturition (Liggins and Thornburn 1993).

In many species, particularly primates, inhibin produced by follicles, CLs and/or the placenta (Knight 1996), is believed to be involved in the establishment and maintenance of pregnancy (Florio et al. 2010). In the elephant, whereas CLs are a major source of inhibin during the estrous cycle (Kaewmanee et al. 2011a; Yamamoto et al. 2012a), they do not appear to produce inhibin during gestation, as overall concentrations are low (Yamamoto et al. 2012a) (Fig. 8.2). On closer inspection, inhibin in fact is increased for the first 8 weeks post-conception, mimicking a normal luteal phase increase. Thus, both progestagens and inhibin are increased immediately post-conception and then decrease at 7–8 weeks. After that, progestagens rebound to even higher concentrations on average, whereas inhibin concentrations continue to decline. This pattern suggests a shift in luteal cell function during early gestation, and a deviation in the secretory ability of CLs between cycling and pregnant elephants. Taken together, a role for inhibin in elephant pregnancy seems unlikely. A related protein, activin A, may be worth investigating as it is secreted by stromal endometrial cells and is involved in implantation in other species (reviewed by Florio et al. 2010). It also enhances cytotrophoblast differentiation indirectly by increasing the expression of other molecules involved in embryo implantation, such as matrix metalloproteinases and leukemia inhibitory factor. A local derangement of the activin A pathway has been implicated in some human pregnancy disorders (incomplete and complete miscarriages, recurrent abortion, and ectopic pregnancy), and so may be worth investigating in older elephants that are more susceptible to poor pregnancy outcomes.

2.3 Reproductive Challenges

Transrectal ultrasonography techniques have become instrumental in monitoring reproductive tract health, including in elephants (Hildebrandt et al. 2003). In the U.S., nearly half of Asian and African elephant females in AZA-accredited zoos have had an ultrasound examination (Dow et al. 2011a). Of these, the majority exhibit one or more reproductive tract pathologies of ovarian or uterine origin. As reviewed by Hildebrandt et al. (2006), vestibular cysts occur in both species, whereas vestibular polyps are observed only in Africans, with an incidence of about 70 % in females >30 years of age. In both species, vaginal cysts and neoplastic formations may be extensive and fill the vaginal lumen, blocking semen flow after mating and causing discomfort during estrus and mating. Periodic vaginal discharge containing mucus and clotted blood is a symptom of this condition. Asian and African elephants both develop endometrial hyperplasia, whereas Asian elephants develop multiple benign uterine leiomyomas. Ovarian cysts also occur more frequently in zoo African (~15 %) than Asian (~5 %) elephants compared to wild females (<1 %). In general, reproductive tract pathologies are more prevalant in older (>30 years of age) nulliparous cows, and those where reproduction has not occurred within 10–15 years. In a recent survey, more than half of zoo females with documented tract pathologies had no previous breeding history, either through natural mating or AI (Dow et al. 2011a). The occurrence of urogenital pathologies in older females is termed 'asymmetric reproductive aging' (Hermes et al. 2004), and believed to be the result of continuous ovarian cyclicity of non-bred females. Repetitive remodelling and exposure of the endometrium to ovarian steriods likely has a negative and cumulative effect on reproductive health (Hermes et al. 2008). In the wild, most females are either pregnant or lactating and thus experience comparatively few reproductive cycles in their lifetime. Consequently, these pathologies are not common in wild elephants (Hildebrandt et al. 2006; Freeman et al. 2008).

As Asian and African elephants in zoos age, the risk of developing pathologies increases (Aupperle et al. 2008), so a new treatment to slow or stop their development based on the use of GnRH vaccines is being explored (Boedeker et al. 2012). These vaccines stimulate the production of anti-GnRH antibodies that block the binding of endogenous GnRH to gonadotrope receptors in the pituitary gland (Conforti et al. 2008). This action inhibits the release of FSH and LH from the anterior pituitary, thereby causing the cessation of ovarian steroidogenic activity and reproductive cyclicity. In one case study, a CpG motif-based adjuvant in a recombinant GnRH vaccine (Repro-BLOC, Amplicon Vaccine, LLC, Pullman, WA) suppressed ovarian cycle activity and resolved hemorrhage and anemia associated with a vascular reproductive tract tumor in a 59-year-old Asian elephant (Boedeker et al. 2012). Synthetic oligodeoxynucleotides (ODNs) containing unmethylated CpG motifs directly stimulate B cells and plasmacytoid dendritic cells, thereby promoting the production of cytokines and the maturation/activation of antigen-presenting cells (Klinman 2006). These activities enable CpG ODNs to act as immune adjuvants, accelerating and boosting antigen-specific immune responses by 5- to 500-fold over traditional Freund's adjuvants. Six years after initial vaccination, the elephant continues to lack distinct ovarian cycles and is healthy. In addition to resolving these problems in older females, there is interest in using GnRH vaccines as contraceptives for wild elephants to modulate population growth and mitigate human-elephant conflict and protect resources in limited habitats (Bertschinger et al. 2007; Botha et al. 2008).

Of considerable concern is the tendency for older, first-time mothers to experience dystocia and/or stillbirths. Over 50 % of dystocias occur in nulliparous females >20 years of age or after prolonged barren periods (Hermes et al. 2008). As a management recommendation, nulliparous females in captivity are considered post-reproductive after 25–35 years of age because of this increased risk (Hermes et al. 2004; Hildebrandt et al. 2006); over two thirds of Asian elephants in U.S. zoos are now above this age. Causes of calving problems include: large calves; malposition or an anterior position; loss of flexibility in the pelvic region; muscle fatigue; hypocalcemia; intact hymen obstructing calf passage; edema of the vestibule narrowing the birth canal; elephant endotheliotropic herpes virus infection; and cysts, polyps and/or tumors of the urogenital tract (Hermes et al. 2008). Many of these can be linked not only to the female being older, but to being overweight and lacking in physical fitness. To date, caesarean section has not been successful in resolving dystocia in elephants, although several have been tried (Hermes et al. 2008). The only surgical option to correct a dystocia is vestibulotomy and if necessary a subsequent fetotomy (Schaftenaar 2013). However, these are only possible if fetal parts have already entered the cervix. These problems make tracking gestation progress with daily progestagen analysis critical to ensure staff is prepared to take proactive steps if needed (e.g., oxytocin to enhance labor after opening of the birth canal has been confirmed by ultrasound; Brown et al. 2004b; Hermes et al. 2008). It is now clear that prolonged non-reproductive periods in elephants are associated with serious reproductive issues, making it imperative to breed females soon after puberty and regularly throughout their reproductive lifespan to avoid compromising health and well-being.

2.4 Ovarian Acyclicity

Factors limiting the number of breeding female elephants vary by species. For Asian elephants, the main issue is advancing age. For African elephants, a primary cause of poor reproduction is a high rate of ovarian acyclicity. Based on a 2008 reproductive survey of elephants in AZA-accredited zoos, 46 % of African elephant females exhibited abnormal ovarian cycles, and 31 % did not cycle at all. Most importantly, the majority of ovarian cycle problems occurred in reproductive age females (Dow et al. 2011a). By contrast, only 11 % of Asian elephants were acyclic, and these were mostly post-reproductive females (>40 years of age).

Despite the recognition of this problem in African elephants for over two decades, the etiology of ovarian acyclicity remains a mystery. There is considerably more known about what does not cause it than what does. Several conditions known to be associated with infertility in other species have been examined, such as 'stress' (e.g., increased cortisol; Brown et al. 2004a), hyperandrogenism (Mouttham et al. 2011), hyperestrogenism (Prado-Oviedo et al. 2013) and thyroid dysfunction (Brown et al. 2004a), but none were found to be related to ovarian cycle problems. Similarly, concentrations of LH, FSH (Brown et al. 2004a) and inhibin (J. Brown and J. Hoffman, unpubl) are within normal baseline ranges in acyclic females, although none fluctuate as in cycling elephants. One theory was that acyclic zoo elephants may exhaust their supply of ovarian follicles; i.e., undergoing a premature 'menopause', due to constant cycling. In most mammals, the ovarian reserve is high at birth and undergoes a steady loss through natural attrition and ovulation during pre- and post-pubertal life (Gosden 1987). Ultimately, the reserves become depleted and reproductive senescence ensues in individuals that reach a maximum lifespan (Cohen 2004). In elephants too, ovarian histology of females culled in southern Africa revealed a significant decline in follicle reserves with age, which in a few individuals were depleted by the 7th decade (African: Stansfield et al. 2012). However, this is well beyond the normal lifespan, so the authors concluded that the elephant ovary is capable of supplying oocytes for ovulation right up to the time of death in most individuals. Still, there are questions about whether the continuous cyclicity of non-bred zoo elephants might accelerate this depletion. Thus, a study was conducted to quantify anti-müllerian hormone (AMH) (Dow et al. 2011b), which is produced by granulosa cells and serves as a marker for the number of morphologically healthy oocytes within the follicular reserve (Rico et al. 2009; Grynnerup et al. 2012). It is used clinically to confirm the state of menopause in women. Results showed that AMH concentrations were not different between cycling and noncycling elephants (Dow et al. 2011b), which can be viewed as good news, as it suggests the ovaries should be responsive to follicular and ovulation induction therapies (e.g., LH, FSH, eCG, hCG). Then again, without detailed histological studies of zoo elephant ovaries similar to those of wild elephants, it will be difficult to eliminate primary hypogonadism as a cause of ovarian acyclicity, at least in some individuals.

Of particular concern is the association between ovarian cycle problems in African elephants (but not Asians) and a hormonal imbalance – hyperprolactinemia (Brown et al. 2004a; Yamamoto et al. 2010; Dow and Brown 2012). In a comprehensive endocrine study (Brown et al. 2004a), a third of noncycling African elephant females had elevated concentrations of prolactin compared to cycling females. In a follow-up study 8 years later, 71 % of acyclic African elephant females were diagnosed with this condition, 45 % of which were of reproductive age (Dow and Brown 2012). This increase was due primarily to elephants switching from a normal to a hyperprolactinemic condition between studies, and so it appears to be a growing problem. Prolactin is produced in lactotroph cells in the anterior pitutary and is under inhibitory control by dopamine (Schuff et al 2002; Melmed 2003; Bachelot and Binart 2007). A common cause of hyperprolactinemia in women is a prolactinoma, the most common type of pituitary tumor (Melmed 2003). Prolactinomas retain intact trophic control, so they may develop in an environment with reduced dopamine concentrations, reduced dopamine sensitivity, or as a result of vasculature isolation that prevents dopamine from reaching the lactotrophs (Schuff et al 2002; Melmed 2003). There are no data on the incidence of prolactin-secreting tumors in African elephants, as pituitary histopathology is rarely performed at necropsy; however this could be a possibility and so should be explored. In other species, hyperprolactinemia is associated with infertility (Aron et al. 1985; Yuen 1992; Zacur 1999), and in women it is the most common disorder of the hypothalamic-pituitary axis; up to 40 % presenting with secondary amenorrhea are hyperprolactinemic (Serri et al. 2003; Wang et al. 2012). The negative effects of chronic elevated prolactin secretion on reproductive function generally involve inhibition of hypothalamic GnRH release and subsequent suppression of pituitary LH and FSH secretion, resulting in anovulation (Bachelot and Binart 2007). Whether that is the case for elephants is not known, nor is it clear if changes in prolactin actually precede acyclicity. Understanding if acyclicity is a direct or indirect effect of hyper prolactin production will be key to the development of more targeted fertility treatments.

There is convincing evidence that prolactin in elephants is controlled through dopamine negative feedback, as in other species, based on findings that a dopamine agonist (cabergoline) decreases and an antagonist (domperidone) increases prolactin secretion (Ball and Brown 2004; J. Brown and T. Dow, unpubl). Thus, it is of interest to develop treatments that modulate prolactin secretion as a means of correcting ovarian dysfunction. Cabergoline, a dopamine agonist, is an effective treatment for hyperprolactinemia-induced infertility in women (Verhelst et al. 1999), so a clinical trial was conducted to treat hyperprolactinemic, noncycling African elephant females (1–2 mg twice weekly oral cabergoline for 4–12 months; $n=8$). Cabergoline resulted in a significant and immediate reduction in prolactin during the treatment period; however, no females resumed cycling and prolactin increased to levels as high or higher after treatment withdrawal (Ball and Brown 2004; Morfeld et al. in press). Perhaps increasing the dose or extending the treatment period would be more effective, but until such trials are conducted, it is not clear if ovarian acyclicity can be corrected merely by reducing prolactin.

In addition to hypothalamic inhibition by dopamine, other factors have been identified as stimulants of prolactin synthesis or have a suppressive effect on dopaminergic tone: vasoactive intestinal polypeptide, estradiol, serotonin, oxytocin, thyrotropin releasing hormone and vasopressin (Freeman et al. 2000). With the exception of estradiol, which showed no relation to elevated prolactin (Prado-Oviedo et al. 2013), none of these potential stimulators of prolactin have been evaluated in elephants. Given the growing problem of hyperprolactinemia in the African species, these investigations appear to be warranted. Another factor that has not been examined is the role of ‘stress’ in excess prolactin secretion and ovarian inactivity, as prolactin is sometimes considered a ‘stress hormone’ (Matteri et al. 2000; La Torre and Falorni 2007). Human studies have shown that stressors such as social conflict, a new job, death of a loved one, divorce, separation from a parent during childhood, and academic pressures increase the secretion of prolactin and may predispose individuals to hyperprolactinemia and infertility (Assies et al. 1992; Sobrinho et al. 1984; Sobrinho 2003; Sonino et al. 2004). A consequence of the stress response can be reduced dopamine secretion and its tonic inhibition of prolactin, or an up regulation of prolactin stimulating factors (Calogero et al. 1998). Thus, it would be beneficial to examine how individual elephant temperaments, social relationships and life events relate to prolactin secretion and reproductive status, and if management or husbandry changes could help elephants better cope with the captive environment, similar to that described for other species (Wielebnowski 1998; Wielebnowski et al. 2002a, b; Mellen 2005; Carlstead 2009).

Whereas over two-thirds of noncycling African elephants exhibit elevated prolactin, the other third has low, baseline levels with no cyclic fluctuations. As described above, in normal cycling African elephants prolactin concentrations increase during the follicular phase and reach maximum levels immediately preceding ovulation (Bechert et al. 1999; Brown et al. 2004a). For elephants with chronically low prolactin, a trial was conducted to stimulate prolactin using domperidone (Equidone®), based on its ability to augment follicular development and improve fertility in mares (Panzani et al. 2011; Paccamonti 2012). However, although domperidone was able to increase prolactin within days of oral treatment ($n=6$), continuous treatment for a 6-month period resulted in no resumption of cyclicity

(J. Brown and T. Dow, unpubl), with the exception of one female that exhibited fluctuating prolactin during treatment and did start cycling. This female subsequently conceived, delivered a healthy calf and continues to cycle normally. In a follow-up trial, to mimic natural fluctuations in prolactin during the cycle, domperidone was administered in a 1 month on, 2 months off regimen for 1 year ($n=5$). However, despite stimulating a cyclic pattern of prolactin, progestagen concentrations remained at baseline in all females (J. Brown and T. Dow, unpubl). Suggested next steps are to further mimic natural hormone patterns by interspersing domperidone with progestagen administration, perhaps including injections of GnRH to simulate the double LH surge; however, such drastic efforts will only be worthwhile if cyclicity continues after treatment withdrawal.

One suggested cause of reproductive problems in zoo elephants is obesity (Clubb and Mason 2002), as a high body mass index (BMI) has been correlated with acyclicity in Africans (Freeman et al. 2009). Compared to wild counterparts, captive African elephants are ~27 % heavier (Ange et al. 2001), and together these observations raise questions about whether reproductive problems may be caused in part by metabolic derangements associated with excessive body fat (Clubb et al. 2009; Mason and Veasey 2010). This is plausible given studies in horses and humans showing obesity can lead to metabolic changes that impair fertility (Vick et al. 2006; Miller et al. 2008; Jungheim and Moley 2010). For example, obese mares experience an extended interval between successive ovulations, and amenorrhea is common in obese women, not unlike the irregular cycles observed in elephants (Brown 2000). Stillbirths and dystocias also are common in obese women and horses, and are a major cause of calf mortality in elephants (Clubb et al. 2009). Such evidence suggests that elephants may be experiencing fertility problems associated with obesity, including ovarian acyclicity. To determine if obesity and related metabolic conditions exist in zoo-managed African elephants, body condition scores (BCS; 5-point scale with 1 = thinnest, 5 = heaviest) (Morfeld et al. 2014), and insulin, glucose, and leptin levels were compared between breeding-aged cycling and non-cycling elephants ($N=23$ each; Morfeld 2013). Overall, 72 % had a BCS of 4 or 5, whereas none had a score of 1. The percentage of cycling females was >90 % for a BCS of 2 or 3, but only ~50 % for a BCS of 4 or 5. Perhaps more significant was the finding that leptin and insulin concentrations were higher in non-cycling as compared to cycling elephants. Using "non-cycling" as the outcome variable in regression models, and BCS, leptin, insulin, and the glucose:insulin (G:I) ratio as predictors, all but leptin were predictive of a non-cycling status, with BCS showing the strongest predictive power. Thus, these screening tools may be clinically useful for identifying at-risk elephants and developing targeted management interventions to improve body condition and insulin sensitivity with the goal of reinitiating ovarian activity. Of comparative interest, zoo Asian elephants also appear to be heavier than wild counterparts, yet ovarian acyclicity is not associated with being overweight in that species. Assessments of metabolic factors have not been conducted in that species yet, so it would be of interest to determine if they are altered in females with higher BCS as is the case for African elephants.

Finally, another factor associated with ovarian cycle problems in zoo African elephants appears to be behavioral, not physiological, with ovarian inactivity being associated with a high social dominance rank (Freeman et al. 2004, 2010). In the wild, the largest, oldest female in a herd is the matriarch and is crucial to their survival. In captivity, dominance is still important for maintaining social harmony; therefore, the energy that goes into peace keeping within a captive herd of unrelated, and sometimes incompatible females may be compromising ovarian function (Freeman et al. 2004). In mammals, reproductive inhibition can occur through the suppressive effects of primer pheromones, such as urinary chemosignals when population densities are too high, or alternatively, a dominant individual may use behavior to induce stress and shut-down reproductive mechanisms in subordinates (Wasser and Barash 1983; Creel and MacDonald 1995). While no two species use the exact same strategy, most use either behavioral or chemical suppressive mechanisms to improve their own reproductive success. Reproductive suppression is a natural strategy for many species in the wild; however, when it occurs in captivity it could be indicative of suboptimal situations (Wielebnowski 1998). Thus, research efforts are focused on determining if there are socio-management factors associated with ovarian suppression, and how it might be related to dominance status. Testosterone is known to affect aggression and dominance behaviors in many species, including bull elephants (Giammanco et al. 2005; Brown et al. 2007; Adamafio 2009), while cortisol can inhibit gonadal function through direct and indirect means (see reviews, Dobson and Smith 2000; Moberg 2000). So far, neither serum cortisol (as an index of stress) nor testosterone (as an index of dominance) have been linked to differences in social or cyclicity status (Proctor et al. 2010; Mouttham et al. 2011), so these do not appear to be driving forces in the apparent socially-mediated suppression. Presence of a bull was associated with slightly higher cyclicity rates (by 11 %) in African females; however, there were many facilities with cycling females that did not house a bull, and vice versa, so it certainly is not an absolute requirement (Dow et al. 2011a). Rather it could be due to the purposeful distribution of viable females to facilities with breeding bulls, rather than bull exposure directly. Last, a relationship between dominance and ovarian cyclicity status has not been observed in Asian elephants (e.g., Glaeser et al., 2012; J. Brown, unpubl), representing a significant species difference in how sociality or other behavioral factors affect reproductive functioning. As stated above, African elephants appear to be more sensitive to environmental and management change than Asian elephants with respect to ovarian activity. This would appear to be true for social interactions as well.

Clearly, we need to understand why so many elephant females are not cycling normally, otherwise the population collapse predicted for the U.S. will be inevitable (Faust and Marti 2011a). Prolonged acyclicity does not appear to occur in wild African elephants based on physiological studies that show females can cycle into their 50's, although capacity declines with age (Freeman et al. 2008, 2011). It is unlikely that any one management factor is responsible, as 52 % of zoos house both cycling and non-cycling females (Freeman et al. 2009). Some elephants even alternate between cyclic and non-cyclic periods (Brown 2006). Rather, there likely are multiple etiologies, so it will be key to ascertain if problems are of physical or behavioral origin, and what is the best approach to ameliorate them (e.g., exercise programs for overweight elephants, altered social groupings, creation of multi-generational herds, increased space, etc.). Obviously, it is important to develop targeted treatments for these conditions, but given the complexity of trying to control endocrine function, it is even more important to identify underlying causes so that mitigating steps can be taken to prevent problems from occurring in the first place.

2.5 Male Physiology

Similar to females, there is a shift in the onset of sexual maturity between captive and wild bulls, especially for Asians. Successful mating has been recorded in captive males as young as 6 years of age (Keele et al. 2010; Olson 2011), whereas wild bulls generally do not breed until they are at least 25 years of age (Sukumar 2003). There also is an age difference between wild and captive bulls in the occurrence of musth—the period of heightened aggressive and sexual behavior associated with increased temporal gland secretions (TGS), urine dribbling (UD) and elevated androgen (e.g., testosterone, dihydrotestosterone, androstenedione) production (Yon et al. 2008; Ganswindt et al. 2002). In the wild, musth occurs annually in sexually mature bulls (Sukumar 2003), whereas in captivity, TGS and increased testosterone secretion have been observed in bulls as young as 7 years of age (Asian: Cooper et al 1990). Musth-like changes often are irregular in captive bulls and can occur several times a year, with or without UD, and in fact some Asians appear to be in a continual state of temporal drainage and hyper testosterone secretion (Brown et al. 2007). These patterns probably do not reflect true musth, however, which by definition refers to the competitive state in sexually active male elephants, with the presence of UD being the defining physical signal (Ganswindt et al. 2005). As with females, captive African elephant bulls appear to reach sexual maturity later and they do not exhibit musth as early as Asian bulls (Rasmussen et al. 1984; Cooper et al. 1990; Brown et al. 2007; J. Brown, unpubl). There also appears to be a species differences in elephant bulls' responses to social factors, not unlike that observed for females. For instance, musth generally occurs in most males in multi-bull Asian groups regardless of dominance status, whereas in African groups, it is pronounced only in the dominant bull (Ganswindt et al. 2005; Brown et al. 2007; J. Brown, unpubl).

Bull elephants advertise their musth status through a variety of chemicals exuded in TGS and urine. In young Asian bulls between ~8 and 13 years of age, TGS consists of sweet odors: acetates, an alcohol (3-hexen-2-ol) that smells of leaves, and pleasant smelling ketones (acetophenone and 2-heptanone) (Rasmussen et al. 2002; Riddle et al. 2006). This sweet musth is referred to as honey or moda musth, and is associated with behaviors that are more erratic and unpredictable. Moda musth also is of a shorter duration and associated with comparatively lower androgen levels. To date, there are no reports of moda musth in juvenile African elephant bulls, which may represent yet another species difference. Older males of both species are more socially and sexually adept and capable of sustaining longer periods of musth at higher androgen concentrations, often for several months each year. In Asian elephants, the pleasant-smelling compounds of moda musth are transitionally replaced by more malodorous compounds: carboxylic acids, which reduce the pH of the TGS to as low as 5.5, and a known chemical signal - an acrid ketal, frontalin [1,5-dimethyl-5,8-dioxabicyclo(3.2.1)octane]. Both enantiomeric forms of frontalin are produced during musth, but the proportion varies from day to day, with the (+) form generally predominating (Greenwood et al. 2005). In older Asian and African elephant bulls, musth urine contains higher levels of alkan-2-ones, alkan- 2-ols, and some aromatic compounds compared to urine of females and non-musth males (African: Riddle et al. 2006; Asian and African: Goodwin et al. 2012). Young African bulls also begin to excrete alkan-2-ones and alkan-2-ols during periods of elevated testosterone, but not near to the degree observed in adults (Davenport et al. 2013). Levels of ketones, alcohols and protein-derived aromatic metabolites also increase as urine ages, likely due to microbial metabolism of fatty acids, suggesting that microbes may play a role in timed release of urinary chemical signals (Goodwin et al. 2012). Asian and African elephants have well-developed primary and secondary (vomeronasal) olfactory systems (Lazar et al. 2004), and both males and females respond to these chemical signals. Estrous females seek out musth bulls, and sub-adult males (especially non-musth) exhibit avoidance behavior when exposed to musth semiochemicals (Riddle et al. 2006; Goodwin et al. 2012). Ultimately, the state of musth confers an advantage to adult bulls, which gain more access to estrous females and experience a higher paternity success, at least in African elephants (Hollister-Smith et al. 2007). Understanding the biochemistry of musth could have practical application, as use of musth-like chemicals is being considered in Asian elephants as a strategy to mitigate human-elephant conflict and discourage encroachment into villages and crop fields (Perera 2009).

The relationship between aggressive behaviors and increases in androgen production during musth suggest the two are linked, although the regulatory mechanisms are less clear. A proposed model for the endocrinology of musth is depicted in Fig. 8.3. In a long-term study (4 years) of one African bull, distinct hormonal relationships were observed in association with musth activity (Kaewmanee et al. 2011b). Serum LH increased about 4 weeks before musth began and was maintained for ~5 weeks, and likely was responsible for triggering the rise in testosterone. A distinct pattern of FSH was less clear, but it too was higher during the pre-musth period. FSH was associated with a subsequent rise in inhibin secretion, which in turn correlated positively with testosterone. In other species, inhibin is produced by Sertoli cells in response to FSH, which then controls pituitary FSH secretion through a negative feedback loop. Another hormone produced by Sertoli cells under FSH control is AMH, which can be used as a marker of FSH and testicular function (Valeri et al. 2013). AMH is higher in prebubertal bulls; however, there were no difference in concentrations between musth and nonmusth males (Dow et al. 2011b). Similar to other species, AMH concentrations in elephants are much higher in males compared to females (over 100-fold), regardless of age or gonadal status (Dow et al. 2011b).

Musth may be partially under thyroid control. In one study of Asian bulls exhibiting normal musth cycles, serum thyrotropin-stimulating hormone (TSH) was positively correlated, and thyroid hormones (T_3, T_4) were negatively correlated to testosterone secretion (Brown et al. 2007; Fig. 8.3). Specifically, increases in thyroid

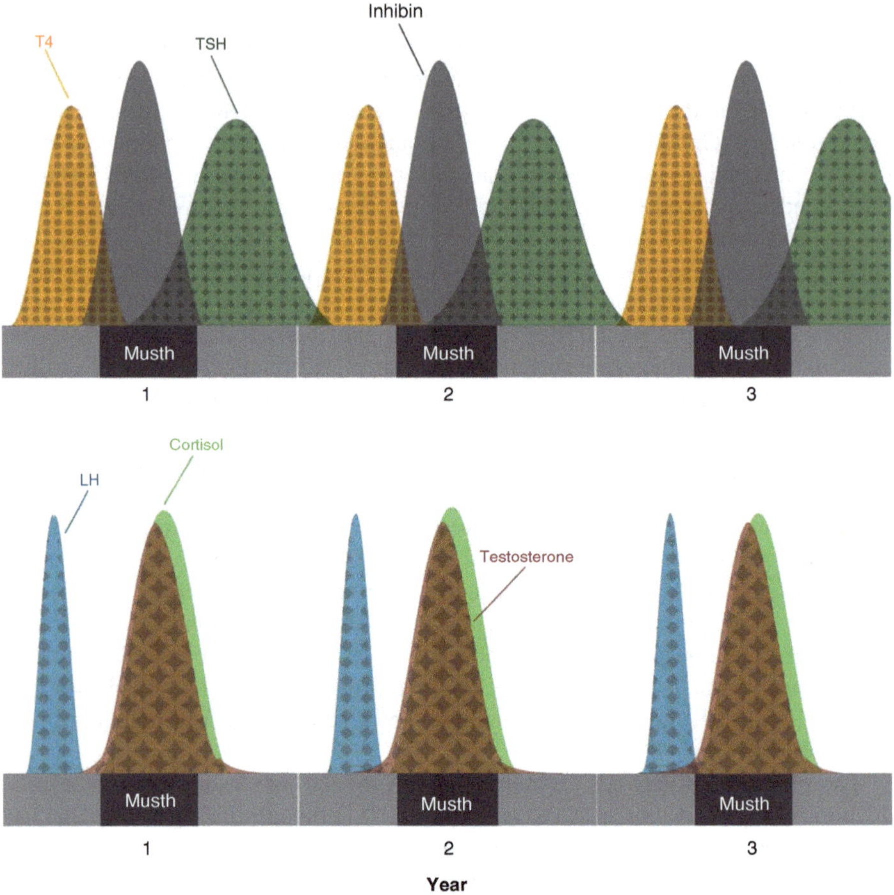

Fig. 8.3 Schematic of the hormonal relationships among testosterone, LH, cortisol, inhibin, TSH and T4 in association with musth in elephant bulls. Adapted from Brown et al. (2007) and Kaewmanee et al. (2011b)

hormones preceded the rise in testosterone associated with musth onset, after which concentrations declined and reached nadir levels in conjunction with the return of testosterone to baseline. A negative relationship between thyroid hormones and TSH is suggestive of negative feedback regulation, as in other species. Similar relationships between T_3 and/or T_4 hormones and testicular function have been reported in other ungulates, as males transition from a breeding to a nonbreeding state (Shi and Barrell 1994). Thus, an increase in thyroid hormones preceding musth may be necessary for increased metabolic activity in preparation for the physical and physiological changes associated with a heightened sexual state.

There are conflicting reports on the role of adrenal activity in the regulation of or in response to musth. Positive correlations have been found between serum cortisol and testosterone, with both being elevated during musth in captive Asian and African elephants (Brown et al. 2007; Yon et al. 2007; see Fig. 8.3). By contrast, measures of fecal androgen and corticoid metabolites did not correlate in wild or captive bulls (Ganswindt et al. 2003, 2005). These differences may be due to using serum vs. feces, although all assays were properly validated. Thus, further studies are warranted to examine how or if adrenal activity is altered in relation to the many physiological and physiological changes associated with musth.

3 Assisted Reproduction

3.1 Artificial Insemination

One the most significant advances in elephant reproductive management has been the development of an artificial insemination (AI) technique. Pioneered by Dr. Thomas Hildebrandt (Institute for Zoo and Wildlife Research, Berlin, Germany) in the 1990s, the non-surgical procedure involves a custom-made balloon catheter (2.5-cm diameter 140 cm length) lubricated with nonspermicidal sterile gel, inserted into the vestibule to slightly distend the reproductive tract for optimal visualization and allow passage of a flexible 3.0-m endoscope containing a customized disposable insemination catheter (3-mm diameter, 300 cm length) (Brown et al. 2004b). Both endoscopic and transrectal ultrasonographic visualizations are used to guide the insemination catheter to the distal vagina in front of the cervical opening or intracervically, where semen deposition can be visualized ultrasonographically to verify placement. The timing of AI is based on identifying LH1 in daily blood samples during the follicular phase and then conducting 2–3 inseminations ~18–21 days later to coincide with LH2 (Brown et al. 2004b). Monitoring LH in elephants can be done only with blood serum or plasma, as immunoactive hormone has not been detected in urine (Brown et al. 2010). The first elephant AI birth was in 1999 (Saragusty et al. 2009a); since then over 40 births have resulted worldwide, with the majority (>70 %) being in African elephants. For a time, there was sex skewing with 83 % of AI calves being male (Saragusty et al. 2009a); however, current studbook data indicate this skew is no longer significant.

3.1.1 Semen Characteristics and Cryopreservation

As a management tool, AI has been in use for nearly 20 years; however, the utility of this technique is hampered by: (1) the inability to consistently obtain good quality ejaculates; (2) poor spermatozoa survival after liquid storage and transport; and (3) poor post-thaw recovery of cryopreserved sperm. To date, successful births have resulted only with fresh or chilled semen, which limits AI effectiveness if good quality ejaculates cannot be collected on peak fertility days. There have been three reported pregnancies from frozen-thawed semen in elephants to date. One was achieved in an Asian elephant, which ended in a stillborn calf after 17 months (Thongtip et al. 2009), one was in an African elephant that terminated after 5 months (Dennis L. Schmitt, personal communication). A third in an African elephant inseminated in 2011 resulted in a live birth in 2013 (Hildebrandt et al. 2012b and unpubl). There is little doubt that use of AI is a significant management tool for captive elephants, but maximal impact will not be attained until good quality samples can be reliably obtained and cryopreserved for use with AI.

In zoo elephants, semen is collected by a transrectal message technique, but sample quality is highly variable within and among individuals, especially for Asians elephants (Schmitt and Hildebrandt 1998). In extensive studies of Asian elephant bulls, less than a third of ejaculates exhibited >60 % motile spermatozoa, and many were contaminated with urine (Kiso et al. 2011, 2012, 2013)—a common problem in elephants that is known to damage cells. Good motility ejaculates (>65 % motility) contained higher proportions of normal morphology spermatozoa with intact acrosomes compared to poor motility ejaculates. Furthermore, ejaculates with higher motility were of a larger volume and lower sperm concentration. To determine if rectal message produces an abnormal complement of seminal components in poor quality ejaculates, seminal plasma analyses were conducted and revealed several correlations between chemistry components and spermatozoal characteristics (Imrat et al. 2013a; Kiso et al. 2013). Differences were found in creatine phosphokinase, alanine aminotransferase, phosphorus, sodium, chloride, magnesium and glucose in seminal plasma from ejaculates exhibiting good versus poor motility, whereas there were no differences in total protein, albumin, lactate dehydrogenase, aspartate aminotransferase, alkaline phosphatase, calcium, potassium, cholesterol, bicarbonate, creatinine, or urea nitrogen. One- and two-dimensional gel electrophoresis revealed similar seminal plasma protein profiles between good and poor motility ejaculates. However, a protein of approximately 80 kDa was abundant in 85–90 % of ejaculates with good motility, and was absent in >90 % of poor motility ejaculates. Mass spectrometry analyses identified the protein as lactotransferrin, which was confirmed by immunoblotting (Kiso et al. 2013) and is considered a potential fertility marker in human semen (Milardi et al. 2012). The ability of lactotransferrin to sequester iron molecules may improve seminal quality by serving as a natural antibiotic and/or an antioxidant in semen (Brock 2002; Sanocka and Kurpisz 2004). Thus, some seminal plasma components correlate with spermatozoa motility in elephants, especially lactotransferrin, which may serve as biomarkers of spermatozoa quality (Kiso et al. 2013).

Immediately after collection, semen is extended for shipment to a recipient zoo. However all too often, even excellent quality samples rapidly decline in motility and variability, and within 12–24 h can be too low in quality for insemination (Kiso et al. 2011; O'Brien et al. 2013). In a study of semen extenders and temperatures, storage at 35 °C resulted in a sharp decline in spermatozoal quality parameters after a few hours in both species, whereas spermatozoa held at 22 °C and 4 °C maintained ~50 % of their initial motility for up to 12 h (Kiso et al. 2011). Even still, the identification of considerable DNA damage and morphological degeneration in ejaculates after only 24 h of chilled storage indicates that sperm ageing may be a primary contributor to inconsistent semen quality (O'Brien et al. 2013). It has been suggested that Asian elephant spermatozoa are particularly susceptible to DNA damage compared that of other mammalian species (Imrat et al. 2012a). For example, in a study of nearly a dozen species, the expression of protamine 2 in sperm significantly enhanced the likelihood of DNA fragmentation, whereas greater numbers of cysteine residues in protamine 1 tended to confer increased sperm DNA stability (Gosalvez et al. 2011). The amount of disulphide bonding and number of arginine–lysine residues in protamines likely influences the relative stability of sperm DNA, and aids in a more efficient chromatin organization. Whereas the number of cysteine residues per molecule of protamine 1 in other species ranges from 6–10, elephant protamine 1 contains only five (Gosalvez et al. 2011). Less cross-linking in elephant sperm likely makes it comparatively more fragile under in vitro conditions.

Type of semen extender affects spermatozoal longevity in Asian elephants, with diluents adding a source of lipoprotein (i.e., skim milk or egg yolk) being the most effective (Kiso et al. 2012), and skim milk showing better post-thaw survival rates than egg yolk in this species (Imrat et al. 2013b). The inclusion of antioxidants to reduce DNA fragmentation (e.g., BullMax) also helps preserve sperm longevity of Asian spermatozoa during storage (Imrat et al. 2012b). By contrast, African elephant spermatozoa maintained viability longer than that of Asian elephants, and there was little difference in spermatozoal quality parameters across the extenders. Altogether, it is recommended that extended elephant semen be stored at temperatures below body temperature. For transport to recipient institutions, Asian elephant spermatozoa should be diluted in extenders containing egg yolk or skim milk, whereas spermatozoa from African elephants have no such requirement (Graham et al. 2004; Saragusty et al. 2005; Hermes et al. 2009; Saragusty et al. 2009b; Kiso et al. 2011). Comparative studies clearly show inherent differences between the species, with Asian elephant spermatozoa being overall more sensitive to storage and culture conditions. Given this, future studies to better understand the physical and biochemical nature of elephant spermatozoa and differential responses to handling procedures certainly are warranted.

Cryopreservation of elephant spermatozoa was first attempted 30–40 years ago in African elephants (Jones 1973; Howard et al. 1986), and about a decade ago in Asian elephants (Hedrick and Schmitt 2001; Thongtip et al. 2004). Initial protocols relied on freezing samples over liquid nitrogen vapor or forming pellets on dry ice; however, post-thaw survival was generally poor (<50 %), more so for Asian than African elephants. As such, cell damage during the freeze-thaw process has been a major limitation to successful cryopreservation. During the process of chilling and freezing, the plasma membrane lipid bilayer of spermatozoa is altered in both composition and structure. The freezing of extracellular water results in hyperosmotic conditions that draw water out of the cells and can lead to altered membrane permeability, cell dehydration and shrinkage, and death. Several studies have attempted to enhance the cryosurvival of elephant spermatozoa by adding various cryoprotectants (i.e. dimethyl sulfoxide, glycerol, ethylene glycol and propylene glycol) before cooling and freezing (Kiso 2004; Thongtip et al. 2004; Saragusty et al. 2009b; Buranaamnuay et al. 2013), with glycerol appearing to be the most effective.

In a preliminary study, fatty acids composition of Asian and African elephant spermatozoa plasma membrane were found to differ (Swain and Miller 2000). A lower proportion of polyunsaturated fatty acids in Asian elephant spermatozoal membranes suggested they might be less fluid. Increasing membrane fluidity by incubating with lipids, such as egg-yolk or egg-phosphatidylcholine liposomes, helps elephant spermatozoa withstand such stresses (Asian: Saragusty et al. 2005). Today, most extenders for elephant semen cryopreservation contain a minimum of 15–20 % egg yolk. In other species, treatment of spermatozoa with membrane stabilizers before cryopreservation has improved cryosurvival by modifying phase transition characteristics and increasing tolerance of the cells to freeze–thawing (Purdy and Graham 2004). Membranes with higher cholesterol or greater cholesterol:phospholipid molar ratios tend to be more tolerant to temperature changes during cryopreservation compared to membranes with lower cholesterol levels (Amann and Pickett 1987). In a recent study in Asian elephants, cholesterol loaded into spermatozoa by co-incubation with cholesterol-loaded cyclodextrins (CLC; 1.5 mg of CLC/120×10^6 spermatozoa) increased membrane cholesterol concentrations and significantly improved post-thaw spermatozoa motility and intact acrosomes; ~50 % compared to ~25 % with no CLC (Kiso et al. 2012). This percentage survival should be adequate for use with AI, as the three reported pregnancies occurred with post-thaw sperm motilities of 45–60 % (Thongtip et al. 2009; Hildebrandt et al. 2012b). A newer method of semen cryopreservation involves the use of directional freezing. In conventional freezing methods, ice forms at an uncontrolled rate and can damage cellular membranes. Using large volume (2.5 or 8 ml) cryogenic tubes (HollowTubes®, IMT Ltd.), semen can be moved at a constant velocity through a linear temperature gradient, which results in better control of ice crystal formation and minimum damage to cells. Post-thaw motility in excess of 50 % has recently been obtained using directional freezing with semen extended in an egg-yolk extender [e.g., Berliner Cryomedium: 2.41 % (w/v) TES, 0.58 % (w/v) Tris, 0.1 % (w/v) fructose and 5.5 % (w/v) lactose, 15.6 % (v/v) egg yolk and 20 IU α-tocopherol/ml] with 7 % glycerol (Asian: Saragusty et al. 2009b; African: Hermes et al. 2013).

Traditionally, zoo elephant populations have been supplemented by the selective importation of females. This is in part due to the increase in husbandry requirements and exhibit costs associated with maintaining bulls, which limits the number of facilities willing to house males. As more zoos have become involved in breeding elephants by both AI and natural mating, the number of males in the captive population has been increasing (Keele et al. 2010; Olson 2011). With the advent of flow cytometric methods to separate sperm based on the DNA difference between X and Y chromosomes (Johnson et al. 1987), the technology now exists to preferentially select females. One obstacle to the practical application of this technique is identifying an extender that will optimize short-term storage of spermatozoa while simultaneously being compatible with flow cytometric sorting. Egg yolk is added to semen extenders because the added lipids enhance sperm survival during in vitro semen storage, especially for Asian elephants. However, it must be removed before sorting because it interferes with the uniform staining of DNA that is critical to separate X- and Y-chromosome bearing populations (Johnson and Welch 1999). Additional processing and handling to remove remnants of egg yolk prior to sex-sorting can cause sperm damage and loss in viability. For these reasons, non-egg yolk or reduced egg yolk-based extenders need to be used. Success has been achieved in sex-sorting of Asian elephant sperm utilizing a MES-HEPES skim milk-based medium (Hermes et al. 2009). Another skim milk diluent supplemented with only 4 % egg yolk (INRA96) sustained sperm as well as media containing 20 % egg yolk and thus can be used to extend ejaculates for shipment to centralized sperm sorting facilities (Kiso et al. 2011; Imrat et al. 2013a). African elephants have no lipid requirement, so based on previous semen storage studies, both TL-Hepes and Modena, which are devoid of egg yolk, should be effective in extending African elephant ejaculates for sorting (Kiso et al. 2011). The next step is to optimize cryopreservation techniques for sex-sorted spermatozoa to be used with AI to favor the selection of females, thus increasing the chances of creating genetically healthy and sustainable populations of captive elephants. Based on these recent successes, we may finally have the means to establish a genome resource bank for elephants, which if it includes sex-sorted samples from captive and wild bulls, could greatly enhance the genetic management of elephants under human care (Hermes et al. 2013).

4 Future Priorities

By better understanding the biology of elephants, we aim to improve breeding management and establish self-sustaining ex situ populations, findings that could potentially have application to the conservation of elephants in situ. Through advancements in endocrine and ultrasound monitoring techniques, many unique aspects of elephant reproduction have been identified. Compared to other mammalian species, female elephants exhibit interesting differences in luteal steroidogenic activity, follicular maturation, pituitary gonadotropin secretion, fetal development and reproductive tract anatomy. However, problems associated with uterine and ovarian pathologies, with or without accompanying ovarian acyclicity, hamper captive propagation efforts. Older, nulliparous cows are particularly susceptible, leading to speculation that continuous ovarian cyclicity of non-bred females in zoos is having a negative and cumulative effect on reproductive health. Most of the ovarian cycle problems, and in some cases delayed puberty, occur in African rather than Asian elephants, and represent significant species differences. New approaches to management of social groupings, modifications in nutrition, and/or medical treatments might avoid some of the problems associated with early asymmetric reproductive aging, gonadal dysfunction and poor pregnancy outcomes. New methods also are needed to identify pregnancy-specific markers, preferably noninvasively in samples of urine and/or feces. Even more effective would be dipstick tests to permit the monitoring of ovarian and pregnancy status in elephants where laboratory capabilities are limited, as is the case in most range countries.

Male elephants, although not as extensively studied as females, exhibit the unique phenomenon of musth and an unusual reproductive anatomy (internal testes, ampullary semen storage). They also appear to be affected by comparable reproductive problem described for females, like social suppression of gonadal function

(Africans), poor gamete quality (Asian and African) and/or decreased libido (Asian and African). Collection of semen by rectal message, although a simple technique that does not require anesthesia and is a behavior that can be easily trained, is not always effective in obtaining good quality samples useful for AI or cryopreservation. Alternative collection approaches should be explored, perhaps using combinations of neurostimulants and transrectal electrical stimulation.

Last, better approaches are needed to assess elephant welfare and stress as it pertains to health and fitness. Cortisol, while capable of identifying adrenal responses to acute stimuli and illness, is less informative when evaluated in the context of understanding chronic stress, such as the impact of social or environmental stressors on reproductive function. Other indicators, physiological and behavioral, may be more revealing, especially if combined, such as measures of general health, heart rate, inflammatory markers, cytokines, hormones related to well-being (e.g., oxytocin, IgA), catacholamines, cognitive bias, stereotypies, or as yet to be identified stress-reactive biomarkers. Rapid test kits to monitor reproductive, stress and nutritional status, especially in the field, would be particularly useful for assessing environmental and/or anthropogenic effects on elephant biology and behavior.

Elephants are not always easy to study, not unlike other wildlife species, and so most existing data are based on investigations of zoo animals. There are both advantages and disadvantages to studying elephants. Prolonged ovarian cycle and gestation lengths hinder the speed at which information can be generated. This likely is one reason it has taken over 20 years to understand the complex dynamics of the hypothalamo-hypophyseal-ovarian axis. On the other hand, the ability to collect serial blood samples combined with ultrasonographic examinations has produced an unprescedented set of reproductive data for female and male elephants. These advances were possible only because of how easily elephants can be trained for nonstressful reproductive assessments. In fact, our knowledge of elephant endocrinology has no match in wildlife biology, making it a model species for reproductive studies of other zoo animals.

Acknowledgments I thank all of the technical, veterinary, zoo and research staff (and elephants) who contributed to the reproductive work described herein. This work was funded or supplied with in-kind donations by the Abbott Fund, AZA Conservation Endowment Fund, FBB Capital Partners, Friends of the National Zoo, Institute of Museum and Library Services, International Elephant Foundation, Morris Animal Foundation, Pfizer Corporation, Siemens Medical Solutions Diagnostics, Shared Earth Foundation, Smithsonian Competetive Grants Program, Smithsonian Scholarly Studies Program, Smithsonian Women's Committee, and the Wolcott Fund.

References

Adamafio NA. The hormone-emotion-behavioural gene-neuromessenger labyrinth: pertinent questions. Afr J Biochem Res. 2009;3:326–31.

Allen WR. Ovulation, pregnancy, placentation and husbandry in the African elephant (*Loxodonta africana*). Phil Trans R Soc B. 2006;361:821–34.

Allen WR, Mathias SS, Wooding FBP, Skidmore JA, van Aarde RJ. Placentation in the African elephant, *Loxodonta africana*. I. Endocrinological aspects. Reprod Suppl. 2002;60:105–16.
Allen WR, Mathias S, Ford M. Placentation in the African elephant, *Loxodonta africana*. IV. Growth and function of the fetal gonads. Reproduction. 2005;130:713–20.
Ange K, Crissey SD, Doyle C, Lance K, Hintz H. A survey of African (*Loxodonta afriana*) and Asian (*Elephas maximus*) elephant diets and measured body dimensions compared to their estimated nutrient requirements. Proc Nutrition Advisory Group Conf. 2001;4:5–14.
Amann RP, Pickett BW. Principles of cryopreservation and a review of cryopreservation of stallion spermatozoa. J Equine Vet Sci. 1987;7:145–73.
Aupperle H, Reischauer A, Bach F, Hildebrandt T, Göritz F, Jäger K, et al. Chronic endometritis in an Asian elephant (*Elephas maximus*). J Zoo Wildl Med. 2008;39:107–10.
Aron DC, Tyrrell JB, Wilson CB. Pituitary tumors: current concepts in diagnosis and management. Western J Med. 1985;162:340–52.
Assies J, Vingerhoets JJM, Poppelaars K. Psychosocial aspects of hyperprolactinemia. Psychoneuroendocrinology. 1992;17:673–9.
Bachelot A, Binart N. Reproductive role of prolactin. Reproduction. 2007;113:361–9.
Baird DT. Factors regulating the growth of the pre-ovulatory follicle in sheep and humans. J Reprod Fertil. 1983;69:343–52.
Ball RL, Brown JL. Treatment of anestrus due to hyperprolactinemia with cabergoline in a captive Asian elephant *(Elephas maximus)*. Proc Int Elephant Res Symp, 2004; Fort Worth, Texas.
Bechert US, Swanson L, Wasser SK, Hess DL, Stormshak F. Serum prolactin concentrations in the captive female African elephant (*Loxodonta africana*): potential effects of season and steroid hormone interactions. Gen Comp Endocrinol. 1999;114:269–78.
Ben-Jonathan N, LaPensee CR, LaPensee EW. What can we learn from rodents about prolactin in humans? Endocr Rev. 2008;29:1–41.
Bertschinger H, Delsink A, van Altena JJ, Kirkpatrick J, Killian H, Ganswindt A, et al. Reproductive control of elephant. In: Scholes RJ, Mennell KG, editors. Assessment of South African Elephant Management. Johannesburg: Wits University Press; 2007. p. 191–245.
Boedeker NC, Hayek LC, Murray S, de Avila DM, Brown JL. Effects of a gonadotropin-releasing hormone vaccine on ovarian cyclicity and uterine morphology of an Asian elephant (*Elephas maximus*). J Zoo Wild Med. 2012;43:603–14.
Botha AE, Schulman ML, Bertschinger HJ, Guthrie AJ, Annandale CH, Hughes SB. The use of a GnRH vaccine to suppress mare ovarian activity in a large group of mares under field conditions. Wildl Res. 2008;35:548–54.
Brock JH. The physiology of lactoferrin. Biochem Cell Biol. 2002;80:1–6.
Brown JL. Reproductive endocrine monitoring of elephants: an essential tool for assisting captive management. Zoo Biol. 2000;19:347–67.
Brown JL. Reproductive endocrinology. In: Fowler ME, Mikota SK, editors. Biology, medicine and surgery of elephants. Ames, IA: Blackwell; 2006. p. 377–88.
Brown JL, Kersey DC, Walker SL. Assessment of luteinizing hormone and prolactin immunoactivity in Asian and African elephant urine using assays validated for serum. Gen Comp Endocrinol. 2010;169:138–43.
Brown JL, Somerville M, Riddle H, Keele M, Duer C, Freeman EW. Comparative endocrinology of testicular, adrenal and thyroid function in Asian and African elephant bulls. Gen Comp Endocrinol. 2007;151:153–62.
Brown JL, Walker SL, Moeller T. Comparative endocrinology of cycling and noncycling Asian (*Elephas maximus*) and African (*Loxodonta africana*) elephants. Gen Comp Endocrinol. 2004a;136:360–70.
Brown JL, Goritz F, Pratt-Hawkes N, Hermes R, Galloway M, Graham LH, et al. Successful artificial insemination of an Asian elephant at the National Zoological Park. Zoo Biol. 2004b; 23:45–63.
Brown JL, Schmitt DL, Bellem A, Graham LH, Lehnhardt J. Hormone secretion in the Asian elephant (*Elephas maximus*): characterization of ovulatory and anovulatory luteinizing hormone surges. Biol Reprod. 1999;61:1294–9.

Brown JL, Lehnhardt J. Serum and urinary hormones during pregnancy and the peri- and postpartum period in an Asian elephant (*Elephas maximus*). Zoo Biol. 1995;14:555–64.
Brown JL, Citino SB, Bush M, Lehnhardt J, Phillips LG. Cyclic patterns of luteinizing hormone, follicle stimulating hormone, inhibin, and progesterone in the Asian elephant (*Elephas maximus*). J Zoo Wildl Med. 1991;22:49–57.
Buranaamnuay K, Mahasawangkul S, Saikhun K. The in vitro quality of frozen-thawed Asian elephant (*Elephas maximus*) spermatozoa in semen supplemented with Equex STM paste and oxytocin during and after cryopreservation. Reprod Biol. 2013;13:169–71.
Calogero AE, Badgy G, D'Agata R. Mechanisms of stress on reproduction – evidence for a complex intra-hypothalamic circuit. Annals NY Acad Sci. 1998;1:364–70.
Carlstead K. A comparative approach to the study of Keeper-Animal Relationships in the zoo. Zoo Biol. 2009;28:589–608.
Campbell BK, McNeilly AS, Mann GE, Baird DT. The effect of stage of estrous cycle and follicular maturation on ovarian inhibin production in sheep. Biol Reprod. 1991;44:483–90.
Castracane VD, Stewart DR, Gimpel T, Overstreet JW, Lasley BL. Maternal serum androgens in human pregnancy: early increases within the cycle of conception. Hum Reprod. 1998;13:460–4.
Clubb R, Mason G. A review of the welfare of zoo elephants in Europe. A report commissioned by the RSPCA. UK: University of Oxford; 2002.
Clubb R, Rowcliffe M, Lee P, Mar KU, Moss C, et al. Fecundity and population viability in female zoo elephants: problems and possible solutions. Anim Welfare. 2009;18:237–47.
Cohen AA. Female post-reproductive lifespan: a general mammalian trait. Biol Rev Camb Philos Soc. 2004;79:733–50.
Conde DA, Flesness N, Colchero F, Jones OR, Scheuerlein A. An emerging role of zoos to conservation biodiversity. Science. 2011;331:1390–1.
Conforti VA, de Avila DM, Cummings NS, Zanella R, Wells KJ, Ulker H, et al. CpG motif-based adjuvant as a replacement for Freund's complete adjuvant in a recombinant LHRH vaccine. Vaccine. 2008;26:907–13.
Cooper KA, Harder JD, Clawson DH, Fredrick DL, Lodge GA, Peachey HC, et al. Serum testosterone and musth in captive male African and Asian elephants. Zoo Biol. 1990;9:297–306.
Creel S, MacDonald D. Sociality, group size, and reproductive suppression among carnivores. Adv Study Behav. 1995;24:203–57.
Czekala NM, MacDonald EA, Steinman K, Walker S, Garrigues II NW, Olson D, et al. Estrogen and LH dynamics during the follicular phase of the estrous cycle in the Asian elephant. Zoo Biol. 2003;22:443–54.
Davenport SA, Yin Q, Mark DB, Harelimana IH, Schulte BA, Brown JL, Goodwin TE. Urinary chemical signals of musth in immature male African elephants. Proceedings of 245th American Chemical Society, 2013; New Orleans, LA, abstract #1158.
Dobson H, Smith RF. What is stress, and how does it affect reproduction? Anim Reprod Sci. 2000;60:743–52.
Dow TL, Brown JL. Evidence that hyperprolactinemia is associated with ovarian acyclicity in female zoo African elephants (*Loxondonta africana*). Reprod Fertil Dev. 2012;24:1019–27.
Dow TL, Holaskova I, Brown JL. Results of the third reproductive assessment survey of North American Asian (*Elephas maximus*) and African (*Loxodonta africana*) female elephants. Zoo Biol. 2011a;30:699–711.
Dow TL, Roudebush W, Parker FN, Brown JL. Influence of age and gender on secretion of anti-müllerian hormone in Asian (*Elephas maximus*) and African (*Loxodonta africana*) elephants. Theriogenology. 2011b;75:620–7.
Downey BR. Regulation of the estrous cycle in domestic animals: a review. Can Vet J. 1980;21: 301–6.
Drews B, Hermes R, Göritz F, Gray C, Kurz J, Lueders I, et al. Early embryo development in the elephant assessed by serial ultrasound examinations. Theriogenology. 2008;69:1120–8.
Duer C, Carden M, Schmitt D, Tomasi T. Utility of maternal serum testosterone analysis for fetal gender determination in Asian elephants (*Elephas maximus*). Anim Reprod Sci. 2002;69:47–52.
Faust L, Marti K. Technical report on Zoo Risk modeling of the North American African elephant population, AZA TAG/SSP. Chicago, IL: Lincoln Park Zoo; 2011a.

Faust L, Marti K. Technical report on Zoo Risk modeling of the North American Asian elephant population, AZA TAG/SSP. Chicago, IL: Lincoln Park Zoo; 2011b.

Fernando P, Kumar MA, Williams AC, Wikramanayake E, Aziz T, Singh SM. Review of human-elephant conflict mitigation measures practiced in south Asia. AREAS Technical Support Document, World Wildl Fund for Nature; 2008. p. 1–45.

Fernando P, Vidya TNC, Payne J, Stuewe M, Davison G, et al. DNA analysis indicates that Asian elephants are native to Borneo and are therefore a high priority for conservation. PLoS Biol. 2003;1:e6. doi:10.1371/journal.pbio.0000006.

Florio P, Gabbanini M, Borges LE, Bonaccorsi L, Pinzauti S, Reis FM, et al. Activins and related proteins in the establishment of pregnancy. Reprod Sci. 2010;17:320–30.

Frasor J, Gibori G. Prolactin regulation of estrogen receptor expression. Trends Endocrinol Metab. 2003;14:118–23.

Freeman EW, Meyer JM, Putman S, Schulte BA, Brown JL. Using a simplified field progestagen method to assess ovarian activity in female African elephants. Biol Conserv. 2011;144:2105–11.

Freeman EW, Schulte BA, Brown JL. Investigating the impact of rank and ovarian activity on the social behavior of captive female African elephants. Zoo Biol. 2010;29:154–67.

Freeman EW, Guagnano G, Olson D, Keele M, Brown JL. Social factors influence ovarian acyclicity in captive African elephants (*Loxodonta africana*). Zoo Biol. 2009;28:1–15.

Freeman EW, Whyte I, Brown JL. Reproductive evaluation of elephants culled in Kruger National Park, South Africa between 1976 and 1995. Afr J Ecol. 2008;47:192–201.

Freeman EW, Weiss E, Brown JL. Examination of the interrelationships of behavior, dominance status, and ovarian activity in captive Asian and African elephants. Zoo Biol. 2004; 23:431–48.

Freeman ME, Kanyicska B, Lerant A, Nagy G. Prolactin: structure, function and regulation of secretion. Physiol Rev. 2000;80:1524–631.

Forsyth IA, Wallis M. Growth hormone and prolactin–molecular and functional evolution. J Mammary Gland Biol Neoplasia. 2002;7:291–312.

Ganswindt A, Rasmussen HB, Heistermann M, Hodges JK. The sexually active states of free-ranging male African elephants (*Loxodonta africana*): defining musth and non-musth using endocrinology, physical signals, and behavior. Horm Behav. 2005;47:83–91.

Ganswindt A, Palme R, Heistermann M, Borragan S, Hodges JK. Non-invasive assessment of adrenocortical function in the male African elephant (*Loxodonta africana*) and its relation to musth. Gen Comp Endocrinol. 2003;134:156–66.

Ganswindt A, Heistermann M, Borragan S, Hodges JK. Assessment of testicular endocrine function in captive African elephants by measurement of urinary and fecal androgens. Zoo Biol. 2002;21:27–36.

Giammanco M, Tabacchi G, Giammanco S, Di Majo D, La Guardia M. Testosterone and aggressiveness. Med Sci Monit. 2005;11:RA136–45.

Glaeser SS, Martin MS, Hunt KE, Finnegan M, Brown JL. Investigation of individual and group variability in estrous cycle characteristics in female Asian elephants (*Elephas maximus*) at the Oregon Zoo. Theriogenology. 2012;78:285–96.

Grynnerup AG, Lindhard A, Sorensen S. The role of anti-mullerian hormone in female fertility and infertility – an overview. Acta Obstet Gynecol Scand. 2012;91:1252–60.

Ginther OJ, editor. Reproductive biology of the mare, basic and applied aspects. 2nd ed. Wisconsin: Equiservices; 1992.

Goodwin TE, Eggert MS, House SJ, Weddell ME, Schulte BA, Rasmussen LE. Insect pheromones and precursors in female African elephant urine. J Chem Ecol. 2006;32:1849–53.

Goodwin TE, Broederdorf LJ, Burkert BA, Hirwa IH, et al. Chemical signals of elephant musth: temporal aspects of microbially-mediated modifications. J Chem Ecol. 2012;38:81–7.

Gosalvez J, Lopez-Fernandez C, Fernandez JL, Gouraud A, Holt WV. Relationships between the dynamics of iatrogenic DNA damage and genomic design in mammalian spermatozoa from eleven species. Mol Reprod Dev. 2011;78:951–61.

Gosden RG. Follicular status at menopause. Hum Reprod. 1987;2:617–21.

Graham LH, Bando J, Gray C, Buhr MM. Liquid storage of Asian elephant (*Elephas maximus*) sperm at 4 °C. Anim Reprod Sci. 2004;80:329–40.

Greenwood DR, Comeskey D, Hunt M, Rasmussen LEL. Chirality in elephant pheromones. Nature. 2005;438:1097–8.

Hedrick H, Schmitt DL. Comparison of semen extenders and cryoprotectants in an Asian elephant. Springfield, MO: College of Natural and Applied Science, Agriculture, Master Thesis, Natural and Applied Science, Southwest Missouri State University; 2001. p. 1–47.

Hermes R, Sargusty J, Goeritz F, Bartels P, Potier R, Baker B, et al. Freezing African elephant semen as a new population management tool. PLoS ONE. 2013;8:e57616.

Hermes R, Behr B, Hildebrandt TB, Blottner S, Sieg B, Frenzel A, et al. Sperm sex-sorting in the Asian elephant (*Elephas maximus*). Anim Reprod Sci. 2009;112:390–6.

Hermes R, Sargusty J, Schaftenaar W, Goeritz F, Schmitt DL, Hildebrandt TB. Obstetrics in elephants. Theriogenology. 2008;70:131–44.

Hermes R, Hildebrandt TB, Göritz F. Reproductive problems directly attributable to long-term captivity: asymmetric reproductive aging. Anim Reprod Sci. 2004;82–83:49–60.

Hermes R, Olson D, Göritz F, Brown JL, Schmitt DL, Hagan D, et al. Ultrasonography of the estrous cycle of female African elephants (*Loxodonta africana*). Zoo Biol. 2000;19:369–82.

Hildebrandt TB, Lueders I, Hermes R, Göritz F, Sargusty J. Anatomical, physiological, behavioral and pathological aspects of the estrous cycle in elephants. Jpn J Zoo Wildl Med. 2012a;17:97–109.

Hildebrandt TB, Hermes R, Saragusty J, Potier R, Schwammer HM, Balfanz F, et al. Enriching the captive elephant population genetic pool through artificial insemination with frozen-thawed semen collected in the wild. Theriogenology. 2012b;78:1398–404.

Hildebrandt TB, Lueders I, Hermes R, Göritz F, Sargusty J. Reproductive cycle of the elephant. Anim Reprod Sci. 2011;124:176–83.

Hildebrandt TB, Göritz F, Hermes R, Reid C, Dehnhard M, Brown JL. Aspects of the reproductive biology and breeding management of Asian and African elephants. Int Zoo Yrbk. 2006;40:20–40.

Hildebrandt TB, Brown JL, Hermes R, Göritz F. Ultrasound for the analysis of reproductive function in wildlife. In: Wildt DE, Holt WV, Pickard AR, Roger JC editors. Reproduction and Integrated Conservation Science, Cambridge University Press: UK; 2003. p. 166–82.

Hoffman M, Hilton-Taylor C, Angulo A, et al. The impact of conservation on the status of the world's vertebrates. Science. 2010;330:1503–9.

Hollister-Smith JA, Poole JH, Archie DA, Vance EA, Georgiadis NJ, Moss CJ, et al. Age, musth and paternity success in wild male African elephants, *Loxodonta africana*. Anim Behav. 2007;74:287–96.

Howard JG, Bush M, de Vos V, Schiewe MC, Pursel VG, Wildt DE. Influence of cryoprotective diluent on post-thaw viability and acrosomal integrity of spermatozoa of the African elephant (*Loxodonta africana*). J Reprod Fertil. 1986;78:295–306.

Imrat P, Suthanmapinanth P, Saikhun K, Mahasawangkul S, Sostaric E, Sombutputorn P, et al. Effect of pre-freeze semen quality, extender and cryoprotectant on the post-thaw quality of Asian elephant (*Elephas maximus indicus*) semen. Cryobiology. 2013;66:52–9.

Imrat P, Hernandez M, Ritem S, Thongtip N, Mahasawangkul S, Gosalvez J, et al. The dynamics of sperm DNA stability in Asian elephant (*Elephas maximus*) spermatozoa before and after cryopreservation. Theriogenology. 2012a;77:998–1007.

Imrat P, Mahasawangkul S, Gosalvez J, Suthanmapinanth P, Sombutputorn P, Jansittiwate S, et al. Effect of cooled storage on quality and DNA integrity of Asian elephant (*Elephas maximus*) spermatozoa. Reprod Fertil Dev. 2012b;4:1105–16.

Jackson T. Ivory apocalypse. Africa geographic. April Issue; 2013. p. 32–57.

Johnson LA, Welch GR. Sex preselection: High-speed flow cytometric sorting of X and Y sperm for maximum efficiency. Theriogenology. 1999;52:1323–41.

Johnson LA, Flook JP, Look MV, Pinkel D. Flow sorting of X and Y chromosome-bearing spermatozoa into two populations. Gamete Res. 1987;16:1–9.

Jones RC. Collection, motility and storage of spermatozoa from the African elephant *Loxodonta africana*. Nature. 1973;243:38–9.

Jungheim ES, Moley KH. Current knowledge of obesity's effects in the pre- and periconceptional periods and avenues for future research. Am J Obstet Gynecol. 2010;203:525–30.

Kaewmanee S, Watanabe G, Kishimoto M, Zhu Jin W, Yamamoto Y, Yamamoto T, et al. Secretion of inhibin during the estrous cycle in the female Asian elephant (*Elephas maximus*). J Vet Med Sci. 2011a;73:77–82.

Kaewmanee S, Watanabe G, Keio M, Yamamoto Y, Yamamoto T, Kishimoto M, et al. A surge-like increase in luteinizing hormone preceding musth in a captive bull African elephant (*Loxodonta africana*). J Vet Med Sci. 2011b;73:379–83.

Kaplowitz PB. Link between body fat and the timing of puberty. Pediatrics. 2008;121:S208–17.

Kapustin N, Critser JK, Olson D, Malven PV. Nonluteal estrous cycles of 3-week duration are initiated by anovulatory luteinizing hormone peaks in African elephants. Biol Reprod. 1996;55:1147–54.

Keele M, Lewis K, Owens T. Asian elephant (*Elephas maximus*) North American Regional Studbook. Portland, Oregon: Oregon Zoo; 2010.

Kiso WK. Cryopreservation of Asian elephant (*Elephas maximus*) semen in conjunction with the use of Spermac® stain to evaluate acrosome status in both Asian and African (*Loxodonta africana*) elephants. Masters Thesis, Missouri State University, Springfield, Missouri; 2004.

Kiso WK, Selvaraj V, Nagashima J, Asano A, Brown JL, Schmitt DL, et al. Lactotransferrin in Asian elephant (*Elephas maximus*) seminal plasma correlates with semen quality. PLOS One. 2013;8:e71033.

Kiso W, Asano A, Travis AJ, Schmitt DL, Brown JL, Pukazhenthi B. Pretreatment of Asian elephant (*Elephas maximus*) spermatozoa with cholesterol-loaded cyclodextrins and glycerol addition at 4 °C improves cryosurvival. Reprod Fertil Dev. 2012;24:1134–42.

Kiso W, Brown JL, Siewerdt SDL, Olson D, Crichton EG, Pukazhenthi BS. Liquid semen storage in elephants (*Elephas maximus* and *Loxodonta africana*): Species differences and storage optimization. J Androl. 2011;21:420–31.

Klinman DM. Adjuvant activity of CpG oligodeoxynucleotides. Int Rev Immunol. 2006;25: 135–54.

Knight PG. Roles of inhibins, activins, and follistatin in the female reproductive system. Front Neuroendocrinol. 1996;17:476–509.

La Torre D, Falorni A. Pharmacological causes of hyperprolactinaemia. Ther Clin Risk Manag. 2007;3:929–51.

Lazar J, Rasmussen LEL, Greenwood DR, Bang IS, Prestwich GD. Elephant albumin: A multipurpose pheromone shuttle. Chem Biol. 2004;11:1093–100.

Lee PC, Graham MD. African elephants and human-elephant interactions: Implications for conservation. Int Zoo Yb. 2006;40:9–19.

Lejeune H, Habert R, Saez JM. Origin, proliferation and differentiation of Leydig cells. J Mol Endocrinol. 1998;20:1–25.

Liggins GC, Thornburn GD. Initiation of parturition. In: Lamming GE, editor. Marshall's physiology of reproduction, vol. 3. 4th ed. London: Chapman & Hall; 1993.

Lueders I, Niemuller C, Rich P, Gray C, Hermes R, Göritz F, et al. Gestation for 22 months: Luteal development and pregnancy maintenance in elephants. Proc Royal Soc B. 2012;279:3687–96.

Lueders I, Taya K, Watanabe G, Yamamoto Y, Yamamoto T, Kaewmanee S, et al. Role of the double luteinizing hormone peak, luteinizing follicles, and the secretion of inhibin for dominant follicle selection in Asian elephants (*Elephas maximus*). Biol Reprod. 2011;85:714–20.

Lueders I, Niemuller C, Gray C, Rich P, Hildebrandt TB. Luteogenesis during the estrous cycle in Asian elephants (*Elephas maximus*). Reproduction. 2010;140:777–86.

MacLennan AH. Relaxin – a review. Aust N Z J Obstet Gynaecol. 1981;21:195–202.

Maisels F, Strindberg S, Blake S, Wittemyer G, Hart J, et al. Devastating decline of forest elephants in Central Africa. PLoS ONE. 2013;8(3):e59469.

Mason GJ, Veasey JS. What do population-level welfare indices suggest about the well-being of zoo elephants? Zoo Biol. 2010;29:256–73.

Matteri RL, Carroll JA, Dyer CJ. Neuroendocrine responses to stress. In: Mench JA, Moberg GP, editors. The biology of animal stress. Wallingford, UK: CABI Publishing; 2000. p. 43–76.

Medan MS, Arai KY, Watanabe G, Taya K. Inhibin: Regulation of reproductive function and practical use in females. Anim Sci J. 2007;78:16–27.

Mellen J. Factors influencing reproductive success in small captive exotic felids (*Felis spp.*): A multiple regression analysis. Zoo Biol. 2005;10:95–110.

Melmed S. Mechanisms for pituitary tumorgensis: The plastic pituitary. J Clin Invest. 2003;112: 1603–18.

Meyer JM, Walker SL, Freeman EW, Steinetz BG, Brown JL. Species and fetal gender effects on the endocrinology of pregnancy in elephants. Gen Comp Endocrinol. 2004;138:263–70.

Milardi D, Grande G, Vincenzoni F, Messana I, Pontecorvi A, De Marinis L, et al. Proteomic approach in the identification of fertility pattern in seminal plasma of fertile men. Fertil Steril. 2012;97:67–73.

Miller PB, Forstein DA, Styles S. Effect of short-term diet and exercise on hormone levels and menses in obese, infertile women. J Reprod Med. 2008;53:315–9.

Moberg GP. Biological response to stress: Implications for animal welfare. In: Moberg GP, Mench JA, editors. The biology of animal stress: Basic principles and implications for animal welfare. UK: CABI Publishing; 2000. p. 1–21.

Morfeld K, Ball RL, Brown JL. Recurrence of hyperprolactinemia and continuation of ovarian acyclicity in captive African elephants (*Loxodonta africana*) treated with cabergoline. J. Zoo Wildl. Med. (in press).

Morfeld K. Investigating body condition and metabolic hormones in relationship to reproductive cyclicity in zoo female African elephants, *Loxodonta africana*. PhD Dissertation, George Mason University, Fairfax, Virginia; 2013.

Morfeld KA, Lehnhardt J, Alligood C, Bolling J, Brown JL. Development of a body condition scoring index for female African elephants validated by ultrasound measures of subcutaneous fat. PLoS ONE 2014:94(4). DOI: 10.1371/journal.pone.0093802.

Mouttham L, Buhr M, Brown JL. Ovarian acyclicity and dominance status of African elephants in zoos is not associated with increased serum testosterone production. Zoo Biol. 2011;126: 115–21.

Niemuller CA, Gray C, Cummings E, Liptrap RM. Plasma concentrations of immunoreactive relaxin activity and progesterone in the pregnant Asian elephant (*Elephas maximus*). Anim Reprod Sci. 1998;53:119–31.

O'Brien JK, Steinman KJ, Montano GA, Love CC, Robeck TR. Sperm DNA fragmentation and morphological degeneration in chilled elephant (*Elephas maximus* and *Loxodonta africana*) semen collected by transrectal massage. Andrology. 2013;1:387–400.

Olson D. North American Region Studbook for the African Elephant (*Loxodonta africana*). Azle, TX: International Elephant Foundation; 2011. p. 1–71.

Paccamonti D. Inducing earlier cyclicity in mares. Bologna, Italy: Proc Ann Meeting Italian Assoc Equine Vet; 2012. p. 75–9.

Panzani D, Zicchino I, Taras A, Marmorini P, Crisci A, Rota A, et al. Clinical use of dopamine antagonist sulpiride to advance first ovulation in transitional mares. Theriogenology. 2011;75:138–43.

Perera BMAO. The human-elephant conflict: A review of current status and mitigation methods. Gajah. 2009;30:41–52.

Prado-Oviedo NA, Mallow EJ, Deng X, Brown JL. Hyperprolactinemia is not associated with hyperestrogenism in noncycling African elephants (*Loxodonta africana*). Gen Comp Endocrinol. 2013;89C:7–14. doi:10.1016/j.ygcen.2013.03.035.

Proctor CM, Freeman EF, Brown JL. Influence of dominance status on adrenal activity and ovarian cyclicity status in captive African elephants. Zoo Biol. 2010;29:168–78.

Purdy PH, Graham JK. Effect of cholesterol-loaded cyclodextrin on the cryosurvival of bull sperm. Cryobiology. 2004;48:36–45.

Rasmussen LEL. Source and cyclic release pattern of (*Z*)-7-dodecenyl acetate, the pre-ovulatory pheromone of the female Asian elephant. Chem Senses. 2001;26:611–23.

Rasmussen LEL, Riddle HS, Krishnamurthy V. Mellifluous matures to malodorous in musth. Nature. 2002;415:975–6.

Rasmussen LEL, Lee TD, Zhang A, Roelofs WL, Daves Jr GD. Purification, identification, concentration and bioactivity of (Z)-7-dodecen-1-yl acetate: sex pheromone of the female Asian elephant, *Elephas maximus*. Chem Senses. 1997;22:417–37.

Rasmussen LE, Buss IO, Hess DL, Schmidt MJ. Testosterone and dihydrotestosterone concentrations in elephant serum and temporal gland secretions. Biol Reprod. 1984;30:352–62.
Rico C, Fabre S, Medigue C, di Clemente N, Clement F, Bontoux M, et al. Anti-müllerian hormone is an endocrine marker of ovarian gonadotropin-responsive follicles and can help to predice superovulatory responses in the cow. Biol Reprod. 2009;80:50–9.
Riddle HS, Greenwood DR, Rasmussen LEL. The shifting chemical signals of musth. Gajah. 2006;24:39–44.
Robinson MR, Mar KU, Lummaa V. Senescence and age-specific trade-offs between reproduction and survival in female Asian elephants. Ecol Lett. 2012;15:260–6.
Sanocka D, Kurpisz M. Reactive oxygen species and sperm cells. Reprod Biol Endocrinol. 2004;2:12–9.
Saragusty J, Hermes R, Göritz F, Schmitt DL, Hildebrandt TB. Skewed birth sex ratio and premature mortality in elephants. Anim Reprod Sci. 2009a;115:247–54.
Saragusty J, Hildebrandt TB, Behr B, Knieriem A, Kruse J, Hermes R. Successful cryopreservation of Asian elephant (*Elephas maximus*) spermatozoa. Anim Reprod Sci. 2009b;115: 255–66.
Saragusty J, Hildebrandt TB, Natan Y, Hermes R, Yavin S, Göritz F, et al. Effect of egg-phosphatidylcholine on the chilling sensitivity and lipid phase transition of Asian elephant (*Elephas maximus*) spermatozoa. Zoo Biol. 2005;24:233–45.
Schaftenaar W. Delayed postpartum fetotomy in an Asian elephant (*Elephas maximus*). J Zoo Wildl Med. 2013;44:130–5.
Schmitt DL, Hildebrandt TB. Manual collection and characterization of semen from Asian elephants (Elephas maximus). Anim Reprod Sci. 1998;53:309–14.
Schuff KG, Hentges ST, Kelly MA, Binart N, Kelly PA, Iuvone PM, et al. Lack of prolactin receptor signaling in mice results in lactotroph proliferation and prolactinomas by dopamine dependent and independent mechanisms. J Clin Invest. 2002;110:973–81.
Schulte BA, Feldman E, Lambert R, Oliver R, Hess DL. Temporary ovarian inactivity in elephants: Relationship to status and time outside. Physiol Beh. 2000;71:123–31.
Serri O, Chik CL, Ur E, Ezzat S. Diagnosis and management of hyperprolactinemia. Canadian Med Assoc J. 2003;169:575–81.
Shi ZD, Barrell GK. Thyroid hormones are required for the expression of seasonal changes in red deer (*Cervus elaphus*) stags. Reprod Fertil Dev. 1994;6:187–92.
Sobrinho LG. Prolactin, physchological stress and environment in humans: Adaptation and maladaptation. Pituitary. 2003;6:35–9.
Sobrinho LG, Nunes MCP, Calhaz-Jorge C, Afonso AM, Pereira MC, Santos MA. Hyperprolactinemia in women with paternal deprivation during childhood. Obstet Gynec. 1984;64:65–468.
Sonino N, Navarrini C, Ruini C, Fallo F, Boscaro M, Fava GA. Life events in the pathogenesis of hyperprolactinemia. Europ J Endocrinol. 2004;151:61–5.
Squires EL, Ginther OJ. Follicular and luteal development in pregnant mares. J Reprod Fertil Suppl. 1975;23:429–33.
Stansfield FJ, Allen WR. Luteal maintenance of pregnancy in the African elephant (*Loxodonta africana*). Reproduction. 2012;143:845–54.
Stansfield FJ, Nothling JO, Allen WR. The progression of small-follicle reserves in the ovaries of wild African elephants (*Loxodonta africana*) from puberty to reproductive senescence. Reprod Fertil Dev Online, http://dx.doi.org/10.1071/RD12296 (2012).
Stiles D. The elephant and ivory trade in Thailand. Petaling Jaya, Selangor, Malaysia: TRAFFIC Southeast Asia; 2009.
Sukumar R, editor. The living elephants: evolution, ecology, behavior and conservation. UK: Oxford University Press; 2003. p. 478.
Swain JE, Miller RR. A postcryogenic comparison of membrane fatty acids of elephant spermatozoa. Zoo Biol. 2000;19:461–73.
Takahashi T. Biology of the prolactin family in bovine placenta. I. Bovine placental lactogen: Expression, structure and proposed roles. Anim Sci Journal. 2006;77:10–7.

Thitaram C. Breeding management of captive Asian elephant (*Elephas maximus*) in range countries and zoos. Jpn J Zoo Wildl Med. 2012;17:91–6.

Thitaram C, Pongsopawijit P, Chansitthiwet S, Brown JL, Nimtragul N, Boonprasert K, et al. Induction of the ovulatory LH surge in Asian elephants (*Elephas maximus*): A novel aid in captive breeding management of an endangered species. Reprod Fertil Dev. 2009;21:672–8.

Thitaram C, Brown JL, Pongsopawijit P, Chansitthiwet S, Wongkalasin W, Daram P, et al. Seasonal effects on the endocrine pattern of the semi-captive female Asian elephant (*Elephas maximus*): timing of the luteinizing hormone surge determines the length of the estrous cycle. Theriogenology. 2008;69:237–44.

Thongtip N, Mahasawangkul S, Thitaram C, Pongsopavijitr P, Kornkaewrat K, Pinyopummin A, et al. Successful artificial insemination in the Asian elephant (*Elephas maximus*) using chilled and frozen-thawed semen. Reprod Biol Endocrinol. 2009;7:75–83.

Thongtip N, Saikhun J, Damyang M, Mahasawangkul S, Suthunmapinata P, et al. Evaluation of post-thaw Asian elephant (*Elephas maximus*) spermatozoa of extender and cryoprotectant. Theriogenology. 2004;62:748–60.

Urwin VE, Allen WR. Pituitary and chorionic gonadotrophin control of ovarian function during early pregnancy in equids. J Reprod Fertil Suppl. 1982;32:371–82.

Verhelst J, Abs R, Maiter D, van den Bruel A, Vandeweghe M, Velkeniers B, et al. Cabergoline in the treatment of hyperprolactinemia: a study in 455 patients. J Clin Endocrinol Metab. 1999;84:2518–22.

Valeri C, Schteingart HF, Rey RA. The prepubertal testis: Biomarkers and functions. Curr Opin Endocrinol Diabetes Obes. 2013;20:224–33.

Vick MM, Sessions DR, Murphy BA, Kennedy EL, Reedy SE, Fitzgerald BP. Obesity is associated with altered metabolic and reproductive activity in the mare: effects of metformin on insulin sensitivity and reproductive cyclicity. Reprod Fertil Dev. 2006;18:609–17.

Wang AT, Mullan RJ, Lane MA, Hazem A, Prasad C, Gathaiya NW, et al. Treatment of hyperprolactinemia: A systematic review and meta-analysis. Syst Rev. 2012. doi:10.1186/2046-4053-1-33.

Wasser SL, Clark B, Laurie C. The ivory trail. Sci Am. 2009;301:68–74.

Wasser SL, Barash DP. Reproductive suppression among female mammals: Implications for biomedicine and sexual selection theory. Quart Rev Biol. 1983;58:513–38.

Wasser SL, Papageorge S, Foley C, Brown JL. Excretory fate of estradiol and progesterone in the African elephant (*Loxodonta africana*) and pattern of fecal steroid concentrations throughout the estrous cycle. Gen Comp Endocrinol. 1996;102:255–62.

Wielebnowski N. Contribution of behavioral studies to captive management and breeding of rare and endangered animals. In: Caro T, editor. Behavioral ecology and conservation biology. New York, NY: Oxford University Press; 1998. p. 130–62.

Wielebnowski NC, Fletchall N, Carlstead K, Busso JM, Brown JL. Non-invasive assessment of adrenal activity associated with husbandry and behavioral factors in the North American clouded leopard population. Zoo Biol. 2002a;21:77–98.

Wielebnowski NC, Ziegler K, Wildt DE, Lukas J, Brown JL. Impact of social management on reproductive, adrenal and behavioural activity in the cheetah (*Acinonyx jubatus*). Anim Conserv. 2002b;5:291–301.

Yamamoto Y, Yuto N, Yamamoto T, Kaewmanee S, Shiina O, Mouri Y, et al. Secretory pattern of inhibin during the estrous cycle and pregnancy in African (*Loxodonta africana*) and Asian (*Elephas maximus*) elephants. Zoo Biol. 2012a;31:511–22.

Yamamoto Y, Yamamoto T, Yuto N, Hildebrandt TB, Lueders I, et al. The secretory pattern and source of immunoreactive prolactin in pregnant African (*Loxodonta africana*) and Asian (*Elephas maximus*) elephants. J Reprod Dev. 2012b;58:105–11.

Yamamoto Y, Yamamoto T, Taya K, Watanabe G, Stansfield FJ, Allen WR. Placentation in the African elephant (*Loxodonta africana*). V. The trophoblast secretes placental lactogen. Placenta. 2011;32:506–10.

Yamamoto Y, Yamamoto T, Watanabe G, Yuto N, Keio M, Narushima E, et al. Prolactin secretion and ovarian function in cycling and non-cycling African female elephants (*Loxodonta africana*). J Vet Med Sci. 2010;72:845–52.

Yamoto M, Minami S, Nakano R, Kobayashi M. Immunolocalization of inhibin/activin subunits in human ovarian follicles during the menstrual cycle. J Clin Endocrinol Metab. 1992;74: 989–93.

Yamoto M, Minami S, Nakano R. Immunohistochemical localization of inhibin subunits in human corpora lutea during menstrual cycle and pregnancy. J Clin Endocrinol Metab. 1991;73: 470–7.

Yon L, Chen J, Moran P, Lasley B. An analysis of the androgens of musth in the Asian bull elephant. Gen Comp Endocrinol. 2008;155:109–15.

Yon L, Kanchanapangka S, Chaiyabutr N, Meepan S, Stanczyk FZ, Dahl N, et al. A longitudinal study of LH, gonadal and adrenal steroids in four intact Asian bull elephants (*Elephas maximus*) and one castrate African bull (*Loxodonta africana*) during musth and non-musth periods. Gen Comp Endocrinol. 2007;151:241–5.

Yuen BH. Hyperprolactinemia in women of reproductive age: Etiology, diagnosis and management. Canadian Family Physician. 1992;38:367–79.

Zacur HA. Hyperprolactinemia. In: Knobil E, Neill J, editors. Encyclopedia of reproduction, vol. 2. New York, NY: Academic Press; 1999. p. 725–35.

Chapter 9
The Koala (*Phascolarctos cinereus*): A Case Study in the Development of Reproductive Technology in a Marsupial

Stephen D. Johnston and William V. Holt

Abstract The successful development and application of an assisted breeding program in any animal relies primarily on a thorough understanding of the fundamental reproductive biology (anatomy, physiology and behaviour) of the species in question. Surely, the ultimate goal and greatest hallmark of such a program is the efficacious establishment of a series of reliable techniques that facilitate the reproductive and genetic management of fragmented populations, both in captivity and in the wild. Such an achievement is all the more challenging when the reproductive biology of that species is essentially rudimentary and without adequate reproductive models to compare to. Using the koala (*Phascolarctos cinereus*) as a case study, this chapter provides some personal insights into the evolution of a concept that began as a small undergraduate student project but that subsequently evolved into the first-ever successful artificial insemination of a marsupial. Apart from this historical perspective, we also provide a brief review of the current reproductive biology of the koala, discuss technical elements of current assisted breeding technology of this species, its potential application to the wombat, and the future role it might play in helping to conserve wild koala populations. There is little doubt that the unique reproductive biology and tractability of the koala has in this case been a benefit rather than a hindrance to the success of artificial breeding in this species.

Keywords Koala • Phascolarctos cinereus • Assisted breeding technology • Reproductive biology • Semen collection • Semen cryopreservation • Oestrus detection • Induction of ovulation • Artificial insemination

---

S.D. Johnston, B.Sc.(Hons.), Ph.D. (✉)
Wildlife Biology Unit, School of Agriculture and Food Science,
The University of Queensland, Gatton 4343, QLD, Australia
e-mail: s.johnston1@uq.edu.au

W.V. Holt, Ph.D. (✉)
Academic Department of Reproductive and Developmental Medicine,
University of Sheffield, Jessop Wing, Tree Root Walk, Sheffield S10 2SF, UK
e-mail: Bill.holt@sheffield.ac.uk

W.V. Holt et al. (eds.), *Reproductive Sciences in Animal Conservation*,
Advances in Experimental Medicine and Biology 753, DOI 10.1007/978-1-4939-0820-2_9,

1 Introduction

Reproductive technology offers significant advantages for the genetic management and propagation of captive wildlife but only in a small proportion of species is it currently making a significant contribution to conservation (e.g. Holt and Lloyd 2009). The application of reproductive technology to marsupial species is even less well developed and has the added complication of being restricted by a lack of fundamental knowledge with respect to their unusual mode of reproduction (Johnston and Holt 2001). Only in the koala is the use of artificial insemination significantly advanced, that it is regularly being used for the production of pouch young (Johnston et al. 2003; Rodger et al. 2009). Nonetheless and despite significant advances in the establishment of koala reproductive technologies, there are still some significant technical problems that need to be solved with regard to the use of cryopreservation of the spermatozoa and the actual implementation of the technology into koala conservation more generally. The koala represents an excellent case study of what it ultimately takes to develop an assisted breeding program in a species with only limited reproductive information available. This chapter provides a unique opportunity to document the development of reproductive technology in a novel wildlife species over a 20 year period; a project that began (1992) as a modest undergraduate investigation but which expanded to the most successful artificial insemination program of any marsupial.

2 Conservation Status of the Koala

The conservation status of the koala is a vexed issue because it depends largely upon which populations you are referring to in Australia and the degree of human intervention with respect to their genetic management. The Natural Resource Management Ministerial Council (NRMMC 2009) has reported on the regional conservation status of the koala and has indicated that it varies from secure in some areas to vulnerable or extinct in others. In Queensland, koalas occur throughout most of their natural range, although the overall population numbers continue to decline as a result of clearing and habitat fragmentation, drought and climate change. There is also significant pressure from urban development in southeast Queensland, where the koala is currently listed as vulnerable by the state government. In 2009, the Queensland Department of Environment and Resource Management released a report on the decline of the Koala Coast koala population in SE Queensland and noted that koala numbers had fallen by 64 % from 1996 to 2008. In New South Wales (NSW), trends in koala populations vary across the state; some populations on the NSW coast are declining, while some populations west of the Great Dividing Range have actually expanded (NRMMC 2009). Koalas in Victoria occur over much of the southern and eastern lowlands and population densities are artificially high in some areas (Raymond Island, Snake Island, French

Island, parts of the Otway Ranges, Framlingham, Mount Eccles National Park, Tower Hill Game Reserve; NRMMC 2009). While the densities of koalas are lower in the dry forests and woodlands in northern parts of Victoria where the habitat is of lower quality, much of the habitat remaining in the state is fragmented and many populations are isolated (NRMMC 2009). Koalas in South Australia can now be found in a greater range and abundance than at the time of European settlement, because of introductions both within its natural range and areas where it did not occur naturally (NRMMC 2009).

3 Why Develop Assisted Reproductive Technology in the Koala?

The management and conservation benefits of assisted reproductive technology (ART) to wildlife species have been promoted heavily (e.g. Pukazhenthi et al. 2006) and in theory at least, are worthy of pursuit. However, given the current conservation status of the koala discussed above, questions still remain as to the specific uses of such tools in this marsupial and how they may be implemented both in captive and wild populations. As the application of ART requires a detailed understanding of reproductive biology, it is likely that the most significant, but probably most overlooked contribution that ART can make to conservation, is the acquisition of novel fundamental biology (Johnston and Holt 2001; Wildt et al. 2003; Holt and Lloyd 2009); this includes basic information such as reproductive seasonality, oestrous cycle characterisation and mating behaviour and choice, all of which impact on behavioural ecology (Johnston et al. 2013b).

A major concern of captive koalas is the maintenance of appropriate genetic diversity within the closed population, so that techniques such as artificial insemination are likely to play an important role here. Semen cannot only be transported across borders and physical boundaries between zoos but also, with the use of cryopreservation technology, it can also be stored through time, prolonging the generation interval of genetically important sires. Discussion on use of this technology for the genetic management of koalas is less well developed; we have recently advocated the use of what we refer to as "live" and "frozen" genome banks for koala populations in SE Queensland that are likely to need urgent human intervention in order to survive (Johnston et al. 2013b)—these specific concepts will be discussed later in the chapter. There are also significant animal welfare benefits to the koala associated with the use of ART; while some of the procedures such as electroejaculation and artificial insemination are mildly invasive, the transportation of semen, rather than whole animals between institutions (national and international) is likely to be substantially less stressful in terms of animal welfare. Associated with the transport of semen is the lower comparative cost of this approach and the fact that many semen samples can be sent and stored in the one shipment container. Thus far we have only used chilled koala semen transported between zoological institutions on the Gold Coast to successfully produce a joey by artificial insemination (Allen et al. 2008a).

ART techniques are also very useful for identifying and overcoming reproductive problems or specific anatomical, physiological and behavioural blocks to reproduction (Johnston et al. 1999). Any knowledge that contributes to the fundamental reproductive biology of the koala, adds to a better understanding of what is "normal" and what is "abnormal". This has always been a major issue for those of us working in the area of wildlife or exotic species reproduction, with this type of database information difficult to find or accumulate, especially when compared to domestic animals or humans. The assessment of what represents a normal semen sample in the koala is a classical example of this phenomenon; the standard set of seminal characteristics in the koala, boar, bull or human are likely to be completely different to those animals with very divergent reproductive strategies.

Artificial insemination (AI) is an example of an ART that could be used to overcome mating problems in koalas with physical injuries or behavioural incompatibilities. Recent observations by Dief (2011) have shown that male koalas positive for chlamydial prostatitis or urethritis, can still possess epididymidal tissue free of the organism—these animals would be ideal for gamete recovery and AI, as would non-infected koalas that come into koala hospitals that need to be euthanasia because of dog attack or car accidents. ART has the potential to recover important alleles from individuals that would otherwise not make a contribution.

Another important reason for developing ART in the koala was because at the time of commencing our project, there were still significant numbers of animals in the wild and in captivity that could be utilised to develop the technology. Too often the development of ART is implemented as an afterthought or at a stage in conservation management where there are insufficient animals to conduct the fundamental science. Instead, what is required, is a proactive approach to the development of ART - in the koala, we have largely been able to achieve this. Now that AI is a relatively routine procedure in the koala, it will be necessary to see how the technology can be further refined and improved so that it has an important role in the conservation of wild populations.

4 The Koala as a Research Animal

Irrespective of its iconic status, the successful and rapid development of assisted breeding in the koala has been very much dependent on the unique biology of the species. Although perhaps regarded as a difficult animal to maintain in captivity, in our hands, we have found the koala to be an ideal research animal for ART. Its relatively passive nature, size, tractability and adaptability to captivity, mean that handling the animal for blood collection and reproductive procedures can occur without need for anaesthesia or sedation. While the koala is no doubt a "fussy" and expensive animal to feed, their ability to habituate to human presence while in captivity, has allowed us to develop and conduct procedures that would not normally be possible in other marsupials—for example, semen collection by means of an artificial vagina (Johnston et al. 1997) and artificial insemination (Johnston et al. 2003; Allen et al. 2008a) can both be accomplished successfully without the need for anaesthesia. In addition, and probably most importantly, the koala possesses an extremely overt and prolonged (10 days) behavioural oestrus (Johnston et al. 2000b) and an induced ovulatory response (Johnston et al. 2004), so that this makes predicting the timing of insemination a relatively straightforward process.

5 How to Make a Koala Pouch Young

There are three fundamental components to the development of any artificial insemination program (Johnston and Holt 2001). These include (1) an understanding of male reproductive biology in order to collect viable semen for evaluation, preservation and insemination, (2) characterisation of the oestrous cycle, with particular focus on expression of oestrus and the type and timing of ovulation and (3) identification of the most appropriate site and mechanism of semen deposition into the female's reproductive tract. Each of these components are of course based on building a framework of koala reproductive biology composed of structural and functional anatomy, endocrinology and gamete physiology and natural and captive reproductive behavioural observations, all of which are continually refined over many years. The following summary of koala reproductive biology will be used as the basis on which to discuss the development of ART in this species and will primarily focus on captive animals.

6 The Reproductive Biology of the Female Koala

The female reproductive tract of the koala (Fig. 9.1) is like that of all marsupials being small sized and possessing a urogenital sinus that opens into a cloaca alongside a vascular clitoris and the rectum. Johnston (1999) noted the presence of vestibular or Bartholin's glands opening into the urogenital sinus. The urogenital sinus gives rise to the urethra ventrally and dorsally into two small poorly defined ostia that open into the muscular lateral vaginae. The marsupial purposes two vaginae because the mullerian ducts are prevented from fusing on the mid-line due to the medial migration of the ureters during organogenesis. The lateral vaginae open into a vaginal cul-de-sac that receives the simple but muscular cervices. The vaginal cul-de-sac in the koala is partitioned by a medial septum that completely separates the left and right sides and which means that for semen to enter both uteri it must travel up through both lateral vaginae separately; this has obvious implications for the koala artificial insemination procedure. The duplex uteri each open into a convoluted oviduct that terminates in the infundibulum. Depending on the stage of the oestrous cycle the ovary is ovoid and typically 10–12 mm long, 7–9 mm wide and

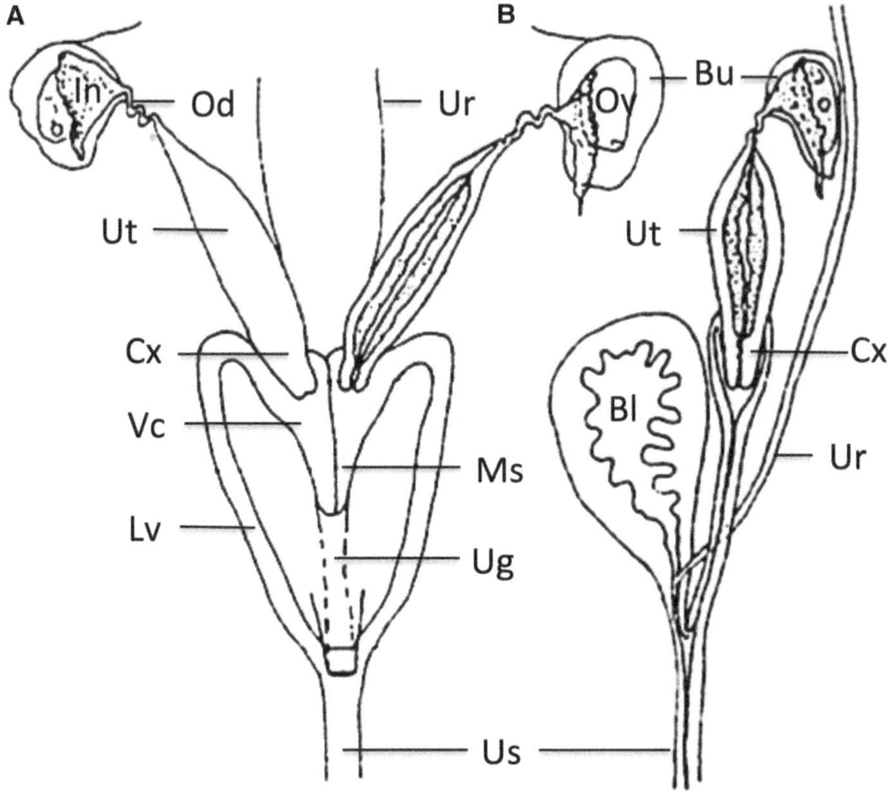

Fig. 9.1 Schematic drawing of female koala reproductive tract; (A) Ventral aspect and (B) Lateral aspect. Bl—bladder; Bu—bursa; Cx—Cervix; Fi—Fimbria; In—Infundibulum; Lv—Lateral vagina; Ms—Medial septum; Od—Oviduct; Ug—Urogenital strand; Ur—Ureter; Us—Urogenital sinus; Vc—Vaginal cul-de-sac (Modified from Obendorf 1988)

3–5 mm deep. The morphology of folliculogenesis in the Koala ovary is unremarkable apart from the large size (up to 7 mm) of the mature Graafian follicles, which represent some of the largest recorded follicles in any marsupial; the active corpora luteum are of similar size or larger (Johnston 1999). Of particular interest is the thickness of the Koala zona pellucida (14.2 μm) which is over twice that recorded for any marsupial, being more typical in thickness to that of the zonae pellucidae of eutherian species. Johnston (1999) has commented that a lack of coitus during oestrus in the koala presumably leads to an "over-ripening" of the pre-ovulatory follicle, subsequent atresia and formation of haemorrhagic follicles, similar to what has been observed in the rabbit ovary and describes large numbers of haemorrhagic and/or atretic follicles in varying stages of formation in the koala ovary at all phases of the reproductive cycle. Interestingly, part of the remaining theca interna and granulosal tissue also undergoes hyperplasia, with cells of the theca interna ultimately being transformed into what appear to be ovarian "interstitial-like" tissue; the functional significance of this tissue is presently unknown. The koala ovary is surrounded

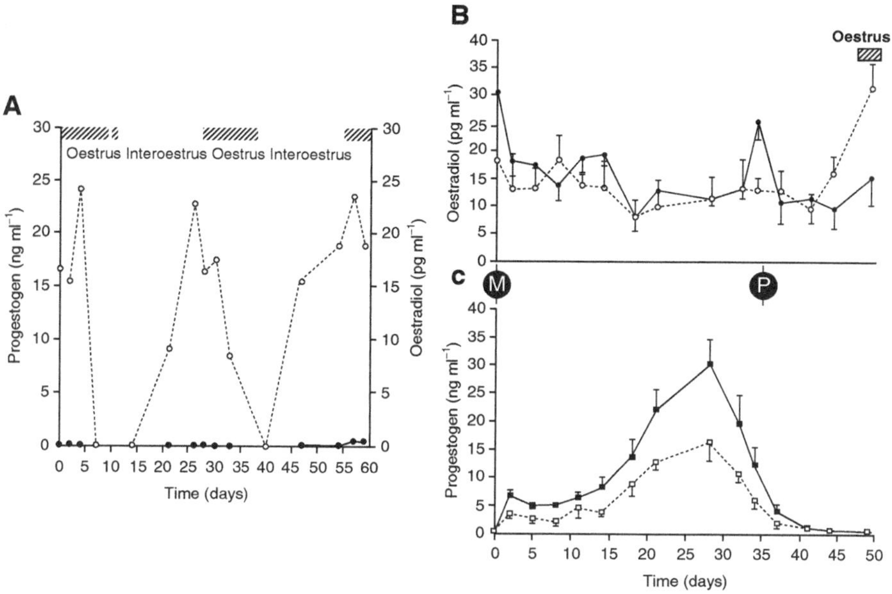

Fig. 9.2 Reproductive endocrinology of the koala. (A) Oestradiol (open circles) and progestogen (closed circles) secretion during an anovulatory oestrous cycle; (B) Oestradiol secretion post-mating during pregnancy (closed circle) and a non-sterile oestrous cycle (including luteal phase; open circle); (C) Progestogen secretion post-mating during pregnancy and a non-sterile oestrous cycle (including luteal phase). Cross-hatching—behavioural oestrus; M—mating; P—parturition (From Johnston et al. 2000b)

by a translucent ovarian bursa; the oviduct and bursa are often sites of inflammation associated with chlamydia infection, leading to infertility resulting from fluid pressure atrophy on the ovary and/or stenosis of the oviduct. Although no direct observations have been documented, histological observations by Johnston (1999) have reported that birth of the koala foetus most likely occurs through the terminal portion of the vaginal cul-de-sac, into the softened oedemic tissue of the urogenital strand, with the neonate exiting to the exterior through the urogenital sinus and cloaca. The pouch young then climbs unaided into the pouch where it firmly attaches to the teat, suckles and continues to develop. Presumably, the reproductive tract heals over after each parturition event.

The reproductive cycle of the koala is reasonably well understood (Johnston et al. 2000b, c, 2004); endocrinology of the pregnant and non-pregnant koala cycle is illustrated in Fig. 9.2. While ovulation appears to be induced by coitus, the exact mechanism of induction is still unresolved as there is some evidence that semen may possess some type of ovulation-inducing biochemical, similar to what has been described in the camel (Johnston et al. 2004). Like the felidae, there are three possible types of koala oestrous cycle. The first involves the follicular phase without induction of ovulation and a luteal phase and is referred to as interoestrus (Fig. 9.2a; Johnston et al. 2000b). The female displays oestrus for approximately 10 days and if mating does not occur, oestradiol secretion subsides and the female takes approximately 33 days to come back into oestrus again. If the female mates with a male, an LH surge occurs approximately 28 h after mating and plasma progesterone levels commence to rise and peak about 28d after mating (Fig. 9.2b). If pregnancy is not successful, the female will come back into oestrus approximately 50 days from the previous oestrus (Fig. 9.2b). If a pregnancy does result, gestation will occur over a period of 35 days (Fig. 9.2b) and following parturition, the female will fail to return to oestrus, most probably because of the suckling stimulus of the pouch young preventing further ovarian activity. This detailed understanding of the physiology of the koala oestrous cycle and ovulatory pattern has been fundamental to the success of the artificial insemination program, in terms of the most appropriate timing for insemination, but also in developing procedures to induce ovulation. We are currently in the process of sequencing Koala GnRH, FSH and LH for the purposes of developing specific anti-sera so that these protein hormones can be monitored throughout the cycle, pregnancy and early lactation; we are also sequencing the respective receptors of these hormones. This information will also help us design or select the most appropriate GnRH antagonists for work oestrous synchronisation.

Oestrus in the captive koala is one of the most visually and auditory overt behaviours reported for any marsupial (Johnston et al. 2000a; Feige et al. 2006). In a procedure known as "teasing", an adult male is brought into an all female enclosure. The male is initially presented to each female by the zookeeper by holding the male at eye-level with each female so that all females are aware of the male's presence. Those females that are not in oestrus will typically reject the male with an aggressive vocalisation and or by physically striking out at the male with their forelimb. Females in oestrus will show interest in the male and some will respond by immediately initiating oestrus-related behaviours. The male is then placed on the enclosure floor where he undergoes prescriptive pre-copulatory behaviours including urination, scent marking a tree pole within the enclosure with his scent gland, and most characteristically, vocalising with deep guttural bellows. The bellow of the male is typically the trigger that initiates the expression of oestrus in the female and on hearing the bellow the oestrus female's excitement level rapidly increases. Interestingly, the female will not necessarily approach the bellowing male but more typically, locates other females in the enclosure with whom she attempts to mount and copulate. The oestrous female will normally take on the male role during pseudo-copulation, establishing a neck bite to the nap of the neck to stabilise the female that is being mounted and orientating herself as if she were the male performing the copulatory act (Feige et al. 2006; Fig. 9.3). The mounting female will display the full range of male copulatory behaviours, including pelvic thrusting, a period of stillness, which is similar to the period when the male would be ejaculating, and a finally, a neck bite to the shoulder as a signal of disengagement; some females will even bellow at the end of the pseudo-mating attempt. Remarkably, the female that is being mounted will take on the normal female role during this process and demonstrate lordosis (arching of the backbone and backward tilt of the head), a period of stillness while the female is mimicking male thrusting behaviour and then a period where she shows convulsive jerking of her body associated what is presumably a mimic of ejaculatory behaviour (Feige et al. 2006). Typically, the oestrous

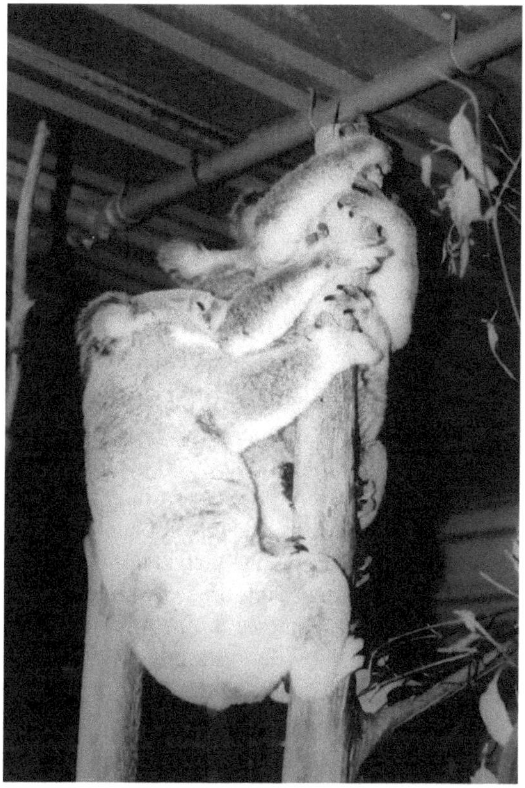

Fig. 9.3 Pseudomale copulatory behavior in three oestrous koalas. Note the homosexual behaviour of the middle and lower females (From Johnston 1999)

female seeks out other females that are either in oestrus or coming into or out of oestrus, as females that have been mounted will then proceed to serve the female who has just finished mounting. Given that oestrus in the koala is on average 10 days of a 33 day cycle, it is not all that surprising that their will be more than one or two oestrous females in the enclosure at the same time. Other behaviours that are associated with koala oestrus include, bellowing behaviour by the female, convulsive jerking of the body in what has been likened to uncontrollable "hiccoughing", urination and increased agitation or restlessness (Johnston et al. 2000b). Koalas are normally very sedentary in captivity but there is a notable change in movements and activity during oestrus. As in a range of domestic species, koalas also appear to demonstrate a standing oestrus in which the female will stand still (receptivity) while the male ensues a copulatory position. In some cases, the female may even back down "rump first" in the face of the male on the same pole in what appears to be a form of "cloacal presentation". All of these behaviours are unmistakable indicators of oestrus for the zookeeper, who along with help from the teaser male koala can determine the daily reproductive status of the female from behavioural observations. Some zookeepers have even learnt to mimic the bellow of the male koala or play recordings of the male bellow to the female and this has been used for effective oestrus detection. All these "oestrus" behaviours have been linked to an elevated secretion of oestradiol (Johnston et al. 2000b). The ecological significance of captive koala homosexual behaviour is difficult to comprehend but is probably an artefact of captivity, as the behaviour has never been observed in the wild. Feige et al. (2006) have suggested that it may be associated with sexual excitement linked to high levels of oestradiol in the systemic circulation and that this may be the physiological mechanism that causes wild koalas to seek out males outside of their normal home range.

7 The Reproductive Biology of the Male Koala

The reproductive anatomy of the koala has been well described by Temple-Smith and Taggart (1990) so that only a brief overview will be presented here. As for most marsupials, the male Koala has a pre-penile scrotum; the scrotal; skin being non-pigmented and covered with short hairs. The importance of the scrotum for thermoregulation of the testes has not yet been investigated. The flaccid penis is maintained within the prepuce in an s-shaped configuration. The glans penis is covered proximally with prominent keratinised spines but the precise role of these spines has yet to be clearly determined—we have speculated that they are likely to have role in stimulating reflex ovulation along with ovulation factors in the semen (Johnston et al. 2004). In the centre of the chest of the male is a sternal gland which produces an oily sebaceous and odorous secretion; the scent gland appears to be androgen dependent and its activity increases leading up to and during the breeding season (Allen et al. 2010).

The testis of the koala is supplied by a rete mirabile of over 100 blood vessels and five large lymphatics. The testis is ellipsoidal and small in comparison to that of other marsupials, weighing only approximately 0.07 % of total body weight (Johnston 1994); the significance of small testicular size on the reproductive strategy and behavioural ecology of the koala requires further investigation. Oishi et al. (2013) has recently described the quantitative testicular histology and the dynamics of the seminiferous cycle in the koala and wombat in which we report both species possessing eight stages of cellular associations. Interestingly, the koala (≈33 %) differs from the Southern Hairy-nosed Wombat in typically having a greater proportion of interstitial testicular parenchyma. Temple-Smith and Taggart (1990) refer to the koala as possessing a Type 3 pattern of organisation with large tracts of Leydig cells completely isolating adjacent seminiferous tubules. The Sertoli cells of the koala contain nuclei that are 5X larger than that those in eutherian species and which possess unusual crystalloid inclusions, the precise function of which is still unknown. Testicular volume of koalas in SE Queensland changes throughout the year in both wild and captive populations with an increase over spring and summer and a decrease in autumn and winter (Allen et al. 2010).

Spermiogenesis in the koala has a number of unusual features including; the formation of a proacrosomal granule within the acrosomal vacuole, an uneven condensation of chromatin and a unique flattening of the sperm nucleus which results in the wide variation observed in sperm head morphology (Temple-Smith and Taggart 1990). The acrosome is located in a dorsal nuclear cleft and there is an unusual ventral neck insertion of the mid-piece into the sperm head (Harding et al. 1979; 1987; Harding and Aplin 1990). Harding and Aplin (1990) have used these unusual features of the spermatozoon to review the phylogenetic position of the koala (Harding et al. 1987; Harding and Aplin 1990) confirming a close relationship with the wombat.

The gross anatomy of the Koala epididymis and associated vasculature is similar to that described for other diprotodontid marsupials (Temple-Smith and Taggart 1990). The caput epididymidis is expanded in shape and connected via a corpus segment to a bulbous *cauda epididymidis*. The epithelium of Koala epididymis consists of four basic cell types; principal, basal, mitochondrial rich and electron-translucent, but they have no specialised regions of phagocytic principal cells. As for most mammals, Koala sperm appear to gain the capacity to become motile as they move into the *cauda epididymidis*. Fluid absorption in the *cauda epididymidis* results in a higher sperm concentration than that found in the *caput epididymidis* (Temple-Smith and Taggart 1990). While there is a marked increase in the curvature of the sperm head and a folding of the acrosomal surface and condensation of the accessory cytoplasmic droplet to within the hook of the sperm head (Hughes 1977), there appear to be no other morphological changes to the Koala spermatozoa during epididymidal transit; this is in sharp contrast to changes in sperm morphology that occur during epididymidal transit in other marsupials (Harding et al. 1983; Rodger and Mate 1993).

The *ductus deferentia* run from the neck of the scrotum through the inguinal canal and into the abdominal cavity, emptying finally into the cranial portion of the prostatic urethra. The *ductus deferens* is non-specialised, non-glandular, with no seminal vesicles or ampullae (sperm storage organs in eutherian mammals). The Koala prostate is pyriform and can be divided into three histologically distinct regions, the functional significance of which have yet to determined. A short membranous urethra connects the prostate to the penile urethra. At the distal extremity of the membranous urethra and in close association with the penile crura and bulbs are three pairs of bulbourethral glands. Glands BII and BIII are filled with a thick, clear, mucin-like secretion that is strongly eosinophilic (Temple-Smith and Taggart 1990). Although Mitchell (1990) noted the presence of paracloacal glands in koalas, he gave no anatomical details or references. It is likely that the prostate provides the bulk of seminal plasma in koala semen, while the bulbo-urethral gland secretions have an important role in producing a seminal plug. In SE Queensland, a study of post-mortem specimens entering koala hospitals revealed no seasonal change in the size of the prostate but an increase in bulbo-urethral gland volume over spring, a decrease over summer and autumn and an increase towards the end of winter (Allen et al. 2010).

Like most mammals, plasma testosterone secretion in the Koala is highly episodic (Johnston 1999). The highest concentration of testosterone appears to occur in the koala during periods of male dispersal just prior to the breeding season, and not as may have been expected, during the breeding season (McFarlane 1990; Handasyde et al. 1990; Allen et al. 2010). A rise in testosterone concentration is coincident with the onset of bellowing, prior to the commencement of breeding activity (Handasyde et al. 1990). In order to obtain a more reliable measure of the testosterone secretory capacity of the koala testis, a GnRH or hCG stimulation test can be employed (Allen et al. 2006). This approach was used to demonstrate a seasonal change in testosterone secretion of both captive and wild koalas in southeast Queensland with a peak in spring and nadir in autumn (Allen et al. 2010).

8 Semen Collection

Our first major step towards the development of a successful AI program in the koala was the ability to reliably collect semen in sufficient volumes to be used for insemination. Semen collection in the koala using electro-ejaculation was first described by Wildt et al. (1991) and later by Johnston et al. (1994); the procedure has been extremely effective with greater than 90 % of semen collection attempts resulting in spermatozoa. The procedure is field applicable (Fig. 9.4a) and a recent study by Allen et al. (2010) showed that it was possible to repeatedly collect semen from both captive (monthly) and wild koalas (every 6 weeks) from the same individuals. However, there are also disadvantages with respect to the use of the procedure that need to be acknowledged, including the requirement for the koala to be anaesthetised (which can be problematic), production of semen with variable compositions of seminal plasma, lower sperm concentration, urine contamination and in rare cases rectal trauma through improper placement of or overstimulation from the rectal probe. We have used the electro-ejaculation procedure to determine semen quality and provide semen for studies of sperm physiology, preservation and artificial insemination.

A major break through in determining the appropriate parameters of koala semen for AI was the ability to collect semen using an artificial vagina (Johnston et al. 1997; Fig. 9.4b, c). Before attempting the procedure we carefully observed the natural mating behaviour of the koala to inform us of how we might engage the animal to serve the artificial vagina, for a major problem that we needed to overcome, was the fact that the male koala thrusts his penis into the female urogenital sinus in a vertical direction and that semen was presumably ejaculated in a similar trajectory against "gravity". Contributing to the success of the technique was the tractability of both male and female koala to tolerate human presence during mating activity. In fact, it was not uncommon for zookeepers at Lone Pine Koala Sanctuary to assist the male to direct his penis into the urogenital sinus of the female during natural mating. Prior to attempting semen collection with the AV, zookeepers had reported to us that during movement of koalas from one enclosure to another that males held against the abdomen of the zookeeper would often ejaculate during transit. Captive koalas are handled from an early age and there is no doubt this early exposure to the zookeeper resulted in animals with a high tolerance to intervention and manipulation. We had observed that during copulation that the male was very focussed on securing the female in an appropriate position by establishing a neck bite and using his forearms but that his hind limbs were rarely used to support his hold. Based on our observations of natural mating behaviour, we constructed a koala AV from a ram artificial vagina that was adjusted to the length of the koala penis. After preparing

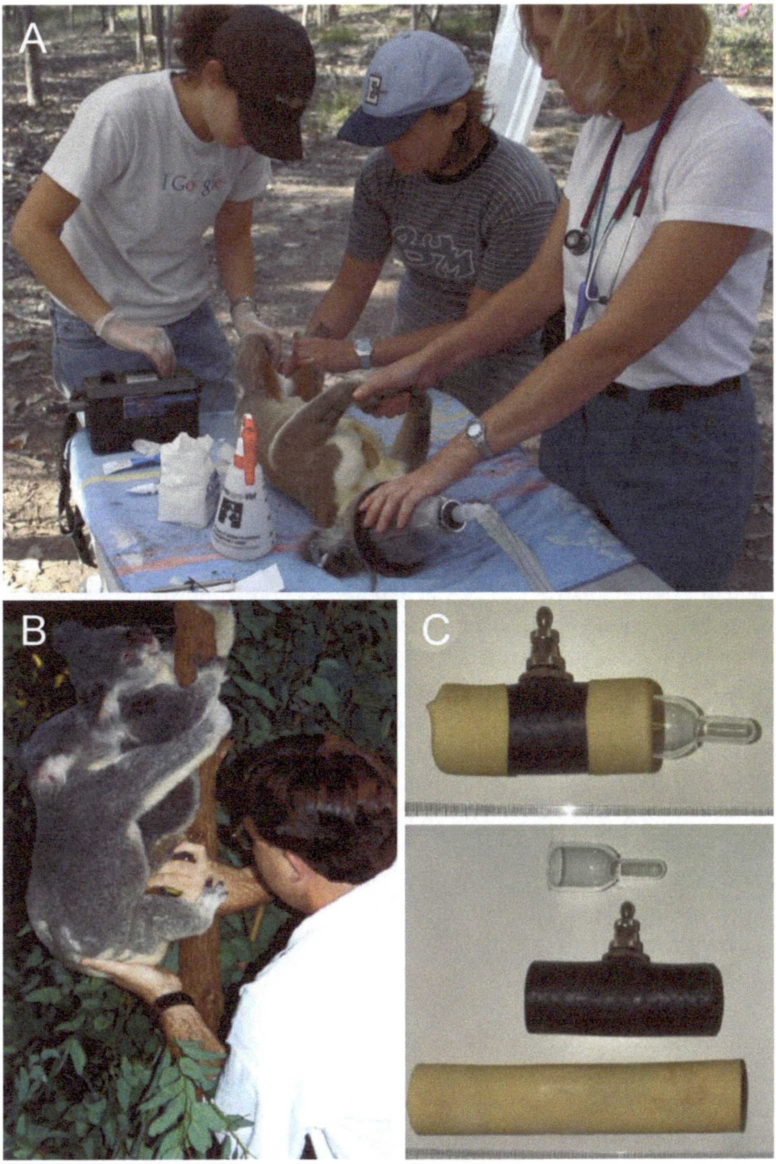

Fig. 9.4 Semen collection in the Koala. (A) Electro-ejaculation can be used in the field; (B and C) Use of a modified sheep artificial vagina for captive koalas (From Johnston et al. 1997 and Johnston 1999)

the AV with hot water (42–45 °C) and inflating it with air, the female is placed on a single tree-pole so that she was secure for mating at approximately eye height of the collector. The male koala is then brought into the female's enclosure, placed on the floor and allowed to conduct his normal pre-copulatory behaviour, including sternal gland scent marking, urinating and bellowing, all of which stimulates oestrus behaviour in the female. The male is then placed on the tree pole directly below an oestrus female and allowed to establish his mating position. The male typically manipulates the female into position while at the same time establishes penile erection. Once the penis is fully erect the collector would step in and direct the male's penis into the warm AV; the male in most cases appeared to be unaware that he was serving the AV rather than the female and the female continues to display her normal copulatory behaviour. The male then thrusts vigorously into the AV and completes this behaviour with two strong final ejaculatory thrusts, which appear to signal the commencement of ejaculation. At this point, the collector lifts the rump of the male slightly horizontally so as to direct the flow of semen into the collection vial; this manipulation was possible because the hind limb of the male is not used for support during copulation. Once ejaculation is completed the male disengages the female by biting her shoulder or arm and which results in an aggressive response by the female and separation. During disengagement the collector removes the AV from the male's penis. Semen collected by the AV provides a means of determining a more natural estimate of seminal characteristics, including semen volume ($\approx$1 mL), pH and sperm concentration; all these seminal parameters are crucial for developing appropriate methods of artificial insemination. While this method of semen collection (58 %) is less reliable than electro-ejaculation (96 %) it does offer the additional advantage of observing structure physical defects in breeding soundness and assessment of male sexual drive. To date the koala remains the only marsupial for which the AV has been used successfully.

It is also possible to recover spermatozoa from post-mortem specimens collected from koala hospitals and veterinary clinics and use these gametes in assisted breeding programs (Johnston and Holt 2001; Johnston et al. 2013b). Each year in SE Queensland, 100 s of koalas are euthanased because of disease or trauma; this is such an appalling waste of genetic material that could potentially be utilised for genetic exchange programs. While this procedure has been described and proven in the Common Wombat (MacCallum and Johnston 2005) and Southern Hairy-nosed Wombat (Johnston unpublished observations) there are currently no published studies of gamete recovery in the koala. Although the yield of recovered epididymidal spermatozoa is likely to be less than that of the wombat, we nevertheless, have preliminary data that indicates that this approach is feasible. Dief (2011) has recently noted that while ascending chlamydia infection can cause orchitis and epididymitis in the koala, some animals have prostatic infection only. These animals can be used for sperm collection and used directly for AI or for cryopreservation. We are also exploring the use of semen "clean-up" technology to remove or destroy viable chlamydial elementary bodies.

9 Semen Evaluation

The koala ejaculate collected by electro-ejaculation offers up some challenges for seminal analysis as the sample is typically very viscous and contains a high proportion of prostatic and bulbo-urethral secretions; on rare occasions the semen may even coagulate preventing further manipulation. In order to assess semen quality, the raw semen is typically diluted 1:10 in a Tris-citrate glucose or fructose diluent at room temperature. The raw semen sample has little mass activity, the pH is typically neutral and osmolality is in the order of 300–350 mOsm kg^{-1}. Sperm concentration of koala semen collected by AV was typically higher (165×10^6/ml) than that collected by electro-ejaculation (83×10^6/ml).

The assessment of sperm motility is conducted on a diluted sample on a warm stage set to 35 °C (approximate body temperature of the koala) and the percentage of sperm swimming in a progressively forward manner is determined along with an assessment of the rate of sperm movement (0—no movement to 5—extremely rapid movement). In early studies and without the use of fluorescent microscope, viability (intactness of the plasma membrane) was assessed using nigrosin-eosin (Johnston et al. 1994; 1997) but this was later substituted for the use of SYBR-14 (live) and propidium (dead) stains; we found the use of nigrosin-eosin somewhat problematic when using diluents containing egg yolk. More recently we have turned to the use of the JC-1 stain (5,5',6,6'-tetrachloro-1,1',3,3'-tetraethylbezimidazolyl carbocyanine iodide) for the assessment of mitochondrial membrane potential, which provides additional insight to quality of sperm motility. Sperm motility of semen collected by electro-ejaculation or by artificial vagina is usually in the order of 70–80 % progressive motility. The live-dead and JC-1 stains can be used in combination to provide simultaneous assessment of both parameters (Fig. 9.5A). Interestingly, sperm motility of wild and captive koalas was highest in winter, as was the post-thaw survival of cryopreserved spermatozoa from the same animals (Allen et al. 2010).

One of most striking feature of the Koala spermatozoon is the extent of pleiomorphy in the head morphotype. Temple-Smith and Taggart (1990) identified two extreme and four common intermediate head morphologies from both testicular and epididymal sperm, including a structural abnormality of the neck-midpiece region. Wildt et al. (1991) have identified ten nuclear morphotypes, while Johnston et al. (1994) identified 11 (Fig. 9.5B); mid-piece and principal piece structural abnormalities have also been described (Johnston et al. 1994); as to what specifically represents normal sperm head morphology in the koala has yet to be defined as is the exact mechanism of fertilisation given the unusual location of the acrosome within the curvature of the sperm head and thickened zona pellucida (Johnston 1999).

Recently we developed an assay for the assessment of sperm DNA fragmentation in the koala (Johnston et al. 2007; Zee et al. 2009a; Fig. 9.5C). This has been a first for a marsupial and was initially developed to explore reasons for post-thaw decondensation of koala sperm DNA. The assay has been appropriately validated with in situ nick translation and comet assay and is based on sperm chromatin dispersion in a microgel. Koala spermatozoa are loaded into a microgel on a microscope slide and treated with protein lysing agent to expose loops of DNA. Single stranded and double stranded DNA is then allowed to disperse in the microgel and DNA identified by fluorescence microscopy. Our data thus far suggests that DNA fragmentation occurs at a relatively low incidence (6.7 %; Johnston et al. 2013a) in captive population of koalas and that the DNA molecule is able tolerate prolonged periods of chilled

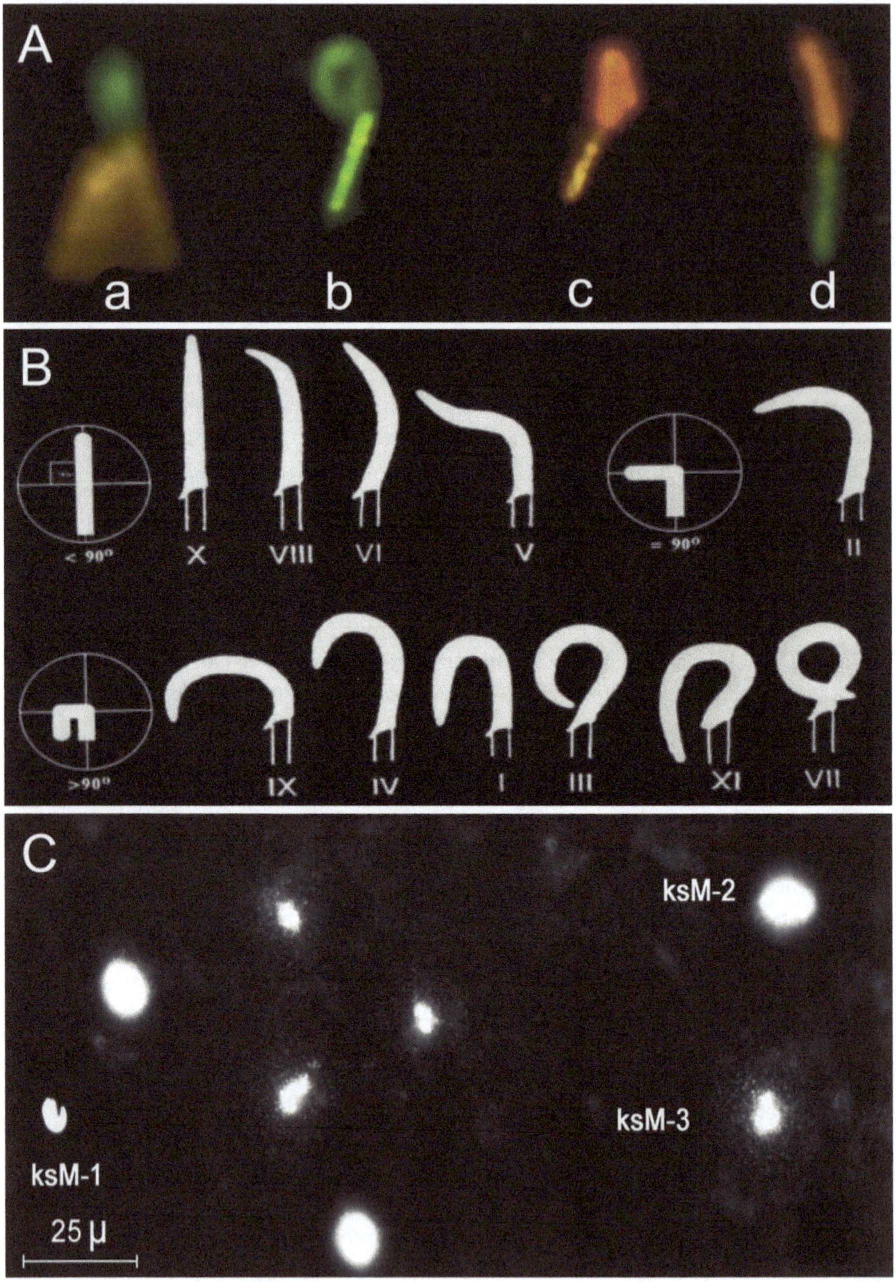

Fig. 9.5 Evaluation of Koala spermatozoa; A. Assessment of live (Sybr14—green) and dead (Propidium Iodide - red) sperm cells in combination with the assessment of mitochondrial membrane potential using the JC-1 stain (orange—high MMP; green—low MMP) (From Zee et al. 2007); B. Koala sperm nuclear morphotypes (I–XI); C. Assessment of koala sperm DNA fragmentation - KSM-1, rod-shaped nuclei without a halo of chromatin dispersion; KSM-2, round and rod-shaped nuclei showing a compact halo of chromatin dispersion about a nuclear core; KSM-3, sperm nuclei with enlarged halo and stellar chromatin corresponding to DNA fragments diffusing from the central core. (From Johnston et al. 2007)

storage. While our analysis revealed individual koalas with high levels of DNA fragmentation (up to 15.3 %) the significance of this phenomenon on fertility has yet to be determined. Given the relationship between chlamydia infection and high levels of DNA fragmentation in human spermatozoa (Gallegos et al. 2008), future studies are planned to investigate this relationship in the koala.

10 Semen Manipulation and Liquid Preservation

The practicality of an assisted breeding program based on artificial insemination is dependent on the ability to prepare and extend semen for insemination and the period of time that semen can be successfully preserved. Raw spermatozoa typically die rapidly ex vivo so that the semen must be rapidly diluted in a medium that is compatible for its evaluation and survival; this requires an analysis of the how spermatozoa tolerate physicochemical conditions such as pH, temperature and osmolality. Early in the development of our artificial insemination program we showed that koala sperm prefer a pH of 7–8, an osmolality of approximately 300 mOsm kg^{-1} and can tolerate a temperature range without a loss of sperm motility from 15 to 35 °C (Johnston et al. 2000a).

Contrary to evidence in most other marsupials, there is a small but significant decrease in sperm motility after rapid cooling of diluted semen from 35 to 5 °C, but compared to most eutherian species, koala spermatozoa would still be regarded as cold-shock tolerant and can be readily stored and manipulated at room temperature. In an attempt to explain the cold tolerance of marsupial spermatozoa we investigated the sperm membrane fatty acid composition in the koala and compared this to the common wombat and the eastern grey kangaroo (Miller et al. 2004). We discovered that koala sperm membranes had a high ratio of unsaturated/saturated membrane fatty acids compared to wombats and that sterol levels in marsupial sperm generally were very low. These marsupials were first to be examined for lipid membrane composition and highlight the dearth of information that exists for this taxon. We also employed a cryomicroscope to examine the effect of chilling on the sperm membrane directly (Zee et al. 2007). Using this technique we were able to view the effect of cooling to 5 °C on the plasma membrane and mitochondrial membrane potential simultaneously but there was no significant effect on the proportion of spermatozoa with high MMP or intact plasma membranes. Given the tolerance of koala sperm to cope with chilled temperatures, we are somewhat equivocal regards the effectiveness of egg yolk in the protection of the plasma membrane.

Tolerance of spermatozoa to changes in osmotic pressure can provide a guide to the sperm's ability to cope with osmotic flux during cryopreservation. We originally determined that koala spermatozoa had a relatively narrow osmotic tolerance and were particularly susceptible to hypo-osmotic media < than 250 mOsm kg^{-1}. Later we compared koala spermatozoa to that of wombat spermatozoa and found that koala spermatozoa were more susceptible to hyperosmotic damage than wombat spermatozoa (Johnston et al. 2006) both in terms of sperm motility and plasma membrane integrity. More recently, we have also examined the effect of osmolality on mitochondrial membrane potential, relaxed (swollen) chromatin and DNA fragmentation (Johnston et al. 2012). Plasma membrane integrity, chromatin relaxation and SDF appeared particularly susceptible to extreme hypotonic environments, whereas mitochondrial membrane potential (MMP), while susceptible to extreme hypo- and hypertonic environments, showed an ability to rebound from hypertonic stress when returned to isotonic conditions. The problem of chromatin relaxation is a major impediment to the successful cryopreservation of koala sperm.

Another important, but often ignored component in the development of semen preservation technology and artificial insemination, is an understanding of the normal flora and pathogens associated with the external genitalia, prepuce and semen. Prior to determining what antibiotics are required for semen diluents, it is first necessary to conduct culture and sensitivity tests of what organisms are present and what antibiotics they are sensitive to. We conducted a screening procedure with koala semen and were surprised by the diversity of microflora, describing a new *Corynebacterium* spp. in the process (Johnston et al. 1998). Once we understood what organisms were present (9 bacteria and two yeasts) and what antibiotics they were sensitive to (Penicillin G 1,000 IU/mL and gentamicin 100 μg/mL), we then had to ensure that the antibiotics were not spermicidal; this included antibiotic dose response studies on sperm parameters (motility) and validation of antibiotic action to prevent bacterial growth (Johnston et al. 1998).

We then extended this approach to examine the possibility of using antibiotic therapy to successfully kill chlamydial elementary bodies following experimental inoculation of cultured EB in semen samples. These results revealed that penicillin at 25 iu/mL, erythromycin at 1,000 μg/mL and tetracycline at 200 μg/mL were highly effective at rendering the chlamydiae non-viable, but streptomycin showed no antichlamydial activity (Bodetti et al. 2003). There was a significant reduction of the motility of spermatozoa extended in diluents containing erythromycin but spermatozoa incubated with tetracycline up to concentrations of 200 μg/mL were not affected. The use of "clean up" semen using antibiotic therapy or other approaches could prove to be an important mechanism of genetic recovery for animals actively shedding elementary bodies, from post-mortem tissue of animals with clinical signs of the disease or that prove to PCR positive to chlamydia. We currently have a PCR screen test for chlamydia in koala semen (Bodetti et al. 2002). If there is no adverse effect on sperm viability from these techniques then they should be incorporated as routine therapies prior to artificial insemination.

Short-term preservation of spermatozoa offers an alternative for temporary sperm storage in those species where the sperm are difficult to cryopreserve; the koala is certainly one of those species. Other domestic species that fit into this category include the stallion and boar, and in both species, there is a robust and commercially viable industry based on chilled or liquid sperm preservation. We conducted the first short-term preservation studies in 1992 where we naively used phosphate buffered saline to maintain sperm survival for 2 days (Johnston et al. 1992). Later we included antibiotics in a Tris-citrate-glucose diluent to successfully store the sperm at 5 °C for 8 days (Johnston et al. 2000a). Remarkably, the sperm still had 46 % progressive motility after 8 days but on leaving the sperm in the refrigerator for a total of 22, 35 and 42 days there was still approximately 47 %, 31 % and 16 % motility, respectively. More recently, we examined the effect of chilled storage for 16 days on sperm motility, membrane integrity, the percentage of sperm with relaxed chromatin and sperm DNA fragmentation. After 16 days storage at 5 °C, the progressive motility, plasma membrane integrity and nuclei relaxed chromatin were 50 %, 57 % and 25 % respectively (Johnston et al. 2013a). We also examined the DNA fragmentation dynamics of the spermatozoa over a 16-day storage period. DNA dynamics is a concept in which DNA of the sperm at defined time periods of storage is examined following incubation of sperm at body temperature in order to mimic in vivo conditions with the female reproductive system. For the koala, we examined sperm DNA dynamics immediately after ejaculation, 4 h, 1d, 2d, 4d, 8d and 16 d after chilled storage at 5 °C. At each of these time periods the sperm was then incubated at 35 °C and DNA fragmentation assessed at 0, 2, 6, 24 and 48 h. Remarkably, these results revealed that koala sperm DNA fragmentation after 16d storage at 5 °C, followed by 48 h incubation at 35 °C was still only 15 %. These results suggest that koala sperm DNA is very stable when stored in a chilled state for up to 16 days and based on this evidence and the motility and viability data we expect that sperm is potentially fertile and could be used for AI. We have already used chilled (5 °C) semen stored for 24 and 72 h to produce two and four koala pouch young, respectively (Allen et al. 2008a). Chilled storage and transport of semen not only can be used for national genetic transfers but also has the potential to be used for international exchange. As the koala has a fertile oestrus of approximately 8–10 days (Allen et al. 2008a), it possible that given notification of oestrus on day 1 of an animal located in the USA or Japan, to collect semen from males in Australia the following day. The semen from those males can then be loaded in a chilled liquid preservation diluent containing antibiotics and antichlamydials; in addition, the semen could be screened for chlamydia using PCR. The sample is then loaded into a commercial shipper that would maintain the sperm at 5 °C for up to 2–4 days during transportation; even after 5 days, there is a high likely-hood that the female will still be in heat and ready for insemination. The semen is examined on arrival and the sperm with the best post-chilled parameters is used for insemination. Based on our most recent observations (Johnston et al. 2013a), it might even be possible that semen stored chilled for up to 16 days could be used for insemination.

11 Sperm Cryopreservation

There is no doubt that sperm cryopreservation for use in successful artificial insemination currently remains one of the greatest challenges and limitations to the broad scale use and application of ART in the koala. The ability to freeze-thaw koala spermatozoa allows genetic management to occur through both time and space and would facilitate the global shipment of gametes and the use of genetics recovered from post-mortem animals. We first attempted to freeze koala spermatozoa in 1993 and at that stage there had been only one other attempt to cryopreserve marsupial sperm by Rodger et al. (1991) on the brush-tail possum (*Trichosurus vulpecula*). Remarkably, this study found possum spermatozoa had the best post-thaw survival when frozen in 17.5 % glycerol. Based on the possum methodology, we conducted a similar protocol for koala spermatozoa using a range of glycerol concentrations (4–18 % w/w) in a Tris-citrate 20 % egg yolk extender (Johnston et al. 1993). The results of these investigations revealed that glycerol had a detrimental effect on sperm motility when stored for 1 h at room temperature but more importantly, koala spermatozoa, similar to Brush-tail possum, required 14 % glycerol in order to produce post-thaw motility of 30 %.

Some 13 years later we returned to investigate sperm cryopreservation in the koala and this time compared the cryopreservation success of koala spermatozoa with that of common wombat spermatozoa (Johnston et al. 2006). The use of wombat sperm was significant as both species share a close phylogeny and similar morphology but wombat spermatozoa showed a remarkable tolerance to cryopreservation (Taggart et al. 1996; 1998; MacCallum and Johnston 2005) compared to that of koala spermatozoa. In our comparative study, we confirmed that common wombat spermatozoa showed greater post-thaw survival than koala sperm in terms of motility, plasma membrane integrity and decondensed (relaxed) chromatin. We also showed that both koala and wombat prefer to be frozen at slow rate of freezing (−6 °C/min) compared to rapid freezing (3 cm above the liquid N_2 interface) and that both species show greatest post-thaw survival when frozen in 14 % rather 8 % (w/w) glycerol. Koala spermatozoa showed a large increase in the proportion of spermatozoa with relaxed chromatin compared to wombats following 2 h incubation post-thaw. The lower tolerance of koala sperm to cope with hyperosmotic media may be contributing to lower post-thaw survival. We also examined both spermatozoa for the presence of F-actin in an attempt to relate poor cryopreservation success with F-actin induced plasma membrane inflexibility—we discounted this hypothesis when we found F-actin in the wombat sperm but not in the koala.

Following some success at using a cryomicroscope for investigating freeze-thaw methodologies in kangaroo spermatozoa (Holt et al. 1999), we subsequently turned our attention to this instrument to investigate the cryopathology of koala spermatozoa (Zee et al. 2007). This work represented a shift away from the traditional empirical approach to investigating cryopreservation to more hypotheses driven sperm organelle focus. Using a combination of fluorescent probes to directly observe plasma membrane integrity, MMP, lipid raft stability and phosphatidylserine translocation, during the freeze-thaw protocol, we were able to gain a better understanding of koala sperm specific cryopathology. These observations revealed that high MMP declined significantly during rewarming, especially as the temperature increased from 5 to 35 °C. We also concluded that chilling and freezing had no effect on the distribution of ganglioside GM1 in sperm membrane or on plasma membrane lipid asymmetry.

We have also examined the use of cryoprotectants other than glycerol for use with koala spermatozoa, including dimethyl sulfoxide (DMSO), methanol, propylene glycol and dimethylacetamide (DMA) (Zee et al. 2008). These experiments revealed that 10–15 % DMA may be a useful alternative to glycerol for cryopreservation.

We have also examined individual variability in post-thaw survival with a large captive koala population (Zee et al. 2009b). There were significant differences in post-thaw survival from different animals that were independent of pre-freeze semen quality. We showed that glycerol (14 %) was a better cryoprotectant than DMA (12.5 %) in terms of maintaining motility, plasma membrane integrity and high MMP, but there was no difference between the two compounds regards their ability to prevent chromatin relaxation. We also noted the efficiency of energy generation by the mitochondria was lowered by cryopreservation. From this study we postulated that the unpredictability of assessing post-thaw survival from pre-freeze koala semen parameters is likely to be associated with variation in ejaculate composition or inherent genetic differences between animals.

The issue of sperm chromatin relaxation following cryopreservation led us to work in Spain and a very fruitful collaboration with the laboratory of Prof Jaime Gosalvez at the Autonomous University of Madrid (See Chap. 15 of this volume). We were initially concerned that koala sperm were showing evidence of fragmented DNA after the freeze-thaw procedure, so we developed and validated an assay for the assessment of koala sperm DNA fragmentation based on the sperm chromatin dispersion test (SCDt; Johnston et al. 2007). In this early work, we reported three koala nuclear morphotypes following the assay (Fig. 9.5C); KSM1 and KSM2 morphotypes showed no or limited sperm DNA fragmentation. We proposed that KSM2 result from DNA that is damaged as part of the normal processing of the spermatozoa in the assay and is primarily a consequence of the lack of cysteine residues and associated stabilising disulphide bonds in marsupial DNA generally. We also concluded that "True" DNA fragmentation was represented by the KSM3 morphotype that showed massive dispersion of chromatin following the SCDt. This work also revealed that the proportion of KSM3 actually increased and the KSM2 decreased following incubation of frozen thawed spermatozoa. The next step in this investigation was to determine whether the sperm DNA fragmentation we were visualising was a result of single stranded or double stranded DNA breaks; our prediction was that KSM2 were being caused as a result of single stranded breaks, whereas KSM3 were associated with more severe double stranded breaks. This was achieved by developing a double comet assay to assess sperm chromatin damage (Zee et al. 2009a) and comparing these results with SCDt. Following SCDt, we discovered a continuum of nuclear morphotypes, ranging from no apparent DNA fragmentation to those with highly dispersed chromatin; these morphotypes were mirrored by a similar diversity of comet morphologies that could be further differentiated by double comet assay. Spermatozoa with "true" DNA fragmentation exhibited a continuum of comet morphologies, ranging from a more severe form of alkaline-susceptible DNA with a diffuse single tail to nuclei that exhibited both single and double-stranded breaks with two comet tails (See Chap. 15 this volume). In our most recent papers on this topic, we explored the effect of cryopreservation on koala sperm DNA fragmentation. As for our studies on chilled koala spermatozoa, we used a different variation of the SCDt, which examined the DNA fragmentation dynamics of frozen-thawed koala spermatozoa over a 48 h period of incubation at 35 °C (Johnston et al. 2013a) to compare sperm frozen in 14 % glycerol and 10 % DMA.

This study revealed that while the survivorship of pre-freeze sperm DNA fragmentation was not different when compared with sperm frozen in DMA or between sperm frozen in DMA and glycerol, spermatozoa frozen in glycerol showed a higher rate of DNA fragmentation than pre-freeze spermatozoa. This result differed from that of observations of progressive motility, plasma membrane integrity and relaxed chromatin, which were all adversely affected after cryopreservation in both glycerol and DMA. We also found that following thawing, koala sperm chromatin tended to relax but interestingly, the incidence of sperm DNA fragmentation was not correlated with the incidence of sperm chromatin relaxation after cryopreservation. These results suggested that chromatin relaxation was not necessarily associated with DNA fragmentation.

In attempt to explore the underlying aetiology of sperm chromatin relaxation associated with koala sperm cryopreservation and post-thaw incubation, we explored an experimental model that mimicked the structural and physiological effects of osmotic flux on the sperm cell (Johnston et al. 2012). Similar to our previous findings, DNA labelling after in situ nick translation of thawed cryopreserved spermatozoa revealed a positive correlation between area of relaxed chromatin in the nucleus and the degree of nuclear labelling and while the chromatin of some spermatozoa increased more than eight times its normal size, not all sperm nuclei with relaxed chromatin showed evidence of nucleotide incorporation. Preferential staining associated with DNA fragmentation was typically located in the peri-acrosomal and peripheral regions of the sperm head and at the base of the sperm nucleus where it appeared as "hot spots" of damage following cryopreservation. We also compared the effect of exposure to anisotonic media and cryopreservation on the integrity of koala spermatozoa and found that injury induced by exposure to osmotic flux essentially imitated the results found following cryopreservation. Plasma membrane integrity, chromatin relaxation and SDF appeared very susceptible to extreme hypotonic exposure. Although susceptible to extreme hypo- and hypertonic media, MMP showed an ability to rebound from hypertonic stress when return to isotonic conditions. Koala spermatozoa exposed to 64 mOsm kg^{-1} media showed an equivalent, or more severe, degree of structural and physiological injury to that of frozen-thawed spermatozoa, supporting the hypothesis that cryoinjury is principally associated with a hypo-osmotic effect. A direct comparison of SDF of thawed cryopreserved spermatozoa and those exposed to a 64 mOsm kg^{-1} excursion showed a significant correlation but no correlation was found when the percentage of sperm with relaxed chromatin was compared. We concluded that while a cryo-induced osmotic injury model appears to explain post-thaw changes in koala SDF, the mechanisms resulting in relaxed chromatin require further study and will be essential to understand before genome resource banking becomes a reality for the koala. A lack of correlation between the percentage of sperm with relaxed chromatin and SDF might indicate that the timing of these pathologies are asynchronous; perhaps cryo-induced injury involves a combination of structural damage (rupture of the membrane) and oxidative stress, that first leads to the reduction of MMP and the relaxation of chromatin, which is then ultimately followed by an increase in DNA fragmentation.

12 Oestrus Detection and Induction of Ovulation

In most eutherian mammals it is possible to synchronise oestrus by intervening in the endocrine control of the corpus luteum (CL). The inhibitory effects on GnRH secretion imposed by progesterone can be removed by the use of prostaglandin and its analogues to cause regression of the CL or by administration and timed removal of progestogen implants (Allen et al. 2008b). This approach is not possible in marsupials as uterine prostaglandin does not result in luteolysis in non-pregnant females and exogenous progestogens do not suppress GnRH secretion. This means that other methods of oestrus and ovulation control are required in marsupials. As LH and FSH are considered to be important hormones in the control of the marsupial reproduction, we have considered that an alternative approach for taking control of the oestrous cycle without the benefit of being able to manipulate prostaglandin and progesterone, is the use of GnRH antagonists. GnRH antagonists bind to the GnRH receptor on the gonadotroph and typically outcompete natural sequence GnRH but they do not illicit activation of gonadotrophin secretion. This inhibition of FSH and LH secretion results in a suppression of the steroidogenesis and prevents the female from cycling. The female will continue in anoestrus while the GnRH antagonist is administered but once the molecule has been metabolised, the receptors are free to bind to natural GnRH and the cycle commences again. The consistency on which the cycle can be interrupted or recommenced and synchronised depends on the dosage of antagonist administered, its biological half-life and its binding affinity. In a preliminary study using the male koala as our model (Allen et al. 2008b) we used a single injection of the GnRH antagonist acyline (100 μg = 14.3 μg/kg or 500 μg = 71.4 μg/kg) but this had no effect on suppressing pituitary or testicular steroidogenesis. The inability of acyline to suppress the reproductive axis of male koalas maybe associated with a low affinity for the GnRH receptor, although it is also possible that acyline is cleared quickly from circulation. Further studies are required using females to investigate other possible GnRH antagonists as this may still be the most productive and direct approach to synchronising oestrus. It is in fact possible to induce and potentially synchronise oestrus in lactating female marsupials by removing the sucking pouch young. Removal of the suckling stimulus results in reactivation of the ovarian cycle but such an approach is hardly appropriate or ethical, unless the "pulled" pouch young can be successfully transferred to another lactating mother.

A very significant advantage in the reproductive physiology of the koala that has contributed to the success of artificial insemination is that ovulation is induced via coitus (Johnston et al. 2000b). Our initial investigations lead us to believe that ovulation was induced by a complete duration of penile thrusting and that the mechanism was essentially a copulo-receptive reflex (Johnston et al. 2000c) similar to which occurs in the felidae. However, further study revealed that mechanical stimulation of the urogenital sinus with a glass rod alone did not appear to induce a luteal phase (0/9), that semen alone with stimulation induced a luteal phase in some koalas (4/9) but that a combination of semen and glass rod simulation was most effective (7/9) at inducing a luteal phase. These results suggested that there might be factors in the semen that promote induction of the luteal phase. We have speculated that the presence of specific ovulating factors in the semen and the mechanical stimulation of the urogenital sinus during coitus, work synergistically to induce ovulation; perhaps the thrusting of the male's penis in the urogenital sinus in someway prepares the urogenital epithelium for exposure to the ovulating factors? We talk about a luteal phase, rather than ovulation because it is difficult to observe the ovary directly in this species. The ability to effectively document ovulation by ultrasound would be an important advance that needs to be perfected in this species and is part of our ongoing research.

While we currently use glass rod stimulation of the urogenital sinus as part of our standard artificial insemination technique, it is also technically possible to use pharmacological induction of the ovulation using either GnRH and human chronic gonadotrophin (hCG; Johnston et al. 2000c; 2003; Allen et al. 2008b). To date, we have only produced one pouch young following administration of hCG (Johnston et al. 2003). We have also induced what appear to be normal luteal phases in the koala oestrus using the GnRH agonist buserelin (Allen et al. 2008b). If hCG or GnRH agonists are going to be employed to routinely induce ovulation as part of the koala AI protocol, then it might be prudent to inseminate the female approximately 24–32 before injection of these pharmaceuticals; clearly, there needs to be further studies to refine their respective applications. The use of pharmaceutical induction of ovulation is particularly relevant when using cryopreserved or heavily diluted spermatozoa, as dilution with semen extenders or the washing out of cryoprotectants could potentially remove the ovulating factors in the semen. Nevertheless, we regularly use diluted semen 1:1 with Tris-citrate extender as part of AI program and have been successful; it would be interesting to see how far the semen could be extended before the concentration of ovulating factor is insufficient to induce ovulation.

While it is currently not possible to synchronise oestrus in the koala to the same extent of domestic animals, it is possible to make a case that this form of oestrus control may not be that necessary in order to develop an effective artificial insemination program. For example, given that the non-mated koala has an oestrous cycle of 33 days, an oestrus of typically 10 days (Johnston et al. 2000b) and assuming that the non-lactating females are cycling normally, all koalas should come into oestrus and be artificially inseminated in a 43 day period. Females that fail to conceive and do not ovulate will come back into oestrus again 33 days later while those who do ovulate but failed to conceive will return approximately 50 days. Given that the koala is polyoestrus there is approximately 5–6 AI opportunities each breeding season.

13 Artificial Insemination

We have developed two different successful techniques for artificial insemination in the koala. Our most commonly used protocol involves the conscious female and insemination using a custom designed Foley insemination catheter (Johnston et al. 2003; Fig. 9.6a). Typically we will choose a female for AI who is in day 2–5 of

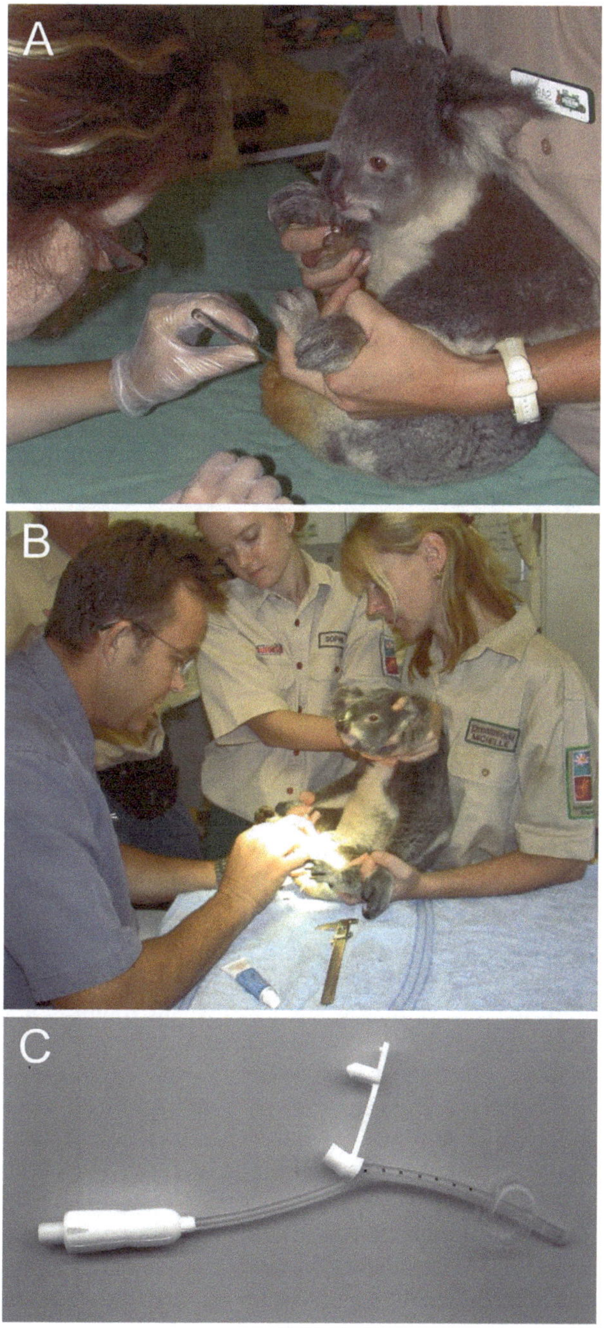

Fig. 9.6 Artificial insemination of the koala. A. Glass rod stimulation of the koala urogenital sinus; B. Artificial insemination of the conscious koala; C. Customised designed koala insemination Foley catheter. (From Johnston et al. 2003)

oestrus. She is brought into the veterinary surgery and held in a full restraint position by the zookeeper (Fig. 9.6b). The koala's urogenital sinus is positioned slightly upwards and towards the inseminator. A sterile glass rod that mimics the glans penis of the koala is then gently inserted into the urogenital sinus and stroked back and forth with a slight twisting motion to a depth of 40–60 mm; stimulation of the urogenital sinus is based on previous descriptions of natural koala coitus (40 penile thrusts per 20 s; Johnston et al. 2000b, c). The koala AI catheter is then inserted the full depth of the urogenital sinus and the cuff of the catheter inflated so to produce a seal to prevent retrograde flow of semen; the basis of the insemination technique is one of positive displacement of seminal volume into the ostia of the lateral vaginae. We typically use a seminal volume of approximately 1 ml and this can be diluted or undiluted semen. We have been successful using semen with a sperm concentration as low as 3×10^6/mL but we would suggest insemination sperm concentration of approximately $20–40 \times 10^6$/mL as standard (Allen et al. 2008a). The semen is drawn up into the catheter with a 1 mL syringe and inseminated slowly; following deposition of semen; a second 1 mL syringe of air is used to displace any semen remaining in the bore of the catheter. The bore of the insemination catheter is then capped and the female koala restrained with the catheter in place for another 5 min, after which the catheter cuff is deflated and removed. As the female is conscious she can then be immediately returned to her enclosure. We also developed another method of insemination, which we refer to as the urogenitoscopic method in which the female is anaesthetised, placed in ventral recumbency on a tilt table, the ostia of the vaginae visualised using an otoscope and speculum and a tom-cat catheter directed into the vaginal ostia (Johnston et al. 2003). Apart from the inconvenience and risk associated with anaesthesia, we found that proliferation of the urogenital epithelium made visualisation of the ostia virtually impossible. Reliable insemination into both ostia is important as the left and right sides of the koala reproductive tract are separated by a medial septum of the vaginal cul-de-sac, so that semen needs to be deposited into each ostium. As it is extremely difficult to determine which ovary has ovulated without ultrasonography or laparoscopy, we concluded that the use of the catheter had inherently less procedural risk and a greater probability of successful semen delivery into both sides of the tract. Nevertheless, the use of small volumes of semen or the use of cryopreserved semen will probably require us to rethink new methods of insemination higher up in the reproductive such as laparoscopic intrauterine insemination.

The most immediate and current use of koala AI has been within and between zoological institutions for the purposes of genetic exchange, animal pairings with behaviour incompatibilities or structural soundness problems that prevent or impede physical mating. However, there is also a greater potential for use of artificial insemination, especially if it is combined with cryopreservation and genome banking principles. For example, we have already demonstrated that it possible to collect semen from wild donors (Allen et al. 2010), so that this semen can be screened for disease and then used after chilled or cryopreservation for AI into captive reservoirs of female koalas or females in wild isolated genetically restricted habitat fragments. Semen can also be collected, used or stored in genome banks, from animals that require euthanasia or even post-mortem specimens associated with dog and car accident trauma and disease. While any genetic exchange should be carefully monitored with a detailed understanding of genetic history and relatedness of individuals in the population, we are getting to the point in certain parts of SE Queensland where intensive genetic management is going to be necessary to prevent localised extinction (Johnston et al. 2013b).

To date we have produced a total of 34 koala pouch by means of artificial insemination but much of this work has been experimental rather than specifically for genetic management. Consequently, our results have been limited by the constraints of our own experimental approach, so that we have not yet been able to reliably determine the overall efficiency of the procedure, as might be the case for commercial AI in the cattle industry. Nevertheless, the number of offspring produced (32) significantly outnumbers what has been produced in other marsupials (Rodger et al. 2009) and rivals that of successful ABT programs in other captive wildlife.

It is one thing to develop efficient and reliable protocols for koala AI but another to implement these technologies within commercial zoo practices and governmental genetic management of wild animals. That is why we were particularly satisfied with an excerpt in the latest Queensland koala Nature Conservation (Koala) Conservation Plan (2006–2016) (Department of Environment Heritage and Protection, Queensland Government) which mentioned our work in Sect. 4.8 of the policy document regards the use of artificial insemination "*To reduce the need to export live koalas in the future, the EPA is also working with the University of Queensland, a world leader on koala biological research, on reproductive technologies for koalas. In the future, artificial insemination will provide for the introduction of new genetic diversity to local and overseas zoos and provide more effective, less costly measures for maintaining viable breeding colonies*". It is only with government and public acceptance that we will be able to implement and evaluate all the potential benefits that ABT can deliver.

14 The Transfer of Koala Art to the Wombat

From day 1 when the plan for a koala artificial insemination program was first conceived it has always been our goal to apply what we learnt from the koala to other species. The establishment of model species is a sound principle for developing an artificial insemination technique. The problem that we faced working in the area of marsupial reproduction was that were very few well-studied species for direct comparison. The characterisation of all marsupials into a generalised mode of reproduction has been a major limitation to the development of ART; this point is well illustrated in the koala and wombat. While the close phylogeny of these two species is reflected in their respective reproductive anatomies and gamete ultrastructure, the reproductive physiology and behaviour of these two species is frustratingly, very different. For example, the female reproductive tract of the wombat is similar enough that a slightly larger version of the koala insemination catheter would be ideal for the wombat, but unlike the koala, the wombat ovulates spontaneously and appears to have a relatively short oestrus of less than 24 h. These differences, combined with the cryptic, non-tractile and fossorial nature of the wombat, make the application of ART a much more "hands-off" and technically difficult challenge (Hogan et al. 2013). Another interesting difference is that wombat spermatozoa are very tolerant of cryopreservation and yet it is this element of the ART program in the koala that is currently the rate-limiting step for the use of frozen-thawed semen and establishment of a genome resource bank. To date there have been no marsupial young born following AI with frozen-thawed semen.

15 Application of Art to Behavioural Ecology of the Koala

Key physiological and behavioural information accumulated in the development of an ART program can also have direct impact on understanding behavioural ecology of the wild animal; an example of this is the observation that ovulation in the koala is induced by coitus. This information is fundamental to understanding and interpreting female koala movement and home range observations. Although yet to be proven, we would postulate that female koalas in the wild that come into oestrus will show increased levels of physical activity and seek out males for mating, motivated by the high concentration of oestradiol from the pre-ovulatory follicle. Once mated, the follicle would collapse, the female losses sexual interest and returns to her home range. It is only when we understand the physiological parameters of the reproductive biology that we can interpret what we see in the wild. We are currently examining other fundamental aspects of koala biology that are relevant to refining ART but also have a role in understanding ecology (e.g. refractory period of the male).

16 Conclusion and Future Directions

Given that we have produced 34 koala young by means of artificial insemination, there is no doubt that we currently have a reliable means of genetic management for captive koalas based on ART. It would also be safe to say that the koala program has been the most successful for any marsupial (Rodger et al. 2009). Despite this success, there are still a number of missing pieces to the complete ART puzzle; the two most significant being the development of a reliable method of sperm cryopreservation and the application of the technology to wild populations. The establishment of a frozen genome bank of koala spermatozoa from representative males from every wild population in Australia is long-term objective of our group. We also recommend developing AI protocols that allow the use of extended chilled semen, as we are confident that combination of chilled liquid preserved semen (2–3 weeks) could be useful for the international exchange of semen between zoos and facilitating the timing of artificial insemination.

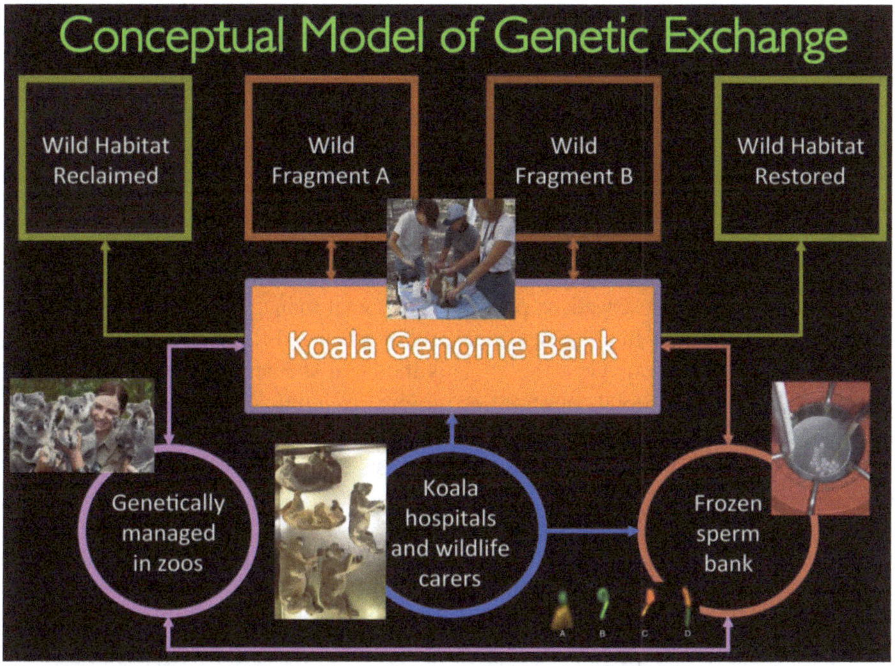

Fig. 9.7 Conceptual model of genetic exchange for koala conservation through the use of live and frozen genome banks (From Johnston et al. 2013b)

In close association with koala ART, we would also like to develop genetic exchange programs in the form of a live genome bank. This concept is similar to frozen genome bank except that live female koalas are used as the storage units. In such a program we would use koalas that are genetically selected to be representative of the population under investigation and then manage them closely. These concepts include (Johnston et al. 2013b; Fig. 9.7); (1) *Genetic connectivity* where koalas spend short periods of time in captivity to facilitate genetic exchange between fragmented populations such that animals from isolated fragments could be brought into captivity bred and released back into their respective fragments (Holt et al. 1996). In this we have ensured greater genetic diversity and gene flow into both fragments; (2) *Genetic capture* where koalas destined for translocation because of sudden habitat loss or destruction could be brought into captivity into a live genome bank to make a quick genetic deposit; males for example could spend a short period of time in a captive facility to breed, or perhaps semen could be collected from wild males and inseminated to dedicated genetically selected females. In this way their genes would be captured and stored both in live and frozen genome banks; (3) *Genetic recovery*—where gametes from diseased, trauma or post-mortem koalas could be recovered and used for germplasm storage or artificial insemination. This source of genetic material is currently being wasted in SE Queensland hospitals—

between 1997 to 2009, 6581 koalas were euthanased in koala hospitals (DERM 2009); (4) *Genetic propagation*—where we propose the establishment of purpose koala breeding centre that could be used for the generation of genetically scripted koalas for release into either reclaimed or restored habitat. SE Queensland zoos have world's best practice at producing koalas in captivity and a skill set ideally suited for this type or conservation (Johnston et al. 2013b).

Some of these ideas are quite unconventional and might even be regarded as controversial, but there are certain regions in Queensland where intensive management may be the only way that koala populations can ultimately survive. In combination with relevant habitat preservation, predator and social management and policies, we are convinced that ART in its most simple and/or advanced forms has a major role in koala conservation. In our view the koala is the ideal species for demonstrating the benefits of ART to marsupial conservation biology; its iconic status and public appeal, tractility, unique and amenable reproductive biology and its ability to adapt well to temporary captivity and release, all point towards a high probability for success.

Acknowledgements The success of the koala ART program has largely been achieved through the dedication and support of a wide range of scientific collaborators, post-graduate students and zoological institutions. In alphabetic order, we are extremely grateful for the contribution and support of Dr Camryn Allen, Dr Rosie Booth, Ms Michelle Burridge, Mr Kevin Bradley, Prof Randal Cameron, Prof Frank Carrick, Dr Ron Cox, Dr Jon Curlewis, Mr Robert Douglas, Dr Bill Ellis, Prof Jaime Gosalvez, Dr Carmen Lopez-Fernandez, Mr Alan Lisle, Prof Michael McGowan, Dr Allan McKinnon, Mr Al Mucci, Mr Paul O'Callaghan, Dr Vere Nicolson, Ms K Nilsson, Dr Nancy Phillips, Dr Michael Pyne, Mr Peter Theilman, Dr Andrew Tribe and Dr Zee Yeng Peng.

References

Allen CD, McKinnon AJ, Lisle AT, D'Occhio MJ, Johnston SD. Testosterone secretion in the koala (Phascolarctos cinereus). J Androl. 2006;27:720–4.

Allen CD, Burridge M, Mulhall S, Chafer M, Nicolson V, Pyne M, et al. Successful artificial insemination in the koala (*Phascolarctos cinereus*) using extended and extended-chilled semen collected by electroejaculation. Biol Reprod. 2008a;78:661–6.

Allen CD, Burridge M, Chafer ML, Nicolson VN, Jago SC, Booth RJ, et al. Control of the koala (*Phascolarctos cinereus*) anterior pituitary-gonadal axis with analogues of GnRH. Reprod Fertil Dev. 2008b;20:598–605.

Allen CD, de Villiers DL, Manning BD, Dique DS, Burridge M, Chafer ML, et al. Seasonal reproduction in wild and captive male koala (*Phascolarctos cinereus*) populations in south-east Queensland. Reprod Fertil Dev. 2010;22:695–709.

Bodetti T, Johnston SD, Pospischil A, Knox C, Timms P. PCR detection of *Chlamydia* in koala (*Phascolarctos cinereus*) semen. Vet Rec. 2002;151:147–9.

Bodetti TJ, Hengstberger K, Johnston SD, Timms P. Evaluation of four antibiotics, tetracycline, erythromycin, penicillin and streptomycin for their ability to decontaminate *Chlamydia*-spiked koala (*Phascolarctos cinereus*) semen *in vitro*. Veterinary Record. 2003;155:588–91.

Dief HHA. Chlamydial disease of the male koala (*Phascolarctos cinereus*) reproductive tract. Gatton: The University of Queensland; 2011.

Department of Environment and Resource Management QLD. Decline of the koala coast koala population: Population status in 2008. Queensland Government: DERM; 2009.

Feige S, Nilsson K, Phillips CJC, Johnston SD. Heterosexual and homosexual behaviour and vocalisations in captive female koalas (Phascolarctos cinereus). Appl Animal Behav Sci. 2006;103:131–45.

Gallegos G, Ramos B, Santiso R, Goyanes V, Gosalvez J, Fernandez JL. Sperm DNA fragmentation in infertile men with genitourinary infection by *Chlamydia trachomatis* and mycoplasma. Fertil Steril. 2008;90:328–34.

Handasyde KA, McDonald IR, Than KA, Michaelides J, Martin RW. Reproductive hormones in the koala. In: Lee AK, Handasyde KA, Sanson GD, editors. Biology of the koala. Chipping North, New South Wales: Surrey Beatty & Sons Pty Ltd.; 1990. p. 203–10.

Harding HR, Carrick FN, Shorey CD. Special features of sperm structure and function in marsupials. In: Fawcett DW, Bedford JM, editors. The Spermatozoon. Baltimore, MD/Munich: Urban & Schwarzenburg; 1979. p. 289–303.

Harding HR, Carrick FN, Shorey CD. Acrosome development during spermiogenesis and epididymal sperm maturation in Australian marsupials. In: Andrè J, editor. The sperm cell. The Hague: Martinus Nijhoff; 1983. p. 411–4.

Harding HR, Carrick FN, Shorey CD. The affinities of the Koala, *Phascolarctos cinereus* on the basis of sperm ultrastructure and development. In: Archer M, editor. Possums and opossums; studies in evolution. Sydney: Surrey Beatty & Sons Pty Ltd., and Royal Zoological Society of New South Wales; 1987. p. 353–64.

Harding HR, Aplin KP. Phylogenetic affinities of the Koala (Phascolarctidae : Marsupialia): A reassessment of the spermatozoal evidence. In: Lee AK, Handasyde KA, Sanson GD, editors. Biology of the KOALA. Chipping North, New South Wales: Surrey Beatty & Sons; 1990. p. 1–31.

Hogan LA, Janssen T, Johnston SD. Wombat reproduction (Marsupialia; Vombatidae): An update and future directions for the development of artificial breeding technology. Reproduction. 2013;145(6):R157–73.

Holt WV, Bennett PM, Volobouev V, Watson PF. Genetic resource banks in wildlife conservation. J Zool. 1996;238:531–44.

Holt WV, Penfold LM, Johnston SD, Temple-Smith P, McCallum C, Shaw J, et al. Cryopreservation of macropodid spermatozoa: new insights from the cryomicroscope. Reprod Fertil Dev. 1999;11:345–54.

Holt WV, Lloyd RE. Artificial insemination for the propagation of CANDES: The reality! Theriogenology. 2009;71:228–35.

Hughes RL. Light and electron microscope studies on the spermatozoa of the Koala, *Phascolarctos cinereus* (Marsupialia). J Anat, London. 1977;124:153.

Johnston SD, McGowan MR, Carrick FN, Cameron RDA. Towards the development of an artificial insemination programme for koalas (*Phascolarctos cinereus*). *Proceedings of 12th International Congress on Animal Reproduction*, vol. 4. Netherlands: The Hague; 1992. 1954–1957.

Johnston SD, McGowan MR, Carrick FN, Tribe A, Douglas R. Preliminary investigations into the feasibility of freezing koala semen. Aust Vet J. 1993;70:424–5.

Johnston SD. *The reproductive biology of the queensland koala*. Honours Thesis, Department of Zoology, The University of Queensland, Brisbane; 1994.

Johnston SD, McGowan MR, Carrick FN, Cameron RDA. Seminal characteristics and spermatozoa morphology of captive Queensland Koalas, (*Phascolarctos cinereus adustus*). Theriogenology. 1994;42:501–11.

Johnston SD, O'Callaghan P, McGowan MR. Seminal characteristics of koala ejaculates collected using an artificial vagina. J Reprod Fertil. 1997;109:319–23.

Johnston SD, O'Boyle D, Frost AJ, McGowan MR, Tribe A, Higgins D. Antibiotics for the preservation of koala (*Phascolarctos cinereus*) semen. Aust Vet J. 1998;76:335–8.

Johnston SD. *Studies towards the development of an artificial insemination program in the koala* (*Phascolarctos cinereus*), PhD Thesis, The University of Queensland; 1999.

Johnston SD, McGowan MR, O'Callaghan P. Assisted breeding technology for the conservation and propagation of *Phascolarctos cinereus* or How to make a koala pouch young. *Proceedings of the Postgraduate Committee in Veterinary Science*, University of Sydney, Sydney; 1999. pp 199–217.

Johnston SD, McGowan MR, Phillips NJ, O'Callaghan P. Optimal physico-chemical conditions for the manipulation and short-term preservation of koala (*Phascolarctos cinereus*) spermatozoa. J Reprod Fertil. 2000a;118:273–81.

Johnston SD, McGowan MR, O'Callaghan P, Cox R, Nicolson V. Studies of the Koala (*Phascolarctos cinereus*) oestrous cycle, oestrus and pregnancy. J Reprod Fertil. 2000b;120: 49–57.

Johnston SD, McGowan MR, O'Callaghan P, Cox R, Nicolson V. Natural and artificial methods of inducing the luteal phase in the Koala (Phascolarctos cinereus). J Reprod Fertil. 2000c; 120:59–64.

Johnston SD, Holt WV. Germplasm conservation in marsupials, Chapter 12. In: Watson PF, Holt WV, editors. Cryobanking the genetic resource. Wildlife conservation for the future? London: Taylor and Francis; 2001. p. 201–27.

Johnston SD, McGowan MR, O'Callaghan P, Cox R, Houlden B, Haig S, et al. Birth of Koalas *Phascolarctos cinereus* at Lone Pine Koala Sanctuary following artificial insemination. Int Zoo Yearbook. 2003;38:160–72.

Johnston SD, O'Callaghan P, Nilsson K, Curlewis JD. Semen-induced luteal phase and identification of a LH surge in the koala (*Phascolarctos cinereus*). Reproduction. 2004;129:629–34.

Johnston SD, MacCallum C, Blyde D, McClean R, Lisle A, Holt WV. An investigation into the similarities and differences governing the cryopreservation success of koala (*Phascolarctos cinereus*: Goldfuss) and wombat (*Vombatus ursinus*: Shaw) spermatozoa. Cryobiology. 2006;53:218–28.

Johnston SD, Lopez-Fernandez C, Gosalbez A, Zee Y, Holt WV, Allen CA, et al. The relationship between sperm morphology and chromatin integrity in the koala (*Phascolarctos cinereus*) as assessed by the sperm chromatin dispersion test (SCDt). J Androl. 2007;28:891–9.

Johnston SD, Satake N, Zee YP, Lopez-Fernandez C, Holt WV, Gosalvez J. Osmotic stress and cryoinjury of koala sperm: an integrative study of the plasma membrane, chromatin stability and mitochondrial function. Reproduction. 2012;143:789–97.

Johnston SD, Zee YP, López-Fernández C, Gosálvez J. The effect of chilled storage and cryopreservation on the sperm DNA fragmentation dynamics of a captive population of koalas. J Androl. 2013a;33:1007–15.

Johnston SD, Mucci A, Ellis W. The role of captive koalas in koala conservation. In: Conserving central Queensland's koalas. Rockhampton: Koala Research Centre of Central Queensland, Central Queensland University; 2013b.

MacCallum C, Johnston SD. Studies on the cryopreservation of common wombat (*Vombatus ursinus*) spermatozoa. Reprod Fertil Dev. 2005;17:727–32.

McFarlane JR. *Endocrine factors influencing reproduction in male non-eutherian mammals.* Unpublished PhD Thesis, The University of Queensland, Brisbane; 1990.

Miller Jr RR, Sheffer CJ, Johnston SD. Sperm membrane fatty acid composition in the Eastern Grey kangaroo (*Macropus giganteus*), Koala (*Phascolarctos cinereus*) and Common Wombat (*Vombatus ursinus*) and its relationship to cold shock injury and cryopreservation success. Cryobiology. 2004;49:137–48.

Mitchell P. Social behaviour and communication of koalas. In: Lee AK, Handasyde KA, Sanson GD, editors. Biology of the koala. Chipping North, New South Wales: Surrey Beatty & Sons Pty Ltd.; 1990. p. 151–70.

Natural Resource Management Ministerial Council. National Koala Conservation and Management Strategy, 2009–2014. Canberra: Department of the Environment, Water, Heritage and the Arts; 2009.

Obendorf DL. The pathogenesis of urogenital tract disease in the koala. In: Australian Wildlife, Proceedings Number 104, The University of Sydney Post-graduate Committee in Veterinary Science now the Centre for Veterinary Education, Sydney; 1988. p. 649–655.

Oishi M, Takahashi M, Amasaki H, Janssen T, Johnston SD. The seminiferous epithelial cycle and microanatomy of the koala (*Phascolarctos cinereus*) and southern hairy-nosed wombat (*Lasiorhinus latifrons*). J Anat. 2013;222:380–9.

Pukazhenthi B, Comizzoli P, Travis AJ, Wildt DE. Applications of emerging technologies to the study and conservation of threatened and endangered species. Reprod Fertil Dev. 2006;18:77–90.

Rodger JC, Cousins SJ, Mate KE. A simple glycerol-based freezing protocol for the semen of a marsupial *Trichosurus vulpecula*, the Common Brushtail Possum. Reprod Fertil Dev. 1991;3:119–25.

Rodger JC, Mate KE. Marsupial gametes and fertilisation. Today's Life Science. 1993;5:28–33.

Rodger JC, Paris DBBP, Czarny NA, Harris MS, Molinia FC, Taggart DA, et al. Artificial insemination in marsupials. Theriogenology. 2009;71:176–89.

Taggart DA, Leigh CM, Steele VR, Breed WG, Temple-Smith PD, Phelan J. Effect of cooling and cryopreservation on sperm motility and morphology of several species of marsupial. Reprod Fertil Dev. 1996;8:673–9.

Taggart D, Steele VR, Schultz D, Dibben J, Temple-Smith PD. Semen collection and cryopreservation in the southern hairy-nosed wombat *Lasiorhinus latifrons*: implications for conservation of the northern hairy-nosed wombat *Lasiorhinus krefftii*. In: Wells RT, Pridmore PA, editors. Wombats. Chipping Norton: Surrey Beatty and Sons; 1998. p. 180–91.

Temple-Smith PD, Taggart DA. On the male generative organs of the koala (*Phascolarctos cinereus*), an update. In: Lee AK, Handasyde KA, Sanson GD, editors. Biology of the koala. Chipping North, New South Wales: Surrey Beatty and Sons Pty Ltd.; 1990. p. 33–54.

Wildt DE, Bush M, O'Brien SJ, Murray ND, Taylor A, Marshall Graves JA. Semen characteristics in free-living Koalas (*Phascolarctos cinereus*). J Reprod Fertil. 1991;92:99–107.

Wildt DE, Ellis D, Jannsen D, Buff J. Toward more effective reproductive science for conservation. In: Holt WV, Pickard AR, Rodger JC, Wildt DE, editors. Reproductive science and integrated conservation. Cambridge: Cambridge University Press; 2003. p. 2–20.

Zee YP, Holt WV, Allen CA, Nicolson V, Burridge M, Lisle A, et al. Effects of cryopreservation on mitochondrial function and heterogeneity, lipid raft stability and phosphatidylserine translocation in koala (*Phascolarctos cinereus*) spermatozoa. Reprod Fertil Dev. 2007;19:850–60.

Zee Y, Holt WV, Gosalvez J, Allen CD, Nicolson V, Pyne M, et al. Dimethylacetamide (DMA) can be used as an alternative to glycerol for the successful cryopreservation of koala (*Phascolarctos cinereus*) spermatozoa. Reprod Fertil Dev. 2008;20:724–33.

Zee Y, Lopez-Fernandez C, Arroyo F, Johnston SD, Holt WV, Gosalvez J. Evidence that single-stranded DNA breaks are a normal feature of koala sperm chromatin, while double-stranded DNA breaks are indicative of DNA damage. Reproduction. 2009a;38:267–78.

Zee Y, Holt WV, Nicolson V, Pyne M, Johnston SD. Individual variability in post-thaw sperm survival in a captive koala population. Cryobiology. 2009b;59:69–74.

Chapter 10
Reproduction and Advances in Reproductive Studies in Carnivores

Katarina Jewgenow and Nucharin Songsasen

Abstract Reproductive mechanisms are extraordinarily diverse among species, even within the same phylogenetic clade. Due to this, it has been difficult to directly apply reproductive technologies developed in human and livestock to genetically manage *ex situ* wildlife, including carnivores. To date, more common, closely related species, e.g., domestic cats, dogs and ferrets have served as valuable models for developing reproductive technologies for managing rare, endangered carnivores. Artificial insemination and sperm cryopreservation have already been successfully used to manage *ex situ* populations in some carnivore species, such as the black-footed ferret, cheetah and giant panda. However, technologies aiming at preserving genetics of valuable females have not been fully developed in carnivores, due to the lack of fundamental knowledge about reproductive anatomy and physiology, gamete development, embryogenesis and cryopreservation. The present chapter is divided into two parts. The first part focuses on current knowledge about carnivore reproduction, with emphasis on species diversity in reproductive mechanisms. The second part highlights the progress in reproductive science and related technologies made during the last decade. In addition, we provide examples of how reproductive technologies can contribute to carnivore management and conservation. Although carnivores are comprised of 19 families, we will only focus our attention on four taxonomic groups, including felids, canids, ursids and mustelids.

Keywords Carnivores • Reproductive mechanisms • Genome resource bank • Ovarian tissue cryopreservation • Non-invasive endocrine monitoring

---

K. Jewgenow
Department for Reproduction Biology, Leibniz-Institute for Zoo and Wildlife Research, Alfred-Kowalke-Str 17, Berlin, Germany

N. Songsasen, D.V.M., Ph.D. (✉)
Center for Species Survival, Smithsonian Conservation Biology Institute, National Zoological Park, Front Royal, VA 22630, USA
e-mail: songsasenn@si.edu

W.V. Holt et al. (eds.), *Reproductive Sciences in Animal Conservation*, Advances in Experimental Medicine and Biology 753, DOI 10.1007/978-1-4939-0820-2_10,

1 Introduction

To-date, global biodiversity has been increasingly threatened due to continued habitat loss and degradation, overexploitation, pollution, invasive species, human overpopulation and climate change (www.iucn.org). For carnivores, the magnitude of threats associated with human pressures increases exponentially, as these species require large space and are highly vulnerable to environmental change and human persecution (Comizzoli et al. 2009). Historically, approaches for preserving biodiversity have focused on saving habitat and protecting species living in their natural environments. However, the extent of the species crisis now means that all options, including intensive management in zoos and breeding centers, should be considered (Comizzoli et al. 2010; Wildt et al. 2010). The establishment of a ‘hedge’ collection of wildlife *ex situ* is largely for three purposes (Holt et al. 2003). First, *ex situ* animals are a ‘genetic repository’ for retaining all existing heterozygosity, especially important in the case of an unexpected catastrophic event impacting wild counterparts. Secondly, these animals serve as an invaluable resource for systematic and controlled basic and applied research studies, including collection of bio-information difficult (if not impossible) to collect from free-ranging individuals. Lastly, animals living in *ex situ* collections have proven useful as ambassadors for raising public awareness and significant funding to address the plight of their wild counterparts. Nevertheless, *ex situ* programs are expensive and complex as well as demanding space, human resources and they often require specialized reproductive knowledge and technologies to ensure that all existing gene diversity is retained for at least the next century (Wildt et al. 2010).

It is well established that reproductive mechanisms are extraordinarily diverse among species, even within the same taxonomic group (Comizzoli et al. 2010; Wildt et al. 2010). Due to this, it has been difficult to apply reproductive technologies developed in human and livestock directly to the genetic management of *ex situ* wildlife, including carnivores (Comizzoli et al. 2010; Pukazhenthi and Wildt 2004). To-date, only artificial insemination has contributed to genetic management and species recovery programs of carnivore species (Howard and Wildt 2009; Pukazhenthi and Wildt 2004). For other modern reproductive technologies to be applied successfully and reliably to wildlife species, resulting in live offspring, there certainly are needs for species-specific research to generate fundamental knowledge on reproductive anatomy and physiology, gamete development, embryogenesis and cryopreservation.

The present chapter is divided into two parts. The first part focuses on current knowledge of carnivore reproduction with emphasis on the way that species diversity impacts reproductive mechanisms. The second part highlights the progress made during the last decade in reproductive science and technologies. In addition, we provide examples of how reproductive technologies can contribute to carnivore management and conservation programs. Although carnivores are comprised of 19 families, we will only focus our attention on four taxonomic groups, including felids, canids, ursids and mustelids.

2 Current Knowledge on Carnivore Reproductive Biology

2.1 *Felidae*

Within the family *Felidae*, a total of 37 species is currently recognized (Johnson et al. 2006), although recent morphological and molecular data have suggested that the species number may increase (Johnson et al. 2006). Almost 90 % of the cat family is included in the 2013 International Union for Conservation of Nature (IUCN) Red List, and due to ongoing habitat destruction and persecution, close to 50 % of wild felids are listed as vulnerable, endangered or critically-endangered (IUCN 2013). Felids occur in every continent except Australasia and Antarctica, and inhabit a huge range of climate zones (Sunquist and Sunquist 2009). Around three-quarters of cat species live in forested terrain, and they are generally agile climbers.

According to molecular (and fossil) data, the family Felidae diverges into eight lineages, with two New World groups (ocelot and puma lineages) and seven Old World lineages (domestic cat, leopard cat, lynx, caracal, bay cat and panthera lineages) (Johnson et al. 2006). Felids vary greatly in size, from the black-footed cat (*Felis nigripes*), measuring 35–40 cm in length, to the tiger (*Panthera tigris*), which can attain up to 350 cm in length and weigh 300 kg. However, most felids share similar morphology: lithe and flexible bodies with muscular limbs, retractable claws, domed head with a short muzzle, distinct senses and vocalization. Reproductive mechanisms are also quite similar among felid species, although there are some differences in certain characteristics dependent on species size. Sexual dimorphism is limited but ubiquitous, with males being about 5–10 % larger than females (Sunquist and Sunquist 2009).

Small cats normally reach sexual maturity between the ages of 7 and 9 months, whereas lions (*Panthera leo*) and tigers (*Panthera tigris*) reproduce after 3–4 years of age. Females of most species are polyestrous with induced ovulation (Pelican et al. 2006), although some felids, such as lions, clouded leopards (*Neofelis nebulosa*), margay cats (*Leopardus wiedii*) and lynxes, spontaneously ovulate (Brown and Wildt 1997; Goeritz et al. 2009; Moreira et al. 2001; Schramm et al. 1994). However, the latter phenomenon was observed mainly in captivity when females were kept separate from males. Some domestic cats, especially young individuals that are housed in groups have been shown to ovulate spontaneously (Gudermuth et al. 1997). The feline reproductive cycle is characterized by repeated estrus until mating (or ovulation), followed by a pregnant cycle characterized by increased circulating progesterone that remains elevated until parturition (Brown and Wildt 1997). In addition to progesterone, relaxin and prolactin levels increase in pregnant females (Braun et al. 2009; de Haas van Dorsser et al. 2007; Harris et al. 2008; Stewart and Stabenfeldt 1985). Relaxin concentration rises at the middle of gestation and declines shortly prior to parturition, whereas prolactin elevates toward the end of pregnancy and remains high throughout lactation (Tsutsui and Stabenfeldt 1993).

One reproductive peculiarity of felids is an obligatory pseudo-pregnant cycle after an infertile mating or in some rare cases after a spontaneous ovulation. In these females, the corpus luteum forms and produces progesterone that remains elevated for approximately 3 weeks. However, relaxin and prolactin remain at the baseline concentration in pseudo-pregnant females, indicating that these hormones play important roles in placenta development and the maintenance of pregnancy (Tsutsui and Stabenfeldt 1993). A pseudo-pregnant cycle is usually about 45 days, after which the corpus luteum regresses and ceases its function. However, in some felid species, complete luteolysis does not occur even after parturition or weaning. As shown for Eurasian (*Lynx lynx*) (Carnaby et al. 2012), Iberian (*L. pardinus*) and Canada (*L. canadensis*) lynxes (Dehnhard, et al. 2010), progesterone level remains relatively elevated throughout the year and that, in turn, may induce a negative feedback to inactivate folliculogenesis and turn the lynx into a monoestrous species (Goeritz et al. 2009). It has been suggested that this prolonged elevation of progesterone concentration may in part be the species' adaptability to a harsh environment to ensure that only one litter is produced per year (Goeritz et al. 2009).

In general, reproductive seasonality in felids seems to be associated with the animal's size, the length of time it takes the young to mature and the habitat it lives in. For example, the domestic cat may have up to three litters per year and other small cats (*Felis chaus, Felis silvestris* and *Felis serval*; Hayssen et al. 1993) may reproduce twice, while the larger cats normally have only one litter every 2 years. Nevertheless, most wild felids have one litter per year, and express a period of anestrus (no hormone activity) after the annual breeding period, which can last several months. The males of these species also express seasonal reproductive activity, with decreased sperm production and quality during the annual anestrous periods (Goeritz et al. 2006; Jewgenow et al. 2006).

Gestation length of felids varies between 56 and 115 days depending upon the species (Hayssen et al. 1993), and litter size is usually between one and six young. The potential longevity for most cats is probably at least 15 years, and some individuals have lived over 30 years. Nevertheless, most cats die, or are killed, before they reach sexual maturity in the wild (Sunquist and Sunquist 2009).

2.2 Canidae

The canid family consists of 36 species, six of which are listed as 'threatened' or 'endangered' by the IUCN (2013). Canids are diverse in morphometrics and natural distributions. Ranging from the smallest, the fennec fox (*Vulpes zerda*) weighing less than 1 kg, to the largest grey wolf (*Canis lupus*) exceeding 50 kg (Sillero-Zubiri et al. 2004), at least one species of canid lives on every continent except Antarctica. To date, with the exception of the domestic dog (*Canis familiaris*), there is little information about the details of the reproductive physiology of this family. Yet, there is growing information on the intriguing diversity in reproductive mechanisms among species. For example, within South America, social structure ranges from solitary in the maned wolf (*Chrysocyon brachyurus*) (Dietz 1984) to monogamous pairings in the hoary fox (*Pseudalopex vetulus*; Dalponte and Courtenay 2004) to cohesive social units in the bush dog (*Speothos venaticus*; Zuercher et al. 2004).

While the domestic dog exhibits non-seasonal, sporadic monestrus occurring once or twice a year (Concannon 2009), most wild canids are seasonal breeders with onset apparently dependent on species, latitudinal location and/or variety of environment factors (e.g., rainfall; Asa and Valdespino 1998). For examples, canids living in North America, such as the red wolf (*Canis rufus*; Walker et al. 2002), gray wolf (*Canis lupus*; Asa and Valdespino 1998), coyote (*Canis latran*; Green et al. 1984; Minter and DeLiberto 2008), red fox (*Vulpes, Vulpes*; Mondain-Monval et al. 1984) and island fox (*Urocyon littoralis*; Asa et al. 2007b) become reproductively active as day-length begins to increase. However, in South America, the time of breeding season varies among species. For example, the maned wolf breeds as day-length decreases (i.e., fall to early winter; Rodden et al. 2004); the crab-eating fox (*Cerdocyon thous*) becomes reproductively active in winter (e.g., June–July in Southern hemisphere; Souza et al. 2012), while the bush dog shows no reproductive seasonality and can breed year-round (DeMatteo et al. 2006). Wild African canids, such as side-striped jackal (*Canis adutus*; Asikainen et al. 2003) and Ethiopian wolf (*Canis simensis*; Sillero-Zubiri et al. 1998) normally breed in summer to early fall. Yet, only female African wild dogs (*Lycaon pictus*; van der Horst et al. 2009) exhibit strict seasonality in reproductive activity, while males can produce sperm year-round, albeit poor quality during non-breeding season (Asa and Valdespino 1998; Johnston et al. 2007; van der Horst et al. 2009). For canids that range across regions or continents, such as the golden jackal (*Canis aureus*; Jhala and Moehlman 2004) and Asiatic wild dogs (*Cuon alpinus*; Durbin et al. 2004), the time of breeding season appears to vary between regions. Finally, some small canid species, such as fennec fox (*Vulpes zurda*), bush dog and crab-eating fox may exhibit variations in the number of reproductive cycles per year between animals living *ex situ* and *in situ*. Specifically, under a controlled environment in captivity, females of these canid species can cycle twice per year compared to once per year for counterparts living in the wild (Valdespino et al. 2002).

Generally, the female canid reproductive cycle is characterized by an extended period of proestrus and then estrus (~1 week each). The estrous period is characterized by an estrogen peak along with rising progesterone concentration, even before ovulation (Concannon 2009; Songsasen et al. 2006; Souza et al. 2012; Valdespino et al. 2002; Van den Berghe et al. 2012; Velloso et al. 1998). Estrus is followed by diestrus (metestrus), a luteal phase averaging 2 months in duration irrespective of pregnancy (Asa and Valdespino 1998; Concannon 2009). Diestrus is succeeded by anestrus, an extended (2–10 month) interval of ovarian quiescence (Concannon 2009). However, recent studies utilizing non-invasive endocrine monitoring demonstrate significant deviations in this pattern in certain wild canids. For example, female maned wolves (Reiter 2012; Songsasen et al. 2006) and island foxes (Asa et al. 2007a, b) only ovulate in the presence of a male, implicating the existence of some form of induced ovulation. The Asian wild dog or dhole exhibits seasonal polyestrus with periods of sexual receptivity occurring every 4–6 weeks (Durbin et al. 2004; Khonme, unpublished data). Pup production among African wild dogs (*Lycaon pictus*) and Ethiopian wolf is highly regulated by pack social structure with the dominant female capable of suppressing reproduction in subordinate counterparts through a combination of increased aggression and higher estrogen and progesterone output (Creel et al. 1997; Van den Berghe et al. 2012; van Kesteren et al. 2012). Nevertheless, subordinate individuals are reproductively fertile. Because of the lack of opportunity to breed, the subordinate females become pseudo-pregnant and this, in turn, enables them to be behaviorally and hormonally receptive as caregivers to pups produced by the dominant female (allosuckling) (Van den Berghe et al. 2012; van Kesteren et al. 2012). For the bush dog, peri-pubertal females require a novel male to be present for the initiation of estrus, and can undergo sequential estrus cycles without the interruption of ovarian quiescence of the anestrous period (Porton et al. 1987). Although the presence of a male significantly shortens inter-estrous interval, it is not required for adult females to enter estrus and ovulate (DeMatteo et al. 2006).

The female gamete of canids is unique compared to that of other carnivore species. The canid ovary, especially in the domestic dog, contains an unusually high proportion (11 %) of polyovular follicles (more than one oocyte/follicle) compared to other species (e.g., 4 % in the domestic cat; Telfer and Gosden 1987). Although polyovular follicles can release multiple oocytes, it is likely that only one gamete is of good quality and capable of undergoing maturation and fertilization (Chastant-Maillard et al. 2011; Reynaud et al. 2009). Domestic dog and arctic fox (and perhaps wild canid) oocytes contain a much higher amount of cytoplasmic lipids than gametes of other mammalian species with the same opacity (i.e., cat and pig; Chastant-Maillard et al. 2011). Canid oocytes ovulate at an immature stage and require up to 72 h to complete nuclear maturation within the oviduct (Chastant-Maillard et al. 2011; Pearson and Enders 1943; Hyttel et al. 1990; Songsasen and Wildt 2007). However, there is circumstantial evidence suggesting that there may be species variation in the time required for nuclear maturation among canid species. Specifically, nuclear maturation may be shorter in silver foxes (*Vulpes vulpes*) because breeding within 24 hours after ovulation results in a high conception rate (Farstad 1998). It has also been shown that there are ultrastructural differences between oocytes from the dog and fox. Specifically, dog preovulatory oocytes are abundant with large concentric strands of smooth endoplasmic reticulum and mitochondrial clouds which are not observed in the fox gamete (Chastant-Maillard et al. 2011; Hyttel et al. 1990).

Canid sperm can survive in the female reproductive tract for up to 7 days (Doak et al. 1967; Tsutsui 1989), whereas the oocyte remains fertile 4–5 days after nuclear maturation (i.e., 6–7 days post ovulation; Tsutsui et al. 2009). Although dog oocytes complete nuclear maturation 48–72 hours within the oviduct, fertilization does not occurs until 83 hours post-ovulation even in the presence of sperm (Reynaud et al. 2005). Unlike other species, fertilization occurs at the distal part of the oviduct in dogs and embryos with two pronuclei can be observed 92 hours after ovulation (Reynaud et al. 2005). The duration during which canid early stage embryos remain within the oviduct varies among species (Farstad 2000; Farstad et al. 1993; Reynaud et al. 2005). Specifically, domestic dog embryos stay in the oviduct until the morula stage before entering the uterine horn 5–6 days after fertilization (Tsutsui 1989). In the blue (or Arctic) fox (*Alopex vulpes*), embryos remain in the oviduct for 6–8 days and enter the uterus at the morula stage (Farstad 1998). However, silver (or red) fox embryos have a faster oviductal transport, as they enter the uterus at the 14–16 cell stage, 4–6 days post mating (Farstad 1998).

Spermatogenesis occurs year round in aseasonal canids. However, spermatogenesis only takes place during the breeding season in canids that are strict seasonal breeders (Asa and Valdespino 1998; Songsasen et al. 2013). In these canids, there is also temporal variation in testosterone concentration (Asikainen et al. 2003; Rudert et al. 2011) with the hormone level beginning to rise prior to the breeding season, with the peak coinciding with maximum sperm production (Minter and DeLiberto 2008; Walker et al. 2002; Weng et al. 2006).

2.3 *Ursidae*

The family Ursidae is comprised of eight extant species in five genera: brown bears (*Ursus arctos*), polar bears (*Ursus maritimus*), American black bears (*Ursus americanus*), Asian black bears (*Ursus thibetanus*), sun bears (*Ursus malayanus*), sloth bears (*Ursus ursinus*), spectacled bears (*Tremarctos ornatus*) and giant pandas (*Ailuropoda melanoleuca*). Six of eight extant bear species are currently at risk of extinction and the remainders face significant risks to their future survival (IUCN 2013; Spady et al. 2007). Members of this family are found in a variety of habitats ranging from the Arctic coasts to tropical jungles. The brown bear is the most widespread of all ursids. Because this species occurs in different geographical locations, the brown bear can be further divided into three separate subspecies: the North American grizzly bear (*U. arctos horribilis*), the Kodiak bear (*U. arctos middendorffi*) from Kodiak Island in Alaska and the Eurasian brown bear (*U. arctos arctos*). The giant panda has recently been shown to represent an early divergence from the bear family (Agnarsson et al. 2010; Mayr 1986; Nakagome et al. 2008; Peng et al. 2007). The panda, which lives in central and west China, is unique among the bear species in that it relies on bamboo as the main diet.

The bear family also ranges in size ranging from the smallest, sun bear (male: 65 kg, female: 45 kg (Garshelis 2009) to the largest, polar bear (reaches 3 m in height and 650 kg in weight). Male polar bears accumulate a large amount of body fat and reach a weight of 800 kg (Garshelis 2009). Reproductive biology among bear species is very similar and characterized by the following main traits: reproductive seasonality (Spady et al. 2007), delayed implantation (Renfree and Shaw 2000; Zhang et al. 2009) and the occurrence of pseudo-pregnancy (Goeritz et al. 1997). Bears in temperate regions have a seasonal cycle of reproduction [exception: bears living in zones without seasonal changes in light and/or food supply [for review: (Spady et al. 2007)]. Mating season lasts several months during spring and summer with several receptive phases of 2–3 days [exception: Giant panda expressing mono-estrous ovarian cycle (Goeritz et al. 1997; Goeritz et al. 2001)]. In the males, there are changes in testicular size, androgen production and seminal characteristics with increased reproductive activities during reproductive seasonality (Aitken-Palmer et al. 2012; Knauf et al. 2003; Spady et al. 2007). However, male spectacled bears produce sperm year-round, although births mainly occur between December and March (Spady et al. 2007).

Following fertilization, bears exhibit another peculiar evolutionary adaptation known as delayed implantation or diapause. Bear embryos develop to the blastocyst stage after entering a state of suspended animation and then 'float' around in the female reproductive tract for varying intervals (Renfree and Shaw 2000). During this state, the blastocyst grows at a very slow rate (Renfree and Shaw 2000). All ursids, except sun bears, are believed to experience delayed implantation (Zhang et al. 2009). Because the length of diapause mainly influences pregnancy length, gestation period varies greatly even within the same females. To date, the mechanisms regulating embryonic diapause are poorly understood. It has been demonstrated in the Japanese black bear (*U. thibetanus japonicas*) that a rise of circulating prolactin is closely associated with increased progesterone concentration in both pregnant and pseudo-pregnant females, but not in individuals that have not bred (Sato et al. 2001). The authors suggested that prolactin may play roles in reactivating dormant corpora lutea, a prerequisite process preceding implantation (Sato et al. 2001).

The phenomenon of pseudo-pregnancy is characterized by the prolonged elevation of progesterone with profiles that are indistinguishable from the pregnant conspecific (Goeritz et al. 1997; Kersey et al. 2010; Onuma et al. 2001). This presents as a significant challenge in determining pregnancy in ursids (via hormone metabolites in fecal or urine samples), especially during the pre-implantation period (Kersey et al. 2010).

2.4 Mustelidae

The family Mustelidae is the largest and most diverse carnivore family, comprised of 59 species. Recent classification recognizes up to eight subfamilies: Mustelinae, Galictinae, Helictidinae, Martinae, Melinae, Lutrinae, Mellovorinae and Taxidiinae (Yu et al. 2011; IUCN 2013). To-date, there is limited information on the reproductive biology of mustelids with knowledge on reproductive characteristics available in only seven of the 59 species. These include the sea otter (*Enhydra lutris;* Da Silva and Larson 2005; Jameson 1993; Sinha et al. 1966), North American river otter (*Lontra Canadensis;* Bateman et al. 2009; Reed-Smith 2008), giant otter (*Pteronura brasiliensis*; Carter and Rosas 1997; Lariviere 1999; Londono and Munoz 2006), wolverine (*Gulo gulo*; Mead et al. 1993; Persson et al. 2006), Siberian polecat (*Mustela eversmanii;* Mead et al. 1990; Williams et al. 1992), long-tailed weasel (*Mustela frenata;* Nowak 1999) and black-footed ferret (*Mustela nigripes;* Williams et al. 1991). Despite the major knowledge gap, studies, conducted to-date, have revealed that reproductive characteristics of mustelids vary greatly between species. Most mustelids reach sexual maturity at 12 months of age. However, female short-tailed weasels (or Stoat; *Mustela erminea*), long-tailed weasels, African striped weasels (*Poecilogale alginucha*), American badgers (*Taxidea taxus*), least weasel (*Mustela nivlis*) and European marbled polecats (*Vormela pergusna*) can breed at an early age (1–3 months) and give birth in the following year due to delayed implantation (Amstislavsky and Ternovskaya 2000; Nowak 1999).

However, males of these species do not reach sexual maturity until they are at least 1 year old (Amstislavsky and Ternovskaya 2000). Large sized mustelids, including otters, marten and wolverines do not breed until 2 years of age or older (Amstislavsky and Ternovskaya 2000).

There are also variations in reproductive strategies among species, even in those within the same subfamily. Specifically, Asian small-clawed otter (Bateman et al. 2009), sea otter (Jameson 1993; Sinha et al. 1966) and giant otter (Londono and Munoz 2006) are non-seasonal, polyestrus, whereas North American river otter (*Lontra Canadensis*) are seasonal, monestrus (Bateman et al. 2009). The majority of the subfamily Mustelidae are seasonal breeders with the time and duration of breeding season depending on species and location. Long-tailed weasel (Nowak 1999), black-footed ferret (Williams et al. 1991) and wolverine (Persson et al. 2006) breed once per year (monestrus), whereas females of other species, including the Siberian polecat (Mead et al. 1990) cycle multiple times during a breeding season. For the Bornean ferret badger (*Melogale everetti*), only males exhibit reproductive seasonality where spermatogenesis occurs in January through August (Edmison 2003). It has been shown that there are seasonal changes in testicular size in the Siberian polecat (Mead et al. 1990) and North American river otter (Bateman et al. 2009). Furthermore, seasonal variations in testosterone concentration and sperm production are also observed in the North American river otter (Bateman et al. 2009). The timing of seasonal increase in testosterone concentration is coincident with increasing day-length, and the onset of peak testosterone level is dependent on the latitude of animal location; as latitude increases, hormone peaks appear to occur later in the calendar year (Bateman et al. 2009). In this species, peak sperm production is observed in spring (Bateman et al. 2009).

Ovulation in most mustelids species, such as lesser grisón (*Galictis cuja*), wolverine, sea otter, Siberian polecat, long-tailed weasel and black-footed ferret is induced by copulation that, in turn, stimulates the release of luteinizing hormone from the anterior pituitary (Amstislavsky and Ternovskaya 2000). However, some species, such as Asian small-clawed otters, North American river otters and giant otters ovulate spontaneously (Amstislavsky and Ternovskaya 2000). It has been shown that oocytes can be recovered 72–96 hours post coitum in Stoat and 33–72 hours in American mink (*Neovison vison*; Amstislavsky and Ternovskaya 2000).

A recent study has shown that three distinct chromatin configurations can be identified in ferret oocytes, including fibrillar chromatin (FC), intermediate condensed chromatin (ICC) and condensed chromatin (CC; Sun et al. 2009). The degree of chromatin condensation has been found to be associated with the degree of interaction between the oocyte and surrounding cumulus cells, oocyte diameter and meiotic competence (Sun et al. 2009). Specifically, oocytes with condensed chromatin are often surrounded by compact layers of cumulus cells. Furthermore, the diameter of CC oocyte is significantly larger than that of their FC and ICC counterparts, and the former gametes complete nuclear maturation to a higher percentage (Sun et al. 2009).

Three types of pregnancy have been described in mustelids (Amstislavsky and Ternovskaya 2000). Specifically, the black-footed ferret, Siberian pole cat and least weasel have short gestation (4–6 weeks) and do not exhibit delayed implantation

(Amstislavsky and Ternovskaya 2000; Nowak 1999; Williams et al. 1991). The gestation of the American mink is relatively short and variable, depending on the date of mating; a brief period of delayed implantation occurs only if the females are mated early in the season (pregnancy with facultative diapause; Amstislavsky and Ternovskaya 2000). However, it is worth noting that there are differences in reproductive physiology between the two species, American mink (*Neovison vison*) and European mink (*Mustela lutreola*), as embryonic diapauses does not occur in the latter. The third type of pregnancy involves obligatory delayed implantation (7–10 months) which has been characterized in the majority of mustelid species (Amstislavsky and Ternovskaya 2000). In these species, the blastocyst gradually increases in size through the accumulation of fluid within the blastocoel and through blastomere proliferation, with few morphological changes (Amstislavsky and Ternovskaya 2000). During this period, progesterone remains at a baseline level (Bateman et al. 2009; Dalerum et al. 2005). Shortly prior to implantation, blastocysts rapidly increase in size and undergo differentiation. Implantation is accompanied by a significant rise in progesterone concentration which remains elevated until shortly before parturition (Bateman et al. 2009; Dalerum et al. 2005).

3 Recent Advances in Reproductive Studies

3.1 Non-invasive Endocrine Monitoring

The advance in understanding reproductive biology, as well as physiological responses of wild carnivores to environmental changes, has been made possible by noninvasive assessment of hormonal metabolites in urine or feces (Brown and Wildt 1997; Brown et al. 2001). The non-invasive endocrine monitoring technology was initially pioneered in *ex situ* wildlife population with the goal of better understanding species-specific reproductive biology for enhancing captive management (Monfort 2003; Pickard 2003). This technology has been later adapted to free-ranging individuals for studying reproductive seasonality, gonadal and adrenal status, pregnancy rate and age-specific fecundity as well as the endocrine mechanisms controlling reproductive fitness in social mammals (Creel et al. 1993, 1997; Spercoski et al. 2012). The usefulness of this technology can be amplified when combined with the utilization of "scat detection dogs" and molecular technologies. Specifically, the scat detection dogs increase the likelihood of locating feces in the field, and the development of DNA technologies allows the genetic identification of individual samples collected opportunistically. Thus, from a single fecal sample, it is now possible to identify not only the species, but the individual and its sex, population of origin, reproductive and social statuses by combining molecular and endocrine approaches (Schwartz and Monfort 2008).

In *felids*, noninvasive fecal hormone monitoring was initially developed in the domestic cat, and later validated in many non-domestic felid species. Currently, patterns of ovarian steroid hormones have been reported for more than one half of the

37 felid species (Brown 2006). The utilization of this technology has led to the discovery of a high degree of variability in estrous cycle characteristics among species (Brown 2006). For example, there is species specificity in estrogen secretion pattern during gestation; estrogen increases after mid-gestation in the domestic cat, cheetah, Pallas's cat and fishing cat, while remaining constant in the clouded leopard and tiger (Brown 2006). With some exceptions (Fanson et al. 2010b; Goeritz et al. 2009; Pelican et al. 2006), fecal estrogen and progestagen metabolites reflect female reproductive cyclicity and pregnancy status, whereas androgen metabolites are suitable for monitoring testicular activity in males (Fanson et al. 2010a; Jewgenow et al. 2006; Morais et al. 2002). However, pregnant and pseudo-pregnant cycles cannot be differentiated based on progestagen metabolites alone unless long term sampling over the entire pregnancy period is performed. Pregnant felids exhibit elevated progestagen past the normal pseudo-pregnant luteal phase (i.e., 2/3 of gestation) (Brown 2006). In addition to steroids, the placenta-related hormone relaxin has been indicated as a useful marker of pregnancy in felids. The presence of relaxin in serum or urine provides a clear indication of an ongoing pregnancy in several felid species (Braun et al. 2009; de Haas van Dorsser et al. 2006, 2007). Recently, elevation of 13,14-dihydro-15-keto-prostaglandin F2α (PGFM), a metabolite of prostaglandin F2α, was shown to be specific for the last trimester of pregnancy in several felids (Dehnhard et al. 2012). This hormone metabolite can be detected in urine and feces (Finkenwirth et al. 2010), thus allowing this technology to be applied to free-ranging felids. However, in other carnivores, PGFM is detected in feces only during the peri-partum period, and therefore limits its application for pregnancy diagnosis in canids, ursids and mustelids (Dehnhard, personal communication).

Non-invasive hormone monitoring has also been used to study reproductive mechanisms of various *canid* species, including the red wolf (Walker et al. 2002), maned wolf (Songsasen et al. 2006; Velloso et al. 1998), Ethiopian wolf (van Kesteren et al. 2011, 2012), African wild dog (Creel et al. 1997), bush dog (DeMatteo et al. 2006), fennec fox (Valdespino et al. 2002), crab-eating fox (Souza et al. 2012), island fox (Asa et al. 2007b) and arctic fox (Sanson et al. 2005). Similar to felids, the use of non-invasive hormone monitoring has revealed several species-specific reproductive mechanisms in canids. Specifically, it has been demonstrated that the island fox (Asa et al. 2007b) and maned wolf (Reiter 2012; Songsasen et al. 2006) are unique compared to other canids in that ovulation only occurs in the presence of a male. Non-invasive hormone monitoring has also been used to study the responses of female maned wolves to an estrus induction protocol using a gonadotropin releasing hormone (GnRH) agonist, Ovuplant® (Johnson 2012). Although it has been shown in the domestic dog (Gudermuth et al. 1998), maned wolf (Songsasen et al. 2006) and red wolf (Walker et al. 2002) that fecal progestagen metabolites in pregnant females are higher than those of non-pregnant individuals, the use of this hormone assay for pregnancy diagnosis has been very limited. This is due to individual variations in the excretion of hormone metabolites and the need for longitudinal endocrine monitoring throughout the luteal period. While the use of commercial canine relaxin kits has proven to be reliable for detecting circulating hormone concentration for pregnancy diagnosis in wild canids (Bauman et al. 2008), the use of urinary relaxin for this purpose is rather limited (Steinetz et al. 2009).

Table 10.1 Non-invasive endocrine monitoring assays developed in bear species

Species	Hormone assays	Reproductive stages	Samples	Citations
Giant panda	Estrogens Estrone sulfate	Estrus	Urine	Dehnhard et al. (2006), Durrant et al. (2006),
	Estrone-3-glucorinade Pregnandiol	Luteal phase		Hodges et al. (1984), Hama et al. (2008), Dehnhard et al. (2006)
	Progesterone	Luteal phase	Feces	Kersey et al. (2010)
Spectacled bear	Progesterone	Luteal phase	Urine, Feces	Dehnhard et al. (2006)
Asian black bear	Estrogens	Estrus	Feces	Chang et al. (2011)
	Progesterone	Luteal phase		
Sun bear	Epi-androsterone	Estrus	Feces	Schwarzenberger et al. (2004)
	Pregnanediol	Luteal phase		
Brown bear	Estrogens	Estrus	Feces	Dehnhard et al. (2006), Ishikawa et al. (2003)
	Progesterone	Luteal phase		Ishikawa et al. (2003), Goeritz et al. (2001)

In *bears*, interspecies comparison of fecal endocrine data from several species demonstrates the importance of testing several steroid assays for reproductive investigations due to species-specific differences in gonadal hormone metabolites excreted in urine and feces (Table 10.1). As for other carnivores, the differentiation between pregnant and pseudo-pregnant cycles is especially important for breeding management in captivity, but urinary and fecal steroids are unable to discriminate between non-pregnant, pregnant or pseudo-pregnant states (Kersey et al. 2010). Recently, a potential marker of pregnancy, the acute phase protein ceruloplasmin, has been described (Willis et al. 2011). In the giant panda, the levels of active urinary ceruloplasmin increase during the first week of pregnancy and remain elevated until 20–24 days prior to parturition, while no increase was observed in pseudo-pregnant females (Willis et al. 2011).

Reproductive hormone metabolites have been determined to characterize reproductive cyclicity and seasonality as well as understand causes of reproductive failure in some *mustelid* species, including the North American river otter (Bateman et al. 2009), Asian small-clawed otter (Bateman et al. 2009), sea otter (Larson et al. 2003), black-footed ferret (Young et al. 2001) and wolverine (Dalerum et al. 2005). Specifically, analyses of gonadal hormone metabolites have revealed species differences in reproductive seasonality between the Asian small-clawed otter and North American river otter, with the latter being a seasonal breeder and females entering estrus once per year (Bateman et al. 2009). Furthermore, it has been demonstrated that the solitary North American river otter is an induced ovulator, whereas the social Asian small-clawed otter appears to be primarily a spontaneous ovulator (Bateman et al. 2009). A study in female wolverines has shown that reproductive failure may be related to low social rank and likely to be due to implantation failure independent of elevated glucocorticoid metabolites (Dalerum et al. 2005).

3.2 Semen Collection and Artificial Insemination

Semen collection in wild felids is usually performed by electroejaculation (Howard and Wildt 2009). Recently, a field-friendly technology utilizing urethral catheterization has been developed in the domestic cat (Zambelli and Cunto 2006; Zambelli et al. 2008). When compared with electroejaculation, urethral catheterization yields smaller seminal volume (67.1 ± 25.9 μl vs. 10.5 ± 5.3 μl) with high sperm cell concentration ($542.9 \pm 577.9 \times 10^6$/ml vs. $1{,}868.4 \pm 999.8 \times 10^6$/ml; Zambelli et al. 2008). Furthermore, cryopreserved sperm obtained by urethral catheterization are able to fertilize conspecific oocytes at a similar percentage to electroejaculated samples (Zambelli et al. 2008). This newly established technique has been used in the African lion following medetomidine administration to yield high quality semen samples (volume [mean ± SD]: 422.9 ± 296.1 μl; motility [mean ± SD]: 88.8 ± 13.2 %; sperm concentration of 1.9×10^9/ml) (Lueders et al. 2012).

Recovery of sperm cells from the epididymis *post castrationem* or *post mortem* has also been reported in wild carnivores (Anel et al. 2011; Jewgenow et al. 1997). This technology allows the rescue of gametes from genetically under-represented animals that undergo castration for medical reasons or die unexpectedly (Jewgenow et al. 1997; Johnston et al. 1991). In the domestic cat, seminal characteristics and fertilizing capacity of epididymal samples are comparable to those of ejaculated samples (Filliers et al. 2010). Due to the rarity of conspecific gametes, fertility of epididymal sperm obtained from wild felids has only been examined *in vitro* using heterologous fertilization with domestic cat oocytes (Ganan et al. 2009; Jewgenow et al. 2011). It is worth mentioned that heterologous *in vitro* fertilization using domestic cat oocytes has also been used to assess the fertility of sperm obtained from other carnivore species including the giant panda (Spindler et al. 2006) and to differentiate binding/fertilizing capacity between individual males and species (Baudi et al. 2008; Niu et al. 2006). To-date, there are no reports on collection of epididymal sperm from wild canids (Silva et al. 2004; Thomassen and Farstad 2009). However, a study in the domestic dog has reported birth of puppies born from a female inseminated with fresh epididymal sperm from a benign prostatic hyperplasia dog (Klinc et al. 2005). Furthermore, it has been shown that *in vivo* fertility of cryopreserved dog epididymal spermatozoa is lower than that of ejaculated samples, although no differences in sample quality post-thaw are observed (Thomassen and Farstad 2009). Recently, it has been reported that epididymal sperm from a Cantabrian brown bear (*Ursus arctos*) cryopreserved in a 430 mOsm Test-Tris Fructose + 4 % glycerol + 15 % egg yolk exhibited close to 70 % motility after freezing and thawing (Anel et al. 2011).

Artificial insemination, especially in combination with sperm cryopreservation, is a valuable tool for managing threatened wildlife populations because this strategy eliminates transportation of animals from different locations for breeding, and allows re-infusion and dissemination of valuable genes even after death of the sperm donor. Despite these great potentials, the practical application of AI to the conservation of carnivores is still quite limited (Swanson 2006). Successful AI has been

Table 10.2 List of canids, ursids and mustelids that offspring have been produced by artificial insemination

Species	Gonadotropin	Sperm deposition	Sperm type	No pregnancies	References
Canids					
Gray wolf	GNRH agonist	TVI/TCI	Fresh	1/3 (33.3 %)	Asa et al. (2006)
Gray wolf			Frozen	1	Seager et al. (1975)
Mexican gray wolf	N/A	IUI	Fresh	3/3	Thomassen and Farstad (2009)
Red wolf	natural	LUI	Fresh	1 L produced	Goodrowe et al. (1998)
Red fox	N/A	N/A	Fresh/Frozen	80 % (fresh sperm)	Farstad (1998)
Blue fox	N/A	N/A	Fresh/Frozen	80 %	Farstad (1998, 1992)
Ursids					
Giant panda	Natural	TVI	Frozen	5/14	Huang et al. (2012)
Giant panda	Natural		Fresh		Masui et al. (1989)
Mustelids					
European ferret	Natural estrus/hCG	LUI	Fresh	17/24 (70.8 %)	Wildt et al. (1989)
European ferret	Natural estrus/hCG	LUI	Frozen	7/10 (70.0 %)	Howard et al. (1991)
Black-footed ferret	Natural	LUI	Fresh/Frozen		Howard and Wildt (2009)

LUI laparoscopic uterine insemination, *TV* transvaginal insemination, *TC* transcervical insemination, *TU* intrauterine insemination

reported in only a few wild carnivore species to-date using both surgical and non-surgical techniques, with some progress during the past 10 years. Summary of felid species in which live offspring have been produced by AI (fresh or frozen semen) has recently been reported elsewhere (Howard and Wildt 2009). Table 10.2 summarizes successful AI in canids, ursids and mustelids.

Most recently, laparoscopic oviductal AI has been developed in felids (Swanson 2012). The utilization of this technique overcomes challenges associated with the complexity of the female reproductive tract and poor seminal quality observed in many felid species. To-date, offspring have been produced after laparoscopic oviductal AI in the ocelot, Pallas' cat and domestic cat (Swanson 2012). Furthermore, an ultrasound guided trans-cervical insemination in which sperm are non-surgically deposited deep into the cervix has also been developed and successfully applied in the lion, cheetah and Amur leopard (Goeritz et al. unpublished data; Goeritz et al. 2012).

For *canids*, with the exception of the domestic dog and the farmed fox species, semen collection is normally performed by electroejaculation (Asa et al. 2007a, b; Goodrowe et al. 1998; Johnston et al. 2007; Minter and DeLiberto 2008; Songsasen et al. 2013). A major challenge in semen collection in canids is contamination of urine in semen samples obtained from electroejaculation (Platz et al. 2001). The presence of urine alters osmolarity and pH of the seminal sample and increases the proportions of sperm with bent and coiled tail as well as decreasing motility (Platz et al. 2001) and may affect the susceptibility of sperm to osmotic stress during

Table 10.3 Seminal traits of the wild canids

Species	Volume (ml)	Concentration ($\times 10^6$ sperm/ml)	Motility (%)	Morphologically normal sperm (%)	Citations
Gray wolf[a]	1.7±0.2	290.8±53.5	91.7±1.5	N/A	Mitsuzuka (1987)
Coyote	1.7±0.4	549.2±297.7	90.4±4.5	78.0±13.5	Minter and DeLiberto (2008)
Red wolf	4.7±0.7	146.5±25.7	71.2	80 %	Goodrowe et al. (1998)
African wild dog	0.6±0.1	212.3±87.3	69.5±3.3	N/A	Johnston et al. (2007)
Maned wolf	2.0±0.6	29.5±9.3	65.0±6.1	50.1±8.1	Comizzoli et al. (2009), Johnson (2012), Songsasen et al. (2013)
Maned wolf[a]	1.3±0.14	56.8±7.8	76.1±2.8	36.5±3.4	Teodoro et al. (2012)
Blue fox	0.4±0.3	491.8±372.1	N/A	89.9±4.4	Stasiak et al. (2008)

[a]Samples were collected by digital stimulation

the cryopreservation process (Johnson 2012). A recent study has demonstrated that maned wolf ejaculates can be obtained using digital stimulation (Teodoro et al. 2012) with seminal characteristics similar to those obtained by electroejaculation (Johnson 2012; Songsasen et al. 2013).

Seminal characteristics vary among wild canid species (Table 10.3), with the maned wolf having the lowest seminal quality (low sperm output and high percentage of structurally abnormal cells (Comizzoli et al. 2009; Johnson 2012; Teodoro et al. et al. 2012). Ejaculate traits of free ranging maned wolves have also been investigated, showing similar comparable seminal traits to those of captive individuals with a high proportion of structurally abnormal spermatozoa (Songsasen et al. 2013). The high proportions of abnormal spermatozoa observed in both captive and wild wolves may be linked to low genetic diversity (nucleotide diversity [π] = 0.0013) as the result of genetic bottle neck during or at the end of the last glacial maximum (Prates Junior 2008). Low gene diversity has been shown to be associated with poor seminal quality in other carnivore species, including the Florida panther (*Felis concolor coryi*; Facemire et al. 1995), cheetah (Fitzpatrick and Evans 2009), black-footed ferret (Fitzpatrick and Evans 2009) and gray wolf (Asa et al. 2007a). Unlike wild felids, artificial insemination has not been widely used in wild canids, due to the lack of knowledge on species' reproductive biology and the challenges in predicting ovulation onset and effectively manipulating the female reproductive cycle (Thomassen and Farstad 2009).

With the exception of the giant panda, little progress has been made in establishing AI in *ursids*, although this technology will greatly benefit captive management of Malaysian sun bear and spectacled bear populations. Successful pregnancies of the giant panda following AI was achieved in the 1980s (Masui et al. 1989; Moore et al. 1984). Because the conception rate after AI with fresh semen is similar to that after natural mating, this technology is now routinely used in the captive management of this species (Hori et al. 2006; Huang et al. 2012). Furthermore, it has been shown that there are no differences in the ability of fresh and cryopreserved giant panda sperm to undergo decondensation after exposure to cat oocyte ooplasm (Spindler et al. 2006). To-date, live offspring have been produced after AI with cooled and frozen-thawed sperm (Pukazhenthi and Wildt 2004). Due to similarities in morphology of the reproductive organs, it may be possible that instruments used for giant panda AI can be adapted to other bear species (Knauf 2006). Semen collection, seminal analysis and sperm cryopreservation procedures have been reported in other bear species (Chen et al. 2007; de Paz et al. 2012; Ishikawa et al. 2002).

Semen samples have been collected by electroejaculation in some *mustelids*, including North American river otter (Bateman et al. 2009), black-footed ferret, Siberian polecat and domestic ferret (Howard et al. 1991; van der Horst et al. 2009). The timing of the breeding season has been shown to influence seminal quality in ferrets (van der Horst et al. 2009). Specifically, there is a higher percentage of structurally abnormal sperm in samples collected at the beginning of a breeding season than those collected during the peak of the reproductive season. Although low seminal output is observed in the black-footed ferret compared with the domestic ferret and Siberian polecat, fresh sperm from the three species exhibit similar percentages of normal morphology, sperm motion characteristics and responses to cryopreservation (van der Horst et al. 2009), indicating that the two common ferret species are suitable models for developing reproductive technologies for their endangered counterpart. To-date, offspring have been produced from artificial inseminations with fresh and frozen semen in two mustelid species, including the black-footed ferret (Table 10.2; Howard and Wildt 2009; Howard et al. 1991).

3.3 In Vitro *Oocyte Maturation/Fertilization and Embryo Transfer*

The abundant supply of domestic cat ovaries from veterinary and spay clinics have made this species a practical and valuable model for reproductive studies in wild counterparts (Pope et al. 2006; Comizzoli et al. 2010). Extrapolation of reproductive technologies developed in the domestic cat has yielded encouraging outcomes in several wild *felids*. Currently, *in vitro* fertilization of ovarian oocytes (*in vivo* or *in vitro* matured) obtained from live animals (ovum pick-up or ovariohysterectomy) or after death of a female has been reported in several wild felid species. These include the leopard cat (*Felis bengalensis*; Goodrowe et al. 1989), tiger (Donoghue et al. 1990), Siberian tigers (*P. tigris altaica*; Crichton et al. 2003), cheetah (Crosier et al. 2011; Donoghue et al. 1992), Indian desert cat (*Felis silvestris ornate*; Pope et al. 1993), jungle cat (*Felis chaus*; Pope et al. 1993), black-footed cat (*Felis nigripes*; Pope et al. 1993), fishing cat (*Felis viverrinus*; Pope et al. 1993), lion (Armstrong et al. 2004), African wild cat (Pope et al. 2006), ocelot (Swanson 2006), sand cat (Swanson 2012) and caracal (Pope et al. 2006). Among these, live offspring have been produced after transferring fresh or frozen embryos in the Siberian tiger (Donoghue et al. 1993), Indian desert cat (Pope et al. 1993), African wildcat, Caracal

(Pope et al. 2006), ocelot (Swanson 2006), sand cat (Swanson 2006, 2012) and black-footed cat (Pope et al. 2012a). Intracytoplasmic sperm injection (ICSI) has also been attempted in wild felids, and high cleavage rates were reported (lion: 60 %, fishing cat: 70 %), but transfer of pre-implantation stage embryos did not result in a pregnancy (Pope et al. 2006). Although live cubs have been produced in numbers of wild felids, *in vitro* embryo production has not been widely used into captive breeding program mainly due to several reasons. First, there is limited information on species-specific reproductive endocrinology, gamete biology and embryogenesis. The second is the complexity of the procedure and the need for specialized equipment and facilities to recover oocyte, perform *in vitro* fertilization and culture embryos. The third reason is limited availability of developmentally competent oocytes, especially of poor health/aging females. Because ovaries are required to be transported to a laboratory, it is essential to optimize storage condition that allows the maintenance of gamete quality for at least 24 h. Unlike the male, where ICSI can be applied to overcome poor quality samples (low motility; Ringleb et al. 2010), there is no alternative method to circumvent this challenge for the female gamete.

In the domestic dog, *in vitro* maturation and fertilization has been far from being successful, due to the unique reproductive (prolonged anestrus) and oocyte (protracted oocyte maturation within the oviduct) biology (Songsasen and Wildt 2007). Therefore, no or little progress has been made regarding *in vitro* production of embryos in wild *canids*. The first report on *in vitro* fertilization of *in vitro* matured oocytes was in the domestic dog by Mahi and Yanagimachi (1976), although embryonic development was not reported. Since then, several studies have been conducted to examine the variety of micro-environmental factors (e.g., hormones, energy substrates and culture media) influencing developmental competence of dog oocytes (Songsasen and Wildt 2007), but still with limited success in oocyte and embryo development outcomes (Luvoni et al. 2005; Songsasen and Wildt 2007). To date, there is only one report demonstrating the production of a single blastocyst from *in vitro* fertilization (Otoi et al. 2000), and one non-full term pregnancy after transferring close to 100 *in vitro* derived presumptive zygotes in the domestic dog (England et al. 2001). *In vitro* maturation and fertilization have also been conducted in silver (red) fox with minimal success (Feng et al. 1994). In that study, only 19 % of oocytes reached the metaphase I/metaphase II after 48 h *in vitro* culture, and sperm penetration was observed in 18 % of cultured oocytes (Feng et al. 1994); no embryonic development was recorded in this study. *In vitro* fertilization of *in vivo* matured oocytes has been reported in the blue fox (Farstad et al. 1993). In that study, 5 of 13 oocytes (38.4 %) obtained 6 days post LH developed into 2-cell stage post-insemination, and one embryo reached the morula stage.

In *bears*, very little efforts have been made to establish protocols for *in vitro* production of embryos, although some sporadic attempts have been performed. Ovarian oocytes recovered from American black bear, sun bear, sloth bear (Johnston et al. 1994) and giant pandas (Zhang et al. 1998) are able to complete nuclear maturation *in vitro*. More recently, it has been shown that 50 % of brown bear oocytes complete nuclear maturation after 48 h *in vitro* culture (Yin et al. 2007). Electrical activation of the *in vitro* cultured oocytes has resulted 31 % of activated gametes developing into 2–4 cell stage (Yin et al. 2007). Recovery of *in vivo* produced embryos during diapause has been reported in Brown bear (Tsubota et al. 1991). Furthermore, Boone et al. (1999) reported the live birth of an American black bear cub after non-surgical embryo collection following by laparoscopic embryo transfer. In that study, the delayed implantation allowed successful development of the embryos despite a substantial asynchrony between the donor and the recipient (Boone et al. 1999).

In vitro maturation of *mustelid* oocytes has been conducted only in the domestic ferret. Supplementing culture media with gonadotropins has been shown to enhance nuclear maturation capacity (Li et al. 2002; 70 % after 24 hours IVM compared with 30 % of no hormone treatment). Oocytes (43 %) matured under this condition were able to develop to the blastocyst stage after *in vitro* activation using cycloheximide and 6-dimethylaminopurine followed by electrical stimulation (Li et al. 2002).

To-date, *in vitro* fertilization has not been reported in mustelids. However, there have been a few reports on *in vitro* culture of *in vivo* produced embryos in the domestic ferret (Li et al. 2001), European polecat (Lindeberg and Jarvinen 2003), stoat (Amstislavsky et al. 2012) and American mink (Moreau et al. 1995). For the ferret and European polecat that do not exhibit delayed implantation, blastocyst development after *in vitro* culture has been observed (Li et al. 2001; Lindeberg and Jarvinen 2003), and the production of live offspring after embryo transfer has been reported in the former (Li et al. 2001). For stoat, a species with obligate delayed implantation, 1 to 4 cells stage embryos are able to develop to the blastocyst stage; however, diapausing embryos fail to develop in culture (Amstislavsky et al. 2012). Co-culture with Buffalo rat liver cell has been shown to enable American mink diapausing embryos to develop into hatch blastocysts (Moreau et al. 1995).

3.4 Preservation of Female Germplasm

3.4.1 Oocyte and Embryo Cryopreservation

Genetic resource banking is defined as the storage of gametes (sperm and oocytes) and embryos with the deliberate intention to use these valuable biomaterials to maintain gene diversity of threatened populations (Holt and Pickard 1999). In combination with reproductive technologies, it allows re-infusion and dissemination of valuable genes independently of time and geographical locations. In addition to gametes and embryos, somatic cells and tissues have been considered as valuable biological resource because of their broad applications in genetics, toxicology and epidemiology (Leon-Quinto et al. 2008). Furthermore, with the rapid progress in somatic cell nuclear transfer technologies (Gomez et al. 2009; Kim et al. 2007), preservation of somatic cell lines may offer future reproductive opportunities for animals whose viable gametes cannot be obtained (Leon-Quinto et al. 2008; Lermen et al. 2009).

Cryopreservation of *feline* oocytes is still considered to be an experimental technique (Luvoni 2006), although births of live kittens produced from vitrified domestic cat oocytes have been reported (Pope et al. 2012b; Tharasanit et al. 2011). Despite this success, cryopreservation of immature and mature cat oocytes is not proven to be a reliable tool; the cleavage rate of frozen-thawed domestic cat oocytes is less than 25 % (Cocchia et al. 2010; Comizzoli et al. 2006; Merlo et al. 2008; Tharasanit et al. 2011). Yet, a recent study has offered a novel alternative through germinal vesicle preservation (Comizzoli et al. 2011). In that study, germinal vesicles recovered from fresh or vitrified oocytes were transferred into the cytoplasm of fresh Grade I gametes; the reconstructed oocytes from both groups were able to complete nuclear maturation (80 %) and developed to blastocyst after IVF at a similar capacity (15 %). To date, there are no reports on oocyte cryopreservation in wild felids.

For *canids*, advances in oocyte cryopreservation have been reported during the past 5 years (Abe et al. 2008; Boutelle et al. 2011; Turatham et al. 2010; Zhou et al. 2009). By using the open-pulled straw technique, the percentage of vitrified-warmed dog oocytes completing nuclear maturation was similar to that of fresh controls, although more cryopreserved gametes were arrested at the GV stage than those of the fresh counterparts (Turatham et al. 2010). Blue fox oocytes vitrified using the two-step open-pulled straw method also developed to the MII at the similar percentage to those of the fresh control (Zhou et al. 2009). Recently, it has been shown that dog and Mexican gray wolf oocytes maintained viability (based on propidium iodide staining) after vitrification using the cryotop technique (Abe et al. 2008; Boutelle et al. 2011).

Successful embryo cryopreservation in canids either by vitrification (Abe et al. 2011) or slow freezing method (Guaitolini et al. 2012) has recently been reported in the domestic dog. It has been shown that there appears to be a stage-dependency in the susceptibility to vitrification (Abe et al. 2011). Specifically, blastocysts are more sensitive to vitrification than those at the earlier stages of development (one-cell to morula stages) (Abe et al. 2011). Transferring of vitrified-warmed dog embryos has resulted in the birth of live offspring (7 pups/77 transferred embryos; 9.1 %; Abe et al. 2011). Dog blastocysts also have been cryopreserved using a slow freezing method (Guaitolini et al. 2012; Kim et al. 2002). Although cryopreserved blastocysts maintain viability *in vitro* for 6 days, transfer of frozen-thawed embryos results in no offspring (Guaitolini et al. 2012; Kim et al. 2002).

To date, there have been no reports on cryopreservation of *mustelid* oocytes. However, successful embryo cryopreservation resulting in live births after embryo transfer has been reported in stoats, European polecats and ferrets (Piltti et al. 2004; review in Amstislavsky et al. 2012).

3.4.2 Ovarian Tissue Cryopreservation

The ability to successfully cryopreserve ovarian tissues has potential importance for the genetic management of rare species, especially carnivores, as the majority of these animals are strictly seasonal breeders (i.e., developmentally competent gametes can be recovered only during reproductively active period; Comizzoli et al. 2012). Viable banks of tissues can be used in combination with *in vitro* culture, allo- or xenografting followed by IVM/IVF to produce developmentally competent embryos (Agca et al. 2009; Cleary et al. 2003; Comizzoli et al. 2012; Fassbender et al. 2007; Jewgenow and Paris 2006; Kim et al. 2009; Newton et al. 1996; Paris et al. 2004). This technology is also useful in cases of unexpected death of pre-pubertal individuals. The challenges in establishing cryopreservation procedures for ovarian tissue are the complexity of gonadal tissue structure, cell heterogeneity and the lack of basic cryobiological and developmental knowledge. Within the ovarian cortex, there are thousands of primordial follicles containing oocytes at their least differentiated stage, which appear to be relatively resistant to vitrification procedure (Comizzoli et al. 2012). Ovarian tissue cryopreservation has been conducted in the domestic cat (Bosch et al. 2004; Comizzoli et al. 2012). It has been demonstrated that more cat follicles within ovarian cortices maintain structural integrity after vitrification in a solution containing 15 % ethylene glycol + 15 % DMSO + 0.5 M sucrose than after being cryopreserved using a standard slow-cooling method (Comizzoli et al. 2012). To-date, ovarian cortices of the black-footed ferret, cheetah and clouded leopard have been vitrified using the method developed in the domestic cat; the vitrified tissues maintained structural integrity, although the developmental potential of cryopreserved follicles has not been examined (Comizzoli et al. 2012). Recently, Wiedemann et al. (2012) have demonstrated follicular development in lions after ovarian tissue cryopreservation using a slow cryopreservation protocol following xenografting into immunodeficient mice. Despite some progress in the cryopreservation of ovarian tissue, it has been demonstrated that isolated small preantral cat follicles are rather susceptible to cryopreservation with only 10 % remaining structurally intact and physiologically active after thawing (Jewgenow et al. 1998).

3.5 Somatic Cell Nuclear Transfer

Since the birth of the sheep Dolly in the late 1990s (Campbell et al. 1996), much progress has been made in establishing somatic cell nuclear transfer (SCNT) technology in a wide-variety of species, including carnivores. This reproductive technology is often suggested to have potential for conservation of endangered wildlife, especially when it aims at recovering genes from under-represented individuals (Holt et al. 2004). To-date, SCNT studies focusing on establishing cell line and cell cycle synchronization have been conducted in both domestic and wild carnivores (de Barros et al. 2010; Koo et al. 2009; Tao et al. 2009; Verma et al. 2012; Wittayarat et al. 2012). Live births also have been reported in the domestic cat (Yin et al. 2005; Yin et al. 2008), dog (Hossein et al. 2009a, b; Jang et al. 2010; Jang et al. 2008) and gray wolf (Kim et al. 2007; Oh et al. 2008). Cats and dogs produced by SCNT have been shown to exhibit normal health and reproductive potential (Choi et al. 2010; Park et al. 2010). However, the practical application of SCNT to wildlife conservation is still limited because of its low success rate of <0.1–6 % of reconstructed embryos developing to live offspring, due to abnormal nuclear reprogramming of the transplanted somatic cells (Holt et al. 2004; Loi et al. 2011).

An additional limiting factor in the implementation of SCNT technology for endangered species is the scarcity of conspecific oocytes and embryo transfer recipients. It has been suggested that interspecific (or intergeneric) SCNT which involves transferring cell of one species into enucleated oocytes of another species, and then establishing pregnancy by interspecific embryo transfer, can circumvent this challenge (Loi et al. 2011). While transferring wild felid embryos into domestic cat recipients has resulted in births of live offspring (Pope et al. 1993, 2012a, b), embryos produced by interspecific SCNT have limited developmental capacity, due to abnormal reprogramming of donor cells (Gomez et al. 2012; Imsoonthornruksa et al. 2012; Lee et al. 2010; Thongphakdee et al. 2010). Epigenetic modifications of donor somatic cells have been shown to enhance *in vitro* development of reconstructed embryos, and regulated the expression of pluripotent genes, but this strategy does not result in full-term pregnancy in the leopard cat (Lee et al. 2010) and black-footed cat (Gomez et al. 2011, 2012). To-date, births of live offspring produced by interspecific SCNT have been reported in the African wild cat (1 % success rate; Gomez et al. 2004) and sand cat (0.9 %; Gomez et al. 2008). Intergeneric SCNT also has been attempted in the giant panda (Chen et al. 2002). Giant panda cell nuclei were transplanted into rabbit oocytes and development of reconstructed embryos up to blastocyst stages was demonstrated (Chen et al. 2002). Furthermore, co-transferring of giant panda-rabbit reconstructed embryos with cat-rabbit cloned counterparts resulted in implantation of the former in the cat uterus (Chen et al. 2002).

Recently, alternative approaches have been explored to overcome the challenge associated with the scarcity of conspecific oocytes for SCNT (or IVM/IVF/ICSI). These include recovery of gametes from ovarian tissues transplanted to immunodeficient hosts (Bosch et al. 2004; Fassbender et al. 2007; Wiedemann et al. 2012) and generating *in vitro*-derived gametes from gonadal stem cells (Gomez et al. 2010; Kashir et al. 2012; Ko et al. 2012). Although both technologies are in a very early stage in carnivores (Gomez et al. 2010), the feasibility of these approaches has already been demonstrated in other mammalian species, including the mouse and human (Hayashi, et al. 2012; Li et al. 2010). Similar approaches have also been investigated for *in vitro* production of male gametes in many mammalian species, including the domestic cat and dog (Kim et al. 2006; Kim et al. 2008; Powell et al. 2012; Travis et al. 2009).

4 Captive Breeding Programs for Conservation, Linking *Ex Situ: In Situ* Conservation in Critically Endangered Carnivores

For many critically endangered carnivore species (or subspecies), e.g. the Iberian lynx, Amur leopard, black-footed ferret, the giant panda, or red wolf, captive breeding programs have been considered as an integral part of the conservation action plans. As mentioned in the Introduction, *ex situ* populations mainly serve as (1) a genetic repository stock for reintroduction to bolster the numbers of the wild counterpart projects (Calzada et al. 2009; Christie 2009; Vargas et al. 2008) and (2) research resources to better understand species' biology. Although reintroduction programs have been launched in many carnivores, including the black-footed ferret, gray wolf, African lion, African wild dog, tiger, snow leopard, leopards, jaguars and cheetah (for review; Hayward and Somers 2009), captive breeding programs that take advantage of advanced reproductive sciences and technologies to genetically manage the population are rarely found.

The Conservation programs for the black-footed ferret (Comizzoli et al. 2009), Iberian lynxes (Vargas et al. 2008, 2009) and giant panda may serve as examples of how reproductive science contributes to the conservation of a species. Because there is a chapter in this book dedicated specifically to the Black-footed ferret, here, we will only focus on how reproductive sciences assist in the conservation program of the Iberian lynx and the giant panda.

Since the inauguration of the Iberian lynx captive breeding program in Spain (ILCBP) in 2002, research has been conducted to advance the understanding of reproductive biology of this species. So far, the reproductive biology of this critically endangered species has been fully characterized (Goeritz, et al. 2006; Goeritz et al. 2009; Jewgenow et al. 2006; Pelican et al. 2006), semen and somatic cells regularly obtained for genome banking (Ganan et al. 2009; Leon-Quinto, et al. 2008), and some important breeding management tools, including pregnancy diagnosis have been developed and implemented (Braun et al. 2009; Finkenwirth et al. 2010). Because of this monumental effort, ILCBP has successfully propagated lynxes *ex-situ* and reintroduced the first captive born individuals to the wild in 2010, only 8 years after the inauguration of the program. Currently, ILCBP is continuing captive breeding efforts to produce healthy individuals for reintroduction programs (Vargas et al. 2008).

The giant panda, a bear that rely mainly on bamboos as a diet is listed as endangered by the IUCN and <2,000 individuals are estimated to remain in the wild (IUCN 2013). Because of the unstable status of the wild giant panda, tremendous efforts have been placed on establishing a self-sustaining captive population this species. Since the beginning of this century, reproductive and health studies have been conducted with the goal of overcoming poor reproduction and health of *ex situ* individuals to eliminate the need for removing wild counterparts (Wildt et al. 2003). Reproductive endocrinology of male and female giant pandas has been extensively characterized (Aitken-Palmer et al. 2012; Kersey et al. 2010). Furthermore, reproductive technologies such as artificial insemination and semen cryopreservation have been successfully established, and are often incorporated into breeding programs (Huang et al. 2012; Wildt et al. 2003). This multidisciplinary effort has resulted in the tremendous success in breeding and maintaining giant panda ex situ; the ex situ population has now increased from 104 in 1996 to >300 adults during the past 15 years (Wildt et al. 2007). During the same period, efforts at protecting species' habitat have also been emphasized, and the number of giant panda reserves have increased from 13 in 1989 to >40 today (Wildt et al. 2007). Due to these successes, there is a plan to re-introduce captive born individuals to increase the numbers of wild population in the near future (Gong et al. 2012).

5 Summary

Advances in reproductive studies have improved our understanding about reproductive mechanisms that in turn allows us to establish tools to genetically manage carnivore populations ex situ (Comizzoli et al. 2009; Howard and Wildt 2009; Pukazhenthi and Wildt 2004; Vargas et al. 2009). While reproductive science has greatly contributed to the management of captive breeding programs of some carnivore species, including the Iberian lynx, black-footed ferret and giant panda, several reproductive technologies (especially for preserving female genetics) are still considered experimental and far from being included to the routine practices (Wildt et al. 2010), mainly due to the lack of basic knowledge on species' reproduction. As of the beginning of the twenty-first century, only 2 % of mammals in the planet are considered 'well studied' (Wildt et al. 2010). Obviously, domestic species (i.e., cat, dog and ferret) have served as valuable models for establishing reproductive technologies in wild carnivores. Yet, there are still needs for species-specific research due to enormous diversity in reproductive biology among species, even within the same phylogenetic clade (Wildt et al. 2010). Specifically, high priority research should include characterizing reproductive traits (cyclicity and seasonality) especially in under-studied species. Furthermore, it is crucial for us to enhance the ability to manipulate female reproductive cycles, especially of embryo recipients. Finally, fundamental research should be conducted in parallel to advance our understanding of molecular and cellular mechanisms regulating gamete and embryo development, information that is crucial for successful implementation of 'high-tech' reproductive technologies to wild carnivores.

References

Abe Y, Lee D-S, Kim S-K, et al. Vitrification of canine oocytes. J Mamm Ova Res. 2008;28:32–6.

Abe Y, Suwa Y, Asano T, et al. Cryopreservation of canine embryos. Biol Reprod. 2011;84:363–8.

Agca C, Lucy MC, Agca Y. Gene expression profile of rat ovarian tissue following xenotransplantation into immune-deficient mice. Reproduction. 2009;137:957–67.

Agnarsson I, Kuntner M, May-Collado LJ. Dogs, cats, and kin: a molecular species-level phylogeny of Carnivora. Mol Phylogenet Evol. 2010;54:726–45.

Aitken-Palmer C, Hou R, Burrell C, et al. Protracted reproductive seasonality in the male giant panda (*Ailuropoda melanoleuca*) reflected by patterns in androgen profiles, ejaculate characteristics, and selected behaviors. Biol Reprod. 2012;86:195,1–10.

Amstislavsky S, Ternovskaya Y. Reproduction in mustelids. Anim Reprod Sci. 2000; 60–61:571–81.

Amstislavsky S, Lindeberg H, Luvoni GC. Reproductive technologies relevant to the genome resource bank in carnivora. Reprod Domest Anim. 2012;47:164–75.

Anel L, Alvarez M, Anel E, et al. Evaluation of three different extenders for use in emergency salvaging of epididymal spermatozoa from a Cantabric brown bear. Reprod Domest Anim. 2011;46:85–90.

Armstrong DL, Chrichton EG, Dankof SM, et al. Ovarian stimulation, laparoscopic retrieval, IVF and blastocyst production using sequential media in the African lion (*Panthera leo*). J Reprod Fertil. 2004;16:221.

Asa C, Valdespino C. Canid reproductive biology: an integration of proximate mechanisms and ultimate causes. Amer Zool. 1998;38:251–9.
Asa C, Bauman K, Callahan P, et al. Induction of fertile estrus with either natural mating or artificial insemination followed by birth of pups in gray wolves (*Canis lupus*). Theriogenology. 2006;66:1778–82.
Asa C, Miller P, Agnew M, et al. Relationship of inbreeding with sperm quality and reproductive success in Mexican gray wolves. Anim Conserv. 2007a;10:326–31.
Asa CS, Bauman JE, Coonan TJ, et al. Evidence for induced estrus or ovulation in a canid, the Island fox (*Urocyon littoralis*). J Mammal. 2007b;88:436–40.
Asikainen J, Mustonen A-M, Hyvarinen H, et al. Seasonal reproductive endocrine profile of the raccoon dog (*Nyctereutes procyonoides*)—Effects of melatonin and food deprivation. J Exp Zool. 2003;229A:180–7.
Bateman HL, Bond JB, Campbell M, et al. Characterization of basal seminal traits and reproductive endocrine profiles in North American river otters and Asian small-clawed otters. Zoo Biol. 2009;28:107–26.
Baudi DL, Jewgenow K, Pukazhenthi BS, et al. Influence of cooling rate on the ability of frozen-thawed sperm to bind to heterologous zona pellucida, as assessed by competitive *in vitro* binding assays in the ocelot (*Leopardus pardalis*) and tigrina (*Leopardus tigrinus*). Theriogenology. 2008;69:204–11.
Bauman JE, Clifford DL, Asa CS. Pregnancy diagnosis in wild canids using a commercial available relaxin assay. Zoo Biol. 2008;27:406–13.
Boone WR, Catlin JC, Casey KJ, et al. Live birth of a bear cub following nonsurgical embryo collection. Theriogenology. 1999;51:519–29.
Bosch P, Hernandez-Fonseca HJ, Miller DM, et al. Development of antral follicles in cryopreserved cat ovarian tissue transplanted to immunodeficient mice. Theriogenology. 2004;61:581–94.
Boutelle S, Lenahan K, Krisher R, et al. Vitrification of oocytes from endangered Mexican gray wolves (*Canis lupus baileyi*). Theriogenology. 2011;75:647–54.
Braun BC, Frank A, Dehnhard M, et al. Pregnancy diagnosis in urine of Iberian lynx (*Lynx pardinus*). Theriogenology. 2009;71:754–61.
Brown JL, Wildt DE. Assessing reproductive status in wild felids by non-invasive faecal steroid monitoring. Int Zoo Yearb. 1997;35:173–91.
Brown JL, Graham LH, Wielebnowski N, et al. Understanding basic reproductive biology of wild felids by monitoring of faecal steroids. J Reprod Fertil Suppl. 2001;57:71–82.
Brown JL. Comparative endocrinology of domestic and nondomestic felids. Theriogenology. 2006;66:25–36.
Calzada J, González LM, Guzmán JN, et al. A new strategy for the conservation of the Iberian lynx. In: Vargas A, Breitenmoser C, Breitenmose U, editors. Iberian Lynx ex-situ conservation: an interdisciplinary approach. Madrid: Fundación Biodiversidad; 2009. p. 23–31.
Campbell KH, McWhir J, Ritchie WA, et al. Sheep cloned by nuclear transfer from a cultured cell line. Nature. 1996;380:64–6.
Carnaby K, Painer J, Soderberg A, et al. Histological and endocrine characterisation of the annual luteal activity in Eurasian lynx (*Lynx lynx*). Reproduction. 2012;144:477–84.
Carter SK, Rosas FCW. Biology and conservation of the giant otter *Pteronura brasiliensis*. Mammal Rev. 1997;27:1–26.
Chang GR, Yang CC, Hsu SH, et al. Fecal reproductive steroid profiles for monitoring reproductive patterns in female Formosan black bears (*Ursus thibetanus formosanus*). Ann Zool Fenn. 2011;48:275–86.
Chastant-Maillard S, de Lesegno CV, Chebrout M, et al. The canine oocyte: uncommon features of *in vivo* and *in vitro* maturation. Reprod Fertil Dev. 2011;23:391–402.
Chen DY, Wen DC, Zhang YP, et al. Interspecies implantation and mitochondria fate of panda-rabbit cloned embryos. Biol Reprod. 2002;67:637–42.
Chen LM, Hou R, Zhang ZH, et al. Electroejaculation and semen characteristics of Asiatic black bears (*Ursus thibetanus*). Anim Reprod Sci. 2007;101:358–64.

Choi EG, Lee YS, Cho SJ, et al. Semen characteristics of genetically identical male cats cloned via somatic cell nucleus transfer. Theriogenology. 2010;73:638–44.
Christie S. Breeding far Eastern leopards for reintroduction: the zoo programme perspective. In: Vargas A, Breitenmoser C, Breitenmoser U, editors. Iberian lyxn *ex situ* conservation: an interdisciplinary approach. Madrid: Fundacion Biodersidad; 2009. p. 462–77.
Cleary M, Paris MC, Shaw J, et al. Effect of ovariectomy and graft position on cryopreserved common wombat (*Vombatus ursinus*) ovarian tissue following xenografting to nude mice. Reprod Fertil Dev. 2003;15:333–42.
Cocchia N, Ciani F, Russo M, et al. Immature cat oocyte vitrification in open pulled straws (OPSs) using a cryoprotectant mixture. Cryobiology. 2010;60:229–34.
Comizzoli P, Wildt DE, Pukazhenthi BS. *In vitro* development of domestic cat embryos following intracytoplasmic sperm injection with testicular spermatozoa. Theriogenology. 2006;66:1659–63.
Comizzoli P, Crosier AE, Songsasen N, et al. Advances in reproductive science for wild carnivore conservation. Reprod Domest Anim. 2009;44 Suppl 2:47–52.
Comizzoli P, Songsasen N, Wildt DE. Protecting and extending fertility options for females of wild and endangered species. Cancer Treat Res. 2010;156:87–100.
Comizzoli P, Pukazhenthi B, Wildt DE. The competence of germinal vesicle oocytes is unrelated to nuclear chromatin configuration and strictly depends on cytoplasmic quantity and quality in the cat model. Hum Reprod. 2011;26:2165–77.
Comizzoli P, Songsasen N, Hagedorn M, et al. Comparative cryobiological traits and requirements for gametes and gonadal tissues collected from wildlife species. Theriogenology. 2012;78:1666–81.
Concannon PW. Endocrinolgic control of normal canine ovarian function. Reprod Domest Anim. 2009;44 Suppl 2:3–15.
Creel S, Wildt DE, Monfort SL. Aggression, reproduction, and androgen in wild dwarf mongooses: a test of the challenge hypothesis. Am Nat. 1993;141:816–25.
Creel S, Creel NM, Mills MGL, et al. Rank and reproduction in cooperatively breeding African wild dogs: behavioral and endocrine correlates. Behav Ecol. 1997;8:298–306.
Crichton EG, Bedows E, Miller-Lindholm AK, et al. Efficacy of porcine gonadotropins for repeated stimulation of ovarian activity for oocyte retrieval and *in vitro* embryo production and cryopreservation in Siberian tigers (*Panthera tigris altaica*). Biol Reprod. 2003;68:105–13.
Crosier AE, Comizzoli P, Baker T, et al. Increasing age influences uterine integrity, but not ovarian function or oocyte quality, in the cheetah (*Acinonyx jubatus*). Biol Reprod. 2011;85:243–53.
Da Silva IM, Larson S. Predicting reproduction in captive sea otters (*Enhydra lutris*). Zoo Biol. 2005;24:73–81.
Dalerum F, Creel S, Hall SB. Behavioral and endocrine correlates of reproductive failure in social aggregations of captive wolverine (*Gulo gulo*). J Zool. 2005;269:527–36.
Dalponte J, Courtenay O. Hoary fox (*Pseudalopex vetulus*). In: Sillero-Zubiri C, Hoffman M, Macdonald DW, editors. Canids: foxes, wolves, jackals and dogs. status survey and conservation action plan. Gland: IUCN/SSC Canid Specialist Group; 2004. p. 72–5.
de Barros FR, Goissis MD, Caetano HV, et al. Serum starvation and full confluency for cell cycle synchronization of domestic cat (*felis catus*) foetal fibroblast. Reprod Domest Anim. 2010;45:38–41.
de Haas van Dorsser FJ, Swanson WF, Lasano S, et al. Development, validation, and application of a urinary relaxin radioimmunoassay for the diagnosis and monitoring of pregnancy in felids. Biol Reprod. 2006;74:1090–5.
de Haas van Dorsser FJ, Lasano S, Steinetz BG. Pregnancy diagnosis in cats using a rapid, benchtop kit to detect relaxin in urine. Reprod Domest Anim. 2007;42:111–2.
de Paz P, Alvarez-Rodriguez M, Nicolas M, et al. Optimization of glycerol concentration and freezing rate in the cryopreservation of ejaculate from brown bear (*Ursus arctos*). Reprod Domest Anim. 2012;47:105–12.
Dehnhard M, Hildebrandt TB, Knauf T, et al. Comparative endocrine investigations in three bear species based on urinary steroid metabolites and volatiles. Theriogenology. 2006;66:1755–61.

Dehnhard M, Fanson K, Frank A, et al. Comparative metabolism of progestagens and estrogens in the four lynx species, the Eurasian (*Lynx lynx*), the Iberian (L. pardinus), the Canada lynx (*L. canadensis*) and the bobcat (*L. rufus*). Gen Comp Endocrinol. 2010;167:287–96.

Dehnhard M, Finkenwirth C, Crosier A, et al. Using PGFM (13,14-dihydro-15-keto-prostaglandin F2alpha) as a non-invasive pregnancy marker for felids. Theriogenology. 2012;77:1088–99.

DeMatteo KE, Porton IJ, Kleiman DG, et al. The effect of the male bush dog (*Speothos venaticus*) on the female reproductive cycle. J Mammal. 2006;87:723–32.

Dietz JM. Ecology and social organization of the maned wolf (*Chrysocyon brachyurus*). Washington, DC.: Smithsonian Institution Press; 1984. p. 51.

Doak RL, Hall A, Dale HE. Longevity of spermatozoa in the reproductive tract of the bitch. J Reprod Fertil. 1967;13:51–8.

Donoghue AM, Johnston LA, Seal US, et al. *In vitro* fertilization and embryo development *in vitro* and *in vivo* in the tiger (*Panthera tigris*). Biol Reprod. 1990;43:733–44.

Donoghue AM, Howard JG, Byers AP, et al. Correlation of sperm viability with gamete interaction and fertilization *in vitro* in the cheetah (*Acinonyx jubatus*). Biol Reprod. 1992;46:1047–56.

Donoghue AM, Johnston LA, Armstrong DL, et al. Birth of a Siberian tiger cub (*Panthera tigris altaica*) following laparoscopic intrauterine artificial insemination. J Zoo Wildl Med. 1993;24:185–9.

Durbin LS, Venkataraman A, Hedges S, et al. Dhole (*Cuon alpinus*). In: Sillero-Zubiri C, Hoffman M, Macdonald DW, editors. Canids: foxes, wolves, jackals and dogs. Gland, Switzerland: IUCN; 2004. p. 201–18.

Durrant BS, Ravida N, Spady T, et al. New technologies for the study of carnivore reproduction. Theriogenology. 2006;66:1729–36.

Edmison N (2003) Melogale everetti (On-line). Animal Diversity Web, http://animaldiversity.ummz.umich.edu/accounts/Melogale_everetti. Access 30 Oct 2012.

England GWW, Verstegen JP, Hewitt DA. Pregnancy following in vitro fertilisation of canine oocytes. Vet Rec. 2001;148:20–2.

Facemire CF, Gross TS, Guillette LJ. Reproductive impairment in the Florida panther—nature or nurture. Environ Health Perspect. 1995;103:79–86.

Fanson KV, Wielebnowski NC, Shenk TM, et al. Patterns of testicular activity in captive and wild Canada lynx (*Lynx canadensis*). Gen Comp Endocrinol. 2010a;169:210–6.

Fanson KV, Wielebnowski NC, Shenk TM, et al. Patterns of ovarian and luteal activity in captive and wild Canada lynx (*Lynx canadensis*). Gen Comp Endocrinol. 2010b;169:217–24.

Farstad W, Fougner JA, Andersen Berg K. Species differences in fertility after artificial insemination with frozen semen in fox pure breeding. Norw J Agric Sci (Suppl). 1992;9:115–211.

Farstad W, Hyttel P, Grondahl C, et al. Fertilization *in vitro* of oocytes matured *in vivo* in the blue fox (*Alopex lagopus*). J Reprod Fertil Suppl. 1993;47:219–26.

Farstad W. Reproduction in foxes: current research and future challenges. Anim Reprod Sci. 1998;53:35–42.

Farstad W. Assisted reproductive technology in canid species. Theriogenology. 2000;53:175–86.

Fassbender M, Hildebrandt TB, Paris MC, et al. High-resolution ultrasonography of xenografted domestic cat ovarian cortex. J Reprod Dev. 2007;53:1023–34.

Feng HL, Liu JM, Wen XH, et al. Follicular oocyte maturation and sperm penetration *in vitro* in the silver fox (*Vulpes vulpes*). Anim Reprod Sci. 1994;36:163–70.

Filliers M, Rijsselaere T, Bossaert P, et al. *In vitro* evaluation of fresh sperm quality in tomcats: a comparison of two collection techniques. Theriogenology. 2010;74:31–9.

Finkenwirth C, Jewgenow K, Meyer HHD, et al. PGFM (13,14-dihydro-15-keto-PGF2α) in pregnant and pseudo-pregnant Iberian Lynx: a new non-invasive pregnancy marker for felid species. Theriogenology. 2010;73:530–40.

Fitzpatrick JL, Evans JP. Reduced heterozygosity impairs sperm quality in endangered mammals. Biol Lett. 2009;5:320–3.

Ganan N, Gonzalez R, Sestelo A, et al. Male reproductive traits, semen cryopreservation, and heterologous *in vitro* fertilization in the bobcat (*Lynx rufus*). Theriogenology. 2009;72: 341–52.

Garshelis DL. Family Ursidae (Bears). In: Mittermeier RA, Wilson DE, editors. Handbook of the mammals of the world. Barcelona: Lynx Edicions; 2009. p. 448–98.

Goeritz F, Hildebrandt T, Jewgenow K, et al. Transrectal ultrasonographic examination of the female urogenital tract in nonpregnant and pregnant captive bears (Ursidae). J Reprod Fertil Suppl. 1997;51:303–12.

Goeritz F, Quest M, Hildebrandt T, et al. Control of reproduction with anti-progestin and oestrogens in captive bears. J Reprod Fertil Suppl. 2001;57:249–54.

Goeritz F, Neubauer K, Naidenko SV, et al. Investigations on reproductive physiology in the male Eurasian lynx (*Lynx lynx*). Theriogenology. 2006;66:1751–4.

Goeritz F, Dehnhard M, Hildebrandt TB, et al. Non cat-like ovarian cycle in the Eurasian and the Iberian lynx—ultrasonographical and endocrinological analysis. Reprod Domest Anim. 2009;44:87–91.

Goeritz F, Painer J, Jewgenow K, et al. Embryo retrieval after hormonal treatment to control ovarian function and non-surgical artificial insemination in African lions (*Panthera leo*). Reprod Domest Anim. 2012;47 Suppl 6:156–60.

Gomez MC, Pope CE, Giraldo A, et al. Birth of African wildcat cloned kittens born from domestic cats. Cloning Stem Cells. 2004;6:247–58.

Gomez MC, Pope CE, Kutner RH, et al. Nuclear transfer of sand cat cells into enucleated domestic cat oocytes is affected by cryopreservation of donor cells. Cloning Stem Cells. 2008;10:469–83.

Gomez MC, Pope CE, Ricks DM, et al. Cloning endangered felids using heterospecific donor oocytes and interspecies embryo transfer. Reprod Fertil Dev. 2009;21:76–82.

Gomez MC, Serrano MA, Pope CE, et al. Derivation of cat embryonic stem-like cells from *in vitro*-produced blastocysts on homologous and heterologous feeder cells. Theriogenology. 2010;74:498–515.

Gomez MC, Pope CE, Biancardi MN, et al. Trichostatin A modified histone covalent pattern and enhanced expression of pluripotent genes in interspecies black-footed cat cloned embryos but did not improve *in vitro* and *in vivo* viability. Cell Reprogram. 2011;13:315–29.

Gomez MC, Biancardi MN, Jenkins JA, et al. Scriptaid and 5-aza-2′ deoxycytidine enhanced expression of pluripotent genes and *in vitro* developmental competence in interspecies black-footed cat cloned embryos. Reprod Domest Anim. 2012;47 Suppl 6:130–5.

Gong MH, Song YL, Yang ZS, et al. Important population viability analysis parameters for giant pandas (*Aliuropoda melanoleuca*). Dongwuxue Yanjiu. 2012;33:01–02e18.

Goodrowe KL, Miller AM, Wildt DE. *In vitro* fertilization of gonadotropin-stimulated leopard cat (*Felis bengalensis*) follicular oocytes. J Exp Zool. 1989;252:89–95.

Goodrowe KL, Hay MA, Platz CC, et al. Characteristics of fresh and frozen-thawed red wolf (*Canis rufus*) spermatozoa. Anim Reprod Sci. 1998;53:299–308.

Green JS, Adair RA, Woodruff RA. Seasonal variation in semen production by captive coyotes. J Mammal. 1984;65:506–9.

Guaitolini CRF, Taffarel MO, Teixeira NS, et al. Post-thaw viability of *in vivo*-produced canine blastocysts cryopreserved by slow freezing. Theriogenology. 2012;78:576–82.

Gudermuth DF, Concannon PW, Daels PF, et al. Pregnancy-specific elevations in fecal concentrations of estradiol, testosterone and progesterone in the domestic dog (*Canis familiaris*). Theriogenology. 1998;50:237–48.

Gudermuth DF, Newton L, Daels P, et al. Incidence of spontaneous ovulation in young, group-housed cats based on serum and faecal concentrations of progesterone. J Reprod Fertil Suppl. 1997;51:177–84.

Hama N, Kanemitsu H, Sakamoto K, et al. Development and evaluation of a rapid enzyme-immunoassay system for measurement of the urinary concentration of estrone-3-glucuronide in a female giant panda (*Ailuropoda melanoleuca*). J Reprod Dev. 2008;54:281–5.

Harris LA, Steinetz BG, Bond JB, et al. Refinement of a commercial bench-top relaxin assay for pregnancy diagnosis using urine from domestic and nondomestic felids. J Zoo Wildl Med. 2008;39:170–9.

Hayashi K, Ogushi S, Kurimoto K, et al. Offspring from oocytes derived from *in vitro* primordial germ cell-like cells in mice. Science. 2012;338:971–5.

Hayward MW, Somers M. Reintroduction of top-order predators (conservation scinece and practice). West Sussex: Wiley-Backwell; 2009. 480 pp.
Hayssen V, Tienhoven VA, Tienhoven VA. Asdell's pattern of mammalian reproduction. a compendium of species-specific data. Ithaca, NY: Cornell University Press; 1993. p. 1027.
Hodges JK, Bevan DJ, Celma M, et al. Aspects of the reproductive endocrinology of the female giant panda (*Ailuropoda-Melanoleuca*) in captivity with special reference to the detection of ovulation and pregnancy. J Zool. 1984;203:253–67.
Holt WV, Pickard AR. Role of reproductive technologies and genetic resource banks in animal conservation. Rev Reprod. 1999;4:143–50.
Holt WV, Pickard AR, Prather RS. Wildlife conservation and reproductive cloning. Reproduction. 2004;127:317–24.
Holt WV, Pickard AR, Rodger JC, et al. Reproductive sciences and integrated conservation. Cambridge: Cambridge University Press; 2003. p. 409.
Hori T, Hashizaki F, Narushima E, et al. A trial of intrauterine insemination using a fiberscope in the giant panda (*Ailuropoda melanoleuca*). J Vet Med Sci. 2006;68:987–90.
Hossein MS, Jeong YW, Park SW, et al. Birth of Beagle dogs by somatic cell nuclear transfer. Anim Reprod Sci. 2009a;114:404–14.
Hossein MS, Jeong YW, Park SW, et al. Cloning missy: obtaining multiple offspring of a specific canine genotype by somatic cell nuclear transfer. Cloning Stem Cells. 2009b;11:123–30.
Howard J, Wildt DE. Approaches and efficacy of artificial insemination in felids and mustelids. Theriogenology. 2009;71:130–48.
Howard JG, Bush M, Morton C, et al. Comparative semen cryopreservation in ferrets (*Mustela putorius furo*) and pregnancies after laparoscopic intrauterine insemination with frozen-thawed spermatozoa. J Reprod Fertil. 1991;92:109–18.
Huang Y, Li D, Zhou Y, et al. Factors affecting the outcome of artififcial insemination using cryopreserved spermatozoa in the giant panda (*Ailuropoda melanoleuca*). Zoo Biol. 2012;31:561–73.
Hyttel P, Farstad W, Mondain-Monval M, et al. Structural aspects of oocyte maturation in the blue fox (*Alopex lagopus*). Anal Embryol (Berl). 1990;181:325–31.
Imsoonthornruksa S, Sangmalee A, Srirattana K, et al. Development of intergeneric and intrageneric somatic cell nuclear transfer (SCNT) cat embryos and the determination of telomere length in cloned offspring. Cell Reprogram. 2012;14:79–87.
Ishikawa A, Matsui M, Sakamoto H, et al. Cryopreservation of the semen collected by electroejaculation from the Hokkaido brown bear (*Ursus arctos yesoensis*). J Vet Med Sci. 2002;64:373–6.
Ishikawa A, Sakamoto H, Katagiri S, et al. Changes in sexual behavior and fecal steroid hormone concentrations during the breeding season in female Hokkaido brown bears (*Ursus arctos yesoensis*) under captive condition. J Vet Med Sci. 2003;65:99–102.
IUCN: IUCN Red List of Threatened Species. Version 2013.2, www.iucnredlist.org. (2013)
Jameson RJ. Reproductive characteristics of female sea otters. Marine Mammal Sci. 1993;9:156–67.
Jang G, Hong SG, Oh HJ, et al. A cloned toy poodle produced from somatic cells derived from an aged female dog. Theriogenology. 2008;69:556–63.
Jang G, Kim MK, Lee BC. Current status and applications of somatic cell nuclear transfer in dogs. Theriogenology. 2010;74:1311–20.
Jewgenow K, Blottner S, Lengwinat T, et al. New methods for gamete rescue from gonads of non-domestic felids. J Reprod Fertil Suppl. 1997;51:33–9.
Jewgenow K, Penfold LM, Meyer HH, et al. Viability of small preantral ovarian follicles from domestic cats after cryoprotectant exposure and cryopreservation. J Reprod Fertil. 1998;112:39–47.
Jewgenow K, Naidenko SV, Goeritz F, et al. Monitoring testicular activity of male Eurasian (*Lynx lynx*) and Iberian (*Lynx pardinus*) lynx by fecal testosterone metabolite measurement. Gen Comp Endocrinol. 2006;149:151–8.

Jewgenow K, Paris MC. Preservation of female germ cells from ovaries of cat species. Theriogenology. 2006;66:93–100.

Jewgenow K, Wiedemann C, Bertelsen MF, et al. Cryopreservation of mammalian ovaries and oocytes. Inter Zoo Yearb. 2011;45:124–32.

Jhala YV, Moehlman PD. Golden jackal (*Canis aureus*). In: Sillero-Zubiri C, Hoffman M, Macdonald DW, editors. Canids: foxes, wolves, jackals and dogs. Status survey and conservation action plan. Gland: IUCN/SSC Canid Specialist Group; 2004. p. 156–60.

Johnson AEM (2012) The development of assisted reproductive techniques for managing maned wolves (*Chrysocyon brachyurus*) ex situ. Master Thesis. George Mason University, Fairfax, Virginia, USA. 116

Johnson WE, Eizirik E, Pecon-Slattery J, et al. The late Miocene radiation of modern Felidae: a genetic assessment. Science. 2006;311:73–7.

Johnston LA, Donoghue AM, O'Brien SJ, et al. Rescue and maturation *in vitro* of follicular oocytes collected from nondomestic felid species. Biol Reprod. 1991;45:898–906.

Johnston LA, Donoghue AM, Igo W, et al. Oocyte recovery and maturation in the American black bear (*Ursus americanus*): a model for endangered ursids. J Exp Zool. 1994;269:53–61.

Johnston SD, Ward D, Lemon J, et al. Studies of male reproduction in captive African wild dogs (*Lycaon pictus*). Anim Reprod Sci. 2007;100:338–55.

Kashir J, Jones C, Child T, et al. Viability assessment for artificial gametes: the need for biomarkers of functional competency. Biol Reprod. 2012;87:114,1–11.

Kersey DC, Wildt DE, Brown JL, et al. Unique biphasic progestagen profile in parturient and non-parturient giant pandas (*Ailuropoda melanoleuca*) as determined by faecal hormone monitoring. Reproduction. 2010;140:183–93.

Kim MK, Jang G, Oh HJ, et al. Endangered wolves cloned from adult somatic cells. Cloning Stem Cells. 2007;9:130–7.

Kim SS, Lee WS, Chung MK, et al. Long-term ovarian function and fertility after heterotopic autotransplantation of cryobanked human ovarian tissue: 8-year experience in cancer patients. Fertil Steril. 2009;91:2349–54.

Kim Y, Selvaraj V, Dobrinski I, et al. Recipient preparation and mixed germ cell isolation for spermatogonial stem cell transplantation in domestic cats. J Androl. 2006;27:248–56.

Kim Y, Turner D, Nelson J, et al. Production of donor-derived sperm after spermatogonial stem cell transplantation in the dog. Reproduction. 2008;136:823–31.

Kim YK, Kim BJ, Yu IJ. Embryo transfer with frozen embryos. J Vet Clin. 2002;19:73–9.

Klinc P, Majdic G, Sterbenc N, et al. Establishment of a pregnancy following intravaginal insemination with epididymal semen from a dog castrated due to benign prostatic hyperplasia. Reprod Domest Anim. 2005;40:559–61.

Knauf T, Jewgenow K, Dehnhard M, et al. Comparative investigations on reproduction in different bear species—anatomical, ultrasonographic and endocrine investigations. Verhber Erkrf Zootiere. 2003;41:287–96.

Knauf T (2006) Vergleichende Studien zur Reproduktionsbiologie bei Großbären (Ursidae) [Comparative investigations on the reproduction biology of bears (Ursidae)], Thesis, Free University Berlin

Ko K, Wu G, Arauzo-Bravo MJ, et al. Autologous pluripotent stem cells generated from adult mouse testicular biopsy. Stem Cell Rev. 2012;8:435–44.

Koo OJ, Hossein MS, Hong SG, et al. Cell cycle synchronization of canine ear fibroblasts for somatic cell nuclear transfer. Zygote. 2009;17:37–43.

Lariviere S. *Lontra provocax*. Mamm Species. 1999;610:1–4.

Larson S, Casson CJ, Wasser S. Noninvasive reproductive steroid hormone estimates from fecal samples of captive female sea otters. J Mammal. 2003;134:18–25.

Lee HS, Yu XF, Bang JI, et al. Enhanced histone acetylation in somatic cells induced by a histone deacetylase inhibitor improved inter-generic cloned leopard cat blastocysts. Theriogenology. 2010;74:1439–49.

Leon-Quinto T, Simon MA, Cadenas R, et al. Developing biological resource banks as a supporting tool for wildlife reproduction and conservation the Iberian lynx bank as a model for other endangered species. Anim Reprod Sci. 2008;112:347–61.
Lermen D, Blomeke B, Browne R, et al. Cryobanking of viable biomaterials: implementation of new strategies for conservation purposes. Mol Ecol. 2009;18:1030–3.
Li J, Kawamura K, Cheng Y, et al. Activation of dormant ovarian follicles to generate mature eggs. Proc Natl Acad Sci U S A. 2010;107:10280–4.
Li Z, Jiang Q, Sabet M, et al. Conditions for *in vitro* maturation and artificial activation of ferret oocytes. Biol Reprod. 2002;66:1380–6.
Li ZY, Jiang QS, Zhang YL, et al. Successful production of offspring after superovulation and *in vitro* culture of embryos from domestic ferrets (*Mustela putorius furos*). Reproduction. 2001;122:611–8.
Lindeberg H, Jarvinen M. Early embryonic development and *in vitro* culture of *in vivo* produced embryos in the farmed European polecat (*Mustela putorius*). Theriogenology. 2003;60:965–75.
Loi P, Modlinski JA, Ptak G. Interspecies somatic cell nuclear transfer: a salvage tool seeking first aid. Theriogenology. 2011;76:217–28.
Londono GC, Munoz NT. Reproduction, behaviour and biology of the giant river otter *Pteronura brasiliensis*. Int Zoo Yearb. 2006;40:360–71.
Lueders I, Luther I, Scheepers G, et al. Improved semen collection method for wild felids: urethral catheterization yields high sperm quality in African lions (*Panthera leo*). Theriogenology. 2012;78:696–701.
Luvoni GC, Chigioni S, Allievi E, et al. Factors involved *in vivo* and *in vitro* maturation of canine oocytes. Theriogenology. 2005;63:41–59.
Luvoni GC. Gamete cryopreservation in the domestic cat. Theriogenology. 2006;66:101–11.
Mahi CA, Yanagimachi R. Maturation and sperm penetration of canine ovarian oocytes *in vitro*. J Exp Zool. 1976;196:189–95.
Masui M, Hiramatsu H, Nose N, et al. Successful artificial insemination in the giant panda (*Ailuropoda melanoleuca*) at Ueno zoo. Zoo Biol. 1989;8:17–26.
Mayr E. Uncertainty in science: is the giant panda a bear or a raccoon? Nature. 1986;323:769–71.
Mead RA, Neirinchx S, Czekala NM. Reproductive cylce of the steppe polecat. J Reprod Fertil. 1990;88:353–60.
Mead RA, Bowles M, Starypan G, et al. Evidence for pseudopregnancy and induced ovulation in captive wolverines (*Gulo gulo*). Zoo Biol. 1993;12:353–8.
Merlo B, Iacono E, Regazzini M, et al. Cat blastocysts produced *in vitro* from oocytes vitrified using the cryoloop technique and cryopreserved electroejaculated semen. Theriogenology. 2008;70:126–30.
Minter LJ, DeLiberto TJ. Seasonal variation in serum testosterone, testicular volume, and semen characteristics in the coyote (*Canis latrans*). Theriogenology. 2008;69:946–52.
Mitsuzuka M. Collection, evaluation and freezing of wolf semen. In: Frank H, editor. Man and Wolf. Dordrecht: Dr. W. Junk Publishers; 1987. p. 127–41.
Monfort SL. Non-invasive endocrine measures of reproduction and stress in wild populations. In: Holt WV, Pickard AR, Rodger JC, Wildt DE, editors. Reproductive science and integrated conservation. Cambridge: Cambridge University Press; 2003. p. 147–65.
Mondain-Monval M, Bonnin M, Canivence R, et al. Heterologous radioimmunoassay of fox LH: levels during reproductive season and the anoestrus of the red fox (*Vulpes vulpes L.*). Gen Comp Endocrinol. 1984;55:125–32.
Moore HDM, Bush M, Celma M, et al. Artificial insemination in the Giant panda (*Ailuropoda melanoleaca*). J Zool. 1984;203:269–78.
Morais RN, Mucciolo RG, Gomes ML, et al. Seasonal analysis of semen characteristics, serum testosterone and fecal androgens in the ocelot (*Leopardus pardalis*), margay (*L. wiedii*) and tigrina (*L. tigrinus*). Theriogenology. 2002;57:2027–41.
Moreau GM, Arslan A, Douglas DA, et al. Development of immortalized endometrial epithelial and stromal cell lines from the mink (*Mustela vison*) uterus and their effects on the survival *in vitro* of mink blastocysts in obligate diapause. Biol Reprod. 1995;53:511–8.

Moreira N, Monteiro-Filho EL, Moraes W, et al. Reproductive steroid hormones and ovarian activity in felids of the *Leopardus* genus. Zoo Biol. 2001;20:103–16.
Nakagome S, Pecon-Slattery J, Masuda R. Unequal rates of Y chromosome gene divergence during speciation of the family Ursidae. Mol Biol Evol. 2008;25:1344–56.
Newton H, Aubard Y, Rutherford A, et al. Low temperature storage and grafting of human ovarian tissue. Hum Reprod. 1996;11:1487–91.
Niu Y, Greube A, Ji W, et al. The application of *in vitro* sperm competition test to evaluate the impact of ZP-derived peptides on fertilization capacity of cat sperm. Theriogenology. 2006;66:989–95.
Nowak RM. Walker's mammals of the world, vol. 1. Baltimore: The John Hopskin Press; 1999. p. 836.
Oh HJ, Kim MK, Jang G, et al. Cloning endangered gray wolves (*Canis lupus*) from somatic cells collected postmortem. Theriogenology. 2008;70:638–47.
Onuma M, Suzuki M, Ohtaishi N. Reproductive pattern of the sun bear (*Helarctos malayanus*) in Sarawak, Malaysia. J Vet Med Sci. 2001;63:293–7.
Otoi T, Murakami M, Fujii M, et al. Development of canine oocytes matured and fertilised in *vitro*. Vet Rec. 2000;146:52–3.
Paris MC, Snow M, Cox SL, et al. Xenotransplantation: a tool for reproductive biology and animal conservation? Theriogenology. 2004;61:277–91.
Park JE, Kim MK, Kang JT, et al. Growth and hematologic characteristics of cloned dogs derived from adult somatic cell nuclear transfer. Cell Reprogram. 2010;12:141–50.
Pearson OP, Enders RK. Ovulation, maturation and fertilization in the fox. Anat Rec. 1943;85:69–83.
Pelican KM, Wildt DE, Pukazhenthi B, et al. Ovarian control for assisted reproduction in the domestic cat and wild felids. Theriogenology. 2006;66:37–48.
Peng R, Zeng B, Meng X, et al. The complete mitochondrial genome and phylogenetic analysis of the giant panda (*Ailuropoda melanoleuca*). Gene. 2007;397:76–83.
Persson J, Landa A, Andersen R, et al. Reproductive characteristics of female wolverines (*Gulo gulo*) on Scandinavia. J Mammal. 2006;87:75–9.
Pickard AR. Reproductive and welfare monitoring for the management of ex situ populations. In: Holt WV, Pickard AR, Rodger JC, Wildt DE, editors. Reproductive science and integrated conservation. Cambridge: Cambridge University Press; 2003. p. 133–46.
Piltti K, Lindeberg H, Aalto J, Korhonen H. Live cubs born after transfer of OPS vitrified-warmed embryos in the farmed European polecat (*Mustela putorius*). Theriogenology. 2004;61: 811–20.
Platz CC, Waddell W, Smith R, et al. Semen collection in wild canidae using electroejaculation. In: Loskutiff NM, Chrichton EG, editors. The first international symposium on assisted reproductive technology for conservation and genetic management of wildlife. 9th ed. Omaha: Henry Doorly Zoo; 2001. p. 134–5.
Pope CE, Keller GL, Dresser BL. *In vitro* fertilization in domestic and nondomestic cats including sequences of early nuclear events, development *in-vitro*, cryopreservation and successful intraspecies and interspecies embryo-transfer. J Reprod Fertil Suppl. 1993;47:189–201.
Pope CE, Gomez MC, Dresser BL. *In vitro* production and transfer of cat embryos in the 21st century. Theriogenology. 2006;66:59–71.
Pope CE, Gomez MC, Galiguis J, et al. Applying embryo cryopreservation technologies to the production of domestic and black-footed cats. Reprod Domest Anim. 2012a;47 Suppl 6:125–9.
Pope CE, Gomez MC, Kagawa N, et al. *In vivo* survival of domestic cat oocytes after vitrification, intracytoplasmic sperm injection and embryo transfer. Theriogenology. 2012b; 77:531–8.
Porton IJ, Kleiman DG, Rodden MD. Aseasonality of bush dog reproduction and the influence of social factors on the estrous cycle. J Mammal. 1987;68:867–71.
Powell RH, Biancardi MN, Galiguis J, et al. 286 expression of pluripotent stem cell markers in domestic cat spermatogonial cells. Reprod Fertil Dev. 2012;25:290–1.

Prates Junior PHS (2008) Diversidade Genetica e Historia Evolutiva do Lobo Guara (*Chrysocyon brachyurus*). PhD Thesis. Pontificia Universidade Catolica do Rio Grande Do Sul, Porto Alegre, Brasil. 125 pp.

Pukazhenthi BS, Wildt DE. Which reproductive technologies are most relevant to studying, managing and conserving wildlife? Reprod Fertil Dev. 2004;16:33–46.

Reed-Smith J. North American river otter: husbandry notebook. Grand Rapid: John Ball Zoo; 2008. p. 307.

Reiter LE. Understanding poor reproduction in the maned wolf (*Chrysocyon brachyurus*) held ex situ. Fair Fax, VA: Master Thesis. George Mason University; 2012. p. 137.

Renfree MB, Shaw G. Diapause. Annu Rev Physiol. 2000;62:353–75.

Reynaud K, Fontbonne A, Marseloo N, et al. *In vivo* meiotic resumption, fertilization and early embryonic development in the bitch. Reproduction. 2005;130:193–201.

Reynaud K, Viaris de Lesegno C, Chebrout M, et al. Follicle population, cumulus mucification, and oocyte chromatin configuration during the periovulatory period in the female dog. Theriogenology. 2009;72:1120–31.

Ringleb J, Waurich R, Wibbelt G, et al. Save the cats! Challenges and changes of a felid-gamete-rescue project. Proc Int Conf Dis Zoo Wild Anim. 2010;45:86–91.

Rodden M, Rodrigues F, Bestelmeyer S. Maned Wolf (*Chrysocyon brachyurus*). In: Sillero-Zubiri C, Hoffman M, Macdonald DW, editors. Foxes, wolves, jackals and dogs. Status survey and conservation action plan. Cambridge: IUCN/SSC Canid Specialist Group; 2004. p. 38–43.

Rudert S, Brown JL, Ganslober U, et al. Activity pattern, reproductive behaviors and gonadal hormones in the raccoon dog (*Nyceteutes procyonoides*). Zoo Biol. 2011;30:134–48.

Sanson G, Brown JL, Farstad W. Non-invasive faecal steroid monitoring of ovarian and adrenal activity in farmed blue fox (*Alopex lagopus*) females during late pregnancy, parturition and lactation onset. Anim Reprod Sci. 2005;87:309–19.

Sato M, Tsubota T, Komatsu T, et al. Change in sex steroids, gonadotropins, prolactin, and inhibin in pregnant and nonpregnant Japanese black bears (*Ursus thibetanus japonicus*). Biol Reprod. 2001;65:1006–13.

Schramm RD, Briggs MB, Reeves J. Spontaneous and induced ovulation in the lion (*Panthera-Leo*). Zoo Biol. 1994;13:301–7.

Schwartz MK, Monfort SL. Genetic and endocrine tools for carnivore surveys. In: Long R, Mackay P, Justina R, Zielinski W, editors. Noninvasive survey methods for North American carnivores. Washington, DC: Island Press; 2008. p. 228–50.

Schwarzenberger F, Fredriksson G, Schaller K, et al. Fecal steroid analysis for monitoring reproduction in the sun bear (*Helarctos malayanus*). Theriogenology. 2004;62:1677–92.

Seager SWJ, Platz CC, Hodge W. Successful pregnancy using frozen semen in the wolf. Int Zoo Yearb. 1975;15:140–3.

Sillero-Zubiri C, Hoffman M, Macdonald DW. Canids: foxes, wolves, jackals and dogs: status survey and conservation action plan. Gland Switzerland and Cambridge, UK: IUCN/SSC Canid Specialist Group; 2004. 430 pp.

Sillero-Zubiri C, Johnson PJ, Macdonald DW. A hypothesis for breeding synchrony in Ethiopian wolves (*Canis simensis*). J Mammal. 1998;79:853–8.

Silva AR, Morato RG, Silva LDM. The potential for gamete recovery from non-domestic canids and felids. Anim Reprod Sci. 2004;81:159–75.

Sinha AA, Conaway CH, Kenyon KW. Reproduction in the female sea otter. J Wildl Manag. 1966;30:121–30.

Songsasen N, Rodden M, Brown JL, et al. Patterns of fecal gonadal hormone metabolites in the maned wolf (*Chrysocyon brachyurus*). Theriogenology. 2006;66:1743–50.

Songsasen N, Wildt DE. Oocyte biology and challenges in developing *in vitro* maturation systems in the domestic dog. Anim Reprod Sci. 2007;98:2–22.

Songsasen N, Johnson AE, Reiter LE, et al. Reproduction and factors impacting reproductive success in the maned wolf. In: Corsorte-McCrea A, Ferraz E, editors. Ecology and conservation of the maned wolf. Boca Raton, FL: Tay & Francis Group; 2013.

Souza NP, Furtado PV, da Paz RCR. Non-invasive monitoring of the estrous cycle in captive crab-eating foxes (*Cerdocyon thos*). Theriogenology. 2012;77:233–9.
Spady TJ, Lindburg DG, Durrant BS. Evolution of reproductive seasonality in bears. Mammal Rev. 2007;37:21–53.
Spercoski KM, Morais RN, Morato RG, et al. Adrenal activity in maned wolves is higher on farmlands and park boundaries than within protected areas. Gen Comp Endocrinol. 2012;179:232–40.
Spindler RE, Huang Y, Howard JG, et al. Giant panda (*Ailuropoda melanoleuca*) spermatozoon decondensation *in vitro* is not compromised by cryopreservation. Reprod Fertil Dev. 2006;18:767–75.
Stasiak K, Janicki B, Kupcewicz B. Biological parameters of polar fox (*Alopex lagopus L.*) semen during the breeding season. Turk J Vet Anim Sci. 2008;34:327–31.
Steinetz B, Lasano S, van Dorsser FD, et al. Relaxin concentrations in serum and urine endangered and crazy mixed-up species new methods, uses and findings. Ann N Y Acad Sci. 2009;1160: 179–85.
Stewart DR, Stabenfeldt GH. Relaxin activity in the pregnant cat. Biol Reprod. 1985;32:848–54.
Sun X, Li Z, Ding W, et al. Chromatin configurations in the ferret germinal vesicle that reflect developmental competence for *in vitro* maturation. Reprod Domest Anim. 2009;44:320–5.
Sunquist ME, Sunquist FC. Family felidae (cats). In: Mittermeier RA, Wilson DE, editors. Handbook of the mammals of the world. Barcelona: Lynx Edicions; 2009. p. 54–170.
Swanson WF. Application of assisted reproduction for population management in felids: the potential and reality for conservation of small cats. Theriogenology. 2006;66:49–58.
Swanson WF. Laparoscopic oviduct embryo transfer and artificial insemination in felids—challenges, strategies and successes. Reprod Domest Anim. 2012;47 Suppl 6:136–40.
Tao Y, Liu J, Zhang Y, et al. Fibroblast cell line establishment, cryopreservation and interspecies embryos reconstruction in red panda (*Ailurus fugens*). Zygote. 2009;17:117–24.
Telfer E, Gosden RG. A quantitative cytological study of polyovular follicles in mammalian ovaries with particular reference to the domestic bitch (*Canis familiaris*). J Reprod Fertil. 1987;81:137–47.
Teodoro LO, Melo-Junior AA, Spercoski KM, et al. Seasonal aspects of reproductive physiology in captive male maned wolves (*Chrysocyon brachyurus*, Illiger 1815). Reprod Domest Anim. 2012;47 Suppl 6:250–5.
Tharasanit T, Manee-In S, Buarpung S, et al. Successful pregnancy following transfer of feline embryos derived from vitrified immature cat oocytes using 'stepwise' cryoprotectant exposure technique. Theriogenology. 2011;76:1442–9.
Thomassen R, Farstad W. Artificial insemination in canids: a useful tool in breeding and conservation. Theriogenology. 2009;71:190–9.
Thongphakdee A, Siriaroonrat B, Manee-in S, et al. Intergeneric somatic cell nucleus transfer in marbled cat and flat-headed cat. Theriogenology. 2010;73:120–8.
Travis AJ, Kim Y, Meyers-Wallen V. Development of new stem cell-based technologies for carnivore reproduction research. Reprod Domest Anim. 2009;2:22–8.
Tsubota T, Takahashi Y, Kanagawa H, et al. Embryo recovery during delayed implantation in the captive Hokkaido brown bear, *Ursus arctos yesoensis*. J Vet Med Sci. 1991;53:141–2.
Tsutsui T. Gamete physiology and timing of ovulation and fertilization in dogs. J Reprod Fertil Suppl. 1989;39:269–75.
Tsutsui T, Stabenfeldt GH. Biology of ovarian cycles, pregnancy and pseudopregnancy in the domestic cat. J Reprod Fertil Suppl. 1993;47:29–35.
Tsutsui T, Takahashi F, Hori T, et al. Prolonged duration of fertility of dog ova. Reprod Domest Anim. 2009;44 Suppl 2:230–3.
Turatham B, Saikhun K, Sangsuwan P, et al. Effects of vitrification on nuclear maturation, ultrastructural changes and gene expression of canine oocytes. Reprod Biol Endocrinol. 2010;8:70.
Valdespino C, Asa CS, Bauman JE. Estrous cycle, copulation, and pregnancy in the fennce fox (*Vulpes zerda*). J Mammal. 2002;83:99–109.

Van den Berghe F, Paris DBBP, Van Soom A, et al. Reproduction in the endangered African wild dog: Basic physiology, reproductive suppression and possible benefits of artificial insemination. Anim Reprod Sci. 2012;133:1–9.

van der Horst G, Kitchin RM, van der Horst M, et al. The effect of the breeding season, cryopreservation and physiological extender on selected sperm and semen parameters of four ferret species: implications for captive breeding in the endangered black-footed ferret. Reprod Fertil Dev. 2009;21:351–63.

van Kesteren F, Paris M, Macdonald DW, et al. (2011) The physiology of cooperative breeding in a rare social canid; sex, suppression and pseudopregnancy in female Ethiopian wolves. In: The physiology of cooperative breeding in a rare social canid; sex, suppression and pseudopregnancy in female Ethiopian wolves. Thesis, Ms.science, University of Oxford.

van Kesteren F, Sillero-Zubiri C, Millar R, Farstad W, Macdonald DW, Paris M. (2012) Reproductive physiology of the world's rarest canid: patterns in estradiol, progesterone, glucocorticoid and testosterone metabolites in Ethiopian wolves, *Canis simensis*, 3rd Annual conference of The International Society for wildlife endocrinology (ISWE) 2012, September 23-26, Vienna, Austria (abstr)

Vargas A, Sánchez I, Martínez F, et al. Interdisciplinary methods in the Iberian lynx (*Lynx pardinus*) conservation breeding programme. In: Vargas A, Breitenmoser C, Breitenmoser U, editors. Iberian lynx *ex situ* conservation: an interdisciplinary approach. Madrid: Fundación Biodiversidad; 2009. p. 42–55.

Vargas A, Sánchez I, Martínez F, et al. The Iberian lynx (*Lynx pardinus*) conservation breeding program. Int Zoo Yearb. 2008;42:190–8.

Velloso A, Wasser SK, Monfort SL, et al. Longitudinal fecal steroid excretion in maned wolves (*Chrysocyon brachyurus*). Gen Comp Endocrinol. 1998;112:96–107.

Verma R, Holland MK, Temple-Smith P, et al. Inducing pluripotency in somatic cells from the snow leopard (*Panthera uncia*), an endangered felid. Theriogenology. 2012;77:220–8.

Walker SL, Waddell WT, Goodrowe KL. Reproductive endocrine patterns in captive female and male red wolves (*Canis rufus*) assessed by fecal and serum hormone analysis. Zoo Biol. 2002;21:321–35.

Weng Q, Medan MS, Xu M, et al. Seasonal changes in immunolocalization of inhibin/activin subunits and testicular activity in wild male raccoon dogs (*Nyctereutes procyonoides*). J Reprod Dev. 2006;52:503–10.

Wiedemann C, Hribal R, Ringleb J, et al. Preservation of primordial follicles from lions by slow freezing and xenotransplantation of ovarian cortex into an immunodeficient mouse. Reprod Domest Anim. 2012;47 Suppl 6:300–4.

Wildt DE, Bush M, Moron C, et al. Semen characteristics and testosterone profiles in ferrets kept in long-day photoperiod and the influence of hCG timing and sperm dilution on pregnancy rate after laparoscopic insemination. J Reprod Fertil. 1989;86:349–58.

Wildt DE, Ellis S, Janssen D, et al. Toward more effective reproductive science for conservation. In: Holt WV, Pickard AR, Rodger JC, Wildt DE, editors. Reproductive science and integrated conservation. Cambridge: Cambridge University Press; 2003. p. 2–20.

Wildt DE, Comizzoli P, Pukazhenthi B, et al. Lessons from biodiversity-the value of nontraditional species to advance reproductive sciences, conservation, and human health. Mol Reprod Dev. 2010;77:397–409.

Wildt DE, Zhang A, Zhang H, Jenssen DL, Ellis S. Giant pandas. Cambridge: Cambridge University Press; 2007. p. 610.

Williams ES, Thorne ET, Kiwiatkowski DR, et al. Reproductive biology and management of captive black-footed ferrets (*Mustela nigripes*). Zoo Biol. 1991;10:383–98.

Williams ES, Thorne ET, Kwiatkowski DR, et al. Comparative vaginal cytology of the estrous cycle of black-footed ferrets *(Mustela nigripes*), Siberian polecats (*M. eversmanni*), and domestic ferrets (*M. putorius furo*). J Vet Diagn Invest. 1992;4:38–44.

Willis EL, Kersey DC, Durrant BS, et al. The acute phase protein ceruloplasmin as a non-invasive marker of pseudopregnancy, pregnancy, and pregnancy loss in the giant panda. PLoS One. 2011;6:e21159.

Wittayarat M, Thongphakdee A, Saikhun K, et al. Cell cycle synchronization of skin fibroblast cells in four species of family felidae. Reprod Domest Anim. 2012;27:1439–0531.

Yin XJ, Lee HS, Lee YH, et al. Cats cloned from fetal and adult somatic cells by nuclear transfer. Reproduction. 2005;129:245–9.

Yin XJ, Lee HS, Choi EG, et al. *In vitro* maturation of oocytes derived from the brown bear (*Ursus arctos*). J Reprod Dev. 2007;53:685–90.

Yin XJ, Lee HS, Yu XF, et al. Generation of cloned transgenic cats expressing red fluorescence protein. Biol Reprod. 2008;78:425–31.

Young KM, Brown JL, Goodrowe KL. Characterization of reproductive cycles and adrenal activity in the black-footed ferret (*Mustela nigripes*) by fecal hormone analysis. Zoo Biol. 2001;20:517–36.

Yu L, Peng, D, Liu J, et al. On the phylogy of Mustelidae subfamilies: analysis of seventeen nuclear non-coding loci and mitochondrial complete genomes. BMC Evol Biol 11.92: (2011) http://www.biomedcentral/1471-2148/11/52.

Zambelli D, Cunto M. Semen collection in cats: techniques and analysis. Theriogenology. 2006;66:159–65.

Zambelli D, Prati F, Cunto M, et al. Quality and *in vitro* fertilizing ability of cryopreserved cat spermatozoa obtained by urethral catheterization after medetomidine administration. Theriogenology. 2008;69:485–90.

Zhang H, Li D, Wang C, et al. Delayed implantation in giant pandas: the first comprehensive empirical evidence. Reproduction. 2009;138:979–86.

Zhang MJ, Hou R, Zhang AJ, et al. *In vitro* maturation of follicular oocytes of the Giant Panda (*Ailuropoda melanoleuca*): a case report. Theriogenology. 1998;49:1251–5.

Zhou G-B, Ma C-B, Liu G-S, et al. Vitrification of farmed blue fox oocytes in ethylene glycol and DMSO-based solutions using open-pulled straw (OPS). CryoLetters. 2009;30:112–8.

Zuercher GL, Swarner M, Silveira L, et al. Bush dog (*Speothos venaticus*). In: Sillero-Zubiri C, Hoffman M, Macdonald DW, editors. Canids: foxes, wolves, jackals and dogs. Status survey and conservation action plan. Gland: IUCN/SSC Canid Specialist Group; 2004. p. 76–80.

Chapter 11
Methods to Examine Reproductive Biology in Free-Ranging, Fully-Marine Mammals

Janet M. Lanyon and Elizabeth A. Burgess

Abstract Historical overexploitation of marine mammals, combined with present-day pressures, has resulted in severely depleted populations, with many species listed as threatened or endangered. Understanding breeding patterns of threatened marine mammals is crucial to assessing population viability, potential recovery and conservation actions. However, determining reproductive parameters of wild fully-marine mammals (cetaceans and sirenians) is challenging due to their wide distributions, high mobility, inaccessible habitats, cryptic lifestyles and in many cases, large body size and intractability. Consequently, reproductive biologists employ an innovative suite of methods to collect useful information from these species. This chapter reviews historic, recent and state-of-the-art methods to examine diverse aspects of reproduction in fully-aquatic mammals.

Keywords Reproduction • Marine mammals • Techniques • Cetaceans • Sirenians

1 Introduction

Many marine mammal species have severely depleted populations from historical exploitation including commercial harvest and/or subsistence hunting, and other anthropogenic activities (e.g., incidental drowning through fisheries by-catch, vessel strike, pollution) (Fowler 1981, 1984; Clapham et al. 1999). An estimated 36 % of marine mammal species are now threatened *cf.* 25 % of terrestrial mammal taxa (Schipper et al. 2008). Recovery rates of these impacted populations are dependent on their capacity to survive and reproduce in the face of contemporary threats.

J.M. Lanyon, B.Sc. (Hons), Ph.D. Zoology (✉) • E.A. Burgess, B.Sc. (Hons), M.Sc., Ph.D.
School of Biological Sciences, The University of Queensland,
Brisbane, QLD 4072, Australia
e-mail: j.lanyon@uq.edu.au

W.V. Holt et al. (eds.), *Reproductive Sciences in Animal Conservation*, Advances in Experimental Medicine and Biology 753, DOI 10.1007/978-1-4939-0820-2_11,

Reproductive potential is the maximum reproductive capacity of a population when resources are unlimited, and this is probably rarely achieved by wild mammalian populations, and possibly never realized by marine mammals. More realistically, the potential of a population to maintain or increase its size through reproductive growth is contingent on a suite of factors including life history parameters (e.g., growth rate, age at maturity), reproductive cycles (e.g., timing of estrous, mating, gestation, birth, lactation, resting periods) and population size and structure (e.g., sex ratio, number of reproductively active adults), which is then tempered by how well individuals reproduce under various levels of resource limitation and external stressors, i.e., their reproductive fitness. Understanding reproductive patterns of marine mammals and the factors that influence their reproductive potential is critical if we are to identify and manage threats to population survival and growth.

Marine mammals are K-selected species meaning that most are long-lived and slow to mature with protracted reproductive parameters, low birth rates and relatively large investment in their young (Fowler 1981). These life history traits make reproductive studies of marine mammal species challenging to conduct (i.e., requiring long-term datasets) but paramount to conservation practice because most populations are vulnerable to extrinsic mortality. For instance, the past heavy exploitation and incidental by-catch of many cetacean species has caused marked changes to many reproductive parameters (Fowler 1981, 1984). Slow breeding (including increased age at first calving, reduced number of reproductively active females, low calf production, longer inter-birth intervals) is a common response of long-lived iteroparous species such as marine mammals to adverse environmental conditions (DeMaster 1981; Kraus et al. 2001; Hadley et al. 2006). Interestingly though, in some baleen whales, enhanced reproductive capacity (i.e., decreased age at sexual maturation, increased proportion of mature females in ovulation, and increased pregnancy rates) appears to have occurred, possibly in response to increased prey availability following reduction in whale numbers after commercial whaling (Lockyer 1984; Lockyer and Smellie 1985; Boyd et al. 1999).

Furthermore, many marine mammal species are distributed over wide latitudinal ranges, across multiple oceans or biogeographic regions (Schipper et al. 2008), so that reproductive parameters of demographically-isolated populations are likely to vary across environmental gradients. Intraspecific populations of marine mammals may therefore exhibit regional differences in asymptotic body size, pre-reproductive period, breeding frequency or season, length of reproductive cycles, and ultimately, reproductive potential and resilience. For example, eastern and northern spinner dolphin (*Stenella longirostris*) populations vary in body length, breeding seasonality, and ovulation rates (Perrin and Reilly 1984; Perrin et al. 1985). Populations of pantropical spotted dolphins (*Stenella attenuata*) north and south of the equator show differences in each of timing of breeding seasons, proportion of pregnant females (Barlow 1985), and the average body length/age at puberty (Chivers and Myrick 1993). For some cetacean species, differences in growth and reproductive parameters have been sufficiently distinctive to identify discrete management units (e.g., Botta et al. 2010). Population life history studies have led to a better understanding of the basic biology of marine mammal species, their relationships to local environments (Börjesson and Read 2003; Danil and Chivers 2006), and to their stock structure (Dizon et al. 1994), behavior (Whitehead and Mann 2000), mating systems (Murphy et al. 2005), and resilience to anthropogenic threats (Chivers and Myrick 1993; Wells et al. 2005). When working to conserve marine mammals, it is therefore imperative that life history parameters appropriate to the focal populations are understood and applied to population models. As a fundamental step, reproductive scientists need reliable and useful methodologies to assess reproductive processes of marine mammals.

Determining reproductive parameters of wild marine mammals offers particular challenges. Species that breed on land, such as pinnipeds (seals, sea lions, walrus), tend to be better understood because their terrestrial life stage(s) present opportunities to directly measure reproductive timing, pregnancy rates and recruitment into a population via sampling of individuals of different life stages including pups at haul-out sites (e.g., Boyd 2000; McKenzie et al. 2005; Gibbens et al. 2010; Oosthuizen et al. 2012). In contrast, fully aquatic marine mammals, i.e., cetaceans (dolphins, porpoises, whales) and sirenians (manatees and dugongs), offer relatively limited opportunities for direct observation because of their offshore and/or remote habitats, great mobility (including long breeding migrations), patchy distribution and/or cryptic natures that arise from their spending large proportions of their lives underwater and out of sight (Costa 1993). For these reasons, the reproductive scientist must employ a series of non-traditional, often indirect, methods to elucidate reproductive parameters. These differences in approach are reflected in the type and amount of reproductive information available for the major groups of marine mammals: information for the semi-aquatic pinnipeds is markedly more extensive (Atkinson 1997; Boyd et al. 1999), possibly because of greater access to individuals at particular life stages. For the entirely-aquatic cetaceans, the bottlenose dolphin's reproductive biology is the best known due to its accessibility to researchers, i.e., it is the species most commonly held in aquaria and is found in coastal waters worldwide, often in close proximity to humans (Connor et al. 2000). In contrast, reproductive parameters of most other cetacean species are relatively unknown. This chapter reviews the historical, contemporary and state-of-the-art techniques that are applied to examine various aspects of reproductive biology in fully-aquatic marine mammals, i.e., cetaceans and sirenians (Table 11.1).

2 State of the Art: How to Study Reproduction in Fully Marine Mammals

2.1 Captive Fully-Marine Mammals

Marine mammals in captivity provide a unique opportunity to investigate species-specific behavior (e.g., vocalizations: Edds et al. 1993; Vergara et al. 2010; Ridgway et al. 2012; cognitive function: Mercado and deLong 2010; post-natal development:

Table 11.1 Summary of reproductive techniques in both female and male marine mammals with referenced applications in free-ranging populations (where possible)

Technique	Sex	Significance	Example studies
Captive studies	F	Longitudinal monitoring, estrous cycling, gestational changes, growth rate	[a]Biancani et al. (2009), [a]Steinman et al. (2012)
	M	Longitudinal monitoring, sexual maturity and activity, ejaculate characteristics, growth rate	[a]Robeck and Monfort (2006), [a]Yuen et al. (2009)
Carcass dissection and histology, immunohistology	F	Ovarian follicular activity, previous ovulations and birth, absolute age	Marsh and Kasuya (1986), Foote (2008)
	M	Testicular development, stage of spermatogenesis, absolute age	Desportes et al. (1993), Holt et al. (2004), Murphy et al. (2005)
Direct observation	M/F	Mating behavior, social associations, calf abundance, length of nursing, intercalving period, growth and maturity rates	Würsig and Jefferson (1990), Parks et al. (2007), de Bruyn et al. (2011)
Sexual dimorphism	M/F	Mating system, indicator of maturity	Whitehead (1994), MacLeod (1998), Burgess et al. (2012b), Fitzpatrick et al. (2012)
Body scarring	M/F	Aggressive social interactions: sexual coercion, male-male competition, energetics	Scott et al. (2005), Burgess et al. (2013a, b)
Endocrinology	F	Estrous cycling, pregnancy diagnosis, lactation, fertility potential	Rolland et al. (2005), Kellar et al. (2006), Wilson et al. (2011), Burgess et al. (2012a)
	M	Sexual maturity and activity, mating behaviour, development of secondary sexual characters, fertility potential	Rolland et al. (2005), Kellar et al. (2009), Wilson et al. (2011), Burgess et al. (2012b)
Ultrasound	F	Reproductive tract activity, fetal development	[a]Brook (2001), [a]Lacave et al. (2004), Burgess et al. (2012a)
	M	Gonadal activity	[a]Brook et al. (2000)
Spermatozoa analysis	M	Sperm morphology and morphometrics, ultrastructure, phylogenetic comparisons	Plön and Bernard (2006), [a]Yuen et al. (2009)
Gamete cryopreservation	F	Oocyte preservation	Fujihira et al. (2006), Bhuiyan et al. (2009)
	M	Sex sorting, artificial insemination (AI)	[a]O'Brien et al. (2008), [a]O'Brien et al. (2009), [a]O'Brien and Robeck (2012b)

[a]Technique primarily applied in captive population of marine mammals

Lyamin et al. 2005; Favaro et al. 2013) and physiology (e.g., dive response: Kooyman et al. 1981) that normally would be challenging to study in the wild. Extrapolation of information gained from captive individuals to wild populations can be complicated (see Lambrechts et al. 1999; Jenssen et al. 2001), because extended periods spent in artificial environment may induce variation in physiological traits (e.g., Mellish et al. 2006), including reproduction. Nonetheless, significant insights into reproductive cycles of marine mammals have been made through captive breeding programs, where individuals' reproductive status have been monitored in longitudinal studies of behavior and internal physiological state (e.g., Atkinson 1997; Boyd et al. 1999; Robeck et al. 2001).

Captive marine mammals for which reproductive parameters have been obtained, including within artificial breeding programs, comprise delphinids (dolphins, killer whales, false killer whales; e.g., O'Brien and Robeck 2012a; Robeck and Monfort 2006; Atkinson et al. 1999), monodontids (beluga; e.g., Steinman et al. 2012), sirenians (manatees, dugongs; e.g., Amaral et al. 2009; Burgess et al. 2013a) and pinnipeds (seals, sea lions, walrus; e.g., Atkinson and Gilmartin 1992; Myers et al. 2010; Kinoshita et al. 2012). Due to limitations of space and provision of adequate nutrition, marine mammals housed in zoological settings are typically small (<10 m long) and mostly tractable. Larger marine mammals, such as all mysticete cetaceans (Edds et al. 1993; Sumich et al. 2001), have not been held in captivity past reproductive age. A number of factors may affect the quality and robustness of reproductive data obtained from captive animals. First, numbers of captive individuals are usually small. Second, individuals are often segregated by sex and age so that natural social groupings may be fragmented or nonexistent. Furthermore, the effects of highly artificial physical surroundings on an animal's reproductive behavior and functioning are not well understood.

Despite these limitations, observations of mating and subsequent parturition in captive animals have enabled at least initial estimates of age at maturity, reproductive seasonality, and length of gestation for many smaller species (McBride and Hebb 1948; McBride and Kritzler 1951; Essapian 1962; Cornell et al. 1987; Odell et al. 1995; Blanchet et al. 2006). More recently, training programs using operant conditioning along with advanced monitoring technologies (see Sects. 2.4 and 2.5) have enabled collection of serial biological samples and/or physical examinations under temporal regimes that are adequate to characterize reproductive cycles in live females and males (reviewed by O'Brien and Robeck 2012b).

A great advantage of captive marine mammals is that most can be trained to present voluntarily for collections of body fluids including blood (Robeck and Monfort 2006; Bauer et al. 2010), saliva (Pietraszek and Atkinson 1994; Hogg et al. 2005), respiratory mucus (Hogg et al. 2005), ocular secretions (Amaral et al. 2009), urine (Walker et al. 1988; Robeck et al. 1993, 2005a; Wakai et al. 2002), feces (Biancani et al. 2009), and milk (West et al. 2000), all of which may be useful for longitudinal monitoring of hormone profiles. The relatively non-invasive techniques of fecal sampling (and urinary sampling with training) can be conducted daily and offer the best prospect for examining endocrine-reproductive relationships. Such investigations of reproductive cycling have been enhanced by monitoring body temperature (Terasawa et al. 1999; Katsumata et al. 2006a, b), examining the reproductive tract using trans-abdominal ultrasound (Brook et al. 2000, 2001, 2004; Lacave et al. 2004), conducting vaginal cytology of females (Pietraszek and Atkinson 1994), collecting semen from males (Schroeder and Keller 1989; Robeck and Monfort 2006; O'Brien et al. 2008; Robeck et al. 2009; Yuen et al. 2009) and simultaneously documenting reproductive behavior (e.g., Tavolga and Essapian 1957; Shane et al. 1986).

Understanding the reproductive cycles of captive individuals has advanced husbandry practices including management of pregnant females and fetal health (Williamson et al. 1990; Lacave et al. 2004; Katsumata et al. 2006b), establishment of socially cohesive groupings (Shane et al. 1986; Waples and Gales 2002; O'Brien and Robeck 2006), contraception (Atkinson et al. 1993; Briggs 2000; Calle 2005) and captive breeding programs (Robeck et al. 2005b). It has enabled the development of assisted reproductive technologies including artificial insemination (Robeck et al. 2004, 2005b, 2009, 2010; O'Brien and Robeck 2006, 2010; O'Brien et al. 2008), estrous synchronization (Robeck et al. 2009), sperm sexing (O'Brien and Robeck 2006; O'Brien et al. 2009), and semen preservation (Robeck and O'Brien 2004; O'Brien et al. 2008; Yuen et al. 2009). Understanding the relationships between reproductive behavior and physiological events for marine mammals in captivity has enabled facilities to determine and/or manipulate readiness for breeding, and to ensure successful conception for several cetacean taxa (Cornell et al. 1987; Robeck et al. 2005b; O'Brien and Robeck 2012a, b). With the increasing need for aquaria to display animals born in captivity rather than wild-caught individuals, the success of captive breeding programs for marine mammals is necessary to sustain *ex situ* populations and manage genetic bottlenecks that can occur in reproductively-isolated facilities (reviewed by O'Brien and Robeck 2012b).

2.2 Dissection of Carcasses

Almost all of the information regarding basic reproductive patterns of wild cetaceans and sirenians has been derived from the examination of reproductive tracts recovered from carcasses obtained from hunted (e.g., Best 1969; Miyazaki 1984; Desportes et al. 1993; Dawbin 1997; Kwan 2002) or stranded animals (e.g., Calzada et al. 1996; Evans and Hindell 2004) or from those incidentally caught and killed in fishing nets (e.g., Read 1990; Van Waerebeek and Read 1994; Hohn et al. 1996) or shark nets (e.g., Marsh et al. 1984a, b; Cockcroft and Ross 1990). Early studies using post-mortem material provided information on gross morphology and anatomy of the reproductive tract (e.g., Harrison and Ridgway 1971). Functional morphology of reproductive tracts, histology of gonads and mammary glands, and embryology (van der Schoot 1995; Thewissen and Heyning 2007) have provided information on the timing of reproductive activity, conception, gestation, birth and lactation, and also information concerning age and/or body size-related life history parameters including first reproduction and numbers of offspring produced over lifetimes (e.g., cetaceans: Laws 1961; Ohsumi and Masaki 1975; Lockyer and Martin 1983; Marsh and Kasuya 1984, 1986; Kato and Sakuramoto 1991; Read and Hohn 1995; sirenians: Marsh et al. 1984a, b; Hernandez et al. 1995; Marmontel 1995; Kwan 2002). For example, sexual maturity in males can be determined through assessment of testicular development (including seminiferous tubule diameter, Sertoli and Leydig cell activity), relative weight and length of testes, stage of spermatogenesis and presence/abundance of sperm (e.g., Perrin and Henderson 1984; Perrin and Reilly 1984). More recently, immunocyto-chemical methods have been applied to examine maturation of gonads: these include screening for distribution of cytoskeletal proteins (e.g., smooth muscle actin and vimentin) as markers of testis development (Holt et al. 2004). For females, ovarian follicular activity is indicative of maturity, and previous ovulations have been confirmed by presence of *corpus luteum* and/or *corpus albicans*, since in many marine mammals, ovarian scars appear to be persistent (e.g., Marsh and Kasuya 1984; Perrin and Reilly 1984). Pregnancy rate and calving intervals can be determined though counts of placental scars (e.g., dugongs, Marsh et al. 1984a) or ovulatory scars (in cetaceans, most are thought to represent pregnancies, Boyd et al. 1999) regressed against age or body length at death. Timing of sexual activity is suggested by the presence of mature spermatozoa in males and ovarian activity in females (Perrin and Reilly 1984) and reproductive senescence through their absence (Marsh and Kasuya 1986).

Carcass analysis, for all of its value, is prone to both temporal and/or spatial biases (Boyd et al. 1999). Carcasses are usually sourced from places where hunting, harvest or other causes of mortality are most likely, or from areas where reportage or recovery of carcasses are most common (e.g., Chittleborough 1954; Marsh et al. 1984a, b; Marsh and Kasuya 1986; Desportes et al. 1993; Kwan 2002). In some cases, certain individuals or life stages are more likely to be sampled, e.g., the targeted fishery for dugongs where hunters preferentially take pregnant rather than resting females (Boyd et al. 1999); or samples obtained at certain times of the year, e.g., from baleen whales during the hunting season (Chittleborough 1954; Laws 1961). Further, the utility of a carcass depends on its degree of decomposition, so that if carcasses from warmer climates decompose faster, less information may be available for these than for animals from higher latitudes. However, the value of carcasses for studying reproduction and other facets of marine mammal biology cannot be understated. Indeed, for some rare marine mammal species, carcass recovery has been the only avenue for direct examination (e.g., Reyes et al. 1991; Thompson et al. 2012). In many cases, however, recoveries are too few to obtain useful information regarding reproductive biology. In contrast, for those species that strand *en masse*, recovery of multiple carcasses has yielded reproductive profiles of social groupings (e.g., Mignucci-Giannoni et al. 2000). When using carcass analysis to obtain information, there is significant value in establishing coordinated responses to marine mammal strandings (e.g., Thompson et al. 2012) at regional or national levels to ensure that, when rescue attempts fail, there is timely collection of tissues (including from reproductive tracts) and other vital data.

2.3 Direct Observations

Direct observations of reproductive activity and behavior at sea of free-ranging, fully marine mammals vary from single or opportunistic sightings (e.g., courtship: Preen 1989; birth: Stacey and Baird 1997; copulation: Adulyanukosol et al. 2007; naso-suckling: Gero and Whitehead 2007) through to longitudinal sightings (Glockner-Ferrari and Ferrari 1990; Scott et al. 1990; Whitehead 1993; Langtimm et al. 1998; Rathbun et al. 1995; Wells 2000; Rowntree et al. 2001; Burgess et al. 2012a, b). Unsurprisingly, the quality of these data varies markedly depending on approachability of the species and the observational methods applied. The observation methods should be selected carefully to meet the objectives and scale of the research project.

Distinguishing sexes of marine mammals through visual means can be challenging for species that exhibit no obvious sexual dimorphism in body size or morphology, e.g., most cetaceans and sirenians. Male marine mammals are testicond (ascrotal) with their testes permanently concealed in the abdomen (excluding otariid seals which have scrotal testes; Atkinson 1997) and the flaccid penis can be retracted into the body wall (Marsh et al. 1984b; Rommel et al. 1992) facilitating streamlining. Thus, for the majority of sexually-monomorphic species, the sex of most observed individuals remains unknown unless genetic sexing is accomplished. For most cetaceans and sirenians, mating has never been confirmed and courtship is difficult to interpret or recognize.

In fact, detailed information on patterns of courtship, mating and parturition in the wild only exists for a small number of fully-marine mammal species (e.g., Karczmarski et al. 1997; Stacey and Baird 1997; Hückstadt and Antezana 2001; Parks et al. 2007) and is mostly documented for pinnipeds that breed and give birth on land (e.g., Le Boeuf 1991; Cassini 1999; Acevedo et al. 2008; Karamanlidis et al. 2010). A general understanding of reproductive patterns for most entirely-aquatic species has largely been compiled from limited, often opportunistic, data and then analogized according to our knowledge of other mammalian species, e.g., primates (Wells 2003; Connor 2007). For some large odontocetes including beaked and bottlenose whales, their rarity, remote habitats and secretive habits ensure that they are rarely, if ever, observed alive (Heyning 1984; Thompson et al. 2012) so that their reproductive behavior remains a mystery. For many of the mysticete whales, direct observation is often fleeting and opportunistic, and rarely extensive or enlightening in terms of reproductive behavior. Exemplifying this point is the little we know about the reproductive habits of the largest animals that have ever existed, blue whales, *Balaenoptera musculus* (Yochem and Leatherwood 1985) despite having documented extensively various aspects of their ecology. Their elusive nature, deep-water habits and open geographic ranges (i.e., the only clear geographic boundaries to baleen whale movement are large continental land masses) have meant that breeding grounds of Southern Hemisphere populations remain unverified (Attard et al. 2012). However, for other large baleen whales (e.g., Kraus et al. 2001; Clapham 2008), some observations regarding reproductive behavior have been conducted at particular locations along predictable migration routes between feeding and breeding grounds or in breeding grounds themselves, particularly for species that travel along topographic features such as coastlines or continental shelves. Amongst the best known of the marine mammals are those that spend significant time in coastal habitats (including migratory whales, delphinids, manatees) or at known breeding aggregation points, e.g., around islands, sea mounts, reefs and along mid ocean ridges.

Direct observation methods usually include distant sightings from elevated platforms: vessels, aircraft or land-based vantage points. Even though some fully-marine mammals may be approached closely by boat at times (e.g., manatees, Hartman 1979; dugongs, Preen 1992), most of their reproductive activities probably occur underwater and are unavailable to observers, so that some inference may be required (e.g., sex of observed individuals). In contrast, indirect 'observations' rely on telemetry to track animals, and these methods call for capture and attachment of radio/satellite/GPS/acoustic tags or recoverable data loggers (e.g., Mate et al. 1998; Noad and Cato 2001; Deutsch et al. 2003). Information on location and movements of tagged individuals have been particularly useful for tracking breeding migrations, determining timing of reproductive cycles, or for examining smaller scale movements and social associations within breeding grounds (Mate et al. 1998, 2003; Russell et al. 2013).

Direct visual observations may also be conducted at the population level and may yield information regarding spatial distribution of animals, density and timing of breeding migrations, adult aggregations, breeding associations, and presence of neonates. In these cases, individual identity of group members is not always necessary. Examples of this approach include aerial surveys of humpback whale adults aggregating within breeding grounds (Andriolo et al. 2006), surveys of right whale nursery areas and adult aggregations (Rowntree et al. 2001; Parks et al. 2007) and calf counts for various taxa as an index of annual recruitment of breeding success (Perryman et al. 2002). Compilation of these observations over extended temporal scales gives information regarding reproductive timing and breeding patterns.

Behavioral observations are greatly enhanced through recognition of individuals. If individual animals can be distinguished and identified through discriminate marks (permanent pigmentation, dorsal fin shape and/or body scarring, genetic tags) or artificially-applied marks (physical tags), there is the opportunity for longitudinal studies of life history. Recognizable individuals allow for descriptions of basic activity patterns (resting, socializing, travelling, feeding) as well as interpretation of inter-specific behaviors, especially if sex and reproductive condition are known (e.g., Connor et al. 1992; Smolker et al. 1992; Brown et al. 1995). For example, long-term observations of known identified mother-calf pairs of Florida manatees at winter aggregation sites have provided insight into frequency of calving, intercalving period, growth rates of calves, duration of nursing, time to maturity and other reproductive measures critical to population modeling (Langtimm et al. 2004). The value of longitudinal sightings data lies not only in compilation of life history parameters relevant to reproductive biology (including reproductive and total life span if individuals are recognized from an early age), but in elucidation of social behaviors including social associations, mate selection, nurturing of young and other details about a species' breeding behavior (Mann et al. 2000). Individual identification has become a staple of many field studies of cetaceans (reviewed by Würsig and Jefferson 1990).

Despite the technical difficulties in directly observing and identifying mammals in the pelagic environment, our understanding of reproductive behavior and social interactions is improving. New techniques that allow us to track individuals through time and space, such as the amplification of molecular markers from tissue biopsies for tagging as well as advanced telemetry and biologging (Block 2005; Bograd et al. 2010), are providing superior insights into social structure and social associations related to reproduction.

2.4 Endocrinology

Since most reproductive processes are mediated through hormones circulated in the bloodstream, and hormonal functions tend to be conserved across mammalian taxa, endocrine analysis is a useful means of assessing reproductive status (e.g., sexual maturity, pregnancy diagnosis) and reproductive activity (e.g., estrous cycles, seasonality), particularly for live individuals.

The most commonly measured hormone in marine mammals is progesterone (Table 11.2) since its diagnostic use indicates the onset of female sexual maturity or pregnancy. Understanding reproductive status of females provides vital information for the management of both wild (e.g., Gardiner et al. 1996; Atkinson et al. 1999; Rolland et al. 2005; Kellar et al. 2006; Tripp et al. 2008; Villegas-Amtmann et al. 2009; Burgess et al. 2012a) and captive populations (e.g., Pietraszek and Atkinson 1994; Atkinson et al. 1999; Biancani et al. 2009; O'Brien and Robeck 2012a). The other sex steroid hormones (Table 11.2) that are routinely measured are testosterone (Atkinson and Gilmartin 1992; Desportes et al. 1994; Kellar et al. 2009; Burgess et al. 2012b) and various forms of estrogen (e.g., Francis-Floyd et al. 1991; Pietraszek and Atkinson 1994; Robeck et al. 2005b). Gonadotropins including luteinizing hormone (LH) and follicle stimulating hormone (FSH), and other protein hormones (Table 11.2) have primarily been measured in cases of assisted reproduction (i.e., artificial insemination or gamete harvesting), often requiring multiple samples collected daily (Robeck et al. 2004, 2009; Muraco et al. 2010); although some studies exploiting opportunistic sample collection have been reported (Suzuki et al. 2001; Watanabe et al. 2004; Hao et al. 2007)

In early studies, endocrine analysis relied on collecting blood samples to determine circulating hormone concentrations so that marine mammals held in captivity were sampled more frequently than those in the wild (Yoshioka et al. 1986; Cornell et al. 1987; Duffield et al. 1995). The first ground-breaking hormonal study on marine mammals investigated serum testosterone in two captive male bottlenose dolphins (Harrison and Ridgway 1971). Broad-scale blood sampling for endocrine analysis in field studies has been possible in few circumstances such as during

Table 11.2 Summary of endocrine glands and hormones investigated in marine mammals with referenced applications in free-ranging populations (where possible)

Gland	Hormone	Chemical class	Principal functions	Example studies
Ovary Graafian follicle	Estrogens (e.g. estradiol)	Steroid	Female mating behavior, secondary sex characteristics, maintenance of female duct system, mammary growth	[a]Robeck et al. (2005a, b), Rolland et al. (2005), [a]Steinman et al. (2012)
	Inhibin	Protein	Regulates release of FSH from anterior pituitary	Wetzel et al. (2009)
	Anti-Müllerian hormone (AMH)	Protein	Sophisticated biomarker of reproductive potential	Wilson et al. (2011)
Ovary Corpus luteum	Progestins (e.g. progesterone)	Steroid	Maintenance of pregnancy, mammary growth and secretion	Rolland et al. (2005), Burgess et al. (2012a), [a]O'Brien and Robeck (2012a)
	Relaxin	Polypeptide	Expansion of pelvis, dilation of cervix	Schwabe et al. (1989), [a]Bergfelt et al. (2011)
Testis Leydig cells	Androgens (e.g. testosterone)	Steroid	Male mating behavior, spermatogenesis, maintenance of male duct system and accessory glands	Rolland et al. (2005), [a]Robeck and Monfort (2006), Burgess et al. (2012b)
Sertoli cells	Inhibin	Protein	Regulates release of FSH	Miller et al. (2002a, b), [a]Katsumata et al. (2012)
	Anti-Müllerian hormone (AMH)	Protein	Sophisticated biomarker of reproductive potential	Wilson et al. (2011)
Adrenal cortex	Glucocorticoids (e.g. cortisol, corticosterone)	Steroid	Stress response, parturition induction, milk synthesis	[a]Hunt et al. (2004, 2006), Myers et al. (2010), Burgess et al. (2013a, b)
Pineal gland	Melatonin	Biogenic amine	Control of seasonal reproduction	Barrell and Montgomery (1989), Aarseth et al. (2003), [a]Funasaka et al. (2011), [a]Panin et al. (2012)
Posterior Pituitary	Oxytocin	Octapeptide	Parturition and milk ejection in females	Archer et al. (1964), Eisert et al. (2013)
Anterior pituitary	Follicle stimulating hormone (FSH)	Glycoprotein	Stimulate follicle growth and estrogen production in females, spermiogenesis in males	[a]Walker et al. (1988), Gardiner et al. (1999), Suzuki et al. (2001), Watanabe et al. (2004), Hao et al. (2007)
	Luteinizing hormone (LH)	Glycoprotein	Stimulate ovulation, support corpus luteum formation and progesterone secretion, stimulate testosterone synthesis by Leydig cells of testis	Gardiner et al. (1999), Suzuki et al. (2001), Watanabe et al. (2004), [a]Robeck et al. (2005a, b), Hao et al. (2007)
	Prolactin	Protein	Development of mammary function, maintain lactation, effects on maternal behavior, post-partum estrous cycling	Boyd (1991), [a]Steinman et al. (2012)
	Adrenocorticotrophic hormone (ACTH)	Protein	Release of corticosteroids and glucocorticoids, initiate parturition	[a]Schmitt et al. (2010), Tripp et al. (2010), Schaefer et al. (2011)
Hypothalamus	Gonadotrophic releasing hormone (GnRH)	Decapeptide	Stimulate release of FSH and LH from anterior pituitary	Atkinson et al. (1998), [a]Robeck et al. (2010), [a]O'Brien and Robeck (2012a)

[a]Technique primarily applied in captive population of marine mammals

haul-out of the semi-aquatic pinnipeds (e.g., Atkinson and Gilmartin 1992; Gardiner et al. 1996; Harcourt et al. 2010), or when sampling relatively small numbers of fully aquatic mammals during health assessments (e.g., Tripp et al. 2008) or even during lethal harvests (e.g., Suzuki et al. 2001; Kjeld et al. 2004, 2006). The need for physical restraint out-of-water to ensure collection of blood samples uncontaminated by seawater has hampered endocrine evaluations in most species, particularly of large or rare cetaceans. More recently however, sampling of media other than blood for hormone profiles has been increasingly recognized as a useful approach for free-ranging aquatic mammals (reviewed by Amaral 2010; Hunt et al. 2013).

Hormones are removed from circulation through metabolic processes and eventually excreted from the body through various pathways (depending on the species), providing alternative opportunities for measuring hormones and their byproducts. Hormone concentrations have been measured in a variety of body tissues and excreta from marine mammals, including saliva (Pietraszek and Atkinson 1994; Atkinson et al. 1999; Hogg et al. 2005; Amaral et al. 2009), milk (West et al. 2000), ocular secretion (Amaral et al. 2009), respiratory exudate (Hogg et al. 2005, 2009), adipose tissue (Mansour et al. 2002; Kellar et al. 2006, 2009; Pérez et al. 2011), muscle (Yoshioka et al. 1994), urine (Walker et al. 1988; Robeck et al. 1993; Wakai et al. 2002; Muraco et al. 2010) and feces (Larkin et al. 2005; Rolland et al. 2005; Biancani et al. 2009; Burgess et al. 2012a, b). These alternative endocrine approaches have already been widely applied to monitoring reproductive processes in captive individuals (see Sect. 2.1) and show promise for live, free-ranging populations. In fact, measuring levels of excreted hormones may offer advantages over blood levels because these represent averaged values pooled over time (i.e., integrated over the gut passage time for fecal samples), rather than a single point-in-time measure. However, it is important that for each new species under study, careful biochemical and biological validation of hormones in each sampled medium is conducted (Lasley and Kirkpatrick 1991). Furthermore, researchers must carefully consider the biology, ecology, and habits of the species, as well as the project's scale, when deciding on which media are most appropriate (see Hunt et al. 2013).

To date, a few field studies have successfully sampled feces (Lanyon et al. 2005; Larkin et al. 2005; Rolland et al. 2005; Burgess et al. 2012a, b) or blubber (Kellar et al. 2006, 2009; Pérez et al. 2011) to assess reproductive hormone profiles of both sexes of live cetaceans and sirenians. Obviously, logistical difficulties arise with collection of such samples from elusive and large marine mammals (reviewed by Hunt et al. 2013). In some cases, floating feces voided by large whales have been sampled by dip-net during focal follows of, for example, right whales (Rolland et al. 2005; Hunt et al. 2006), sperm whales (Smith and Whitehead 2006; Marcoux et al. 2007), gray whales (Newell and Cowles 2006), blue and humpback whales (Lefebvre et al. 2002) either in feeding grounds or along migration routes. Sinking feces have also been collected from smaller cetaceans by snorkelers towed by boats traveling in close proximity to dolphin groups, although a large proportion of fecal material disperses too rapidly for collection (Parsons et al. 1999, 2003). One innovation to facilitate collection has been the use of 'sniffer' dogs trained to detect floating whale feces (Rolland et al. 2006). Another approach has been to collect feces of uncertain origin, either floating or as stools washed ashore, and then conduct molecular analysis for taxon and/or sex identity (Rolland et al. 2005). In live-capture studies of smaller marine mammals, fecal samples have been collected either opportunistically as they are voided during restraint (Lanyon et al. 2005, 2010b; Larkin et al. 2005) or directly by inserting a soft latex tube into the distal rectum (Biancani et al. 2009; Burgess et al. 2012a, b).

For blubber hormone analysis, remote biopsy darting has been used to obtain small cores of fatty hypodermis from free-swimming cetaceans (Kellar et al. 2006, 2009) and this also has potential application in pinnipeds (e.g., Hoberecht et al. 2006). One unique study on Weddell seals (*Leptonychotes weddellii*) analyzed hormones in urine samples that had become preserved in ice (Constable et al. 2006). However, the routine collection of fluid samples is logistically difficult when collecting in an aquatic environment, unless the mammal is first removed from the water.

Perhaps the most innovative approach for measuring hormone levels in marine mammals has been the use of respiratory exudate or 'blow', which is currently in the early stages of development (see Hunt et al. 2013). Cetaceans ventilate a larger percentage of the respiratory system with greater force than other mammals (Ridgway et al. 1969) and their lungs are heavily vascularized (Pabst et al. 1999), causing expulsion of a substantial volume of lung exudate droplets (along with respiratory gases) with each exhalation. Preliminary research on captive bottlenose dolphins and large free-ranging baleen whales (humpback whales *Megaptera novaeangliae* and North Atlantic right whales *Eubalaena glacialis*) has demonstrated that at least some exhaled blow samples (collected into a sampling device at the end of a cantilevered pole) contain detectable steroid hormones (Hogg et al. 2005, 2009). However, current analytical techniques only detect presence/absence and not quantitative concentrations of steroid hormones (see Trout 2008), suggesting that important methodological issues need to be addressed before hormone analysis of exhaled blow can be applied more widely to assess physiological state of free-ranging marine mammals (Hunt et al. 2013). The relative ease with which blow samples can be collected and the promise that these may be chemically analyzed to answer a number of biological questions (Acevedo-Whitehouse et al. 2010; Frère et al. 2010) have already led several researchers to begin collecting these samples (e.g., Gero and Whitehead 2007). Ultimately, less-invasive hormone monitoring (i.e., using non-blood samples) provides marine mammal scientists with the capability to conduct both basic and applied physiological-endocrine research that can be integrated with other disciplines including genetics, behavior, nutrition, animal health, ecology and evolution.

The discipline of conservation endocrinology is still evolving for free-ranging marine mammals, and will undoubtedly continue to provide new and valuable information to ensure the survival of viable marine mammal populations into the future. Exciting developments in wildlife endocrine studies are enabling us to move beyond assessing an individual's reproductive status to being able to understand an individual's reproductive potential (i.e., fertility status). Managers of marine mammal populations need to know not only how many individuals are present and/or how many are of breeding age (i.e., effective population size), but also understand an individual's contribution to population growth. Most recently, sophisticated biomarkers (including anti-Müllerian hormone, AMH, inhibin A, inhibin B) have been developed to provide direct measures of gonadal function as well as fertility potential (ovarian reserve) in both female and male mammals (Table 11.2) (Lee and Donahoe 1993; Tremellen et al. 2006; Kumar et al. 2010). The first reports of detection of AMH and inhibin A and B in marine mammals, including Florida manatees (Wilson et al. 2011), bottlenose dolphins (Schwierzke-Wade et al. unpubl. data), beluga whales and dugongs (Wetzel et al. unpubl. data) suggest that these biomarkers offer promise as a means of evaluating the 'reproductive quality' of individuals comprising a population. With careful development and validation, the use of such biomarkers may provide insight into the effects of stressors on critical biological functions such as reproduction, which can then be used to inform effective and focused conservation efforts.

2.5 Diagnostic Imaging

Diagnostic imaging (ultrasonography and endoscopy) has been used to examine the urogenital tracts and implement advanced reproductive technologies (including artificial insemination) of terrestrial wildlife (Hildebrandt et al. 2003). These techniques have been applied to fully marine mammals, but in relatively limited ways due to the need for restraint and immobility during the procedures. Ultrasonography is a useful technique for examination and differentiation of internal soft tissues and compared to other reproductive imaging techniques is arguably the simplest, safest, less invasive and most cost effective method. It has been useful for obtaining vital information regarding reproductive activity and fetal development in mostly captive cetaceans (Brook 1997, 1999; Brook et al. 2001, 2004). Recent advances in ultrasound technologies have enabled researchers to examine gonadal activity through development of testes and ovarian follicles, monitor ovulatory cycles, implement artificial insemination, and estimate gestational age of fetuses and therefore parturition date for small captive marine mammals (Robeck et al. 1998, 2001, 2005b; Brook 1997; Brook et al. 2001, 2004; Lacave et al. 2004). Some captive cetaceans have been trained to station for ultrasound examination (Brook et al. 2001), whilst in a field situation, ultrasound may be applied to individuals removed from the water (Wells et al. 2004; Goldstein et al. 2006) or even restrained alongside a vessel if a portable device with waterproof transducer is deployed (Nichols 2005). For larger, but still tractable marine mammals, e.g., big males or individuals with thick blubber layers, ultrasonography may be relatively challenging and the outcome of examinations doubtful, whilst it is not practical for the largest free-ranging cetaceans.

Perhaps the greatest value of diagnostic sonography with respect to reproduction has been as an adjunct tool to physical examination and endocrinology. The use of ultrasonography for validation of reproductive state has been crucial in developing hormone-based diagnostic tests for pregnancy (e.g., in wild sirenians: Burgess et al. 2012a, Fig. 11.1) and for confirming pregnancy in captive cetaceans. Pseudopregnancy is possible in captive marine mammals (i.e., elevated hormone levels in

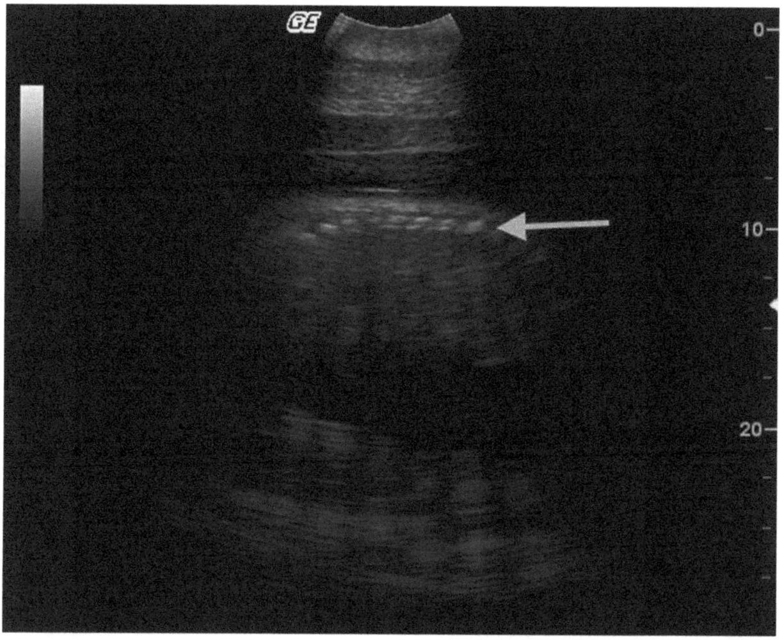

Fig. 11.1 Ultrasound image showing a sagittal cross-section of a fetus of a free-ranging dugong. An *arrow* marks the bony axial components of the fetus (Burgess et al. 2012a)

animals that are not carrying a fetus) (Atkinson 1997; West et al. 2000) and it is only through the use of ultrasound examination, that true pregnancy can be discriminated from this condition (Robeck et al. 2001). Furthermore, ultrasound combined with endocrine monitoring offers insight into how hormonal changes directly relate to ovulation (Robeck et al. 2001). Interestingly, the first evidence of facultative-induced ovulation in a cetacean, the beluga, was determined using serial urinary hormone monitoring combined with ovarian ultrasound (Steinman et al. 2012). Reproductive cycles of both males and females of some cetacean taxa have been determined through routine repeat ultrasounds of the gonads (e.g., inshore bottlenose dolphins: Brook 1997; Indo-Pacific dolphins: Brook et al. 2004). As ultrasound devices become smaller, more robust and affordable, and their value apparent, their use in marine field situations is only likely to increase.

Endoscopy is a minor surgical procedure and, although more invasive than ultrasonography, allows direct visual internal examination through minimal tissue trauma. Laparoscopy of the abdominal cavity of marine mammals has been conducted for some captive cetaceans and pinnipeds, mostly as a diagnostic tool for disease or for removal of swallowed foreign objects, with limited application in reproductive studies (Dover and van Bonn 2001). Flexible endoscopes have also been used to guide practitioners during artificial insemination operations of captive cetaceans (Robeck et al. 1994; Robeck 2000; Dover and van Bonn 2001).

2.6 Gamete Collection and Cryopreservation

Spermatogenesis is an important indicator of the reproductive status of the male mammal, including the onset of sexual maturity, the start of breeding in seasonally reproducing species, and gamete quality. Studies of the morphology of spermatozoa have been extensive for land mammals but remain in their infancy for marine mammals because of limited access to gametes. Detailed studies of the gross morphology of spermatozoa from wild marine mammals have been mostly limited to samples collected post-mortem from the epididymis and/or vas deferens, e.g., in cetaceans (reviewed by Plön and Bernard 2006; Kita et al. 2001), pinnipeds (Cummins and Woodall 1985) and sirenians (Marsh et al. 1984b; Miller et al. 2001). Sperm rescued postmortem from urogenital tracts of variable states of decomposition (dependent on timely collection and preservation, e.g., Mogoe et al. 1998; Miller et al. 2002a; Hiwasa et al. 2009), often comprise small samples of variable quality and viability, which makes comparisons between both taxa and ontogenetic stages within a taxon challenging (see Plön and Bernard 2006). Semen ejaculates from live, free-ranging marine mammals can be obtained using electro-ejaculation techniques (e.g., gray seals *Halichoerus grypus*, Lawson et al. 1996) or passively through opportunistic collection of voluntary ejaculate (e.g., dugongs, Burgess et al. 2012b; JM Lanyon unpubl. data).

Live captive cetaceans have provided the most detailed knowledge of spermatozoa (in terms of morphology, ultrastructure, motility, viability) in fully marine mammals, with specimens collected via electro-ejaculation (Fleming et al. 1981; Lawson et al. 1996) or conditioned for voluntary semen collection after manual external stimulation (Keller 1986; Miller et al. 2002b; Robeck and O'Brien 2004; O'Brien et al. 2008). The latter method potentially produces higher quality samples more indicative of ejaculates produced during natural matings. Such knowledge has helped elucidate aspects of sperm development and maturation and the fertilization process in marine mammals, and has provided fundamental insights for the development of assisted reproductive technologies.

Assisted reproductive technologies, including semen cryopreservation (Robeck and O'Brien 2004), artificial insemination (Robeck et al. 2005b) and more recently, sex-sorting of spermatozoa (O'Brien and Robeck 2006; Montano et al. 2010), provide a number of benefits to captive breeding programs such as permitting easier genetic exchange between facilities without the need for translocation of animals (which can be risky, expensive and disruptive to the stability of social groups) as well as sex-ratio management. Thus, research into gamete collection and preservation is considered high priority for *in situ* management of captive marine mammals, particularly cetaceans (O'Brien and Robeck 2012b). Due to species-specific ejaculate characteristics and sperm biology, the composition of diluents used to preserve spermatozoa *in vitro* varies considerably across species (e.g., slightly higher osmolarity of diluents for cetaceans than for terrestrial animals) (reviewed by Fukui et al. 2007), and requires careful experimentation for each new species under study (e.g., Fukui et al. 1996; Robeck and O'Brien 2004; Robeck et al. 2004, 2009; O'Brien and Robeck 2010).

Cryopreservation of ejaculated semen and formation of a genome (sperm) resource bank for captive populations has been achieved for bottlenose dolphins (Robeck and O'Brien 2004), killer whales (Robeck et al. 2004), Pacific white-sided dolphins (Robeck et al. 2009), belugas (O'Brien and Robeck 2010) and dugongs (T Keeley and R Bathgate unpubl. data). For wild populations, cryopreservation techniques have also been attempted for rescued spermatozoa from harvested common minke whales *Balaenoptera acutorostrata* (Mogoe et al. 1998) and Bryde's whales *Balaenoptera edeni* (Hiwasa et al. 2009) as well as voluntary ejaculates from live dugongs (JM Lanyon unpubl. data).

The potential use of genome banks can be further enhanced by the development of viable methods for oocyte and embryo preservation. To date, all significant research on *in vitro* oocyte maturation, fertilization and embryo culture has been performed using salvaged post-mortem tissues from freshly harvested whale (mysticete) species (Asada et al. 2001a; Iwayama et al. 2005; Fukui et al. 2007; Watanabe et al. 2007; Bhuiyan et al. 2009). Advanced studies have experimented with frozen-thawed oocytes (Asada et al. 2001a) and vitrified oocytes (Asada et al. 2001b; Fujihira et al. 2006) and have examined the feasibility of using interspecies somatic cell nuclear transfer to produce whale embryos (Lee et al. 2009; Bhuiyan et al. 2010). However, successful embryo preservation has not yet been reported for a marine mammal species.

It is easy to foresee that, as with terrestrial mammals, the collection and cryopreservation of viable gametes, combined with studies to ensure fertility following thawing, will enable the indefinite storage of valuable genetic material from threatened and endangered marine mammals. These approaches are already useful for captive breeding of marine mammals (Robeck et al. 2004, 2009, 2010; O'Brien and Robeck 2006; O'Brien et al. 2008, 2009), and in the future, may also be of value for conservation of wild populations (reviewed by O'Brien and Robeck 2012b).

3 Case Study

3.1 Investigating Reproductive Biology of a Fully Aquatic Cryptic Species: The Dugong

Understanding reproductive status and potential is important for the effective management of vulnerable marine mammal species such as the dugong. Dugongs are challenging to study because they are fully aquatic, live in turbid coastal waters, spend almost all of their lives underwater foraging on benthic seagrasses (Marsh et al. 2011), are shy and elusive, and have few distinguishing features to discriminate individuals (Lanyon et al. 2002 *cf.* Anderson 1995). Until recently, all knowledge of the reproductive biology of dugongs (e.g., sexual maturation, gestation, reproductive activity, seasonality) was obtained from analysis of dugong carcasses recovered from tropical north Queensland, Australia (Marsh et al. 2011). Tissue samples obtained from dead dugongs of both sexes, all ages and reproductive states have formed the basis of descriptions of functional anatomy and histology of the reproductive system and timing of breeding cycles (Marsh et al. 1984a, b, c). Direct observations of social associations in wild dugongs including presumed reproductive behaviors (courtship, mating) are few (Preen 1989; Anderson 1997) and remain largely anecdotal because the sexes of individuals involved could not be confirmed. No dugongs have been bred in captivity, and no wild births have been witnessed.

Examination of known-aged reproductive tracts suggests that dugongs may be amongst the slowest reproducers of all the marine mammals, with life history parameters broadly similar to humans. Female dugongs in parts of north Queensland mature at 13–17 years, have one calf at a time and then a long intercalving period of 3–7 years (Marsh et al. 1984a). Male dugongs similarly mature late (>9 years) and have been described as discontinuous breeders due to asynchronous spermatogenesis (Marsh et al. 1984b, c). However, there is evidence of marked inter- and even intra-population variation in reproductive parameters (Marsh et al. 2011) so that dugongs elsewhere in the tropics or at different times, possibly under more favorable nutritional circumstances, appear to grow faster, mature earlier (4–7 years) and reproduce more frequently (Kwan 2002; Marsh and Kwan 2008). The geographically-biased and diverse nature of life history and reproductive parameters suggests that a regionally-specific approach to determining reproductive parameters may be particularly appropriate for this species. Dugongs have a wide geographic distribution between 26° and 27° north and south of the equator, yet until now, reproductive parameters of dugongs have not been determined in non-tropical regions (i.e., beyond 23° 27′ latitude) where seasons may be more pronounced than in the tropics. Such gaps in our knowledge of dugong populations existed due to a previous lack of non-lethal methodologies to collect quantitative data on reproductive status and activity. Therefore, novel research approaches were warranted to investigate reproductive processes in live wild populations.

Recently, the first long-term capture-mark-recapture population study of dugongs has offered an opportunity to examine reproductive biology in live, free-ranging animals (Lanyon et al. 2002, 2006, 2012). During this program, >1,500 sampling events of dugongs of both sexes and all body sizes, in a population of ~1,000 dugongs, were conducted over a decade. Dugongs were captured using an open water 'rodeo' technique (Lanyon et al. 2006) and sampled at the water surface for a short period (5–6 min) (Lanyon et al. 2010a, b). Life history and reproductive parameters for individual dugongs including growth rate, body size at reproductive maturity and reproductive status have been elucidated through an innovative sampling approach that includes integration of information on molecular identity (Broderick et al. 2007; McHale et al. 2008; Lanyon et al. 2009; Kellogg Hunter et al. 2011), body morphometrics (Lanyon et al. 2010b), body scarring (Athousis 2012) and fecal hormone analysis (Burgess et al. 2012a, b).

Since most reproductive processes are hormone-dependent, endocrine analysis is an effective and reliable non-lethal method for assessment of reproductive status (Schwarzenberger et al. 1996; Touma and Palme 2005). Concentrations of the steroid hormones testosterone, estrogen and progesterone were measured in fecal samples collected (via rectal tube) from dugongs of both sexes, all size classes and in all months over 10 years, using enzyme-immuno-assay (EIA) (Burgess et al. 2012a, b). All EIAs were biologically validated for this new species under study. Fecal hormone concentrations were examined against serum levels of a subsample of 50 dugongs sampled out-of-water to ensure that circulating concentrations in the blood were reliably reflected in the feces (Burgess et al. 2012a, b).

A pregnancy test for dugongs based on fecal progesterone in combination with body morphometrics (body length, maximum girth and teat length) was developed, validated through ultrasonography and then applied across the female population (Burgess et al. 2012a). Body size to first reproduction in females in this wild population could then be determined: compared to dugongs in tropical locations, these subtropical females grew slower, matured later but achieved greater asymptotic body size (Burgess et al. 2012a). For males, increases in fecal testosterone levels and tusk eruption (a secondary sex trait) characterized puberty, with further ontogenetic increases in testosterone indicating late onset of mature sexual activity (Burgess et al. 2012b). Temporal profiles of fecal hormone levels indicated discrete seasonality in reproductive activity, mating and pregnancy (spring-summer months), with synchronous peaks in male testosterone occurring in spring (Burgess et al. 2012b), coincident with heightened stress (fecal glucocorticoids), loss of body condition (Burgess et al. 2013b) and increased body scarring indicative of aggressive conspecific conflict (Athousis 2012; Burgess et al. 2013b): all consistent with competitive scramble mating that has been suggested for this species (Preen 1989). Interestingly, hormone profiles for recaptured individuals has confirmed discontinuous breeding in males but has also indicated the possibility of male senescence with advanced age (Burgess et al. 2012b).

Further longitudinal monitoring of individuals in this population will provide information on breeding cycles, interbreeding intervals and lifetime reproductive output in this long-lived species. This new integrated approach for investigating reproductive parameters in dugongs has wider application to other live populations of sirenians, and to cryptic marine mammals in general. This case study provides an example of how perseverance and evolving state-of-the-art science can make headway in understanding reproduction and associated processes in difficult-to-study marine mammals. Moreover, such vital data acquired on species' and population life history can be integrated across scientific disciplines in order to answer more complicated questions (e.g., genealogy (kinship, paternity), reproductive fitness, evolutionary processes), and ultimately lead to better knowledge to assess conservation status and management of vulnerable species.

4 Future Priorities

Our knowledge of the reproductive biology of fully marine mammals is very uneven. For those few species that are commonly held captive or for which carcasses have been recovered and analyzed, we have a basic understanding of reproductive processes. However, for the vast majority of fully aquatic marine mammals we know very little. The rarest, largest and/or most elusive of whales will always offer extreme challenges to the reproductive practitioner and we may never discover a great deal about their reproductive behavior or details of reproductive patterns at a population level. However, for many other species, recent advances in reproductive technologies with more 'remote' collection procedures (e.g., molecular and endocrine analysis) present the opportunity to initiate focused and integrated programs. Of particular importance to this field has been the development of wildlife endocrinology over the past 20–30 years. Although initially developed with terrestrial wildlife in mind, the advent of technologies to collect and analyze body tissues other than blood has been nothing short of revolutionary. Now, marine mammal field biologists have opportunities to sample tissues in non-invasive (feces, sloughed skin, exhaled blow) or minimally invasive ways (biopsied skin, blubber) and develop reproductive hormonal profiles for live individuals or across free-ranging populations. With some ingenuity, creativity and a little perseverance, marine mammal biologists are applying these new technologies in innovative and sometimes remarkable ways. Furthermore, the expanding suite of parameters that can be measured (e.g., steroid and non-steroid hormones, and other sophisticated reproductive biomarkers in combination with genetics) permits ever more profound and meaningful questions in marine mammal reproductive science. For instance, fine-scale genetic tools are now being used in association with morphometric and endocrine data to reconstruct genealogies for dugong populations along the Australian coast (Cope et al. 2011). Such ground-breaking approaches will not only inform us as to social and reproductive connectedness (movements and mating events) within and between populations, but have the capacity to elucidate mating strategies (relatedness and mate choice) and other evolutionary processes in elusive, cryptic marine mammals.

With increasing human usage of the world's oceans for commercial, recreational and strategic purposes, many populations of marine mammals are in a parlous state, and their survival depends on their ability to reproduce in degrading aquatic environments. Reproductive parameters, if these can be determined, provide vital information on natural population fecundity and reproductive capacity, potential population growth rate and minimum viable population size. Having a solid understanding of the way in which marine mammals reproduce will inform us as to what levels of mortality and disturbance, populations or even species can tolerate and help us identify those most at risk. We can only hope that the opportunity to collect and integrate reproductive data will not be a case of 'too little, too late' to save our marine mammal biota.

References

Aarseth JJ, Van't Hof TJ, Stokkan K-A. Melatonin is rhythmic in newborn seals exposed to continuous light. J Comp Physiol B. 2003;173:37–42.

Acevedo J, Aguayo-Lobo A, Torres D. Fetus presentation and time taken for parturition in Antarctic fur seal, *Arctocephalus gazelle*, at Cape Shirreff, Antarctica. Polar Biol. 2008;31:1137–41.

Acevedo-Whitehouse K, Rocha-Gosselin A, Gendron D. A novel non-invasive tool for disease surveillance of free-ranging whales and its relevance to conservation programs. Anim Conserv. 2010;13:217–25.

Adulyanukosol K, Thongsukdee S, Hara T, Arai N, Tsuchiya M. Observations of dugong reproductive behavior in Trang Province, Thailand: further evidence of intraspecific variation in dugong behavior. Mar Biol. 2007;151:1887–91.

Archer R, Chauvet J, Chauvet MT. Isolation of finback whale oxytocin and vasopressin. Nature. 1964;201:191–2.

Amaral RS. Use of alternative matrices to monitor steroid hormones in aquatic mammals: a review. Aquat Mamm. 2010;36:162–71.

Amaral RS, Rosas FCW, Viau P, d'Affonse ca Neto JA, da Silva VMF, de Oliveira CA. Noninvasive monitoring of androgens in male Amazonian manatee (*Trichechus inunguis*): biologic validation. J Zoo Wildl Med. 2009;40:458–65.

Anderson PK. Shark Bay dugongs in summer. I: Lek mating. Behaviour. 1997;134:433–62.

Anderson PK. Scarring and photo identification of dugongs in Shark Bay, Western Australia. Aquat Mamm. 1995;21:205–11.

Andriolo A, Martins CCA, Engel MH, Pizzorno JL, Mas-Rosa S, Freitas AC, Morete ME, Kinas PG. The first aerial survey to estimate abundance of humpback whales (*Megaptera novaeangliae)* in the breeding ground off Brazil (Breeding Stock A). J Cetac Res Manage. 2006;8:307–11.

Asada M, Tetsuka M, Ishikawa H, Ohsumi S, Fukui Y. Improvement on *in vitro* maturation, fertilization and development of minke whale (*Balaenoptera acutorostrata*) oocytes. Theriogenology. 2001a;56:521–33.

Asada M, Wei H, Nagayama R, Tetsuka M, Ishikawa H, Ohsumi S, Fukui Y. An attempt at intracytoplasmic sperm injection of frozen-thawed minke whale (*Balaenoptera bonaerensis*) oocytes. Zygote. 2001b;9:299–307.

Athousis C. Body scarring as an indicator of tusk function in intraspecific social interactions in the dugong (*Dugong dugon*). Unpublished Honours Thesis, The University of Queensland, Brisbane, Australia; 2012. 69 pages.

Atkinson S. Reproductive biology of seals. Rev Reprod. 1997;2:175–94.

Atkinson S, Combelles C, Vincent D, Nachtogall P, Pawloski J, Breese M. Monitoring of progesterone in captive female false killer whales, *Pseudorca crassidens*. Gen Comp Endocrinol. 1999;115:323–32.

Atkinson S, Gilmartin WG. Seasonal testosterone pattern in Hawaiian monk seals (*Monachus schauinslandi*). J Reprod Fertil. 1992;96:35–9.

Atkinson S, Gilmartin WG, Lasley BL. Testosterone response to a gonadotrophin-releasing hormone agonist in Hawaiian monk seals (*Monachus schauinslandi*). J Reprod Fertil. 1993;97:35–8.

Atkinson S, Ragen TJ, Gilmartin WG, Becker BL, Johanos TC. Use of a GnRH agonist to suppress testosterone in wild male Hawaiian monk seals (*Monachus schauinslandi*). Gen Comp Endocrinol. 1998;112:178–82.

Attard CRM, Beheregaray LB, Jenner KCS, Gill PC, Jenner M-N, Morrice MG, Robertson KM, Möller LM. Hybridization of Southern Hemisphere blue whale subspecies and a sympatric area off Antarctica: impacts of whaling or climate change? Mol Ecol. 2012;21:5715–27.

Barlow J. Variability, trends, and biases in reproductive rates of spotted dolphins, *Stenella attenuata*. Fish Bull. 1985;83:657–69.

Barrell GK, Montgomery GW. Absence of circadian patterns of secretion of melatonin or cortisol in Weddell seals under continuous natural daylight. J Endocrinol. 1989;122:445–9.

Bauer GB, Colbert DE, Gaspard III JC. Learning about manatees: a collaborative program between New College of Florida and Mote Marine Laboratory to conduct laboratory research for manatee conservation. Int J Comp Psychol. 2010;23:811–25.

Bergfelt DR, Steinetz BG, Lasano S, West KL, Campbell M, Adams GP. Relaxin and progesterone during pregnancy and the post-partum period in association with live and stillborn calves in bottlenose dolphins (*Tursiops truncatus*). Gen Comp Endocrinol. 2011;170:650–6.

Best PB. The sperm whale (*Physeter catodon*) off the west coast of South Africa. 3. Reproduction in the male. Investig Rep Div Sea Fish South Afr. 1969;72:1–20.

Bhuiyan MM, Suzuki Y, Watanabe H, Lee E, Hirayama H, Matsuoka K, Fujise Y, Ishikawa H, Ohsumi S, Fukui Y. Production of sei whale (*Balaenoptera borealis*) cloned embryos by inter- and intra-species somatic cells nuclear transfer. J Reprod Dev. 2010;56:131–9.

Bhuiyan MMU, Suzuki Y, Watanabe H, Matsuoka K, Fujise Y, Ishikawa H, Ohsumi S, Fukui Y. Attempts at *in vitro* fertilization and culture of *in vitro* matured oocytes in sei (*Balaenoptera borealis*) and Bryde's (*B. edeni*) whales. Zygote. 2009;17:19–28.

Biancani B, Da Dalt L, Lacave G, Romagnoli S, Gabai G. Measuring fecal progestogens as a tool to monitor reproductive activity in captive female bottlenose dolphins (*Tursiops truncatus*). Theriogenology. 2009;72:1282–92.

Blanchet M-A, Desportes G, Nance T, Vanman C. The description of the pregnancies, labours and pre- and post-partum events of two harbour seals in a zoological environment. Aquat Mamm. 2006;32:145–51.

Block BA. Physiological ecology in the 21st century: advancements in biologging science. Integr Comp Biol. 2005;45:305–20.

Bograd SJ, Block BA, Costa DP, Godley BJ. Biologging technologies: new tools for conservation. Introduction. Endanger Species Res. 2010;10:1–7.

Börjesson P, Read AJ. Variation in the timing of conception between populations of harbor porpoise. J Mammal. 2003;84:948–55.

Botta S, Secchi ER, Muelbert MMC, Danilewicz D, Negri MF, Cappozzo HL, Hohn AA. Age and growth of franciscana dolphins, *Pontoporia blainvillei* (Cetacea: Pontoporiidae) incidentally caught off southern Brazil and northern Argentina. J Mar Biol Assoc U K. 2010;90:1493–500.

Boyd IL. Changes in plasma progesterone and prolactin concentrations during the annual cycle, and role of prolactin in the maintenance of lactation and luteal development in the Antarctic fur seal (*Arctocephalus gazella*). J Reprod Fertil. 1991;91:637–47.

Boyd IL. State-dependent fertility in pinnipeds: contrasting capital and income breeders. Funct Ecol. 2000;14:623–30.

Boyd IL, Lockyer C, Marsh HD. Reproduction in marine mammals. In: Reynolds III JE, Rommel SA, editors. Biology of marine mammals. Washington, DC: Smithsonian Institution Press; 1999. p. 218–86.

Briggs M. Contraception in bottlenose dolphins (*Tursiops truncatus*). In: Duffield DA, Robeck TR, editors. The bottlenose dolphin breeding workshop. Silver Springs, MD: American Zoological Association, Marine Mammal Taxon Advisory Group; 2000. p. 201–4.

Broderick D, Ovenden J, Slade R, Lanyon JM. Characterisation of 26 new microsatellite loci in the dugong (*Dugong dugon*). Mol Ecol Notes. 2007;7:1275–7.

Brook FM. The use of diagnostic ultrasound in assessment of the reproductive status of the bottlenose dolphin, *Tursiops aduncus*, in captivity and applications in management of a controlled breeding program. Unpublished PhD Thesis, Hong Kong Polytechnic University, Hong Kong; 1997. 339 pages.

Brook FM. Sonographic ovarian and testicular evaluation in bottlenose dolphins, *Tursiops truncatus aduncus*. In: Duffield DA, Robeck TR, editors. Proceedings Bottlenose Dolphin Reproduction Workshop; June 1999; San Diego, California; 1999. p. 207–22.

Brook FM. Ultrasonographic imaging of the reproductive organs of the female bottlenose dolphin, *Tursiops truncatus aduncus*. Reproduction. 2001;121:419–28.

Brook FM, Kinoshita RE, Brown B, Metreweli C. Ultrasonographic imaging of the testis and epididymis of the bottlenose dolphin, *Tursiops truncatus aduncus*. J Reprod Fertil. 2000;119:233–40.

Brook FM, Lim EHT, Chua FHC, Mackay B. Assessment of the reproductive cycle of the Indo-Pacific humpback dolphin, *Sousa chinensis*, using ultrasonography. Aquat Mamm. 2004;30:137–48.

Brook FM, van Bonn W, Jensen E. Ultrasonography. In: Dierauf LA, Gulland FMD, editors. CRC handbook of marine mammal medicine. 2nd ed. Boca Raton, FL: CRC Press; 2001. p. 590–620.

Brown MR, Corkeron PJ, Hale PT, Schultz KW, Bryden MM. Evidence for a sex-segregated migration in the humpback whale (*Megaptera novaeangliae*). Proc R Soc Lond B. 1995;259:229–34.

Burgess EA, Blanshard WH, Barnes A, Gilchrist S, Chua J, Lanyon JM. Reproductive hormone monitoring of dugongs in captivity: detecting the onset of sexual maturity in a cryptic marine mammal. Anim Reprod Sci. 2013a;140(3–4):255–67.

Burgess EA, Brown JL, Lanyon JM. Sex, scarring and stress: understanding seasonal costs in a cryptic marine mammal. Conserv Physiol. 2013b;1:cot014.

Burgess EA, Lanyon JM, Brown JL, Blyde D, Keeley T. Diagnosing pregnancy in free-ranging dugongs using fecal progesterone metabolite concentrations and body morphometrics: a population application. Gen Comp Endocrinol. 2012a;177:82–92.

Burgess EA, Lanyon JM, Keeley T. Testosterone and tusks: maturation and seasonal reproductive patterns of live free-ranging dugongs (*Dugong dugon*) in a subtropical population. Reproduction. 2012b;143:683–97.

Calle PP. Contraception in pinnipeds and cetaceans. In: Asa CS, Porton IJ, editors. Wildlife contraception: issues, methods, and applications. Baltimore, MD: John Hopkins University Press; 2005. p. 168–76.

Calzada N, Aguilar A, Sorensen TB, Lockyer C. Reproductive biology of female striped dolphin (*Stenella coeruleoalba*) from the western Mediterranean. J Zool. 1996;240:581–91.

Cassini MH. The evolution of reproductive systems in pinnipeds. Behav Ecol. 1999;10:612–6.

Chittleborough RG. Aspects of reproduction in the male humpback whale *Megaptera nodosa* (Bonnaterre). Aust J Mar Freshwat Res. 1954;6:1–29.

Chivers SJ, Myrick Jr AC. Comparisons of age at sexual maturity and other reproductive parameters for two stocks of spotted dolphin, *Stenella attenuata*. Fish Bull. 1993;91:611–8.

Clapham PJ. The social and reproductive biology of humpback whales: an ecological perspective. Mammal Rev. 2008;26:27–49.

Clapham PJ, Young SB, Brownell RL. Baleen whales: conservation issues and the status of the most endangered populations. Mammal Rev. 1999;29:35–60.

Cockcroft VG, Ross GJB. Age, growth and reproduction of bottlenose dolphins *Tursiops truncatus* from the east coast of Southern Africa. Fish Bull. 1990;88:289–302.

Constable S, Parslow A, Dutton G, Rogers T, Hogg C. Urinary cortisol sampling: a non-invasive technique for examining cortisol concentrations in the Weddell seal, *Leptonychotes weddelli*. Zoo Biol. 2006;25:137–44.

Cope RC, Lanyon JM, Seddon JR, Pollett PK. Reconstructing family trees for free-ranging dugongs in southern Queensland, Australia. Published Abstract, International Sirenian Symposium, 19th International Biennial Conference on Marine Mammals; 2011 Nov 27; Tampa Florida; 2011.

Connor RC. Dolphin social intelligence: complex alliance relationships in bottlenose dolphins and a consideration of selective environments for extreme brain size evolution in mammals. Philos Trans R Soc Lond B Biol Sci. 2007;362:587–602.

Connor RC, Smolker RA, Richards AF. Two levels of alliance formation among male bottlenose dolphins (*Tursiops* sp.). Proc Natl Acad Sci. 1992;89:987–90.

Connor RC, Wells R, Mann J, Read A. The bottlenose dolphin: social relationships in a fission-fusion society. In: Mann J, Connor RC, Tyack PJ, Whitehead H, editors. Cetacean societies: field studies of whales and dolphins. Chicago, IL: University of Chicago Press; 2000. p. 91–126.

Cornell LH, Asper ED, Antrim JE, Searles SS, Young WG, Goff T. Progress report: Results of a long-range captive breeding program for the bottlenose dolphin, *Tursiops truncatus* and *Tursiops truncatus gilli*. Zoo Biol. 1987;6:41–53.

Costa DP. The secret lives of marine mammals: novel tools for studying their behaviour and biology at sea. Oceanography. 1993;6:120–8.

Cummins JM, Woodall PF. On mammalian sperm dimensions. J Reprod Fertil. 1985;75:153–75.

Danil K, Chivers SJ. Habitat based spatial and temporal variability in life history characteristics of female common dolphins (*Delphinus delphis*) in the eastern tropical Pacific. Mar Ecol Prog Ser. 2006;318:277–86.

Dawbin WH. Temporal segregation of humpback whales during migrations in Southern Hemisphere waters. Mem Queensl Mus. 1997;42:105–38.

de Bruyn PJN, Tosh CA, Bester MN, Cameron EZ, McIntyre T, Wilkinson IS. Sex at sea: alternative mating system in an extremely polygynous mammal. Anim Behav. 2011;82:445–51.

DeMaster DP. Estimating the average age of first birth in marine mammals. Can J Fish Aquat Sci. 1981;38:237–9.

Desportes G, Saboureau M, Lacroix A. Reproductive maturity and seasonality of male long-finned pilot whales off the Faroe Islands. Rep Int Whaling Comm. 1993;14:233–62.

Desportes G, Saboureau M, Lacroix A. Growth-related changes in testicular mass and plasma testosterone concentrations in long-finned pilot whales, *Globicephala melas*. Reproduction. 1994;102:237–44.

Deutsch CJ, Reid JP, Bonde RK, Easton DE, Kochman HI, O'Shea TJ. Seasonal movements, migratory behavior, and site fidelity of West Indian manatees along the Atlantic coast of the United States. Wildl Monogr. 2003;151:1–77.

Dizon AE, Perrin WF, Akin PA. Stocks of dolphins (*Stenella* spp. and *Delphinus delphis*) in the Eastern Tropical Pacific: a phylogeographic classification. NOAA Technical Report NMFS-SWFSC-TM-119; 1994.

Dover SR, van Bonn W. Flexible and rigid endoscopy in marine mammals. In: Dierauf LA, Gulland FMD, editors. CRC handbook of marine mammal medicine. 2nd ed. Boca Raton, FL: CRC Press; 2001. p. 621–42.

Duffield DA, Odell DK, McBain JF, Andrews B. Killer whale (*Orcinus orca*) reproduction at SeaWorld. Zoo Biol. 1995;14:417–30.

Edds PL, Odell S, Tershy BR. Calls of a captive juvenile and free-ranging adult-calf pairs of Bryde's whales, *Balaenoptera edeni*. Mar Mamm Sci. 1993;9:269–84.

Eisert R, Oftedal OT, Barrell GK. Milk composition in the Weddell seal *Leptonychotes weddellii:* evidence for a functional role of milk carbohydrates in pinnipeds. Physiol Biochem Zool. 2013;86:159–75.

Essapian FS. Observations of abnormalities of parturition in captive bottlenose dolphins, *Tursiops truncatus*, and concurrent behavior of other porpoises. J Mammal. 1962;44:405–14.

Evans K, Hindell MA. The age structure and growth of female sperm whales (*Physeter macrocephalus*) in southern Australian waters. J Zool. 2004;263:237–50.

Favaro L, Gnone G, Pessani D. Postnatal development of echolocation abilities in a bottlenose dolphin (*Tursiops truncatus*). Zoo Biol. 2013;32:210–5.

Fitzpatrick JL, Almbro M, Gonzalez-Voyer A, Kolm N, Simmons LW. Male contest competition and the coevolution of weaponry and testes in pinnipeds. Evolution. 2012;66:3595–604.

Fleming AD, Yanagimachi R, Yanagimachi H. Spermatozoa of the Atlantic bottlenose dolphin, *Tursiops truncatus*. J Reprod Fertil. 1981;63:509–14.

Foote AD. Mortality rate acceleration and post-reproductive lifespan in matrilineal whale species. Biol Lett. 2008;4:189–91.

Fowler CW. Density dependence as related to life history strategy. Ecology. 1981;62:602–10.

Fowler CW. Density dependence in cetacean populations. Rep Int Whaling Comm. 1984;6: 373–9.

Francis-Floyd R, White JR, Chen CL, Cardeilhac PT, Cichra CE. Serum progesterone and estradiol concentrations in captive manatess *Trichechus manatus*. J Aquat Anim Health. 1991;3:70–3.

Frère CH, Krzyszczyk E, Patterson EM, Hunter S, Ginsburg A, Mann J. Thar she blows! A novel method for DNA collection from cetacean blow. PLoS One. 2010;5:e12299.

Fujihira T, Kobayashi M, Hochi S, Hirabayashi M, Ishikawa H, Ohsumi S, Fukui Y. Developmental capacity of Antarctic minke whale (*Balaenoptera bonaerensis*) vitrified oocytes following *in vitro* maturation, and parthenogenetic activation or intracytoplasmic sperm injection. Zygote. 2006;14:89–95.

Fukui Y, Iwayama H, Matsuoka T, Nagai H, Koma N, Mogoe T, Ishikawa H, Fujise Y, Hirabayashi M, Hochi S, Kato H, Ohsumi S. Attempt at intracytoplasmic sperm injection of *in vitro* matured oocytes in common minke whales (*Balaenoptera acutorostrata*) captured during the Kushiro Coast Survey. J Reprod Dev. 2007;53:945–52.

Fukui Y, Mogoe T, Jung YG, Terawaki Y, Miyamoto A, Ishikawa H, Fujise Y, Ohsumi S. Relationships among morphological status, steroid hormones, and post-thawing viability of frozen spermatozoa of male minke whales (*Balaenoptera acutorostrata*). Mar Mamm Sci. 1996;12:28–37.

Funasaka N, Yoshioka M, Suzuki M, Ueda K, Miyahara H, Uchida S. Seasonal difference of diurnal variations in serum melatonin, cortisol, testosterone, and rectal temperature in Indo-Pacific bottlenose dolphins (*Tursiops aduncus*). Aquat Mamm. 2011;37:433–42.

Gardiner KJ, Boyd IL, Racey PA, Reijnders PJH, Thompson PM. Plasma progesterone concentrations measured using an enzyme-linked immunosorbent assay useful for diagnosing pregnancy in harbor seals (*Phoca vitulina*). Mar Mamm Sci. 1996;12:265–73.

Gardiner KJ, Boyd IL, Follett BK, Racey PA, Reijnders PJH. Changes in pituitary, ovarian, and testicular activity in harbour seals (*Phoca vitulina*) in relation to season and sexual maturity. Can J Zool. 1999;77:211–21.

Gero S, Whitehead H. Suckling behavior in sperm whale calves: observations and hypotheses. Mar Mamm Sci. 2007;23:398–413.

Gibbens J, Parry LJ, Arnould JPY. Influences on fecundity in Australian fur seals (*Arctocephalus pusillus doriferus*). J Mammal. 2010;91:510–8.

Glockner-Ferrari DA, Ferrari MJ. Reproduction in the humpback whale (*Megaptera novaeangliae*) in Hawaiian waters, 1975–1988: the life history, reproductive rates and behavior of known individuals identified through surface and underwater photography. Rep Int Whaling Comm. 1990;12:161–9.

Goldstein J, Reese E, Reif JS, Varela RA, McCulloch SD, Defran RH, Fair PA, Bossart GD. Hematologic, biochemical, and cytologic findings from apparently healthy Atlantic Bottlenose dolphins (*Tursiops truncatus*) inhabiting the Indian River lagoon, Florida, USA. J Wildl Dis. 2006;42:447–54.

Hadley GL, Rotella JJ, Garrot RA, Nichols JD. Variation in probability of first reproduction of Weddell seals. J Anim Ecol. 2006;75:1058–70.

Hao YJ, Chen DQ, Zhao QZ, Wang D. Serum concentrations of gonadotropins and steroid hormones of *Neophocaena phocaenoides asiaeorientalis* in middle and lower regions of the Yangtze River. Theriogenology. 2007;67:673–80.

Harcourt RG, Turner E, Hall A, Waas JR, Hindell M. Effects of capture stress on free-ranging, reproductively active male Weddell seals. J Comp Physiol A. 2010;196:147–54.

Harrison RJ, Ridgway SH. Gonadal activity in some bottlenose dolphins (*Tursiops truncatus*). J Zool. 1971;165:355–66.

Hartman DS. Ecology and behavior of the manatee (*Trichechus manatus*) in Florida. Am Soc Mammalogists. 1979;5:1–153.

Hernandez P, Reynolds III JE, Marsh H, Marmontel M. Age and seasonality of spermatogenesis of Florida manatees. In: O'Shea TJ, Ackerman BB, Percival HF, editors. Population biology of the Florida manatee. Information and Technology Report 1. USA: National Biological Service; 1995. p. 84–97.

Heyning JE. Functional morphology involved in intraspecific fighting of the beaked whale, *Mesoplodon carlhubbsi*. Can J Zool. 1984;62:1645–54.

Hildebrandt TB, Brown JL, Hermes R, Göritz F. Ultrasound analysis of reproductive function in wildlife species. Conserv Biol. 2003;8:166–82.

Hiwasa M, Suzuki Y, Watanabe H, Bhuiyan MMU, Matsuoka K, Fujise Y, Ishikawa H, Ohsumi S, Fukui Y. Effects of semen extenders and storage temperatures on characteristics of frozen-thawed Bryde's (*Balaenoptera edeni*) whale spermatozoa. J Reprod Dev. 2009;55: 599–606.

Hoberecht LK, Vos DJ, VanBlaricom GR. A remote biopsy system used to sample Steller sea lion (*Eumetopias jubatus*) blubber. Mar Mamm Sci. 2006;22:683–9.

Hogg C, Rogers T, Shorter A, Barton K, Miller P, Nowacek D. Determination of steroid hormones in whale blow: it is possible. Mar Mamm Sci. 2009;25:605–18.

Hogg C, Vickers E, Rogers T. Determination of testosterone in saliva and blow of bottlenose dolphins (*Tursiops truncatus*) using liquid chromatography-mass spectrometry. J Chromatogr B. 2005;814:339–46.

Hohn AA, Read AJ, Fernândez S, Vidâl O, Findley LT. Life history of the vaquita, *Phocoena sinus* (Phocoenidae, Cetacea). J Zool (Lond). 1996;239:235–51.

Holt WV, Waller J, Moore A, Jepson PD, Deaville R, Bennett PM. Smooth muscle actin and vimentin as markers of testis development in the harbour porpoise (*Phocoena phocoena*). J Anat. 2004;205:201–11.
Hückstadt L, Antezana T. An observation of parturition in a stranded *Kogia breviceps*. Mar Mamm Sci. 2001;17:362–5.
Hunt KE, Moore MJ, Rolland RM, Kellar NM, Hall AJ, Kershaw J, Raverty SA, Davis CE, Yeates LC, Fauquier DA, Rowles TK, Kraus SD. Overcoming the challenges of studying conservation physiology in large whales: a review of available methods. Conserv Physiol. 2013;1:1–24.
Hunt KE, Rolland RM, Kraus SD, Wasser SK. Analysis of fecal glucocorticoids in the North Atlantic right whale (*Eubalaena glacialis*). Gen Comp Endocrinol. 2006;148:260–72.
Hunt KE, Trites AW, Wasser SK. Validation of a fecal glucocorticoid assay for Steller sea lions (*Eumetopias jubatus*). Physiol Behav. 2004;80:595–601.
Iwayama H, Ishikawa H, Ohsumi S, Fukui Y. Attempt at *in vitro* maturation of minke whale (*Balaenoptera bonaerensis*) oocytes using a portable CO_2 incubator. J Reprod Dev. 2005;51:69–75.
Jenssen TA, Lovern MB, Congdo JD. Field-testing the protandry-based mating system for the lizard, *Anolis carolinesis*: does the model organism have the right model? Behav Ecol Sociobiol. 2001;50:162–72.
Karamanlidis AA, Paravas V, Trillmich F, Dendrinos P. First observations of parturition and postpartum behavior in the Mediterranean monk seal (*Monachus monachus*) in the Eastern Mediterranean. Aquat Mamm. 2010;36:27–32.
Karczmarski L, Thornton M, Cockroft VG. Description of selected behaviours of humpback dolphins, *Sousa chinensis*. Aquat Mamm. 1997;23:127–34.
Kato H, Sakuramoto K. Age at sexual maturity of Southern minke whales: a review and some additional analyses. Rep Int Whaling Comm. 1991;41:331–7.
Katsumata E, Furuta C, Katsumata H, Watanabe G, Taya K. Basal body temperature method for detecting ovarian cycle in the captive beluga (*Delphinapterus leucas*). J Reprod Dev. 2006a;52:59–63.
Katsumata E, Jaroenporn S, Katsumata H, Konno S, Maeda Y, Watanabe G, Taya K. Body temperature and circulating progesterone levels before and after parturition in killer whales (*Orcinus orca*). J Reprod Dev. 2006b;52:65–71.
Katsumata E, Ueda Y, Arai K, Katsumata H, Kishimoto M, Watanabe G, Taya K. Circulating testosterone and inhibin levels at different ages in the male beluga (*Delphinapterus leucas*). J Vet Med Sci. 2012;74:292–300.
Kellar NM, Trego ML, Marks CI, Chivers SJ, Danil K, Archer FI. Blubber testosterone: a potential marker of male reproductive status in short beaked common dolphins. Mar Mamm Sci. 2009;25:507–22.
Kellar NM, Trego ML, Marks CI, Dizon AE. Determining pregnancy from blubber in three species of delphinids. Mar Mamm Sci. 2006;22:1–16.
Keller KV. Training of the Atlantic bottlenose dolphins (*Tursiops truncatus*) for artificial insemination. Int Assoc Aquat Anim Med. 1986;14:22–4.
Kellogg Hunter M, Broderick D, Ovenden JR, Tucker KP, Bonde RK, McGuire PM, Lanyon JM. Characterization of highly informative cross-species microsatellite panels for the Australian dugong (*Dugong dugon*) and Florida manatee (*Trichechus manatus latirostris*) including five novel primers. Mol Ecol Resour. 2011;10:368–77.
Kinoshita K, Kiwata M, Kuwano R, Sato N, Tanaka T, Nagata M, Taira H, Kusunoki H. Temporal association of serum progesterone concentrations and vaginal cytology in walruses (*Odobenus rosmarus*). Theriogenology. 2012;77:933–9.
Kita S, Yoshioka M, Kashiwagi M, Ogawa S, Tobayama T. Comparative external morphology of cetacean spermatozoa. Fish Sci. 2001;67:482–92.
Kjeld M, Alfredsson A, Ólafsson O, Tryland M, Christensen I, Stuen S, Árnason A. Changes in blood testosterone and progesterone concentrations of the North Atlantic minke whale (*Balaenoptera acutorostrata*) during the feeding season. Can J Fish Aquat Sci. 2004;61:230–7.

Kjeld M, Ólafsson O, Víkingsson GA, Sigurjónsson J. Sex hormones and reproductive status of the North Atlantic fin whales (*Balaenoptera physalus*) during the feeding season. Aquat Mamm. 2006;32:75–84.
Kooyman GL, Castellini MA, Davis RW. Physiology of diving in marine mammals. Annu Rev Physiol. 1981;43:343–56.
Kraus SD, Hamilton PK, Kenney RD, Knowlton AR, Slay CK. Reproductive parameters of the North Atlantic right whale. J Cetac Res Manage. 2001;2:231–6.
Kumar A, Kalra B, Patel A, McDavid L, Roudebush WE. Development of a second generation anti-Müllerian hormone (AMH) ELISA. J Immunol Methods. 2010;362:51–9.
Kwan D. Towards a sustainable indigenous fishery for dugongs in Torres Strait: a contribution of empirical data analysis and process. Unpublished PhD thesis, James Cook University, Australia; 2002. 282 pages.
Lacave G, Eggermont M, Verslycke T, Brook FM, Salbany A, Roque L, Kinoshita R. Prediction from ultrasonographic measurements of the expected delivery date in two species of bottlenose dolphin (*Tursiops truncatus* and *Tursiops aduncus*). Vet Rec. 2004;154:228–33.
Lambrechts MM, Perret P, Maistre M, Blondel J. Do experiments with captive non-domesticated animals make sense without population field studies? A case study in blue tit's breeding time. Proc R Soc Lond B. 1999;266:1311–5.
Langtimm CA, O'Shea TJ, Pradel R, Beck CA. Estimates of annual survival probabilities for adult Florida manatees (*Trichechus manatus latirostris*). Ecology. 1998;79:981–97.
Langtimm CA, Beck CA, Edwards H, Fick-Child K, Ackerman BB, Barton S, Hartley WC. Survival estimates for Florida manatees from the photo-identification of individuals. Mar Mamm Sci. 2004;20:438–63.
Lanyon JM, Slade RW, Sneath HL, Broderick D, Kirkwood JM, Limpus D, Limpus CJ, Jessop T. A method for capturing dugongs (*Dugong dugon*) in open water. Aquat Mamm. 2006;32:196–201.
Lanyon JM, Smith K, Carrick FN. Reproductive steroids are detectable in the faeces of dugongs. Aust Zool. 2005;33:247–50.
Lanyon JM, Sneath HL, Kirkwood JM, Slade RW. Establishing a mark-recapture program for dugongs in Moreton Bay, south-east Queensland. Aust Mammal. 2002;24:51–6.
Lanyon JM, Sneath HL, Long T. Three skin sampling methods for molecular characterisation of free-ranging dugong (*Dugong dugon*) populations. Aquat Mamm. 2010a;36:298–306.
Lanyon JM, Sneath HL, Long T. Evaluation of exertion and capture stress in serum of wild dugongs, *Dugong dugon*. J Zoo Wildl Med. 2012;43:20–32.
Lanyon JM, Sneath HL, Long T, Bonde RK. Physiological response of wild dugongs (*Dugong dugon*) to out-of-water sampling for health assessment. Aquat Mamm. 2010b;36:46–58.
Lanyon JM, Sneath HL, Ovenden JR, Broderick D, Bonde RK. Sexing sirenians: validation of visual and molecular sex determination in both wild dugongs (*Dugong dugon*) and Florida manatees (*Trichechus manatus latirostris*). Aquat Mamm. 2009;35:187–92.
Larkin ILV, Gross TS, Reep RL. Use of faecal testosterone concentrations to monitor male Florida manatee (*Trichechus manatus latirostris*) reproductive status. Aquat Mamm. 2005;31: 52–61.
Lasley BL, Kirkpatrick JF. Monitoring ovarian function in captive and free-ranging wildlife by means of urinary and fecal steroids. J Zoo Wildl Med. 1991;22:23–31.
Laws RM. Reproduction, growth, and age of southern fin whales. Discov Rep. 1961;31:327–486.
Lawson JW, Parsons JL, Craig SJ, Eddingotn JD, Kimmins WC. Use of electroejaculation to collect semen samples from wild seals. J Am Vet Med Assoc. 1996;209:1615–7.
Le Boeuf BJ. Pinniped mating systems on land, ice and in the water: emphasis on the Phocidae. In: Renouf D, editor. Behavior of pinnipeds. London, UK: Chapman and Hall; 1991. p. 45–65.
Lee E, Bhuiyan MMU, Watanabe H, Matsuoka K, Fujise Y, Ishikawa H, Fukui Y. Production of cloned sei whale (*Balaenoptera borealis*) embryos by interspecies somatic cell nuclear transfer using enucleated pig oocytes. J Vet Sci. 2009;10:285–92.
Lee MM, Donahoe PK. Müllerian inhibiting substance: a gonadal hormone with multiple functions. Endocr Rev. 1993;14:152–64.

Lefebvre KA, Bargu S, Kieckhefer T, Silver MW. From sanddads to blue whales: the pervasiveness of domoic acid. Toxicon. 2002;40:971–7.
Lockyer C. Review of baleen whale (Mysticeti) reproduction and implications for management. Rep Int Whaling Comm. 1984;6:27–50.
Lockyer C, Martin AR. The sei whale off western Iceland: II Age, growth, and reproduction. Rep Int Whaling Comm. 1983;33:465–76.
Lockyer C, Smellie CG. Assessment of reproductive status of female fin and sei whales taken off Iceland, from a histological examination of the uterine mucosa. Rep Int Whaling Comm. 1985;35:343–8.
Lyamin O, Pryaslova J, Lance V, Siegel J. Continuous activity in cetaceans after birth. Nature. 2005;435:1177.
MacLeod CD. Intraspecific scarring in odontocete cetaceans: an indicator of male 'quality' in aggressive social interactions. J Zool. 1998;244:71–7.
Mann J, Connor RC, Tyack PL, Whitehead H. Cetacean Societies: field studies of dolphins and whales. Chicago, IL: University of Chicago Press; 2000. 448 pages.
Mansour AAH, McKay DW, Lien J, Orr JC, Banoub JH, ØIen N, Stenson G. Determination of pregnancy status from blubber samples in minke whales (*Balaenoptera acutorostrata*). Mar Mamm Sci. 2002;18:112–20.
Marcoux M, Whitehead H, Rendell L. Sperm whale feeding variation by location, year, social group and clan: evidence from stable isotopes. Mar Ecol Prog Ser. 2007;333:309–14.
Marmontel M. Age and reproduction in female Florida manatees. In: O'Shea TJ, Ackerman BB, Percival HF, editors. Population biology of the Florida manatee. Information and Technology Report 1. USA: National Biological Service; 1995. p. 98–119.
Marsh H, Heinsohn GE, Channells PW. Changes in the ovaries and uterus of the dugong, *Dugong dugon* (Sirenia: Dugongidae), with age and reproductive activity. Aust J Zool. 1984a;32:743–66.
Marsh H, Heinsohn GE, Glover TD. Changes in the male reproductive organs of the dugong, *Dugong dugon* (Sirenia: Dugondidae) with age and reproductive activity. Aust J Zool. 1984b;32:721–42.
Marsh H, Heinsohn GE, Marsh LM. Breeding cycle, life history and population dynamics of the dugong, *Dugon dugon* (Sirenia: Dugongidae). Aust J Zool. 1984c;32:767–88.
Marsh H, Kasuya T. Changes in the ovaries of the short-finned pilot whale, *Globicephala macrorhynchus*, with age and reproductive activity. Rep Int Whaling Comm. 1984;6:331–5.
Marsh H, Kasuya T. Evidence for reproductive senescence in female cetaceans. Rep Int Whaling Comm. 1986;8:57–74.
Marsh H, Kwan D. Temporal variability in the life history and reproductive biology of female dugongs in Torres Strait: the likely role of seagrass dieback. Cont Shelf Res. 2008;28:2152–9.
Marsh H, O'Shea TJ, Reynolds III JE. Ecology and conservation of the sirenia. Dugongs and manatees. Cambridge, UK: Cambridge University Press; 2011. 521 pages.
Mate BR, Gisiner R, Mobley J. Local and migratory movements of Hawaiian humpback whales tracked by satellite telemetry. Can J Zool. 1998;76:863–8.
Mate BR, Lagerquist BA, Urban-Ramirez J. A note on using satellite telemetry to document the use of San Ignacio Lagoon by gray whales (*Eschrichtius robustus*) during their reproductive season. J Cetac Res Manage. 2003;5:149–54.
McBride AF, Hebb DO. Behavior of the captive bottlenose dolphin *Tursiops truncatus*. J Comp Physiol Psychol. 1948;41:111–23.
McBride AF, Kritzler H. Observations on pregnancy, parturition, and postnatal behavior in the bottlenose dolphin. J Mammal. 1951;32:251–66.
McHale M, Broderick D, Ovenden JR, Lanyon JM. A multiplexed PCR assay for gender assignment in dugong (*Dugong dugon*) and West Indian manatee (*Trichechus manatus*). Mol Ecol Notes. 2008;8:669–70.
McKenzie J, Parry LJ, Page B, Goldsworthy SD. Estimation of pregnancy rates and reproductive failure in New Zealand fur seals (*Arctocephalus forsteri*). J Mammal. 2005;86:1237–46.

Mellish JE, Calkins DG, Christen DR, Horning M, Rea LD, Atkinson SK. Temporary captivity as a research tool: comprehensive study of wild pinnipeds under controlled conditions. Aquat Mamm. 2006;32:58–65.

Mercado III E, deLong CM. Dolphin cognition: representations and processes in memory and perception. Int J Comp Psychol. 2010;23:344–78.

Mignucci-Giannoni AA, Toyos-Gonzalez GM, Perez-Padilla J, Rodriguez-Lopez MA, Overing J. Mass stranding of pygmy killer whales (*Feresa attenuata*) in the British Virgin Islands. J Mar Biol Assoc U K. 2000;80:759–60.

Miller DL, Bossart GD, Nadji M, Tarpley R, Roberts B, O'Hara TM. A note on the possibility of identifying Leydig and Sertoli cells by immunohistochemistry in bowhead whales (*Balaena mysticetus*). J Cetac Res Manage. 2002a;4:149–53.

Miller DL, Dougherty MM, Decker SJ, Bossart GD. Ultrastructure of the spermatozoa from a Florida manatee (*Trichechus manatus latirostris*). Anat Histol Embryol. 2001;30:253–6.

Miller DL, Styer EL, Decker SJ, Robeck T. Ultrastructure of the spermatozoa from three odonocetes: a killer whale (*Orcinus orca*), a Pacific white-sided dolphin (*Lagenorhynchus obliquidens*) and a beluga (*Delphinapterus leucas*). Anat Histol Embryol. 2002b;31:158–68.

Miyazaki N. Further analyses of reproduction in the striped dolphin, *Stenella coeruleoalba*, off the Pacific coast of Japan. Rep Int Whaling Comm. 1984;6:343–53.

Mogoe T, Fukui Y, Ishikawa H, Ohsumi S. Effects of diluent composition and temperature on motility and viability after liquid storage and precryopreservation of minke whale (*Balaenoptera acutorostrata*) spermatozoa. Mar Mamm Sci. 1998;14:854–60.

Montano GA, Kraemer DC, Love CC, Robeck TR, O'Brien JK. Evaluation of motility, membrane status and DNA integrity of frozen-thawed bottlenose dolphin (*Tursiops truncatus*) spermatozoa after sex-sorting and recyropreservation. Reproduction. 2010;143:799–813.

Muraco H, Clough P, Teets V, Arn D, Muraco M. Ovarian follicular dynamics during the luteinizing hormone surge in the bottlenose dolphin (*Tursiops truncatus*). Int J Comp Psychol. 2010;23:723–33.

Murphy S, Collet A, Rogan E. Mating strategy in the male common dolphin (*Delphinus delphis*): what gonadal development tells us. J Mammal. 2005;86:1247–58.

Myers MJ, Litz B, Atkinson S. The effects of age, sex, season and geographic region on circulating serum cortisol concentrations in threatened and endangered Steller sea lions (*Eumetopias jubatus*). Gen Comp Endocrinol. 2010;165:72–7.

Newell CL, Cowles TJ. Unusual gray whale *Eschrichtius robustus* feeding in the summer of 2005 off the central Oregon Coast. Geophys Res Lett. 2006;33:L22S11.

Nichols C. Temporal change in dugong body condition and seagrass in Moreton Bay, a temperate environment. Unpublished Honours Thesis, The University of Queensland, Australia; 2005. 55 pages.

Noad M, Cato DH. A combined acoustic and visual survey of humpback whales off southeast Queensland. Mem Queensl Mus. 2001;47:507–23.

O'Brien JK, Robeck TR. Development of sperm sexing and associated assisted reproductive technology for sex pre-selection of captive bottlenose dolphins (*Tursiops truncatus*). Reprod Fertil Dev. 2006;18:319–29.

O'Brien JK, Robeck TR. Preservation of beluga (*Delphinapterus leucas*) spermatozoa using a trehalose-based cryodiluent and directional freezing technology. Reprod Fertil Dev. 2010;22: 653–63.

O'Brien JK, Robeck TR. The relationship of maternal characteristics and circulating progesterone concentrations with reproductive outcome in the bottlenose dolphin (*Tursiops truncatus*) after artificial insemination, with and without ovulation induction, and natural breeding. Theriogenology. 2012a;78:469–82.

O'Brien JK, Robeck TR. The value of *ex situ* cetacean populations in understanding reproductive physiology and developing assisted reproductive technology for *ex situ* and *in situ* species management and conservation efforts. Int J Comp Psychol. 2012b;23:227–48.

O'Brien JK, Steinman KJ, Robeck TR. Application of sperm sorting and associated reproductive technology for wildlife management and conservation. Theriogenology. 2009;71:98–101.

O'Brien JK, Steinman KJ, Schmitt TL, Robeck TR. Semen collection, characterisation and artificial insemination in the beluga (*Delphinapterus leucas*) using liquid-stored spermatozoa. Reprod Fertil Dev. 2008;20:770–83.

Odell DK, Bossart GD, Lowe MT, Hopkins TD. Reproduction of the West Indian manatee in captivity. In: O'Shea TJ, Ackerman BB, Percival HF, editors. Population biology of the Florida manatee. Information and Technology Report 1. USA: National Biological Service; 1995. p. 192–3.

Ohsumi S, Masaki Y. Biological parameters of the Antarctic minke whale at the virginal population level. J Fish Res Board Can. 1975;32:995–1004.

Oosthuizen WC, de Bruyn PJN, Bester MN. Unmarked individuals in mark-recapture studies: comparisons of marked and unmarked southern elephant seals at Marion Island. Austral Ecol. 2012;37:556–68.

Pabst DA, Rommel SA, McLellan WA. The functional morphology of marine mammals. In: Reynolds III JE, Rommel SA, editors. Biology of marine mammals. Washington, DC: Smithsonian Institution Press; 1999. p. 15–72.

Panin M, Gabai G, Ballarin C, Peruffo A, Cozzi B. Evidence of melatonin secretion in cetaceans: plasma concentration and extrapineal HIOMT-like presence in the bottlenose dolphin *Tursiops truncatus*. Gen Comp Endocrinol. 2012;177:238–45.

Parks SE, Brown MW, Conger LA, Hamilton PK, Knowlton AR, Kraus SD, Slay CK, Tyack PL. Occurrence, composition, and potential functions of North Atlantic right whale (*Eubalaena glacialis*) surface active groups. Mar Mamm Sci. 2007;23:868–87.

Parsons KM, Dallas JF, Claridge DE, Durban JW, Balcomb III KC, Thompson PM, Noble LR. Amplifying dolphin mitochondrial DNA from faecal plumes. Mol Ecol. 1999;8:1776–78.

Parsons KM, Durban JW, Claridge DE. Comparing two alternative methods for sampling small cetaceans for molecular analysis. Mar Mamm Sci. 2003;19:224–31.

Pérez S, García-López A, De Stephanis R, Giménez J, García-Tiscar S, Verborgh P, Mancera JM, Martínez-Rodriguez G. Use of blubber levels of progesterone to determine pregnancy in free-ranging live cetaceans. Mar Biol. 2011;158:1677–80.

Perrin WF, Henderson JR. Growth and reproductive rates in two populations of spinner dolphins, *Stenella longirostris*, with different histories of exploitation. Rep Int Whaling Comm. 1984;6:417–30.

Perrin WF, Reilly SB. Reproductive parameters of dolphins and small whales of the family Delphinidae. Rep Int Whaling Comm. 1984;6:97–133.

Perrin WF, Scott MD, Walker GJ, Cass VL. Review of geographical stocks of tropical dolphins (*Stenella* spp. and *Delphinus delphis*) in the eastern Pacific. In: NOAA Technical Report NMFS 28, US Department of Commerce, Washington, D.C.; 1985.

Perryman WL, Donahue MA, Perkins PC, Reilly SB. Gray whale calf production 1994–2000: are observed fluctuations related to changes in seasonal ice cover? Mar Mamm Sci. 2002;18:121–44.

Pietraszek J, Atkinson S. Concentrations of estrone sulfate and progesterone in plasma and saliva, vaginal cytology, and bioelectric impedance during the estrous cycle of the Hawaiian monk seal (*Monachus schauinslandi*). Mar Mamm Sci. 1994;10:430–41.

Plön S, Bernard RTF. A review of spermatozoan morphology in Cetacea with new data for the genus *Kogia*. J Zool. 2006;2006:466–73.

Preen AR. Observations of mating behaviour in dugongs *(Dugong dugon)*. Mar Mamm Sci. 1989;5:382.

Preen AR. Interactions between dugongs and seagrasses in a sub-tropical environment. Unpublished PhD Thesis, James Cook University; Townsville, Queensland, Australia; 1992.

Rathbun GB, Reid JP, Bonde RK, Powell JA. Reproduction in free-ranging Florida manatees. In: O'Shea TJ, Ackerman BB, Percival HF, editors. Population biology of the Florida manatee. Information and Technology Report 1. USA: National Biological Service; 1995. p. 135–56.

Read AJ. Estimation of body condition in harbour porpoises, *Phocoena phocoena*. Can J Zool. 1990;68:1962–6.

Read AJ, Hohn AA. Life in the fast lane: the life history of harbor porpoises from the gulf of Maine. Mar Mamm Sci. 1995;11:423–40.
Reyes JC, Mead JG, Van Waerebeek K. A new species of beaked whale *Mesoplodon peruvianus* sp. N (Cetacea: Ziphiidae) from Peru. Mar Mamm Sci. 1991;7:1–24.
Ridgway S, Carder D, Jeffries M, Todd M. Spontaneous human speech mimicry by a cetacean. Curr Biol. 2012;22:860–1.
Ridgway SH, Scronce BL, Kanwisher J. Respiration and deep diving in the bottlenose porpoise. Science. 1969;166:1651–4.
Robeck TR. Advances in the understanding and manipulation of bottlenose dolphin reproduction. In: Duffield DA, Robeck TR, editors. Proceedings Bottlenose Dolphin Reproduction Workshop, San Diego, California; June 1999; 2000. p. 109–31.
Robeck TR, Atkinson SKC, Brook F. Reproduction. In: Dierauf LA, Gulland FMD, editors. CRC handbook of marine mammal medicine. 2nd ed. Boca Raton, FL: CRC Press; 2001. p. 193–236.
Robeck TR, Curry BE, McBain JF, Kraemer DC. Reproductive biology of the bottlenose dolphin (*Tursiops truncatus*) and the potential application of advanced reproductive technologies. J Zoo Wildl Med. 1994;25:321–36.
Robeck TR, McBain JF, Mathey S, Kraemer DC. Sonographic evaluation of the effects of exogenous gonadotrophin on follicular recruitment and ovulation induction in the Atlantic bottlenose dolphin, *Tursiops truncatus*. J Zoo Wildl Med. 1998;29:6–13.
Robeck TR, Monfort S. Characterization of male killer whale (*Orcinus orca*) sexual maturation and reproductive seasonality. Theriogenology. 2006;66:242–50.
Robeck TR, Monfort SL, Calle PP, Dunn JL, Jensen E, Boehm JR, Young S, Clark ST. Reproduction, growth and development in captive beluga (*Delphinapterus leucus*). Zoo Biol. 2005a;24:29–49.
Robeck TR, O'Brien JK. Effect of cryopreservation methods and precryopreservation storage on bottlenose dolphin (*Tursiops truncatus*) spermatozoa. Biol Reprod. 2004;70:1340–8.
Robeck TR, Schneyer ÀL, McBain JF, Dalton LM, Walsh MT, Czekala NM, Kraemer DC. Analysis of urinary immunoreactive steroid metabolites and gonadotrophins for characterization of the estrous cycle, breeding period, and seasonal estrous activity of captive killer whales (*Orcinus orca*). Zoo Biol. 1993;12:173–87.
Robeck TR, Steinman KJ, Gearhart S, Reidarson TR, McBain JF, Monfort SL. Reproductive physiology and development of artificial insemination technology in killer whales (*Orcinus orca*). Biol Reprod. 2004;71:650–60.
Robeck TR, Steinman KJ, Montano GA, Katsumata E, Osborn S, Dalton L, Dunn JL, Schmidt TL, Reidarson TR, O'Brien JK. Deep intra-uterine artificial inseminations using cryopreserved spermatozoa in beluga (*Delphinapterus leucas*). J Theriogenol. 2010;74:989–1001.
Robeck TR, Steinman KJ, Ramirez K, Greenwell M, Van Bonn W, Yoshioka M, Katsumata E, Dalton L, Osborn S, O'Brien JK. Seasonality, estrous cycle characterization, estrus synchronization, semen cryopreservation and artificial insemination in the Pacific white-sided dolphin (*Lagenorhynchus obliquidens*). Reproduction. 2009;138:391–405.
Robeck TR, Steinman KJ, Yoshioka M, Jensen E, O'Brien JK, Katsumata E, Gili C, McBain JF, Sweeney J, Monfort SL. Estrous cycle characterization and artificial insemination using frozen-thawed spermatozoa in the bottlenose dolphin (*Tursiops truncatus*). Reproduction. 2005b;129:659–74.
Rolland RM, Hamilton PK, Kraus SD, Davenport B, Gillett RM, Wasser SK. Faecal sampling using detection dogs to study reproduction and health in North Atlantic right whales (*Eubalaena glacialis*). J Cetac Res Manage. 2006;8:121–5.
Rolland RM, Hunt KE, Kraus SD, Wasser SK. Assessing reproductive status of right whales (*Eubalaena glacialis*) using fecal hormone metabolites. Gen Comp Endocrinol. 2005;142: 308–17.
Rommel SA, Pabst DA, McLellan WA, Mead JG, Potter CW. Anatomical evidence for a countercurrent heat exchanger associated with dolphin testes. Anat Rec. 1992;232:150–6.

Rowntree VJ, Payne RS, Schell DM. Changing patterns of habitat use by southern right whales (*Eubalaena australis*) on their nursery ground at Península Valdés, Argentina, and in their long-range movements. J Cetacean Res Manag. 2001;2:133–43.

Russell DJF, McConnell B, Thompson D, Duck C, Morris C, Harwood J, Matthiopoulos J. Uncovering the links between foraging and breeding regions in a highly mobile mammal. J Appl Ecol. 2013;50:499–509.

Schaefer AM, Stavros H-CW, Bossart GD, Fair PA, Goldstein JD, Reif JS. Associations between mercury and hepatic, renal, endocrine, and hematological parameters in Atlantic bottlenose dolphins (*Tursiops truncatus*) along the eastern coast of Florida and South Carolina. Arch Environ Contam Toxicol. 2011;61:688–95.

Schipper J, Chanson JS, Chiozza F, et al. The status of the world's land and marine mammals: diversity, threat, and knowledge. Science. 2008;322:225–30.

Schmitt TL, St. Aubin DJ, Schaeffer AM, Dunn JL. Baseline, diurnal variations and stress-induced changes of stress hormones in three captive beluga whales, *Delphinapterus leucas*. Mar Mamm Sci. 2010;26:635–47.

Schroeder JP, Keller KV. Seasonality of serum testosterone levels and sperm density in *Tursiops truncatus*. J Exp Zool. 1989;249:316–21.

Schwabe C, Büllesbach EE, Heyn H, Toshioka M. Cetacean relaxin: isolation and sequence of relaxins from *Balaenoptera acutorostrata* and *Balaenoptera edeni*. J Biol Chem. 1989;264: 940–3.

Shane SH, Wells RS, Würsig B. Ecology, behavior and social organization of the bottlenose dolphin: a review. Mar Mamm Sci. 1986;2:34–63.

Schwarzenberger F, Möstl E, Palme R, Bamberg E. Faecal steroid analysis for non-invasive monitoring of reproductive status in farm, wild and zoo animals. Anim Reprod Sci. 1996;42:515–26.

Scott EM, Mann J, Watson-Capps JJ, Sargeant BL, Connor RC. Aggression in bottlenose dolphins: evidence for sexual coercion, male-male competition, and female tolerance through analysis of tooth-rake marks and behaviour. Behaviour. 2005;142:21–44.

Scott MD, Wells RS, Irvine AB. A long-term study of bottlenose dolphins on the west coast of Florida. In: Leatherwood SJ, Reeves RR, editors. The bottlenose dolphin. San Diego, CA: Academic; 1990. p. 235–44.

Smith SC, Whitehead H. The diet of Galapagos sperm whales *Physeter macrocephalus* as indicated by fecal sample analysis. Mar Mamm Sci. 2006;16:315–25.

Smolker RA, Richards AF, Connor RC, Pepper JW. Sex differences in patterns of association among Indian Ocean bottlenose dolphins. Behaviour. 1992;123:38–69.

Stacey PJ, Baird RW. Birth of a 'resident' killer whale off Victoria, British Columbia, Canada. Mar Mamm Sci. 1997;13:504–8.

Steinman KJ, O'Brien JK, Monfort SL, Robeck TR. Characterization of the estrous cycle in female beluga (*Delphinapterus leucas*) using urinary endocrine monitoring and transabdominal ultrasound: evidence of facultative induced ovulation. Gen Comp Endocrinol. 2012;175:389–97.

Sumich J, Goff T, Perryman WL. Growth of two captive gray whale calves. Aquat Mamm. 2001;27:231–3.

Suzuki T, Mogoe T, Asada M, Miyamoto A, Tetsuka M, Ishikawa H, Ohsumi S, Fukui Y. Plasma and pituitary concentrations of gonadotropins (FSH and LH) in minke whale (*Balaenoptera acutorostrata*) during the feeding season. Theriogenology. 2001;55:1127–41.

Tavolga MC, Essapian FS. The behavior of the bottle-nosed dolphin (*Tursiops truncatus*): mating, pregnancy, parturition and mother-infant behavior. Zoologica. 1957;42:11–31.

Terasawa F, Yokoyama Y, Kitamure M. Rectal temperature before and after parturition in bottlenose dolphins. Zoo Biol. 1999;18:153–6.

Thewissen JGM, Heyning J. Embryogenesis and development in *Stenella attenuata* and other cetaceans. In: Miller D, editor. Reproductive biology and phylogeny of Cetacea: whales, dolphin and porpoises. Enfield, NH: Science Publishers, Inc.; 2007. p. 307–29.

Thompson KV, Baker CS, van Helden A, Patel S, Millar C, Constantine R. The world's rarest whale. Curr Biol. 2012;22:905–6.

Touma C, Palme R. Measuring fecal glucocorticoid metabolites in mammals and birds: the importance of validation. Ann N Y Acad Sci. 2005;1046:54–74.

Tremellen KP, Kolo M, Gilmore A, Laekamge DN. Anti-Müllerian hormone as a marker of ovarian reserve. Aust N Z J Obstet Gynaecol. 2006;45:20–4.

Tripp KM, Verstegen JP, Deutsch CJ, Bonde RK, Rodriguez M, Morales B, Schmitt DL, Harr KE. Validation of a serum immunoassay to measure progesterone and diagnose pregnancy in the West Indian manatee (*Trichechus manatus*). Theriogenology. 2008;70:1030–40.

Tripp KM, Verstegen JP, Deutsch CJ, Bonde RK, de Wit M, Manire CA, Gaspard J, Harr KE. Evaluation of adrenocortical function in Florida manatees (*Trichechus manatus latirostris*). Zoo Biol. 2010;29:1–15.

Trout T. Comment on 'Determination of testosterone in saliva and blow of bottlenose dolphins (*Tursiops truncatus*) using liquid chromatography-mass spectrometry'. J Chromatogr B. 2008;863:192–3.

Van der Schoot P. Studies on the fetal development of the gubernaclum in cetacea. Anat Rec. 1995;243:449–60.

Van Waerebeek K, Read AJ. Reproduction of dusky dolphins, *Lagenorhynchus obscurus*, from coastal Peru. J Mammal. 1994;75:1054–62.

Vergara V, Michaud R, Barrett-Lennard L. What can captive whales tell us about their wild counterparts? Identification, usage, and ontogeny of contact calls in belugas (*Delphinapterus leucas*). Int J Comp Psychol. 2010;23:278–309.

Villegas-Amtmann S, Atkinson S, Costa DP. Low synchrony in the breeding cycle of Galapagos sea lions revealed by seasonal progesterone concentrations. J Mammal. 2009;90: 1232–7.

Wakai Y, Hasegawa K, Sakamoto S, Asano S, Watanabe G, Taya K. Annual changes of urinary progesterone and estradiol-17β of the dugong (*Dugong dugon*) in captivity. Zool Sci. 2002;19:679–82.

Walker L, Cornell L, Dahl K, Czekala N, Dargen C, Joseph B, Hsueh A, Lasley B. Urinary concentrations of ovarian steroid hormone metabolites and bioactive follicle-stimulating hormone in killer whales (*Orcinus orcus*) during ovarian cycles and pregnancy. Biol Reprod. 1988;39:1013.

Waples KA, Gales NJ. Evaluating and minimising social stress in the care of captive bottlenose dolphins (*Tursiops aduncus*). Zoo Biol. 2002;21:5–26.

Watanabe H, Mogoe T, Asada M, Hayashi K, Fujise Y, Ishikawa H, Ohsumi S, Miyamoto A, Fukui Y. Relationship between serum sex hormone concentrations and histology of seminiferous tubules of captured baleen whales in the western North Pacific during the feeding season. J Reprod Dev. 2004;50:419–27.

Watanabe H, Tateno H, Kusakbe H, Matsuoka T, Kamiguchi Y, Fujise Y, Ishikawa H, Ohsumi S, Fukui Y. Fertilizability and chromosomal integrity of frozen-thawed Bryde's whale (*Balaenoptera edeni*) spermatozoa intracytoplasmically injected into mouse oocytes. Zygote. 2007;15:9–14.

Wells RS. Reproduction in wild bottlenose dolphins: overview of the patterns observed during a long-term study. In: Duffield DA, Robeck TR, editors. Proceedings Bottlenose Dolphin Reproduction Workshop; June 1999; San Diego, California; 2000. p. 57–74.

Wells RS. Dolphin social complexity: lessons from long-term study and life history. In: de Waal FBM, Tyack PL, editors. Animal social complexity: intelligence, culture, and individualized societies. Cambridge, MA: Harvard University Press; 2003. p. 32–56.

Wells RS, Rhinehart HL, Hansen LJ, Sweeney JC, Townsend FI, Stone R, Casper DR, Scott MD, Hohn AA, Rowles TK. Bottlenose dolphins as marine ecosystem sentinels: developing a health monitoring system. Ecohealth. 2004;1:246–54.

Wells RS, Tornero V, Borrell A, Aguilar A, Rowles TK, Rhinehart HL, Hofmann S, Jarman WM, Hohn AA, Sweeney JC. Integrating potential life-history and reproductive success data to examine relationships with organochlorine compounds for bottlenose dolphins (*Tursiops truncatus*) in Sarasota Bay, Florida. Sci Total Environ. 2005;349:106–19.

West K, Atkinson S, Carmichael M, Sweeney J, Krames B, Krames J. Concentrations of progesterone in milk from bottlenose dolphins during different reproductive states. Gen Comp Endocrinol. 2000;117:218–24.

Wetzel DL, Reynolds III JE, Breuel KF, Roudebush WE. Validation of an inhibin-B ELISA for use in an endangered marine mammal, the West Indian manatee (*Trichechus manatus*). Biol Reprod Suppl. 2009;215:97–7.

Whitehead H. The behaviour of mature male sperm whales on the Galapagos Islands breeding grounds. Can J Zool. 1993;71:689–99.

Whitehead H. Delayed competitive breeding in roving males. J Theor Biol. 1994;166:127–33.

Whitehead H, Mann J. Female reproductive strategies of cetaceans. In: Mann J, Connor RC, Tyack PL, Whitehead H, editors. Cetacean Societies: field studies of dolphins and whales. Chicago, IL: University of Chicago Press; 2000. p. 219–46.

Williamson P, Gales NJ, Lister S. Use of real-time B-mode ultrasound for pregnancy diagnosis and measurement of fetal growth in captive bottlenose dolphins (*Tursiops truncatus*). J Reprod Fertil. 1990;88:543–8.

Wilson RC, Reynolds III JE, Wetzel DL, Schwierzke-Wade L, Bonde RK, Breuel KF, Roudebush WE. Secretion of anti-Müllerian hormone in the Florida manatee *Trichechus manatus latirostris*, with implications for assessing conservation status. Endanger Species Res. 2011;14:107–12.

Würsig B, Jefferson TA. Methods of photo-identification for small cetaceans. Rep Int Whaling Comm. 1990;12:43–52.

Yoshioka M, Aida K, Hanyu I. Annual changes in serum reproductive hormone levels in captive bottlenose dolphins. Bull Jpn Soc Sci Fish. 1986;52:1939–46.

Yoshioka M, Okumura T, Aida K, Fujise Y. A proposed technique for quantifying muscle progesterone content in minke whales (*Balaenoptera acutorostrata*). Can J Zool. 1994;72:368–70.

Yochem PK, Leatherwood S. Blue whale *Balaenoptera musculus* (Linnaeus, 1758). In: Ridgway SH, Harrison R, editors. Handbook of marine mammals, The sirenians and baleen whales, vol. 3. London: Academic; 1985. p. 193–240.

Yuen QWH, Brook FM, Kinoshita RE, Ying MTC. Semen collection and ejaculate characteristics in the Indo-Pacific bottlenose dolphin (*Tursiops truncatus*). J Androl. 2009;30:432–9.

Chapter 12
Amphibian Declines in the Twenty-First Century: Why We Need Assisted Reproductive Technologies

John Clulow, Vance L. Trudeau, and Andrew J. Kouba

Abstract Each amphibian species is evolutionarily distinct, having developed highly specialized and diverse reproductive strategies in both terrestrial and aquatic environments. These unique reproductive patterns and mechanisms, key to species propagation, have only been explored in a limited number of laboratory models. Although the development of applied reproductive technologies for amphibians has proven useful for a few threatened species, the real benefit of this technology has been new insights into the reproductive adaptations, behavior, endocrinology, and physiological mechanisms that have evolved over millions of years. As the basic fundamental database on amphibian reproductive physiology has grown, so has the applied benefit for species conservation. In particular, technologies such as non-invasive fecal and urinary hormone assays, hormone treatments for induced breeding or gamete collection, *in vitro* fertilization, and the ability to establish genome resource banks have all played important roles in monitoring or managing small populations of captive species. Amphibians have the ability to produce a large excess of germplasm (up to 10,000 ovulated eggs in a single reproductive event) that if not collected and preserved, would represent a wasted valuable resource. We discuss the current state of knowledge in assisted reproductive technologies for

J. Clulow
School of Environmental and Life Sciences, University of Newcastle,
Newcastle, NSW, Australia
e-mail: John.Clulow@newcastle.edu.au

V.L. Trudeau (✉)
Department of Biology, Centre for Advance Research in Environmental Genomics,
University of Ottawa, Ottawa, ON, Canada K1N 6N5
e-mail: trudeauv@uottawa.ca

A.J. Kouba
Conservation and Research Department, Memphis Zoo, Memphis, TN, USA
e-mail: akouba@memphiszoo.org

W.V. Holt et al. (eds.), *Reproductive Sciences in Animal Conservation*, 275
Advances in Experimental Medicine and Biology 753, DOI 10.1007/978-1-4939-0820-2_12,

amphibians and why their extinction crisis means these available tools can no longer be implemented as small-scale, last-ditch efforts. The reproductive technologies must be established early as a key component of large-scale species recovery.

Keywords Captive breeding • Cryopreservation • De-extinction • Frog • Hormones • In vitro fertilization • Oocyte • Sperm • Stem cells

1 The Amphibian Extinction Crisis

Recognition of global amphibian population declines initially manifested at the First World Congress of Herpetology in 1989 (Bishop et al. 2012). Since this initial warning alarm was sounded, the once heralded declines have now become an extinction crisis with no sign of abating (Alford and Richards 1999; Houlahan et al. 2000; Stuart et al. 2004; Bishop et al. 2012; Stuart 2012). Analysis of the data from the 2004 Global Amphibian Assessment showed that 34 amphibians have disappeared (1 % of 5,915 described species) and 1,893 species are threatened with extinction (32.0 % of listed species). However, the proportion threatened with extinction may be closer to 40 % when the proportion of data-deficient species is taken into account (Bishop et al. 2012). In comparison, the number of birds (12 %) and mammals (25 %) threatened with extinction is much lower (Bishop et al. 2012), making amphibians the most threatened vertebrate class. It is also probable that the actual number of extinct amphibians is much higher than 34, given that many of the species listed as data-deficient have not been seen for a decade or longer.

The rate of decline and extinction has made it incredibly difficult for conservation organizations to respond in an effective manner and there is not enough holding space or resources in captive facilities to provide a safety net for all imperilled species. The situation is compounded by the fact that one of the primary drivers of the extinction process, a rapidly spreading disease known as chytridiomycosis, remains largely a problem without an effective solution (Bishop et al. 2012). Despite major progress in understanding the pathogenicity, virulence, etiology, and species susceptibility to the chytrid fungus *Batrachochytrium dendrobatidis* (Bd) (Berger et al. 1998; Longcore et al. 1999; Lips et al. 2006; Skerratt et al. 2007, 2010; Murray et al. 2010), we are no closer to halting or even reversing its deadly impact on amphibian populations. Another primary driver of population decline is habitat alteration and loss, which is estimated to affect 63 % of all amphibian species (Bishop et al. 2012; Chanson et al. 2008), suggesting the number of threatened species will escalate in the near future (Rowley et al. 2010).

In cases where habitat loss, over-exploitation or pollution is the clear driver of extinction (e.g., Kihansi spray toad, dusky gopher frog, hellbender, giant salamanders), urgent action is needed to establish captive assurance colonies. Habitat restoration efforts and reintroductions can then be a feasible conservation strategy in the future. Although some aspects of population sustainability have been addressed by setting aside nature reserves or protected habitats, these conservation efforts have proven insufficient when the primary driver of losses are diseases. For example, many of the extinctions caused by Bd infection have occurred in pristine, unaltered ecosystems for which further formal protection would have had little direct impact on the eventual outcome. In Australia, four species recognised by the Federal Government and IUCN as extinct, including two species of unique gastric brooding frogs (*Rheobatrachus* spp.), disappeared from relatively pristine rain forest habitats, some of which are now World Heritage sites. Furthermore, of the 29 Australian species officially listed as threatened, the decline of only a small proportion can reasonably be linked to habitat threats, while it is widely believed that most of the declines should be attributed to disease (Skerratt et al. 2007; Murray et al. 2010). In addition, many neo-tropical amphibian species in Latin America have declined due to the epizootic spread of Bd (Lips et al. 2005, 2006).

In 2006, the Amphibian Ark was founded and is a consortium of zoos, governmental organizations and other non-profit conservation organizations committed to captive breeding and maintenance of assurance colonies (Bishop et al. 2012). After evaluating less than half (~42 %), of all the amphibian species, it was concluded that 360 require captive breeding assistance, which extrapolates to ~950 species when assumptions on all threatened and data deficient species are made (Bishop et al. 2012). These 950 species equate to 16 % of all described species (Stuart et al. 2008; Bishop et al. 2012). Since the analysis of the Global Amphibian Population Assessment (Stuart et al. 2008), approximately 100 new species have been reported and global numbers are now approaching 7,000. This implies that the number of candidate species for captive populations will likely increase. The scale of the challenge can be put into perspective when one considers the current global capacity for managing viable captive amphibian populations is in the order of 50 species (Bishop et al. 2012). It is doubtful that there will be a surge in resources and funding proportionate to the scale of the amphibian crisis (Bishop et al. 2012; Stuart 2012). Hence, the challenge facing the amphibian conservation community is the coordinated optimization of available resources and approaches.

We focus on various amphibian assisted reproductive technologies (ART) for captive breeding that are the foundation for research, maintaining genetic diversity, translocations, and reintroductions. Many amphibian species reproduce very poorly or not at all in captivity, so it will be important to prioritize and systematically apply technological advances as required. Varying levels of population status (e.g., critically endangered, endangered, vulnerable, etc.) and knowledge about a species under consideration for captive breeding dictates the level of intervention required, ranging from simple bioassays to more complicated procedures such as animal cloning. Yet, given the speed with which the herpetological community was overtaken by the current crisis, it is clear that techniques such as cryopreservation of genetic resources (Kouba et al. 2013) should be used to mitigate future risks. One example of a response hierarchy for extinction mitigation according to threat status is shown in Table 12.1.

Table 12.1 Application and prioritization of assisted reproductive technologies according to IUCN threat category in order to address the amphibian extinction crisis

IUCN category	Number of species[a]	Response	Urgency
Extinct	34	Explore de-extinction; (cloning)	Not urgent, as long as governance and curation of frozen specimens is in place
Extinct in the wild	2	Implement ART; (cryobanking, hormone therapy, IVF)	Highest priority
Critically endangered	519	Implement ART; (cryobanking, hormone therapy, IVF)	Highest priority
Endangered	773	Implement ART; (cryobanking, hormone therapy, IVF)	High priority
Vulnerable	656	Develop ART procedures as contingency; (implement cryobanking)	Medium priority
Near threatened	390	Implement cryobanking	Low priority
Least concern	2,404	Implement cryobanking; select a few as model species for more advanced technologies	Low priority
Data deficient	1,633	Implement cryobanking	More information required
Total	6,411		

Abbreviations: *ART* assisted reproductive technologies, *IVF in vitro* fertilisation
[a]IUCN Red List Version 2013.1 Table 3a, updated 8 July, 2013

2 Benefits of ART to Amphibian Conservation

Amphibian ART has gained considerable interest in the last decade but still lags behind technological developments for mammalian species. Although amphibian ART is a relatively new field of study, several recent advancements in protocol development have led to significant conservation achievements (Table 12.2) such as the production of endangered tadpoles from frozen-thawed sperm or the release of thousands of tadpoles produced by *in vitro* fertilization (IVF) (Kouba et al. 2009, 2013; Kouba and Vance 2009). When applied effectively, ART could result in increased breeding efficiency, reduced costs, halt loss of genetic diversity, reduce mate-pairing issues, and possibly even reject extinction as the only scenario for critically endangered species. Amphibian ART can achieve these outputs through the following mechanisms:

2.1 Increased Efficiency of Captive Breeding and Release Programmes

Amphibian ART offers the potential to increase the efficiency of captive breeding by induced spawning or controlled release of gametes for IVF. Induced ovulation and spawning may remove the requirement for challenging behavioural and phermonal cues in recalcitrant species that may not otherwise breed readily in captivity, if at all (Table 12.2).

Table 12.2 Amphibian conservation achievements using assisted reproductive technologies

Application	Detail	References
Hormone-induced spawning in frogs and toads	Production of anurans in and out of season for research should reduce pressures on wild populations	Trudeau et al. (2010, 2012, 2013), Germano et al. (2011), Kouba et al. (2011, 2012a), Silla (2011, 2012), Clulow et al. (2012)
Hormone-induced spawning and IVF in salamanders	Production of hellbenders (*Cryptobranchus alleganiensis*) for captive breeding program; production of tiger salamanders (*Ambystoma tigrinum*) as a model species for other ambystomids	Trudeau et al. (2012), Marcec et al. (unpublished)
Use of IVF in captive breeding and release	2,000 Wyoming Toads (*Anaxyrus baxteri*) produced by IVF; >10,000 boreal toads (*Anaxyrus boreas boreas*) produced by IVF; release of all tadpoles to the wild IVF performed in critically endangered Corroboree frog (*Pseudophyrne corroboree*)	Kouba et al. (2013), Browne et al. (2006a), Kouba (unpublished data), Byrne and Silla (2010)
Use of gamete transport between institutions, prior to successful fertilisation by IVF	Dusky gopher frog (*Lithobates sevosa*) sperm transported from Memphis Zoo to Omaha Henry Doorly Zoo, where IVF resulted in 200 eggs fertilised	Kouba et al. (2011, 2012a, b, 2013)
Collection of sperm from wild males followed by cryopreservation, and IVF producing mixed wild-captive produced offspring	Sperm was collected from wild boreal toads (*Anaxyrus boreas boreas*), frozen in the field and used to fertilize eggs from hormone-induced females in captivity	Langhorne et al. (unpublished)
Sexing of frogs using fecal or urinary steroids: NZ captive leiopelmatid frogs (urinary steroids); Australian captive bred *Geocrinia* juveniles (faecal steroids); N. American bufonids	Leiopelmatids are monoecious; sexing is required for mate pairing in captive breeding; Captive bred *Geocrinia* juveniles cannot be sexed visually; determination of captive bred juveniles prior to release to ensure even sex ratios at release sites	Germano and Molinia (unpublished), Hogan et al. (2013), Szymanski et al. (2006)

2.2 Reduce the Costs Associated with Captive Assurance Colonies by Limiting the Number of Live Animals Required to Sustain an Outbred Colony

By the combined use of small live populations with cryobanked gametes for breeding, heterozygosity and allelic diversity can be maintained with a fraction of the number of live animals. The savings per species are potentially enormous, with no reduction in effectiveness of the programme. Financial and logistical resources could be redirected to support more species and thereby contribute to the management of extinction risk.

2.3 Decrease the Rate of Genetic Diversity Loss in Captive and Small Wild Populations

The capacity of ART to improve genetic outcomes for species is under-recognised and under-utilised. Small wild populations, such as those managed on islands or remnant mainland populations, risk extinction from inbreeding depression (Frankham and Ralls 1998; Weeks et al. 2011). Captive populations in zoos and fenced sanctuaries risk loss of genetic fitness through selection for domestication (Williams and Hoffman 2009) that would reduce chances of later successful reintroduction to the wild. Storing genomes as gametes or embryos in the first or second generation after establishment of captive populations can prevent the loss of genetic diversity (Schad 2008). Long-term storage in gene banks offers many advantages to captive breeding programs (Table 12.3) and means that genetic diversity can be restored in small populations at any stage in the future. If diversity can be captured early in the process, the options for later restoring genetic fitness in wild populations are enhanced. Such genetic rescue can be very effective (Madsen et al. 1999; Westemeir et al. 1998).

2.4 Genetic Management Through the Control of Mate Pairing

This can be accomplished by selective breeding through induced spawning or IVF (Table 12.2). Genetic management of small populations requires the capacity to control the pairing of males and females; ART can overcome problems with mate choice where pairs show behavioural incompatibility, and thus slow the rate of loss of genetic diversity through more intensive management of captive animals (Schad 2008). For example, a female may produce thousands of eggs from hormone-induced spawning, which can be fertilized by sperm from multiple males during IVF.

2.5 Restoration of Extinct Species

A number of species that are extinct persist as frozen tissues (not cryopreserved) in various museums around the world. At one level, it is possible to argue that these lost species are not extinct because there is a high probability that the cell nuclei in

Table 12.3 Justification for establishing Genetic Resource Banks for amphibians

Justification	Outcome for amphibian conservation
Reproductive failure	Incompatible breeding in amphibians could be overcome through the use of IVF
Increased security	Provides some protection against disease outbreaks causing local extinctions to amphibian populations
Unlimited space	Cryobanking offers a large amount of space to conserve diversity versus limited space and resources for live colonies
Increased gene flow	Transportation of frozen gametes between zoos or wild populations (or between zoos and wild populations) has advantages over moving live amphibians
Minimize introgression	Secures the integrity of a gene pool against the threat of hybridisation
Extend generation times	The genetic lifespan is extended thereby reducing loss of alleles (genetic drift)
Maximize genetic diversity	Storage of unrepresented founder amphibians, under-represented descendants and deceased animals
Minimize inbreeding	Restoring germplasm to unrelated or more distantly related amphibians
Manage effective population size	Equalize family size by manipulating age-specific fertility rates and sex ratios
Minimize selection	Detailed pedigree analysis combined with GRBs can reduce genetic drift and increase genetic diversity for small assurance colonies
Mutation	Extending generation lengths assists in decreasing the load of harmful mutations in small amphibian populations
Safeguarding existing resources	Preservation of cell lines for future research and technological advances in ART
Long-term benefits	Possibilities of restoring lost genes; discovering medicinal compounds for curing illnesses; pathogenic studies for disease resistance; nuclear transfer or parthenogenesis (gynogenesis and androgenesis) experiments
Extinction risk	Reduce the risk of extinction
De-extinction	Provide cells and genetic resource for the potential recovery of extinct populations and species with future technological advances

Adapted from Kouba and Vance (2009), Holt et al. (1996) and Bennett (2001)

the frozen specimens are still functional (in the same way that cells and gametes in cryostorage for human IVF labs, medical disease models or in agricultural industries are still functional and retrievable). Some consideration might be given to the possibility of retrieving at least some species as technology advances.

The ease of performing IVF in a taxon exhibiting external fertilization and development has resulted in the successful production of many amphibian species (Hollinger and Corton 1980; Browne et al. 1998; Edwards et al. 2004; Kouba et al. 2009, 2011). This includes the capacity to perform intra-cytoplasmic sperm injection (Kroll and Amaya 1996). The work relevant to reproductive technology has been concentrated in a relatively few laboratory species, especially of the genera *Xenopus*, *Silurana* and *Lithobates* (Rugh 1962; Lofts 1974; McKinnell 1978; Di Berardino 1997; Gurdon and Hopwood 2000; Schultz and Dawson 2003; Ogawa et al. 2011) and bufonids (Cabada 1975; Browne et al. 1998, 2001, 2002a, b, c, d, 2006a, b; Fitzsimmons et al. 2007). Yet there is a large body of published work on reproductive biology across many amphibian families (for comprehensive sources see Jamieson 2007; Ogielska 2009; Norris and Lopez 2011) that lay the foundations to develop captive breeding strategies for a wide range of species.

The impetus for research in amphibian developmental and reproductive biology has been due to the technical advantages of working with amphibian models in basic research, phylogeny and evolution. The focus therefore has not been on the applied aspects of amphibian reproduction or how this knowledge might assist conservation efforts. This area has developed slowly over the last two decades (Clulow et al. 1999; Kouba et al. 2009, 2013; Kouba and Vance 2009). Consequently, while some aspects of developmental and reproductive biology in amphibians are highly advanced, others relevant to their practical application and value in conservation biology are not. While reproductive endocrinology, nuclear transfer, and IVF are well developed in a few model laboratory species, other aspects such as genome storage via sperm, egg, embryo and somatic cells are well behind the advances made with other vertebrate taxa. In particular, very little has been studied in urodeles (salamanders and newts) or caecilians and the reader will note that most of the literature presented here is related to anurans (frogs and toads). Before exploring these topics in greater detail it is relevant to set the stage for understanding how the study of endocrinology has advanced our knowledge of amphibian reproductive mechanisms and allowed biologist to exploit the use of exogenous hormones for applied conservation.

3 Neuroendocrine Control of Reproduction

While amphibians share the main common reproductive neuropeptides, pituitary hormones and gonadal steroids with other vertebrates, very little is actually known about how the hypothalamic-pituitary-gonadal (HPG) axis is regulated, and therefore how it can be manipulated experimentally or for endangered species propagation. Here we review major aspects of the HPG axis that are directly relevant to induction of spawning for captive breeding. This review is pertinent given that obtaining gametes and fertilizing them (either through hormone-induced breeding or IVF) is central to the success of captive breeding programs where animals fail to breed naturally. Specific applications of injected hormones and agonists will be presented in later sections. The hypothalamic decapeptide gonadotropin-releasing hormone (GnRH) stimulates the synthesis and release of the gonadotropins from the anterior pituitary (Kim et al. 2011; Kah et al. 2007). The gonadotropins are luteinizing hormone (LH) and follicle stimulating hormone (FSH). These dimeric proteins are composed of a common alpha subunit and a differing beta subunit that confers specificity of binding to either the LH or FSH receptors, and thus also confers specificity of biological action at the level of the gonads. In vertebrates, natural ovulation in females and sperm release in males results from a surge release of LH (Fernandez and Ramos 2003; Trudeau 1997), so most ART research focuses on inducing LH release or stimulating LH receptors in the gonads. Little is known about the neuroendocrine regulation of FSH in amphibia.

3.1 Gonadotropin-Releasing Hormone

The neuropeptide GnRH exists in multiple forms across the vertebrate families (Kim et al. 2011; Kah et al. 2007). They are usually named for the species or animal group where they are first discovered. There are two GnRH forms in most amphibians that are the products of two different genes. These are mammalian GnRH (mGnRH; pGlu-His-Trp-Ser-Tyr-Gly-Leu-Arg-Pro-Gly-amide) and chicken GnRH-II (cGnRH-II; pGlu-His-Trp-Ser-His-Gly-Trp-Tyr-Pro-Gly-amide), also called GnRH1 and GnRH2, respectively. The hypophysiotropic neurons expressing mGnRH are restricted to the anterior preoptic area in the anuran brain and project to the median eminence (Collin et al. 1995). This indicates that mGnRH will be the predominant endogenous neuropeptide released into the portal blood vessels and delivered to the anterior pituitary to regulate gonadotropin synthesis and release in frogs. Neurons producing cGnRH-II are widely distributed throughout the frog brain (Collin et al. 1995). Thus, it could be speculated that cGnRH-II would exert central control over GnRH-regulated sexual behaviours rather than being the main regulator of gonadotropin release. It has recently been reported that mGnRH also regulates aspects of neurosteroid synthesis in the frog brain, implicating this GnRH form in the control of behaviour as well (Burel et al. 2013).

Elegant ligand–receptor interaction studies in several ranid species indicate high GnRH receptor (GnRH-R) selectivity for endogenous and synthetic analogs of GnRH (Wang et al. 2001; Seong et al. 2003). Relevant here is the observation that mGnRH exhibits high potency on the type 1 GnRH-R, the dominant receptor in the anterior pituitary of frogs. Therefore, efforts to evaluate the use of GnRH for spawning induction should focus on mammalian GnRH super-agonists because they will likely target the GnRH-R in the amphibian pituitary. To date the main GnRH agonists that are highly biologically active in anurans are des-Gly^{10}-His(Bzl)6-GnRH-ethylamide and des-Gly^{10}, D-Ala^{6}, Pro^{9}-GnRH-ethylamide (Kouba et al. 2009; Michael et al. 2004; Minucci et al. 1989; Waggener and Carroll 1998a, b; Browne et al. 2006a; Trudeau et al. 2010, 2013). The most promising to date is des-Gly^{10}, D-Ala^{6}, Pro^{9}-GnRH-ethylamide because of high activity, longer half-life, availability, and low cost. It is a synthetically modified form of mGnRH. The modified alanine at position 6 is particularly important for increasing GnRH activity to stimulate LH release (Millar et al. 2004).

3.2 Control of Pituitary Gonadotropin Synthesis and Release

The distribution of gonadotrophs in the pars distalis (anterior) of the pituitary gland is similar to the situation in mammals (Gracia-Navarro and Licht 1987; Mizutani et al. 1994; Tanaka et al. 1990). Immunolocalization studies in several ranids indicate that the majority of gonadotrophs (~50–80 %) produce both LH and FSH, whereas the remainder produce only one of the gonadotropins. Very little is known about the factors controlling LH or FSH synthesis and release in amphibians. There are many studies exploring the role of a vast array of neuropeptides, neurotransmitters and sex steroids controlling gonadotroph function in fish (Peter et al. 1986; Trudeau 1997; Trudeau et al. 2000; Popesku et al. 2008; Zohar et al. 2010) and mammals (Barraclough et al. 1984; Herbison 1998; Smith and Clarke 2010). In marked contrast, most studies in frogs focus only on the stimulatory effects of GnRH on LH synthesis and release (Stamper and Licht 1990, 1993, 1994), which indicate a similar mode of action as found in other vertebrates. This includes upregulation of LH by GnRH self-priming (Porter and Licht 1985), a principle that has been demonstrated to enhance spermiation, ovulation and fertility in frogs (Silla 2011; Kouba et al. 2012a; Trudeau et al. 2013). Low doses of the neuropeptide GnRH can upregulate its own receptor, thereby enhancing the LH response to a specifically timed second injection (e.g., Trudeau et al. 2013). On the other hand, overstimulation by excessive GnRH treatments can have the opposite effect because GnRH receptors in the pituitary can be slowly down-regulated or desensitized (Pawson et al. 2008). Therefore, from a practical sense, the timing and frequency of GnRH treatments for induced breeding in a given species can have a significant impact on the outcome (i.e., success or failure).

Other data indicate that the pituitary hormone prolactin (PRL) can enhance GnRH-stimulated LH release from bullfrog pituitary cells *in vitro*; PRL cells are in close proximity to gonadotropin-producing cells, so it is probable the PRL is a paracrine factor (Oguchi et al. 1997). Prolactin can also act within the brain to enhance courtship behaviour in the male newts, *Cynops pyrrhogaster* (Toyoda et al. 2005). Therefore, environmental and hormonal factors modulating prolactin production could impact reproductive success in amphibia.

There are also few studies on the role of sex steroid feedback control of LH release in amphibians. Pavgi and Licht (1993) showed that estradiol (E2) has a direct inhibitory effect on FSH and LH secretion at the level of the bullfrog pituitary, indicating the existence of a negative feedback loop. In contrast, the androgen 5α-dihydrotestosterone (DHT) significantly elevated the GnRH-induced secretion of LH and FSH (Stamper and Licht 1994), which is evidence for positive feedback. On the other hand, *in vivo* studies indicate that both E2 and DHT suppress LH, perhaps indicative of decreased GnRH release in male leopard frogs (Tsai et al. 2005; Tsai and Jones 2005). Regardless, these data suggest the existence of both positive and negative feedback effects of sex steroids in frogs as reported in fish and mammals (Trudeau 1997; Trudeau et al. 2000; Zohar et al. 2010; Smith and Clarke 2010).

In leopard frogs it is known that estrogens and androgens can influence the number of forebrain neurons immunoreactive for tyrosine hydroxylase, a key enzyme in the catecholamine dopamine (DA) synthesis pathway (Chu and Wilczynski 2002; Wilczynski et al. 2003). However, the role that catecholamine neurons play in the steroidal feedback control of GnRH and LH is unknown. In goldfish it is clear that sex steroids modulate the synthesis of DA, and this is part of a negative feedback loop at the level of the brain (Trudeau 1997). Based on what is known in amphibians, we propose a model that can serve as a framework for future research on the

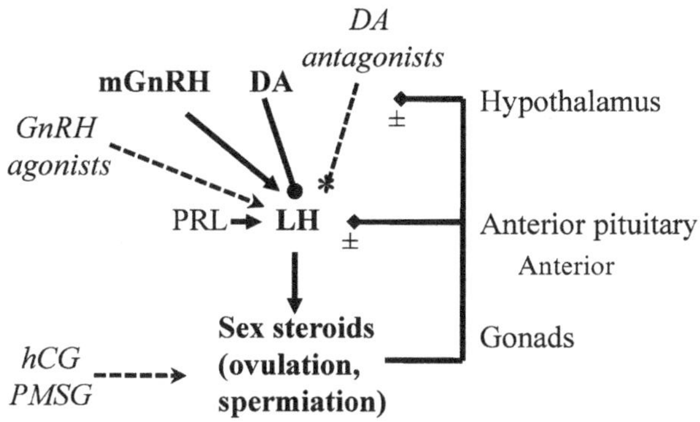

Fig. 12.1 Model for the neuroendocrine control of luteinizing hormone release in amphibians. The decapeptide gonadotropin-releasing hormone (GnRH) is produced in specific hypothalamic neurons in the hypothalamus. In amphibians the main endogenous hypophysiotropic form delivered to the anterior pituitary is mammalian GnRH (mGnRH). The GnRH is released into portal blood vessels at the median eminence. The GnRH is delivered to the anterior pituitary and acts on specific membrane-bound GnRH receptors on gonadotrophs to stimulate the production and release of luteinizing hormone (LH). The glycoprotein hormone LH is released from the pituitary into the general circulation and acts on specific LH receptors. For the ovary, LH receptors are found on theca and granulosa cells, and for the testes LH receptors are on the Leydig cells. The sex steroids (e.g., testosterone, estradiol, progesterone) are produced and released into the blood and act on specific receptors throughout the body to regulate numerous physiological processes. Shown here are the positive or negative feedback effects on LH release. Sex steroids can also act at the level of the hypothalamus to regulate GnRH and dopamine (DA), and sexual behaviors. The catecholamine DA is produced in specific hypothalamic neurons and is a potent inhibitor of LH release. It acts at the pituitary by binding to DA receptors. The pituitary hormone prolactin (PRL) can enhance GnRH-stimulated LH release. Also shown (*italics*, *dashed lines*) are the main hormone preparations currently in use for induced spawning in amphibia. Numerous natural and synthetic agonists of GnRH have been injected to stimulate LH release, leading to gamete release or spawning. They can be injected to stimulate gonadal LH receptors to induce sex steroid production, ovulation and spermiation. In amphibians, the mammalian gonadotropins human chorionic gonadotropin (hCG) and pregnant mare's serum gonadotropin (PMSG) may have follicle stimulating hormone-like activities and stimulate FSH receptors on granulosa cells in the ovary to regulate follicular development or on FSH receptors on Sertoli cells in the testes to regulate spermatogenesis. Antagonists of DA have been used to potentiate GnRH action to super-stimulate LH release from the pituitary, which leads to amplexus and spawning. In this case, a GnRH agonist is co-administered with the DA antagonist. The DA antagonist blocks DA receptors, thus removing the inhibitory effects of DA on LH release (*dashed line* with *asterisk*). See text for further details

hormonal control of spawning in amphibians (Fig. 12.1). While it is known that many other factors play a role in LH and FSH release in other vertebrates, they have been largely unexplored in the context of reproduction in amphibians.

Similar to other vertebrates, seasonal profiles of amphibian hormones in relation to the timing of the breeding period indicate the importance of LH and FSH in controlling gonadal steroidogenesis and gonadal growth (Licht et al. 1983; Polzonetti-Magni et al. 1998; Itoh et al. 1990). Highly purified bullfrog LH and FSH exert season-dependent effects on ovarian production of E2 and progesterone in *Pelophylax esculentus* (Polzonetti-Magni et al. 1998). Interestingly, both LH and

FSH were also shown to directly stimulate hepatic vitellogenin production in females (Polzonetti-Magni et al. 1998). Vitellogenin is the yolk glycoprotein under the stimulatory control of E2. Once produced in the liver, vitellogenin is released into the circulation, taken up by growing oocytes through receptor-mediated endocytosis, and serves to nourish developing embryos in oviparous vertebrates (Polzonetti-Magni et al. 2004).

3.3 The Role of the Catecholamine Dopamine

While it is known that GnRH and GnRH agonists activate recombinant GnRH receptors *in vitro* (Wang et al. 2001; Seong et al. 2003), and in some cases have been demonstrated to stimulate LH release, they may not effectively induce ovulation and spawning in frogs without *in vivo* co-treatments with other hormones or neuroactive agents (Browne et al. 2006a, b; Kouba et al. 2012a). This is highly suggestive of the existence of an inhibitory neuroendocrine mechanism controlling the surge release of LH. It has been known for more than two decades (Peter et al. 1986) that there are specific regions in the teleost forebrain that inhibits spawning by inhibition of LH release. In fish, a multitude of studies have clearly established that the main inhibitory substance is DA (Peter et al. 1986; Trudeau 1997; Dufour et al. 2005; Popesku et al. 2008). In marked contrast, central inhibitory control mechanisms have not been considered important in frog reproduction, even though there is convincing but sparse evidence. Electrolytic lesions in the hypothalamus and infundibular regions of the European frog *Rana temporaria* increased GnRH and LH release and advanced spawning times (Sotowska-Brochocka 1988; Sotowska-Brochocka and Licht 1992). Immunocytochemical visualization of DA neuronal fibres in the hypothalamus and median eminence of *Pelophylax ridibundus* and *Pleurodeles waltl* indicates that DA can be delivered to the amphibian pituitary (Gonzalez and Smeets 1991). In the túngara frog, *Engystomops pustulosus*, mapping of the catecholamine synthesis enzyme tyrosine hydroxylase in relation to the DA signaling protein DARPP32 indicated that these neuroendocrine regions contain DA and respond to DA (O'Connell et al. 2010). Moreover, the DA type 2 receptor is expressed in the frog pituitary (Nakano et al. 2010a, b). This neuroanatomical and biochemical evidence indicates that DA is potentially involved in the control of LH in amphibians.

More direct evidence for the involvement of DA also comes from studies of *Rana temporaria* (Sotowska-Brochocka et al. 1994). *In vivo* treatments with the DA type 2 receptor agonist bromocriptine, inhibits LH release and ovulation in *Rana temporaria* in some situations. In contrast, long-term treatment with a slow release implant containing the DA type 2 receptor antagonist metoclopramide (MET) induced ovulation in hibernating *Rana temporaria* (Sotowska-Brochocka et al. 1994). These data indicate that DA is an important inhibitor of LH release in frogs as it is in numerous fish species, birds and some mammals, including sheep and humans (Dufour et al. 2005). Browne et al. (2006a) explored the effects of combinations of hormones on spawning in *Anaxyrus fowleri*. In that study, they used the DA antagonist pimozide (PIM) and concluded that PIM may increase spawning in some situations and hormone combinations. Following this, Trudeau et al. (2010) proposed the use of combinations of GnRH agonists with DA receptor antagonists (PIM and MET) to induce spawning. It was reported that the combination of MET with the mammalian GnRH agonist des-Gly^{10}, D-Ala^{6}, Pro-GnRH-ethylamide was the most effective in *Lithobates pipiens*. Both PIM and MET cross the blood–brain barrier, and thus their sites of action can be both in the brain and pituitary and may be antagonist to endogenous DA action. It should be noted that PIM is not specific to DA receptors, and is known to act on adrenergic and serotoninergic receptors in addition to DA receptors (Bezchlibnyk-Butler and Jefferies 2005); therefore the use of PIM is not recommended.

The specific DA type 2 receptor antagonist domperidone (DOM) does not cross the blood–brain barrier, and acts at the level of the pituitary to potentiate GnRH action on LH release in goldfish (Omeljaniuk et al. 1987). In leopard frogs, the combination of des-Gly^{10}, D-Ala^{6}, Pro^{9}-GnRH-ethylamide with either MET or DOM are equally effective at inducing spawning (Trudeau et al. 2013), so current evidence suggests that the major site of action of DA to potentiate LH release is the pituitary. Dopamine may have additional roles in anuran reproduction. Creighton et al. (2013) recently reported that activation of D2-like receptors by injection of a specific agonist has an inhibitory effect on vocalization in breeding male green tree frog, *Hyla cinerea*. It is therefore possible that DA acts in the brain to modulate sexual vocalizations and at the pituitary to control LH release to ensure that amplexus, oviposition, sperm release and fertilization are coordinated. Further experimentation is required to substantiate this hypothesis. A clearer understanding of the physiology, biochemistry, receptor signaling pathways and molecular mechanisms leading to the control of release of LH and FSH is critical for the development of specialized hormone treatments for applied conservation.

4 Collection of Gametes Following Injection with Exogenous Hormones

It is essential to understand the basic neuroendocrine control mechanisms and the site of hormone action in the body (Fig. 12.1). A wide array of hormones has been used in attempts to induce ovulation or sperm release in amphibians. These include synthetic GnRH agonists, purified pituitary gonadotropins (LH and FSH), human chorionic gonadotropin (hCG), pregnant mare serum gonadotropin (PMSG), and pituitary homogenates/extracts, alone or in combination with other agents. There are specific challenges with each hormone preparation, and lack of success with one method does not preclude success with another approach or a combination of therapies. Much remains to be discovered about the hormonal control of reproduction, and how to induce breeding in amphibians.

We recommend that new protocols applied to the induction of gamete release in any species for the first time should be initially tested on a subset of animals. This is particularly important if the protocols are being applied to rare or endangered animals that are critical to a captive breeding or reintroduction programme. Fatalities are extremely rare using the aforementioned hormones (<1 %), when administered properly (Kouba et al. 2012a). Potentially, these hormones could cause physiological upsets or even death if given too frequently (e.g., improper handling or multiple injections given within a 1 week span) or accidentally over-dosed due to human error. In the unlikely situation that an animal dies and it was a female, death may occur because of egg retention. In this scenario, the female would not have died as a direct result of the hormone therapy. Ovulation may have occurred but oviposition did not happen, so the retained eggs may lead to problems in such cases (Green et al. 2007).

It is clear that ART is having a positive impact on the reintroduction and restoration of wild amphibian populations. For example, to date, more than 100,000 endangered Wyoming toads have been produced collectively and released into the wild through the use of hormones for induced spawning. Similarly, over 250,000 threatened Puerto Rican crested toads and 50,000 boreal toads have been produced and released because of hormone therapy. In this context, it will be important to carefully monitor the health of amphibians produced via ART to ensure long-term success of re-established populations. The rest of this section will provide in detail information on several hormones used to induce egg maturation, ovulation, spermiation, and amplexus.

4.1 Use of GnRH Agonists for Spermiation and Ovulation

In amphibian literature, the natural decapeptide GnRH, and the many potent synthetic agonists, are often referred to incorrectly as luteinizing hormone-releasing hormone (LHRH). This is a misnomer because it does not specify the correct structure or source of the peptide. For consistency in this review we will refer to the peptide's more appropriate classification as GnRH. There are several advantages of using GnRH agonists over other hormone therapies because it acts at the level of the pituitary (Fig. 12.1) to stimulate endogenous LH production and release. Although there are dozens of GnRH compounds available commercially, the most common analog used by the amphibian community to date is des-Gly^{10}, D-Ala^{6}, Pro^{9}-GnRH-ethylamide (Bachem, catalog # H4070; also called des-Gly^{10}, D-Ala^{6}, LHRH ethylamide, Sigma-Aldrich, catalog #L4513). While limited comparative trials have been conducted in amphibians, most other GnRH agonists have not performed as well as des-Gly^{10}, D-Ala^{6}, Pro^{9}-GnRH-ethylamide for the stimulation of spermiation and ovulation (Mann et al. 2010; Michael et al. 2004; Trudeau et al. 2010). This is likely because of reduced binding of the agonist to the amphibian GnRH receptor (Wang et al. 2001; Seong et al. 2003). Goncharov et al. (1989) tested this GnRH agonist on 39 different amphibian species and found several thousand-fold differences in apparent concentration potency needed to induce spermiation in males. This could be a true species difference in sensitivity and/or due to differences in the physiological state of the injected animals. It is advisable to consider GnRH plus a DA antagonist for less sensitive species so that endogenous LH release is maximized (Trudeau et al. 2010, 2012).

In males, GnRH has been shown to stimulate spermiation in a diverse array of anuran and some urodele species (Vellano et al. 1974; Trottier and Armstrong 1975; Waggener and Carroll 1998a, b; Goncharov et al. 1989; Minucci et al. 1989; Clulow et al. 1999; Obringer et al. 2000; Rowson et al. 2001; Kouba et al. 2012a; Browne et al. 2006a, b; Pozzi et al. 2006; Kouba et al. 2009; Mansour et al. 2010; Mann et al. 2010; Silla 2010; Byrne and Silla 2010; Silla and Roberts 2012). Moreover, injection of GnRH combined with the DA antagonist MET also stimulated spermiation in male leopard frogs (Trudeau et al. 2010, 2013). In female amphibians GnRH stimulates final egg maturation, ovulation, and spawning in a number of species as well (Waggener and Carroll 1998a; Whitaker 2001; Roth and Obringer 2003; Michael and Jones 2004; Michael et al. 2004; Kouba et al. 2009; Byrne and Silla 2010; Roth et al. 2010; Silla 2011)

4.2 Use of Chorionic Gonadotropins for Spermiation and Ovulation

Given the fundamental importance of gonadotropins to regulate gonadal function, numerous studies in amphibians have used these hormones to induce ovulation, spermiation or spawning, including hCG and PMSG (Clulow et al. 2012; Kouba et al. 2012a). These gonadotropins are produced outside the pituitary in placental tissues of select mammals, and may have both LH-like and FSH-like effects, depending on the species studied. These chorionic gonadotropins are relatively easy to obtain in large quantities from urine of pregnant women or mares and they have been used to partially mimic endogenous LH or FSH in numerous non-mammalian species. In the 1940s to 1960s the standard pregnancy test used ovulation or spermiation in amphibians to determine a positive or negative result (Bellerby 1934; Shapiro and Zwarenstein 1934; Galli-Mainini 1947; Gurdon and Hopwood 2000). After injection of human female urine into *Xenopus laevis* or *Rhinella arenarum*, the presence of eggs or production of sperm was monitored in the female or male anuran, respectively. If a woman was pregnant, there would be enough hCG in the urine to induce these responses in the amphibians. This method became known as the “Bufo test” for pregnancy. Injection of hCG does not have the same efficacy in amphibians as it does in mammals, therefore the effective dose for amphibians being nearly 2,000 times higher than that given to a mammal on a per weight basis (Kouba et al. 2009). Nevertheless, amongst the amphibians, a number of bufonids have been shown to produce gametes following administration of hCG (Clulow et al. 2012). In contrast, Licht (1995) found reduced sensitivity to hCG in *Lithobates pipiens* production of gametes and hypothesized this poor response was related to the low affinity of amphibian LH receptors for hCG in this species. Even though hCG shows reduced specificity, species-specific responses, and is required in high concentrations, it is widely used due to its commercial availability, standardized activity (International Units; IU) that can be adjusted on a per weight basis, and low side effects or reported health problems.

Injection of hCG has been shown to induce spermiation in a wide range of male amphibian species. It has proven effective in most bufonids tested to date including numerous species in the genus *Anaxyrus* and *Rhinella* (Iimori et al. 2005; Browne et al. 2006a; Kouba et al. 2009, 2012b). Moreover, an increasing number of amphibians have been reported to respond to hCG including those in the genus *Lithobates* and *Rana* (McKinnell et al. 1976; Easley et al. 1979; Kurian and Saidapur 1982; Rosemblit et al. 2006; Minucci et al. 1989; Uteshev et al. 2012; Subcommittee of amphibian standards 1996; Kouba et al. 2009, 2011), *Leptodactylus* species (Rosemblit et al. 2006), *Xenopus* species (Subcommittee of amphibian standards 1996) and *Litoria* species (Clulow et al. 1999). Administration of hCG can also induce ovulation in a number of other species, including *Mixophyes fasciolatus* (Clulow et al. 2012), *Eleutherodactylus coqui* (Michael et al. 2004), *Engystomops pustulosus* (Lynch et al. 2006), *Ambystoma mexicanum* (Mansour et al. 2011), several *Bufonidae* (Browne et al. 2006a, b; Kouba et al. 2009, 2013), and *Xenopus laevis* (Hollinger and Corton 1980).

Phylogenetic analysis of the gonadotropin beta subunits has shown that hCG and PMSG are protein products of genes only distantly related to amphibian LH beta and FSH beta (Fig. 12.2). This may help explain the wide range of hCG-sensitivities in amphibians. As illustrated in the phylogenetic analysis (Fig. 12.2), chorionic gonadotropins group most closely with mammalian LH beta subunits, being clearly separated from the amphibian LH beta subunit clade. The purification of sufficient amounts of endogenous amphibian LH and FSH is difficult, as is the production of recombinant gonadotropins that retain full biological activity. The gonadotropins are highly glycosylated, and these carbohydrate additions are critical for full biological activity. Therefore, strategies using homologous gonadotropins, which would be much more effective for amphibian ART, are unlikely to be available in the near future, although recombinant human LH and FSH have shown some promise in inducing spermiation in toads and were more effective than hCG-induced spermiation in *A. arenarum* (Pozzi et al. 2006). Hence, recombinant gonadotropins warrant more investigation for their usefulness to amphibian ART.

In a number of cases, the use of a combination of a GnRH agonist and hCG have proven more effective at inducing spermiation or ovulation than either hormone alone, even when tested at higher individual doses (Browne et al. 2006a; Kouba et al. 2009, 2012a; Germano et al. 2011). These hormone combinations may be more effective because both the pituitary and gonad would be stimulated (Fig. 12.1). Additionally, the combination of hCG with PMSG was highly successful for stimulating ovulation in the barred frog *Mixophyes fasciolatus* (Clulow et al. 2012). In this case, the combinations of heterologous hormones were likely effective at stimulating both LH and FSH receptors in the ovaries (Fig. 12.1).

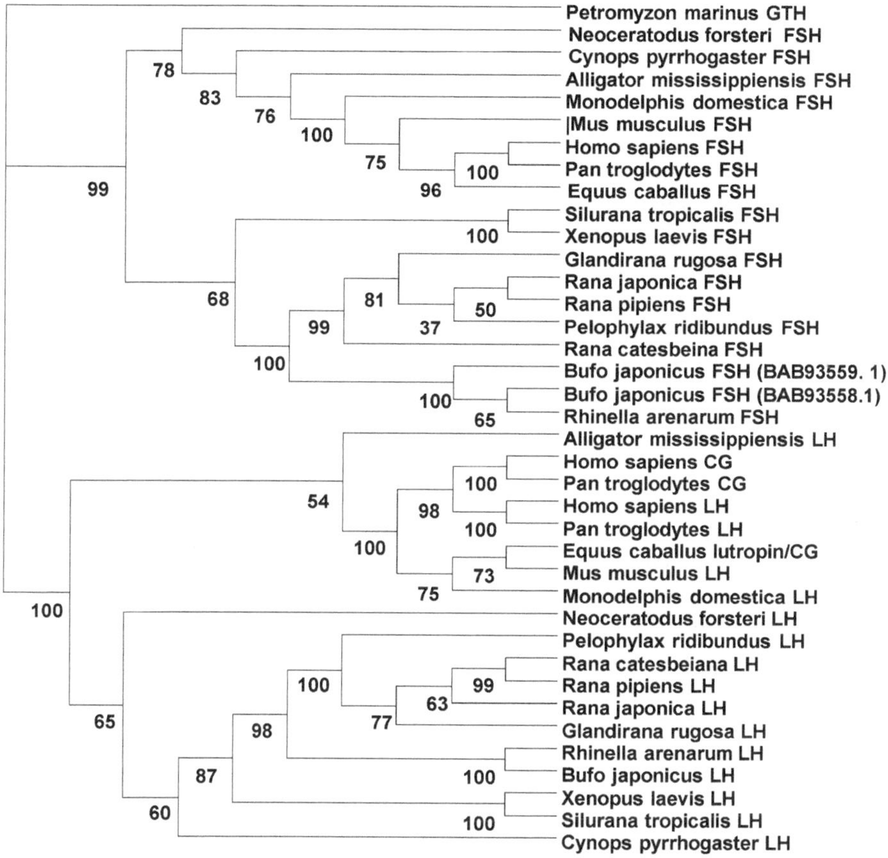

Fig. 12.2 Phylogenetic tree depicting the evolutionary relationships between the beta subunits of the vertebrate gonadotropins. The gonadotropins are composed of a common alpha subunit linked to a unique beta subunit which confers biological specificity. The main gonadotropins are luteinizing hormone (LH), follicle stimulating hormone (FSH), and the chorionic gonadotropins (CG). The gonadotropin (GTH) subunit from the lamprey *Petromyzon mariunus* was used as an outgroup. Shown is the consensus tree generated from DAMBE (Xia 2001; Xia and Xie 2001), with bootstrap support for individual nodes based on the aligned amino acid sequences. The Poisson-corrected distance and neighbor-joining method were used. See text for further details

4.3 Use of Pituitary Homogenates for Inducing Spermiation or Ovulation

Pituitary homogenates are the excised pituitary gland from a sacrificed donor amphibian, with the gland then crushed in a suitable medium and subsequently administered to a recipient for purposes of breeding or obtaining gametes. Pituitary extracts go through a further crude purification step to concentrate the hormones and remove some of the cellular debris. Injections of pituitary homogenates stimulate ovulation and spawning in female amphibians and, to a lesser degree, spermiation in males (Subcommittee on amphibian standards 1996; Edwards et al. 2004). The use of pituitary homogenates was the first method for the collection and study of gametes from live animals. There are numerous disadvantages to using pituitary homogenates; therefore, their use should be avoided. First, a pituitary homogenate/extract may contain dangerous transmissible diseases that could be passed to the recipient. Given the global spread of pathogens, such as chytrid fungus and ranavirus, the passage of these diseases to endangered or threatened species must be avoided (especially if the future goal is reintroduction). Second, animals must be sacrificed to collect pituitaries. Researchers and conservationists must weigh the ethical and acceptable risk of this method. Third, the exact reproductive state and hormonal milieu of the donor animal(s) is typically not known and the active amount of gonadotropin available to the recipient can vary several thousand-fold. Fourth, the homogenate/extract is comprised of other cellular debris and/or pituitary hormones that can have adverse effects on the recipient. Most commercial supply companies have discontinued the sale of frog pituitaries possibly due to disease issues. However, if all other possible routes of hormone stimulation have been exhausted and failed to produce gametes, the ethical sacrifice of a common or invasive species should be considered as a mechanism to save a near extinct species. Although none of the authors endorse the use of pituitary homogenates in practice, there may be future circumstances where the technique may be employed, for example, harvesting pituitaries from an invasive species to use for induced reproduction in an endangered species.

5 *In Vitro* Fertilization and Short-Term Storage of Gametes

In vitro fertilization for amphibians has been performed for more than 60 years, predominantly for studies on early embryonic development or for commercial production of laboratory species. Since the detailed description of IVF in *Xenopus* (Wolf and Hedrick 1971), few IVF protocols have been applied to anurans outside of the common laboratory models. The development of IVF for amphibians is relatively simple compared to mammals because of the many advantages of external fertilization in water. Complex cell culture media and highly specific incubation conditions are all necessary to perform IVF in mammals due to the fact that natural fertilization is internal. For the most part, these complex protocols are not needed for amphibian IVF, which makes this technique highly amenable for captive amphibian breeding programs. Unfortunately, limited studies have been done for internally fertilizing urodeles (salamanders/newts) or for anuran species that do not have an aquatic life-stage (i.e., direct developers).

To conduct IVF for aquatic breeding amphibians, eggs are quickly removed from the water once spawned, placed into a dry Petri dish and mixed with spermatozoa for 5–10 min before flooding the dish with water (Kouba et al. 2009, 2012a). If eggs are stored in buffers containing electrolytes, they need to be rinsed in water several times before IVF and subsequent placement into the dry Petri dishes. Incubating the gametes in any buffer solution with an osmolality higher than 50 mOsmol per kg will inhibit fertilization (Edwards et al. 2004). It is likely that this inhibition to fertilization is due to inactivation of sperm motility at higher osmolalities (Browne et al. 1998; Kouba et al. 2003; Edwards et al. 2004). Typically, fertilization takes place with sperm concentrations ranging from 10^4 to 10^6 spermatozoa per mL (Wolf and Hedrick 1971; Browne et al. 1998; Edwards et al. 2004). Sperm concentrations for fertilization will likely vary between species and investigations should be undertaken to determine the best sperm to egg ratio for new species.

It was not until 1998 that Waggener and Carroll (1998a) conducted the first amphibian IVF in which both sperm and eggs were obtained from live males and females, even though the technique, using sperm collected post-mortem, had been employed in amphibians for more than 40 years prior to this report. These researchers found that both sperm and eggs could be collected from *Lepidobatrachus* species (*L. laevis* and *L. llanensis*) following an injection of a GnRH analog and when mixed together produced high rates of fertilization. This seminal paper demonstrated that ART for endangered amphibians was a real possibility. To date, IVF trials for threatened species have resulted in more than 2,000 *Anaxyrus baxteri* tadpoles produced and released to the wild (Browne et al. 2006a), over 10,000 *Anaxyrus boreas boreas* released to the wild (Kouba et al. 2013) and more than a 1,000 *Lithobates sevosa* produced by IVF but not released (Kouba et al. 2011). The endangered *Pseudophryne corroboree* frog has also been produced by IVF but all the embryos failed to develop and further studies are underway to improve these initial efforts (Byrne and Silla 2010). Methods for IVF have been developed in *Ambystoma mexicanum* (Mansour et al. 2011), *Pelophylax lessonae* (Uteshev et al. 2013), *Xenopus laevis* (Subcommittee on amphibian standards 1996), *Pseudophryne guentheri* (Silla 2011), several ranids (Shishova et al. 2011, 2013), *Crinia georgiana* (Dziminski et al. 2010), *Limnodynastes tasmaniensis* (Edwards et al. 2004), and *Lepidobatrachus* species (Waggener and Carroll 1998a). Toro and Michael (2004) were successfully able to produce offspring in a true direct developing frog (no aquatic life-stage), *Eleutherodactylus coqui*, using IVF. In addition, the first terrestrial salamander, *Ambystoma tigrinum*, was recently produced from IVF and hormone therapy for the collection of gametes (Marcec et al. unpublished; Fig. 12.3).

Gamete production in males versus females is often asynchronous when using ART; hence, the greatest challenge is typically storing gametes until the eggs or spermatozoa are obtained from the other sex. Amphibian sperm from a diversity of species has been shown to survive days or weeks at a time when stored at 4 °C (Browne et al. 2001; Kouba et al. 2009; Silla 2012) and do not generally appear to experience cold shock as in mammalian spermatozoa. The ability to store spermatozoa easily for extended periods in the refrigerator or ice slurry means that although the exact time of ovulation may be unpredictable, spermatozoa for fertilization can be readily available when eggs are eventually procured. Anuran eggs, on the other hand, are much more sensitive to short-term storage than spermatozoa. For the most part, once eggs are oviposited into water, the egg jelly quickly hydrates, resulting in structural changes that prevent sperm penetration after a short period of 30–60 min (Hollinger and Corton 1980; Elinson 1986; Olson and Chandler 1999). This can be circumvented to a limited degree by placing the eggs in a higher osmolality buffer solution (Browne et al. 2001; Edwards et al. 2004).

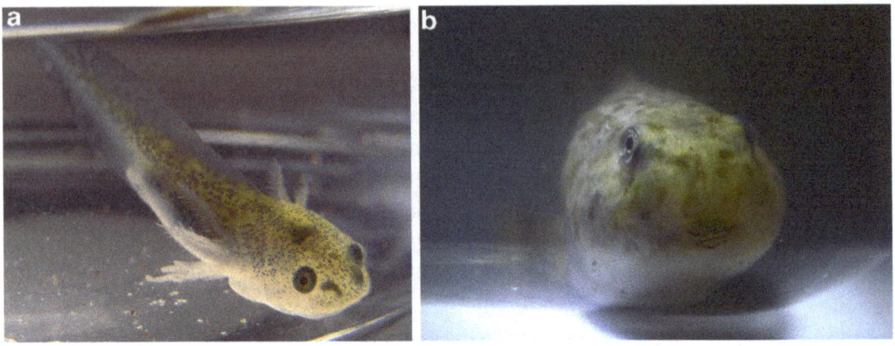

Fig. 12.3 Amphibian larvae produced through the use of hormone treatments and IVF at the Memphis Zoo and Mississippi State University. (**a**) Common tiger salamanders (*Ambystoma tigrinum*) have been produced by injecting hCG and a GnRH agonist to stimulate ovulation and spermatophore production, followed by IVF. (**b**) Critically endangered Dusky gopher frogs (*Lithobates sevosa*) created through a similar hormone therapy and IVF process, except that they were produced using frozen-thawed sperm held for months in cryostorage. Photographs: H. Bement

The short-term storage of gametes at temperatures above 0 °C can be an effective tool in ART where there is a need to transport sperm or eggs from one facility to another for IVF and genetic management. In the USA, cooled but unfrozen sperm of the endangered dusky gopher frog, *Lithobates sevosa*, were transferred from the Memphis Zoo in Tennessee to the Henry Doorly Zoo in Omaha, Nebraska where they were used successfully for IVF, sparing the need to transport live animals (Kouba et al. 2011, 2013). This represents the first time amphibian gametes have been shipped between institutions for the production of a critically endangered amphibian using hormone therapy and IVF technologies. Technological advancements for IVF, along with hormone therapy, are all valuable tools for amphibian conservation programs. In addition, the utility of stored gametes within a genetic resource bank depends on these previously developed techniques to successfully produce offspring for managed captive breeding programs.

6 Cryopreservation in Amphibian ART

6.1 *Cryopreservation of the Male Germ-Line*

Advances in amphibian cryobiology indicate that cryopreservation can play an important part in amphibian conservation (Table 12.2), and the establishment and operation of genome resource banks as a conservation action are as applicable to amphibians as for other vertebrate taxa (Holt et al. 1996; Bennett 2001; Holt 2001; Lermen et al. 2009; Rawson et al. 2011; Kouba et al. 2013). The creation of national amphibian genome resource banks (GRBs) that store somatic cells, tissues, gametes, embryos, and blood in a suspended state (typically in liquid nitrogen) has gathered momentum over the last several years (Kouba et al. 2013). Currently, GRBs have been established in Russia, the U.K., Australia, Germany, and the USA, housing a number of threatened and endangered species. While somatic cells and tissues are valuable for genomic, transcriptomic and proteomic studies, the real conservation value lies in the long-term preservation of gametes and embryos (Rawson et al. 2011). Cryopreservation of amphibian sperm has been shown to be a viable technology for an increasing number of species (Kouba et al. 2009, 2013; Browne and Figiel 2010) (see Table 12.2) since the pioneering work of Barton and Guttman (1972) and the achievement of the first fertilizations using frozen sperm (Browne et al. 1998). Typically, the use of GRB material requires the application of other advanced methods to produce live offspring.

The majority of amphibian species whose spermatozoa have been cryopreserved are all aquatic breeding anurans and there is a desperate need to begin research on how to cryopreserve urodele and caecilian sperm. Cryopreservation of amphibian sperm and the successful retrieval of post-thaw motility has been accomplished in a number of common species including *Xenopus* (Buchholz et al. 2004; Sargent and Mohun 2005; Mansour et al. 2009), *Anaxyrus americanus* (Barton and Guttman 1972; Beesley et al. 1998), *Rhinella marina* (Browne et al. 1998, 2002d), *Lithobates sylvaticus* and *Lithobates pipiens* (Costanzo et al. 1998; Mugnano et al. 1998), *Anaxyrus fowleri* (Kouba and Vance 2009) and *Eleutherodactylus coqui* (Michael and Jones 2004). Although these investigations provided a wealth of information on the practicality of anuran sperm freezing and different cryodiluents, cryoprotectants, and storage mechanisms that afford survival of spermatozoa at low temperatures, they all utilized testis macerates collected from euthanized males. Sacrificing endangered or threatened species for gene banking its hereditary line is not widely accepted in the conservation field, which resulted in numerous studies on the use of exogenous hormone injections for the collection of sperm from live animals. Recently, spermatozoa have been gene banked from two species of live anurans, the common pool frog, *Pelophylax lessonae* (Uteshev et al. 2013) and the common European frog, *Rana temporaria* (Shishova et al. 2011; Uteshev et al. 2012), which resulted in live offspring. Post-thaw sperm motility for these two species was over 40 % with percent fertilizations near 30 % and 80 % for pool frogs and European frogs, respectively. Recently, the first critically endangered anurans, the dusky gopher frog *Lithobates sevosus* and the boreal toad *Anaxyrus boreas boreas*, have been produced from frozen-thawed sperm held in cryostorage (Langhorne et al. 2012, 2013) (Fig. 12.3). These recent studies highlight the rapid development and successful application of an array of advanced techniques and GRBs for threatened species.

6.2 *Cryopreservation of the Female Germ-Line or Diploid Genome*

Although male germ lines are being recovered for an increasing number of species (Kouba et al. 2009), the recovery of cryopreserved female germ lines as either oocytes or post-fertilization diploid genomes, has proven much more elusive.

To recover cryopreserved female germ lines, we need to be able to either: (1) cryopreserve oocytes so that embryos can be generated with fresh or cryopreserved sperm, or (2) generate viable embryos from cryopreserved cells (e.g., diploid embryonic cells, larval cells or adult somatic cells). In practical terms, this means we need to either successfully cryopreserve oocytes that can resume meiosis and develop as viable embryos after fertilization, or successfully cryopreserve diploid cells and generate embryos (by nuclear transfer or as chimeras) that also are capable of completing development and maturation of reproductively competent adults. Here, we discuss the current challenges associated with cryopreserving the female germ line. Moreover, we will compare and contrast what is known in fish in relation to technological advancements that could be accomplished with amphibians.

Currently, there are no reports of the recovery of viable cryopreserved amphibian oocytes or embryos. Although cryopreservation of intact amphibian embryos would be the most efficient and practical way to store amphibian genomes for use, the challenges associated with this approach have proven difficult to overcome. There are relatively few systematic studies involving freezing of amphibian oocytes (Guenther et al. 2006; Kleinhans et al. 2006; Mazur and Kleinhans 2008), and no published studies of cryopreservation attempts on whole embryos. Early embryonic cells are the optimal target for cryopreservation and genome storage due to their smaller size and yolk content. There are only two reports to date of the recovery of cryopreserved cells from early amphibian embryos, but the results are encouraging. Uteshev et al. (2002, 2005) reported the generation of blastulae from vitrified early embryonic cells of *Bufo bufo* and Lawson et al. (2013) reported the recovery of live cells from dissociated and cryopreserved gastrulae and neurulas of the striped marsh frog, *Limnodynastes peronii*. The application of this technology in amphibians is at a very early stage. Nevertheless, the results with fish are also encouraging with reports of the cryopreservation of blastomeres from dissociated embryos of several species (Calvi and Maisse 1998, 1999; Cardona-Costa and García-Ximénez 2007; Dash et al. 2008; Harvey 1983; Kusuda et al. 2002; Lin et al. 2009; Nilsson and Cloud 1993; Routray et al. 2010).

Long-term efforts to cryopreserve embryos of fish have yet to produce viable post-thaw embryos (Harvey 1983; Hagedorn et al. 1996, 1997a, c, 1998; Liu et al. 1999, 2001; Hagedorn and Kleinhans 2011). Fish and amphibian oocytes and embryos have similar biophysical properties, yet fish embryos are not a good proxy for amphibian studies. Gastrulation during early embryogenesis in particular is different, especially with the formation of a syncytium around the yolk in fish but not amphibians. Permeability of oocytes and embryos to water and cryoprotectants seems to be a major problem for successful cryopreservation in fish and amphibians. The impermeability of the syncytium that forms around the highly condensed yolk mass during the early phase of fish embryogenesis is a major failed contributor to recovery of the embryonic genome after cryopreservation (Hagedorn et al. 1996, 1997a, 1998; Liu et al. 2001). The syncytium is present in fish during gastrulation (Balinsky 1975) but not in amphibians. In amphibians, the yolk is widely distributed through the cytoplasm of the dividing cells, especially in the vegetal pole (Balinsky

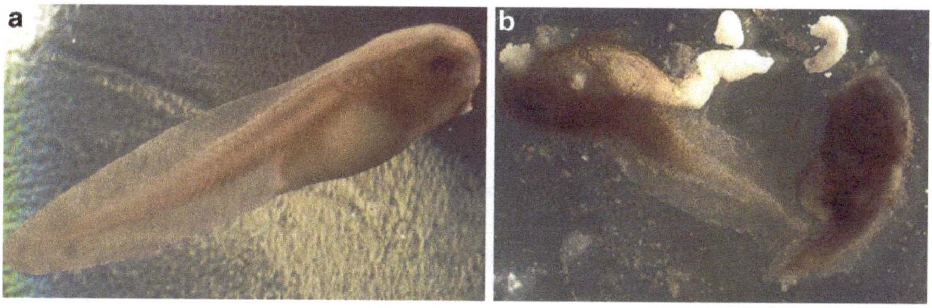

Fig. 12.4 Striped Marsh Frog, *Limnodynastes peronii* larva, Gosner Stage 20. (**a**) Control larva. (**b**) Vitrified and thawed larvae. Curphey et al. unpublished data. Photographs: L. Curphey

1975), but not in a condensed mass in the centre of the embryo, as seen in fish. The yolk syncytial layer of fish has been shown to be impermeable to water and cryoprotectants (Hagedorn et al. 1996, 1997b, 1998; Liu et al. 2001; Hagedorn and Kleinhans 2011). This indicates the yolk compartment is largely inaccessible to cryoprotectants during cryopreservation and unable to dehydrate to prevent lethal intracellular ice formation. While amphibians are not burdened with the yolk syncytial layer, they do not express aquaporin channels in their oocyte plasmalemma (Kleinhans et al. 2006). Hence, the measured permeability of the mature amphibian oocyte to water is very low (equivalent to a lipid bilayer) (Zhang and Werkman 1991; Kleinhans et al. 2005), suggesting poor permeability to water and cryoprotectants during cryopreservation.

High yolk content has implications for the dehydration and ice formation processes during cryopreservation. Yolk content in amphibian embryos may be proportionally higher than in fish (Wallace 1963; Guenther et al. 2006; Hagedorn et al. 1997c; Lawson et al. 2013). Ice crystallization in zebrafish embryos has been reduced by removing some of the yolk (Liu et al. 1999); yet, the optimal freezing stage for larvae of the polychaete *Nereis* occurs at the point at which the total larval yolk content is almost exhausted (Olive and Wang 1997). Yolk may interfere with cryopreservation because of high lipid content (Liu et al. 1999) and/or by upsetting water and cryoprotectant efflux (Hagedorn et al. 1997c). The spread of ice formation (the ice flash) as *Xenopus* oocytes freeze starts at the edge of the cell (Guenther et al. 2006), suggesting that yolk particles may not be the source of ice nucleation during the freezing process. Vitrification, or the rapid controlled freezing of tissues to prevent ice crystal formation, may offer hope for cryopreservation of amphibian oocytes and embryos (Hagedorn and Kleinhans 2011). However, our initial attempts to cryopreserve striped marsh frog embryos (*Limnodynastes peronii*) by vitrification were not successful (Curphey et al. unpublished, Fig. 12.4). Whether yolk is inherently detrimental to vitrification procedures for amphibians has not been studied, although the injection of cryoprotectant directly into the yolk compartment of zebrafish embryos did not protect the embryos from cryodamage during vitrification

(Janik et al. 2000). While the role of yolk in freezing damage during slow cooling and vitrification is not well understood at this stage, it is likely that it plays a significant role. Thus, the major biophysical similarities of high yolk content (Kouba et al. 2013; Lawson et al. 2013) appear detrimental to cryopreservation of the oocytes and embryos as intact structures in both fish and amphibians.

Fish and amphibians share another similar biophysical property of large oocyte and embryo size. Mature oocytes and egg diameters are approximately 0.8–1.0 mm at the lower end of the range in aquatic breeding frogs and marine fish, but can be as large as 10–20 mm in freshwater fish, direct developing frogs and urodeles (Hagedorn and Kleinhans 2011; Kouba et al. 2013). Useful comparisons in key laboratory species in which oocytes and/or embryo cryopreservation have been investigated include 0.8 mm for zebrafish egg, 1.2 mm for mature *Xenopus* oocytes, compared to only 75 μm for mouse oocytes (Guenther et al. 2006; Mazur and Kleinhans 2008; Hagedorn and Kleinhans 2011). Large size *per se* appears to be inherently unfavourable to cryopreservation (Mazur and Kleinhans 2008). Empirical data indicates that intracellular freezing in the large oocytes of both fish (Liu et al. 2001; Hagedorn et al. 2004) and amphibians (Guenther et al. 2006; Kleinhans et al. 2006; Mazur and Kleinhans 2008) and at least one marine invertebrate (Köseoglu et al. 2001), occurs at relatively high temperatures. These high temperatures are close to the temperature of extracellular ice formation: between −14 and −18 °C for zebrafish (Hagedorn et al. 2004), −8 to −10 °C for most *Xenopus* stage I and II oocytes and all mature stage V and VI oocytes (Guenther et al. 2006; Kleinhans et al. 2006), compared to −41 °C for mouse oocytes (Mazur et al. 2005). The evidence (including direct observations of ice formation starting at the periphery of *Xenopus* oocytes (Guenther et al. 2006)) suggest that external ice acts as the ice nucleator penetrating the oocyte through membrane pores or discontinuities to initiate intracellular ice formation, which moves rapidly from the periphery to the core of the oocyte.

There are two major consequences of large oocyte and embryo size. First, the high temperature of freezing means that intracellular ice formation occurs before the oocyte or embryo is sufficiently dehydrated to avoid formation of lethal ice crystals, the primary source of cryoinjury during cryopreservation (Mazur 2004). Second, the low surface area to volume ratio means that the rate of water efflux (i.e., dehydration) is greatly reduced in comparison to smaller structures, such as mouse oocytes. These biophysical parameters leading to lethal intracellular ice formation may be sufficient to preclude slow cooling as an approach to the cryopreservation of full size, mature amphibian oocytes and embryos of fish (Liu et al. 2001; Hagedorn et al. 2004) and amphibians (Mazur and Kleinhans 2008). Unfortunately, incorporation of aquaporin channels into embryos of fish (Hagedorn et al. 2002) and frogs (Yamaji et al. 2006) through the use of new osmotic and chemical treatments (Rahman et al. 2011) to increase permeability to water and cryoprotectants and the physical removal of yolk (Liu et al. 1999) have not yet led to reports of improved cryopreservation outcomes. Thus, the full potential of cryobanking (Table 12.2) as a tool to address the amphibian biodiversity crisis of the twenty-first century will only be reached if procedures to generate live offspring from maternal haploid and diploid somatic/embryonic genomes can be established.

6.3 *Cryopreservation of Amphibian Differentiated Somatic Diploid Cells*

Without the ability to cryopreserve the amphibian embryo or oocyte, other techniques must be developed and implemented if maternal and diploid genomes are to be stored and recovered. There is considerable interest in indirect approaches to this goal in amphibians (Clulow et al. 1999; Kouba et al. 2013; Lawson et al. 2013) as in fish (Thorgaard et al. 2005). This section focuses on the progress to date in this area, and suggests future research directions.

Somatic cells carry the diploid nuclear and mitochondrial genomes and should be a target for genome storage. The generation of amphibian cell lines is challenging, yet has been achieved in a limited number of cases (Kouba et al. 2013). The cryopreservation of isolated cells from primary tissues, tissue explants, and cultured cell lines appears to be achievable at similar recovery rates to mammalian somatic cells and tissues (Kouba et al. 2013). One of the issues limiting progress in this area is that much of the expertise and knowledge is encapsulated in an older literature base from the 1960s and 1970s when amphibians were predominant models for developmental and cancer biology (Mizell 1969). Currently, the field is not very active, although studies continue to be published on this topic (Okumoto 2001). This field of investigation could benefit from the application of advances in molecular and cell biology, particularly in stem cell biology, with the potential to identify new growth and other regulatory factors involved in reprogramming cells and initiating proliferation.

There are important challenges for the use of cryopreserved somatic cells in the generation of viable, reproductively competent adults. The first successful nuclear transfers in vertebrates were achieved with amphibians (Briggs and King 1952), and amphibian models contributed significantly to the advancement of nuclear transfer research for several decades (McKinnell 1978; Di Berardino 1997; Gurdon and Byrne 2003). However, there are no reports of adult amphibians being generated from post-metamorphic or adult somatic cells or tissues. The most advanced differentiated cells from which viable adults have been generated were those from tadpole intestine (Gurdon and Uehlinger 1966) and epidermis (Kobel et al. 1973), achievements that have yet to be repeated (Di Berardino 1997). On the other hand, numerous reports of viable fertile adults generated from adult mammalian somatic cells are evident (Wilmut et al. 1997, 2002). It is not clear whether the failure to generate adult amphibians from differentiated adult somatic cells is an inherent limitation of the biology of those cells or because research activity shifted to mammalian models following their success. Tadpoles have been generated from adult tissues, but these offspring invariably fail before metamorphosis (Di Berardino 1997; Gurdon et al. 1975; Gurdon and Byrne 2003). Irrespective of the ultimate developmental potential of larval and adult cells, the experience to date suggests that there should be a focus on cryopreserving somatic cells from early developmental stages when the goal is to generate fertile adults from nuclear transfer in amphibians using cryopreserved materials.

6.4 Cryopreservation of Early Stage Ovarian Follicles and Regeneration Through Xeno-transplantation

The recovery of viable oocytes is a primary objective in the development of amphibian genome storage methods. This is unlikely to be achieved by direct cryopreservation of mature stage V/VI oocytes for the reasons discussed in the previous sections. However, since the size of the frozen follicle seems to determine the temperature of intracellular ice formation (see above), there is the potential for smaller, earlier stage follicles to be successfully cryopreserved. In *Xenopus laevis* (Guenther et al. 2006; Kleinhans et al. 2006), the smaller stage I and II oocytes, which are 300 μm in diameter, fall into two groups, one that undergoes intracellular ice formation at low temperatures (−30 to −40 °C) and another group that undergoes intracellular ice formation at high temperatures (−8 to −10 °C) close to that of extracellular ice formation (−6 to −9 °C); the group freezing at the lower temperature may have some potential for recovery after cryopreservation. This potential has not been extensively investigated to date, although some preliminary data (Wooi et al. unpublished) from *Rhinella marina* suggests that isolated follicles with attached thecal cells can be recovered after cryopreservation (Fig. 12.5). Even if such early stage ovarian follicles were cryopreserved successfully, they would presumably need to resume growth (vitellogenesis) before undergoing final maturation and ovulation. One possibility is transplantation of cryopreserved follicles into host females to allow the completion of oogenesis and ovulation. This technology is certainly feasible as the generation of young after transplantation of unfrozen ovarian tissue has already been reported in amphibians (Dournon et al. 1997) and the generation of young from transplanted cryopreserved ovarian fragments has been achieved in mammals (Paris et al. 2004). This is a line of investigation that has the potential to significantly advance amphibian ART.

7 Advanced ART for Amphibian Conservation

7.1 Embryonic and Other Stem Cells as Sources of Viable Offspring from Stored Genomes

The probability of successful generation of amphibian embryos by nuclear transfer that develop, complete metamorphosis, and become reproductively competent adults is highest with the earliest embryonic stages (McKinnell 1978; Di Berardino 1997). Nuclear transfers generated from cells of blastulae, have the highest success rate, but gastrulae and neurulas are also capable of reprogramming to produce fertile adults (McKinnell 1978; Di Berardino 1997), although at a lower rate.

The potential to generate viable amphibian embryos using stem cell types other than early embryonic cells is largely unexplored. These include primordial germ cells (PGCs) and induced pluripotent stem cells. The PGCs from later stage embryos

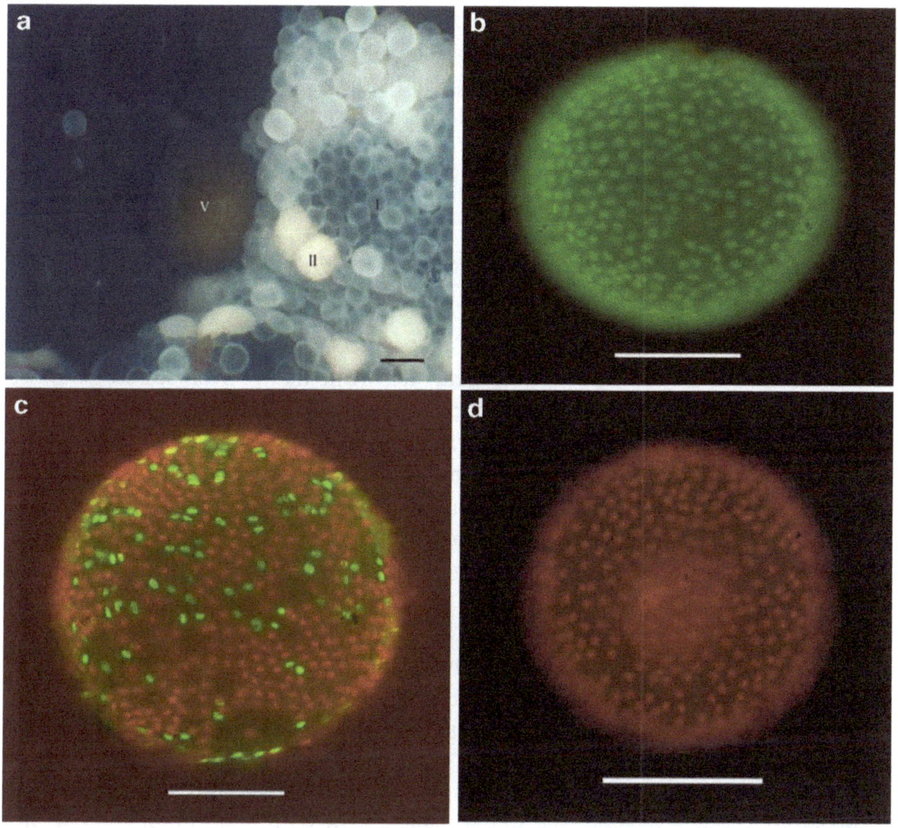

Fig. 12.5 Early stage follicles (stage I and II) of *Rhinella marina* using viability of attached follicular cells as an assay of oocyte recovery after exposure to cryoprotectants (pre-freezing), or post-cryopreservation. (**a**) Ovary with stage I and II follicles; bar = 550 μm. (**b**) Unfrozen control oocyte (stage I), live follicular cells stain green with Sybr14; 100 % viable; bar = 110 μm. (**c**) Oocyte stage (I) with 25 % viable cells (*green*), red stain is propidium iodide, indicating non-viable cells; bar = 110 μm. (**d**) Oocyte (Stage I) with 0 % recovery (all follicular cells stain red with propidium iodide); bar = 110 μm. Wooi et al. unpublished data. Photomicrographs: K. Wooi

of zebrafish have been isolated, cryopreserved and transferred into other zebrafish embryos in which the host embryo PGCs have been sterilised (Higaki et al. 2010). Fertile offspring resulted from the PGC transferred embryos, and their offspring expressed the genotypes of the transferred PGC's, indicating that the PGC's were viable. Viable stem cell lines that may support embryonic development after nuclear transfer may also potentially be derived from adult germ cells (spermatogonia and oogonia) (Ogawa et al. 2004). Moreover, there is also the possibility of the generation of induced pluripotent stem cells from post-metamorphic and adult somatic tissues using reprogramming factors to generate stem cells with the potential to produce viable embryos after nuclear transfer. This approach is now widely applied in mammalian stem cell biology (Takahashi and Yamanaka 2006; Takahashi et al. 2007), and stem cells expressing markers of induced pluripotent cells have been generated in endangered mammalian species including the drill, an endangered primate, and the nearly extinct Northern white rhinoceros (Ben-Nun et al. 2011). Similar strategies may ultimately lead to the generation of fertile adult amphibians derived from post-metamorphic somatic tissues.

7.2 Recovery of Stored Genomes by Nuclear Transfer or Chimeras

The two routes that are available to regenerate fertile adults from stored genomes are: (1) nuclear transfer, and (2) chimeras. Nuclear transfer may generate fertile adults if the sources of the donor nuclei are from early embryos. The limitation of nuclear transfer for recovery of stored genomes is that the mitochondrial genome of the donor is likely to be lost since it is the host oocyte mitochondria that normally persist (Sumida 1997), although this is not always the case (Meirelles et al. 2001). This might not matter in same-species nuclear transfers. However, in inter-species nuclear transfers, the resulting offspring would be nuclear-mitochondrial genomic hybrids. This would not be the optimal outcome for maintenance of the genetic integrity of the recovered species. The other limitation of nuclear transfers in which cross-species nuclear-cytoplasmic hybrids are generated is the lower probability of success. Nuclear-cytoplasmic hybrids between species often do not produce viable embryos (McKinnell 1978).

To date, the generation of amphibian chimeras remains unexplored. Chimeras are generated from embryos that derive their cells from two different sources, the host embryo and foreign donor cells that are incorporated into the developing host embryo. Chimeras from frozen or unfrozen cells have not been reported for amphibians. However, fish chimeras have been generated from unfrozen (Hong et al. 2012; Lin et al. 1992; Nilsson and Cloud 1992; Yamaha et al. 1997) and cryopreserved blastomeres (Kusuda et al. 2004; Yasui et al. 2011) and PGCs (Higaki et al. 2010). Chimeras may be expected to produce sperm and eggs that are derived from both the donor and host embryo lineages; however, the genetic lineage of each would remain separate, allowing recovery of pure donor lines by normal fertilisation mechanisms. The main advantage of using chimeras is that the whole donor species genome is retrieved (nuclear and mitochondrial), since it is the intact cell that is incorporated into the chimera. Should the donor cell differentiate into the germ cell lineage in the chimera, it will produce oocytes whose mitochondrial and nuclear genomes is entirely donor cell-derived. This would be the optimal outcome when recovering stored genomes from diploid cells stored as insurance against loss of all live females from both wild and captive assurance populations. The generation of amphibian chimeras should be a major line of investigation in the development of amphibian genome resource banking and ART.

7.3 Cloning and De-extinction

De-extinction is a new term that has been coined to describe the recovery of extinct species from preserved genomes. Most notable is the proposal to recover mammoths from frozen tissues in the tundra (Loi et al. 2011). This concept must now be considered in light of the amphibian extinction crisis. There are frozen tissues from recently extinct amphibians that are held in various institutions throughout the world (e.g., tissues from extinct Australian frogs in the South Australian Museum (Mahony and Clulow 2011). There is the potential to recover species by this route if the technology continues to develop for amphibians. Unfortunately, most of the currently frozen tissues from extinct frogs have been accessioned into collections by freezing without cryoprotectants. Can live offspring be generated from such tissues? For mammalian tissues the answer is yes; both mice (Wakayama et al. 2008) and cattle (Hoshino et al. 2009) have been cloned from somatic tissues frozen without cryoprotectants and stored for many years in conventional freezers (−20 °C for 16 years, mouse; −80 °C for 10 years, bovine). Recently, nuclear transfer studies have resulted in the production of embryos of the extinct gastric brooding frog *Rheobatrachus vitellinus*, although these failed to develop after a few days (French et al. unpublished data).

7.4 Androgenesis

When cryopreserved sperm are the only genomic resource available, androgenesis has the potential to be a useful procedure for the recovery of populations and species (Corley-Smith and Brandhorst 1999; Thorgaard et al. 2005). Androgenetic (double haploid) offspring are uniparental and their nuclear genome is entirely derived from the sperm pronucleus following inactivation of the maternal pronucleus, usually by ultra-violet radiation. Androgenesis has been used to generate fertile adults in a number of fish species (Parsons and Thorgaard 1984; Corley-Smith et al. 1996; Thorgaard et al. 2005) and fertile androgenetic adults have been generated in at least one amphibian, the axolotl *Ambystoma mexicanum* (Gillespie and Armstrong 1980). The generation of haploid zygotes has been reported for various amphibians including *Xenopus laevis* (Gurdon 1960) and *Lithobates pipiens* (Porter 1939). Development of the haploid embryo (Fig. 12.6) fails at the early larval phase (Porter 1939; Gurdon 1960) if it is not converted to the diploid (doubled haploid) state by the inhibition of first cleavage. This approach to amphibian genome storage does have two potential disadvantages: (1) depending on the sex-determining mechanism, only one sex might be recovered by the process, requiring back-crossing to nearest relatives to regenerate a lineage or species; and (2) the mitochondrial lineage is lost, as mitochondria are not inherited through the male germ line.

Fig. 12.6 Putative haploid, androgenetic larva of the Striped Marsh Frog, *Limnodynastes peronii. Left larva*, haploid; *right larva*, diploid sibling. Clulow et al. unpublished data. Photograph: J. Clulow

8 Conclusions

Assisted reproduction technologies are beginning to be recognized for their potential to contribute to amphibian conservation in the face of an unprecedented decline in amphibian biodiversity. At the base of ART is the induction of gamete release, often under suboptimal conditions in captivity. We no longer have the time to spend years exploring the basic environmental conditions required for captive breeding. We have, in hand, numerous inexpensive, easy-to-use hormone therapies that have been demonstrated to work in all amphibian groups except caecilians.

The hormonal control of reproduction in amphibians has hardly been researched in comparison to fish and mammals. Nevertheless, there are clear directions for future research efforts. These include the mode of delivery of GnRH agonists, including frequency and duration of treatments. There are several inexpensive GnRH agonists that have already proven highly effective in amphibians. The role of neurotransmitters such as DA, a known potent inhibitor of LH release, is pivotal to our understanding of the control of spawning. A combination of a GnRH agonist with a DA antagonist to induce simultaneous surge release of LH in both sexes shows promise because it induces the full complement of spawning behaviors, including amplexus and oviposition. This has now been documented for seven anurans and one salamander (Trudeau et al. 2010, 2012, 2013). The use of injectable gonadotropin preparations, especially when coupled with egg and sperm collection for IVF, is effective. Currently administered to a wide variety of anuran species are hCG and PMSG, sometimes given in combination to maximally stimulate gonadal function. The use of recombinant frog LH and FSH has hardly been considered but all the methods required for their production are available. Regardless of the gonadotropin preparation chosen, much work remains to determine optimal species-specific injection protocols, given the large variations in responses to gonadotropins (Clulow et al. 2012; Kouba et al. 2012a). Optimization of egg and sperm collection methods for IVF is also highly species-specific and requires considerable effort. While good progress is being made for anurans, IVF methods in urodeles and caecilians is lagging.

Excellent progress has been made in developing sperm cryopreservation methods for anuran species with embryos or more advanced offspring generated from frozen sperm in several species, including sperm successfully cryopreserved after non-invasive collection by hormonal induction. There seems no major impediment to the widespread use of sperm cryopreservation in amphibian conservation, and biobanking cryopreserved sperm should be implemented using current technologies and capabilities.

The potential for cryopreservation of the female and embryonic genomes is less optimistic, with no offspring reported to date from either cryopreserved oocytes or whole embryos. Egg size and structure, and yolk composition appear to create technical barriers to cryopreservation. Nevertheless, circumventing this block is likely to be achieved using nuclear transfer and the generation of chimeras using dissociated, cryopreserved embryonic cells, thus effectively achieving cryopreservation of the diploid genome. While cryopreserved somatic cells can be recovered, these provide less favourable targets for the generation of embryos by nuclear transfer, given that no complete development to sexual maturity has been reported from amphibian nuclear transfer using post-metamorphic somatic cells. Advances in molecular and cellular techniques may overcome this problem. The direct cryopreservation of immature ovarian follicles holds promise, but would need to be combined with procedures such as xeno-transplantation to generate mature, ovulated oocytes. Cryopreservation of primordial germ cells also holds promise, but would likely need to be combined with the generation of chimeras to obtain adults that can produce viable gametes. When only cryopreserved sperm from a threatened population or species are available, the use of androgenesis may have the potential to generate viable, live offspring. As a last resort with species that have become extinct in the wild and in captivity, future technological advances may generate viable offspring from frozen tissues currently held in collections.

Model species are important for all aspects of biological research. Given the diversity of reproductive strategies in the Amphibia, it remains challenging to have only a few "model species". Nevertheless, work on bufonids, ranids and pipids in particular have advanced most aspects of ART in the last two decades. Yet, a wider more collaborative effort between ecologists, molecular and cellular biologists, and reproductive physiologists must be established in order to address some of the more challenging aspects relative to conserving the female genome.

Nearly 2,000 amphibians are threatened with extinction, which represents ~32 % of currently listed species. This extinction crisis shows no sign of abating. Therefore, all available knowledge, tools and resources must be used efficiently and immediately as a foundation for large-scale species recovery. Here we have outlined important success stories and future research directions for assisted reproductive technologies that will have paramount importance for the conservation of threatened amphibians.

Acknowledgements The authors acknowledge funding support from the Australian Research Council, the University of Newcastle and WWF (J.C.); University of Ottawa Research Chair Program and Environment Canada (V.L.T.); the U.S. Institute of Museum and Library Services National Leadership Grants(LG-25-09-0064-09 and LG-25-11-0186-11), and Morris Animal Foundation grant (D08ZO-037) (A.J.K.). Dr. Xuhua Xia (University of Ottawa) performed the phylogenetic analysis presented in Fig. 12.2 and this is acknowledged with appreciation. The help of Maria Vu, Michael Mahony, Simon Clulow and Carrie Vance for editing and helpful comments is gratefully acknowledged.

References

Alford RA, Richards SJ. Global amphibian declines: a problem in applied ecology. Annu Rev Ecol Syst. 1999;30(1):133–65.

Balinsky BI. An introduction to embryology. 4th ed. Philadelphia, PA: WB Saunders; 1975. p. 161–75.

Barraclough CA, Wise PM, Selmanoff MK. A role for hypothalamic catecholamines in the regulation of gonadotropin secretion. Recent Prog Horm Res. 1984;40:487–529.

Barton HL, Guttman SI. Low temperature preservation of toad spermatozoa (Genus *Bufo*). Tex J Sci. 1972;23:363–70.

Beesley SG, Costanzo JP, Richard Jr EL. Cryopreservation of spermatozoa from freeze-tolerant and -intolerant anurans. Cryobiology. 1998;37:155–62.

Bellerby CW. A rapid test for the diagnosis of pregnancy. Nature. 1934;133:494–5.

Bennett PM. Establishing animal germplasm resource banks for wildlife conservation: genetic, population and evolutionary aspects. In: Watson PF, Holt WV, editors. Cryobanking the genetic resource. London: Taylor and Francis; 2001. p. 47–68.

Ben-Nun IF, Montague SC, Houck ML, Tran HT, Garitaonandia I, Leonardo TR, Wang Y-C, Charter SJ, Laurent LC, Ryder OA, Loring JF. Induced pluripotent stem cells from highly endangered species. Nat Methods. 2011;8:829–31.

Berger L, Speare R, Daszak P, Green DE, Cunningham AA, Goggin CL, et al. Chytridiomycosis causes amphibian mortality associated with population declines in the rain forests of Australia and Central America. Proc Natl Acad Sci U S A. 1998;95:9031–6.

Bezchlibnyk-Butler Z, Jefferies J. Clinical handbook of psychotropic drugs. 15th ed. Cambridge, MA: Hogrefe and Huber Publishing Company; 2005. p. 348.

Bishop PJ, Angulo A, Lewis JP, Moore RD, Rabb GB, Garcia Moreno J. The amphibian extinction crisis—what will it take to put the action into the Amphibian Conservation Action Plan? Surveys and Perspectives Integrating Environment and Society (SAPIENS). 2012;5.2. http://sapiens.revues.org/1406.

Briggs RK, King TJ. Transplantation of living nuclei from blastula cells into enucleated frogs' eggs. Proc Natl Acad Sci U S A. 1952;38:455–63.

Browne RK, Figiel C. Cryopreservation in amphibians. In: Tiersch T, Mazie P, editors. Cryopreservation of aquatic species, vol. 8. Baton Rouge, LA: World Aquaculture Society; 2010. p. 450.

Browne RK, Clulow J, Mahony M, Clark A. Successful recovery of motility and fertility of cryopreserved cane toad (*Bufo marinus*) sperm. Cryobiology. 1998;37:339–45.

Browne RK, Clulow J, Mahony M. Short-term storage of cane toad (*Bufo marinus*) gametes. Reproduction. 2001;121:167–73.

Browne RK, Clulow J, Mahony M. The effect of saccharides on the post-thaw recovery of cane toad (*Bufo marinus*) spermatozoa. Cryo Letters. 2002a;23:121–8.

Browne RK, Clulow J, Mahony M. The short-term storage and cryopreservation of spermatozoa from Hylid and Myobatrachid frogs. Cryo Letters. 2002b;23:129–36.

Browne RK, Clulow J, Mahony M. A comparison of sucrose, saline, and saline with egg-yolk diluents on the cryopreservation of cane toad (*Bufo marinus*) sperm. Cryobiology. 2002c;44: 251–7.

Browne RK, Davis J, Pomering M, Clulow J. Storage of cane toad (*Bufo marinus*) sperm for 6 days at 0 deg C with subsequent cryopreservation. Reprod Fertil Dev. 2002d;14:267–73.

Browne RK, Seratt J, Vance C, Kouba A. Hormonal priming, induction of ovulation and in-vitro fertilization of the endangered Wyoming toad (*Bufo baxteri*). Reprod Biol Endocrinol. 2006a;4:34.

Browne RK, Li H, Seratt J, Kouba A. Progesterone improves the number and quality of hormone induced Fowler toad (*Bufo fowleri*) oocytes. Reprod Biol Endocrinol. 2006b;4:3.

Buchholz DR, Fu L, Shi YB. Cryopreservation of *Xenopus* transgenic lines. Mol Reprod Dev. 2004;67(1):65–9.

Burel D, Li JH, Do-Rego JL, Wang AF, Luu-The V, Pelletier G, et al. Gonadotropin-releasing hormone stimulates the biosynthesis of pregnenolone sulfate and dehydroepiandrosterone sulfate in the hypothalamus. Endocrinology. 2013;154(6):2114–28.

Byrne PG, Silla AJ. Hormonal induction of gamete release, and in vitro fertilisation, in the critically endangered Southern Corroboree frog, *Pseudophryne corroboree*. Reprod Biol Endocrinol. 2010;8:144.

Cabada, MO. Sperm concentration and fertilization rate in *bufo arenarum* (amphibian: anura). J Exp Biol. 1975;62:481–6.

Calvi SL, Maisse G. Cryopreservation of Rainbow Trout (*Oncorhynchus mykiss*) blastomeres: influence of embryo stage on postthaw survival rate. Cryobiology. 1998;36:255–62.

Calvi SL, Maisse G. Cryopreservation of carp (*Cyprinus carpio*) blastomeres. Aquat Living Resour. 1999;12:71–4.

Cardona-Costa J, García-Ximénez F. Vitrification of zebrafish embryo blastomeres in microvolumes. Cryo Letters. 2007;28:303–9.

Chanson J, et al. The state of the world's amphibians. In: Stuart SN, Hoffmann M, Chanson JS, Cox NA, Berridge RJ, Ramani P, Young BE, editors. Threatened amphibians of the world. Barcelona, Spain: Lynx Edicions; 2008. p. 33–52.

Chu J, Wilczynski W. Androgen effects on tyrosine hydroxylase cells in the northern leopard frog, *Rana pipiens*. Neuroendocrinology. 2002;76(1):18–27.

Clulow J, Mahony M, Browne R, Pomering M, Clark A. Applications of assisted reproductive technologies (ART) to endangered anuran amphibians. In: Campbell A, editor. Declines and disappearances of Australian frogs. Canberra: Environment Australia; 1999. p. 219–25.

Clulow J, Clulow S, Guo J, French AJ, Mahony MJ, Archer M. Optimisation of an oviposition protocol employing human chorionic and pregnant mare serum gonadotropins in the Barred Frog *Mixophyes fasciolatus* (Myobatrachidae). Reprod Biol Endocrinol. 2012;10(1):60.

Collin F, Chartrel N, Fasolo A, Conlon JM, Vandesande F, Vaudry H. Distribution of two molecular forms of gonadotropin-releasing hormone (GnRH) in the central nervous system of the frog *Rana ridibunda*. Brain Res. 1995;703:111–28.

Corley-Smith GE, Brandhorst BP. Preservation of endangered species and populations: a role for genome banking, somatic cell cloning, and androgenesis? Mol Reprod Dev. 1999;53:363–7.

Corley-Smith GE, Lim CJ, Brandhorst BP. Production of androgenetic zebrafish (*Danio rerio*). Genetics. 1996;142:1265–76.

Costanzo JP, Mugnano JA, Wehrheim HM, Richard Jr EL. Osmotic and freezing tolerance in spermatozoa of freeze-tolerant and -intolerant frogs. Am J Physiol. 1998;275:713–9.

Creighton A, Satterfield D, Chu J. Effects of dopamine agonists on calling behavior in the green tree frog, *Hyla cinerea*. Physiol Behav. 2013;116–117C:54–9.

Dash SN, Routray P, Dash C, Guru BC, Swain P, Sarangi N. Use of the nontoxic cryoprotectant trehalose enhances recovery and function of fish embryonic stem cells following cryogenic storage. Curr Stem Cell Res Ther. 2008;3:277–87.

Di Berardino MA. Genomic potential of differentiated cells. New York, NY: Columbia University Press; 1997.

Dournon C, Aimar C, Bautz A, Ropars A, Houillon C. Grafts of ovaries in males and females of *Pleurodeles waltl* (urodele amphibian): evidence of a long-term tolerance of allografts; application following a space biology experiment. Reprod Nutr Dev. 1997;37:637–49.

Dufour S, Weltzien FA, Sebert ME, Le Belle N, Vidal B, Vernier P, Pasqualini C. Dopaminergic inhibition of reproduction in teleost fishes: ecophysiological and evolutionary implications. Ann N Y Acad Sci. 2005;1040:9–21.

Dziminski MA, Roberts JD, Simmons LW. Sperm morphology, motility, and fertilization capacity in the myobatrachid frog *Crinia georgiana*. Reprod Fertil Dev. 2010;22:516–22.

Easley KA, Culley Jr DD, Horseman ND, Penkala JE. Environmental influences on hormonally induced spermiation of the bullfrog, *Rana catesbeiana*. J Exp Zool. 1979;207:407–16.

Edwards DL, Mahony MJ, Clulow J. Effect of sperm concentration, medium osmolality, and oocyte storage on artificial fertilisation success in a myobatrachid frog (*Limnodynastes tasmaniensis*). Reprod Fertil Dev. 2004;16:347–54.

Elinson RP. Fertilization in amphibians: the ancestry of the block to polyspermy. Int Rev Cytol. 1986;101:59–100.

Fernandez S, Ramos I. Endocrinology of reproduction. In: Jamieson BGM, editor. Reproductive biology and phylogeny of anura, vol. 2. Enfield, NH: Science Publisher, Inc.; 2003. p. 73–118.

Fitzsimmons C, McLaughlin EA, Mahony MJ, Clulow J. Optimisation of handling, activation and assessment procedures for *Bufo marinus* spermatozoa. Reprod Fertil Dev. 2007;19:594–601.

Frankham R, Ralls K. Inbreeding leads to extinction. Nature. 1998;392:441–2.

Galli-Mainini C. Pregnancy test using the male toad. J Clin Endocrinol Metab. 1947;7:653–8.

Germano J, Drake M, Dowling M, Martin M, Hasenstab S, Kouba A. Hormone-induced spermiation in the Northern leopard frog (*Rana pipiens*) and endangered dusky gopher frog (*Rana sevosa*). Joint meeting of Icthyologists and herpetologists (JMIH) associations; 2011 Jul 6–11; Minneapolis, MN, USA. (abstract).

Gillespie LL, Armstrong JB. Production of androgenetic diploid axolotls by suppression of first cleavage. J Exp Zool. 1980;213:423–5.

Goncharov BF, Shubravy OI, Serbinova IA, Uteshev VK. The USSR programme for breeding amphibians, including rare and endangered species. Int Zoo Yearbook. 1989;28:10–21.

Gonzalez A, Smeets WJAJ. Comparative analysis of dopamine and tyrosine hydroxylase immunoreactivities in the brain of two amphibians, the anuran *Rana ridibunda* and the urodele *Pleurodeles waltlii*. J Comp Neurol. 1991;303(3):457–77.

Gracia-Navarro F, Licht P. Subcellular localization of gonadotrophic hormones LH and FSH in frog adenohypophysis using double-staining immunocytochemistry. J Histochem Cytochem. 1987;35(7):763–9.

Green SL, Parker J, Davis C, Bouley DM. Ovarian hyperstimulation syndrome in gonadotropin-treated laboratory South African clawed frogs (*Xenopus laevis*). J Am Assoc Lab Anim Sci. 2007;46(3):64–7.

Guenther JF, Seki S, Kleinhans FW, Edashige K, Roberts DM, Mazur P. Extra- and intra-cellular ice formation in Stage I and II *Xenopus laevis* oocytes. Cryobiology. 2006;52:401–16.

Gurdon JB. The effects of ultraviolet irradiation on uncleaved eggs of *Xenopus laevis*. Q J Microsc Sci. 1960;101:299–311.

Gurdon J, Byrne J. The first half-century of nuclear transplantation. Proc Natl Acad Sci U S A. 2003;100:8048–52.

Gurdon JB, Hopwood N. The introduction of *Xenopus laevis* into developmental biology: of empire, pregnancy testing and ribosomal genes. Int J Dev Biol. 2000;44:43–50.

Gurdon JB, Uehlinger V. "Fertile" intestine nuclei. Nature. 1966;210:1240–1.

Gurdon JB, Laskey RA, Reeves OR. The developmental capacity of nuclei transplanted from keratinized skin cells of adult frogs. J Embryol Exp Morphol. 1975;34:93–112.

Hagedorn M, Kleinhans FW. Problems and prospects in cryopreservation of fish embryos. In: Tierch TR, Green CC, editors. Cryopreservation in aquatic species. 2nd ed. Baton Rouge, LA: World Aquaculture Society; 2011. p. 483–502.

Hagedorn M, Hsu EW, Pilatus U, Wildt DE, Rall WF, Blackband SJ. Magnetic resonance microscopy and spectroscopy reveal kinetics of cryoprotectant permeation in a multicompartmental biological system. Proc Natl Acad Sci U S A. 1996;93:7454–9.

Hagedorn M, Hsu E, Kleinhans FW, Wildt DE. New approaches for studying the permeability of fish embryos: toward successful cryopreservation. Cryobiology. 1997a;34:335–47.

Hagedorn M, Kleinhans FW, Freitas R, Liu J, Hsu EW, Wildt DE, et al. Water distribution and permeability of zebrafish embryos, *Brachydanio rerio*. J Exp Zool. 1997b;278:356–71.

Hagedorn M, Kleinhans FW, Wildt DE, Rall WF. Chill sensitivity and cryoprotectant permeability of dechorionated zebrafish embryos, *Brachydanio rerio*. Cryobiology. 1997c;34:251–63.

Hagedorn M, Kleinhans FW, Artemov D, Pilatus U. Characterization of a major permeability barrier in the zebrafish embryo. Biol Reprod. 1998;59:1240–50.

Hagedorn M, Lance SL, Fonseca DM, Kleinhans FW, Artimov D, Fleischer R, et al. Altering fish embryos with aquaporin-3: an essential step toward successful cryopreservation. Biol Reprod. 2002;67:961–6.

Hagedorn M, Peterson A, Mazur P, Kleinhans FW. High ice nucleation temperature of zebrafish embryos: slow-freezing is not an option. Cryobiology. 2004;49:181–9.

Harvey B. Cooling of embryonic cells, isolated blastoderms, and intact embryos of the zebra fish *Brachydanio rerio* to −196 °C. Cryobiology. 1983;20:440–7.

Herbison AE. Multimodal influence of estrogen upon gonadotropin-releasing hormone neurons. Endocr Rev. 1998;19(3):302–30.

Higaki S, Eto Y, Kawakami Y, Yamaha E, Kagawa N, Kuwayama M, et al. Production of fertile zebrafish (*Danio rerio*) possessing germ cells (gametes) originated from primordial germ cells recovered from vitrified embryos. Reproduction. 2010;139:733–40.

Hogan LA, Lisle AT, Johnson SD, Goad T, Robertson H. Adult and juvenile sex identification in threatened monomorphic *Geocrinia* frogs using fecal steroid analysis. J Herpetol. 2013;47(1):112–8.

Hollinger T, Corton G. Artificial fertilization of gametes from the South African clawed Frog, *Xenopus laevis*. Gamete Res. 1980;3:45–7.

Holt WV. Genetic resource banking and maintaining biodiversity. In: Watson PF, Holt WV, editors. Cryobanking the genetic resource. London: Taylor and Francis; 2001. p. 9–20.

Holt WV, Bennett PM, Volobouev V. Genetic resource banks in wildlife conservation. J Zool (London). 1996;238:531–44.

Hong N, Chen S, Ge R, Song J, Yi M, Hong Y. Interordinal chimera formation between medaka and zebrafish for analyzing stem cell differentiation. Stem Cells Dev. 2012;21:2333–41.

Hoshino Y, Hayashi N, Taniguchi S, Kobayashi N, Sakai K, Otani T, et al. Resurrection of a bull by cloning from organs frozen without cryoprotectant in a −80 °C freezer for a decade. PLoS One. 2009;4(1):e4142.

Houlahan JE, Findlay CS, Schmidt BR, Meyer AH, Kuzmin SL. Quantitative evidence for global amphibian population declines. Nature. 2000;404(6779):752–5.

Iimori E, D'Occhio MJ, Lisle AT, Johnston SD. Testosterone secretion and pharmacological spermatozoal recovery in the cane toad. Anim Reprod Sci. 2005;90:163–73.

Itoh M, Inoue M, Ishii S. Annual cycle of pituitary and plasma gonadotropins and plasma sex steroids in a wild population of the toad, *Bufo japonicus*. Gen Comp Endocrinol. 1990;78(2):242–53.

Jamieson BGM. Reproductive biology and phylogeny of anura, Reproductive biology and phylogeny, vol. 2. Enfield, NH: Science Publishers; 2007.

Janik M, Kleinhans FW, Hagedorn M. Overcoming a permeability barrier by microinjecting cryoprotectants into zebrafish embryos (*Brachydanio rerio*). Cryobiology. 2000;41:25–34.

Kah O, Lethimonier C, Somoza G, Guilgur LG, Vaillant C, Lareyre JJ. GnRH and GnRH receptors in metazoa: a historical, comparative, and evolutive perspective. Gen Comp Endocrinol. 2007;153(1–3):346–64.

Kim DK, Cho EB, Moon MJ, Park S, Hwang JI, Kah O, et al. Revisiting the evolution of gonadotropin-releasing hormones and their receptors in vertebrates: secrets hidden in genomes. Gen Comp Endocrinol. 2011;170(1):68–78.

Kleinhans FW, Guenther JF, Seki S, Edashige K, Roberts DM, Mazur P. Water P_f of Xenopus (AQP +/–) Stage I and II oocytes. Cryobiology. 2005;51:389.

Kleinhans FW, Guenther JF, Roberts DM, Mazur P. Analysis of intracellular ice nucleation in Xenopus oocytes by differential scanning calorimetry. Cryobiology. 2006;52:128–38.

Kobel HR, Brun RB, Fischberg M. Nuclear transplantation with melanophores, ciliated epidermal cells, and the established cell-line A-8 in *Xenopus laevis*. J Embryol Exp Morphol. 1973;29:539–47.

Köseoglu M, Eroglu A, Toner M, Sadler KC. Starfish oocytes form intracellular ice at unusually high temperatures. Cryobiology. 2001;43:248–59.

Kouba AJ, Vance CK. Applied reproductive technologies and genetic resource banking for amphibian conservation. Reprod Fertil Dev. 2009;21:719–37.

Kouba AJ, Vance CK, Frommeyer MA, Roth TL. Structural and functional aspects of *Bufo americanus* spermatozoa: effects of inactivation and reactivation. J Exp Zool. 2003;295A:172–82.

Kouba AJ, Vance CK, Willis EL. Artificial fertilization for amphibian conservation: current knowledge and future considerations. Theriogenology. 2009;71:214–27.

Kouba AJ, Willis E, Vance CK, Hasenstab S, Reichling S, Krebs J, et al. Development of assisted reproduction technologies for the endangered Mississippi gopher frog (*Rana sevosa*) and sperm transfer for in-vitro fertilization. Reprod Fertil Dev. 2011;24(1):170.

Kouba AJ, Delbarco-Trillo J, Vance CK, Milam C, Carr M. A comparison of human chorionic gonadotropin and luteinizing hormone releasing hormone on the induction of spermiation and amplexus in the American toad (*Anaxyrus americanus*). Reprod Biol Endocrinol. 2012a;10:59.

Kouba AJ, Vance CK, Calatayud N, Rowlison T, Langhorne C, Willard S. Assisted reproduction technologies (ART) for amphibians; Chapter 2. In: Poole VA, Grow S, editors. Amphibian husbandry resource guide. 20th ed. Silver Springs, MD: Amphibian Taxon Advisory Group, Association of Zoos and Aquariums (AZA); 2012b. p. 60–118.

Kouba AJ, Lloyd RE, Houck ML, Silla AJ, Calatayud N, Trudeau VL, et al. Emerging trends for biobanking amphibian genetic resources: the hope, reality and challenges for the next decade. Biol Conserv. 2013;164:10–21.

Kroll K, Amaya E. Transgenic Xenopus embryos from sperm nuclear transplantations reveal FGF signaling requirements during gastrulation. Development. 1996;122:3173–83.

Kurian T, Saidapur SK. Comparative effects of hCG and PMSG on spermatogenesis, Leydig cells and thump pads of the frog, *Rana tigrina*, during post-breeding regression. Indian J Exp Biol. 1982;20:577–81.

Kusuda S, Teranishi T, Koide N. Cryopreservation of chum salmon blastomeres by the straw method. Cryobiology. 2002;45:60–7.

Kusuda S, Teranishi T, Koide N, Nagai T, Arai K, Yamaha E. Pluripotency of cryopreserved blastomeres of the goldfish. J Exp Zool A Comp Exp Biol. 2004;301(2):131–8.

Langhorne C, Calatayud N, Rowlison T, Kouba A, Vance C, Willard S. Short-term cold storage and cryopreservation of boreal toad (*Bufo boreas boreas*) sperm. World Congress of Herpetology; 2012 Aug 8–11; Vancouver, BC; p. 405. (abstract).

Langhorne C, Calatayud NE, Vance CK, Willard ST, Kouba AJ. Cryoconservation: developing sperm cryopreservation protocols for endangered North American amphibians; 2013 Jul 28–31. Bethesda, MD: Society for Cryobiology. (abstract).

Lawson B, Clulow S, Mahony MJ, Clulow J. Towards gene banking amphibian maternal germ lines: short-term incubation, cryoprotectant tolerance and cryopreservation of embryonic cells of the frog, *Limnodynastes peronii*. PLoS One. 2013;8(4):e60760.

Lermen D, Blomeke B, Browne R, Clarke A, Dyce PW, Fixemer T, Fuhr GR, Holt WV, Jewgenow K, Lloyd RE, Lotters S, Paulus M, Reid GM, Rapoport DH, Rawson D, Ringleb J, Ryder OA, Sporl G, Schmitt T, Veith M, Muller P. Cryobanking of viable biomaterials: implementation of new strategies for conservation purposes. Mol Ecol. 2009;18(6):1030–3.

Licht P. Conservation of endangered species in captivity: an interdisciplinary approach. In: Gibbons Jr E, Durrant B, Desmarest J, editors. Reproductive physiology and reptiles and amphibians. New York, NY: State University of New York Press; 1995. p. 169–86.

Licht P, McCreery BR, Barnes R, Pang R. Seasonal and stress related changes in plasma gonadotropins, sex steroids, and corticosterone in the bullfrog, *Rana catesbeiana*. Gen Comp Endocrinol. 1983;50(1):124–45.

Lin S, Long W, Chen J, Hopkins N. Production of germ-line chimeras in zebrafish by cell transplants from genetically pigmented to albino embryos. Proc Natl Acad Sci U S A. 1992;89:4519–23.

Lin C, Zhang T, Rawson DM. Cryopreservation of zebrafish (*Danio rerio*) blastomeres by controlled slow cooling. Cryo Letters. 2009;30:132–41.

Lips KR, Burrowes PA, Mendelson JR, Parra-Olea G. Amphibian declines in Latin America: widespread population declines, extinctions, and impacts. Biotropica. 2005;37:163–5.

Lips KR, Brem F, Brenes R, Reeve JD, Alford RA, Voyles J, et al. Emerging infectious disease and the loss of biodiversity in a Neotropical amphibian community. Proc Natl Acad Sci U S A. 2006;103:3165–70.

Liu X-H, Zhang T, Rawson DM. The effect of partial removal of yolk on the chilling sensitivity of zebrafish (*Danio rerio*) embryos. Cryobiology. 1999;39:236–42.

Liu X-H, Zhang T, Rawson DM. Differential scanning calorimetry studies of intramembrane freezing and cryoprotectant penetration in zebrafish (*Danio rerio*) embryos. J Exp Zool. 2001;290:299–310.

Lofts B. Reproduction, Chapter 4. In: Lofts B, editor. Physiology of the amphibia, vol. 2. New York, NY: Academic; 1974. p. 107–218.

Loi P, Wakayama T, Saragustry J, Fulka J, Ptak G. Biological time machines: a realistic approach for cloning an extinct mammal. Endanger Species Res. 2011;14:227–33.

Longcore JE, Pessier AP, Nichols DK. *Batrachochytrium dendrobatidis* gen. et sp. nov., a chytrid pathogenic to amphibians. Mycologia. 1999;91:219–27.

Lynch KS, Crews D, Ryan MJ, Wilczynski W. Hormonal state influences aspects of female mate choice in the Túngara Frog (*Physalaemus pustulosus*). Horm Behav. 2006;49:450–7.

Madsen T, Shine R, Olsson M, Wittzell H. Restoration of an inbred adder population. Nature. 1999;402:34–5.

Mahony MJ, Clulow J. Appendix 2. Cryopreservation and reconstitution technologies: a proposal to establish a genome resource bank for threatened Australian amphibians. In: Murray K, Skerratt LF, Marantelli G, Berger L, Hunter D, et al., editors. Guidelines for minimising disease risks associated with captive breeding, raising and restocking programs for Australian frogs. Canberra: Australian Government, Department of Sustainability, Environment, Water, Population and Communities; 2011. http://www.environment.gov.au/biodiversity/invasive/publications/pubs/frogs-captive-breeding-appendix2.pdf.

Mann RM, Hyne RV, Choung CB. Hormonal induction of spermiation, courting behavior and spawning in the southern bell frog, *Litoria raniformis*. Zoo Biol. 2010;29:1–9.

Mansour N, Lahnsteiner F, Patzner RA. Optimization of the cryopreservation of African clawed frog (*Xenopus laevis*) sperm. Theriogenology. 2009;72(9):1221–8.

Mansour N, Lahnsteiner F, Patzner RA. Motility and cryopreservation of spermatozoa of European common frog, *Rana temporaria*. Theriogenology. 2010;74:724–32.

Mansour N, Lahnsteiner F, Patzner RA. Collection of gametes from live axolotl, *Ambystoma mexicanum*, and standardization of in vitro fertilization. Theriogenology. 2011;75:354–61.

Mazur P. Principles of cryobiology. In: Fuller BJ, Lane N, Benson EE, editors. Life in the frozen state. Boca Rotan, FL: CRC Press; 2004. p. 3–65.

Mazur P, Kleinhans FW. Relationship between intracellular ice formation in oocytes of the mouse and Xenopus and the physical state of the external medium—a revisit. Cryobiology. 2008;56:22–7.

Mazur P, Seki S, Pinn IL, Kleinhans FW, Edashige K. Extra- and intracellular ice formation in mouse oocytes. Cryobiology. 2005;51:29–53.

McKinnell RG. Cloning: nuclear transplantation in amphibia. Minneapolis, MN: University of Minnesota Press; 1978.

McKinnell RG, Picciaano DJ, Kreig RE. Fertilization and development of frog eggs after repeated spermiation induced by human chorionic gonadotrophin. Lab Anim Sci. 1976;26:932–5.

Meirelles FV, Bordignon V, Watanabe Y, Watanabe M, Dayan A, Lobo RB, et al. Complete replacement of the mitochondrial genotype in a *Bos indicus* in a calf reconstructed by nuclear transfer to a *Bos taurus* oocycte. Genetics. 2001;158:351–6.

Michael SF, Jones C. Cryopreservation of spermatozoa of the terrestrial Puerto Rican frog, *Eleutherodactylus coqui*. Cryobiology. 2004;48:90–4.

Michael SF, Buckley C, Toro E, Estrada AR, Vincent S. Induced ovulation and egg deposition in the direct developing anuran *Eleutherodactylus coqui*. Reprod Biol Endocrinol. 2004; 2(6):1–5.

Millar RP, Lu ZL, Pawson AJ, Flanagan CA, Morgan K, Maudsley SR. Gonadotropin-releasing hormone receptors. Endocr Rev. 2004;25(2):235–75.

Minucci S, Di Matteo L, Chieffi Baccari G, Pierantoni R. A gonadotropin releasing hormone analog induces spermiation in intact and hypophysectomized frogs, *Rana esculenta*. Experientia. 1989;45(11–12):1118–21.

Mizell M. Biology of amphibian tumours. In: Mizel M, editor. Recent results in cancer research. New York, NY: Springer; 1969. Special Supplement.

Mizutani F, Iwasawa H, Tanaka S. A morphometric analysis of the subcellular distribution of LH beta and FSH beta in secretory granules in the pituitary gonadotrophs of the frog (*Rana japonica*). Cell Tissue Res. 1994;277(3):417–26.

Mugnano JA, Costanzo JP, Beesley SG, Richard Jr ER. Evaluation of glycerol and dimethyl sulphoxide for the cryopreservation of spermatozoa from the wood frog (*Rana sylvatica*). Cryo Letters. 1998;19:249–54.

Murray K, Retallick R, McDonald KR, Mendez D, Aplin K, Kirkpatrick P, et al. The distribution and host range of the pandemic disease chytridiomycosis in Australia spanning surveys from 1956 to 2007. Ecology. 2010;91:1557.

Nakano M, Hasunuma I, Okada R, Yamamoto K, Kikuyama S, Machida T, et al. Molecular cloning of bullfrog D2 dopamine receptor cDNA: tissue distribution of three isoforms of D2 dopamine receptor mRNA. Gen Comp Endocrinol. 2010a;168(1):143–8.

Nakano M, Minagawa A, Hasunuma I, Okada R, Tonon MC, Vaudry H, et al. D2 Dopamine receptor subtype mediates the inhibitory effect of dopamine on TRH-induced prolactin release from the bullfrog pituitary. Gen Comp Endocrinol. 2010b;168(2):287–92.

Nilsson EE, Cloud JG. Rainbow trout chimeras produced by injection of blastomeres into recipient blastulae. Proc Natl Acad Sci U S A. 1992;89:9425–8.

Nilsson EE, Cloud JG. Cryopreservation of rainbow trout (*Oncorhynchus mykiss*) blastomeres. Aquat Living Resour. 1993;6:77–80.

Norris DO, Lopez KH. Hormones and reproduction of vertebrates, Amphibians, vol. 2. London: Academic; 2011.

O'Connell LA, Matthews BJ, Ryan MJ, Hofmann HA. Characterization of the dopamine system in the brain of the túngara frog, *Physalaemus pustulosus*. Brain Behav Evol. 2010;76(3–4): 211–25.

Obringer AR, O'Brien JK, Saunders RL, Yamamoto K, Kikuyama S, Roth TL. Characterisation of the spermiation response, luteinizing hormone release and sperm quality in the American toad *(Bufo americanus)* and the endangered Wyoming toad *(Bufo baxteri)*. Reprod Fertil Dev. 2000;12:51–8.

Ogawa T, Ohmura M, Tamura Y, Kita K, Ohbo K, Suda T, et al. Derivation and morphological characterization of mouse spermatogonial stem cell lines. Arch Histol Cytol. 2004;67: 297–306.

Ogawa A, Dake J, Iwashina Y, Tokumoto T. Induction of ovulation in *Xenopus* without hCG injection: the effect of adding steroids into the aquatic environment. Reprod Biol Endocrinol. 2011;9:11.

Ogielska M. Reproduction of amphibians. Enfield, NH: Science Publishers; 2009.

Oguchi A, Tanaka S, Aida T, Yamamoto K, Kikuyama S. Enhancement by prolactin of the GnRH-induced release of LH from dispersed anterior pituitary cells of the bullfrog (*Rana catesbeiana*). Gen Comp Endocrinol. 1997;107(1):128–35.

Okumoto H. Establishment of three cell lines derived from frog melanophores. Zoolog Sci. 2001;18:483–96.

Olive PJW, Wang WB. Cryopreservation of *Nereis virens* (polychaeta, annelida) larvae: the mechanism of cryopreservation of a differentiated metazoan. Cryobiology. 1997;34:284–94.

Olson JH, Chandler DE. X*enopus laevis* egg jelly contains small proteins that are essential to fertilisation. Dev Biol. 1999;210:401–10.

Omeljaniuk RJ, Shih SH, Peter RE. In-vivo evaluation of dopamine receptor-mediated inhibition of gonadotrophin secretion from the pituitary gland of the goldfish, *Carassius auratus*. J Endocrinol. 1987;114(3):449–58.

Paris MCJ, Snow M, Cox S-L, Shaw JM. Xenotransplantation: a tool for reproductive biology and animal conservation? Theriogenology. 2004;61:277–91.

Parsons JE, Thorgaard GH. Production of androgenetic diploid rainbow trout. J Hered. 1984;76: 177–81.

Pavgi S, Licht P. Inhibition of in vitro pituitary gonadotropin secretion by 17 beta-estradiol in the frog *Rana pipiens*. Gen Comp Endocrinol. 1993;89(1):132–7.

Pawson AJ, Faccenda E, Maudsley S, Lu ZL, Naor Z, Millar RP. Mammalian type I gonadotropin-releasing hormone receptors undergo slow, constitutive, agonist-independent internalization. Endocrinology. 2008;49(3):1415–22.

Peter RE, Chang JP, Nahorniak CS, Omeljaniuk RJ, Sokolowska M, Shih SH, et al. Interactions of catecholamines and GnRH in regulation of gonadotropin secretion in teleost fish. Recent Prog Horm Res. 1986;42:513–48.

Polzonetti-Magni AM, Mosconi G, Carnevali O, Yamamoto K, Hanaoka Y, Kikuyama S. Gonadotropins and reproductive function in the anuran amphibian, *Rana esculenta*. Biol Reprod. 1998;58(1):88–93.

Polzonetti-Magni AM, Mosconi G, Soverchia L, Kikuyama S, Carnevali O. Multihormonal control of vitellogenesis in lower vertebrates. Int Rev Cytol. 2004;239:1–46.

Popesku JT, Martyniuk CJ, Mennigen J, Xiong H, Zhang D, Xia X, et al. The goldfish (*Carassius auratus*) as a model for neuroendocrine signaling. Mol Cell Endocrinol. 2008;293(1–2):43–56.

Porter KR. Androgenetic development of the egg of *Rana pipiens*. Biol Bull. 1939;77:233–57.

Porter DA, Licht P. Pituitary responsiveness to superfused GnRH in two species of ranid frogs. Gen Comp Endocrinol. 1985;59(2):308–15.

Pozzi AG, Rosemblit C, Ceballos NR. Effect of human gonadotropins on spermiation and androgen biosynthesis in the testis of the toad *Bufo arenarum* (Amphibia, Anura). J Exp Zool. 2006;305A(1):96–102.

Rahman SM, Strüssmann CA, Majhi SK, Suzuki T, Watanabe M. Efficiency of osmotic and chemical treatments to improve the permeation of the cryoprotectant dimethyl sulfoxide to Japanese whiting (*Sillago japonica*) embryos. Theriogenology. 2011;75:248–55.

Rawson DM, McGregor RG, Lloyd RE. Conservation rationale, research applications and techniques in the cryopreservation of lower vertebrate biodiversity from marine and freshwater environments. Int Zoo Yearbook. 2011;45:1–16.

Rosemblit C, Pozzi AG, Ceballos NR. Relationship between steroidogenesis and spermiation in *Rana catesbeiana* and *Leptodactylus ocellatus*. J Comp Physiol B. 2006;176:559–66.

Roth TL, Obringer AR. Reproductive research and the worldwide amphibian extinction crisis. In: Holt WV, Pickard AR, Rodger JC, Wildt DE, editors. Reproductive science and integrated conservation. Cambridge: Cambridge University Press; 2003. p. 359–74.

Roth T, Szymanski DC, Keyster ED. Effects of age, weight, hormones, and hibernation on breeding success in boreal toads (*Bufo boreas boreas*). Theriogenology. 2010;73:501–11.

Routray P, Dash C, Dash SN, Tripathy S, Verma DK, Swain SK, et al. Cryopreservation of isolated blastomeres and embryonic stem-like cells of Leopard danio, *Brachydanio frankei*. Aquacult Res. 2010;41:579–89.

Rowley J, Brown R, Bain R, Kusrini M, Inger R, Stuart B, et al. Impending conservation crisis for Southeast Asian amphibians. Biol Lett. 2010;6:336–8.

Rowson AD, Obringer AR, Roth TL. Non-invasive treatments of luteinizing hormone-releasing hormone for inducing spermiation in American (*Bufo americanus*) and Gulf coast (*Bufo valliceps*) toads. Zoo Biol. 2001;20:63–74.

Rugh R. Experimental embryology techniques and procedures. Minneapolis, MN: Burgess Publishing Company; 1962.

Sargent MG, Mohun TJ. Cryopreservation of sperm of *Xenopus laevis* and *Xenopus tropicalis*. Genesis. 2005;41:41–6.

Schad K, editor. Amphibian population management guidelines. Amphibian Ark Amphibian Population Management Workshop; 2007 December 10–11; San Diego, CA, USA; Amphibian Ark. www.amphibianark.org. 2008. 31 p.

Schultz TW, Dawson DA. Housing and husbandry of *Xenopus* for oocyte production. Lab Anim. 2003;32:34–9.

Seong JY, Wang L, Oh DY, Yun O, Maiti K, Li JH, et al. Ala/Thr(201) in extracellular loop 2 and Leu/Phe(290) in transmembrane domain 6 of type 1 frog gonadotropin-releasing hormone receptor confer differential ligand sensitivity and signal transduction. Endocrinology. 2003;144:454–66.

Shapiro HA, Zwarenstein H. A rapid test for pregnancy on *Xenopus laevis*. Nature. 1934;133:762.

Shishova NR, Uteshev VK, Kaurova SA, Browne RK, Gakhova EN. Cryopreservation of hormonally induced sperm for the conservation of threatened amphibians with *Rana temporaria* as a model research species. Theriogenology. 2011;75(2):220–32.

Shishova NR, Uteshev VK, Sirota NP, Kuznetsova EA, Kaurova SA, Browne RK, Gakhova EN. The quality and fertility of sperm collected from European common frog (*Rana temporaria*) carcasses refrigerated for up to seven days. Zoo Biol. 2013;32:400–6.

Silla AJ. Effects of luteinizing hormone-releasing hormone and arginine-vasotocin on the sperm-release response of Gunther's toadlet, *Pseudophyrne guentheri*. Reprod Biol Endocrinol. 2010;8:139.

Silla AJ. Effect of priming injections of luteinizing hormone-releasing hormone on spermiation and ovulation in Gunther's toadlet, *Pseudophryne guentheri*. Reprod Biol Endocrinol. 2011;9(1):68.

Silla AJ. Artificial fertilisation in a terrestrial toadlet (*Pseudophryne guentherii*): effect of medium osmolality, sperm concentration and gamete storage. Reprod Fertil Dev. 2012;25(8):1134–41. doi:10.1071/RD12223.

Silla AJ, Roberts JD. Investigating patterns in the spermiation response of eight Australian frogs administered human chorionic gonadotropin (hCG) and luteinizing hormone-releasing hormone (LHRHa). Gen Comp Endocrinol. 2012;79(1):128–36.

Skerratt LF, Berger L, Speare R, Cashins S, McDonald KR, Phillott AD, et al. Spread of chytridiomycosis has caused the rapid global decline and extinction of frogs. Ecohealth. 2007;4:125–34.

Skerratt LF, McDonald KR, Hines HB, Berger L, Mendez D, Phillott AD, et al. Application of the survey protocol for chytridiomycosis to Queensland, Australia. Dis Aquat Organ. 2010;92:117–29.

Smith JT, Clarke IJ. Seasonal breeding as a neuroendocrine model for puberty in sheep. Mol Cell Endocrinol. 2010;324(1–2):102–9.

Sotowska-Brochocka J. The stimulatory and inhibitory role of the hypothalamus in the regulation of ovulation in grass frog, *Rana temporaria* L. Gen Comp Endocrinol. 1988;70(1):83–90.

Sotowska-Brochocka J, Licht P. Effect of infundibular lesions on GnRH and LH release in the frog, *Rana temporaria*, during hibernation. Gen Comp Endocrinol. 1992;85(1):43–54.

Sotowska-Brochocka J, Martyńska L, Licht P. Dopaminergic inhibition of gonadotropic release in hibernating frogs, *Rana temporaria*. Gen Comp Endocrinol. 1994;93(2):192–6.

Stamper DL, Licht P. Effect of gonadotropin-releasing hormone on gonadotropin biosynthesis in pituitaries of the frog, *Rana pipiens*. Biol Reprod. 1990;43(3):420–6.

Stamper DL, Licht P. In vitro exposure of frog (*Rana pipiens*) pituitaries to gonadotropin-releasing hormone results in a persistent stimulation of gonadotropin biosynthesis. J Exp Zool. 1993;265(6):646–52.

Stamper DL, Licht P. Influence of androgen on the GnRH-stimulated secretion and biosynthesis of gonadotropins in the pituitary of juvenile female bullfrogs, *Rana catesbeiana*. Gen Comp Endocrinol. 1994;93(1):93–102.

Stuart SN. Responding to the amphibian crisis: too little, too late? Alytes. 2012;29:9–12.

Stuart SN, Chanson JS, Cox NA, Young BE, Rodrigues AS, Fischman DL, et al. Status and trends of amphibian declines and extinctions worldwide. Science. 2004;306:1783–6.

Stuart SN, Hoffmann M, Chanson JS, Cox NA, Berridge RJ, Ramani P, Young BE. Threatened amphibians of the world. Barcelona, Spain: Lynx Edicions; 2008.

Subcommittee on Amphibian Standards, Committee on Standards, National Research Council, editor. Amphibians: guidelines for the breeding, care, and management of laboratory animals. Washington, DC: National Academy Press Inc; 1996. http://books.nap.edu/html/amphibian.

Sumida M. Inheritance of mitochondrial DNAs and allozymes in the female hybrid lineage of two Japanese pond frog species. Zoolog Sci. 1997;14:277–86.

Szymanski DC, Gist DH, Roth TL. Anuran gender identification by faecal steroid analysis. Zoo Biol. 2006;25:35–46.

Takahashi K, Yamanaka S. Induction of pluripotent stem cells from mouse embryonic and adult fibroblast cultures by defined factors. Cell. 2006;126:663–76.

Takahashi K, Tanabe K, Ohnuki M, Narita M, Ichisaka T, Tomoda K, et al. Induction of pluripotent stem cells from adult human fibroblasts by defined factors. Cell. 2007;131:861–72.

Tanaka S, Park MK, Hayashi H, Hanaoka Y, Wakabayashi K, Kurosumi K. Immunocytochemical localization of the subunits of glycoprotein hormones (LH, FSH, and TSH) in the bullfrog pituitary gland using monoclonal antibodies and polyclonal antiserum. Gen Comp Endocrinol. 1990;77(1):88–97.

Thorgaard G, Wheeler P, Fields R. Utilization of androgenesis for strain recovery from cryopreserved sperm. In: Tiersch TR, Green CC, editors. Cryopreservation in aquatic species. 2nd ed. Louisiana, USA: World Aquaculture Society; 2005. p. 732–7.

Toro E, Michael SF. In vitro fertilisation and artificial activation of eggs of the direct developing anuran *Eleutherodactylus coqui*. Reprod Biol Endocrinol. 2004;2:60.

Toyoda F, Hasunuma I, Yamamoto K, Yamashita M, Kikuyama S. Prolactin acts centrally to enhance newt courtship behavior. Gen Comp Endocrinol. 2005;141(2):172–7.

Trottier TM, Armstrong JB. Hormonal stimulation as an aid to artificial insemination in *Ambystoma mexicanum*. Can J Zool. 1975;53(2):171–3.

Trudeau VL. Neuroendocrine regulation of gonadotrophin II release and gonadal growth in the goldfish, *Carassius auratus*. Rev Reprod. 1997;2(1):55–68.

Trudeau VL, Spanswick D, Fraser EJ, Larivière K, Crump D, Chiu S, et al. The role of amino acid neurotransmitters in the regulation of pituitary gonadotropin release in fish. Biochem Cell Biol. 2000;78(3):241–59.

Trudeau VL, Somoza GM, Natale GS, Pauli B, Wignall J, Jackman P, Doe K, Schueler FW. Hormonal induction of spawning in 4 species of frogs by coinjection with a gonadotropin-releasing hormone agonist and a dopamine antagonist. Reprod Biol Endocrinol. 2010;8:36.

Trudeau VL, Schueler FW, Somoza GM, Natale GS, Salgado Costa C, Thoney D, McGinnity D. Hormonal control of spawning in amphibians. World Congress of Herpetology; 2012 Aug 8–12; Vancouver BC, Canada; p. 730. (abstract).

Trudeau VL, Schueler FW, Navarro-Martin L, Hamilton CK, Bulaeva E, Bennett A, Fletcher W, Taylor L. Efficient induction of spawning of Northern leopard frogs (*Lithobates pipiens*) during and outside the natural breeding season. Reprod Biol Endocrinol. 2013;11:14.

Tsai PS, Jones JT. Steroid-induced changes in the morphology of GnRH neurons in the male leopard frog, *Rana pipiens*: correlation with plasma gonadotropin and gonadal size. Gen Comp Endocrinol. 2005;141(2):152–60.

Tsai PS, Kessler AE, Jones JT, Wahr KB. Alteration of the hypothalamic-pituitary-gonadal axis in estrogen- and androgen-treated adult male leopard frog, *Rana pipiens*. Reprod Biol Endocrinol. 2005;3:2.

Uteshev V, Gakhova E. Gene cryobanks for conservation of endangered amphibian species. Russ J Herpetol. 2005;12(Suppl):233–4.

Uteshev VK, Melnikova EV, Kaurova SA, Nikitin VA, Gakhova EN, Karnaukhov VN. Fluorescent analysis of cryopreserved totipotent cells of amphibian embryos. Biofizika. 2002;47:539–45.
Uteshev VK, Shishova NV, Kaurova SA, Browne RK, Gakhova EN. Hormonal induction of spermatozoa from amphibians with *Rana temporaria* and *Bufo bufo* as anuran models. Reprod Fertil Dev. 2012;24(4):599–607.
Uteshev VK, Shishova NV, Kaurova SA, Manokhin AA, Gakhova EN. Collection and cryopreservation of hormonally induced sperm of pool frog (*Pelophylax lessonae*). Russ J Herpetol. 2013;20(2):105–9.
Vellano C, Bona A, Mazzi V, Colucci D. The effect of synthetic luteinizing hormone-releasing hormone on ovulation in the crested newt. Gen Comp Endocrinol. 1974;24:338–40.
Waggener WL, Carroll Jr EJ. A method for hormonal induction of sperm release in anurans (eight species) and *in vitro* fertilization in lepidobatrachus species. Dev Growth Differ. 1998a;40(1):19–25.
Waggener WL, Carroll Jr EJ. Spermatozoon structure and motility in the anuran *Lepidobatrachus*. Dev Growth Differ. 1998b;40:27–34.
Wakayama S, Ohta H, Hikichi T, Mizutani E, Iwaki T, et al. Production of healthy cloned mice from bodies frozen at −20 degrees C for 16 years. Proc Natl Acad Sci U S A. 2008;105: 17318–22.
Wallace RA. Studies on amphibian yolk III. A resolution of yolk platelet components. Biochim Biophys Acta. 1963;74:494–504.
Wang L, Bogerd J, Choi HS, Seong JY, Soh JM, Chun SY, et al. Three distinct types of GnRH receptor characterized in the bullfrog. Proc Natl Acad Sci U S A. 2001;98:361–6.
Weeks AR, Sgro CM, Young AG, Frankham R, Mitchell NJ, Miller KA, et al. Assessing the benefits and risks of translocations in changing environments: a genetic perspective. Evol Appl. 2011;4:709–25.
Westemeir RL, Brawn JD, Simpson SA, Esker TL, Jansen RW, Walk JW, et al. Tracking the long-term decline and recovery of an isolated population. Science. 1998;282:1695–8.
Whitaker BR. Reproduction. In: Whitaker BR, Wright KM, editors. Amphibian medicine and captive husbandry. Malabar, FL: Krieger Publishing Co; 2001. p. 285–99.
Wilczynski W, Yang EJ, Simmons D. Sex differences and hormone influences on tyrosine hydroxylase immunoreactive cells in the leopard frog. J Neurobiol. 2003;56(1):54–65.
Williams SE, Hoffman EA. Minimizing genetic adaptation in captive breeding programmes: a review. Biol Conserv. 2009;142(11):2388–400.
Wilmut I, Schnieke AE, McWhir J, Kind AJ, Campbell KH. Viable offspring derived from fetal and adult mammalian cells. Nature. 1997;385:810–3.
Wilmut I, Beaujean N, de Sousa PA, Dinnyes A, King TJ, et al. Somatic cell nuclear transfer. Nature. 2002;419:583–6.
Wolf DP, Hedrick JL. A molecular approach to fertilization: II. Viability and artificial fertilization of *Xenopus laevis* gametes. Dev Biol. 1971;25:348–59.
Xia X. Data analysis in molecular biology and evolution. Boston, MA: Kluwer Academic Publishers; 2001. p. 2001.
Xia X, Xie Z. DAMBE: software package for data analysis in molecular biology and evolution. J Hered. 2001;92:371–3.
Yamaha E, Mizuno T, Hasebe Y, Yamazaki F. Chimeric fish produced by exchanging upper parts of blastoderms in goldfish blastulae. Fish Sci. 1997;63:514–9.
Yamaji Y, Valdez DM, Seki S, Yazawa K, Urakawa C, Jin B, et al. Cryoprotectant permeability of aquaporin-3 expressed in Xenopus oocytes. Cryobiology. 2006;53:258–67.
Yasui GS, Fujimoto T, Sakao S, Yamaha E, Arai K. Production of loach (*Misgurnus anguillicaudatus*) germ-line chimera using transplantation of primordial germ cells isolated from cryopreserved blastomeres. J Anim Sci. 2011;89:2380–8.
Zhang RB, Werkman AS. Water and urea permeability properties of Xenopus oocytes: expression of mRNA from toad urinary bladder. Am J Physiol Cell Physiol. 1991;260:C26–34.
Zohar Y, Muñoz-Cueto JA, Elizur A, Kah O. Neuroendocrinology of reproduction in teleost fish. Gen Comp Endocrinol. 2010;165(3):438–55.

Chapter 13
The Reality, Use and Potential for Cryopreservation of Coral Reefs

Mary Hagedorn and Rebecca Spindler

Abstract Throughout the world coral reefs are being degraded at unprecedented rates. Locally, reefs are damaged by pollution, nutrient overload and sedimentation from out-dated land-use, fishing and mining practices. Globally, increased greenhouse gases are warming and acidifying oceans, making corals more susceptible to stress, bleaching and newly emerging diseases. The coupling of climate change impacts and local anthropogenic stressors has caused a widespread and well-recognized reef crisis. Although *in situ* conservation practices, such as the establishment and enforcement of marine protected areas, reduce these stressors and may help slow the loss of genetic diversity on reefs, the global effects of climate change will continue to cause population declines. Gamete cryopreservation has already acted as an effective insurance policy to maintain the genetic diversity of many wildlife species, but has only just begun to be explored for coral. Already we have had a great deal of success with cryopreserving sperm and larval cells from a variety of coral species. Building on this success, we have now begun to establish genetic banks using frozen samples, to help offset these threats to the Great Barrier Reef and other areas.

Keywords Coral • Reef • Cryobiology • Cryobanking • Assisted reproduction • Invertebrate

M. Hagedorn, Ph.D. (✉)
Smithsonian Conservation Biology Institute, Smithsonian Institution, Washington, DC, USA

Hawaii Institute of Marine Biology, University of Hawaii, 46-007 Lilipuna Rd, Kaneohe, HI 96744, USA
e-mail: hagedornm@si.edu

R. Spindler, Ph.D.
Taronga Conservation Society Australia, Mosman, NSW, Australia

W.V. Holt et al. (eds.), *Reproductive Sciences in Animal Conservation*, Advances in Experimental Medicine and Biology 753, DOI 10.1007/978-1-4939-0820-2_13,

1 Introduction

Coral reefs are natural superstructures created by hard coral species that provide economic, ecological and social services. Reefs as a tourism magnet, with the Great Barrier Reef of Australia drawing 1.6 million visitors each year, contributing $1.5 billion each year (CRC Reef Research Centre 2003) in marine tourism on the Great Barrier Reef (http://www.reef.crc.org.au/publications/brochures/marine%20tourism_web.pdf). Globally, the combined value of marine tourism, fishing industry, ecological services and pharmaceutical development is estimated at $375 billion each year (National Oceanic and Atmospheric Administration (2010) Heat Stress to Caribbean Corals in 2005 Worst on Record www.noaanews.noaa.gov/stories2010/20101115_coralbleaching.html).

Reefs provide invaluable ecological services, such as acting as nursery grounds for marine fish and invertebrates, providing natural storm barriers for coastlines, promoting mangroves and sea grass beds, reefs facilitate nitrogen fixation, and carbon/calcium regulation, waste assimilation and provide potential sources for undiscovered pharmaceuticals. These ecosystem services, in turn have significant social impact by providing food security, supporting recreation, supporting spiritual, and cultural practises and aesthetic values (Moberg and Folke 1999; Cesar et al. 2003).

Despite their value, reefs are being degraded at unprecedented rates. Locally, reefs are damaged by pollution, nutrient overload and sedimentation from inappropriate land-use, fishing and mining practices. Globally, increased greenhouse gases are warming and acidifying oceans, making corals more susceptible to stress, bleaching and newly emerging diseases (Hoegh-Guldberg 1999; Goreau et al. 2000; Hughes et al. 2003). The coupling of climate change impacts and local anthropogenic stressors has caused a widespread and well-recognized reef crisis (Glynn and D'Croz 1990; Glynn 1996; Hoegh-Guldberg 1999; Goreau et al. 2000; Hughes et al. 2003; Shearer et al. 2009). Urgent and effective conservation action is now required to address this widespread crisis facing coral reefs. To emphasize this point, the Chair of the IUCN's Species Survival Commission, Simon Stuart, lists corals as one of the planet's three major species extinction crises. Most importantly, the proportion of corals threatened with extinction has increased dramatically in recent decades, exceeding most terrestrial groups.

The full impact of the ecosystems and economies is still unknown. This is because we do not understand enough about the inter-relationships between reefs and other marine ecosystems. Coral may be a keystone ecosystem within the marine environment. For example, coral reefs only encompass 0.2 % of the Earth's surface, but over one quarter of all marine life lives on a coral reef at some point in their life cycle. Moreover, they are some of the oldest and most diverse ecosystems on our planet. Approximately 50 % of all the Earth's oxygen is produced in our oceans by green algae, so if reefs failed, would it affect this most important element for life on Earth? We don't really know, but it is fair to assume that the web of ocean life is closely tied together, and the reef failures around the world will impact most aspects of ocean life, such as the availability of seafood. According to Cassandra de Young of the UN

Food and Agricultural Organization, with over one billion people on the planet already hungry, the disappearance of seafood may cause great instability in food security around the world.

Of all the reefs, Caribbean reefs are suffering the most severe declines, and their fate may predict the future of the corals throughout the world. For example, *Acropora palmata* (elkhorn coral) and *Acropora cervicornis* (staghorn coral), critical Caribbean reef-building species have declined 80–99 % from their historical population levels (Bellwood et al. 2004; Bruckner 2002), resulting in a loss of structure and function of reefs throughout the Caribbean. Recent studies (Gardner et al. 2003; Buddemeier et al. 2003; Pandolfi et al. 2003) have identified similar patterns of ecosystems degradation and species loss in all the oceans, including the Great Barrier Reef. According to the Great Barrier Reef Marine Park Authority (GBRMPA), in their Great Barrier Reef Outlook Report (2009), "Given the strong management of the Great Barrier Reef, it is likely that the ecosystem will survive better … than most reef ecosystems around the world. However … the overall outlook for the Great Barrier Reef is poor and catastrophic damage to the ecosystem may not be averted. Ultimately, if changes in the world's climate become too severe, no management actions will be able to climate-proof the Great Barrier Reef ecosystem".

The solutions to coral conservation must be multivariant and incorporate a sound knowledge of the complex coral biology. Each coral individual (holobiont) consists of a complex of coral cells, algal symbionts (for many coral species) and bacteria. All of these cells play an important role in the growth, development and health of mature coral colonies. This complex relationship requires that all elements of these living systems be adequately preserved as part of a global coral conservation program. Cryopreservation of gametes and small coral fragments would achieve these goals and this is currently under investigation, as part of a multi-disciplinary recovery process.

Critical to any kind of restoration or conservation strategy for coral is a consideration of how much genetic diversity remains in our wild populations. Unfortunately, this diversity is not well-described for most coral species (Shearer et al. 2009). These authors suggest that "coral restoration strategies using 10–35 randomly selected local donor colonies would retain at least 50–90 % of the genetic diversity of the original population". But some populations, such as *Pocillopora damicornis*, that fragment easily, suffer from inbreeding (Combosch and Vollmer 2011). So, clearly conservation and restoration strategies must be carefully tailored to the populations they will serve.

Although *in situ* conservation practices, such as marine protected areas may help slow the loss of genetic diversity on reefs, the global effects of climate change will continue to cause population declines (Pandolfi et al. 2011). Cryopreservation has already acted as an effective insurance policy to maintain the genetic diversity of many wildlife species (Wolf et al. 2001; Wildt et al. 2010; Combosch and Vollmer 2011). We have had a great deal of success with cryopreserving sperm and pluripotent embryonic cells from a variety of coral species (Hagedorn et al. 2012). Building on this success, we have now begun to establish genetic banks using frozen samples, to help offset these threats to the Great Barrier Reef and other areas and have applied these techniques to eight coral species worldwide (Fig. 13.1).

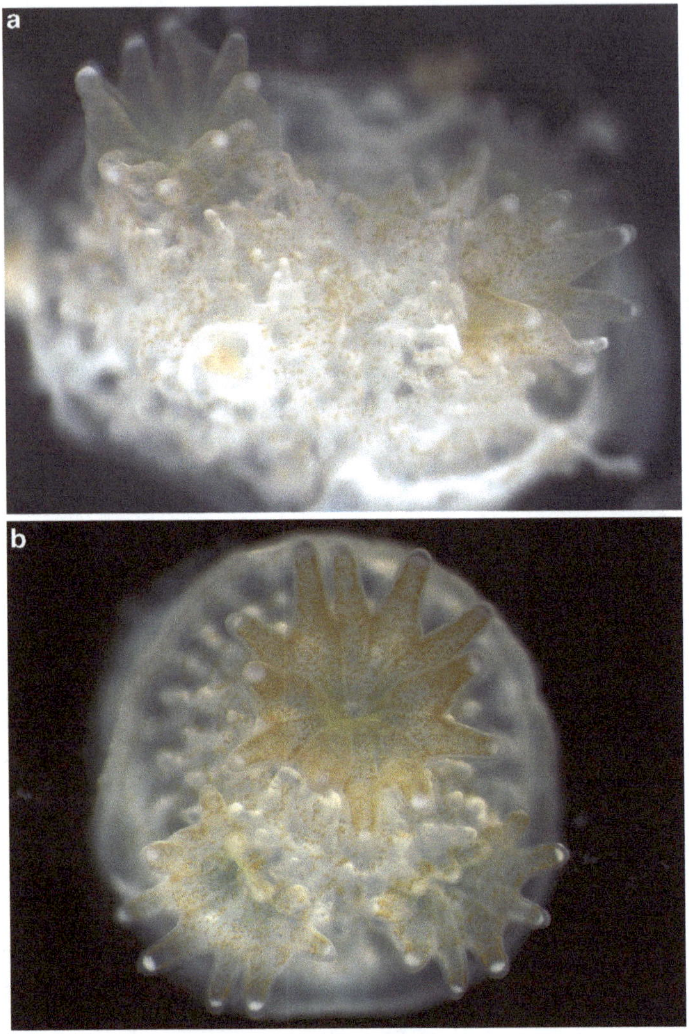

Fig. 13.1 Developing embryos. (**a**) Fresh and (**b**) cryopreserved sperm were used to produce developing coral larvae. Regardless of whether fresh or frozen sperm was used to fertilize fresh eggs, both groups developed, grew, settled and absorbed their symbiotic algae (*brown spots* in the tentacles) in a similar manner. Photo Credit: Emily Howells

Cells that are cryopreserved and banked properly can retain viability for years, or even centuries, without DNA damage. The greatest challenges facing this critical conservation effort are the time and resources to train individuals to form their own banks. However, the successful accomplishment of worldwide capacity building would create an insurance population for reefs, securing their biodiversity and helping to maintain or bolster their related economies.

2 Assisted Reproductive Technologies (ART)

2.1 *The Fundamentals of Cryobiology*

Cryopreservation (the study of cells under cold conditions) is an extremely effective conservation tool for maintaining genetic diversity. In this approach (see Box 13.1), cells are frozen in sugar-like compounds called cryoprotectants, frozen to −80 °C and placed into liquid nitrogen where they can remain frozen, but alive, for decades in a genetic bank. Most technological innovations in the field of germplasm cryopreservation arose from a sound understanding of the mechanisms of cryodamage and cryoprotection (Mazur 1970; 1984). Successful cryopreservation of cells, germplasm and tissue must address intrinsic biophysical properties (e.g., water and cryoprotectant permeability, osmotic tolerance limits, intracellular ice nucleation, etc.) to maximize survival (Rall 1993). A similar systematic approach is vital to improving post-thaw survival of coral and its associated organisms.

Conventional cryopreservation of many types of cells relies upon cryoprotectants and slow freezing to dehydrate and shrink the cell. Cryoprotectants that enter the cell, such as dimethyl sulfoxide, propylene glycol, or glycerol, are effective, yet their mechanisms of action are not completely understood. They depress the freezing point of solutions in and around the cells and may directly alter membrane bilayers or interact with bound proteins on the external cell surface (Hammerstedt et al. 1990). Too little entering the cell before cooling reduces effectiveness and may lead to damaging intracellular ice formation (Taylor et al. 1974); too much entering the cell causes osmotic swelling and rupture during thawing and dilution (Levin and Miller 1981). Often, these procedures must be tailored for each type of cell, based upon a thorough understanding of its properties.

Box 13.1 Cryopreservation Primer

1. *Slow Freezing Cryopreservation*: Uses extracellular ice to dehydrate cells, slowly dehydrating and freezing cells over minutes to hours.
   Advantage: amenable to most cells.
   Disadvantage: some cells are damaged by a slow reduction in temperature.
2. *Vitrification*: Uses high concentrations of cryoprotectants and ultrafast freezing temperatures to form a glass instead of ice.
   Advantage: good for chill sensitive cells or organisms.
   Disadvantage: solutions can be toxic, thawing must use very fast warming temperatures to prevent ice formation.
3. *Slow Vitrification*: Increases cryoprotectant concentration slowly over time to prevent ice crystal formation.
   Advantage: good for chill sensitive cells or organisms, no ice crystals form, no need for rapid thawing.
   Disadvantage: more complicated handling process.

Preventing intracellular ice formation is essential to successful cryopreservation. During slow cooling, extracellular fluid freezes before intracellular fluid, pulling pure water out of the cell, leading to osmotic dehydration of the cells as they supercool. If ~90 % of the intracellular water can be removed before lethal intracellular ice forms, then many cells will survive thawing and dilution (Mazur 1984).

However, certain cells can be damaged by the slow-freezing process because a sudden reduction in temperature can cause cold shock injury (or chilling sensitivity), often resulting in severe membrane damage. It is common in some mammalian sperm cells, such as in pigs (He et al. 2001) and aquatic oocytes, embryos and larvae (Hagedorn et al. 1997). Vitrification, whereby cell water is converted to a glass rather than undergoing a damaging phase transition to ice, may prove to be a more viable technique for aquatic cells. Vitrification entails the use of: (1) highly concentrated cryoprotectants (5–6 M), which cause dehydration before cooling; and (2) rapid cooling of the cell suspension, forming a transparent glass-state. Vitrification permits rapid cryopreservation with improved survival in some cells (Rall and Fahy 1985).

If a tissue is chill sensitive, yet too large or too sensitive to the toxic cryoprotectants used for vitrification, a "slow vitrification" method can be used (Farrant 1965; Pegg et al. 2006). Generally, cytotoxicity of the cryoprotectant decreases with temperature, because of the reduced permeability and metabolism of the cryoprotectant. During slow vitrification the concentration of cryoprotectant is slowly increased at sub-zero temperatures (instead of at room temperature for vitrification). Slow vitrification reduces toxicity and the necessity for fast cooling and thawing rates.

2.2 *Current Status of the Cryobiology of Reef Organisms*

Storage of important coral and related cells through cryopreservation will profoundly advance basic research in embryology, genetics, systematics and molecular biology, as well as enhance management strategies for reef restoration. Although cryopreservation is a proven method for long-term maintenance of genetic material, current protocols for coral and associated organisms are not fully developed, and so the associated programs that could employ these important genetic resources have not reached their full potential. An important point, however, is that once the material is frozen, a great deal of research can be done to determine how it might best be used in the future. In the past 10 years, we have characterised some of the fundamental cryobiology for coral sperm, larvae and associated symbionts (Hagedorn et al. 2006a, b, 2010, 2012).

2.2.1 Coral Sperm Cryopreservation (Successful)

The sperm from eight different coral species (Caribbean: *Acropora palamata* (threatened), Hawaii: *Fungia scutaria*, Great Barrier Reef: *Acropora millepora, Acropora tenuis, Acropora loripes, Platygyra lamolina, Platygyra daedalea, Goniastrea aspera* Fig. 13.2) has been successfully cryopreserved, using the same

Fig. 13.2 Examples of coral currently cryopreserved and stored in banks around the world. Photo Credits: *Acropora palmata*, Raphael Ritson-Williams, Smithsonian Institution; *Fungia scutaria*, Ginnie Carter, Smithsonian Institution; *Acropora cervicornis*, Eric Borneman, University of Houston; *Acropora tenuis*, Andrew Heyward, Australian Institute of Marine Sciences

standardised cryopreservation protocol and preserved in banks around the world (Hagedorn et al. 2012).

The general cryopreservation method for coral germplasm and embryonic cells has been described in detail in Hagedorn et al. (2012). Briefly, the sperm was collected and held in a concentrated form (approximately 2×10^9 cells/ml). Sub-samples were diluted either 1:10 or 1:100 in filtered seawater, counted with a hemocytometer and their motility assessed on a phase microscope approximately 30–45 min after collection. This standardized process was important because some Acroporid species only reach full motility 20–30 min after they have been released from their bundle (Hagedorn et al., unpublished data). Sperm samples with 50 % motility or greater were pooled across males and prepared for cryopreservation. The sample was diluted 1:1 with 20 % dimethyl sulfoxide in filtered seawater. Aliquots (1 ml) were loaded into 2 ml cryovials held at 26–28 °C. After a 10 min exposure to the cryoprotectant, the cells in the vials were frozen at 20 °C/min, quenched in liquid nitrogen, and then placed into a dry shipper for transport to permanent storage. A single sample from each freezing trial was thawed to examine post-thaw motility and fertilization success with fresh eggs.

2.2.2 Assessment of Sperm Viability and Use of the Frozen Bank

Frozen-thawed sperm have been used to fertilize conspecific eggs released in the same spawn and from successive spawns (Hagedorn et al. 2012). While variability remains across species and even within individuals on different nights of the same spawn these sperm have reached fertilization success of 60 % (Hagedorn et al. 2012). These small-scaled *in vitro* experiments demonstrated how to improve the cryopreservation process in developing larvae up to 12 h. In recent preliminary experiments, however, larvae produced from frozen/thawed *Acropora clathrata* sperm have developed, settled and assimilated symbionts over a 8 week period (Hagedorn et al, unpublished; Fig. 13.1). These longer-duration small-scaled experiments suggested that larger-scale grow outs would be possible to examine the effects of cryopreservation on the growth and maturity of coral over several years. In 2013, tens of thousands of *Acropora tenuis* embryos were generated with (1) sperm collected immediately after spawning; (2) this same sperm frozen for 1 h and then thawed, and; (3) sperm that been frozen for 1 year and then thawed. These coral are growing and maturing in the SeaSim facility at the Australian Institute of Marine Science. These studies will help guide future usage of these invaluable frozen resources, because one day they may be needed to help expand and diversify shrinking coral populations worldwide.

2.2.3 Coral Larvae and Oocytes (Not Yet Successful)

No coral larvae have yet been successfully cryopreserved because of their chilling sensitivity. With less than 1 min of exposure to 0 °C, 100 % of all tested *F. scutaria* larvae disintegrated (Hagedorn et al. 2006a). Oocytes have never been cryopreserved either because of chilling sensitivity (Lin et al. 2011, 2012).

2.2.4 Dissociated Coral Embryonic Cells (Successful)

Using modified embryonic stem cells protocols, dissociated larval cells were successfully cryopreserved from eight different species (Caribbean: *Acropora cervicornis* (threatened), Hawaii: *Fungia scutaria*, Great Barrier Reef: *Acropora millepora, Acropora tenuis, Acropora loripes, Platygyra lamolina, Platygyra daedalea, Goniastrea aspera*) and demonstrated 50–90 % post-thaw viability (Hagedorn et al. 2012) and Hagedorn et al. (unpublished data).

The pluripotent nature of 8-cell coral cells has been clearly defined by Heyward and Negri (2012). We have concentrated our embryo cell cryopreservation efforts on this stage of embryo to maximize the potential of the bank. To preserve embryonic cells, approximately 1 ml of 8-cell embryos was placed in a 15-ml tube with 0.1 % Bovine Serum Albumin in filtered seawater. This was diluted 1:1 with 20 % DMSO in filtered seawater. This sample was placed into a glass tissue homogenizer to create a homogenous cell suspension with a targeted cell concentration of approximately 5×10^6 cells/ml. Aliquots (1 ml) of the cell suspension were loaded into 2 ml cryovials, placed into a passive freezing device, such as the Biocision

Coolcell®, placed in a 80° freezer for at least 4 h, and then quenched in liquid nitrogen prior to being loaded into a dry-shipper for shipment to the repository. At least one sample in each group of samples was stained with the Live/Dead Viability Stain (Invitrogen) assessed on a fluorescent microscope or run on a flow cytometer to determine cell integrity post-thaw. The viability of these cells in culture, and the potential for them to develop to maturity has not been measured due to the lack of robust, well-defined culture methods for coral, but given the steady advancement of human stemcell culture, these cells have enormous future potential for conservation and coral disease work.

2.2.5 Coral Fragment Cryopreservation (Cryostudies Underway in Hawaii)

The cryopreservation of small fragments (1 cm × 0.5 cm) containing ~20 to 30 polyps would rapidly advance the *ex situ* conservation of reef species. Once thawed and placed back on the reef or simulated conditions, these small fragments quickly would become reproductive adults. Small fragments from the coral *Pocillopora damicornis* have been cryopreserved and survive to 72 h post-thaw (Hagedorn et al., unpublished data). However after this time, all of the polyps on the fragments died, most likely due to slow-acting lethal damage during the cryopreservation process and or stress. New methods must be developed to overcome this damage and preserve this important tissue.

2.2.6 Symbiodinium Cryopreservation (Cryostudies Underway in Hawaii)

The algae in the genus *Symbiodinium* (often referred to as symbionts) live within some coral cells and produce energy-rich compounds in exchange for the carbon substrates needed for photosynthesis. The cryobiology of three species of *Symbiodinium* algae has been described (Hagedorn et al. 2010). The different subtypes studied demonstrated remarkable similarities in their morphology, sensitivity to cryoprotectants and permeability characteristics; however, they differed greatly in their sensitivity to hypo- and hyperosmotic challenges and sensitivity to chilling, suggesting that standard slow freezing cryopreservation may not work well for *Symbiodinium*. Methods for vitrifying zooxanthellae using thin film technology that results in ultrarapid freezing (>15,000 °C/min) have shown promising results (Hagedorn and Carter, in prep.).

2.2.7 Crustose Coralline Algae (Cryostudies Underway in Hawaii)

Certain species of crustose coralline, such as *Hydrolithon* spp. and *Titanoderma* spp. (Heyward and Negri 1999; Harrington et al. 2004; Ritson-Williams et al. 2009), and components from their associated bacterial films have demonstrated effects on promoting coral larval settlement (Tebben et al. 2011). Experiments focused on the basic cryobiology are underway for multispecies communities of crustose coralline algae (Hagedorn and Carter, unpublished data). Crustose coralline algae are important reef components, but little is known about their cryobiology. Using multispecies-communities, studies are underway in our laboratory to understand their sensitivity to chilling temperatures and to cryoprotectants.

3 The Future

Today, there are three main sites in the world where coral cells are stored long-term, in the U.S. at the Smithsonian Institution at the Hawaii Institute of Marine Biology, the U.S. Department of Agriculture's Animal Germplasm Program and in Australia at the Taronga Western Plains Zoo. Due to the high recruitment rate that is curtailed in the wild by predation, wave motion, lack of available settling sites etc., the billions of cells we have banked might represent only a modest number of individuals. Restoring all of these cells to the source reef would result in a limited gene pool for future selective agents to act upon and possibly result in a less adaptive reef post restoration. Clearly, our cryobanks must be expanded to include more species and more individuals banked within each species, but for now, future use of this frozen material, will more than likely, include three separate but complementary streams:

1. A proportion of cells (approximately 10 %) remain banked for future generations, cloistered in a long-term repository for large scale restoration.
2. A larger proportion (approximately 30 %) could be used to effect local restoration efforts in response to a specific event such as dredging, disease, silting etc. The threat must be mitigated before cells will be thawed and the resources allocated to the growth and care of developing coral colonies. If local threats can be mitigated, there is great hope that we can restore the function of these much needed global drivers.
3. The remaining cells will be used to advance our understanding of biology (i.e., in systematics, genetic, development and disease and pharmaceutical explorations). For example, these cells will provide source material for innovative work by our colleagues on the exploration, selection and nurturing of coral most likely to maintain resilience of the system in warmer and more acidic conditions.

The potential outcomes of these genome resource banks would: (1) preserve gene diversity; (2) prevent extinctions; (3) store the entire genome, including as yet unknown but critically valuable epigenetic factors; (4) create opportunities for diversifying shrinking populations by avoiding natural losses in heterozygosity due to genetic drift; and, (5) advance the science of coral biology. In particular, coral developmental biology which is typically limited to a period of 3 days a year during the spawn event.

In order for the coral cryobank to fulfill its potential, there is an immediate need to capture a comprehensive representation of genetic diversity in each selected species. Our progress to date has been promising and ultimately, this is easily accomplished as only 35 coral adults are needed to maintain 50–90 % of the allelic diversity in the population (Shearer et al. 2009). Further, the bank must be expanded in terms of coral morphological diversity, and their functional roles and geographic locations on the reef. We will include priority species (as defined by the worlds coral experts) to ensure ecosystem function and system resilience are maximised by the species maintained in the bank.

In addition to the biological benefits that the coral cryobanks would support, the banks may provide cultural benefits, as well. For example, both Western Australia and the Great Barrier Reef encompass World Heritage Sites important to the history and culture of the Aboriginal and Torres Strait Islander groups. Some of these sites, such as Shark Bay in Western Australia have evidence of continual occupation by Aboriginal groups for over 22,000 years. The loss of these biomes to the social and cultural fabric within Australia would be incalculable. Traditional local knowledge must be incorporated in decision-making and prioritization of corals, as well as understanding reef function and cultural roles.

Looking to the future, cryobanks have the potential to contribute to many aspects of fundamental and applied coral science, but may also provide an avenue for local organizations to expand coral nurseries to include sexually reproduced corals. This opportunity provides expanded income for these groups and maintains the skills required for rapid generation of many corals for local re-seeding of reefs. The potential of these banks is therefore relevant across the spectrum of economic, cultural and ecological realms. Overall, the early success in coral cryo banking is promising and should provide some avenues to further research and ways to protect the reefs of the world. Both research and conservation are fundamentally impacted by the level of cooperation within present and future collaborations and the now rapidly growing interest in the value and potential uses of the cells in the cryo banks.

References

Bellwood DR, Hughes TP, Folke C, Nystrom M. Confronting the coral reef crisis. Nature. 2004;429:827–33.

Bruckner AW. Proceedings of the Caribbean *Acropora* workshop: potential application of the U.S. endangered species act as a conservation strategy. Silver Spring, MD: NOAA Tech Memo NMFS-OPR-24; 2002. p. 1–23.

Buddemeier RW, Ware JR, Gardner TA, Côté IM, Gill JA, Grant A, Watkinson AR. Coral reef decline in the Caribbean. Science. 2003;302:391–3.

Cesar H, Burke L, Pet-Soede L. The economics of worldwide coral reef degradation. Arnhem, The Netherlands: Cesar Environmental Economics Consulting; 2003. p. 1–23.

Combosch DJ, Vollmer SV. Population genetics of an ecosystem-defining reef coral *Pocillopora damicornis* in the Tropical Eastern Pacific. PLoS One. 2011;6(8):e21200.

Farrant J. Mechanism of cell damage during freezing and thawing and its prevention. Nature. 1965;205:1284–7.

Gardner TA, Côté IM, Gill JA, Grant A, Watkinson AR. Long-term region-wide declines in Caribbean corals. Science. 2003;301:958–60.

Glynn PW, D'Croz L. Experimental evidence for high temperature stress as the cause of El Niño-coincident coral mortality. Coral Reefs. 1990;8:181–91.

Glynn PW. Coral reef bleaching: facts, hypotheses and implications. Glob Change Biol. 1996;2(6):495–509.

Goreau TJ, McClanahan T, Hayes RL, Strong AE. Conservation of coral reefs after the1998 global bleaching event. Conserv Biol. 2000;14:5–15.

Great Barrier Reef Marine Park Great Barrier Reef outlook report. http://www.gbrmpa.gov.au/outlook-for-the-reef/great-barrier-reef-outlook-report. (2009).

Hagedorn M, Kleinhans FW, Wildt DE, Rall WF. Chill sensitivity and cryoprotectant permeability of dechorionated zebrafish embryos, *Brachydanio rerio*. Cryobiology. 1997;34(3):251–63.

Hagedorn M, Carter VL, Steyn RA, Krupp D, Leong J, Lang RP, Tiersch TR. Preliminary studies of sperm cryopreservation in the mushroom coral, *Fungia scutaria*. Cryobiology. 2006a;52(3):454–8.

Hagedorn M, Pan R, Cox E, Hollingsworth L, Lewis TD, Leong JC, Krupp DA, Mazur P, Rall WF, MacFarlane D, Fahy G, Kleinhans FW. Coral larvae conservation: physiology and reproduction. Cryobiology. 2006b;52:33–47.

Hagedorn M, Carter VL, Leong JC, Kleinhans FW. Physiology and cryosensitivity of coral symbiotic algae (*Symbiodinium sp.*). Cryobiology. 2010;60:147–58.

Hagedorn M, Carter VL, Martorana K, Paresa MK, Acker J, Baums IB, Borneman EH, Brittsan M, Byers M, Henley M, Laterveer M, Leong JC, McCarthy M, Meyers S, Nelson BD, Petersen D, Tiersch T, Cuevas Uribe R, Wildt D. Preserving and using germplasm and dissociated embryonic cells for conserving Caribbean and Pacific coral. PLoS One. 2012;7(3):e33354. doi:10.1371/journal.pone.0033354.

Hammerstedt RH, Graham JK, Nolan JP. Cryopreservation of mammalian sperm: what we ask them to survive. J Androl. 1990;111(1):73–88.

Harrington LM, Fabricius K, Dea'th G, Negri AP. Habitat selection of settlement substrata determines post-settlement survival in corals. Ecology. 2004;85:3428–37.

He L, Bailey JL, Buhr MM. Incorporating lipids into boar sperm decreases chilling sensitivity but not capacitation potential. Biol Reprod. 2001;64(1):69–79.

Heyward AJ, Negri AP. Natural inducers for coral larval metamorphosis. Coral Reefs. 1999;18:273–9.

Heyward AJ, Negri AP. Turbulence, cleavage, and the naked embryo: a case for coral clones. Science. 2012;335(6072):1064.

Hoegh-Guldberg O. Climate change, coral bleaching and the future of the world's coral reefs. Mar Freshw Res. 1999;50:839–66.

Hughes TP, Baird AH, Bellwood DR, Card M, Connolly SR, Folke C, Grosberg R, Hoegh-Guldberg O, Jackson JB, Kleypas J, Lough JM, Marshall P, Nyström M, Palumbi SR, Pandolfi JM, Rosen B, Roughgarden J. Climate change, human impacts, and the resilience of coral reefs. Science. 2003;301(5635):929–33.

Lin C, Zhang T, Kuo FW, Tsai S. Gorgonian coral (*Junceella juncea* and *Junceella fragilis*) oocyte chilling sensitivity in the context of adenosine triphosphate response (ATP). Cryo Letters. 2011;32(2):141–7.

Lin C, Wang L-H, Fan T-Y, Kuo F-W. Lipid content and composition during the oocyte development of two gorgonian coral species in relation to low temperature preservation. PLoS One. 2012;7(7):e38689.

Levin RL, Miller TW. An optimum method for the introduction or removal of permeable cryoprotectants: isolated cells. Cryobiology. 1981;18:32–48.

Mazur P. Cryobiology: the freezing of biological systems. Science. 1970;168(934):939–49.

Mazur P. Freezing of living cells: mechanisms and implications. Am J Physiol. 1984;247 (3 Pt 1):C125–42.

Moberg F, Folke C. Ecological goods and services of coral reef ecosystems. Ecol Econ. 1999;29:215–33.

Pandolfi JM, Bradbury RH, Sala E, Hughes TP, Bjorndal KA, Cooke RG, McArdle D, McClenachan L, Newman MJ, Paredes G, Warner RR, Jackson JB. Global trajectories of the long-term decline of coral reef ecosystems. Science. 2003;301(5635):955–8.

Pandolfi JM, Connolly SR, Marshall DJ, Cohen AL. Projecting coral reef futures under global warming and ocean acidification. Science. 2011;22:418–22.

Pegg DE, Wang L, Vaughan D. Cryopreservation of articular cartilage. Part 3: the liquidus-tracking method. Cryobiology. 2006;52(3):360–8.

Rall WF. Advances in the cryopreservation of embryos and prospects for the application to the conservation of salmonid fishes. In: Thorgaard GH, Cloud JG, editors. Genetic conservation of salmonid fishes. New York, NY: Plenum Press; 1993. p. 137–58.

Rall WF, Fahy GM. Ice-free cryopreservation of mouse embryos at −196 °C by vitrification. Nature. 1985;313:573–5.

Ritson-Williams R, Arnold SN, Fogarty ND, Steneck RS, Vermeij MJA, Paul VJ. New perspectives on ecological mechanisms affecting coral recruitment on reefs. Smithson Contrib Mar Sci. 2009;38:437–57.

Taylor R, Adams GD, Boardman CF, Wallis RG. Cryoprotection—permeant vs. nonpermeant additives. Cryobiology. 1974;11(5):430–8.

Tebben J, Tapiolas DM, Motti CA, Abrego D, Negri AP, Blackall LL, Steinberg PD, Harder T. Induction of larval metamorphosis of the coral Acropora millepora by tetrabromopyrrole isolated from a Pseudoalteromonas bacterium. PLoS One. 2011;6(4):e19082.

Shearer TL, Porto I, Zubillaga AL. Restoration of coral populations in light of genetic diversity estimates. Coral Reefs. 2009;28(3):727–33.

Wildt DE, Comizzoli P, Pukazhenthi B, Songsasen N. Lessons from biodiversity—the value of nontraditional species to advance reproductive science, conservation, and human health. Mol Reprod Dev. 2010;77(397–409):2010.

Wolf KN, Wildt DE, Vargas A, Marinari PE, Ottinger MA, Howard JG. Reproductive inefficiency in male black-footed ferrets (*Mustela nigripes*). Zoo Biol. 2001;19:517–28.

Chapter 14
Recent Advances and Prospects in Germplasm Preservation of Rare and Endangered Species

Pierre Comizzoli and William V. Holt

Abstract Fertility preservation strategies using cryopreservation have enormous potential for helping sustain and protect rare and endangered species, especially to assist managing or 'rescuing' the genomes of genetically valuable individuals. However, wide-scale applications are still limited by significant physiological variations among species and a sheer lack of fundamental knowledge about basic reproductive traits as well as in germplasm cryobiology. Cryo-studies have been conducted in more species (mainly vertebrates) in the recent years but a vast majority still remains un-studied. Semen cryopreservation represents the most extensive effort with live births reported in more and more species after artificial insemination. Oocyte freezing remains challenging and unsuccessful in wild species and will require more research before becoming a standard procedure. As an alternative to fully grown gametes, gonadal tissue preservation has become a promising option in vertebrates. Yet, more fertility preservation options are necessary to save species so a change in strategy might be required. It is worthwhile thinking beyond systematic characterizations and considering the application of cutting edge approaches to universally preserve the fertility of a vast array of species.

Keywords Cryopreservation • Spermatozoa • Oocytes • Embryos • Testis • Ovary

---

P. Comizzoli (✉)
Smithsonian Conservation Biology Institute, National Zoological Park,
3001 Connecticut Avenue NW, Washington, DC 20008, USA
e-mail: comizzolip@si.edu

W.V. Holt, Ph.D.
Academic Department of Reproductive and Developmental Medicine,
University of Sheffield, Jessop Wing, Tree Root Walk, Sheffield S10 2SF, UK
e-mail: Bill.holt@sheffield.ac.uk

W.V. Holt et al. (eds.), *Reproductive Sciences in Animal Conservation*,
Advances in Experimental Medicine and Biology 753, DOI 10.1007/978-1-4939-0820-2_14,

1 Introduction

Reproduction is essential to the continuation and evolution of life but what is actually known about animal reproduction is relatively limited; especially when considering how wild species amazingly self-perpetuate (Wildt et al. 2010). Species-specificities in reproductive form and function have long been recognized, but have been vastly undervalued. While being of fascinating scholarly interest, such data have practical use in helping better manage endangered species and rare genotypes, in some cases to avoid extinction (Wildt et al. 2003, 2010). It is well established that the successful application of assisted reproductive techniques such as artificial insemination (AI), in vitro fertilization (IVF) or embryo transfer (ET) for enhancing propagation is directly related to the amount of basic reproductive information available from each species. This is why, of course, assisted reproductive technologies have become so well incorporated into certain domestic animal breeding and human infertility programs—because so much is known about the whole organism, cellular and molecular biology of livestock and people (Comizzoli et al. 2010; Wildt et al. 2010; Sunderam et al. 2012). Despite much effort over the last 50 years, the routine use of reproductive technologies as a way of supporting conservation breeding programs for endangered species is still largely unachievable, in contrast to the situation with domestic cattle. Almost the entire global dairy industry now depends on the routine use of frozen semen samples for AI, where the breeder can choose the most desirable genetic traits. Similarly, the physiological coincidence that allows pig spermatozoa to survive for 1–2 weeks without freezing means that pig breeders can be supplied with the semen of their choice from males throughout an entire country (Johnson et al. 2000). Obviously, when a calf or piglet is born after AI, the event does not make news headlines because this is the expected outcome. The situation is completely different when a zoo species is born after AI. Typically, the event merits a news article or a TV interview, especially if the species involved is large and charismatic, such as a giant panda (*Ailuropoda melanoleuca*) or an Asian elephant (*Elephas maximus*). Developing assisted reproduction methods for wild species to the point where they become successful but unremarkable is, however, rarely undertaken. Developing appropriate techniques and gaining experience is easier with wild species that are not rare and endangered and where it is possible to establish studies with statistically meaningful group sizes. Current trends suggest, however, that those species that are not yet endangered may one day be threatened with population decline and extinction. If none of the appropriate research has ever been undertaken, applying reproductive technologies in support of species survival will not be possible when they are needed. More studies in wildlife therefore are needed (1) to identify the wondrous ways of how diverse animals naturally reproduce and (2) how such fundamental information can be applied to enhance population and genetic management, including by assisted reproduction (Wildt et al. 2010).

The need for comparative/systematic approaches extends to cryopreservation studies of gametes, embryos and gonadal tissues as well as stem cells (Comizzoli et al. 2012). The practical benefits of the freezing, storing and thawing of such biomaterials are well established for improving breeding efficiency in livestock, sustaining specific laboratory animal genotypes and for addressing certain subpar fertility conditions in humans (Mazur et al. 2008). Less well known is that these same cryo-strategies have enormous implications for developing sustainable populations of rare species and genotypes (Wildt et al. 1997; Holt et al. 2003). The benefits for wildlife include preserving genetic vigor, transporting valuable genes without the stress/expense of moving sensitive, fractious animals and 'insuring' all existing genetic diversity that protects fitness and species integrity (Wildt et al. 1997; Lermen et al. 2009). Yet, most of the specific details that will optimize cryo-storage of biomaterials from people, livestock and laboratory animals sometimes have marginal relevance to wildlife species (Lermen et al. 2009; Comizzoli et al. 2012). This largely is due to remarkable variations in germplasm structure and function across species, especially elements that regulate tolerance to osmotic and toxic effects of cryoprotectants as well as resistance to chilling injuries (Gilmore et al. 1998; Critser et al. 2002; Woods et al. 2004). Although most of the focus has been on the preservation of spermatozoa and oocytes (to a lesser extent), embryo cryo-preservation has not been widely developed during the past 10 years because of persisting limitations in IVF success and embryo transfer (Saragusty 2012). Live births have been reported after the transfer of frozen-thawed embryos in the European polecat (*Mustela putorius*; Lindeberg et al. 2003), sika deer (*Cervus nippon* ssp.; Locatelli et al. 2012), or ocelot (*Leopardus pardalis*; Swanson 2012); however, these encouraging results are still anecdotal as there is no routine use of ET to manage captive populations. On the other hand, research activities have increasingly been oriented towards freeze-storage of ovarian and testicular tissues that may be useful for rescuing early developmental stages of gametes in mammals (Jewgenow et al. 2011; Honaramooz 2012) as well as in birds (Liu et al. 2013a). These efforts have been inspired by the development of new fertility preservation approaches in human medicine for protecting reproductive potential of cancer patients who may lose fertility due to chemical or radiation treatments (Waimey et al. 2013). Specifically, most research has been conducted on the cryopreservation and the *in vitro* culture of gonadal tissues that represent invaluable sources of gametes, especially for prepubertal patients (Waimey et al. 2013). Interestingly, reproductive biologists studying domestic and wild species now benefit from these advances in human infertility treatments (Lermen et al. 2009; Comizzoli et al. 2010; Wiedemann et al. 2013).

There are many reasons for including more options to extend fertility potential in conservation breeding programs, largely for animals that have not yet produced sufficient numbers of descendants to ensure the passing on of their genes. The specific targets include individuals that (1) are living but fail to reproduce naturally, (2) unexpectedly die, (3) are nearing reproductive senescence or (4) have been long-dead, but there is value in rescuing and re-infusing their genome into the modern population. Collective efforts also have been enhanced by significant development of Genome Resource Banks, which are organized repositories of biomaterials to be used for managing both heterozygosity and conducting basic as well as applied research (Wildt et al. 1997; Lermen et al. 2009).The concept of biodiverse *frozen and living* repositories is no longer futuristic, but rather a contemporary collection and use strategy to better address conservation challenges. There now are many projects all around the world such as Frozen Zoo at the San Diego Zoo, the Frozen Ark, or the Pan-Smithsonian Cryo-Initiative involving hundreds of thousands of cryopreserved germplasm, embryo, blood product, tissue, DNA and fecal/urine samples (see Chap. 16 by Mastromonaco et al.). These biomaterials are archived and proactively managed in ways that increase collection and the diffusion of knowledge as well as help sustain genetically diverse, sustainable populations of rare species and genotypes. In addition to a DNA bank, or a traditional museum collection, researchers interested in topics from evolution to infectious diseases can access, through genome resource banks, the entire complement of cellular machinery (with any accompanying pathogens or microbes), for studies of proteins, RNA, mitochondrial DNA, and epigenetics. This does not only address conservation issues, but also unique and invaluable needs for scholarly investigations. Even though many zoos and institutions have started to gather biomaterials, there is still a need for good professional training, education, communication and outreach. The emerging concept of biobanking science in wildlife conservation (associated with environmental monitoring and repositories) still has to be widely spread as it enables us to preserve samples from wild or captive populations and allows the establishment of linkages between the *in situ* and the *ex situ* efforts (storing and transferring genes between captive and wild populations for a better sustainability).

The objective of this chapter is to review the progress made over the last decade and to discuss what strategies will be soon relevant in germplasm preservation and biobanking for rare and endangered animal species.

2 A Steady Progress in Semen Cryopreservation and Banking

The idea of integrating semen freezing and banking into conservation projects is not new. The earliest records of semen freezing can be traced back to the Italian scientist Lazzaro Spallanzani (1776) who, among many other research activities, conducted experiments with frog, human and canine spermatozoa and also described how he carried out a successful artificial insemination of a bitch. These pioneering activities were described in more detail in a wide ranging review (Watson 1990) of artificial insemination and semen freezing, mainly in domestic animals. More recently a comprehensive review about comparative semen freezing technologies across different vertebrate groups was published in book form in 2001 (Watson and Holt 2001), with chapters devoted to several groups of mammals, e.g. lagomorphs (Holt et al. 2001), artiodactyla (Holt 2001), marsupials (Johnston and Holt 2001), but also some non-mammalian groups such as fish (Billard and Zhang 2001) and birds (Wishart 2001). This book represented one output of a European Union-funded Concerted Action project on genetic resource conservation, in which a series of meetings was held between 1994 and 1996 to discuss not only technical aspects of germplasm cryobiology, but also organizational aspects of establishing cryobank, including how to prioritize species, how to organize samples and also how to consider the potential disease risks associated with cryobanks. In a review of germplasm cryopreservation published almost a decade ago, Pickard and Holt (2004) presented a literature search comparing the number of mammalian species appearing in the 2000 edition of the IUCN Red list with the number for which successful inseminations with fresh and frozen semen had been reported. Only four carnivores, four primates and 14 ungulates belonged to this category, compared with a total of 35 species which had been successfully inseminated with fresh semen. A more recent review (Fickel et al. 2007) also provided a table of about 50 (mainly) wild species in which sperm cryopreservation had been studied, but only 11 of these had been translated into successful artificial inseminations. Births in new species after AI with frozen semen have increased only slightly over the past years (see below). Also, when these reports are examined more closely, it is clear that they mostly represent small insemination trials, and few present evidence that the frozen semen is suitable for reliable/routine use to support conservation breeding programs. The black-footed ferret (*Mustela nigripes*) is a notable exception here (Howard et al. 2003; Howard and Wildt 2009), a situation in which semen that was originally frozen more than two decades ago now represents a working genetic bank. The intensive efforts in China to breed giant pandas using artificial insemination have met with some success, but as most of the breeding attempts involved the combined use of natural mating and artificial insemination, with both fresh and frozen semen, the value of the cryopreserved semen is rather difficult to estimate (Huang et al. 2012a, b). Nevertheless, it is clear that banked collections of giant panda semen have been established and are now used. Thus, to some extent they can now reasonably be regarded as constituting working genetic resource banks for the giant panda.

Although semen is relatively simple to recover from many species, so much more is to be learned about taxon-inherent seminal traits and sensitivity of spermatozoa to freeze-thawing. While it appears obvious that small species usually produce minute ejaculate volumes (e.g., 10–50 μl for a black-footed ferret; Santymire et al. 2006) and gigantic animals produce prodigious volumes (e.g., >100 ml for the African elephant, *Loxodonta africana*; Kiso et al. 2011), it is well-established that sperm concentration and total sperm output are unrelated to body mass (Comizzoli et al. 2012). For sperm processing, seminal plasma osmolarity and pH dictate the composition of the required seminal extenders as well as dilution processes to retain sperm structure and function during freezing, storage and thawing (Rossato et al. 2002). While generally seminal fluid osmolarity remains slightly higher (350 mOsmol/l) than that of conspecific serum (~300 mOsmol/l; Comizzoli et al. 2012), there are notable exceptions; for example, the value in black-footed ferret semen can reach 790 mOsmol/l (Santymire et al. 2006). By contrast and based on evaluation of hundreds of species, pH remains the least variable metric, generally remaining near neutral or only slightly alkaline (Comizzoli et al. 2012).

Initial quality of the recovered spermatozoa influences the subsequent ability of these cells to endure freezing and thawing stress. A useful example is the condition of teratospermia, the production of >40 % malformed spermatozoa per ejaculate that is common to certain (but not all) species in the Felidae family (Pukazhenthi et al. 2002). These cells start out with a disadvantage by being challenged not only in form, but also in function, and rarely can withstand freeze-thawing or even cooling to 5 °C (Pukazhenthi et al. 2002).

While it is relatively easy to define sperm structure, there are few data on membrane biophysical properties, even in common domestic and laboratory species. Yet this information is what allows understanding species-specific osmotic tolerances and permeability that ultimately allow formulating science-based protocols for cellular freezing and thawing (Leibo and Songsasen 2002; Gosalvez et al. 2011). In the absence of specific biophysical data, the approach for developing sperm cryomethods has been largely empirical, that is, adapting a satisfactory, 'standard' protocol for the bull, ram, pig or horse (Leibo and Songsasen 2002) to the species of interest. In many cases, a single cryoprotectant can be widely applicable. For example, the use of glycerol has allowed sperm recovery post-thawing in diverse species and at similar volume-to-volume concentrations (4–8 %), ranging from various felid species (Crosier et al. 2006; Stoops et al. 2007) to marine mammals (Robeck and O'Brien 2004; Robeck et al. 2011) to the Asian elephant (Saragusty et al. 2009; Thongtip et al. 2009). More recently, Przewalski horse (*Equus ferus przewalskii*) (Pukazhenthi et al. 2010), Baird's tapir (*Tapirus bairdii*) (Pukazhenthi et al. 2011) and Indian rhinoceros (*Rhinoceros unicornis*) (Stoops et al. 2010) spermatozoa all have been found to respond well to cryo-dilution and freezing protocols originally developed for the domestic horse (all members of Perissodactyla). Thus, over the last decade, more individuals from more species and taxa have been studied, collected, and banked including wolves, primates, equids, tapirs, marine mammals as mentioned above but also other taxa such as corals (Chap. 13 by Hagedorn et al.) and amphibians (Chap. 12 by Clulow et al.). Live births after AI with frozen-thawed semen have been reported in a few new species only (for instance, gerenuk, *Litocranius walleri walleri*, Penfold et al. 2005; Pallas cat, *Otocolobus manul*, Swanson 2006; killer whale, *Orcinus orca*, Robeck et al. 2011; Persian onager, *Equus hemionus onager*, Schook et al. 2013).

Among the mammals, marsupial spermatozoa present interesting but frustrating differences from their eutherian counterparts. Semen cryopreservation research in marsupials (Molinia and Rodger 1996) commenced in the 1990s and, despite the best efforts of several groups of experienced cryobiologists (for review, see Johnston and Holt 2001), the successful insemination of marsupials with frozen semen has remained elusive. This is a pity because conservation management plans need all the input they can muster. For example, the number of macropod species (kangaroos and wallabies) in Australasia currently stands at fifty, and a report published by the World Wildlife Fund in 2011 (Roache 2011) listed twenty one (18 %) as experiencing some level of threat. Unfortunately the current lack of success with semen freezing precludes the inclusion of genetic resource banks and assisted reproductive techniques as working components of the National action plan for macropod conservation (Roache 2011). This report estimated a need for AU $290 million to be spent on macropod conservation over 10 years in order to mitigate their threat level. A family of diverse problems is associated with marsupial sperm cryopreservation; while motility immediately after thawing cryopreserved koala (*Phascolarctos cinereus*) spermatozoa may be relatively high, around 40–50 %, (Zee et al. 2008), plasma membrane integrity decreases drastically after thawing because the sperm heads swell to several times their original volume (Johnston et al. 2012). The reason for this swelling effect has so far resisted all attempts to explain it, but the problem is partly due to the unusual configuration of koala chromatin. Like almost all other marsupials, koala sperm protamine 1 does not contain cysteine residues and is thus precluded from forming chromatin-stabilizing disulphide bonds. At the same time DNA in koala sperm heads naturally contains many single strand breaks, thus further reducing the stability of koala sperm chromatin under adverse conditions (Zee et al. 2009).A considerable variety of approaches to improving the successful cryopreservation of koala spermatozoa have now been investigated, but so far all have failed. As explained in greater detail within the Chap. 9 by Johnston and Holt, the koala AI technique is now highly successful when used with fresh semen, and can be viewed seriously as an option for genetic management. In contrast, cryopreservation of macropodid spermatozoa suffers a different, but equally serious, problem. Extensive studies have revealed that unless cryoprotectant concentration is unusually high (e.g. >15 % v/v glycerol), post-thaw motility is almost never observed in macropod spermatozoa when they are viewed at 35 °C (body temperature). However, if the post-thaw samples are viewed at temperatures below about 20 °C, the post-thaw motility can be as high as 70 % (Holt et al. 2000). This effect is caused by a remarkably rapid destabilization of the plasma membrane when the temperature increases above a narrow threshold, typically around 22 °C. In this case the high glycerol concentration action appears to be twofold: it apparently protects the spermatozoa during cooling and freezing, but induces extensive damage after thawing. Exploration of alternative cryoprotectants showed that dimethylacetamide (DMA) mitigated this damage to some extent (McClean et al. 2008) but not to the extent required for use with artificial insemination.

The comparatively high cryoprotectant concentrations described above, especially with DMA, appear to commonly benefit bird semen, although there are amazing variations among species. Particularly illustrative studies have been conducted by Blanco et al. (2008, 2011) who compared sperm osmotic tolerance among domestic and wild birds. Sandhill crane (*Grus canadensis*) spermatozoa remain viable at 3,000 mOsm/l, whereas turkey spermatozoa are damaged after exposure to 500 mOsm/l. Imperial eagle (*Aquila adalberti*) and Peregrine falcon (*Falco peregrinus*) spermatozoa have higher osmotic tolerance at ~800 mOsm/l than those of poultry (fowl and turkey) and Golden eagle (*Aquila chrysaetos*) and Bonelli's eagle (*Aquila fasciata*). Thus, in this case, species results are not aligned according to expectations for the 'fowl' versus 'birds-of-prey' categories, but unexpectedly to the more distant relatives. Although there are no studies about sperm membrane biophysical properties in birds (except data in the fowl showing clear differences with bull spermatozoa; Watson et al. 1992), variations in cryo-tolerances among species (even among 17 pheasant species; Saint Jalme et al. 2003) likely emanate from differing membrane biophysical properties (Blanco et al. 2011).

3 Progress in Oocyte Cryopreservation Is Still Limited

Oocytes are remarkably different from sperm cells in cryo-sensitivity properties and requirements (Songsasen and Comizzoli 2009). Because the size of a round mammalian oocyte (generally ~120 μm in diameter) is larger than a spermatozoon, there is a smaller surface-to-volume ratio and a correspondingly higher sensitivity to chilling and intracellular ice formation (Songsasen and Comizzoli 2009). The naturally fragile cytoskeleton of eggs also lessens the resistance to volumetric changes (Saragusty and Arav 2011). Adding to the overall challenge is the thick, protective and all encompassing zona pellucida as well as the oocyte's plasma membrane, which has a low permeability coefficient in mammals that impedes or prevents movement of cryoprotectant and water (Songsasen and Comizzoli 2009). Generally, oocytes also have a high water and cytoplasmic lipid content that increases chilling sensitivity. For example, prodigious amounts of lipid exist in the oocytes of canids and felids (Wildt et al. 2010) as well as in bovids (McEvoy et al. 2000) and pigs (Sturmey and Leese 2003).

Oocyte morphology and biophysical characteristics often vary with species that, in turn, can influence cryo-sensitivity and the laboratory protocols needed to successfully store the cell (Comizzoli et al. 2012). Additionally, it is now clearly demonstrated that felid intraovarian oocytes are more highly tolerant to cold temperatures and osmotic changes (Wolfe and Wildt 1996; Pope et al. 2006; Comizzoli et al. 2008, 2010) than counterparts from bovids, cervids or equids that are cold shock sensitive (Comizzoli et al. 2012). Within a given species, there is a growing number of 'markers' suggestive of which oocytes are more likely to survive a freeze-thaw stress (e.g., those having a cytoplasmic homogeneity or a sufficient number of encompassing cumulus cells; Songsasen and Comizzoli 2009). Nonetheless, it still is challenging to predict survivability to low temperatures on the basis of any known oocyte trait, even from common, domesticated species.

Despite a lot of efforts and new discoveries over the past 10 years in human and domestic mammals (Saragusty and Arav 2011), there are still no reports about the successful cryopreservation of oocytes (followed by fertilization, embryo development, and pregnancy) in any wild species. Studies of oocytes and their cold tolerance in wild species are still limited due to cell availability and the difficulty of collecting them in comparison to semen samples (Leibo and Songsasen 2002; Saragusty 2012). And, despite many advances in domestic species (mouse, nonhuman primates and livestock species), optimal oocyte cryopreservation still requires fundamental/biophysical studies (e.g., as done in the rhesus monkey, *Macaca mulatta*; Songsasen et al. 2002) and detailed, comparative studies (Critser et al. 2002). Interestingly, it has been confirmed in carnivores that immature oocytes at the germinal vesicle stage are more cryo-resistant than counterparts at metaphase II, because the former cells do not contain a temperature-sensitive meiotic spindle (Comizzoli et al. 2004, 2008; Songsasen and Comizzoli 2009). There also has been reasonable progress in examining new oocyte cryopreservation methods, beyond the conventional slow-cooling to using ultra-rapid protocols (i.e., vitrification on electron microscope grids and cryo-loops) (Saragusty and Arav 2011). Among the most non-traditional animal models, *in vivo* matured oocytes were successfully vitrified in the domestic cat and embryos obtained after sperm injection developed into healthy offspring (Pope et al. 2012). Regarding the vitrification of cat immature oocytes, several years of basic studies (Comizzoli et al. 2004, 2008) have proved that *in vitro* maturation (IVM) and IVF was possible even though pregnancies were not going to term after ET (Tharasanit et al. 2011). There also is preliminary evidence that vitrification (using high concentrations of ethylene glycol, DMSO and sucrose) of oocytes from the Tasmanian devil (*Sarcophilus harrisii*) (Czarny and Rodger 2010) and Mexican grey wolf (*Canis lupus baileyi*) (Boutelle et al. 2011) can lead to cell survival after warming, although developmental competence has yet to be explored in these species.

Interestingly, the oocytes of amphibian and aquatic species share many of the same complexities as those of terrestrial species. Although these cells are up to 25 times larger than their mammalian counterparts, the presence of (and reliance on) the yolk compartment whose membrane is impermeable to water and cryoprotectant is a limiting factor (Isayeva et al. 2004; Kouba and Vance 2009). As a result, fish oocytes are extremely chill sensitive. Moreover, their large size and small surface area to volume ratio reduce permeability to water and cryoprotectants, thus creating permissive conditions for detrimental ice formation at freezing temperatures (Isayeva et al. 2004). However, as for mammalian oocytes, immature zebrafish (*Danio rerio*) oocytes have a better tolerance to cold temperatures (Seki et al. 2011). Regardless, there have been few oocyte freezing studies involving aquatic invertebrate species, the exceptions being successful cryopreservation of Pacific oyster (*Crassostrea gigas*) (Salinas-Flores et al. 2008) and greenshell mussel (*Perna canaliculus*) (Adams et al. 2009) oocytes.

4 The Preservation of Gonadal Tissue Holds a Lot of Promises

The ovary and testis have a wealth of untapped, arrested, or developing germ cells, most of which never participate in fertilization. The ability to preserve the gonadal tissues and artificially mature early stage oocytes or spermatozoa in culture or by xenografting could provide unlimited germplasm to generate embryos, including from animals that are prepubertal, outside their breeding season, nearing the end of their reproductive lifespan or that die unexpectedly.

However, due to the complexity of gonadal tissue structure, cell heterogeneity and the lack of basic information, there are substantial challenges ahead, including in simply conducting studies on osmotic tolerance, toxicity and chilling sensitivity. Interestingly, more than 35 human babies have been born after grafting of ovarian tissues stored in liquid nitrogen for a long period of time. It will probably take several years of intensive research and development to obtain the first live birth in a wild species using the same approach.

Table 14.1 Specificities in testicular tissue anatomy and cryopreservation methods

Species	Anatomy and pretreatment	Optimal preservation methods
Black-footed ferret (*Mustela nigripes*)	Highly compacted seminiferous tubules 1 h enzymatic digestion	Slow freezing in straws
Cheetah (*Acinonyx jubatus*) Clouded leopard (*Neofelis nebulosa*) Fishing cat (*Prionailurus viverrinus*) Siberian tiger (*Panthera tigris altaica*)	Seminiferous tubules can be easily separated 30 min enzymatic digestion	Rapid freezing in straws
Dama gazelle (*Nanger dama*)	Highly compacted seminiferous tubules 1 h enzymatic digestion	Slow freezing in cryovials

Few studies have been directed at preserving and *in vitro* culturing testicular tissues to produce fully formed spermatozoa that have the capacity to fertilize. Investigations in wild species are being pursued, especially given recent encouraging data from Sato et al. (2011) who demonstrated that mature mouse sperm cells can be produced *in vitro*. Certainly, a next high priority for wild species is to determine the mechanisms related to acquisition of motility and centrosomal maturation in testicular spermatozoa grown *in vitro*, phenomena not yet well understood for any species. As an alternate to *in vitro* culture, the use of xenografting is another option that has been explored. Fresh testis tissue from the common ferret (*Mustela putorius furo*) has been xenografted into the body of immunodeficient mice and then produced mature spermatozoa from the original donor (Gourdon and Travis 2011). While having theoretical relevance to other carnivores, the challenge can be the normally abbreviated life-span of the rodent host (much shorter than for other carnivores) and the protracted (>35 weeks) duration required for gamete maturation from the tissue grafts. Similar observations and limitations have been reported about bison testis xenografting (Honaramooz 2012). Regarding their preservation, testicular pieces (0.5–1.0 mm^3) of human and laboratory species have been cryopreserved successfully by equilibrating them in glycerol or DMSO at room temperature and then transferring into cryovials that are cooled in a programmable unit (Ehmcke and Schlatt 2008). Similar approaches have been successful with goat and bison (*Bison bison*) testicular tissues (Honaramooz 2012). Tissue survival usually is judged on the basis of observing favorable post-thaw histology or by measuring resumption of gametogenesis after grafting. Although there has been some success in the above species, there appears to be significant species variation in tissue cryosensitivity. Tissue structure characteristics (varying from species to species) are exerting an important influence on the pretreatment and the cryopreservation method (Table 14.1) but no real differences have been observed between prepubertal and adult testicular tissues. Specifically, we have explored issues for carnivore and ungulate testes that range from the importance of transport temperature of freshly-excised tissue to the laboratory to the need for seminiferous tubule isolation using collagenase and hyaluronidase to the value of a closed vitrification systems (i.e., tissue sealed into a plastic straws to avoid the contact with liquid nitrogen or vitrification in dry-shipper containers; Comizzoli and Wildt 2012a). We also have found that felid testicular tissue better survives vitrification (based on structural and functional assays; Comizzoli et al. 2010) than in laboratory rodents. In addition, we have demonstrated the feasibility of using collagenase and hyaluronidase to isolate living cat and dog seminiferous tubules for preservation and culture (Comizzoli and Wildt 2012a). Their post-thawing/warming viability is routinely examined after 2 days of *in vitro* culture. Intact histological structures, >65 % of cell viability, and <10 % of apoptosis are good indicators of a correct tissue reanimation after warming. Interestingly, recent results in birds have demonstrated that testicular tissue of Japanese quail (*Coturnix japonica*) can be preserved using vitrification procedures and recovered through transplantation (Liu et al. 2013b).

Based on success with cryopreservation of reproductive tissues from mice and non-human primates used as model species over the past 10 years (Comizzoli et al. 2010), encouraging progress is being made towards producing viable antral ovarian follicles, especially for the cat and the dog, a topic addressed in detail by Songsasen et al. (2011, 2012).Working with partners at other zoological institutions, ovaries are being shipped at 4 °C to our core laboratory, processed by cutting them into 1–2 mm^3 pieces, equilibrated in cryoprotectant and then preserved by comparing of methods. So far, we have clearly demonstrated the value of vitrification (e.g., in 15 % ethylene glycol + 15 % DMSO + 0.5 M sucrose) compared to slow freezing methods for primordial follicles enclosed in the ovarian cortex from prepubertal and adult felids and several ungulate species (Comizzoli et al. 2010, 2012). Optimal techniques now are being used to routinely bank ovarian tissue from various species, including the black-footed ferret, cheetah (*Acinonyx jubatus*), clouded leopard (*Neofelis nebulosa*), Eld's deer (*Rucervus eldii*), scimitar-horned oryx (*Oryx dammah*), tufted deer (*Elaphodus cephalophus*) and Przewalski horse (Comizzoli et al. 2010, 2012). As demonstrated for testicular tissue, preservation of small ovarian biopsies is preferred over the whole gonad to increase the number of frozen samples available from a given individual. The post-warming viability is assessed using different criteria after 2 days of *in vitro* culture (retention of tissue structure and cell communications, >50 % of the cell viability, proliferation of the granulosa as well as the stroma cells, <10 % apoptosis). As mentioned for male gonads, tissue structure characteristics are influencing the pre-treatment and the cryopreservation method (Table 14.2) but no differences between prepubertal (high follicle density) and adult ovarian tissues can usually be observed. We have consistently observed that germ cells tend to be more cryoresistant than the somatic/stroma cells. Also, the viability and the proliferation of the stroma cells surrounding the early follicles are considered as a critical viability indicator for the subsequent follicular growth. High survival of slow frozen-thawed ovarian tissue from some felids or marsupial species (wombat; *Vombatus ursinus*) also has been demonstrated on the basis of cell integrity after culture and grafting success (Jewgenow et al. 2011; Wiedemann et al. 2013; Paris et al. 2004; Cleary et al. 2004). Of course, the ability to grow these early stage follicles and their oocytes *in vitro* to achieve full maturation and fertilization would be more convenient but largely remain unknown, even for common livestock and laboratory species. This requires examining a host of micro-environmental

Table 14.2 Specificities in ovarian tissue anatomy and cryopreservation methods

Species	Anatomy	Optimal preservation methods
Black footed-ferret (*Mustela nigripes*)	Hard tissue	Rapid freezing of whole ovary in cryovial
Cheetah (*Acinonyx jubatus*) Clouded leopard (*Neofelis nebulosa*) Florida panther (*Puma concolor*) Sumatran Tiger (*Panthera tigris sumatrae*)	Soft tissue	Rapid freezing of 1–2 mm biopsies in cryovials
Maned wolf (*Chrysocyon brachyurus*)	Hard tissue	Rapid freezing of 1–2 mm biopsies in cryovials

factors from the hormonal support and oxygen concentration needed to the ability to eliminate wastes in follicular culture systems that will require up to 6 months to produce viable oocytes (Songsasen et al. 2011, 2012).

The preservation of gonadal tissue also is relevant for non-mammal species as indicated by a few studies. For species with eggs containing yolk, there has been interest in cryopreserving primordial germ cells as demonstrated in the rainbow trout (*Oncorhynchus mykiss*). These cells can be frozen in 1.8 M ethylene glycol, thawed and transplanted into the peritoneal cavity of allogenic trout hatchlings where they differentiated into mature spermatozoa and eggs having the genetic constitution of the original donor (Kobayashi et al. 2007). Production of donor-derived offspring also has been reported after transplantation of vitrified ovarian tissue in Japanese quail (*Coturnix japonica*) using the same vitrification solutions as in mammalian species (15 % ethylene glycol + 15 % DMSO + 0.5 M sucrose) (Liu et al. 2010). In that set of studies, vitrification again appears to have been more successful than slow-freezing for bird ovarian tissues. Transplantation of whole frozen ovaries as a means of capturing important genetic quality has been undertaken successfully with silkworms (Mochida et al. 2003; Banno et al. 2013). These authors estimated that there are about 2,000 strains of silkworms in Japan and that these are currently maintained as live cultures. The earlier report used a cryopreservation protocol, which was based on the use of 1.5 M DMSO and suspending the vials containing ovaries in liquid nitrogen vapor for 30 min, resulted in about 22 % of transplanted moths producing fertilized eggs. The more recent study used a slower cooling rate (1 °C/min to −80 °C followed by plunging in liquid nitrogen) and reported improved results (about 70 % of transplanted moths laid fertilized eggs).

5 Emerging Preservation Approaches to Address Current Limitations

5.1 Novel Sources of Germplasm

Stem cell technologies are promising as methods for producing gametes from embryonic stem cells, spermatogonial progenitors, or from differentiated cells. Characterization, isolation, and transfer of spermatogonial stem cells have been attempted in the cat and dog with mixed results (Travis et al. 2009). In brief, this has involved isolating the spermatogonial stem cells followed by transfer into a germ-cell depleted (via radiation) host. On occasion, it has been possible to recover ~20 % of mature sperm cells derived from the donor (Travis et al. 2009). Others have transplanted germ cells from a wild felid (ocelot) into the domestic cat to produce spermatozoa successfully from the donor (Silva et al. 2012). Recent studies on ovarian stem cells in various mammal species could also hold some promises for the production of gametes from endangered species (Dunlop et al. 2013).

The induced Pluripotent Stem (iPS) cell concept also is timely because of the possibility to optimize the use of Genome Resource Banks that include cultured somatic cells like fibroblasts (see Chap. 16 by Mastromonaco et al.). The iPS cells (in appropriate culture conditions) could provide a self-renewing, inexhaustible resource of material from wildlife species (Ben-Nun et al. 2011). Eventually, it will probably be more efficient to differentiate embryonic stem cells or iPS cells *in vitro* for this purpose, the latter being accomplished recently for the snow leopard (*Panthera uncia*) (Verma et al. 2012). The striking potential of these strategies also has been demonstrated in the mouse where *in vitro*-differentiated embryonic stem cells have given rise to sperm-like cells (Nayernia et al. 2006) or oocyte-like cells derived from newborn mouse skin (Dyce et al. 2011).

The limited success achieved with preserving fish oocytes and embryos has stimulated some novel approaches to the problem. One of the most interesting current strategies involves the vitrification of whole embryos at the 22–28 somite stage; this does not result in live post-thaw embryos but the primordial germ cells (PGC) survive the procedure. The PGCs, which are green fluorescent protein-labeled prior to vitrification, are dissected out of the embryos and transplanted into host blastulae from the same or a different species. When the host embryo develops the theoretical outcome is that it will produce gametes that are genetically derived from the original vitrified embryo (Fig. 14.1). One example of this procedure involved transferring common carp (*Cyprinus carpio*) PGCs into goldfish (*Carassius auratus*) embryos (Kawakami et al. 2012). A variant of this method involved isolating and vitrifying loach (*Misgurnus anguillicaudatus*) PGCs, labeling them with a fluorescent protein produced in zebrafish, and transplanting them into fresh embryos (Yasui et al. 2011; Inoue et al. 2012). The transplanted PGCs retained their ability to migrate within the embryo and colonize the genital ridge, an important outcome showing that this method has a realistic chance of resulting in normal sexual differentiation and gamete production. A different and equally ingenious approach to the problem of genome preservation in fish is based on the recovery of Type A spermatogonia (ASG) from slowly frozen rainbow trout (*Oncorhynchus mykiss*) testes (Lee et al. 2013) and their subsequent transfer into the peritoneal cavity of sterile triploid hatchlings of the same species. In this study nearly half of the triploid recipients produced functional eggs or spermatozoa derived from the frozen ASGs. Fertilization of these gametes resulted in the successful production of normal, frozen ASG-derived offspring. From the standpoint of genome conservation one of the most interesting aspects of this study was that the ASGs were derived from testes that had been kept frozen for up to 939 days without significant loss of ASG viability or lowered performance of the derived gametes. The authors state that the isolation and

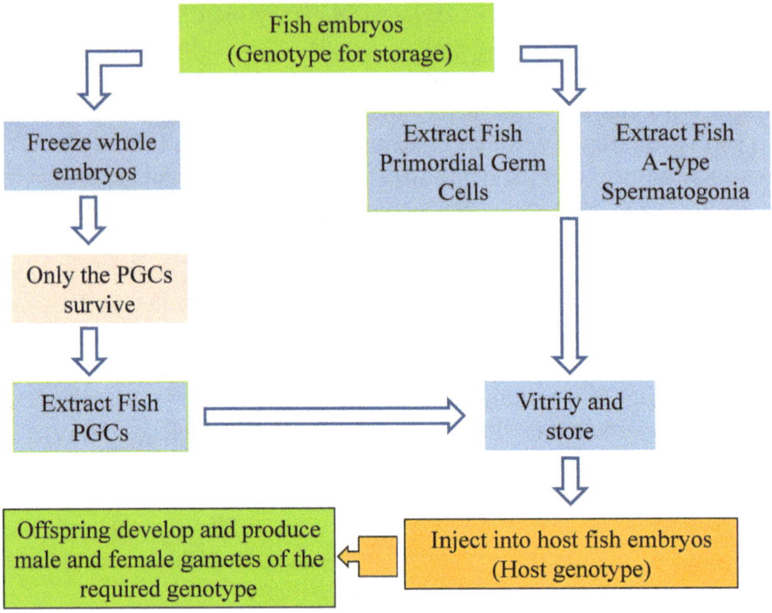

Fig. 14.1 Novel approaches in germplasm production and preservation in fish

transfer of ASGs is very straightforward and practical; however, the ASGs in this study had been labeled with green fluorescent protein prior to isolation and identification of live ASGs was therefore relatively easy. If this method is to be useful for endangered fish species a new approach to the identification of ASGs will be required (Fig. 14.1).

The interest in PGC transplantation combined with the current topicality of research into epigenetics has raised awareness that freezing and thawing might, *per se*, induce epigenetic changes in the materials being preserved. In fact, evidence that this might be the case has recently been published by Riesco and Robles (2013) who studied the cryopreservation of zebrafish genital ridges; these are regarded as useful candidates for genome banking because they contain PGCs. The study detected the cryopreservation-induced downregulation of several zebrafish mRNA transcripts, and the upregulation of two heat shock proteins, Hsp70 and Hsp90. Epigenetic gene downregulation is normally attributed to hypermethylation of gene promoter regions, and the authors found that one specific promoter was indeed hypermethylated as a result of cryopreservation. This study confirmed what had been suspected for several years, namely that cryopreservation methods might have unforeseen epigenetic outcomes that could lead to altered phenotypes and disease. In fact, cryopreservation is not alone in being implicated in the induction of epigenetic changes. Recent concern has been expressed about the negative epigenetic impacts of culture media used for human embryos during assisted reproduction techniques (Nelissen et al. 2004). Similar effects, later attributed to the presence of growth factors in media containing bovine serum albumin (BSA) (Thompson et al. 1998) led to the birth of overweight calves following IVF and embryo transfer, but these effects were eliminated by the expedient of treating the BSA with charcoal. Subtle effects associated with assisted reproductive technologies can occur whenever gametes and embryos are being manipulated (for review, see Yamauchi et al. 2011) and it is clear that this will be an important research area in the future.

5.2 Alternative Preservation Methods at Low Temperatures

Directional freezing is based on a simple thermodynamic principle in which ice crystals are precisely controlled through the sample by regulating the velocity of the sample movement through the predetermined temperature gradient. Directional freezing permits a precise and uniform cooling rate in both small and large volume samples. Directional freezing has been used for slow and rapid freezing, as well as for vitrification of oocytes and embryos using the minimum drop size technique. Sperm samples from a wide range of domestic and wild animals have been successfully cryopreserved using the directional freezing method. The method also has enabled, for the first time, successful freezing of a whole ovary and freeze-drying of mammalian cells followed by thawing and transplantation and rehydration, respectively (Arav and Natan 2012).Post-thaw results showed that within the same sperm type in marine mammals, directional freezing was superior to conventional techniques for maintaining motility and in immediate post-thaw viabilities (O'Brien and Robeck 2006). Another comparative study was carried out on spermatozoa from killer whales, contrasting conventional freezing (using straws) and directional freezing (using hollow tubes) (Robeck et al. 2011). The post-thaw results showed that directional freezing also was superior to conventional freezing in all parameters of motility and viability in gazelle (*Gazella gazelle*) semen (Saragusty et al. 2006), European brown hare (*Lepus europaeus*, Hildebrandt et al. 2009), or rhinoceros (*Ceratotherium simum simum*, Hermes et al. 2009).

Ultra-rapid freezing and vitrification still are regarded by many as 'novel', despite Rall and Fahy's pioneering report more than 25 years ago on its usefulness for preserving mouse embryos (Rall and Fahy 1985). We continue to be enthusiastic about vitrification because of its relative simplicity, low cost and 'field-friendliness' (i.e., the ability to vitrify biomaterials, even in harsh, remote environments using only a liquid nitrogen dry shipper). Interestingly, compared to studies of embryos or oocytes, there continues to be few reports on the efficacy of ultra-rapid freezing, or vitrification, of spermatozoa. However, encouraging results recently have been reported for human (Isachenko et al. 2011) and dog (Kim et al. 2012) spermatozoa. In the latter case, the gametes were exposed to 5 % glycerol and freezing vapors for <1 min before plunging into liquid nitrogen; >50 % of spermatozoa were motile after thawing. Despite its great potential, this approach has not yet been explored in wild species.

Besides the freezing methods, the addition of heat shock protein HSPA8 to cryoprotectant media can improve the survival of spermatozoa post-thawing as demonstrated in the brown bear (*Ursus arctos*) (Alvarez-Rodriguez, et al. 2013).

There also is a trend towards freeze-preservation in lower cryoprotectant concentrations, which appears especially important for dog spermatozoa and cat oocytes that are susceptible to cryoprotectant toxicity (Comizzoli et al. 2012). Another alternative explored in our laboratory is the effectiveness of vitrification solutions that rely only on non-permeating (non-toxic) cryoprotectants, including natural sugars such as sucrose or trehalose. In the case of cat oocytes, we have observed >80 % survival after vitrification in saturated trehalose solutions (Comizzoli and Wildt 2012a).

5.3 *Novel Preservation Methods at Supra-zero Temperatures*

Although isolated cells and tissues can be successfully vitrified and warmed without detrimental formation of ice crystals, the challenge remains that low temperature storage can trigger injury to DNA, membranes, and cell junctions (Yavin and Arav 2007). Dehydration by air-, evaporative-, and vacuum-drying, followed by storage at room temperature is therefore an appealing option for preserving germplasm, because desiccation is similar to natural approaches used by certain small organisms to suspend their life cycle. For instance, tardigrades are protostomal animals well known for their capability of surviving extreme conditions by undergoing anhydrobiosis at ambient temperature for extended periods (Crowe et al. 2002). This phenomenon is possible due to an innate ability to accumulate natural sugars (including the disaccharide trehalose) intracellularly to preserve membrane lipid bilayers and proteins (Crowe et al. 2002, 2005). For this reason, our laboratory is exploring alternative opportunities for preserving carnivore spermatozoa via desiccation in trehalose and storage at supra-zero temperature (Comizzoli and Wildt 2012a). The advantages of both include much simpler sample processing and transport of genetic material and, most impressively, no need for liquid nitrogen. In theory, the latter could markedly reduce the costs and complexity of biomaterials storage. However, results also have revealed a significant loss in sperm motility and the potential of compromised centrosomal function post-rehydration. For example, we observed poor sperm aster formation after injecting dehydrated (at ambient temperature in trehalose) cat spermatozoa into conspecific oocytes (Comizzoli and Wildt 2012a). Centrosomal dysfunction post-freeze drying has been less apparent for nonhuman primate and bull spermatozoa, although rhesus monkey spermatozoa desiccated in trehalose are known to lose fertilizing capacity (Comizzoli and Wildt 2012b). Results from preliminary studies of freeze-drying canine spermatozoa have revealed pronuclear formation after injection into mouse oocytes (Watanabe et al. 2009). Also encouraging are the recent findings of Ringleb et al. (2011) who found early (albeit limited) embryo development after injecting freeze-dried cat spermatozoa into conspecific oocytes. Finally, it is worth noting that desiccation likely has excellent potential for preserving the maternal genome. For example, our laboratory has determined that GVs from cat oocytes that are artificially compacted (with histone deacetylation enhancers), air-dried, and then rehydrated, can resume meiosis after injection into a fresh (enucleated) cytoplast (Graves-Herring et al. 2013). Thus, it may be possible to use this GV rescue approach to salvage the maternal genome from individuals who die early or late in life before reproducing, or who are experiencing cytoplasmic deficits in the oocyte or follicular anomalies. It also has been determined that the chromatin of the cat GV withstands artificial compaction for subsequent injection to reconstruct a viable oocyte, all without encountering the need for the usual complex membrane electrofusion (Graves-Herring et al. 2013). This approach, never reported for other species, may well evolve into a simple, inexpensive, and biologically viable means of storing the female genome (without the cytoplasm) of carnivores as well as other taxa (Holt 2013).

Preservation in a liquid environment at supra-zero or ambient temperatures also is an emerging area in cell or tissue preservation. As an alternate to cat sperm storage in classical extenders at cold temperature (Pope et al. 2006), we have effectively preserved cooled (4 °C) cat spermatozoa for up to 2 weeks in a 2 M trehalose solution while retaining DNA integrity and centrosomal structure (presence of centrin) as well as function (sperm aster formation) (Comizzoli and Wildt, 2012b). Recently, encouraging results also have been obtained in porcine oocytes that were able to retain developmental competence after storage for several days at ambient temperature (Yang et al. 2010). We now are exploring the same strategy for gonadal tissues.

It appears that harmonization of protocols at supra-zero temperatures would be easier since there are no issues related to species-specific cryo-sensitivity. However, in the search for these new and simple preservation protocols, it is critical to thoroughly verify the integrity of the DNA sequence and the multiple epigenetic factors regulating the functionality of the genome. It is expected that newly available tools such as Next Generation Sequencing and other 'omics' in association with bioinformatics will help to accurately control the quality of the germplasm preserved with these new approaches.

5.4 Bridges with Human Fertility Preservation

More interactions between human and animal cryobiology are needed to optimize fertility preservation (Comizzoli et al. 2010). Indeed, the next steps in human and animal fertility preservation are dictated by similar needs for (1) more options in case of complex fertility issues (2) minimal cost, field-friendly methods when lack of resources and limited access to freezing equipment or liquid nitrogen and (3) customized/universal solutions since variations in cryo-sensitivity within animal populations are similar to humans. We argue that human reproductive specialists also could well take advantage of new fundamental knowledge on biological insights from studies of far-from-traditional animal species. Translational fertility preservation could be ensured by promoting more interaction among stakeholders in all areas—whether human, livestock, laboratory animal or wildlife-oriented. For example, there could be significant benefits from the establishment of a fertility preservation network, with benefits ranging from active communication for sharing critical (or simply interesting) information to opportunities for direct collaboration.

As already identified in human reproductive medicine, critical needs for the future regardless of our ability to find 'more-forgiving' technologies are to have wider application across taxonomic groups too. However, there will always be the need for basic, species specific data. For example, we still are going to need to know when an animal cycles, when it ovulates, how many sperm cells it produces, etc. So the shift may come in the timing of when the fundamental information is required. It may become less important *before* a sample is preserved—but it will likely still be crucial to when the sample is used to produce an offspring (from gamete reconstruction or micro-injection to *in vitro* culture to preparation of a female for AI or ET).

6 Conclusions

Most contemporary germplasm preservation research conducted has been concentrated on the cryopreservation of spermatozoa in diverse wild species. There are few studies on oocytes and gonadal tissues (Fig. 14.2). The overall goal needs to be creating the ability to preserve any germplasm from a valuable animal of any age or reproductive state using reasonably simple, cost-effective techniques. Clearly, there are still vast needs in basic cryobiological studies for diverse species, especially fundamental biophysical traits as well as comparative evaluations of permeating

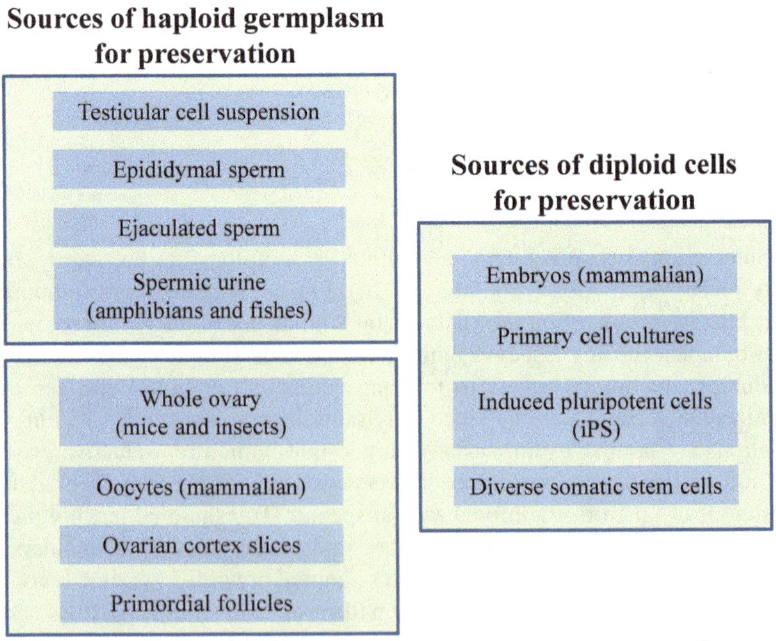

Fig. 14.2 Sources of germplasm, stem cells, and somatic cells for genome preservation

(e.g., DMSO, DMA, propylene glycol) versus non-permeating (e.g. raffinose, trehalose) cryoprotectants as well as freeze/thaw rates. In some cases, progress can be enhanced markedly by relying on data already available in domesticated counterparts (e.g., cattle for antelope, dog for wild canids and cat for wild felids). However, even this approach can fail to overcome within-family physiological variations among species (Wildt et al. 2010). And relying on a domesticated counterpart does not assist the thousands of absolutely unique species (e.g., elephant, killer whale, bats, monotremes, seahorses, Monarch butterfly, among hundreds of others) for which no closely related research model exists. Inter-species (mammals vs. non-mammals, carnivores vs. ungulates) as well as inter-individual variations or sometimes commonalities also can be remarkable. More cryobiological studies will also help to develop customized treatments for some individuals or lead to universal freezing methods valid for a vast array of species. Progress can only be made if we continue to explore basic gamete and gonad biology to identify appropriate quality criteria before and after freezing. There also is a need to be more expansive in our thinking about the priorities in cryobiology and reproductive science, and this certainly includes species that are *not* mammals. The future is exciting because such efforts will continue to demonstrate the amazing reproductive diversity that already exists as well as produce knowledge that actually will be practical—helping to sustain viable populations and even avoid extinctions. Unlike as little as 20 years ago, such statements are no longer hyperbole. The recovery of species as diverse as the whooping crane, black-footed ferret and giant panda owe at least partial credit to modern reproductive science and tools in fertility preservation and assisted breeding.

There is no doubt that the usual methods of preserving wildlife germplasm will continue, with studies involving the conventional cooling, freezing, and storage approaches that rely on liquid nitrogen. But as demonstrated with encouraging data presented here, we assert that it is time to break away from customary practices and to explore novel and likely more cost-effective strategies. We are especially excited about mining the germplasm within the gonads, that is, the premature stage spermatozoa and oocytes that represent an enormous reserve of genetic material normally never used for actual reproduction. In this arena, we believe the priorities should include exploring the developmental potential of early gamete stages, developing *in vitro* culture systems to secure more mature stages, and determining how stem cells can be converted into gametes that can produce viable embryos. In terms of preserving fertility, we believe there is great promise in the simplified storage of genomes without the intricacies and expense associated with liquid nitrogen. Therefore, it seems prudent to invest more research into the areas of desiccation and biostabilization at ambient temperatures associated with optimal reanimation conditions to bring the samples back to life. It also always is wise to monitor the literature for new information on yet-to-be discovered storage and reanimation phenomena that normally are found in nature. For example, how can we take the knowledge that bat and bee spermatozoa remain viable in the female reproductive tract for months (Wildt et al. 2010) and transform it into laboratory techniques to preserve germplasm? Lastly, it is essential that our ability to preserve viable germplasm, embryos, and the entire genome short- and long-term does not surpass our capacity to use it to produce viable young. Therefore, a continued priority for the wildlife science community is to advance assisted reproductive technologies, including developing more consistent artificial insemination, ovulation induction, ovarian cycle synchronization, and embryo transfer protocols. Simultaneous progress with all of these tools will allow the improved production and management of genetically valuable companion animals, models to understand human diseases, and rare wild species. Lastly, biobanking strategies must be coordinated with progresses in preservation. The management and access of preserved germplasm as well as the appending sample data is critical for future use.

References

Adams SL, Tervit HR, McGowan LT, Smith JF, Roberts RD, Salinas-Flores L, Gale SL, Webb SC, Mullen SF, Critser JK. Towards cryopreservation of Greenshell mussel (*Perna canaliculus*) oocytes. Cryobiology. 2009;58:69–74.

Alvarez-Rodriguez M, Alvarez M, Borragan S, Martinez-Pastor F, Holt WV, Fazeli A, de Paz P, Anel L. The addition of heat shock protein HSPA8 to cryoprotective media improves the survival of brown bear (*Ursus arctos*) spermatozoa during chilling and after cryopreservation. Theriogenology. 2013;79:541–50.

Arav A, Natan D. Directional freezing of reproductive cells and organs. Reprod Domest Anim. 2012;47 Suppl 4:193–6.

Banno Y, Nagasaki K, Tsukada M, Minohara Y, Banno J, Nishikawa K, Yamamoto K, Tamura K, Fujii T. Development of a method for long-term preservation of Bombyx mori silkworm strains using frozen ovaries. Cryobiology. 2013;66:283–7.

Ben-Nun IF, Montague SC, Houck ML, Tran HT, Garitaonandia I, Leonardo TR, Wang YC, Charter SJ, Laurent LC, Ryder OA, Loring JF. Induced pluripotent stem cells from highly endangered species. Nat Meth. 2011;8:829–31.

Billard R, Zhang TT. Techniques of genetic resource banking in fish. In: Watson PF, Holt WV, editors. Cryobanking the genetic resource: wildlife conservation for the future? London: Taylor and Francis; 2001. p. 145–70.

Blanco JM, Long JA, Gee G, Donoghue AM, Wildt DE. Osmotic tolerance of avian spermatozoa: influence of time, temperature, cryoprotectant and membrane ion pump function on sperm viability. Cryobiology. 2008;56:8–14.

Blanco JM, Long JA, Gee G, Wildt DE, Donoghue AM. Comparative cryopreservation of avian spermatozoa: benefits of non-permeating osmoprotectants and ATP on turkey. Anim Reprod Sci. 2011;123:242–8.

Boutelle S, Lenahan K, Krisher R, Bauman KL, Asa CS, Silber S. Vitrification of oocytes from endangered Mexican gray wolves (*Canis lupus baileyi*). Theriogenology. 2011;75:647–54.

Cleary M, Shaw JM, Jenkin G, Trounson AO. Influence of hormone environment and donor age on cryopreserved common wombat (*Vombatus ursinus*) ovarian tissue xenografted into nude mice. Reprod Fertil Dev. 2004;16:699–707.

Comizzoli P, Wildt DE. On the horizon for fertility preservation in domestic and wild carnivores. Reprod Domest Anim. 2012a;47 Suppl 6:261–5.

Comizzoli P, Wildt DE. Centrosomal functions and dysfunctions in cat spermatozoa. Centrosome research. In: Schatten H, editor. The centrosome. Totowa, NJ: Humana; 2012b. p. 49–58.

Comizzoli P, Wildt DE, Pukazhenthi BS. Effect of 1,2-propanediol versus 1,2-ethanediol on subsequent oocyte maturation, spindle integrity, fertilization and embryo development *in vitro* in the domestic cat. Biol Reprod. 2004;71:598–604.

Comizzoli P, Wildt DE, Pukazhenthi BS. Impact of anisosmotic conditions on structural and functional integrity of cumulus-oocyte complexes at the germinal vesicle stage in the domestic cat. Mol Reprod Dev. 2008;75:345–54.

Comizzoli P, Songsasen N, Wildt DE. Protecting and extending fertility for females of wild and endangered mammals. Cancer Treat Res. 2010;156:87–100.

Comizzoli P, Songsasen N, Hagedorn M, Wildt DE. Comparative cryobiological traits and requirements for gametes and gonadal tissues collected from wildlife species. Theriogenology. 2012;78:1666–81.

Critser JK, Agca Y, Woods E. Cryopreservation of immature and mature gametes. In: De Jonge CJ, Barratt CLR, editors. Assisted reproductive technology: accomplishment and new horizons. Cambridge: Cambridge University Press; 2002. p. 144–66.

Crosier AE, Pukazhenthi BS, Henghali JN, Howard J, Dickman AJ, Marker L, Wildt DE. Cryopreservation of spermatozoa from wild-born Namibian cheetahs (*Acinonyx jubatus*) and influence of glycerol on cryosurvival. Cryobiology. 2006;5:169–81.

Crowe JH, Oliver AE, Tablin F. Is there a single biochemical adaptation to anydrobiosis? Integ Comp Biol. 2002;42:497–503.

Crowe JH, Crowe LM, Wolkers WF, Oliver AE, Ma X, Auh J-H, Tang M, Zhu S, Norris J, Tablin F. Stabilization of dry mammalian cells: lesions from nature. Integ Comp Biol. 2005;45: 810–20.

Czarny NA, Rodger JC. Vitrification as a method for genome resource banking oocytes from the endangered Tasmanian devil (*Sarcophilus harrisii*). Cryobiology. 2010;60:322–5.

Dunlop CE, Telfer EE, Anderson RA. Ovarian stem cells-potential roles in infertility treatment and fertility preservation. Maturitas. (2013). pii: S0378-5122(13)00128-X. doi:10.1016/j.maturitas.2013.04.017.

Dyce PW, Liu J, Tayade C, Kidder GM, Betts DH, Li J. *In vitro* and *in vivo* germ line potential of stem cells derived from newborn mouse skin. PLoS One. 2011;6:e20339.

Ehmcke J, Schlatt S. Animal models for fertility preservation in the male. Reproduction. 2008;136:717–23.

Fickel J, Wagener A, Ludwig A. Semen cryopreservation and the conservation of endangered species. Eur J Wildlife Res. 2007;53:81–9.

Gilmore JA, McGann LE, Ashworth E, Acker JP, Raath JP, Bush M, Critser JK. Fundamental cryobiology of selected African mammalian spermatozoa and its role in biodiversity preservation through the development of genome resource banking. Anim Reprod Sci. 1998;53: 277–97.

Gosalvez J, Lopez-Fernandez C, Fernandez JL, Gouraud A, Holt WV. Relationships between the dynamics of iatrogenic DNA damage and genomic design in mammalian spermatozoa from eleven species. Mol Reprod Dev. 2011;78:951–61.

Gourdon JC, Travis AJ. Spermatogenesis in ferret testis xenografts: a new model. Comp Med. 2011;61:145–9.

Graves-Herring JE, Wildt DE, Comizzoli P. Retention of structure and function of the cat germinal vesicle after air-drying and storage at suprazero temperature. Biol Reprod. 2013;88:139.

Hermes R, Göritz F, Saragusty J, Sós E, Molnar V, Reid CE, Schwarzenberger F, Hildebrandt TB. First successful artificial insemination with frozen-thawed semen in rhinoceros. Theriogenology. 2009;71:393–9.

Hildebrandt TB, Roellig K, Goeritz F, Fassbender M, Krieg R, Blottner S, Behr B, Hermes R. Artificial insemination of captive European brown hares (Lepus europaeus PALLAS, 1778) with fresh and cryopreserved semen derived from free-ranging males. Theriogenology. 2009;72:1065–72.

Holt WV. Germplasm cryopreservation in elephants and wild ungulates. In: Watson PF, Holt WV, editors. Cryobanking the genetic resource: wildlife conservation for the future? London: Taylor and Francis; 2001. p. 319–48.

Holt WV. Who needs cytoplasm? Genomic preservation for the 21st century. Biol Reprod. 2013;88:1–2.

Holt WV, Penfold LM, Johnston SD, Temple-Smith P, McCallum C, Shaw J, Lindemans W, Blyde D. Cryopreservation of macropodid spermatozoa: new insights from the cryomicroscope. Reprod Fert Dev. 2000;11:345–53.
Holt WV, Penfold LM, Watson PF. Cryopreservation of lagomorphs gametes and embryos. In: Watson PF, Holt WV, editors. Cryobanking the genetic resource: wildlife conservation for the future? London: Taylor and Francis; 2001. p. 319–48.
Holt WV, Abaigar T, Watson PF, Wildt DE. Genetic resource banks for species conservation. In: Holt WV, Pickard A, Rodger JC, Wildt DE, editors. Reproductive sciences and integrated conservation. Cambridge: Cambridge University Press; 2003. p. 2–20.
Honaramooz A. Cryopreservation of testicular tissue in current frontiers in cryobiology edited by Igor I. Katkov. New York: InTech; 2012.
Howard JG, Wildt DE. Approaches and efficacy of artificial insemination in felids and mustelids. Theriogenology. 2009;71:130–48.
Howard JG, Marinari PE, et al. Black-footed ferret: model for assisted reproductive technologies contributing to in situ conservation. In: Holt WV, Pickard AR, Rodger JC, Wildt DE, editors. Reproductive science and integrated conservation, vol. 8. Cambridge: Cambridge University Press; 2003. p. 249–66.
Huang Y, Li DS, et al. Factors affecting the outcome of artificial insemination using cryopreserved Spermatozoa in the Giant Panda (Ailuropoda melanoleuca). Zoo Biol. 2012a;31:561–73.
Huang Y, Zhang H, et al. Relationship of the estrogen surge and multiple mates to cub paternity in the Giant Panda (Ailuropoda melanoleuca): implications for optimal timing of copulation or artificial insemination. Biol Reprod. 2012b;87:111–7.
Inoue D, Fujimoto T, Kawakami Y, Yasui GS, Yamaha E, Arai K. Vitrification of primordial germ cells using whole embryos for gene-banking in loach, *Misgurnus anguillicaudatus*. J Appl Ichthyol. 2012;28:919–24.
Isachenko V, Maettner R, Petrunkina AM, Mallmann P, Rahimi G, Sterzik K, Sanchez R, Risopatron J, Damjanoski I, Isachenko E. Cryoprotectant-free vitrification of human spermatozoa in large (up to 0.5 mL) volume: a novel technology. Clin Lab. 2011;57:643–50.
Isayeva A, Zhang T, Rawson DM. Studies on chilling sensitivity of zebrafish (Danio rerio) oocytes. Cryobiology. 2004;49:114–22.
Jewgenow K, Wiedemann C, Bertelsen MF, Ringleb J. Cryopreservation of mammalian ovaries and oocytes. Intl Zoo Yrbk. 2011;45:124–32.
Johnson LA, Weitze KF, Fiser P, Maxwell WMC. Storage of boar semen. Anim Reprod Sci. 2000;62:143–72.
Johnston SD, Holt WV. Germplasm conservation in marsupials. In: Watson PF, Holt WV, editors. Cryobanking the genetic resource. Wildlife conservation for the future? London: Taylor and Francis; 2001. p. 203–25.
Johnston SD, Satake N, et al. Osmotic stress and cryoinjury of koala sperm: an integrative study of the plasma membrane, chromatin stability and mitochondrial function. Reproduction. 2012;143:787–97.
Kawakami Y, Saito T, Fujimoto T, Goto-Kazeto R, Takahashi E, Adachi S, Arai K, Yamaha E. Technical note: viability and motility of vitrified/thawed primordial germ cell isolated from common carp (Cyprinus carpio) somite embryos. J Anim Sci. 2012;90:495–500.
Kim S, Lee Y, Yang H, Kim YJ. Rapid freezing without cooling equilibration in canine sperm. Anim Reprod Sci. 2012;130:111–8.
Kiso WK, Brown JL, Siewerdt F, Schmitt DL, Olson D, Crichton EG, Pukazhenthi BS. Liquid semen storage in elephants (*Elephas maximus* and *Loxodonta africana*): species differences and storage optimization. J Androl. 2011;32:420–31.
Kobayashi T, Takeuchi Y, Takeuchi T, Yoshizaki G. Generation of viable fish from cryopreserved primordial germ cells. Mol Reprod Dev. 2007;74:207–13.
Kouba AJ, Vance CK. Applied reproductive technologies and genetic resource banking for amphibian conservation. Reprod Fertil Dev. 2009;21:719–37.
Lee S, Iwasaki Y, Shikina S, Yoshizaki G. Generation of functional eggs and sperm from cryopreserved whole testes. Proceedings of the National Academy of Sciences of the United States of America. 2013;110:1640–45.

Leibo SP, Songsasen N. Cryopreservation of gametes and embryos of non-domestic species. Theriogenology. 2002;57:303–26.
Lermen D, Blömeke B, Browne R, Clarke A, Dyce PW, Fixemer T, Fuhr GR, Holt WV, Jewgenow K, Lloyd RE, Lötters S, Paulus M, Reid GM, Rapoport DH, Rawson D, Ringleb J, Ryder OA, Spörl G, Schmitt T, Veith M, Müller P. Cryobanking of viable biomaterials: implementation of new strategies for conservation purposes. Mol Ecol. 2009;18:1030–3.
Lindeberg H, Aalto J, Amstislavsky S, Piltti K, Järvinen M, Valtonen M. Surgical recovery and successful surgical transfer of conventionally frozen-thawed embryos in the farmed European polecat (Mustela putorius). Theriogenology. 2003;60:1515–25.
Liu J, ChengKM SFG. Fundamental principles of cryobiology and application to *ex situ* conservation of avian species. Avian Biol Res. 2013a;6:187–97.
Liu J, Song Y, Cheng KM, Silversides FG. Production of donor-derived offspring from cryopreserved ovarian tissue in Japanese quail (*Coturnix japonica*). Biol Reprod. 2010;83:15–9.
Liu J, Cheng KM, Silversides FG. Production of live offspring from testicular tissue cryopreserved by vitrification procedures in Japanese quail (Coturnix japonica). Biol Reprod. 2013b;88:124.
Locatelli Y, Hendriks A, Vallet JC, Baril G, Duffard N, Bon N, Ortiz K, Scala C, Maurel MC, Mermillod P, Legendre X. Assessment LOPU-IVF in Japanese sika deer (Cervus nippon nippon) and application to Vietnamese sika deer (Cervus nippon pseudaxis) a related subspecies threatened with extinction. Theriogenology. 2012;78:2039–49.
Mazur P, Leibo SP, Seidel Jr GE. Cryopreservation of the germplasm of animals used in biological and medical research: importance, impact, status and future directions. Biol Reprod. 2008;78: 2–12.
McClean R, Zee YP, et al. Cryopreservation of kangaroo spermatozoa using alternative approaches that reduce cytotoxic exposure to glycerol. Cryobiology. 2008;57:304–7.
McEvoy TG, Coull GD, Broadbent PJ, Hutchinson JS, Speake BK. Fatty acid composition of lipids in immature cattle, pig and sheep oocytes with intact zona pellucida. J Reprod Fertil. 2000;118:163–70.
Mochida Y, Takemura Y, Kanda T, Horie Y. Fertilized eggs obtained from transplantation of frozen ovaries and parthenogenesis in combination with artificial insemination of frozen semen of the silkworm, *Bombyx mori*. Cryobiology. 2003;46:153–60.
Molinia FC, Rodger JC. Pellet-freezing spermatozoa of two marsupials: the tammar wallaby, *Macropus eugenii*, and the brushtail possum, *Trichosurus vulpecula*. Reprod Fertil Dev. 1996;8:681–4.
Nayernia K, Nolte J, Michelmann HW, Lee JH, Rathsack K, Drusenheimer N, Dev A, Wulf G, Ehrmann IE, Elliott DJ, Okpanyi V, Zechner U, Haaf T, Meinhardt A, Engel W. *In vitro-differentiated* embryonic stem cells give rise to male gametes that can generate offspring mice. Dev Cell. 2006;11:125–32.
Nelissen EC, Van Montfoort AP, Coonen E, Derhaag JG, Geraedts JP, Smits LJ, Land JA, Evers JL, Dumoulin JC. Further evidence that culture media affect perinatal outcome: findings after transfer of fresh and cryopreserved embryos. Hum Reprod. 2004;27:1966–76.
O'Brien JK, Robeck TR. Development of sperm sexing and associated assisted reproductive technology for sex preselection of captive bottlenose dolphins (*Tursiops truncatus*). Reprod Fertil Dev. 2006;18:319–29.
Paris MC, Snow M, Cox SL, Shaw JM. Xenotransplantation: a tool for reproductive biology and animal conservation? Theriogenology. 2004;61:277–91.
Penfold LM, L Monfort S, Wolfe BA, Citino SB, Wildt DE. Reproductive physiology and artificial insemination studies in wild and captive gerenuk (Litocranius walleri walleri). Reprod Fertil Dev. 2005;17:707–14.
Pickard AR, Holt WV. Cryopreservation as a supporting measure in species conservation. In: Benson E, Fuller B, Lane N, editors. Not the frozen zoo! Life in the frozen state. Baton Rouge, LA: CRC Press; 2004. p. 393–413.
Pope CE, Gómez MC, Dresser BL. *In vitro* production and transfer of cat embryos in the 21st century. Theriogenology. 2006;66:59–71.
Pope CE, Gómez MC, Kagawa N, Kuwayama M, Leibo SP, Dresser BL. *In vivo* survival of domestic cat oocytes after vitrification, intracytoplasmic sperm injection and embryo transfer. Theriogenology. 2012;77:531–8.

Pukazhenthi B, Spindler R, Wildt D, Bush LM, Howard J. Osmotic properties of spermatozoa from felids producing different proportions of pleiomorphisms: influence of adding and removing cryoprotectant. Cryobiology. 2002;44:288–300.

Pukazhenthi BS, Lapiana F, Padilla L, Santiestevan J, Coutinho da Silva M, Alvarenga M, Wildt DE. Improved sperm cryopreservation in the critically endangered Przewalski's horse (Equus ferus przewalskii) using different cryoprotectants. Biol Reprod. 2010;83:675.

Pukazhenthi BS, Togna GD, Padilla L, Smith D, Sanchez C, Pelican K, Sanjur OI. Ejaculate traits and sperm cryopreservation in the endangered Baird's tapir (*Tapirus bairdii*). J Androl. 2011;32:260–70.

Rall WF, Fahy GM. Ice-free cryopreservation of mouse embryos at −196°C by vitrification. Nature. 1985;313:573–5.

Riesco MF, Robles V. Cryopreservation causes genetic and epigenetic changes in Zebrafish genital ridges. PLoS One. 2013;8:e67614.

Ringleb J, Waurich R, Wibbelt G, Streich WJ, Jewgenow K. Prolonged storage of epididymal spermatozoa does not affect their capacity to fertilise *in vitro*-matured domestic cat (*Felis catus*) oocytes when using ICSI. Reprod Fertil Dev. 2011;23:818–25.

Roache M. The action plan for threatened Australian macropods. Sydney, NSW: WWF-Australia – World Wide Fund for Nature (formerly World Wildlife Fund); 2011.

Robeck TR, O'Brien JK. Effect of cryopreservation methods and precryopreservation storage on bottlenose dolphin (*Tursiops truncatus*) spermatozoa. Biol Reprod. 2004;70:1340–8.

Robeck TR, Gearhart SA, Steinman KJ, Katsumata E, Loureiro JD, O'Brien JK. *In vitro* sperm characterization and development of a sperm cryopreservation method using directional solidification in the killer whale (*Orcinus orca*). Theriogenology. 2011;76:267–79.

Rossato M, Balercia G, Lucarelli G, Foresta C, Mantero F. Role of seminal osmolarity in the reduction of human sperm motility. Int J Androl. 2002;25:230–5.

Saint Jalme M, Lecoq R, Seigneurin F, Blesbois E, Plouzeau E. Cryopreservation of semen from endangered pheasants: the first step towards a cryobank for endangered avian species. Theriogenology. 2003;59:875–88.

Salinas-Flores L, Adams SL, Lim MH. Determination of the membrane permeability characteristics of Pacific oyster, Crassostrea gigas, oocytes and development of optimized methods to add and remove ethylene glycol. Cryobiology. 2008;56:43–52.

Santymire RM, Marinari PE, Kreeger JS, Wildt DE, Howard J. Sperm viability in the black-footed ferret (*Mustela nigripes*) is influenced by seminal and medium osmolality. Cryobiology. 2006;53:37–50.

Saragusty J. Genome banking for vertebrates wildlife conservation. In: Katkov I, editor. Current frontiers in cryobiology. New York: InTech; 2012. p. 293–368. 574 p.

Saragusty J, Arav A. Current progress in oocyte and embryo cryopreservation by slow freezing and vitrification. Reproduction. 2011;141:1–19.

Saragusty J, Gacitua H, King R, Arav A. Post-mortem semen cryopreservation and characterization in two different endangered gazelle species (Gazella gazella and Gazella dorcas) and one subspecies (Gazella gazelle acaiae). Theriogenology. 2006;66:775–84.

Saragusty J, Hildebrandt TB, Behr B, Knieriem A, Kruse J, Hermes R. Successful cryopreservation of Asian elephant (*Elephas maximus*) spermatozoa. Anim Reprod Sci. 2009;115:255–66.

Sato T, Katagiri K, Yokonishi T, Kubota Y, Inoue K, Ogonuki N, Matoba S, Ogura A, Ogawa T. *In vitro* production of fertile sperm from murine spermatogonial stem cell lines. Nat Commun. 2011;2:472.

Schook MW, Wildt DE, Weiss RB, Wolfe BA, Archibald KE, Pukazhenthi BS. Fundamental studies on the reproductive biology of the endangered Persian onager (*Equus hemionus onager*) result in first wild equid offspring from artificial insemination. Biol Reprod. 2013;89:41.

Seki S, Kouya T, Tsuchiya R, Valdez Jr DM, Jin B, Koshimoto C, Kasai M, Edashige K. Cryobiological properties of immature zebrafish oocytes assessed by their ability to be fertilized and develop into hatching embryos. Cryobiology. 2011;62:8–14.

Silva RC, Costa GM, Lacerda SM, Batlouni SR, Soares JM, Avelar GF, Böttger KB, Silva Jr SF, Nogueira MS, Andrade LM, França LR. Germ cell transplantation in felids: a potential approach to preserving endangered species. J Androl. 2012;33:264–76.

Songsasen N, Comizzoli P. A historic overview of embryos and oocyte preservation in the world of mammalian *in vitro* fertilization and biotechnology. In: Borini A, Coticchio G, editors. Preservation of human oocytes. London: Informa Healthcare; 2009. p. 1–11.

Songsasen N, Ratterree MS, VandeVoort CA, Pegg DE, Leibo SP. Permeability characteristics and osmotic sensitivity of rhesus monkey (*Macaca mulatta*) oocytes. Hum Reprod. 2002;17: 175–84.

Songsasen N, Woodruff TK, Wildt DE. *In vitro* growth and steroidogenesis of dog follicles are influenced by the physical and hormonal microenvironment. Reproduction. 2011;142:113–22.

Songsasen N, Comizzoli P, Nagashima J, Fujihara M, Wildt DE. The domestic dog and cat as models for understanding the regulation of ovarian follicle development *in vitro*. Reprod Domest Anim. 2012;47 Suppl 6:13–8.

Spallanzani L. Opuscoli de fisice animale e vegetabile opuscola. II. Observatione e sperienze intorno ai vermicelli spermatici dell' homo e degli animali. Modena. (1776).

Stoops MA, Bond JB, Bateman HL, Campbell MK, Levens GP, Bowsher TR, Ferrell ST, Swanson WF. Comparison of different sperm cryopreservation procedures on post-thaw quality and heterologous *in vitro* fertilisation success in the ocelot (*Leopardus pardalis*). Reprod Fertil Dev. 2007;19:685–94.

Stoops MA, Atkinson MW, Blumer ES, Campbell MK, Roth TL. Semen cryopreservation in the Indian rhinoceros (*Rhinoceros unicornis*). Theriogenology. 2010;73:1104–15.

Sturmey RG, Leese HJ. Energy metabolism in pig oocytes and early embryos. Reproduction. 2003;126:197–204.

Sunderam S, Kissin DM, Flowers L, Anderson JE, Folger SG, Jamieson DJ, Barfield WD. Assisted reproductive technology surveillance – United States, 2009. MMWR Surveill Summ. 2012;61:1–23.

Swanson WF. Application of assisted reproduction for population management in felids: the potential and reality for conservation of small cats. Theriogenology. 2006;66:49–58.

Swanson WF. Laparoscopic oviduct embryo transfer and artificial insemination in felids – challenges, strategies and successes. Reprod Dom Anim. 2012;47 Suppl 6:136–40.

Tharasanit T, Manee-In S, Buarpung S, Chatdarong K, Lohachit C, Techakumphu M. Successful pregnancy following transfer of feline embryos derived from vitrified immature cat oocytes using 'stepwise' cryoprotectant exposure technique. Theriogenology. 2011;76:1442–9.

Thompson JG, Allen NW, McGowan LT, Bell AC, Lambert MG, Tervit HR. Effect of delayed supplementation of fetal calf serum to culture medium on bovine embryo development *in vitro* and following transfer. Theriogenology. 1998;49:1239–49.

Thongtip N, Mahasawangkul S, Thitaram C, Pongsopavijitr P, Kornkaewrat K, Pinyopummin A, Angkawanish T, Jansittiwate S, Rungsri R, Boonprasert K, Wongkalasin W, Homkong P, Dejchaisri S, Wajjwalku W, Saikhun K. Successful artificial insemination in the Asian elephant (Elephas maximus) using chilled and frozen-thawed semen. Reprod Biol Endocrinol. 2009;7:75.

Travis AJ, Kim Y, Meyers-Wallen V. Development of new stem cell-based technologies for carnivore reproduction research. Reprod Domest Anim. 2009;44 Suppl 2:22–8.

Verma R, Holland MK, Temple-Smith P, Verma PJ. Inducing pluripotency in somatic cells from the snow leopard (Panthera uncia), an endangered felid. Theriogenology. 2012;77:220–8.

Waimey KE, Duncan FE, Su HI, Smith K, Wallach H, Jona K, Coutifaris C, Gracia CR, Shea LD, Brannigan RE, Chang RJ, Zelinski MB, Stouffer RL, Taylor RL, Woodruff TK. Future directions in oncofertility and fertility preservation: a report from the 2011 oncofertility consortium conference. J Adolesc Young Adult Oncol. 2013;2:25–30.

Watanabe H, Asano T, Abe Y, Fukui Y, Suzuki H. Pronuclear formation of freeze-dried canine spermatozoa microinjected into mouse oocytes. J Assist Reprod Genet. 2009;26:531–6.

Watson PF. Artificial insemination and the preservation of semen. In: Lamming G, editor. Marshall's physiology of reproduction, vol. 2. London: Churchill Livingstone; 1990. p. 747–869.

Watson PF, Holt WV. Cryobanking the genetic resource; wildlife conservation for the future? London: Taylor and Francis; 2001.

Watson PF, Kunze E, Cramer P, Hammerstedt RH. A comparison of critical osmolality and hydraulic conductivity and its activation energy in fowl and bull spermatozoa. J Androl. 1992;13:131–8.

Wiedemann C, Zahmel J, Jewgenow K. Short-term culture of ovarian cortex pieces to assess the cryopreservation outcome in wild felids for genome conservation. BMC Vet Res. 2013;9:37.

Wildt DE, Rall WF, Critser JK, Monfort SL, Seal US. Genome resource banks: 'living collections' for biodiversity conservation. BioScience. 1997;47:689–98.

Wildt DE, Ellis S, Janssen D, Buff J. Toward more effective reproductive science in conservation. In: Holt WV, Pickard A, Rodger JC, Wildt DE, editors. Reproductive sciences and integrated conservation. Cambridge: Cambridge University Press; 2003. p. 2–20.

Wildt DE, Comizzoli P, Pukazhenthi B, Songsasen N. Lessons from biodiversity – the value of nontraditional species to advance reproductive science, conservation and human health. Mol Reprod Dev. 2010;77:397–409.

Wishart GJ. The cryopreservation of germplasm in domestic and non-domestic birds. In: Watson PF, Holt WV, editors. Cryobanking the genetic resource. Wildlife conservation for thye future? London: Taylor and Francis; 2001. p. 179–200.

Wolfe BA, Wildt DE. Development to blastocysts of domestic cat oocytes matured and fertilized *in vitro* after prolonged cold storage. J Reprod Fertil. 1996;106:135–41.

Woods EJ, Benson JD, Agca Y, Critser JK. Fundamental cryobiology of reproductive cells and tissues. Cryobiology. 2004;48:146–56.

Yamauchi Y, Shaman JA, Ward WS. Non-genetic contributions of the sperm nucleus to embryonic development. Asian J Androl. 2011;13:31–5.

Yang CR, Miao DQ, Zhang QH, Guo L, Tong JS, Wei Y, Huang X, Hou Y, Schatten H, Liu Z, Sun QY. Short-term preservation of porcine oocytes in ambient temperature: novel approaches. PLoS One. 2010;5:e14242.

Yasui GS, Fujimoto T, Sakao S, Yamaha E, Arai K. Production of loach (*Misgurnus anguillicaudatus*) germ-line chimera using transplantation of primordial germ cells isolated from cryopreserved blastomeres. J Anim Sci. 2011;89:2380–8.

Yavin S, Arav A. Measurement of essential physical properties of vitrification solutions. Theriogenology. 2007;67:81–9.

Zee YP, Holt WV, Gosalvez J, Allen CD, Nicolson V, Pyne M, Burridge M, Carrick FN, Johnston SD. Dimethylacetamide can be used as an alternative to glycerol for the successful cryopreservation of koala (*Phascolarctos cinereus*) spermatozoa. Reprod Fertil Dev. 2008;20:724–33.

Zee YP, Lopez-Fernandez C, et al. Evidence that single-stranded DNA breaks are a normal feature of koala sperm chromatin, while double-stranded DNA breaks are indicative of DNA damage. Reproduction. 2009;138:267–78.

Chapter 15
Sperm DNA Fragmentation and Its Role in Wildlife Conservation

Jaime Gosálvez, William V. Holt, and Stephen D. Johnston

Abstract Until about 20 years ago, sperm assessment in the laboratory was focused on motility, morphology and acrosomal integrity. Then came the gradual realisation that, because the main objective of a spermatozoon is to deliver an intact genetic payload of DNA to the egg, being able to check DNA quality of spermatozoa would be equally important, if not more so. Research over the last two decades has therefore led to the development of several techniques for reliably detecting DNA strand breaks, and the more recent focus has been directed towards understanding the fertility implications of DNA damage. It is now clear that evolutionary history has played an important role in determining the stability of sperm DNA under stressful conditions, and that the nature of the DNA-protein interactions also influence the extent to which fertility is affected by both technical procedures involved in sperm preservation and the basic biology of the species concerned. Here we present an overview of the principles involved in DNA assessment and also provide some cases studies that illustrate the influences of species diversity.

Keywords Cryopreservation • Spermatozoa • Rhinoceros • Elephant • Koala • Echidna • Donkey • Planigale

---

J. Gosálvez, B.Sc., Ph.D. (✉)
Faculty of Sciences, Department of Biology, University Autónoma de Madrid, Darwin 2, Madrid 28049, Spain
e-mail: Jaime.gosalvez@uam.es

W.V. Holt, Ph.D.
Academic Dept of Reproductive and Developmental Medicine, University of Sheffield, Jessop Wing, Tree Root Walk, Sheffield S10 2SF, UK

S.D. Johnston, B.Sc. (Hons.), Ph.D.
Wildlife Biology Unit, The School of Agriculture and Food Science, The University of Queensland, Gatton, QLD, Australia

W.V. Holt et al. (eds.), *Reproductive Sciences in Animal Conservation*,
Advances in Experimental Medicine and Biology 753, DOI 10.1007/978-1-4939-0820-2_15,

1 Introduction

The mature spermatozoon is an extremely specialized cell exclusively designed to transport a haploid genome to the oocyte for the purposes of fertilisation, resulting in the recuperation of the diploid condition and formation of a new proliferative somatic cell. The manufacture of the spermatozoon is a complex process that comprises the mitotic proliferation of spermatogonia, followed by two meiotic divisions of spermatocytes and differentiation from haploid spermatids in a process known as spermiogenesis. Within human medicine and domestic animal production there has been a current focus on assessing the spermatozoon on its function as "carrier" and less emphasis on actual "content", yet it is the integrity of the DNA molecule that is most critical for syngamy and subsequent embryonic development and pregnancy. This is particularly emphasised in the case of human reproduction where ICSI (injection of single nucleus into the ooplasm) is now a routine procedure so that assessment of sperm motility is becoming somewhat irrelevant for fertilization. Even though sperm motility provides no direct information on "content", many clinics still promote and rely on it as one of primary determinants for sperm selection prior to ICSI; this could be one of the reasons why analysis of standard semen characteristics is typically such an unreliable predictor of pregnancy, providing evidence for (Shulman et al. 1998; Van Waart et al. 2001) and against (Holt 2005; Lewis 2007) a positive relationship.

In most mammalian species for which assisted reproductive techniques (ART) have been derived, standard analysis of sperm quality involves, at the very least, determination of sperm concentration, morphological normality and motility, but typically no assessment of sperm DNA integrity. Given its relevance to fertility (Agarwal and Allamaneni 2004; Zini and Libman 2006; Shafik et al. 2006; Zini and Sigman 2009; Evenson et al. 1991; Bungum et al. 2011), we are of the opinion that sperm DNA integrity should also be included as part of the standard seminogram and to ignore it, is to ignore a major contributor towards male factor infertility. While this argument is no less relevant to semen analysis and fertility prediction in wildlife species than it is to humans or domestic animals, studies of sperm DNA fragmentation in non-domestic species are limited and the field is very much in its infancy.

The development of species-specific DNA fragmentation assays could potentially play a role, in elucidating infertility or sub-fertility that is not obvious (cryptic) using standard sperm parameters, as a bioassay for the assessment of environmental toxicity and reproductive disease (e.g. Chlamydia infection), understanding the negative effects of aetiological damage associated with semen processing and the development of sperm cryopreservation protocols. In addition, and given the structural and biochemical (protamine) differences in chromatin of the different taxa, the development of species-specific sperm DNA fragmentation (SDF) assays will allow us to explore the evolution and functional significance of the different DNA packaging mechanisms of spermatozoa; for example, sub-eutherian mammalian spermatozoa possess no cysteine residues (disulphide bonds) in their protamines yet are still capable of producing spermatozoa with stable and condensed DNA suitable for sperm transport and internal fertilisation.

2 Sperm DNA Fragmentation: What Is It?

Sperm DNA fragmentation (SDF) is a fundamentally simple concept based on a "Watson and Crick" configured deoxyribonucleic acid molecule that has lost its continuity. Mammalian genomes are biologically formed by discrete pieces of DNA called chromosomes, which are the formal (recognized) reflection of a broken genome that has lost its linearity. The presence of telomeric DNA sequences was in fact an elegant solution to turn the problem of broken DNA into an evolutionary advantage, consequently facilitating an increase in size of the genome during cellular division. In this sense, we might consider chromosomes to be stabilized products of a pre-existing and successful DNA damage event. We would suggest that the process of DNA breakage had its genesis when bacterial DNA lost its circularity to become a linear molecule.

The recurrent incidence of stressing factors on living genomes results in the induction of DNA modifications known as mutations that may or may not be assumed by respective daughter cells. Mutations, of course, also have an effect on gametes, with the added disadvantage for spermatozoa, being the lack of a system for DNA repair, so that accumulated mutations can potentially be transmitted to the offspring (González-Marín et al. 2013; Marchetti and Wyrobek 2008). To understand the effect of SDF on fertility it is important to consider whether the spermatozoon has single (ss-DB) or double-stranded DNA breaks (ds-DB). While some ss-DB can be repaired by the oocyte during the pronuclear stage, ds-DB breaks usually lead to chromosomal imbalance after genome reorganization (inversions, translocations, non-centromeric fragments, free atelomeric DNA sites) characterised by the presence of double 5′3′ free DNA ends in the absence of telomeres. While ss-DBs can potentially be repaired using the original strand of DNA acting as a template, it is possible that DNA mutations such as transitions (interchanges of two-ring purines; A ↔ G, or of one-ring pyrimidines; C ↔ T) or transversions (interchanges of purine for pyrimidine bases) may still occur. If these mutations do not affect structural genes, the relative low amount of such DNA sequences will only have a limited impact in embryonic or adult development; if these mutations affects structural genes or standard genome configurations, they cause early abortion during the first steps of embryo development (Marchetti et al. 2007). Where the affected genome sequences play critical roles in gene regulation and epigenetic control, the situation would be very different and embryo viability could be greatly compromised for a longer period time during the ontogeny of the zygote; it is known that epigenetic disturbance may cause problems in the offspring (Kimmins and Sassone-Corsi 2005; Emery and Carrell 2006; Gosden et al. 2003).

Reactive oxygen species (ROS) are active inducers of ss-DB. Although the short-term effects of excessive ROS production are expressed as a loss of overall cell viability, once the DNA molecule is affected the consequences may be long-lasting because they can be inherited. Although some of these mutations could be repaired at the pronuclear stage of the embryo, if the transmitted DNA damage affects constitutive genes or DNA motifs related to gene regulation of epigenetic control, embryo viability could be greatly compromised. Although the complete biological consequences of DNA base modifications by ROS are unknown, attack by •HO free radicals can modify, for example, the C4–C5 double bond of pyrimidine and generate transient 4,8-endoperoxides in purines (Cadet et al. 2003) . This produces an unpredictable spectrum of oxidative stable conformations such as 8-OHdG, 8-OHdA, formamidopyrimidines thymine glycol, uracil glycol, urea residues, 5-OHdU, 5-OHdC or hydantoin that can result in non-repairable DNA lesions. The consequences of this damage may go beyond gene expression; for example, thymine glycol is able to block DNA replication and is therefore potentially lethal to cells. The case of unrepaired 8-oxo-dG mismatching with dA is a well-known phenomenon in the field of mutagenesis, which increases G to T transition mutations (Cooke 2003; Cooke et al. 2003).

Little is known about the potential negative effects of nitric oxide-derived oxidative processes on spermatozoa. Chemical studies suggest that nitric oxide may produce ss-DBs and/or apurinic or apyrimidinic DNA sites (Burney et al. 1999; Caulfield et al. 1998); for example, regions enriched in guanine are prone to produce base modifications such as 8-Oxo-dG and 8-Oxo-nitro-G (Caulfield et al. 1998). In somatic cells, investigations of direct damage of DNA have demonstrated that the interstitial telomeric DNA sequences of Chinese hamster cells are hypersensitive to nitric oxide damage; in this particular case, the DNA-Protein Kinases fulfil a specific local role in repairing such lesions (Burma and Chen 2004; Mosquera et al. 2005). In the germ line, it has been suggested that nitric oxide can decrease sperm motility (Ramya et al. 2011; Hellstrom et al. 1994) but conversely it has also been claimed to increase it (Miraglia et al. 2011). While exposure of spermatozoa to free nitric radicals in ex vivo experiments results in a negative effect on the DNA molecule (Hellstrom et al. 1994), these results showed that not all single stranded breaks can be considered as presenting the same potential for producing genome damage with a direct impact on the offspring. The nature of the DNA sequences affected and the type of DNA-base modification resulting from the stressors are closely related to the fate of the whole genome. Moreover, the differential capacity of the oocyte cytoplasm to repair a pre-existing DNA damage also contributes to whether the sperm DNA damage will impact embryo viability (Meseguer et al. 2011).

The other important issue of interest is that DNA damage should not be regarded as a static value, as reported in most studies. We have found that DNA damage may change rapidly after ejaculation and tends to increase when the samples are handled ex-vivo. By exposing the spermatozoa in semen extenders at a temperature similar to that found in the female reproductive tract it is possible to reveal potential differences in the rate of DNA fragmentation that could not otherwise be detected. Figure 15.1 compares the sperm DNA fragmentation rate of two rams; before incubation (T0), Ram 1 actually has a higher SDF than that of Ram 2; however, after 1 h both rams have similar values and after just 2 h, Ram 1 shows a massive increase in DNA fragmentation. The question then remains as to which ram has the greater sperm DNA quality? Given that fertilization occurs after a period of sperm transport and storage, it is reasonable to assume that the dynamic rate of DNA fragmentation may have a predictive value in terms of fertility. The impact of this DNA behaviour must not be underestimated, especially in those cases where intrauterine insemination is performed. The importance of understanding sperm DNA fragmentation dynamics in endangered

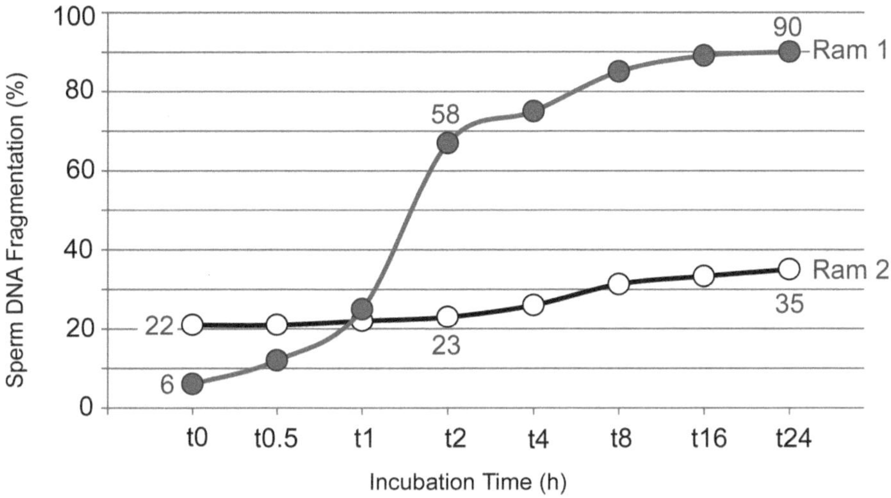

Fig. 15.1 Differential dynamic behaviour of sperm DNA damage in two ejaculates from two different rams after incubation of the sample for 24 h in INRA-84 at 37 °C. *Numbers* inserted in the figure correspond to the level of sperm DNA damage registered during different incubation times

or threatened species is that in general a low number of males are used for insemination purposes. Consequently, the selection of the "best" animal with the lowest level of DNA fragmentation might be critical to achieve the best reproductive outcome. It is also likely that these animals could be genetically restricted, kept in less than ideal captive environments, or be approaching reproductive senescence.

While an understanding of dynamics can reveal intra-specific variation between individuals (López-Fernández et al. 2008a; Gosálvez et al. 2009; Johnston et al. 2012a), we have also discovered significant variation in the rate of sperm DNA degradation between species (Gosálvez et al. 2011a). It is highly likely that this variation is a reflection of the structural and biochemical composition of the sperm chromatin. A dramatic example of major differences in the way that the DNA molecule responds can be seen by comparing frozen-thawed rhinoceros and koala spermatozoa. All three species of rhinoceros that have been examined show a massive degradation of sperm DNA integrity following thawing (Portas et al. 2009), whereas koala spermatozoa show little if any major increase in the rate of DNA fragmentation during post-thaw incubation (Zee et al. 2009a; Johnston et al. 2012b). Rhinoceros spermatozoa possess cysteine residues designed to stabilise the DNA through disulphide bonding and yet show rapid degradation, whereas the koala lacks cysteine residues in its suite of protamines, and thereby appears to maintain its stability using a completely different mechanism. It is therefore fundamental when conducting or interpreting sperm chromatin assays that each individual species be appropriately validated. It stands to reason that differences in chromatin protein chemistry should be reflected in differential responses in the assessment of DNA fragmentation. It seems, for example, that the lack of protamine 2 in some species significantly reduces the likelihood of sperm DNA fragmentation (Gosálvez et al. 2011a). While greater numbers of cysteine residues in protamine 1 tend to confer increased sperm DNA stability, this argument is only likely to be relevant for eutherian sperm. Comparative evolutionary studies involving investigation and validation of SDF assays for different species will provide important insights into the way that the chromatin is packaged and its differential response to stressors, such as environmental toxins or cryopreservation.

3 How Should Sperm DNA Fragmentation Be Assessed?

Several techniques have been developed to assess SDF in humans and other animal species. One of the first experimental approaches performed to measure SDF was dual emission of an altered DNA molecule interacting with fluorescent cationic dye assessed with flow cytometry; this is known as the sperm chromatin structure assay—SCSA—(Evenson 1990). The underlying principle for this method involves subjecting the DNA to mild acid in order to denature ds-DB or ss-DB. This process produces a combination of single stranded DNA stretches originating from the pre-existing DNA breaks, together with orthodox Watson–Crick double stranded DNA molecules. Using the metachromatic characteristics of the fluorochrome, acridine orange, it is possible to produce DNA labelling based on a colour code capable of differentiating damaged from undamaged spermatozoa; those spermatozoa which fluoresce green have mostly double-stranded non-denatured DNA, while red-orange fluorescence is indicative of single stranded DNA motifs. This methodology allows for the objective quantification of spermatozoa with fragmented DNA using a flow cytometer or a standard fluorescence microscope.

Another approach that has been successfully implemented to assess sperm DNA breakage is based upon the enzymatic addition of labelled nucleotides to the ends of pre-existing DNA breaks. These techniques include terminal deoxynucleotidyltransferase (TdT)-mediated nick-end labelling (TUNEL), or *in situ* nick translation (ISNT), using E. coli DNA polymerase (Sharma et al. 2010; Ruvolo et al. 2013). In addition, the comet assay consists of performing single-cell gel electrophoresis on selected cells embedded in microgels (i.e. thin layers of agarose supported by glass). Because of the differential resistance encountered by DNA molecules of different sizes when moving through the gel, a characteristic "comet" distribution is formed after fluorescent staining, with a dense head containing long molecules of DNA and a tail of varying length formed from the shorter fragments. The comet assay can be performed under neutral or alkaline conditions, allowing for the identification of ds-DB or ss-DB respectively. DNA breakage can be evaluated by measuring the number of cells with "comet" tails, as well as the length of the tail and/or percentage of DNA actually contained in the tail (Simon and Carrell 2013). A modification of this technique based on a two dimensional displacement of the DNA fragments offers the possibility of differentiating ds-DB and ss-DB on the same nuclei (Fernández et al. 2001) and has been performed on koala spermatozoa (Zee et al. 2009b).

In our laboratories we have focused on the sperm chromatin dispersion (SCD) test. Improved commercial versions of this test are now available for use in humans and domestic species, and these methodologies can be adapted to a range of wildlife species once the assay has been validated. The SCD is a fast, easy to use method of assessing sperm DNA damage and is based on a controlled DNA denaturation and parallel protein depletion performed on spermatozoa that have been trapped within a microgel. The procedure gives rise to haloes of dispersed chromatin due to the spreading of nuclear DNA loops and/or fragments of DNA; when the sperm nucleus contains fragmented DNA, the size of the halo is directly proportional to the amount of sperm DNA damage. Other methodologies based on the use of specific antibodies against a modified DNA molecule, such as presence of single strand DNA motifs (Zhang et al. 2007), presence of 8-hydroxy-G (Santiso et al. 2010), or indirect markers related to apoptotic processes, are also effective for targeting sperm DNA damage, although at this stage of their development they still have technical and cost constraints for routine use.

While there are some minor differences among the various SDF values produced by the different techniques, in general, the correlation among them is high (Chohan et al. 2006). Despite this agreement, there is nevertheless an unfortunate level of sterile debate on the relative merits of each technique and their respective potential to measure "real" or "potential" sperm DNA damage (Álvarez 2005; Makhlouf and Niederberger 2006; Fernández et al. 2008). It is our view, that such arguments are doctrinaire and fundamentally missing the point when trying to critique the intricate nuances of each technique; rather, they should be utilising the inherent potential differences of each approach to solve the problem.

In terms of selecting the most appropriate technique for the assessment of SDF in wild species, there at least needs to be recognition of the intraspecific differences between sperm structure and chromatin chemistry; for example, it is simply naïve to apply the SCSA using the standardised procedures developed in humans when studying other taxa (e.g. marsupialia or avians) without the appropriate validation. Differential chromatin packaging, closely related to the differential cysteine content in protamine residues, may condition DNA denaturation and subsequently the capacity of acridine orange to bind DNA. In fact it is known that the use of dithiothreitol as an—SS—bond reducer has a direct influence on the capacity to incorporate labelled nucleotides after using the terminal transferase for in situ DNA labelling (Mitchell et al. 2011). In addition to species-specific validation of the assay, it is also important to include appropriate internal controls and to be conscious of how the sperm DNA behaves in conditions that mimic sperm transport and storage in the reproductive tract (Holt and Fazeli 2010); or in the case of those that fertilise oocytes externally (fish and teleosts), the surrounding environment.

4 The Use of SDF in ART: Lessons Learned from Human and Domestic Animals

In general, infertile human males have a higher frequency of spermatozoa with fragmented DNA than fertile controls and there is a certain level of correlation between a poor seminogram and a high level of sperm DNA damage (Liu and Liu 2013). Studies performed on sperm samples used for in vitro fertilization (IVF) demonstrate that the frequency of spermatozoa with fragmented DNA in the sample used for fertilization can influence the reproductive outcome (Bungum et al. 2004). A prospective multi-centre survey analysing 729 couples provided evidence of a correlation between a high incidence of spermatozoa with fragmented DNA in the sample used for insemination and low fertilization rate, poor embryo and blastocyst quality and implantation rate (Vélez de la Calle et al. 2008) . However, in other cases, this correlation is not so clearly established and no consistent relationship between sperm DNA damage and embryo quality and/or development can be detected.

The influence of sperm DNA damage on human embryo quality/development tends to be more significant in ICSI compared to IVF cycles (Zini et al. 2011). One of the possible explanations for some of these discrepancies could be related to the selection of spermatozoa for fertilization and embryos for transfer. Only the most viable spermatozoa selected by visual inspection and those embryos presenting the best prognosis for implantation are normally transferred. While it is likely that the influence of sperm DNA fragmentation on pregnancy would be more demonstrable if all produced embryos were transferred (regardless of quality) such a scenario can only ethically be achieved in animal models. While a large systematic review and meta-analysis of clinical data revealed a small but statistically significant association between sperm DNA integrity and pregnancy following IVF and ICSI cycles (Collins et al. 2008), the complexity of this relationship is no doubt related to the fact that the correlations in these data analysis are not consistent.

In an attempt to try and disentangle this relationship, we have been turning to the use of proven donor oocytes to minimise female factor infertility and thereby provide new insights about the impact of SDF. In a recent study, we reported results obtained in 70 couples assessed for SDF and sperm motility at the time of sperm injection where the sperm samples were assessed and processed for ICSI at the same time (Nuñez-Calonge et al. 2012). In this experimental model, there was no difference in the fertilization rate, cleavage rate or embryo quality, between pregnant and non-pregnant couples. However, the rate of SDF of non-pregnant couples was around 23.9 % and was higher than in those couples who achieved a pregnancy (SDF = 17.0 %; $P=0.002$). The threshold SDF value obtained in this experiment was 17 % and this value could be used to predict pregnancy with a sensitivity and specificity of 78 % and 71 % respectively (Nuñez-Calonge et al. 2012).

The assessment of sperm DNA integrity appears to be an important adjunct to the determination of sperm quality, providing relevant information, not only for reproductive outcomes but for other andrological pathologies. For example, in testicular cancer, the level of DNA damage is very high (Meseguer et al. 2008; Romerius et al. 2010). In these cases, assessment of the level of SDF can provide valuable information about the potential use of cryopreserved sperm samples prior to chemo- or radio-therapy, so that sperm samples from these patients can be cryopreserved for use post-treatment. Patients with a varicocele usually present with an increased level of baseline SDF; two different levels of affected spermatozoa with sperm DNA fragmentation are easily recognized (Enciso et al. 2006; García-Peiró et al. 2012) and fine control of fluctuations in the incidence of these subpopulations can be used to gain information about the effects of antioxidant treatments on final sperm quality

(Abad et al. 2012). Urogenital infections produced by Chlamydia or Mycoplasma infections cause an increased in level of SDF in the ejaculate, but these can be ameliorated using antibiotics at higher doses than those usually recommended (Gallegos et al. 2008). The koala is also known to suffer from chlamydiosis and we are currently examining whether this organism has a similar effect on sperm DNA. The assessment of SDF might also be of relevance in couples with systematic ART failure, or in those cases where the female partner shows poor prognostic conditions such as low ovarian response to stimulation, implantation failure or poor embryo quality; advanced age is also one of the causes that must be taken into account, since the sperm DNA repair capacity of the oocyte in these cases may be compromised.

Associated with strong selection pressure for reproductive performance, a strong correlation between a high level of sperm DNA fragmentation and pregnancy has been established. This is particularly the case for boar or bull (Rybar et al. 2004; Kasimanickam et al. 2006; Fatehi et al. 2006). Strong selection pressure on reproductive performance is likely to be resulting in males with ever decreasing levels of SDF. The most parsimonious hypothesis would be to assume that SDF would have a neutral effect on reproductive outcome such that the distribution of SDF in the population is variable (low and high values), even in those species that are selected for reproductive success. Surprisingly, we actually found this not to be the case, and in those species selected for reproductive characteristics, such as boar, bull or ram, the SDF index was relatively low (López-Fernández et al. 2008a, b; Gosálvez et al. 2011a). Landrace-Large White breed boars, Holstein bulls and Assaf rams, showed values ranging between 0.5 and 10 %, 4.4 % and 4.1 % and 4.1 respectively (Fig. 15.2). Conversely, species that have not been heavily selected for reproductive characteristics, such as human, stallion or donkey, show much more variable values for SDF at 15.8 %, 16.1 % and 16.6 % respectively (Fig. 15.2). This concept obviously has direct relevance for wild populations as the males of rare or endangered, or genetically restricted, populations are purposely often outbred to maintain as much genetic diversity as possible, without any attention to reproductive performance; one would therefore, predict much larger variation of SDF in wild males.

Variable quality in the intra- and inter-specific production of spermatozoa containing a fragmented DNA molecule presents several interesting points that are relevant to the assessment of SDF in wild species. The ejaculates of all species contain spermatozoa with damaged DNA, so that one would perhaps expect continuing directional selection for an important fitness character such as sperm quality, as this result suggests. Further to this, the level of SDF in species selected by man for reproductive purposes is generally lower than those selected for other characteristics. In the case of the boar or ram, all of the animals belong to a random sample from genetic resource centres, where the best animals for particular characteristics demanded by the market are selected. However, all of them also share a common characteristic in that they must be reproductively efficient. For whilst these animals are not deliberately selected by the ART centres to produce low levels of sperm DNA damage in their ejaculate, this is the unintended consequence. This is indicative of stronger selection pressure (whether applied naturally or by human intervention) empirically favouring the presence of spermatozoa with high quality DNA.

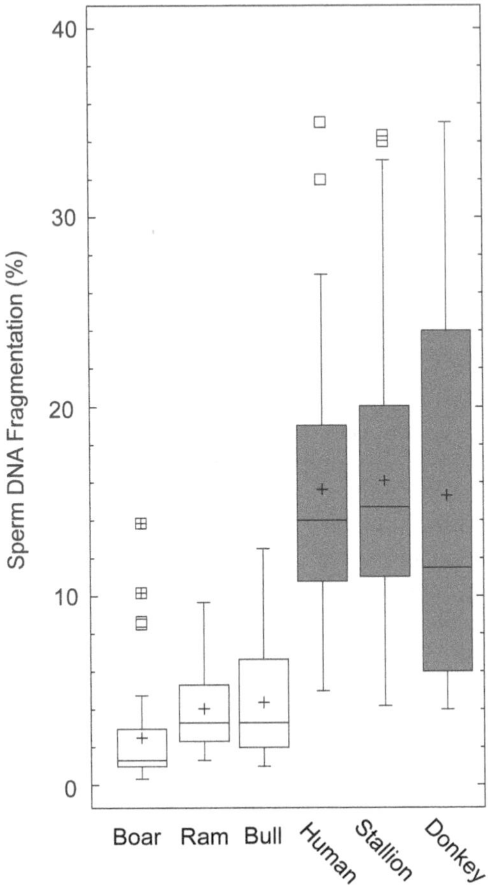

Fig. 15.2 Descriptive statistics for the sperm DNA fragmentation level in two groups of species; in group 1 (*white bars*) males of each species are from populations selected for ART based on reproductive characteristics (boar, ram and bull) whereas those in group 2 (*grey bars*) are from non-selected populations (human, stallion and donkeys)

This mirrors the situation with other sperm parameters where there is considerable evidence that post-copulatory sperm competition is associated with the evolution of sperm features that optimise sperm head and flagellar characteristics and enhance swimming ability (for review, see Birkhead and Immler 2007).

5 The Relevance of Sperm DNA Fragmentation for Wildlife Management and Reproductive Technologies

Environmental destruction, loss of the ecological connectivity, climate change, invasive species or direct predation and hunting are all compromising the survival and conservation of a large number of mammalian species. The Iberian lynx (*Lynx pardinus*) is a wild felid native to the Iberian Peninsula in Southern Europe and is one of the most critically endangered species currently known to man; it is also likely to be

Fig. 15.3 Endangered bovids of Northern Africa (**a**) One of the last free white Oryx herds (*Oryx dammah*) running free in El Krat flats in Western Sahara during the 1940s. (**b**) Indiscriminate hunting of Mohor Gazelles in Tichla-Western Sahara in 1943. Photos courtesy of Prof. José Luís Viejo of Universidad Autónoma de Madrid from the E. Morales-Agacino archives; the original photographs were taken by Dr. Eugenio Morales Agacino during the Sahara campaigns (1943–1956)

the first felid species to become extinct since prehistoric times. Problems emerging from habitat fragmentation and a dependence on rabbits as their primary food prey species, has been a primary driver for population decline in European felids.

The Mohor Gazelle (*Nager dama*) were once running free in north Africa during the 1930s but are now confined to purpose built ex situ captive breeding facilities; hunting was a primary cause of extinction in this species (Fig. 15.3). ART has been presented as adjunct discipline to help overcome some of the genetic issues that are likely to arise from small isolated populations (Abaigar and Holt 2001) and as tools to support efforts to reintroduce the species to its original range (Cano et al. 1993) .

Gamete cryopreservation is one of the most challenging strategies to protect and propagate the species but it also allows precise genetic management both in time and space (Holt et al. 1996; Holt and Moore 1988; Watson and Holt 2001). Many frozen-thawed samples of different individuals and of different species can be stored in the one liquid nitrogen container; this reduces the cost and inconvenience of keeping animals and provides the genetic manager with an opportunity to extend the generation interval of important sires (for review see Holt et al. 2003). The shipment of genetics between populations (zoos or isolated fragments) is also greatly facilitated with the use of frozen semen and leads to a reduction in the animal welfare issues associated with genetic exchange.

Compared to the oocyte, the spermatozoon is a more specialised and differentiated cell, designed to leave and exist in isolation from the soma for hours; consequently the DNA is very well protected and packaged for this exposure. Interestingly, there is large variation amongst different species with respect to viability and DNA integrity when the spermatozoa are handled ex vivo (Gosálvez et al. 2011b). This is partly related to the reproductive strategies of the species, especially with respect to whether it utilises external or internal fertilisation. In external fertilisers such as fish and amphibians the spermatozoa begin activation once exposed to the external environment and remain viable for only minutes. One might expect the DNA integrity of these species to decline rapidly and this has been shown to be the case in teleost fish (López-Fernández et al. 2009), but preliminary results on amphibian (Xenopus sp.) spermatozoa in our laboratory indicate that the sperm DNA remains intact for hours after activation, even after cryopreservation. At the other end of the spectrum, boar or bull sperm DNA is able to remain in the reproductive tract for hours or even days (Holt and Lloyd. 2010; Holt 2011) because specific mechanisms (disulphide bonding between cysteine residues) help to stabilise the DNA. This picture is somewhat complicated by observations on marsupials such as the koala, which do not possess cysteine in the protamines, but are still able to show that DNA fragmentation can remain low following incubation or cryopreservation (Johnston et al. 2012a). Clearly these species have evolved a different mechanism to retain DNA integrity for internal fertilisation. Some avian and reptilian species are known to be able to store sperm in the reproductive tract for long periods of time; in the case of some Squamata and Chelonian reptiles sperm storage has been reported to occur for up 5 years (Birkhead and Moller 1993). Given that these internally fertilising species possess no cysteine in their protamines, it would be fascinating to explore how the sperm DNA has been packaged with respect to the protamines in order to provide such stable DNA integrity. The specifics of how sperm DNA responds once it enters the external environment or the female reproductive tract, has practical application with respect to selecting the best species-specific techniques for sperm preservation and for designing quasi-species specific seminal extenders to provide the spermatozoa with the best media for optimal survival.

Since the first recognized attempt at sperm preservation by Spallanzani (1776), there have been many improvements in preservation technology, including new formulations for semen extenders with various empirical combinations of cryoprotectants, addition of molecules to stabilize membranes, antibiotics to avoid bacterial growth, antioxidant molecules and a range of sophisticated freezing protocols (see reviews by Hammerstedt et al. 1990; Foote 1998; Foote and Parks 1993; Maxwell and Salamon 1993; Watson 1995; Royere et al. 1996; Holt 1997; Vishwanath and Shannon 1997; Woelders 1997). More recently, these developments have been coupled with new technologies for individual sperm selection and injection (ICSI) that in essence do not even require a viable spermatozoon to produce a fertilised embryo. Yoshida (2000) has identified a series of inputs where sperm cryopreservation may have a major impact in wildlife conservation including (1) direct conservation and propagation of threatened or endangered species, (2) international exchange for import-exporting genetic lines, (3) conservation of manipulated sperm for gender selection, (4) conservation of genetic lines having superior genetic traits, (5) rare breed or transgenic lines for establishment of genetic resource banks and (6) gamete reserve to be supplied in response to disease-disaster. Johnston and Holt (2013) have suggested that genome banks can also be established from gametes recovered from post-mortem or wildlife destined for euthanasia. There is also the possibility of recovering gametes from diseased animals and treating or cleaning up these cells prior to cryo-banking (Bielanski 2007; and Bielanski Chap. 17, this book; Bostan et al. 2008). Understanding how sperm DNA integrity is affected by the cryopreservation procedure and maintained following thawing and incubation are important considerations when developing protocols for wild species.

6 Case Study 1: The Rhinoceros

Of the five extant rhinoceros species, the International Union for Conservation of Nature (IUCN) Red List of Threatened Species lists three species as critically endangered, the Javan rhinoceros (*Rhinoceros sondaicus*), the Sumatran rhinoceros (*Dicerorhinus sumatrensis*), and the black rhinoceros (*Diceros bicornis*). Another, the greater one-horned rhinoceros (*Rhinoceros unicornis*) is catalogued as endangered and the white rhinoceros (*Ceratotherium simum*) is threatened. Ex situ reproductive management, including assisted breeding, offers substantial advantages to the conservation of captive rhinoceros populations but the current breeding programs are unfortunately characterized by a low level of reproductive success. Despite investigations into some of the underlying causes of female rhinoceros infertility (Brown et al. 2001; Hermes et al. 2005), there has been comparatively little consideration given to male infertility. Sperm DNA quality has thus far only been assessed in six different animals using cryopreserved semen samples (Portas et al. 2009). The results showed that the baseline level of SDF was relatively low, but that DNA quality rapidly started to decline after only 4 h ex vivo incubation at 37 °C. After 24 h of sperm incubation at 37 °C, SCD of the spermatozoa show large haloes of dispersed chromatin, revealing massive sperm DNA damage (Fig. 15.4). Freshly collected spermatozoa incubated under similar conditions showed no increase in the basal level of DNA fragmentation for up to 48 h. Clearly, cryopreservation of rhinoceros spermatozoa leads to increased levels of sperm DNA fragmentation, either because

Fig. 15.4 (**a**) White rhinoceros (*Ceratotherium simum*) and visualization of sperm DNA damage using the sperm chromatin dispersion test. (**b**) Spermatozoa showing a large halo of dispersed chromatin are those containing a fragmented DNA molecule while the compact sperm head is regarded as normal. (**c**) Visualization of DNA damage using a comet assay test illustrates different levels of chromatin damage according to the size of the comet. (**d**) Electronically filtered image of (**c**) to enhance comet visualization

of an inherent sensitivity of the cell to cryopreservation or an inappropriate cryopreservation procedure. In addition, we have also observed another peculiar condition of the sperm DNA damage in the rhinoceros that is not common in other species; this is the production of double stranded DNA damage after sperm incubation; comet assays of rhinoceros spermatozoa clearly show small comet tails produced under neutral electrophoretic conditions emerging from a compact core (Fig. 15.4c, d). We have not yet investigated this aspect in any detail but certainly the presence of double strand breaks in the sperm DNA is highly unlikely to be repaired by the oocyte, and would lead to embryonic loss or pregnancy failure.

The high incidence of sperm DNA fragmentation following cryopreservation in the rhinoceros provides us an opportunity to show how we could use sperm DNA

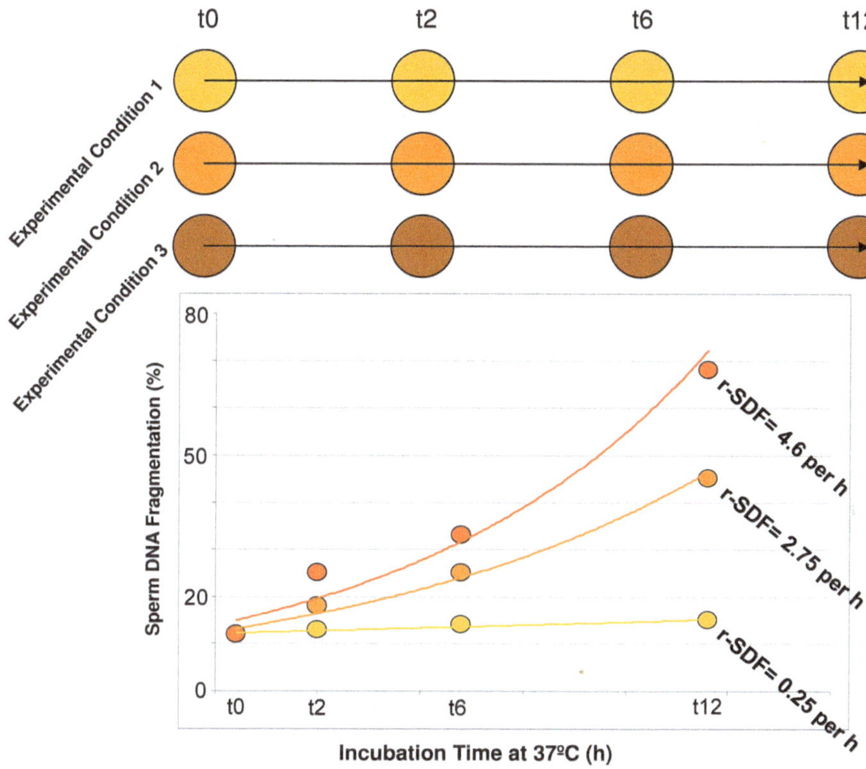

Fig. 15.5 An experimental model to show how the dynamic assessment of sperm DNA damage reveals differences in the cryptic DNA damage, which are not observed after testing for DNA damage values at t0. While the levels of SDF are similar after assessment at t0, these values vary differentially as the incubation time increases, giving rise to different rates of sperm DNA fragmentation (r-SDF)

fragmentation analysis to help us elucidate a more appropriate cryopreservation protocol. The hypothesis to be tested in this investigation would be that the lower the rate of SDF, the better the quality of the frozen-thawed spermatozoa. This model is very simple and could be used to analyze the differential negative impacts of different experimental conditions in maintaining in all sperm characteristics ex vivo. We have exploited this model primarily to assess sperm DNA survival (Gosálvez et al. 2009; Gosálvez et al. 2011b), but it is an analogous to the thermal stress tests (Fiser et al. 1991) used to analyze the dynamic loss of the sperm plasma membrane. Using different experimental conditions (Fig. 15.5), for example, the same sperm sample extended in different diluents, we could calculate and compare the final rate of DNA fragmentation (r-SDF) between these treatments after a series of hours of sperm incubation at 37 °C; in the model depicted in Fig. 15.5, we have proposed 12 h of sperm incubation, but this may vary between species. Based on the dynamic behaviour of SDF observed in Fig. 15.5 we could conclude that treatment 1 is the most beneficial for maintaining sperm DNA integrity, since the rate of SDF (r-SDF) is only 0.25 % per hour, whereas in treatment 3 the r-SDF in an equivalent aliquot is a much higher 4.6 % per hour at equivalent incubation times. Similarly, this experimental model could also be used to analyse the impact of different preservation temperatures for liquid storage and or transportation.

Prior to chilled preservation and transport, the sperm concentration of the inseminate is typically adjusted upwards to account for loss of sperm viability over time. However, we have recently conducted studies on ram spermatozoa that were somewhat counter intuitive to this strategy, and which showed that higher sperm concentration could actually result in a corresponding increase in SDF (López-Fernández et al. 2010). The dynamic assessment of SDF showed that the r-SDF were not only dependent on the inherent sperm DNA fragmentation expressed immediately after thawing, but also on the sperm concentration within the incubated sample. The application of this same model to test the impact of different DNA stressors on human spermatozoa also showed that DNA fragmentation dynamics could be used to assess 'cryptic' sperm damage (Santiso et al. 2012); in this study, increasing acute doses of elevated temperature (41–45 °C), acidic pH and nitric oxide exposure, all resulted in accelerated SDF kinetics following chronic exposure over a 24 h period.

7 Case Study 2: The Elephant

The Asian elephant (*Elephas maximus*) is another seriously endangered mammalian species (IUCN Red List). We have had the opportunity to analyse the spermatozoa of this species at various processing stages before and after cryopreservation (Imrat et al. 2012a; Imrat et al. 2012b). The sperm chromatin dispersion test, specifically adapted to assess SDF in this species, shows large differences in the protein depletion ability among different spermatozoa depending on the level of sperm DNA damage harboured in each sperm (Fig. 15.6). In this species, the overall rate of increase for SDF over 4 h was estimated at about 5 % per hour when the semen was processed in a standard TEST extender. Similar to the situation observed for rhinoceros spermatozoa, no significant changes to this rate were observed at the different processing stages, even including the post-thaw samples. We concluded that Asian elephant spermatozoa were more susceptible to DNA fragmentation than spermatozoa of other mammals (Imrat et al. 2012a). In a second experiment, we redesigned the semen extender to produce an osmotically modified variant of a commercial diluent commonly used for bulls. The results were that this new extender (for commercial reasons the exact composition cannot be specified here) allowed not only more linear sperm movement but also preserved and stabilized the sperm DNA more efficiently than the TEST based extender (Imrat et al. 2012b). The conclusion was that the high post-thaw SDF was not inherent in the unique structure and chemistry of the sperm cell itself but in the conditions used for storage. This is another clear example of how assessment of the SDF using the dynamic concept helped to develop and refine a species-specific preservation protocol.

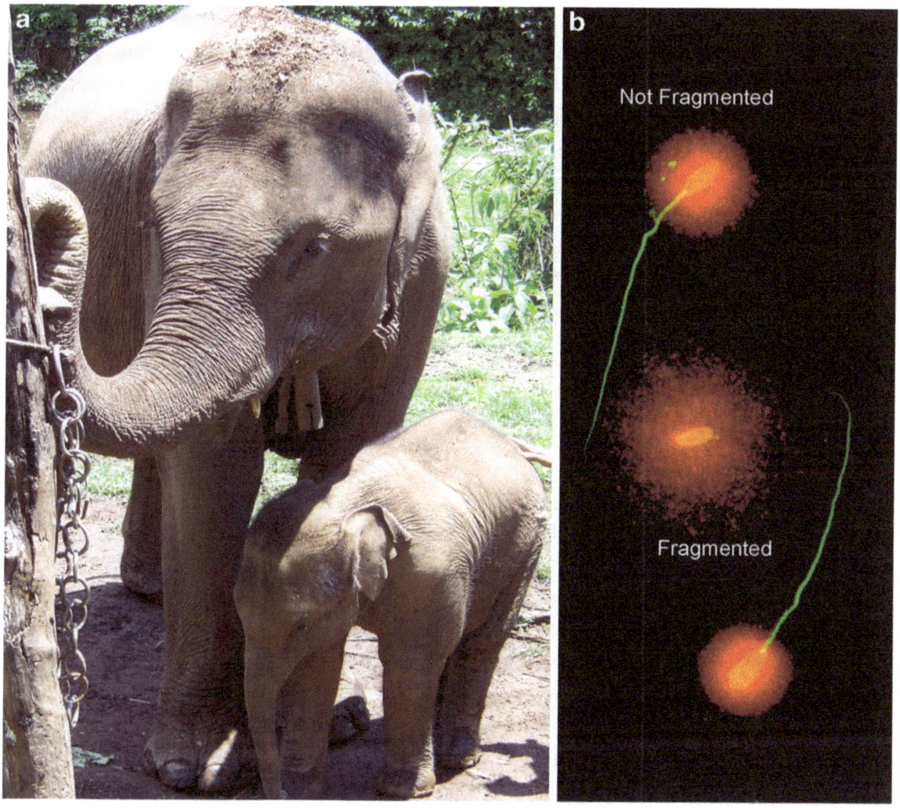

Fig. 15.6 (**a**) Asian elephants (*Elephas maximus*) (photo courtesy of Sittidet Mahasawangku and Podjana Imrat) and (**b**) visualization of the sperm DNA damage after using the sperm chromatin dispersion test

8 Case Study 3: The Iberian Mountain Donkey

Zamorano-Leonés breed (*Equus asinus*) is an Iberian mountain donkey at risk of extinction. Currently, there are only about 50 males in captivity and the population appears to be genetically restricted. Cryopreservation and sperm banking of these species is therefore of critical importance. Frozen-thawed semen samples have been investigated to analyze the sperm DNA longevity under different temperatures after thawing (Cortés-Gutiérrez et al. 2008) and these studies showed that donkey sperm DNA was more sensitive to DNA breakage when incubated at 37 °C than when incubated at 25 or 4 °C. Interestingly, there was large variation in the r-SDF among individuals. We proposed in this study that those individuals with the lowest rate of DNA fragmentation would logically be the most appropriate to use in artificial insemination. We have also investigated stallion sperm DNA in a similar experimental design and found large differences among individuals (López-Fernández et al. 2007). In this case, we found significant differences in the way that the sperm DNA survived during preservation and proposed that such information would be useful when making specific decisions about the most appropriate preservation time for each stallion; i.e. stallions could be chosen to participate in artificial insemination based on the ability of their sperm DNA to withstand chilled preservation. In both the donkey and the stallion, we have developed and suggested preservation strategies based on a screening process for individual animals—"horses for courses", if you will. Such an approach is also applicable to selecting those genetically suitable individuals from within a limited captive population of endangered species that are most likely to produce the best outcome in an artificial breeding program.

9 Case Study 4: The Koala

The koala (*Phascolarctos cinereus*) is an iconic Australian marsupial and although listed as "least concern" on the IUCN red list, it has recently been re-listed as vulnerable by the Queensland State Government; in some local areas of South-east Queensland they are regarded to be close to extinction. The koala is currently the only marsupial for which there is a reliable artificial breeding program with 32 pouch young born by artificial insemination (Rodger et al. 2009; Johnston and Holt 2013 Chap. 9). Despite this success, a major limitation to the koala program, as with other marsupials, has been the lack of a successful sperm cryopreservation method. While the spermatozoon is able to survive the cryopreservation procedure immediately after thawing, further incubation of the spermatozoa at 37 °C results in a form of chromatin swelling akin to de-condensation (Johnston et al. 2006) that presumably renders the cell incapable of fertilisation. In an attempt to investigate this phenomenon, the University of Queensland in Australia, the Zoological Society of London in the UK and the Autónoma University of Madrid in Spain began a series of collaborative studies to determine the reasons for changes in chromatin morphology. In our first paper we described and validated an SCD technique for koala sperm DNA (Johnston et al. 2007). Of all the koala ejaculates that we examined, most showed no evidence of DNA fragmentation, except for one animal called "Beaumont", who possessed a moderate level of DNA damage; Beaumont was essential in helping us to untangle what was a complicated story of sperm DNA fragmentation and which we have outlined in This Book Chap. 9 of the current volume (Johnston and Holt 2013). The koala was also the first sub-eutherian species to be examined using the SCD assay and the response of the spermatozoa to these treatments was particularly fascinating given the lack of cysteine residues in its protamines.

In our initial study, we identified three SCD sperm nuclear morphotypes which we hypothesised represented no DNA damage (morphotype 1; white arrows in Fig. 15.7b), presence of single-stranded DNA damage (yellow arrows in Fig. 15.7b; morphotype 2) and double-stranded DNA plus single strand DNA damage (red arrow in Fig. 15.7b; morphotype 3); in subsequent studies, we have examined this phenomenon using two-way comet assays and further refined our sperm nuclear SCD morphotypes into more of a continuum of DNA fragmentation (Zee et al. 2009b).

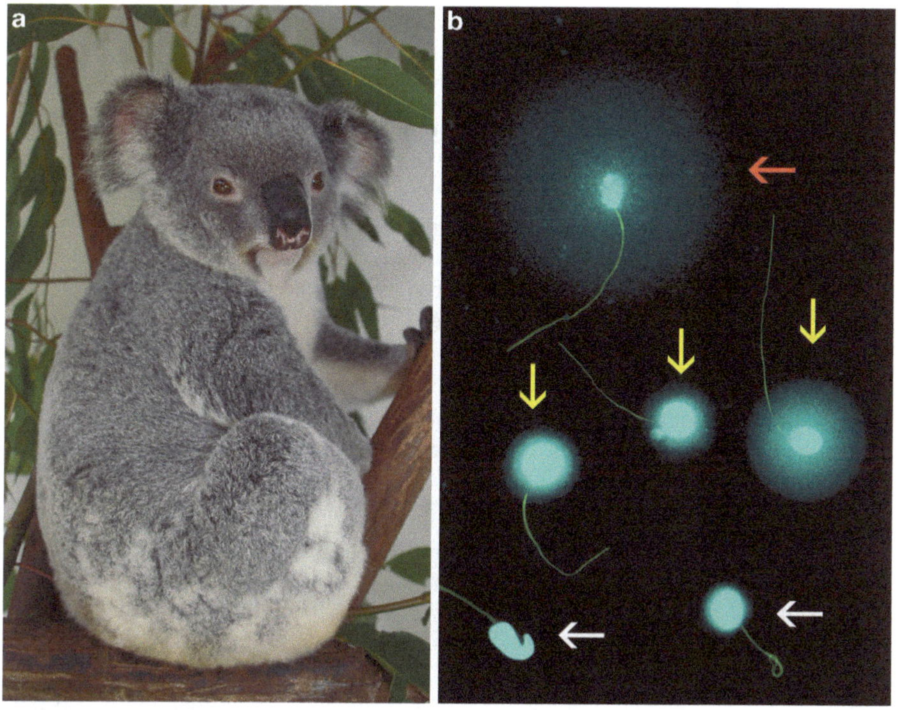

Fig. 15.7 (**a**) Male Koala (*Phascolarctos cinereus*). (**b**) Sperm DNA damage visualization using the sperm chromatin dispersion test. Spermatozoa showing a large halo of dispersed chromatin (morphotype 3; *red arrow*) are those containing a fragmented DNA molecule, while those maintaining the classic shape of the koala sperm or exhibiting a small peripheral and compact halo (morphotype 1; *white arrow*) are considered as devoid of DNA damage. *Yellow arrows* indicate possible intermediate levels of sperm DNA damage (morphotype 2; original published in Reproduction 2012; 143:787–97; Figure 2a in page 789)

More recently we examined the effect of cryopreservation and anisotonic media on sperm DNA fragmentation (Johnston et al. 2012a, b). The major conclusion to our original question about chromatin swelling was that this phenomenon was not necessarily associated with double stranded DNA fragmentation, but rather, was probably a consequence of changes in the tertiary structure resulting from single stranded breaks associated with alkali labile sites on the DNA molecule.

Studies of koala spermatozoa have provided us with the opportunity to re-think how we understand sperm DNA fragmentation, particularly how DNA is packaged into chromatin and condensed in those species that possess no cysteine in the protamines. One of the benefits with working with the spermatozoa of wild species is that the unique biochemistry and structure of these cells has also allowed us to delve into mechanisms that control DNA fragmentation. More recently we have also began to develop SCD protocols for amphibians and reptiles and have plans to extend these studies into avian species. The unique reproductive biology of some taxa provides the basis of natural experiments that can be utilised to explore a deeper understanding of those factors that contribute DNA fragmentation.

10 Case Study 5: Common Planigale

The common planigale (*Planigale maculata*) is a small mouse-size carnivorous marsupial. Although the male reproductive biology of the genus is not well documented, if it is similar to other dasyurid marsupials then it is likely to produce small numbers of spermatozoa and exhibit sperm storage in the female reproductive tract. The sperm DNA of the *Planigale* is extremely fascinating as it is the only known marsupial described to date to possess cysteine residues in its protamines (Retief et al. 1995). Could this phenomenon be considered as an evolutionary strategy to protect the low number of spermatozoa inseminated into the female reproductive tract and hence the need for prolonged sperm survival in the female prior to fertilization. While this is an interesting hypothesis, other dasyurid species possess similar sperm storage strategies but nevertheless lack disulphide bonding. We have never published the results of SDF in this species as we have been unable to induce a detectable level of DNA damage using strategies normally used in other species; these include oxidative stress induction, repeated freezing and thawing and prolonged sperm incubation at 37 °C. Un-processed spermatozoa of this species show a compact nucleus (red in Fig. 15.8a) and a distinctive insertion of the flagellum (green in Fig. 15.8a) In fact, the only morphological change observed in the sperm head after applying the sperm chromatin dispersion test under conditions designed to stress the sperm DNA was the presence of small compact haloes of dispersed chromatin, similar to those considered as containing a non fragmented DNA molecule in other mammalian species (compare Fig. 15.8c with the baseline halos of dispersion showed in Figs. 15.6 and 15.7). It appears as though the *Planigale* sperm nucleus is more resistant to DNA degradation that that present in other mammalian species. We have already discussed the ability of the koala sperm DNA to survive prolong periods of incubation without the use of stabilising disulphide bonds; the source of this stability is still under investigation. Perhaps in the case of the *Planigale*, the sperm is utilising two evolutionary strategies, one that is found in marsupials more generally, and the other the eutherian strategy of stabilising the DNA with disulphide bonds. We are currently in the process of trying to identify other physiological mechanisms that might make the Planigale chromatin vulnerable damage. We are also examining the response of sperm DNA in other closely related dasyurid species that do not possess disulphide bonds such as *Sminthopsis* spp. The stability of the planigale nucleus is likely to present some interesting physiology in terms of the normal nuclear decondensation mechanisms associated with formation of pronuclei in preparation for syngamy.

11 Case Study 6: The Echidna

The short-beaked echidna (*Tachyglos susaculeatus*) (Fig. 15.9a) belongs to an extraordinary group of egg laying mammals known as the monotremes or prototheria. Prototherian spermatozoa are unique amongst the Mammalia, since they

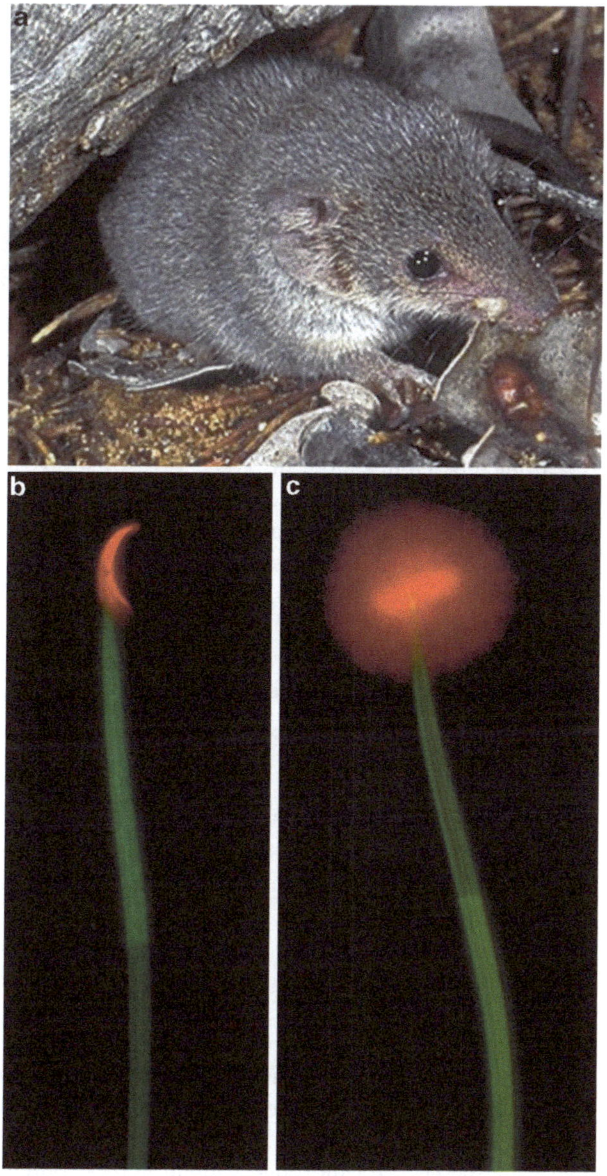

Fig. 15.8 *Planigale maculate* (**a**) and spermatozoa after using the sperm chromatin dispersion test. Spermatozoa do not show the classic halo of dispersed chromatin linked to DNA damage after using this technique. Invariability, all spermatozoa showed a small halo of relaxed chromatin (**c**), different from the untreated cells (**b**) but these haloes do not represent a damaged DNA molecule

present a filiform morphology that is superficially similar in morphology to reptilian spermatozoa (Fig. 15.9b). They also exhibit a tandem arrangement of chromosomes and form sperm bundles during epididymal transit; this is thought to aid sperm transport in a form of co-operative motility. (Jones et al. 2007). Interestingly, sperm DNA damage in this species seems to occur in a directional manner and is co-localised with the presence of highly sensitive alkali sites along the length of the

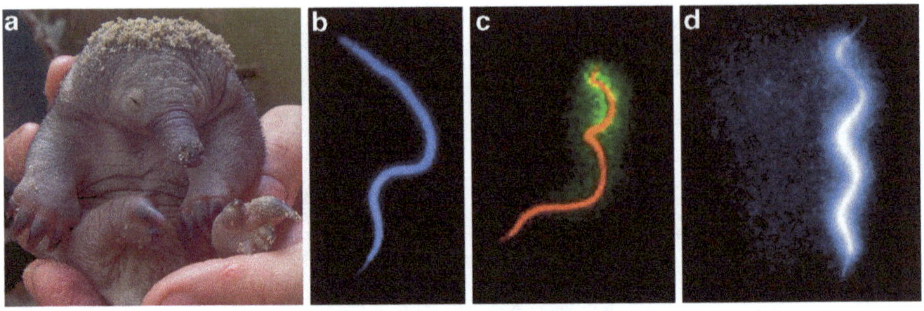

Fig. 15.9 (**a**) Young Echidna (*Tachyglossus aculeatus*) and visualization of sperm DNA damage. (**b**) Individual sperm head with absence of sperm DNA damage. (**c**) Direct DNA labelling breaks—*green* signal—after *in situ* nick translation. (**d**) Visualization of sperm DNA damage using a comet assay

sperm nucleus (Johnston et al. 2009; Fig. 15.9c, d). This directional mapping of sperm DNA damage is observable after in situ nick translation experiments (Fig. 15.9c) and also using a comet assay test (Fig. 15.9d). To date it has not been possible to define whether these alternative DNA configurations are associated with a failure of the sperm nucleus to condense appropriately during spermiogenesis or simply evidence of DNA fragmentation following post-thaw incubation; we have also not ruled out that this sperm DNA behaviour may be a form of sequential structural chromatin rearrangement in preparation for fertilisation.

12 Conclusion and Future Directions

This chapter has introduced the reader to the biology and aetiology of sperm DNA fragmentation, its importance in human reproductive medicine and animal production, how it should be assessed, and its current relevance and potential application to wildlife reproductive and genetic management. Using case study data from 6 diverse species, we have provided examples of how studies of sperm DNA fragmentation in wildlife might be used to investigate and solve, not only issues of fertility (organic and disease) and assisted reproductive technology, but also, how it might contribute to a better fundamental understanding of sperm biology and evolution. Key concepts that we have canvassed include the importance of a species-specific approach, the need for thorough validation procedures when developing the sperm DNA fragmentation assay for "new" wildlife species, the use of a dynamics approach to assessing DNA damage to reveal cryptic sperm DNA fragmentation and the application of sperm DNA fragmentation as a tool for improving methods of sperm preservation (chilled and cryopreservation).

Another potential but under-utilised application of comparative sperm DNA fragmentation analysis is in the development of biomarkers for the assessment of environmental health and reproductive toxicity (Hansen et al. 2010; García-Contreras et al. 2011; Håkonsen et al. 2012). In long-lived species especially, spermatogonial stem cells are likely to be ideal indicators of mutagenesis, and therefore to reflect these cumulative changes in the form of DNA fragmentation of mature sperm cells. Aquatic or semi-aquatic organisms are ideal for this purpose and would represent interesting experimental models; our group is currently exploring these ideas in the salt-water crocodile, amphibians and bivalves. Other experimental model species include corals or other sedentary filter feeding benthic organisms; if sperm fragmentation assays could be developed and validated for these species, they could then be used as bio-sentinels to examine the effects of heavy metals or other environmental contaminates.

While we have already alluded to the phenomenon of selection for reproductive fitness in domestic animals and how this appears to be reflected in lower levels of sperm DNA fragmentation, the effect of genetic restriction or inbreeding depression on DNA quality also requires further study. Recent studies by Petrovic et al. (2013) have suggested that inbred rams show a higher incidence (3X) of sperm DNA fragmentation than that of the outbred group. Similarly, Ruiz-López et al. 2010) have noted a strong positive correlation relationship between inbreeding depression in endangered ungulates and the incidence of sperm DNA fragmentation, highlighting the important role that the integrity of the paternal genome may play when levels of inbreeding are high and which is commonly the situation amongst many genetically restricted endangered species in zoos or fragmented habitat.

Acknowledgements The authors thank Dr C. López-Fernández, F. Arroyo and A. Gosálbez for critical reading and excellent technical assistance. This work was supported by the Ministry of Education and Science, Spain (Grant BFU201016738/BFI).

References

Abad CM, Amengual J, Gosálvez J, Coward K, Hannaoui N, et al. Effects of oral antioxidant treatment upon the dynamics of human sperm DNA fragmentation and subpopulations of sperm with highly degraded DNA. Andrologia. 2012;45(3):211–6. doi:10.1111/and.12003.

Abaigar T, Holt WV. Towards the development of a genetic resource bank for the mohor gazelle: Putting theory into practice. In: Watson PF, Holt WV, editors. Cryobanking the genetic resource. Wildlife conservation for the future? London: Taylor and Francis; 2001. p. 123–39.

Agarwal A, Allamaneni SS. The effect of sperm DNA damage on assisted reproduction outcomes. A review. Minerva Ginecol. 2004;56:235–45.

Álvarez JG. The predictive value of sperm chromatin structure assay. Hum Reprod. 2005;20: 2365–7.

Bielanski A. Disinfection procedures for controlling microorganisms in the semen and embryos of humans and farm animals. Theriogenology. 2007;68:1–22.

Birkhead TR, Immler S. Making sperm: design, quality control and sperm competition. Soc Reprod Fertil Suppl. 2007;65:175–81.

Birkhead TR, Moller AP. Sexual selection and the temporal separation of reproductive events: sperm storage data from reptiles, birds and mammals. Biol J Linn Soc. 1993;50:295–311.

Bostan A, Vannin AS, Emiliani S, Debaisieux L, Liesnard C, et al. Development and evaluation of single sperm washing for risk reduction in artificial reproductive technology for extreme oligospermic HIV positive patients. Curr HIV Res. 2008;6:461–5.

Brown JL, Bellem AC, Fouraker M, Wildt DE, Roth TL. Comparative analysis of gonadal and adrenal activity in the black and white rhinoceros in North America by non-invasive endocrine monitoring. Zoo Biol. 2001;20:463–86.

Bungum M, Bungum L, Giwercman A. Sperm chromatin structure assay (SCSA): a tool in diagnosis and treatment of infertility. Asian J Androl. 2011;13:69–75.

Bungum M, Humaidan P, Spano M, Jepson K, Bungum L, et al. The predictive value of sperm chromatin structure assay (SCSA) parameters for the outcome of intrauterine insemination, IVF and ICSI. Hum Reprod. 2004;19:1401–8.

Burma S, Chen DJ. Role of DNA-PK in the cellular response to DNA double-strand breaks. DNA Repair. 2004;3:909–18.

Burney S, Caulfield JL, Niles C, Whishnok JS, Tannenbaum SR. The chemistry of DNA damage from nitric oxide and peroxynitrite. Mutat Res. 1999;424:37–49.

Cadet J, Douki T, Gasparutto D, Ravanat J-L. Oxidative damage to DNA: Formation, measurement and biochemical features. Mutat Res. 2003;531:5–23.

Cano M, Abaigar T, Vericad JR. Establishment of a group of dama gazelles for reintroduction in Senegal. Int Zoo Yearbook. 1993;32:98–107.

Caulfield JL, Whishnok JS, Tannenbaum SR. Nitric oxide-induced deamination of cytosine and guanine in deoxynucleosides and oligonucleotides. J Biol Chem. 1998;273:12689–95.

Chohan KR, Griffin JT, Lafromboise M, De Jonge CJ, Carrell DT. Comparison of chromatin assays for DNA fragmentation evaluation in human sperm. J Androl. 2006;27:53–9.

Collins JA, Barnhart KT, Schlegel PN. Do sperm DNA fragmentation tests predict pregnancy with in vitro fertilization? Fertil Steril. 2008;89:823–31.

Cooke JP. NO and angiogenesis. Atheroscler Suppl. 2003;4:53–60.

Cooke MS, Evans MD, Dizdaroglu M, Lunec J. Oxidative DNA damage: mechanisms, mutation, and disease. FASEB J. 2003;17:1195–214.

Cortés-Gutiérrez E, Crespo F, Gosálvez J, Dávila-Rodriguez M, López-Fernández C, et al. DNA fragmentation in frozen sperm of Equusasinus: Zamorano-Leonés: a breed at risk of extinction. Theriogenology. 2008;8:1022–32.

Emery BR, Carrell DT. The effect of epigenetic sperm abnormalities on early embryogenesis. Asian J Androl. 2006;8:131–42.

Enciso M, Muriel L, Fernández JL, Goyanes V, Segrelles E, et al. Infertile men with varicocele show a high relative proportion of sperm cells with more intense nuclear damage level in the population with fragmented DNA, evidenced by the sperm chromatin dispersion (SCD) test. J Androl. 2006;27:106–11.

Evenson DP, Jost LK, Baer RK, Turner TW, Schrader SM. Individuality of DNA denaturation patterns in human sperm as measured by the sperm chromatin structure assay. Reprod Toxicol. 1991;5:115–25.

Evenson DP. Flow cytometric analysis of male germ cell quality. Methods Cell Biol. 1990;33:401–10.

Fatehi AN, Bevers MM, Schoevers E, Roelen BA, Colenbrander B, et al. DNA damage in bovine sperm does not block fertilization and early embryonic development but induces apoptosis after the first cleavages. J Androl. 2006;27:176–88.

Fernández JL, Vázquez-Gundín F, Rivero MT, Genescá A, Gosálvez J, et al. DBD-FISH on neutral comets: simultaneous analysis of DNA single- and double-strand breaks in individual cells. Exp Cell Res. 2001;270:102–9.

Fernández JL, Vélez JP, Cajigal D, Gosálvez J. Comments on the clinical value of determination of sperm DNA fragmentation. In: Miura S, Nakano S, editors. Progress in DNA damage research. USA: Nova Science Publishers, Inc.; 2008. p. 341–7. ISBN 978-1-60456-582-9.

Fiser PS, Hansen C, Underhill L, Marcus GJ. New thermal stress test to assess the viability of cryopreserved boar sperm. Cryobiology. 1991;28:454–9.

Foote RH. Artificial insemination to cloning. Ithaca, NY: Cornell University Resource Center; 1998.

Foote RH, Parks JE. Factors affecting preservation and fertility of bull sperm: a brief review. Reprod Fertil Dev. 1993;5:665–73.

Gallegos G, Ramos B, Santiso R, Goyanes V, Gosálvez J, et al. Sperm DNA fragmentation in infertile men with genitourinary infection by Chlamydia trachomatis and Mycoplasma. Fertil Steril. 2008;90:328–34.
González-Marín C, Gosálvez J, Roy R. Types, causes, detection and repair of DNA fragmentation in animal and human sperm cells. Int J Mol Sci. 2013;13:14026–52.
García-Contreras A, De Loera Y, García-Artiga C, Palomo A, Guevara JA, et al. Elevated dietary intake of Zn-methionate is associated with increased sperm DNA fragmentation in the boar. Reprod Toxicol. 2011;31:570–3.
García-Peiró A, Oliver-Bonet M, Navarro J, Abad C, Amengual MJ, et al. Differential clustering of sperm subpopulations in infertile males with clinical varicocele and carriers of rearranged genomes. J Androl. 2012;33:361–7.
Gosálvez J, López-Fernández C, Fernández JL, Gouraud A, Holt WV. Relationships between the dynamics of iatrogenic DNA damage and genomic design in mammalian spermatozoa from eleven species. Mol Reprod Dev. 2011a;78:951–61.
Gosálvez J, Ramirez MA, López-Fernández C, Crespo F, Evans KM, et al. Sex-sorted bovine spermatozoa and DNA damage: I. Static features. Theriogenology. 2011b;75:197–205.
Gosálvez J, Cortés-Gutiérrez E, RocioNuñez R, Fernández JL, Caballero P, et al. A dynamic assessment of sperm DNA fragmentation versus sperm viability in proven fertile human donors. Fertil Steril. 2009;92:1915–9.
Gosden R, Trasler J, Lucifero D, Faddy M. Rare congenital disorders, imprinted genes, and assisted reproductive technology. Lancet. 2003;361:1975–7.
Håkonsen LB, Spano M, Bonde JP, Olsen J, Thulstrup AM, et al. Exposures that may affect sperm DNA integrity: two decades of follow-up in a pregnancy cohort. Reprod Toxicol. 2012;33:316–21.
Hammerstedt RH, Graham JK, Nolan JP. Cryopreservation of mammalian sperm: what we ask them to survive. J Androl. 1990;11:73–88.
Hansen C, Luben TJ, Sacks JD, Olshan A, Jeffay S, et al. The effect of ambient air pollution on sperm quality. Environ Health Perspect. 2010;118:203–9.
Hellstrom WJC, Bell M, Wang R, Sikka SC. Effect of sodium nitroprusside on sperm motility, viability, and lipid peroxidation. Fertil Steril. 1994;61:117–22.
Hermes R, Hildebrandt TB, Blottner S, Walzer C, Schwarzenberger F, et al. Reproductive soundness of captive southern and northern white rhinoceroses (*Ceratotherium simum*): evaluation of male genital tract morphology and semen quality before and after cryopreservation. Theriogenology. 2005;63:219–38.
Holt WV. Alternative strategies for the long-term preservation of spermatozoa. Reprod Fertil Dev. 1997;9:309–19.
Holt WV. Is quality assurance in semen analysis still really necessary? A spermatologist's viewpoint. Hum Reprod. 2005;20:2983–6.
Holt WV. Mechanisms of sperm storage in the female reproductive tract: an interspecies comparison. Reprod Domest Anim. 2011;46:68–74.
Holt WV, Fazeli A. The oviduct as a complex mediator of mammalian sperm function and selection. Mol Reprod Dev. 2010;77:934–43.
Holt WV, Lloyd RE. Sperm storage in the vertebrate female reproductive tract: how does it work so well? Theriogenology. 2010;73:713–22.
Holt WV, Moore HDM. Semen banking—is it now feasible for captive endangered species? Oryx. 1988;22:172–8.
Holt WV, Bennett PM, Volobouev V, Watson PF. Genetic resource banks in wildlife conservation. J Zool. 1996;238:531–44.
Holt WV, Abaigar T, Watson PF, Wildt DE. Genetic resource banks for species conservation. In: Holt WV, Pickard AR, Rodger JC, Wildt DE, editors. Reproductive science and integrated conservation, vol. 8. Cambridge: Cambridge University Press; 2003. p. 267–80.
Imrat P, Hernandez M, Rittem S, Thongtip N, Mahasawangkul S, et al. The dynamics of sperm DNA stability in Asian elephant (*Elephas maximus*) spermatozoa before and after cryopreservation. Theriogenology. 2012a;77:998–1007.

Imrat P, Mahasawangkul S, Gosálvez J, Suthanmapinanth P, Sombutputorn P, et al. Effect of cooled storage on quality and DNA integrity of Asian elephant (*Elephas maximus*) spermatozoa. Reprod Fertil Dev. 2012b;24:1105–16.

Johnston SD, López-Fernández C, Gosálbez A, Holt WV, Gosálvez J. Directional mapping of DNA nicking in ejaculated and caudaepididymidal spermatozoa of the short-beaked echidna (*Tachyglos susaculeatus*: Monotremata). Reprod Fertil Dev. 2009;21:1–7.

Johnston SD, López-Fernández C, Gosálbez A, Zee YP, Holt WV, et al. The relationship between sperm morphology and chromatin integrity in the koala (*Phascolarctos cinereus*) as assessed by the Sperm Chromatin Dispersion (SCD) test. J Androl. 2007;28:891–9.

Johnston SD, MacCallum C, Blyde D, McClean R, Lisle A, Holt WV. An investigation into the similarities and differences governing the cryopreservation success of koala (*Phascolarctos cinereus*: Goldfuss) and common wombat (Vombatusursinus: Shaw) spermatozoa. Cryobiology. 2006;53:218–28.

Johnston SD, Satake N, Zee Y, López-Fernández C, Holt WV, et al. Osmotic stress and cryoinjury of koala sperm: an integrative study of the plasma membrane, chromatin stability and mitochondrial function. Reproduction. 2012a;143:787–97.

Johnston SD, Zee Y, López-Fernández C, Gosálvez J. The effect of chilled storage and cryopreservation on the sperm DNA fragmentation dynamics of a captive population of koalas. J Androl. 2012b;33:1007–15.

Johnston SD, Holt WV. The koala (*Phascolarctos cinereus*): a case study in the development of reproductive technology in a marsupial. In: Holt WV, Brown J, Comizzoli P, editors. Reproductive sciences in animal conservation—progress and prospects. New York, NY: Springer; 2013.

Jones RC, Dacheux JL, Nixon B, Ecroyd HW. Role of the epididymis in sperm competition. Asian J Androl. 2007;9:493–9.

Kasimanickam R, Nebel RL, Peeler ID, Silvia WL, Wolf KT, et al. Breed differences in competitive indices of Holstein and Jersey bulls and their association with sperm DNA fragmentation index and plasma membrane integrity. Theriogenology. 2006;66:1307–15.

Kimmins S, Sassone-Corsi P. Chromatin remodelling and epigenetic features of germ cells. Nature. 2005;434:583–9.

Lewis SEM. Is sperm evaluation useful in predicting human fertility? Reproduction. 2007;134: 31–40.

Liu DY, Liu ML. Clinical value of sperm DNA damage should be assessed in motile sperm fraction rather than whole ejaculated sperm. Fertil Steril. 2013;99:367–71.

López-Fernández C, Crespo F, Arroyo F, Fernández JL, Arana P, et al. Dynamics of sperm DNA fragmentation in domestic animals II. The stallion. Theriogenology. 2007;68:1240–50.

López-Fernández C, Fernández JL, Gosálbez A, Arroyo F, Vázquez JM, et al. Dynamics of sperm DNA fragmentation in domestic animals III Ram. Theriogenology. 2008a;70:898–908.

López-Fernández C, Gage MJ, Arroyo F, Gosálbez A, Larrán AM, et al. Rapid rates of sperm DNA damage after activation in tench (Tincatinca: Teleostei, Cyprinidae) measured using a sperm chromatin dispersion test. Reproduction. 2009;138:257–66.

López-Fernández C, Johnston SD, Fernández JL, Wilson RJ, Gosálvez J. Fragmentation dynamics of frozen-thawed ram sperm DNA is modulated by sperm concentration. Theriogenology. 2010;74:1362–70.

López-Fernández C, Pérez-Llano B, García-Casado P, Sala R, Gosálbez A, et al. Sperm DNA fragmentation in a random sample of the Spanish boar livestock. Anim Reprod Sci. 2008b;103:87–98.

Makhlouf AA, Niederberger C. DNA integrity tests in clinical practice: it is not a simple matter of black and white (or red and green). J Androl. 2006;27:316–23.

Marchetti F, Essers J, Kanaar R, Wyrobek AJ. Disruption of maternal DNA repair increases sperm-derived chromosomal aberrations. Proc Natl Acad Sci U S A. 2007;104:17725–9.

Marchetti F, Wyrobek AJ. DNA repair decline during mouse spermiogenesis results in the accumulation of heritable DNA damage. DNA Repair. 2008;7:572–81.

Maxwell WM, Salamon S. Liquid storage of ram semen: a review. Reprod Fertil Dev. 1993;5: 613–38.

Meseguer M, Santiso R, Garrido N, Fernández JL. The effect of cancer on sperm DNA fragmentation as measured by the sperm chromatin dispersion test. Fertil Steril. 2008;90:225–7.

Meseguer M, Santiso R, Garrido N, García-Herrero S, Remohí J, et al. Effect of sperm DNA fragmentation on pregnancy outcome depends on oocyte quality. Fertil Steril. 2011;95:124–8.

Miraglia E, De Angelis F, Gazzano E, Hassanpour H, Bertagna A, et al. Nitric oxide stimulates human sperm motility via activation of the cyclic GMP/protein kinase G signaling pathway. Reproduction. 2011;141:47–54.

Mitchell LA, De Iuliis GN, Aitken RJ. The TUNEL assay consistently underestimates DNA damage in human spermatozoa and is influenced by DNA compaction and cell vitality: development of an improved methodology. Int J Androl. 2011;34:2–13.

Mosquera A, Gosálvez J, Sabatier L, Fernández JL. Interstitial Telomeric DNA sequences of Chinese hamster cells are hypersensitive to nitric oxide damage, and DNA-PKcs Has a specific local role in its repair. Genes Chromosomes Cancer. 2005;44:76–84.

Nuñez-Calonge R, Caballero P, López-Fernández C, Andrés Guijarro J, Fernández JL, et al. An improved experimental model for understanding the impact of sperm DNA fragmentation on human pregnancy following ICSI. Reprod Sci. 2012;19:1163–8.

Petrovic VC, Maksimovic N, Petrovic MP, Petrovic MM, Ilic ZZ, et al. Effect of inbreeding on body growth traits and sperm DNA fragmentation levels in rams. Anim Sci Pap Rep. 2013;31:27–33.

Portas T, Johnston SD, Hermes R, Arroyo F, López-Fernadez C, et al. Frozen-thawed rhinoceros sperm exhibit DNA damage shortly after thawing when assessed by the sperm chromatin dispersion assay. Theriogenology. 2009;72:711–20.

Ramya T, Misro MM, Sinha D, Nandan D, Mithal S. Altered levels of seminal nitric oxide, nitric oxide synthase, and enzymatic antioxidants and their association with sperm function in infertile subjects. Fertil Steril. 2011;95:135–40.

Retief JD, Rees JS, Westerman M, Dixon GH. Convergent evolution of cysteine residues in sperm protamines of one genus of marsupials, the planigales. Mol Biol Evol. 1995;12:708–12.

Rodger JC, Paris DB, Czarny NA, Harris MS, Molinia FC, et al. Artificial insemination in marsupials. Theriogenology. 2009;71:176–89.

Romerius P, Ståhl O, Moëll C, Relander T, Cavallin-Ståhl E, et al. Sperm DNA integrity in men treated for childhood cancer. Clin Cancer Res. 2010;16:3843–50.

Ruiz-López MJ, Evenson DP, Espeso G, Gomendio M, Roldan ERS. High levels of DNA fragmentation in spermatozoa are associated with inbreeding and poor sperm quality in endangered ungulates. Biol Reprod. 2010;83:332–8.

Ruvolo G, Fattouh RR, Bosco L, Brucculeri AM, Cittadini E. New molecular markers for the evaluation of gamete quality. J Assist Reprod Genet. 2013;30:207–12.

Rybar R, Faldikova L, Faldyna M, Machatkova M, Rubes J. Bull and boar sperm DNA integrity evaluated by sperm chromatin structure assay in the Czech Republic. Vet Med. 2004;49:1–8.

Royere D, Barthelemy C, Hamamah S, Lansac J. Cryopreservation of spermatozoa: a 1996 review. Hum Reprod Update. 1996;2:553–9.

Santiso R, Tamayo M, Gosálvez J, Johnston S, Mariño A, et al. DNA fragmentation dynamics allows the assessment of cryptic sperm damage in human: evaluation of exposure to ionizing radiation, hyperthermia, acidic pH and nitric oxide. Mutat Res. 2012;734:41–9.

Santiso R, Tamayo M, Gosálvez J, Meseguer M, Garrido N, et al. Simultaneous determination in situ of DNA fragmentation and 8-oxoguanine in human sperm. Fertil Steril. 2010;93:314–8.

Sharma RK, Sabanegh E, Mahfouz R, Gupta S, Thiyagarajan A, et al. TUNEL as a test for sperm DNA damage in the evaluation of male infertility. Urology. 2010;76:1380–6.

Shafik A, Shafik AA, Shafik I, El Sibai O. Sperm DNA fragmentation. Arch Androl. 2006;52:197–208.

Shulman A, Hauser R, Lipitz S, Frenkel Y, Dor J, et al. Sperm motility is a major determinant of pregnancy outcome following intrauterine insemination. J Assist Reprod Genet. 1998;15: 381–5.

Simon L, Carrell DT. Sperm DNA damage measured by comet assay. Methods Mol Biol. 2013;927:137–46.

Spallanzani L. Osservazioni e spezienzeinternoai vermicelli spermaticidell'uomo e deglianimali. In Opusculi di FisicaAnimale e Vegetabile. Modena, Italy: Opusculo II; 1776.

Van Waart J, Kruger TF, Lombard CJ, Ombelet W. Predictive value of normal sperm morphology in intrauterine insemination (IUI): a structured literature review. Hum Reprod Update. 2001;7:495–500.

Velez de la Calle JF, Muller A, Walschaerts M, Clavere JL, Jimenez C, et al. Sperm deoxyribonucleic acid fragmentation as assessed by the sperm chromatin dispersion test in assisted reproductive technology programs: results of a large prospective multicenter study. Fertil Steril. 2008;90:1792–9.

Vishwanath R, Shannon P. Do sperm cells age? A review of the physiological changes in sperm during storage at ambient temperature. Reprod Fertil Dev. 1997;9:321–31.

Watson PF. Recent developments and concepts in the cryopreservation of spermatozoa and the assessment of their post-thawing function. Reprod Fertil Dev. 1995;7:871–91.

Watson PF, Holt WV, editors. Cryobanking the genetic resource; wildlife conservation for the future? London: Taylor and Francis; 2001.

Woelders H. Fundamentals and recent development in cryopreservation of bull and boar semen. Vet Q. 1997;19:135–8.

Yoshida M. Conservation of sperms: current status and new trends. Anim Reprod Sci. 2000;60(355):355.

Zee YP, López-Fernández C, Gosálvez J, Holt WV, Johnston SD. DNA fragmentation dynamics of koala spermatozoa. Reprod Fertil Dev. 2009a;21:141–2.

Zee YP, López-Fernández C, Arroyo F, Johnston SD, Holt WV, et al. Evidence that single-stranded DNA breaks are a normal feature of koala sperm chromatin, while double-stranded DNA breaks are indicative of DNA damage. Reproduction. 2009b;138:267–78.

Zhang X, San Gabriel M, Libman J, Phillips S, Courchesne A, et al. Localization of single-stranded DNA in human sperm nuclei. Fertil Steril. 2007;88:1334–8.

Zini A, Jamal W, Cowan L, Al-Hathal N. Is sperm DNA damage associated with IVF embryo quality? A systematic review. J Assist Reprod Genet. 2011;28:391–7.

Zini A, Libman J. Sperm DNA damage: clinical significance in the era of assisted reproduction. CMAJ. 2006;175:495–500.

Zini A, Sigman M. Are Tests of Sperm DNA Damage Clinically Useful? Pros and cons. J Androl. 2009;30:219–29.

Chapter 16
Somatic Cells, Stem Cells, and Induced Pluripotent Stem Cells: How Do They Now Contribute to Conservation?

Gabriela F. Mastromonaco, L. Antonio González-Grajales, Melissa Filice, and Pierre Comizzoli

Abstract More than a decade has now passed since the birth of the first endangered species produced from an adult somatic cell reprogrammed by somatic cell nuclear transfer. At that time, advances made in domestic and laboratory animal species provided the necessary foundation for attempting cutting-edge technologies on threatened and endangered species. In addition to nuclear transfer, spermatogonial stem cell transplantation and induction of pluripotent stem cells have also been explored. Although many basic scientific questions have been answered and more than 30 wild species have been investigated, very few successes have been reported. The majority of studies document numerous obstacles that still need to be overcome to produce viable gametes or embryos for healthy offspring production. This chapter provides an overview of somatic cell and stem cell technologies in different taxa (mammals, fishes, birds, reptiles and amphibians) and evaluates the potential and impact of these approaches for animal species conservation.

Keywords Cell culture • Stem cell • Cloning • Somatic cell nuclear transfer • Biobanking

---

G.F. Mastromonaco, M.Sc, Ph.D. (✉) • L.A. González-Grajales • M. Filice
Reproductive Physiology, Toronto Zoo, Toronto, ON, Canada

Department of Biomedical Sciences, University of Guelph, Guelph, ON, Canada
e-mail: gmastromonaco@torontozoo.ca

P. Comizzoli
Smithsonian Conservation Biology Institute, National Zoological Park, Washington, DC, USA

W.V. Holt et al. (eds.), *Reproductive Sciences in Animal Conservation*, Advances in Experimental Medicine and Biology 753, DOI 10.1007/978-1-4939-0820-2_16,

1 Introduction

Biologists and conservationists have repeatedly warned of the fact that species extinctions are occurring at an alarming rate with no evidence of slowing down within the next few decades. To add to this dilemma, recent assessments of captive populations and their breeding programs have shown that they are not achieving the conditions for sustainability (Lees and Wilcken 2009). Lees and Wilcken (2009) reported that only 48 % of assessed captive populations were breeding to replacement and only 55 % were retaining levels of gene diversity ≥90 %. A key recommendation from these authors was that it is time for global rather than regional population management and any possibility of revitalization will require global audit of populations, list of priority species, target population sizes, investment and commitment (Lees and Wilcken 2009; see also Chap. 2). With this daunting fact facing the world's zoos and aquariums, it is an appropriate time to revisit the role of "frozen zoos" in the sustainable management of captive populations.

Biomaterial banking for threatened or endangered species as a supplemental strategy for genetic management of captive and potentially wild populations, or as insurance against a sudden and extreme loss of genetic diversity has been promoted for over 30 years (Veprintsev and Rott 1979; Holt and Moore 1988; Wildt 1992) and recently reviewed by Saragusty (2012). In most cases, biological or genome resource banks (BRBs or GRBs) were intended to include germplasm, in the form of sperm, oocytes and embryos, as the main goal of these repositories was to enhance future offspring production with the use of assisted reproductive technologies (ARTs). A review of the most recent progress in cryobanking is presented in Chap. 14. Currently, the majority of the stored germplasm consists of frozen sperm samples with minimal contribution from female genetic material due to the difficulties in obtaining and cryopreserving oocytes and embryos. In recent years, there has been a growing interest in the banking of somatic cells for reasons discussed in the present chapter, with a call to action arising from groups such as the Frozen Ark (www.frozenark.org, verified January 20, 2013). Launched in 2004, the Frozen Ark Project's objectives were, and still are, to save the genetic material of threatened animal species (Clarke 2009), preferably in the form of living cells. This global consortium with members from zoos, aquariums, museums and universities located in the United Kingdom, United States, Australia, India, and numerous other countries has outlined and begun to implement a plan for an organized, internationally-linked and properly catalogued repository of genetic material (Clarke 2009). Currently, consortium members hold over 48,000 samples from more than 5,500 animal species (www.frozenark.org, verified January 20, 2013) as frozen tissues, somatic cell cultures and DNA. Some of the newer contributors, such as the German Cell Bank for Wildlife "Alfred Brehm" (Cryo-Brehm), have included stem cells in their sample collection repertoire (www.emb.fraunhofer.de/en/Uebersichtsindex/cellbank_cryo-brehm.html, verified January 20, 2013). This long-term and recently increased investment of time and funds indicates that non-germinal cell banking is being considered as important a genetic resource for the future as sperm, oocytes and embryos (Table 16.1).

Table 16.1 List of the main somatic cell and tissue repositories for threatened and endangered species

Institution/consortium	Country	Cell bank established	Information
North America			
Audubon Nature Institute	USA	INA	>15 feline cell lines[a]
American Museum of Natural History	USA	2001	>28,000 tissue and DNA samples[a, b, c, d, e]
San Diego Zoo Global Wildlife Conservancy	USA	1975	>8,000 cell lines from 800 species[a, b, c, d]
Smithsonian Institution	USA	1970	>1,000,000 tissue and DNA samples and cell lines from >18,000 species[a, b, c, d, e]
Toronto Zoo	Canada	2007	>112 cell lines[a, b, c]
Europe			
Institute of Biomedical and Environmental Science and Technology	UK	2008	19 species[b]
Mediterranean Marine Mammals Tissue Bank	Italy	2002	>11 Mediterranean species[a]
Museo Nacional de Ciencias Naturales[f]	Spain	INA	>250 individual samples of Iberian lynx[a]
Asia			
Kunming Wild Animal Cell Bank	China	1986	1,455 cell lines from 289 species[a, b, c, d, e]
Conservation Genome Resource Bank for Korean Wildlife	Korea	2002	13,475 tissue and DNA samples from 407 species[a, b, d]
National Center for Cell Sciences	India	INA	INA
South America			
Zoologico de Buenos Aires[f]	Argentina	2005	INA
Australia			
Animal Gene Storage and Resource Centre of Australia	Australia	1995	INA
Africa			
Wildlife Biological Resource Centre	South Africa	2003	INA
Global Consortia			
Amphibian ARK	USA	2004	INA
BioBankSA	South Africa	2003	INA
Cryo-Brehm	Germany	INA	Stem cells[a]; INA
The Frozen Ark	UK	1996	48,000 samples from >5, 500 endangered and non-endangered animal species[a, b, c, d, e]

INA Information Not Available
[a]Mammals, [b]Fishes, [c]Reptiles and Amphibians, [d]Birds, [e]Insects
[f]Bi-national study: Genetic Resource Bank for South American Felines

Exciting developments in cell-based technologies in the past 20 years have provided several potential strategies for the use of somatic cells in biomedical applications, specifically reproductive biology. With the advancement of nuclear transfer and stem cell technologies, somatic cells have the potential to be used directly or indirectly for offspring production. The ability to reprogram differentiated somatic cell nuclei into embryonic or germinal cell lineages has suddenly increased the value, or at least interest, in somatic genome banks. The birth of a gaur (*Bos gaurus*) in 2000 following transfer of embryos produced by somatic cell nuclear transfer (SCNT) was the first report of an adult fibroblast being used to create offspring from an endangered species (Lanza et al. 2000). Many years later, in September 2011, a new era in somatic cell manipulation was launched when Ben-Nun et al. (2011) reported the production of embryoid bodies derived from induced pluripotent stem cells (iPSCs): the first cases of induced pluripotency in adult fibroblasts derived from endangered species, the silver-maned drill (*Mandrillus leucophaeus*) and northern white rhinoceros (*Ceratotherium simum cottoni*).

This chapter provides an overview of the successes and challenges of somatic cell banking and its potential application for reproductive purposes today and in the future. Although the most extensive works have been documented in mammal species, an effort has been made to include the current state of technologies in fishes, birds, reptiles and amphibians.

2 Somatic Genome Preservation

The banking of somatic cells, first and foremost, provides valuable genetic and cellular resource material for basic scientific study, including phylogenomics, evolutionary biology, physiology, and pathobiology, to name a few fields of study. The interest in preserving viable cells is their ability to provide a source of renewable material that can be cultured for long periods, and in some cases, indefinitely. Unlike gametes, somatic cells allow the preservation of the entire genome avoiding dilution of valuable alleles as occurs during fertilization, an important factor when considering species with low numbers of remaining individuals. As well as rapidly increasing the contribution of female genetics in biobanks, somatic cells can be used in cases where collection of viable gametes will not be realized, including individuals: (1) who have died unexpectedly, prior to sexual maturity, or outside of the breeding season; (2) in their prime reproductive years experiencing reproductive dysfunction; (3) in their post-reproductive years who failed or lacked the opportunity to breed; (4) who have been castrated; and (5) whose maturational status is unknown (Mastromonaco and King 2007). The technologies that can be implemented to produce gametes or embryos from somatic cells are discussed in Sect. 3 of this chapter.

2.1 *Somatic Cell Collection and Culture*

Fibroblasts are one of the common cell types that are banked and have been used for research purposes for over 50 years since Harry Eagle first identified a chemically-defined culture medium that supported cell growth in vitro (Eagle 1955). They have been utilized as primary cultures or continuous cell lines to better understand cellular physiology, investigate a wide range of normal and pathological molecular and biochemical mechanisms, and evaluate cellular response to exogenous stimuli, such as biological substances, environmental toxins and pharmaceutical agents. Although epithelial cells are also widely used and can be derived from tissue explants along with or instead of fibroblasts, the faster-growing fibroblasts can be more easily cultured and sustained in vitro. The benefits of using fibroblasts is that they generally do not require specialized culture systems or cryopreservation protocols and can be isolated from a variety of sources, including organs, such as lung and kidney, muscle and skin. Furthermore, samples from most taxa can be collected with minimal invasiveness (e.g. ear or skin biopsy), opportunistically (e.g. during hoof trim, health check or surgery), and from recently or long-deceased animals. Therefore, unlike gametes and embryos, fibroblasts provide a diploid nucleus and associated cytoplasmic material without the need for species-specific collection, handling, culturing and freezing techniques. Most importantly for the context of this review, fibroblasts have been identified as being one of the most suitable cell types for SCNT (Campbell et al. 2007). An excellent manual for the preparation and culture of fibroblasts and other cell types is Ian Freshney's Culture of Animal Cells: A Manual of Basic Technique (Freshney 2005). One of the most important criteria for conservation purposes is the establishment of primary cultures from a variety of tissue sources that result in healthy, chromosomally stable and long-lived cell lines.

2.1.1 Mammals

Cell cultures derived from mammals are the most widely used of all the taxa for research and commercial purposes. The American Type Culture Collection (ATCC), the largest biological resource center of its kind in the world, maintains over 3,600 primary and continuous cultures representing over 150 species (www.atcc.org, verified January 20, 2013). In the past 50 years, researchers have established primary cultures from common and endangered breeds or strains of livestock, laboratory animal and companion animal species, including cattle, pig, sheep, rat, guinea pig, horse, dog and cat, but also from a large number of non-domestic and endangered species. San Diego Global Wildlife Conservancy's Frozen Zoo® is one of the oldest and most diverse repositories of mammalian cell cultures from wild species (www.sandiegozooglobal.org, verified January 20, 2013). Samples representing 20 mammal orders, several hundred species and thousands of individuals have been cryobanked since 1975 (Houck 2012). Another equally historical biobank, the National Cancer Institute's repository of over 100,000,000 DNA, tissue and cell line samples, will be integrated into the Smithsonian Institution's collection later this year (Table 16.1). In 2002, the Mediterranean Marine Mammal Tissue Bank (MMMTB) was established by researchers at the University of Padova, Italy, to collect tissues from stranded and captive marine mammals from the Mediterranean region (Peruffo et al. 2010). Due to the body condition of most stranded animals, the majority of the samples are stored in formalin, however, cultures have been derived from bottlenose dolphin (*Tursiops truncatus*), sperm whale (*Physeter macrocephalus*), and striped

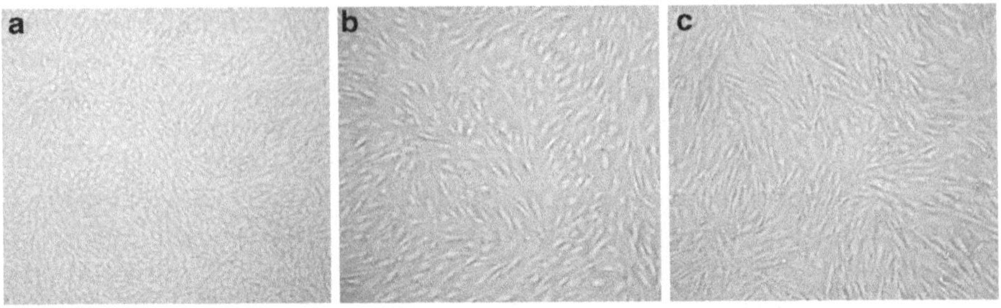

Fig. 16.1 Fibroblast cell cultures at passage 1 derived from different tissue collection techniques. (**a**) piece of ear tissue; (**b**) manual biopsy punch; (**c**) biopsy dart

dolphin (*Stenella coeruleoabla*) (Peruffo et al. 2010). Other attempts to culture and cryopreserve cells from marine mammals include pygmy sperm whale (*Kogia breviceps*) (Mancia et al. 2012), Florida manatee (*Trichechus manatus latirostris*) (Sweat et al. 2001), dolphin (*Lagenorhynchus obliquidens*) and gray seal (*Halichoerus grypus*) (Cecil and Nigrelli 1970).

Fibroblast cell establishment and culture techniques in mammals are not complicated and good plating and growth can be obtained following explant or enzymatic digestion and culture at 37 °C in a 5 % CO_2 environment (Freshney 2005). Basic culture media, such as Eagle's or Dulbecco's Minimal Essential Media (E- or D-MEM), along with the standard supplementation of fetal bovine serum (FBS, 5–20 %) and antibiotics (penicillin/streptomycin or penicillin/streptomycin/fungizone) support cell growth in the primary and passaged cultures (Freshney 2005). Good post-thaw viability of cryopreserved cells can be obtained using a basic freezing medium containing MEM supplemented with 20 % FBS and 10 % dimethyl sulfoxide (DMSO) and a 24 h cooling period at –80 °C in an insulated container prior to immersing in liquid nitrogen. Although researchers have attempted to optimize techniques for specific species or tissue types (León-Quinto et al. 2011), a standardized protocol for mammalian fibroblast culture can be implemented in a laboratory handling a wide variety of species.

Of great importance for wildlife conservation is the ease with which samples can be collected from living mammals without greatly disrupting the animal's function or health. Skin and ear biopsies obtained using a biopsy punch tool (1-8 mm disposable biopsy punches, Integra-Miltex Inc., Plainsboro, NJ, USA) or biopsy dart (biopsy syringe, Palmer Cap-Chur Inc., Powder Springs, GA, USA) have successfully produced fibroblast cultures from living animals in numerous species (León-Quinto et al. 2011; Torvar et al. 2008; Mastromonaco, unpublished data; Fig. 16.1). A recent study using six threatened Chilean species showed that biopsies can be stored in physiological saline solution supplemented with 1 % antibiotic-antimycotic (ABAM) at 4 °C without altering the time required for primary confluence for up to 7 days for ear punch biopsies and 3 days for skin dart biopsies (Torvar et al. 2008). In our laboratory, on-going studies in domestic and non-domestic bovine species are demonstrating that ear biopsies stored in phosphate buffered saline (PBS) supplemented with 1 % ABAM at 4 °C for 0, 3 and 7 days are capable of producing chromosomally stable primary and passaged cultures (Mastromonaco, unpublished data). Benkeddache et al. (2012) demonstrated that viable cells can be retrieved from rabbit ear biopsies stored in a physiological saline solution supplemented with 10 % DMSO at −20 °C for up to 10 days prior to freezing in liquid nitrogen at similar rates to fresh controls (86 versus 82 %). Samples stored up to 20 days prior to freezing in liquid nitrogen had significantly reduced viability (39 %), but still contained living cells (Benkeddache et al. 2012). These temporary storage conditions are ideal for field collections where access to a laboratory or liquid nitrogen can be hours or days away. If liquid nitrogen or nitrogen vapour are available, biopsies can be stored for a longer period of time prior to culture. Many repositories include pieces of tissue submersed in a freezing medium containing cryoprotectant (e.g. DMEM supplemented with 20 % FBS and 10 % DMSO) and frozen in liquid nitrogen. Work done by researchers at San Diego Zoo Global Wildlife Conservancy, USA, has shown that cultures can be obtained upon thawing of these cryopreserved tissues (M. Houck, personal communication). As with the biopsies from living animals, tissues from deceased animals can provide valuable material hours or even days after their death. Ear fibroblasts are consistently obtained from animals that have died unexpectedly and are found typically within 24-48 h (Mastromonaco, unpublished data).

Several studies have attempted to develop novel techniques for cell preservation. In an effort to reduce cellular injury caused by osmotic stress and improve membrane integrity, researchers used methods of microwave heating (dry preservation; Chakraborty et al. 2008) and spin-drying (lyopreservation; Chakraborty et al. 2011) to promote rapid evaporation of water from living cells. Loi et al. (2008) lyophilized sheep granulosa cells and stored them in the dark at room temperature for 3 years. Upon rehydration, the cells were non-viable and exhibited severe membrane damage, but were able to produce blastocysts when used for SCNT (Loi et al. 2008). Similar results were observed with mouse granulosa cells (Ono et al. 2008) and pig fetal fibroblasts (Das et al. 2010).

2.1.2 Fishes

Commercial stakes in the aquaculture industry have been one of the driving forces behind much of the fish cell culture research. In a recent review, Lakra et al. (2011) reported that fish cells had been cultured and banked from more than 74 species with over 283 cell lines originating from various sources, including organs, embryos and fins. The majority of these cell lines were created for toxicology or disease studies. More recently, researchers interested in preserving rare and endangered fish species or strains have identified the importance of banking fish cell cultures as a source of genetic material to compensate for the difficulty with cryopreserving fish sperm, oocytes and embryos (Rawson 2012). A fish cryobank facility, established in 2008 at the Institute of Biomedical and Environmental Science and Technology, University of Bedfordshire, UK, has already banked tissues and cells from 19 IUCN-listed fish species (Rawson 2012). Cells from the Atlantic sturgeon (*Acipenser oxyrhynchus*), a species of conservation interest in North America and Europe, have been cultured and banked (Grunow et al. 2011). However, in these cases, the animals were sacrificed as internal organs or larvae were used for culture establishment.

Although many laboratories use mammal-based techniques for fish cell culture, including culture media (e.g. DMEM supplemented with FBS and antibiotics) and incubation in a 5 % CO_2 environment (Choresca et al. 2012; Han et al. 2011), fish cells are easy to maintain in air or tightly capped flasks at room temperature without the need for excessive feeding (Lakra et al. 2010; Lannan 1994). Studies indicate that cold-water species grow better between 15 and 21 °C, whereas warm-water species grow better between 25 and 35 °C (Lannan 1994). Bols et al. (2005) indicated that optimal growth for most fish cells lines is well above the natural temperature for the fish. The requirements for pH and ability to grow well in air have led many laboratories to use Leibovitz's L-15 medium as the basal medium supplemented with FBS and antibiotics (Mauger et al. 2006; Lakra and Goswami 2011).

Many of the available cell lines have been established from internal organs, which can be collected aseptically and are not exposed to surface or environmental microorganisms. However, it is important to develop protocols for live animal collections, as has been done with mammals. Consequently, the use of fins, a highly regenerative tissue, for culture establishment has become very desirable. Whole fins have been used to grow cultures from glass catfish (*Kryptopterus bicirrhis*) (Han et al. 2011), goldfish (*Carassius auratus*) (Mauger et al. 2006), and pool barb (*Puntius sophore*) (Lakra and Goswami 2011), the latter of which grew to 100 passages. The study in goldfish tested cell growth from all fin types and found comparable results in all except the dorsal fin (Mauger et al. 2006). Caudal fin explants have been used to successfully establish cell cultures in the golden mahseer (*Tor putitora*), an endangered Indian fish species (Prasanna et al. 2000). However, with this tissue source comes an increased risk of microbial contamination and the need to use different or increased concentrations of antibiotics, which often result in increased cytotoxicity and slower or no growth from the tissue explants (Rathore et al. 2007). Fish cultures can be very slow growing (Lannan 1994) sometimes taking up to 3–4 weeks to reach primary confluence (Mauger et al. 2006), which makes it difficult to prevent microbial growth from overcoming cellular growth. Lakra and Bhonde (1996) tested fin culture in eight fish species with unpromising outcomes: only one species attached and grew whereas the other seven either attached but did not grow or were lost to fungal contamination. Studies in our laboratory attempting to optimize culture conditions from fin biopsies of an endangered African cichlid species, ngege (*Oreochromis esculentus*), have been delayed by persistent culture contamination originating from the tissue samples (Filice, unpublished data). Interestingly, fin biopsy cultures of a commercial tilapia species being used as a model are not succumbing to contamination using the same tissue disinfection, sample processing and culture techniques (Filice, unpublished data). Preliminary attempts at short-term storage of fin samples in a variety of chilled media have not yet yielded promising results (Filice, unpublished data).

2.1.3 Birds

Avian cell lines, primarily from domestic chicken, have been extensively used and studied by virologists and toxicologists for the benefits of both human and avian medicine (Moresco et al. 2010; Portz et al. 2008). The majority of cell culture work in bird species has been done using chicken embryos at an early stage of development (day 8–11). To date, there are very few reports in the literature regarding the preservation of avian genomes as somatic cells. Chinese researchers have been producing early embryo-derived cell cultures to preserve nationally protected domestic chicken breeds, including the Qingyuan partridge chicken (Na et al. 2010), Langshan chicken (Guan et al. 2010) and Big Bone chicken (Su et al. 2011). In all these cases, stable cell lines exhibiting good viability and chromosome normality were created. Recently, Kjelland and Kraemer (2012) used semi-mature and mature feathers containing feather pulp and post-hatch egg shells to culture fibroblasts from six domesticated bird species, including emu (*Dromaius novaehollandiae*) and ostrich (*Struthio camelus*). This study is an excellent example of the use of cast-off material and non-invasive sample collection.

The majority of researchers use mammalian-based cell culture systems for the growth of primary and passaged cells as in the studies described above. Minimal essential media supplemented with FBS and antibiotics are generally described. No evidence has been found in the published literature of attempts to optimize culture conditions or cryopreservation techniques in an effort to improve cell viability in other bird species.

2.1.4 Reptiles and Amphibians

Reptiles encompass the least studied taxon when it comes to in vitro techniques, such as cell culture. Lack of pressure from commercial or research interests, particularly health- and food-related industries, has resulted in a smaller number of studies investigating cultured reptile cells. Primary cell cultures have been produced from early embryos of the soft shell turtle (*Pelodiscus sinensis*) (Liu et al. 2012), heart, liver and muscle of the Chinese alligator (*Alligator sinensis*) (Zeng et al. 2011), and venom glands of the South American rattlesnake (*Crotalus durissus terrificus*) (Duarte et al. 1999). Embryos incubated up to 30–40 days were used to establish cultures from green sea turtles (*Chelonia mydas*), a species listed as endangered by the IUCN (Moore et al. 1997). In an attempt to develop non-lethal techniques for culture derivation, tail and toe clippings were used to prepare cell cultures from five species of Australian dragon lizards (Ezaz et al. 2008). The cell lines were grown for ten passages and retained their viability and normal diploidy. In our laboratory, cell cultures were produced from Komodo dragon (*Varanus komodoensis*), another IUCN-listed reptile species, from a thin strip of skin taken from the incision site during a surgical procedure (Mastromonaco, unpublished data).

Reptile cells can be grown in mammalian-based culture systems (e.g. DMEM supplemented with FBS and antibiotics) and a 5 % CO_2 environment (Stephenson 1966). Studies have shown that although the cells can grow at a wide range of temperatures (Stephenson 1966), most species exhibit optimal growth at 28–30 °C (Ezaz et al. 2008; Moore et al. 1997; Clark et al. 1970). Although studies have investigated the influence of temperature on in vitro cell growth rates, very little work has been done, as in birds, on investigating optimal techniques for culture establishment, culture conditions and cryopreservation methods.

Amphibians, on the other hand, have experienced a recent explosion in biobanking activity. The sudden amphibian crisis resulting from both continued habitat loss and the on-going spread of chytridiomycosis have created urgency among zoological and academic professionals to preserve genetic material from any remaining individuals and species. Fortunately, one of the most widely studied laboratory animal species is the African clawed frog (*Xenopus laevis*) in which numerous cell lines, primarily from embryos and tadpoles, have been developed for research purposes and thus, some basic tips for amphibian cell culture are available (Anizet et al. 1981). Tadpole hindlimbs were used to produce cell cultures for transgenic studies in a related species, *Xenopus tropicalis* (Sinzelle et al. 2012). The cell lines were long-lived, surviving over 60 passages, an ideal condition for transgenic studies. There are no descriptions in the scientific literature of cell cultures grown from non-lethal tissue sources or from endangered amphibian species. In November, 2011, Science News Daily reported that researchers at San Diego Zoo Global Wildlife Conservancy had managed to grow cells from frozen biopsy pieces of the endangered Mississippi gopher frog (*Rana sevosa*), increasing the number of banked individuals to 19 (www.sciencenewsdaily.org, verified January 20, 2013).

Similar to the experience with fish tissues, amphibian samples that cannot be collected aseptically, as with internal organs, are highly susceptible to contamination from the external environment. This is especially the case with early embryos in which the jelly coats harbour microbes encountered during passage through the cloaca and from the environment (Laskey 1970). This poses a challenge when attempting to grow uncontaminated cultures and great effort is spent investigating other antibiotics (e.g. gentamycin) or combinations of antibiotics (e.g. kanamycin + polymyxin E) to increase microbicidal activity (Laskey 1970).

2.2 *Ensuring Culture Quality in All Taxa*

Material stored in biobanks must be of the highest quality if it is to be considered in the future, particularly for offspring production. Environmental and genetic factors related to the sample itself (species, donor age, tissue type), collection methods, and culture conditions play a role in cell growth and longevity, thereby influencing the viability and normality of the cultures grown in vitro (Mastromonaco et al. 2007). Very few of the published reports detailing the successful establishment of a new cell line characterize the quality of the culture produced either by examining replicative ability or chromosomal stability. A long-lived cell line able to grow for more than 60 passages but with a chromosomal abnormality as with *Xenopus tropicalis* (trisomic for chromosome 10; Sinzelle et al. 2012) may be adequate for transgenic studies, but not for potential use in offspring production. A similar outcome occurred with common carp (*Cyprinus carpio*) cells, which grew to almost passage 50, but with only 46–48 % of cells containing a normal chromosomal complement (Lakra et al. 2010) and also with iguana (*Iguana iguana*) cells, which exhibited minor chromosome changes between passages 10 and 40 (Clark et al. 1970). Likewise, cells that may not have any whole chromosomal concerns, but grow only to passage 1 (Nel-Themaat et al. 2007) will not have the replicative ability to produce the number of cells required for either nuclear transfer or stem cell technologies. Culture establishment techniques that result in primary confluence occurring after more than double the time expected (i.e. >20 days versus 7–10 days) as in the cold stored dart biopsies from Chilean mammal species (Torvar et al. 2008) should be suspect (Fig. 16.2). Replicative ability is correlated with seeding density, specifically the number of viable cells available or able to initiate the culture. The lower the seeding density, the greater the number of mitotic events, and thus, exhaustion of replicative ability, required by each individual cell in order to achieve confluence. Mastromonaco et al. (2006) found that samples obtained by punch biopsy tool or dart resulted in lower quality cultures compared to larger samples of skin or ear collected post-mortem as indicated by decreased lifespan (<30 population doublings versus >50 population doublings), morphological changes characteristic of senescence and increased percentage of chromosomally abnormal cells (up to 58 % versus <20 %). When used for SCNT, short-lived cell lines or those with high levels of chromosomally abnormal cells result in poor blastocyst development as discussed in further detail below.

When attempting to biobank from endangered species, particularly from living animals, the samples are generally very small and highly valuable pieces of tissue. Little attention has been given to the need for every explant piece to succeed and grow cells (León-Quinto et al. 2011). It is not generally done, but when banking cells from a genetically valuable individual, tissue explants can be re-plated in a new flask following the initial outgrowth and a further growth of cells can be obtained (Mastromonaco, unpublished data). León-Quinto et al. (2011) tested multiple parameters, including tissue explants versus enzymatic digestion, serum and growth factor supplementation, and cryopreservation media for the Iberian lynx. After assessing 20 different culture conditions and 15 different freezing solutions, the authors identified a protocol promoting increased cell growth and post-thaw viability for the lynx. Although these studies are important for other mammal species, a lot more is known about in vitro cell dynamics in mammals than in other taxa. Greater efforts, primarily time and funding, should be spent investigating these basic parameters for fishes, birds, reptiles and amphibians.

It is standard procedure with cryopreserved sperm samples to carry out a post-thaw evaluation to assess basic sperm characteristics, such as motility, morphology and acrosome membrane status. These data provide an estimate of the fertilizing ability of the sample, necessary information when considering how best to use the

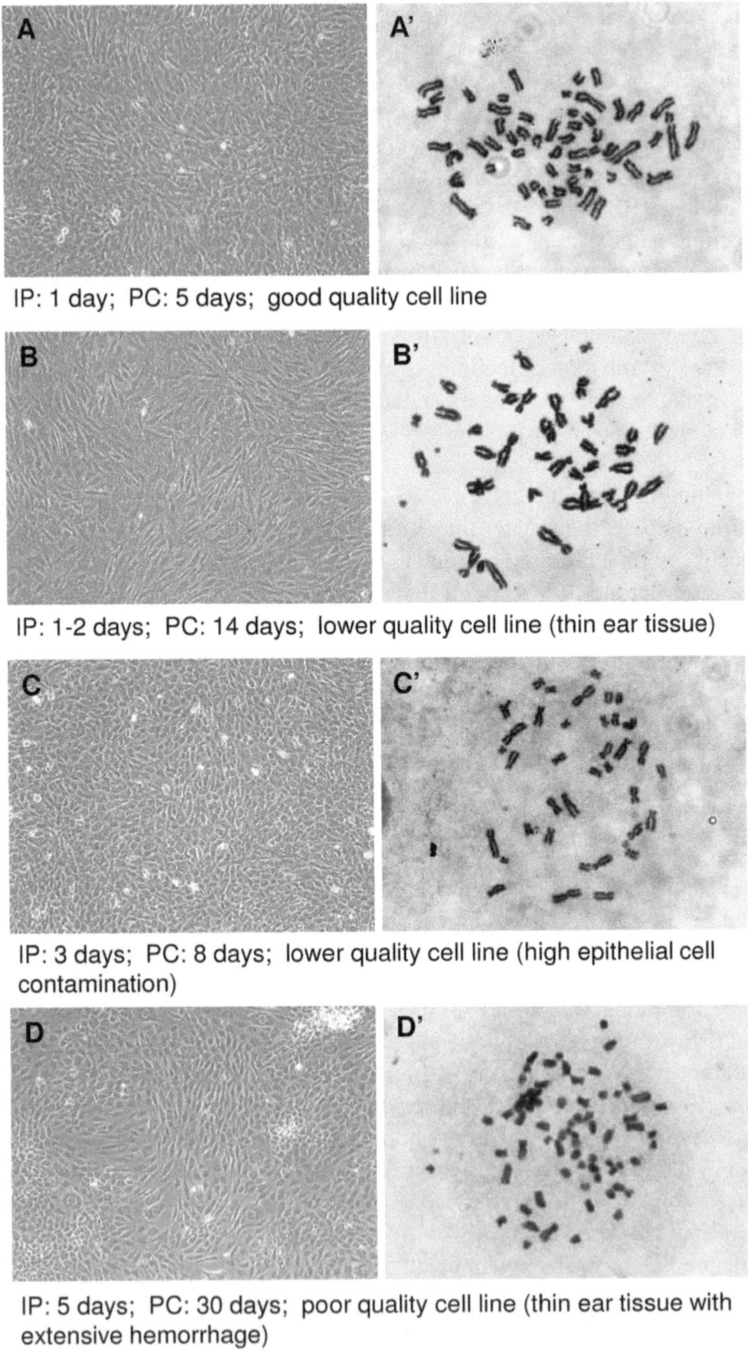

Fig. 16.2 In vitro growth characteristics at passage 1 of cells derived from whole ear tissue. Cell morphology and chromosome spreads from gaur adult (A and A'), Vancouver Island marmot adult (B and B'), cheetah adult (C and C'), polar bear newborn (D and D'). *IP* number of days for initial plating of dissociated cells; *PC* number of days to achieve primary confluence

sample in the future. Is the sample good enough for artificial insemination or will in vitro fertilization or intracytoplasmic sperm injection be required? A similar approach would be beneficial for banked somatic cell samples whereby a standardized list of criteria are assessed, including post-thaw viability, plating efficiency, number of days to confluence, culture to at least passage 10, and chromosomal content at passage 10. A realistic assessment can then be made and the potential use for the banked cells is known. As seen in cases described previously, cells that may be acceptable for genomic or proteomic studies may not be appropriate for use with techniques leading to offspring production.

Of equal importance as cell viability and normality is the proper identification of the species from which the banked tissues and cell lines are derived. Although necessary for all species, it is particularly important for non-mammalian species where identification based on taxonomic knowledge and phenotypic markers may be more difficult and for samples obtained from field collections compared to well-known captive populations. DNA barcoding, as established by Hebert et al. (2003), is a technique that uses short sections of DNA sequence from a standard part of the genome (e.g. mitochondrial cytochrome c oxidase 1 gene in animals) to differentiate samples at the species level. A recent study by Maya-Soriano et al. (2012) confirmed that the mitochondrial DNA 16S ribosomal RNA sequence that is typically used for DNA barcoding in amphibians was used to successfully identify a variety of banked frog and toad, but not newt, samples to the species level. As the number of "barcoded" species grows and the technique becomes more affordable and accessible, it will become important, if not expected, to include DNA barcoding as one of the required pieces of information on catalogued samples.

2.3 *Stem Cell Collection/Production and Culture*

The ground-breaking work of scientists at the University of Toronto in the early 1960s (Till and McCulloch 1961) sparked the new research field of stem cell biology. The very nature of these cells, that is, their ability to self-renew and differentiate to produce mature progeny cells (Wagers and Weissman 2004), is what attracted cell biologists and medical professionals to investigate the potential in these cells for tissue repair and regenerative medicine. As described by Wagers and Weissman (2004), stem cells are classified according to their developmental potential, including totipotent (able to give rise to all embryonic and extra-embryonic cell types), pluripotent (able to give rise to embryonic cell types), and multipotent (able to give rise to a subset of cell lineages). Stem cells can also be categorized according to the tissue from which they arise, including embryonic stem cells (from the inner cell mass of the blastocyst), germline stem cells (from gonadal tissue) and adult stem cells (from various sources in newborns to adults). The field of stem cell research has grown dramatically in the past decade for the study of basic biological processes and for application in therapeutic and reproductive medicine. For the purposes of this chapter and the focus on somatic cell banking, embryonic stem cells will not be discussed further. Embryonic stem cell techniques have been reviewed in companion animals by Tayfur Tecirlioglu and Trounson (2007), in birds by Petitte (2006), and in fishes by Hong et al. (2011). It is important to mention, however, that recent advances using mouse embryonic stem cells has resulted in the production of functional sperm (Hayashi et al. 2011) and oocytes (Hayashi et al. 2012).

2.3.1 Germline Stem Cells

Spermatogonial stem cells (SSCs), the basis for continuous sperm production in males, have been investigated as a source of germline material for sperm production from genetically valuable animals. SSC techniques, originally developed in mice, have now been documented in a variety of species (reviewed by Dobrinski and Travis 2007). In the cat and dog, transplantation of SSCs, which involves characterization, isolation, and transfer of cells, has been attempted with mixed results (Travis et al. 2009). In brief, this involved isolating the SSCs followed by transfer into a germ-cell depleted (via radiation) host. On occasion, it has been possible to recover ~20 % of mature sperm cells derived from the donor (Travis et al. 2009). Others have transplanted germ cells from a wild felid (ocelot; *Leopardus pardalis*) into the domestic cat to produce spermatozoa successfully from the donor (Silva et al. 2012). Recent results in SSC transplantation in primates are encouraging since spermatogenesis can be resumed and lead to the production of fully functional spermatozoa (Hermann et al. 2012). In fishes, fully functional rainbow trout (*Oncorhynchus mykiss*) sperm were produced by masu salmon (*Oncorhynchus masou*) following transplantation of trout spermatogonia into the peritoneal cavity of newly hatched salmon embryos (Yoshizaki et al. 2012). These authors document similar successes in a variety of other fish species (reviewed by Yoshizaki et al. 2012). A recent study in birds showed that Japanese quail (*Coturnix japonica*) spermatogonia were capable of colonizing domestic chicken (*Gallus gallus*) testes, although the production of fully functional sperm was not detected (Pereira et al. 2012).

In a 2009 review, Tilly and Telfer discuss the exciting new findings supporting the existence of proliferative germ cells in the adult ovary, which may soon challenge the age-old belief that females are born with a non-renewable pool of oocytes at birth. These ovarian stem cells have been isolated from multiple species, including the giraffe (*Giraffa camelopardalis*), and successfully cultured in vitro (Telfer, personal communication).

2.3.2 Somatic Stem Cells

Adult or somatic stem cells typically arise in tissues with high regenerative capacity, including blood, skin, intestine, and respiratory tract (Wagers and Weissman 2004). Although some of these tissues are difficult to obtain from living animals, especially non-invasively, the banking of a population of cells with an indefinite lifespan and the capacity to create many other cells within the organism, primarily germ cells, is highly enticing and requires further investment in wildlife species. Furthermore, studies using different cell types as nuclear donors for SCNT indicate that the less differentiated the donor nucleus, the greater the developmental potential of the SCNT embryo both pre- and post-implantation (reviewed by Hochedlinger and Jaenisch 2002), suggesting that some adult stem cell populations may be a more appropriate source of donor material for offspring production by SCNT.

The majority of studies on adult stem cells to date have been carried out in laboratory and livestock species, including mouse (brain: Gritti et al. 1996), rhesus monkey (bone marrow: Chang et al. 2006), goldfish (retina: Wu et al. 2001), zebrafish (melanocyte: Tryon and Johnson 2012), pig (skin: Dyce et al. 2004), cattle (bone marrow: Bosnakovski et al. 2005), red deer (antler: Berg et al. 2007), horse (cord blood: Koch et al. 2007), and chicken (feather follicle: Xi et al. 2003). Isolation of adult stem cells may be more difficult than differentiated somatic cells depending on their source as proportionately smaller numbers of stem cells are available in adult tissue fragments. Furthermore, in vitro culture of stem cells ranges from being as effortless as fibroblast culture (mesenchymal stem cells; Cheng et al. 2012) to requiring specialty media (neural stem cells; Hutton and Pevny 2008) and use of feeder cell layers (epithelial stem cells; Nowak and Fuchs 2009). Most importantly, proper characterization of the isolated and cultured cells is required in order to classify them as stem cells, including extended reproductive lifespan, ability to differentiate into other cell lineages, and expression of a specific array of cell surface markers (Dominici et al. 2006).

Despite the claims that biobanking facilities have started including stem cell cultures from endangered species as part of their cryopreserved inventory, very little information is available in the published literature regarding the isolation, establishment and characterization of adult stem cells from wildlife species. Stem cells derived from adipose tissue biopsies of wild Scandinavian brown bears (*Ursus arctos*) collected during the implantation of tracking devices were shown to differentiate into osteocytes and chondrocytes in vitro (Fink et al. 2011). As with red deer (*Cervus elaphus*), antler-derived stem cells were cultured from fallow deer (*Dama dama*) and characterized using specific mesenchymal stem cell markers, such as STRO-1 (Rolf et al. 2008).

Unlike fibroblast cell culture and banking, which can be easily undertaken by most laboratories, adult stem cell culture and banking requires a greater investment in supplies, equipment, and knowledge (e.g. molecular techniques), as well as more stringent criteria for evaluation, which in turn require greater efforts to complete (e.g. long-term culture, exposure to differentiating factors, cell surface marker identification, chromosome analysis). Although samples for adult stem cell culture can be obtained non-invasively, or opportunistically, there is a greater chance of failure to establish a stem cell line compared to fibroblast culture. Therefore, although fibroblast culture banks can become a widespread movement, stem cell banks may continue to remain in specialized centers, most of which are affiliated with academic institutions.

3 Manipulation of Somatic Cells for Potential Offspring Production

The adult somatic cell is genetically identical to the totipotent blastomere from an early embryo, with the exception that numerous modifications to the genome during embryonic and fetal development determined the adult cell's fate and resulted in a loss in its ability to give rise to multiple cell types, that is loss of pluripotency. However, the presence of the entire genomic code within adult somatic cells makes them an attractive source of genetic material to be considered for propagation of the genetic donor. Researchers have shown that there are multiple strategies for "de-differentiating" somatic cells and inducing an embryonic cell-like state, including nuclear transfer, cell fusion, cell extract exposure, and cell explantation (reviewed by Hochedlinger and Jaenisch 2006). The cytoplasmic machinery within the oocyte has already been proven to reprogram somatic cell nuclei as embryos and living offspring have been produced by SCNT from a wide array of species. Another promising technique still in its infancy is the induction of pluripotency by exposure of cultured somatic cells to cell extracts, virally-mediated transcription factors or other transduction factors. Despite great advances in the past 5–10 years, very little is known about the mechanisms involved in nuclear reprogramming and the potential factors that may influence its efficiency in somatic cells. Many authors agree that an increased understanding of nuclear reprogramming and its control is the only way to enhance the efficiency of techniques, such as SCNT and induction of pluripotent stem cells, in which the goal is to erase all somatic genome modifications and re-establish embryonic genome activity.

3.1 Reproductive Cloning (SCNT)

In the 1950s, nuclear transfer was originally attempted as a method for investigating nuclear equivalence between embryonic and differentiated somatic cells (Briggs and King 1952). Nuclear transfer was, therefore, developed as an experimental tool for the study of specific cellular mechanisms occurring during development. Although it still has great value as an experimental tool, the 1980s brought a shift in the application of nuclear transfer to the propagation of genetically valuable livestock and laboratory animals. The birth of "Dolly" in 1996 came after decades of research by numerous laboratories using a variety of animal models, including frogs, mice, cattle and sheep. Knowing that blastomeres separated from embryos could be reprogrammed to produce viable embryos and offspring, this was the first report of successful reprogramming of an adult somatic cell (from the mammary gland) using SCNT (Wilmut et al. 1997). In the past 16 years since this breakthrough, hundreds of animals from livestock, companion animal and laboratory animal species have been produced using a variety of somatic cell types and sources. However, extensive efforts in technique development and optimization, and in attempting to understand the cellular machinery that drives successful nuclear reprogramming have greatly improved, but not yet overcome, the high rates of developmental loss that are encountered during all phases of SCNT-related reproductive processes (i.e. during embryo reconstruction, pre-implantation development, post-implantation development and neonatal survival). Fetal and neonatal morphological and physiological abnormalities have been linked to placental aberrations (reviewed by Chavatte-Palmer et al. 2012) and epigenetic defects resulting from altered genome methylation patterns (reviewed by Peat and Reik 2012). Despite these challenges, SCNT has the potential to become a valuable tool for the production of: (1) individuals carrying valuable and/or desirable genetic traits; (2) animal models of disease; and (3) samples for the investigation of the fundamental aspects of pre- and post-implantation embryo development and fetal-maternal interactions (Mastromonaco and King 2007). For endangered species, SCNT offers the possibility of creating embryos from individuals where gametes or germline cells are not available. In more extreme circumstances, SCNT has been contemplated as a method for resurrecting extinct species in which undamaged or complete nuclear genomes are no longer available (see Chap. 19).

3.1.1 Mammals

Mammals have been the most extensively studied taxa for SCNT with live offspring being consistently produced from rodents to primates. The application of SCNT as a method of assisted reproduction for endangered species was first done in mammals (gaur (*Bos gaurus*); Lanza et al. 2000) and has yet to be reported in any other taxa. Researchers faced with the task of obtaining large numbers of oocytes required for SCNT from endangered females overcame this challenge by using oocytes from related domestic species as cytoplasmic recipients; a technique called interspecies SCNT (iSCNT). This idea was supported by the study of Dominko et al. (1999), which demonstrated that the bovine cytoplasm was capable of reprogramming, at least partially, diverse donor nuclei (sheep, pig, monkey, rat) past maternal-embryonic transition and, with the exception of the rat, progress to the blastocyst stage. Other than a handful of studies on non-traditional, albeit domesticated or laboratory-based, species using intraspecies SCNT, including dromedary camel (*Camelus dromedarius*; Wani et al. 2010), ferret (*Mustela putorius furo*; Li et al. 2006b), rhesus monkey (*Macaca mulatta*; Zhou et al. 2006), and long-tailed macaque (*Macaca fascicularis*; Ng et al. 2004), all reports in wild species involve iSCNT. To date, published reports show that more than 30 wild species have been attempted with the majority of the species achieving blastocyst development, but poor post-implantation success (Table 16.2; Fig. 16.3).

Bovidae

The domestic cattle (*Bos taurus*) oocyte has been well-studied both in vivo and in vitro for over 20 years. As a result of the volume of knowledge on domestic cattle oocyte maturation and availability of ovaries from local abattoirs, cattle oocytes

Table 16.2 Overview of interspecies SCNT outcomes

Cell donor	Cell type	CITES appendix	Recipient oocyte	Reconst. embryos (n)	Blastocysts (%) [a]	Transferred embryos (n)	No. fetuses/ pregnancies (n) ~D30	~D60	~D120	Off-spring (n)	Outcome	Reference
Bovidae												
Bos gaurus (Gaur)	Skin fibroblast	I	*Bos taurus*	922	33.2–37.5	106		4	3	1	Dead <24 h	Srirattana et al. (2012)
Bos gaurus (Gaur)	Ear fibroblast	I	*Bos taurus*	228	18.1	ND	ND			ND	ND	Mastromonaco et al. (2007)
Bos gaurus (Gaur)	Skin fibroblast	I	*Bos taurus*	692	12.0*	44		6	2	1	Dead <1 week	Lanza et al. (2000)
Budorcas taxicolor (Takin)	Ear fibroblast	II	*Bos taurus*	227	6.6	ND	ND			ND	ND	Li et al. (2006a)
Taurotragus oryx (Common Eland)	Seminal epithelial cell	Not listed	*Bos taurus*	209	0	ND	ND			ND	ND	Nel-Themaat et al. (2008)
Naemorhedus goral (Himalayan Goral)	Skin fibroblast	I	*Bos taurus*	506	0–5.0	ND	ND			ND	ND	Oh et al. (2006)
Bos javanicus (Banteng)	Ear fibroblast	Not listed	*Bos taurus*/ *Bos indicus*	319	20.0–28.0	38	2		0	0	0	Sansinena et al. (2005)
Capra ibex (Alpine Ibex)	Ear fibroblast	Not listed	*Capra hircus*	790	11.0	ND	ND			ND	ND	Wang et al. (2007)
Capra pyrenaica pyrenaica (Pyrenean Ibex)	Ear fibroblast	Extinct	*Capra hircus*	782	NA	184		7		1	Dead <24 h	Folch et al. (2009)
Pantholops hodgsonii (Tibetan Antelope)	Ear fibroblast	I	*Oryctolagus cuniculus*	1097	5.5-20.4	ND	ND			ND	ND	Zhao et al. (2006)

Tragelaphus eurycerus isaaci (Bongo)	Skin fibroblast	III: Ghana	*Bos taurus*	365	14.0-24.2*	ND	ND			ND	ND	Lee et al. (2003)
Ovis canadensis mexicana (Desert Bighorn Sheep)	Skin fibroblast	II: Mexico	*Ovis aries*	265	NA	223	5	0		0	0	Williams et al. (2006)
Ovis orientalis musimon (European Mouflon)	Granulosa cell	I	*Ovis aries*	23	30.4*	7	2	1	1	1	Live	Loi et al. (2001)
Ovis ammon (Argali Sheep)	Skin fibroblast	II	*Ovis aries*	NA	0	28	1			0	0	White et al. (1999)
Ovis orientalis isphahanica (Esfahan Mouflon)	Skin fibroblast	I	*Ovis aries*	667	7.6	12	2			2	Dead <24 h	Hajian et al. (2011)
Pseudoryx nghetinhensis (Saola Antelope)	Skin fibroblast	I	*Bos taurus*	312	35.1	ND	ND			ND	ND	Bui et al. (2002)
Bison bison athabacae (Wood Bison)	Ear fibroblast	II	*Bos taurus*	226	19.2	ND	ND			ND	ND	Kumar et al. (2009)
Cervidae												
Pudu puda (Pudú)	Ear fibroblast	I	*Bos taurus*	89	0–7.4	ND	ND			ND	ND	Venegas et al. (2006)
Felidae												
Pardofelis marmorata (Marbled Cat)	Muscle fibroblast	I	*Felis catus*/ *Oryctolagus cuniculus*	63/56	0–11.5	ND	ND			ND	ND	Thongphakdee et al. (2006)

(continued)

Table 16.2 (continued)

Cell donor	Cell type	CITES appendix	Recipient oocyte	Reconst. embryos (n)	Blastocysts (%) [a]	Transferred embryos (n)	No. fetuses/ pregnancies (n)			Off-spring (n)	Outcome	Reference
							~D30	~D60	~D120			
Pardofelis marmorata (Marbled Cat)	Skin fibroblast	I	*Felis catus*	81	5.4	ND	ND			ND	ND	Thongphakdee et al. (2010)
Felis margarita (Sand Cat)	Skin fibroblast	II	*Felis catus*	485	6.0–43.0	1600	18	14		14	4 Dead 0 h; 5 Dead <24 h; 5 Dead <60 day	Gomez et al. (2008)
Felis nigripes (Black Footed Cat)	Skin fibroblast	I	*Felis catus*	NR	0–3.3*	612	14			0	0	Gomez et al. (2011)
Panthera tigris altaica (Siberian Tiger)	Skin fibroblast	I	*Sus scrofa domesticus*	675	0–1.6	ND	ND			ND	ND	Hashem et al. (2007)
Prionailurus planiceps (Flat-headed Cat)	Muscle/skin fibroblast	I	*Felis catus*	561	8.3–8.6	384	0			0	0	Thongphakdee et al. (2010)
Prionailurus bengalensis (Leopard Cat)	Skin fibroblast	I or II	*Felis catus*	185	5.7–20.7	409	6			0	0	Lee et al. (2010)
Prionailurus bengalensis (Leopard Cat)	Skin fibroblast	I or II	*Felis catus*	412	7.2-7.8*	ND	ND			ND	ND	Yin et al. (2006)
Felis silvestris libica (African Wild Cat)	Skin fibroblast	II	*Felis catus*	484	21.0–41.7	ND	ND			ND	ND	Gomez et al. (2006)

Felis silvestris libica (African Wild Cat)	Skin fibroblast	II	*Felis catus*	1552	NA	1552	24	17	17	7 Dead 0 h; 7 Dead <36 h; 3 Live	Gomez et al. (2004)
Panthera tigris altaica (Siberian Tiger)	Ear fibroblast	I	*Felis catus*	NR	8.8*	NA	0		0	0	Hwang et al. (2001)
Canidae											
Canis lupus (Gray Wolf)	Ear fibroblast	I or II	*Canis familiaris*	251	NA	251	4	2	2	Live	Kim et al. (2007)
Canis lupus (Gray Wolf)	Skin fibroblast	I or II	*Canis familiaris*	372	NA	372	6	6	6	3 Dead <24 h; 3 Live	Oh et al. (2008)
Ursidae											
Ailuropoda melanoleuca (Giant Panda)	Muscle fibroblast	I	*Oryctolagus cuniculus*	612	10.9	ND	ND		ND	ND	Wen et al. (2005)
Ailuropoda melanoleuca (Giant Panda)	Muscle fibroblast	I	*Oryctolagus cuniculus*	2510	NA	2510	2		0	0	Chen et al. (2002)
Ursus thibetanus (Asian Black Bear)	Skin fibroblast	I	*Bos taurus*	270	4.1*	ND	ND		ND	ND	Ty et al. (2003)
Ailuridae											
Ailurus fulgens (Red Panda)	Ear fibroblast	I	*Oryctolagus cuniculus*	194	22.6	ND	ND		ND	ND	Tao et al. (2009)
Balaenopteridae											
Balaenoptera bonaerensis (Minke Whale)	Granulosa cell	I	*Bos taurus*/ *Sus scrofa domesticus*	1003	0/0	ND	ND		ND	ND	Ikumi et al. (2004)

(continued)

Table 16.2 (continued)

Cell donor	Cell type	CITES appendix	Recipient oocyte	Reconst. embryos (n)	Blastocysts (%) [a]	Transferred embryos (n)	No. fetuses/ pregnancies (n)	Off-spring (n)	Outcome	Reference
Balaenoptera borealis (Sei Whale)	Fetal fibroblast	I	*Sus scrofa domesticus*	253	0	ND	ND	ND	ND	Lee et al. (2009)
Balaenoptera borealis (Sei Whale)	Fetal fibroblast	I	*Bos taurus*	397	0	ND	ND	ND	ND	Bhuiyan et al. (2010)
Hominidae										
Pan troglodytes (Chimpanzee)	Skin fibroblast	I	*Bos taurus*	1224	0	ND	ND	ND	ND	Wang et al. (2009)
Macaca mulatta (Rhesus Monkey)	Skin fibroblast	II	*Oryctolagus cuniculus*	557	12.7	ND	ND	ND	ND	Yang et al. (2003)

ND procedure not done, *NA* data not available

[a]Blastocyst development was calculated based on number of cleaved embryos unless otherwise indicated (*)

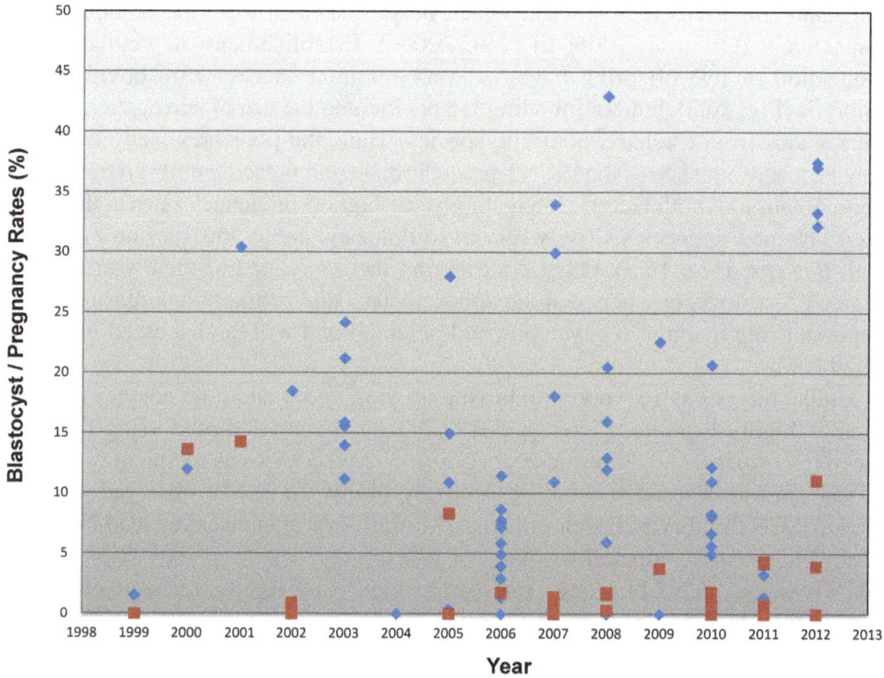

Fig. 16.3 Interspecies SCNT success rates in the past 15 years. (◆) blastocyst development rates; (■) pregnancy rates. Blastocyst development rates were listed as reported in the literature: from cleaved embryos, reconstructed embryos or total embryos. Pregnancy rates were calculated based on number of transferred embryos and number of pregnancies reported on last ultrasound examination

have been used as cytoplasmic recipients for a number of non-domestic species (Table 16.2). Along with closely-related *Bos spp.*, such as gaur (*Bos gaurus*) and banteng (*Bos javanicus*), attempts to produce embryos or offspring have included a wide range of bovid species from the mountain bongo (*Tragelaphus eurycerus isaaci*) to the goral (*Naemorhedus goral*) (Table 16.2). In most cases, development to the blastocyst stage was achieved, however, at a significantly lower rate than that observed in domestic cattle SCNT controls. For instance, only 18.1 % blastocyst development was obtained in the gaur iSCNT treatment group compared to 42.3 % obtained in the cattle SCNT control group (Mastromonaco et al. 2007). It is important to note that embryo cleavage rates are significantly lower or more variable following SCNT than in vitro fertilization (IVF) or intracytoplasmic sperm injection (ICSI) and thus, in this paper blastocyst development rates are reported as a percentage of cleaved embryos rather than total reconstructed embryos, unless otherwise stated. Authors have indicated that the greater the evolutionary divergence between the nuclear donor and cytoplasmic recipient species, the lower the developmental potential of the reconstructed iSCNT embryos past the stage of maternal-embryonic transition. This supports the outcomes observed for the goral (*Naemorhedus goral*) and takin (*Budorcas taxicolor*) in which blastocyst rates were 5.1 % and 6.6 %, respectively (Oh et al. 2006; Li et al. 2006a). Establishment of pregnancy and production of live offspring have had very minimal success with bovid iSCNT embryos (Fig. 16.3), but confounding factors include the use of surrogate or recipient females from a related domestic species. Thus, the poor pregnancy outcomes may be a consequence of the iSCNT procedure, the interspecific embryo transfer, or a combination of both factors. Interestingly, the highest pregnancy rates to date have been obtained in species closely related to domestic cattle, the gaur and banteng, both *Bos spp.* (Fig. 16.3). Other contributing factors related to the donor cell and degree of homogeneity between the donor nucleus and recipient cytoplasm play an important role in embryo developmental potential and will be discussed in further detail below.

Unlike the extensive work with bovine oocytes, sheep and goat oocytes have not been as highly exploited. Interspecies SCNT has been attempted using domestic sheep (*Ovis aries*) and domestic goat (*Capra hircus*) oocytes as recipient cytoplasm in four and three wild species, respectively (Table 16.2). As observed with the bovine iSCNT embryos, development to the blastocyst stage was reduced in iSCNT treatment groups compared to ovine or caprine controls (e.g. ibex (*Capra ibex*) iSCNT vs goat SCNT blastocyst rates were 16.5 % and 40.4 %, respectively; Wang et al. 2007). In the five ovine and caprine species in which embryo transfers were attempted, pregnancies were established in all of the species, but only one mouflon (*Ovis orientalis musimon*; Loi et al. 2001) and one ibex (*Capra pyrenaica pyrenaica*; Folch et al. 2009) have survived.

Felidae

Some of the more successful and reproducible outcomes have been demonstrated in the small wild cat species. The increased success of iSCNT in small felids compared to other wildlife species is partially due to the extensive efforts by the research team at the Audubon Center for Research of Endangered Species, USA. Years of studying domestic cat (*Felis catus*) oocyte maturation and embryo culture laid a strong foundation for the feline iSCNT experiments. To date, embryos from seven species of small wild cats have been produced by iSCNT and living offspring have been achieved from African wild cat (*Felis silvestris libica*) and sand cat (*Felis margarita*). Interestingly, low levels of blastocyst development (<10 %) are observed with domestic cat SCNT (Gómez et al. 2003; Thongphakdee et al. 2010). Therefore, iSCNT blastocyst rates are comparable to, and in some cases, significantly higher than, cat SCNT controls. Interspecies SCNT blastocyst rates for marbled cat (*Pardofelis marmorata*) and flat-headed cat (*Prionailurus planiceps*) were 5.4 % and 8.6 %, respectively (Thongphakdee et al. 2010), whereas Gómez et al. (2003) showed that African wild cat resulted in significantly greater blastocyst rates (24.2 %) than domestic cat controls (3.3 %; percentage of blastocyst development from fused couplets). Attempts to produce embryos from a large wild cat species has only been reported in the Siberian tiger (*Panthera tigris altaica*) using domestic cat, cattle and pig (*Sus scrofa domesticus*) oocytes as recipients for the tiger donor cells (Hwang et al. 2001; Hashem et al. 2007). As with domestic and small wild cats, SCNT blastocyst rates of <10 % were reported with all recipient oocyte species.

Canidae and Other Carnivores

Recently, there have been several studies investigating the use of iSCNT for other species in the order Carnivora. Two studies in gray wolves (*Canis lupus*) using domestic dog (*Canis familiaris*) oocytes as cytoplasmic recipients resulted in the production of live offspring (Kim et al. 2007; Oh et al. 2008). In both cases, embryo development in vitro was not recorded since the embryos were transferred into domestic dog recipients as presumptive zygotes. This is due to the fact that in vitro embryo culture is not well-developed in canids (Luvoni et al. 2006). Furthermore, in vitro maturation of oocytes is also problematic in dogs (Luvoni et al. 2006), therefore, these studies relied on the use of in vivo matured oocytes. An early study in the Asiatic black bear (*Ursus thibetanus*) using domestic cattle oocytes as cytoplasmic recipients resulted in development to the blastocyst stage in a small number of embryos (Ty et al. 2003). The domestic rabbit (*Oryctolagus cuniculus*) oocyte supported good blastocyst development rates (>20 %) for both giant panda (*Ailuropoda melanoleuca*) and red panda (*Ailurus fulgens*) donor cells (Chen et al. 2002; Tao et al. 2009).

Factors Influencing iSCNT Outcomes

Abnormalities affecting SCNT embryos, pregnancies and neonates are well-documented in the literature and are a reminder that the technique is not highly efficient or without its drawbacks. Along with a variety of factors influencing SCNT outcomes, some of which will be discussed below, one of the primary concerns with iSCNT embryos is the presence of cytoplasmic components, specifically mitochondria, from another species. Mitochondria play a primary role in energy production and are involved in such functions as metabolism, cell growth, development, apoptosis, disease and aging (Cummins 2001). The species-specific nature of mitochondrial biogenesis and function (Kenyon and Moraes 1997), which requires active nuclear-mitochondrial communication, makes it an important factor to consider for iSCNT embryos. Incompatibility between the donor nucleus and recipient mitochondria or conflict between the two mitochondrial lineages present within the cytoplasm lead to a disruption in mitochondrial biogenesis and function, and subsequently, to altered ATP levels and incidence of apoptosis (reviewed by Mastromonaco and King 2007). These changes interfere with the basic needs of the early embryo, including proper DNA replication and cell division, activation of the embryonic genome, and initiation of blastulation and cell differentiation. Although it is expected for evolutionarily diverse species to exhibit increased developmental abnormalities related to mitochondrial dysfunction (e.g. sei whale and domestic cattle, 0 % blastocyst; Bhuiyan et al. 2010), nuclear-mitochondrial incompatability can even impact closely-related species, such as the gaur and domestic cattle. Mastromonaco et al. (2007) demonstrated an increase in blastocyst development and quality, similar to cattle controls, in iSCNT embryos reconstructed from hybrid gaur x cattle donor cells compared to gaur donor cells (32.5 % versus 18.1 % blastocyst rate; 186 versus 100 cells). It was hypothesized that the 50 % cattle nuclear genome in the hybrid donor cells was adequately supporting nuclear-cytoplasmic communication and enhancing developmental potential of the embryos. In a more conclusive study, Jiang et al. (2011) injected a mouse somatic cell along with mouse embryonic stem cell extract into a mitochondrial DNA-depleted pig oocyte and significantly increased blastocyst development compared to mouse-pig iSCNT embryos (3.4 % vs 0.5 %; percentage of blastocyst development from fused couplets). Retention of mouse mitochondrial DNA in the later embryo stages along with the increased embryo development demonstrated that addition of compatible cytoplasmic components along with the donor nucleus improves iSCNT outcomes. A greater understanding of the impact of interspecies nuclear-cytoplasmic interaction is required to overcome the developmental problems being observed in iSCNT embryos.

3.1.2 Non-mammalian Species

Nuclear transfer in non-mammalian species has been explored mostly for the study of basic scientific principles, as with amphibians, or the production of commercial or laboratory species, as with fishes. Given that common or non-threatened, species are being used, oocytes are readily accessible and intraspecies SCNT is possible. The original study by Briggs and King (1952) using the frog (*Rana pipiens*) model sparked further studies in nuclear equivalence and differentiation both in this species and *Xenopus laevis* (reviewed by Gurdon and Wilmut 2011). Important milestones in nuclear reprogramming were achieved in the amphibian studies of the 1950s and 1960s, including the ability to induce pluripotency and embryo development in a differentiated somatic cell. Living offspring have been obtained from a wide variety of cell types and experimental systems, and with a number of morphological and physiological abnormalities, as observed in mammals. In 1964, Briggs et al. demonstrated the development of early embryo stages in the Mexican axolotl (*Ambystoma mexicanum*). Unlike mammals, the concerns regarding iSCNT have not been explored to a great extent in amphibians. SCNT between *Xenopus* subspecies was carried out using both *Xenopus laevis laevis* and *X. l. victorianus* as donor cells and as cytoplasmic recipients. Embryos resulted from all nuclear—cytoplasmic combinations (Gurdon 1961). Further attempts at iSCNT included two newt species (*Pleurodeles waltlii* and *Pleurodeles poireti*; Gallien et al. 1973), newt and frog (*P. waltlii* and *X. laevis*; De Robertis and Gurdon 1977), and newt and axolotl (*P. waltlii* and *A. mexicanum*; Signoret et al. 1983). Gallien et al. (1973) noted in their study that increased lethality is observed in interspecies embryos compared to intraspecies ones. Interestingly, *P. poireti* (Edough ribbed newt) is currently listed as endangered on the IUCN Red List (2012). No reports have been found detailing the use of SCNT, experimental or otherwise, in reptile species.

As with amphibians, nuclear transfer in fishes has been on-going since the 1960s, particularly in the laboratory species, zebrafish (*Danio rerio*) and medaka (*Oryzias latipes*). Poor outcomes have been associated with abnormalities that are consistently reported across all taxa, including chromosome aberrations and arrested development in the early embryonic stages (reviewed by Wakamatsu 2008). Until recently, very little success had been obtained with SCNT and donor cells primarily consisted of embryonic blastomeres (reviewed by Wakamatsu 2008). Various attempts with SCNT have now resulted in live offspring (zebrafish: Lee et al. 2002; medaka: Bubenshchikova et al. 2008; goldfish (*Carassius auratus*): Le Bail et al. 2010). SCNT has been used as a method for producing a high-quality breed of goldfish ("Ranchu"; *Carassius auratus auratus*) desired in the ornamental fish culture industry (Tanaka et al. 2010). In this study, gastrula-stage embryos were obtained following the transfer of Ranchu donor cells into non-enucleated oocytes of a common goldfish breed ("Wakin"). Several reports of interspecies nuclear transfer in fishes can be found in the literature; however, the majority of the studies used embryonic blastomeres as donor cells, and not cultured somatic cells. Living offspring were obtained in common carp (*Cyprinus carpio*)—goldfish (Sun et al. 2005) and goldfish—carp (Yan et al. 1984) nucleo-cytoplasmic hybrids following interspecies nuclear transfer. Nuclear transfer hybrids between goldfish and loach (*Paramisgurnus dabryanus*) and tilapia (*Oreochromis nilotica*) and goldfish showed varied developmental potential from gastrula to larva, depending on the donor-recipient combinations (Yan et al. 1990). The authors of this study concluded that the failure to obtain adult fish from any of the interspecies combinations was the result of developmental incompatibilities between the donor nuclei and cytoplasmic recipients due to the evolutionary divergence of the species involved.

Despite the interest in advanced reproductive technologies in birds, such as germ cell transplantation, very little work has been documented in the area of nuclear transfer. In most cases of "nuclear transfer" reported in the literature, electrofusion of two somatic cells or primordial germ cells with somatic cells was attempted and not the standard technique of transferring a somatic or embryonic cell into an enucleated oocyte. For example, Minematsu et al. (2004) used domestic chicken (*Gallus gallus domesticus*) primordial germ cells as cytoplasmic recipients for embryonic blood cells and Bruno et al. (1981) fused dormant erythrocyte nuclei to enucleated fibroblasts. These experiments were mostly geared towards understanding mechanisms of nuclear reprogramming in avian species. Interspecies SCNT was used to develop blastocyst stage embryos from chicken blastodermal cells transferred into enucleated rabbit oocytes (Liu et al. 2004).

To date, there are no reported studies of SCNT use in non-mammalian species as a strategy for the preservation of threatened or endangered wildlife species. It is important to note that there is increased interest in the past few years to implement SCNT in critically endangered amphibian species, which are disappearing at such a rapid pace that in situ conservation efforts are unable to cope. However, further studies are required to better understand the impact of SCNT, and particularly iSCNT, in non-mammalian embryo development and long-term survival and reproductive capacity of clones. Unlike the intensive efforts in a diverse array of mammals to overcome the challenges involved with SCNT, comparatively little work has been carried out in non-mammalian species.

3.1.3 Cell Culture Role in SCNT Success

Low SCNT efficiency reported in all vertebrate taxa does correspond to some extent to the technical requirements of the procedure. Extensive manipulation of the oocyte and donor cells along with the culture needs of the reconstructed embryos and even the status of the donor individual have a definite impact on developmental potential pre- and post-implantation (reviewed by Mastromonaco and King 2007). With iSCNT specifically, reconstructed embryos from a wide variety of non-domestic species are cultured in conditions that have been optimized for related domestic species (typically that of the cytoplasmic recipient), which may not necessarily be the ideal conditions for the nuclear donor species. However, in the context of this review, focus will be placed on the influence of the donor cell, and therefore, somatic cell culture, on SCNT success.

One of the most important requirements for SCNT success is the availability of viable and normal donor cell lines. Although studies have shown that non-viable cells can be used to produce blastocysts following SCNT (sheep granulosa cell: Loi et al. 2008; mouse granulosa cell: Ono et al. 2008; pig fetal fibroblast: Das et al. 2010), the effect on post-implantation development and neonatal survival is not known. Gómez et al. (2008) compared sand cat (*Felis margarita*) iSCNT rates from donor cells used immediately post-thaw or used after 18 h to 5 days of culture post-thaw. Interestingly, no differences in blastocyst development and pregnancy rates were observed between the groups, however, the number of embryos implanted, fetuses at term and expression of the gene POU5F1 were significantly lower when donor cells were used immediately post-thaw. These outcomes can be attributed to the fact that the percentage of necrotic cells immediately post-thaw was 61 % compared to only 6.9 % after 18 h of culture (Gómez et al. 2008). Viability and normality of cells can be affected by culture establishment techniques and long-term culture, which is manifested in the form of genomic instability and telomere shortening. Use of short-lived cell-lines from dart and punch biopsy collections produced significantly lower blastocyst rates (0.8–0.9 %) than long-lived cell lines from ear tissue (11.8 %) in gaur iSCNT trials (Mastromonaco et al. 2006). In the same study, donor cells from high passage cultures resulted in significantly reduced blastocyst development compared to low passage cultures (60 populations doublings, <11 % blastocyst vs 6–10 population doublings, 35 % blastocyst; Fig. 16.4). The association between increased chromosome abnormalities in the cell line and decreased SCNT outcomes has been corroborated by other researchers in cattle (Slimane Bureau et al. 2003), sheep (Loi et al. 2001), and wild cat (Gómez et al. 2006) SCNT embryos.

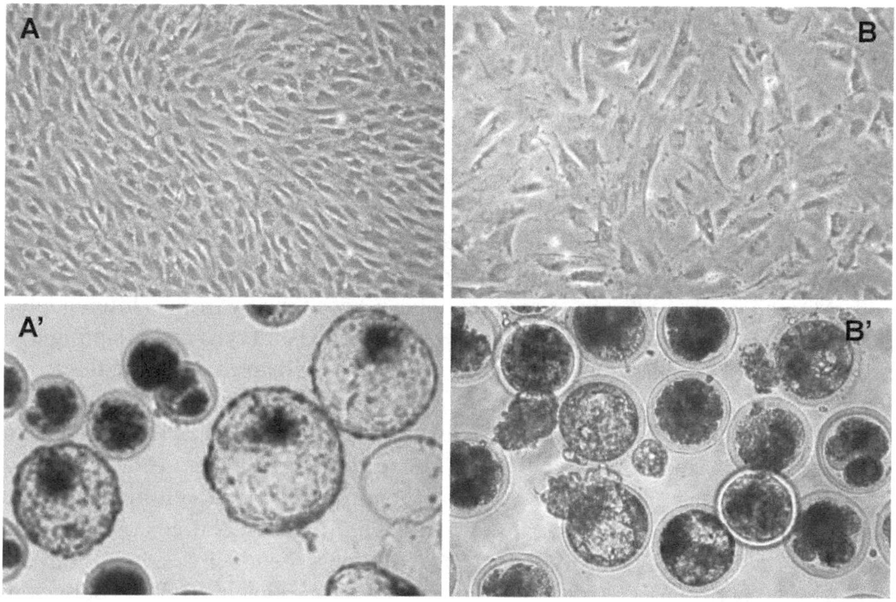

Fig. 16.4 Evidence of the effect of cell culture quality on domestic cattle SCNT outcomes. A and A': fibroblast cells at passage 3–5 and the SCNT blastocysts produced from these cells; B and B': fibroblast cells at passage 30 and the SCNT blastocysts produced from these cells

The influence of the donor cell on SCNT outcome extends beyond cell lifespan and chromosomal normality. Studies have shown that fibroblast cell lines derived from the same tissue but different individuals will have variable outcomes. Zhou et al. (2006) found that only two of four chromosomally-normal fibroblast cell lines derived from the ear skin tissue of four healthy, male, neonatal rhesus monkeys supported routine production of blastocysts following SCNT. In our laboratory, three gaur ear fibroblast cell lines (>85 % normal chromosomes; all donors >15 years of age) were used for iSCNT prior to obtaining successful blastocyst development from one of the cell lines (Mastromonaco, unpublished data). Similarly, two wood bison (*Bison bison athabascae*) ear fibroblast cell lines have been attempted to date for an on-going iSCNT study since the first cell line resulted in 0 % blastocyst development after numerous attempts while the second cell line consistently produces blastocysts at rates of 18-33 % (Mastromonaco, unpublished data).

Aside from genetic and environmental factors influencing SCNT outcome, there are many new studies examining the methylation status of donor cells and embryos in an effort to delineate the role of the epigenome in SCNT success. Epigenetic modifications of the donor chromatin, including DNA methylation and histone modification, play a key role in embryonic processes, such as expression of developmental genes, X-inactivation, and telomere length re-adjustment (Rideout et al. 2001). Recent studies of abnormalities observed in SCNT fetuses and neonates are indicating that they are the result of insufficient reprogramming of the somatic cell nucleus by factors in the oocyte cytoplasm (Loi et al. 2007). Proper reprogramming involves complete elimination of somatic methylation patterns prior to establishment of embryonic methylation patterns. This corroborates decades of work that concluded that increased embryonic development and live offspring from SCNT can be obtained when embryonic blastomeres or undifferentiated cells are used as nuclear donor cells compared to fibroblasts or other differentiated cells (reviewed by Hochedlinger and Jaenisch 2002). The pluripotent state of embryonic blastomeres is thought to support greater development due to the "embryo-ready" state of the epigenome, which requires less reprogramming than a differentiated somatic cell (Morgan et al. 2005). In fact, authors have questioned whether differentiated nuclear donor cells are truly responsible for the cloned embryos and offspring that are produced or whether the small numbers of stem cells present in the tissue are actually the ones capable of initiating long-term development (reviewed by Hochedlinger and Jaenisch 2002). For this reason, several studies have used stem cells as nuclear donor cells for SCNT. Porcine skin-derived stem cells supported significantly greater blastocyst development and quality (total cell number, gene expression profile) than fibroblast cells following SCNT (Zhu et al. 2004) and resulted in live offspring (Hao et al. 2009). Red deer stem cells derived from the antlerogenic periosteum also resulted in live offspring following SCNT (Berg et al. 2007). In this study, however, no differences in pre- and post-implantation development rates were observed between the undifferentiated stem cell group and stem cells that had been chemically induced to initiate osteogenesis or adipogenesis.

In order to increase SCNT efficiency, researchers have looked at techniques for enhancing the reprogramming of donor cells. Xiong et al. (2012) permeabilized yak (*Bos grunniens*) fibroblasts and incubated them with yak or domestic cattle oocyte extracts prior to iSCNT to increase the exposure of the donor nucleus to oocyte-specific reprogramming factors. Embryo development rates and gene expression and methylation patterns were significantly improved in the pre-treated donor cells compared to controls. The use of trichostatin A (TSA; histone deacetylase inhibitor) incubation with donor cells or embryos in an effort to modify histone acetylation has had contradictory outcomes. Some evidence of improved development of iSCNT embryos was observed in the leopard cat (*Prionailurus bengalensis*; Lee et al. 2010), but not in the gaur (Srirattana et al. 2012). In domestic cattle, treatment of the donor cells or reconstructed embryos with a combination of chromatin modifying agents, including 5-aza-2'-deoxycytidine + TSA, resulted in increased blastocyst development and total cell number in a study by Ding et al. (2008), whereas Sangalli et al. (2012) did not find any positive effects on either pre- or post-implantation development of SCNT embryos. Similarly, Jeon et al. (2008) used S-adenosylhomocysteine, a DNA demethylation agent, to treat cattle donor cells prior to SCNT and observed significantly increased telomerase activity, expression of X-chromosome linked genes, and blastocyst development. There is still much to learn about the mechanisms involved in reprogramming and the potential for enhancing reprogramming at various levels: the donor cell, the oocyte or the early embryo.

3.2 *Derivation of Induced Pluripotent Stem Cells*

Reprogramming of a nucleus into a pluripotent state is achievable by the oocyte cytoplasmic machinery as evidenced by SCNT. Likewise, pluripotent cells, such as embryonic germ cells and stem cells, exhibit reprogramming activity when they are fused with a somatic cell. Half a century of work from the pioneers of amphibian nuclear transfer and mammalian stem cell culture has laid down the foundation for a new era of research into pluripotency. Authors have suggested that the cytoplasmic factors responsible for nuclear reprogramming should be identifiable and, therefore, able to induce pluripotency when expressed in somatic cells (Hochedlinger and Plath 2009). Recently, this idea has led to the identification of a defined set of transcription factors that have been used to successfully reprogram somatic cells in culture (reviewed by Hochedlinger and Plath 2009). Takahashi et al. (2007) were the first to demonstrate that retroviral-mediated expression of four human transcription factors, Oct4, Sox2, Klf4 and c-Myc, were sufficient to convert adult mouse fibroblasts into embryonic stem cell-like induced pluripotent stem cells (iPSCs). This and other studies are clearly showing that the mechanisms for nuclear reprogramming appear highly conserved among species. However, as inefficient as SCNT is (<5 % live offspring), the derivation of IPSCs is even less efficient (0.01-0.1 %) due to the fact that reprogramming occurs over several weeks and that expression of the four transcriptions factors alone is not sufficient for complete nuclear reprogramming (reviewed by Hochedlinger and Plath 2009). As with SCNT, the challenges associated with iPSCs may 1 day be outweighed by the considerable benefits that they will provide as a tool for: (1) the study of pluripotency and nuclear reprogramming, and (2) the production of self-renewing cells from genetically valuable individuals. For endangered species, iPSCs offer the possibility of producing germ-line cells from individuals in which these cells are not available.

3.2.1 Mammals

Since the establishment of a feasible technique for reprogramming somatic cells using transcription factors, successful generation of iPSCs for laboratory and domestic animal species (e.g. mouse, human, macaque, rat, pig, dog, horse, sheep and cattle) have all used identical methods, although in some cases iPSCs could be sustained only by continuous expression of the exogenous factors. Characterization of iPSCs is extensive and requires that they show morphological and molecular markers of pluripotent cells, regenerative capacity, and ability to produce all three germ layers following implantation into immunodeficient mice (teratoma formation) or insertion into mouse blastocysts (chimeric mice). A sudden surge of work in this field in the past 5 years has resulted in rapid progress in technique optimization. Inclusion of Nanog expression along with the four transcription factors, Oct4, Sox2, Klf4 and c-Myc, for induction resulted in the production of cells that were more comparable to embryonic stem cells in morphology, proliferation, gene expression and other criteria (Okita et al. 2007). Since expression of c-Myc has been associated with tumour production in offspring, protocols eliminating the use of c-Myc (Nakagawa et al. 2007) or reducing the number of transcription factors from four to two (Kim et al. 2008) have both successfully produced high quality iPSCs. Likewise, the potential for mutagenesis resulting from retroviruses has led researchers to attempt other approaches, including use of adenovirus-mediated induction (Stadtfeld et al. 2008) or virus-free induction (Okita et al. 2008). Studies in mice and pigs have demonstrated the possibility to use extracts of germinal vesicle oocytes to reprogram somatic cells (Bui et al. 2008; Bui et al. 2012).

While reprogramming of differentiated somatic cells has provoked enormous interest for human regenerative medicine, and also certain domestic animals (e.g. race horses), future benefits might extend to helping conserve faunal biodiversity. Recent evidence suggests that it is possible to produce iPSCs from wildlife species. Ben-Nun et al. (2011) were the first to report the successful production of iPSCs from adult fibroblasts of two endangered species: silver-maned drill (*Mandrillus leudophaeus*) and northern white rhinoceros (*Ceratotherium simum cottoni*). This was followed by the production of iPSCs from fibroblasts of snow leopards (*Panthera uncia*; Verma et al. 2012) and prairie voles (*Microtus ochrogaster*; Manoli et al. 2012). In all these cases, induction was carried out by retroviral transfection using the four human transcription factors described previously. Although some minor modifications may have been required, all species attempted resulted in the production of stable iPSC lines.

The striking potential of these strategies has been demonstrated in the mouse where in vitro-differentiated embryonic stem cells have given rise to sperm-like cells (Nayernia et al. 2006) or oocyte-like cells have been derived from newborn mouse skin (Dyce et al. 2011a) and pigs (Dyce et al. 2011b) or from primordial germ cells (Hayashi et al. 2012).

3.2.2 Non-mammalian Species

An extensive search of the literature indicates that very little work is currently being reported on iPSCs in non-mammalian species. The only study reported to date has been in birds where quail iPSCs were derived by retroviral transfection of human transcription factors (Lu et al. 2012). The successful use of mammalian transcription factors in an avian species once again highlights the conserved nature of the mechanisms involved in nuclear programming.

3.2.3 Factors Influencing iPSC Derivation

Factors influencing SCNT success, such as donor cell viability and normality, also play an important role for iPSC production. Pluripotent stem cell induction requires many weeks in culture and extensive passaging of the cell lines, which means that chromosomally-normal cultures with long in vitro lifespan are essential. Researchers have remarked on the difficulties regarding aneuploidy and in vitro passaging of iPSC lines (Manoli et al. 2012). As with SCNT, Okita et al. (2007) suggested that the low efficiency in reprogramming iPSCs may be the result of rare stem cells co-existing in fibroblast cultures that are being properly reprogrammed rather than the differentiated cells. This leads to the same discussion of proper epigenetic modification of iPSCs. Although studies have shown that iPSCs and embryonic stem cells are highly similar in chromatic structure and gene expression, other reports indicate that there are epigenomic differences between these cell types suggestive of incomplete or variable changes in methylation patterns (reviewed by Lister et al. 2011). As with stem cells, the use of iPSCs as donor cells for SCNT may enhance success rates compared to differentiated somatic cells. A better understanding of factors influencing SCNT success will benefit iPSC success, and vice versa.

4 Reality for Animal Species Conservation

At this time, the primary outcome of research on somatic cells and stem cells is the increase of scholarly knowledge by promoting the study of basic cell biology in vitro, including nuclear reprogramming and cell differentiation. There actually are many critical uses for somatic cell and stem cell cultures in relation to species conservation that are not necessarily to produce new individuals. Information stored in cells can be valuable for studies in phylogenomics, evolutionary/developmental biology, pharmacology/toxicology, veterinary regenerative medicine, to name a few areas of interest. The availability of cells representing diverse species and populations for every interested institution will accelerate research progress on analyzing phylogeographic structure, delineating subspecies, tracing paternities, evaluating gene flow, and assessing genetic variation; all of which provide critical information for decision-making in managing both ex situ and in situ wildlife populations. The diminished costs for Next Generation Sequencing of DNA and bioinformatics will boost the knowledge that can be generated from the different cell types (see Chap. 5).

However, one of the most important benefits of reprogramming somatic cells, either by SCNT or derivation of iPSCs, includes the production of embryos and gametes from non-reproductive material. Somatic cells provide an inexhaustible resource compared to banked sperm, oocytes or embryos increasing the potential for experimentation, technique optimization, and sharing of banked resources among facilities and researchers. It is perhaps with these thoughts in mind that a number of "firsts", including the gaur born by iSCNT (Lanza et al. 2000) and the rhinoceros and drill iPSCs derived from adult fibroblasts (Ben-Nun et al. 2011), were supported by Dr. Oliver Ryder and his team at the San Diego Zoo Global Wildlife Conservancy. Recently, the birth of a Pyrenean ibex (*Capra pyrenaica pyrenaica*) by iSCNT years after that subspecies had gone extinct was evidence that, although not perfect, the technique still presents interesting possibilities (Folch et al. 2009). In a November 2012 press release, Brazil's agricultural research agency, Embrapa, stated that they are moving forward with a plan to clone eight species beginning with the maned wolf (*Chrysocyon brachyurus*; www.newscientist.com, verified January 20, 2013).

That leads us to the question "what is the reality for these technologies in the species conservation arena?" Although the potential for in vitro embryo production by SCNT for species conservation has been considered and investigated for more than 10 years with minimal success, the contribution of iPSCs or even germinal stem cells is a recent concept that has, again, more application in the acquisition of scholarly knowledge than in the production of new individuals at this time. In view of the small number of SCNT-produced living offspring in wild species compared to the extensive efforts by many laboratories around the world, the technique still has major challenges that must be overcome. Likewise, the handful of attempts at manipulating stem cells has identified key areas for further development and improvement. And so, it can be agreed, and is well-documented, as noted in a review by Holt et al. (2004), that novel and advanced technologies for assisted reproduction are highly inefficient, overcome with problems and nowhere near the point where they can be applied as consistent and viable strategies for offspring production. In addition, cell culture banks are currently static repositories that are not accessible or known to many individuals, other than those managing or contributing to them. However, what strategies for genetic preservation do we have to compare to at this time? Basic ARTs, primarily artificial insemination, which has long been declared the most powerful tool in the arsenal of reproductive techniques, have made a negligible impact on ex situ conservation programs with repeated successes demonstrated in only a handful of species after more than 30 years of work (e.g. black footed ferrets, koalas; reviewed by Mastromonaco et al. 2011 and discussed in Chap. 7). Gamete and embryo banks currently contain numerous samples which have been inadequately cryopreserved due to a lack of optimized species-specific protocols. Knowing this and the fact that financial resources for more pragmatic conservation programs are not being used to fund research into nuclear transfer and stem cell technologies, the possibilities provided to us by somatic cell culture banks in the near or distant future warrant the continued research and advancement of somatic cell-based techniques for reproductive purposes. The benefits of generating iPSCs from endangered species would be incalculable, especially for the ability to produce an inexhaustible supply of haploid gametes. The production of iPSC-derived sperm and oocytes from long-dead animals could provide an endless resource for fundamental investigations into in vitro oocyte maturation, IVF, ICSI and SCNT-based reproductive strategies and therefore, provide a potential method for infusing much needed genetic diversity. The past decade has taught us that there are no easy paths to success with these technologies, however, the challenges have now been identified and continued progress, albeit slow, can be made.

References

Anizet MP, Huwe B, Pays A, Picard JJ. Characterization of a new cell line, XL2, obtained from *Xenopus laevis* and determination of optimal culture conditions. In Vitro. 1981;17:267–74.

Benkeddache D, Bodinier P, Joly T, Berchiche M, Vignon X. Recovery of viable cells from rabbit skin biopsies after storage at -20°C for up to 10 days. Cell Tissue Bank. 2012;13:479–86.

Ben-Nun IF, Montague SC, Houck ML, Tran HT, Garitaonandia I, Leonardo TR, et al. Induced pluripotent stem cells from highly endangered species. Nat Methods. 2011;8:829–31.

Berg DK, Li C, Asher G, Wells DN, Oback B. Red deer cloned from antler stem cells and their differentiated progeny. Biol Reprod. 2007;77:384–94.

Bhuiyan MMU, Suzuki Y, Watanabe H, Lee E, Hirayama H, Matsuoka K, et al. Production of Sei whale (*Balaenoptera borealis*) cloned embryos by inter- and intra-species somatic cell nuclear transfer. J Reprod Dev. 2010;56:131–9.

Bols NC, Dayeh VR, Lee LEJ, Schirmer K. Use of fish cell lines in the toxicology and ecotoxicology of fish. Piscine cell lines in environmental toxicology. In: Mommsen TP, Moon TW, editors. Biochemistry and molecular biology of fishes, vol. 6. Amsterdam: Elsevier; 2005. p. 43–84.

Bosnakovski D, Mizuno M, Kim G, Takagi S, Okumura M, Fujinaga T. Isolation and multilineage differentiation of bovine bone marrow mesenchymal stem cells. Cell Tissue Res. 2005;319: 243–53.

Briggs R, King TJ. Transplantation of living nuclei from blastula cells into enucleated frogs' eggs. Proc Natl Acad Sci U S A. 1952;38:455–63.

Briggs R, Signoret J, Humphrey RR. Transplantation of nuclei of various cell types from neurulae of the Mexican axolotl (*Ambystoma mexicanum*). Dev Biol. 1964;10:233–46.

Bruno J, Reich N, Lucas JJ. Globin synthesis in hybrid cells constructed by transplantation of dormant avian erythrocyte nuclei into enucleated fibroblasts. Mol Cell Biol. 1981;1:1163–76.

Bubenshchikova E, Kaftanovskaya E, Hattori M, Kinoshita M, Adachi T, Hashimoto H, et al. Nuclear transplants from adult somatic cells generated by a novel method using diploidized eggs as recipients in medaka fish (*Oryzias latipes*). Cloning Stem Cells. 2008;10:443–52.

Bui HT, Wakayama S, Kishigami S, Kim JH, Van Thuan N, Wakayama T. The cytoplasm of mouse germinal vesicle stage oocytes can enhance somatic cell nuclear reprogramming. Development. 2008;135:3935–45.

Bui HT, Dn K, Kang MH, Oh MH, Park MR, Park WJ, et al. Epigenetic reprogramming in somatic cells induced by extract from germincal vesicle stage pig oocytes. Development. 2012;139: 4330–40.

Bui LC, Vignon X, Campion E, Laloy E, Lavergne Y, Ty LV, et al. Use of interspecies nuclear transfer to study the early embryonic development and nuclear activities of the endangered species *Pseudoryx nghetinhensis* (saola). Theriogenology. 2002;57:427 [abstract].

Campbell KHS, Fisher P, Chen WC, Choi I, Kelly RDW, Lee J-H, et al. Somatic cell nuclear transfer: past, present and future perspectives. Theriogenology. 2007;68:S214–31.

Cecil JT, Nigrelli RF. Cell cultures from marine mammals. J Wildl Dis. 1970;6:494–5.

Chakraborty N, Biswas D, Parker W, Moyer P, Elliott GD. A role for microwave processing in the dry preservation of mammalian cells. Biotechnol Bioeng. 2008;100:782–96.

Chakraborty N, Chang A, Elmoazzen H, Menze MA, Hand SC, Toner M. A spin-drying technique for lyopreservation of mammalian cells. Ann Biomed Eng. 2011;39:1582–91.

Chang C, Lee I, Ye F, Tarantal AF. Comparison of growth and differentiation of fetal and adult rhesus monkey mesenchymal stem cells. Stem Cells Dev. 2006;15:209–20.

Chavatte-Palmer P, Camous S, Jammes H, Le Cleac'h N, Guillomot M, Lee RS. Review: placental perturbations induce the developmental abnormalities often observed in bovine somatic cell nuclear transfer. Placenta. 2012;33(Suppl):S99–104.

Chen D-Y, Wen D-C, Zhang Y-P, Sun Q-Y, Han Z-M, Liu Z-H, et al. Interspecies Implantation and mitochondria fate of panda-rabbit cloned embryos. Biol Reprod. 2002;67:637–42.

Cheng C, Lian W, Hsiao F, Liu I, Lin S, Lee Y, et al. Isolation and characterization of novel murine epiphysis derived mesenchymal stem cells. PLoS One. 2012;7:e36085. doi:10.1371/journal.pone.0036085.

Choresca Jr CH, Kang JT, Han JE, Kim JH, Shin SP, Jun JW, et al. Effect of storage media and time on fin explants culture in the goldfish, *Carassius auratus*. Afr J Biotech. 2012;11:6599–602.
Clarke AG. The Frozen Ark Project: the role of zoos and aquariums in preserving the genetic material of threatened animals. Int Zoo Yb. 2009;43:222–30.
Clark HF, Cohen MM, Karzon DT. Characterization of reptilian cell lines established at incubation temperatures of 23 to 36°. Exp Biol Med. 1970;133:1039–47.
Cummins JM. Mitochondria: potential roles in embryogenesis and nucleocytoplasmic transfer. Hum Reprod Update. 2001;7:217–28.
Das ZC, Gupta MK, Uhm SJ, Lee HT. Lyophilized somatic cells direct embryonic development after whole cell intracytoplasmic injection into pig oocytes. Cryobiology. 2010;61:220–4.
De Robertis EM, Gurdon JB. Gene activation in somatic nuclei after injection into amphibian oocytes. Proc Natl Acad Sci U S A. 1977;74:2470–4.
Ding X, Wang Y, Zhang D, Wang Y, Guo Z, Zhang Y. Increased pre-implantation development of cloned bovine embryos treated with 5-aza-2'-deoxycytidine and trichostatin A. Theriogenology. 2008;70:622–30.
Dobrinski I, Travis AJ. Germ cell transplantation for the propagation of companion animals, non-domestic and endangered species. Reprod Fertil Dev. 2007;19:732–9.
Dominici M, Le Blanc K, Mueller I, Slaper-Cortenbach I, Marini FC, Kraus DS, et al. Minimal criteria for defining multipotent mesenchymal stromal cells. The International Society for Cellular Therapy position statement. Cytotherapy. 2006;8:315–7.
Dominko T, Mitalipova M, Haley B, Beyhan Z, Memili E, McKusick B, et al. Bovine oocyte cytoplasm supports development of embryos produced by nuclear transfer of somatic cell nuclei from various mammalian species. Biol Reprod. 1999;60:1496–502.
Duarte MM, Montes de Oca H, Diniz CR, Fortes-Dias CL. Primary culture of venom gland cells from the South American rattlesnake (*Crotalus durissus terrificus*). Toxicon. 1999;37:1673–82.
Dyce PW, Liu J, Tayade C, Kidder GM, Betts DH, Li J. In vitro and in vivo germ line potential of stem cells derived from newborn mouse skin. PLoS One. 2011a;6:e20339. doi:10.1371/journal.pone.0020339.
Dyce PW, Shen W, Huynh E, Shao H, Villagómez DA, Kidder GM, et al. Analysis of oocyte-like cells differentiated from porcine fetal skin-derived stem cells. Stem Cells Dev. 2011b;20:809–19.
Dyce PW, Zhu H, Craig J, Li J. Stem cells with multilineage potential derived from porcine skin. Biochem Biophys Res Commun. 2004;316:651–8.
Eagle H. Nutrition needs of mammalian cells in tissue culture. Science. 1955;122:501–4.
Ezaz T, O'Meally D, Quinn AE, Sarre SD, Georges A, Marshall Graves JA. A simple non-invasive protocol to establish primary cell lines from tail and toe explants for cytogenetic studies in Australian dragon lizards (Squamata: Agamidae). Cytotechnology. 2008;58:135–9.
Fink T, Rasmussen JG, Emmersen J, Fahlman A, Brunberg S, Josefsson J, et al. Adipose-derived stem cells from the brown bear (*Ursus arctos*) spontaneously undergo chondrogenic and osteogenic differentiation. Stem Cell Res. 2011;7:89–95.
Folch J, Cocero MJ, Chesné P, Alabart JL, Domínguez V, Cognié Y, et al. First birth of an animal from an extinct subspecies (*Capra pyrenaica pyrenaica*) by cloning. Theriogenology. 2009;71:1026–34.
Freshney RI. Culture of animal cells: A manual of basic technique. 5th ed. Hoboken, NJ: John Wiley and Sons Inc.; 2005.
Gallien C-L, Aimar C, Guillet F. Nucleocytoplasmic interactions during ontogenesis in individuals obtained by intra- and interspecific nuclear transplantation in the genus *Pleurodels* (Urodele Amphibian): morphology, analysis of two enzymatic systems (LDH and MDH) and immunity reactions. Dev Biol. 1973;33:154–70.
Gomez MC, Pope CE, Biancardi MN, Dumas C, Galiguis J, Morris AC, et al. Trichostatin A modified histone covalent pattern and enhanced expression of pluripotent genes in interspecies black-footed cat cloned embryos but did not improve in vitro and in vivo viability. Cell Reprogram. 2011;13:315–29.

Gómez MC, Pope CE, Kutner RH, Ricks DM, Lyons LA, Ruhe M, et al. Nuclear transfer of sand cat cells into enucleated domestic cat oocytes is affected by cryopreservation of donor cells. Cloning Stem Cells. 2008;4:469–83.

Gómez MC, Pope CE, Lopez M, Dumas C, Giraldo A, Dresser BL. Chromosomal aneuploidy in African wildcat somatic cells and cloned embryos. Cloning Stem Cells. 2006;8:69–78.

Gómez MC, Pope CE, Giraldo A, Lyons LA, Harris RF, King AL, et al. Birth of African wildcat cloned kittens born from domestic cats. Cloning Stem Cells. 2004;6:247–58.

Gómez MC, Jenkins JA, Giraldo A, Harris RF, King A, Dresser BL, et al. Nuclear transfer of synchronized African wild cat somatic cells into enucleated domestic cat oocytes. Biol Reprod. 2003;69:1032–41.

Gritti A, Parati EA, Cova L, Frolichsthal P, Galli R, Wanke E, et al. Multipotential stem cells from adult mouse brain proliferate and self-renew in response to basic fibroblast growth factor. J Neurosci. 1996;16:1091–100.

Grunow B, Noglick S, Kruse C, Gebert M. Isolation of cells from Atlantic sturgeon *Acipenser oxyrinchus oxyrinchus* and optimization of culture conditions. Aquat Biol. 2011;14:67–75.

Guan W, He X, Li L, Liang H, Zhao Q, Pu Y, et al. Establishment and biological characterization of fibroblast cell line from the Langshan chicken. Cell Prolif. 2010;43:157–63.

Gurdon JB. The transplantation of nuclei between two subspecies of *Xenopus laevis*. Heredity. 1961;16:305–15.

Gurdon JB, Wilmut I. Nuclear transfer to eggs and oocytes. Cold Spring Harb Perspect Biol. 2011;3:a002659. doi:10.1101/cshperspect.a002659.

Hajian M, Hosseini SM, Forouzanfar M, Abedi P, Ostadhosseini S, Hosseini L, et al. "Conservation cloning" of vulnerable Esfahan mouflon (*Ovis orientalis isphahanica*): in vitro and in vivo studies. Eur J Wildl Res. 2011;57:959–69.

Han JE, Choresca Jr CH, Koo OJ, Oh HJ, Hong SG, Kin JH, et al. Establishment of glass catfish (*Kryptopterus bicirrhis*) fin-derived cells. Cell Biol Intl Rep. 2011;18(1):e00008. doi:10.1042/CBR20110002.

Hao Y, Wax D, Zhong Z, Murphy C, Ross JW, Rieke A, et al. Porcine skin-derived stem cells can serve as donor cells for nuclear transfer. Cloning Stem Cells. 2009;11:101–9.

Hashem MA, Bhandari BP, Kang SK, Lee BC. Cell cycle analysis and interspecies nuclear transfer of in vitro cultured skin fibroblasts of the Siberian tiger (*Panthera tigris altaica*). Mol Reprod Dev. 2007;74:403–11.

Hayashi K, Ogushi S, Kurimoto K, Shimamoto S, Ohta H, Saitou M. Offspring from oocytes derived from in vitro primordial germ cell-like cells in mice. Science. 2012;338:971–5.

Hayashi K, Ohta H, Kurimoto K, Aramaki S, Saitou M. Reconstitution of the mouse germ cell specification pathway in culture by pluripotent stem cells. Cell. 2011;146:519–32.

Hebert PD, Cywinska A, Ball SL, deWaard JR. Biological identifications through DNA barcodes. Proc Biol Sci. 2003;270:313–21.

Hermann BP, Sukhwani M, Winkler F, Pascarella JN, Peters KA, Sheng Y, et al. Spermatogonial stem cell transplantation into rhesus testes regenerates spermatogenesis producing functional sperm. Cell Stem Cell. 2012;11:715–26.

Hochedlinger K, Jaenisch R. Nuclear reprogramming and pluripotency. Nature. 2006;441:1061–7.

Hochedlinger K, Jaenisch R. Nuclear transplantation: Lessons from frogs and mice. Curr Opin Cell Biol. 2002;14:741–8.

Hochedlinger K, Plath K. Epigenetic reprogramming and induced pluripotency. Development. 2009;136:509–23.

Holt WV, Moore HDM. Semen banking – Is it now feasible for captive endangered species? Oryx. 1988;22:172–8.

Holt WV, Pickard AR, Prather RS. Wildlife conservation and reproductive cloning. Reproduction. 2004;127:317–24.

Hong N, Li Z, Hong Y. Fish stem cell cultures. Int J Biol Sci. 2011;7:392–402.

Houck M. Establishing and cryopreserving cell cultures: procedures developed for the Frozen Zoo®. Biopreserv Biobank. 2012;10:66 [abstract].

Hutton SR, Pevny LH. Isolation, culture, and isolation of progenitor cells from the central nervous system. Cold Spring Harb Protoc. 2008. doi:10.1101/pdb.prot5077.

Hwang W, Kim K, Kim G, Jin Y, Kim Y, Chung H, et al. Interspecies somatic cell nuclear transfer for the production of endangered Korean tiger (*Panthera tigris altaica*). Theriogenology. 2001;55:271 [abstract].

Ikumi S, Sawai K, Takeuchi Y, Iwayama H, Ishikawa H, Ohsumi S, et al. Interspecies somatic cell nuclear transfer for in vitro production of Antarctic minke whale (*Balaenoptera bonaerensis*) embryos. Cloning Stem Cells. 2004;6:284–93.

IUCN 2012. IUCN Red List of Threatened Species. Version 2012.2. (www.iucnredlist.org).

Jeon BG, Coppola G, Perrault SD, Rho GJ, Betts DH, King WA. S-adenosylhomocysteine treatment of adult female fibroblasts alters X-chromosome inactivation and improves in vitro embryo development after somatic cell nuclear transfer. Reproduction. 2008;135:815–26.

Jiang Y, Kelly R, Peters A, Fulka H, Dickinson A, Mitchell DA, et al. Interspecies somatic cell nuclear transfer is dependent on compatible mitochondrial DNA and reprogramming factors. PLoS ONE. 2011;6:e14805. doi:10.1371/journal.pone.0014805.

Kenyon L, Moraes CT. Expanding the functional human mitochondrial DNA database by the establishment of primate xenomitochondrial cybrids. Proc Natl Acad Sci U S A. 1997;94: 9131–5.

Kim JB, Zaehres H, Wu G, Gentile L, Ko K, Sebastiano V, et al. Pluripotent stem cells induced from adult neural stem cells by reprogramming with two factors. Nature. 2008;454:646–50.

Kim MK, Jang G, Oh HJ, Yuda F, Kim HJ, Hwang WS, et al. Endangered wolves cloned from adult somatic cells. Cloning Stem Cells. 2007;9:130–7.

Kjelland ME, Kraemer D. Feathers and post-hatch eggshells: Sources of fibroblast cells for conserving genetic diversity. Avian Biol Res. 2012;5:123–30.

Koch TG, Heerkens T, Thomsen PD, Betts DH. Isolation of mesenchymal stem cells from equine umbilical cord blood. BMC Biotechnol. 2007;7:26. doi:10.1186/1472-6750-7-26.

Kumar BM, St. John E, Mackie PM, King WA, Mastromonaco GF. In vitro development of wood bison (*Bison bison athabascae*) embryos by interspecies somatic cell nuclear transfer. Reprod Fertil Dev. 2009;21:178 [abstract].

Lakra WS, Bhonde RR. Development of a primary cell culture from the caudal fin of an Indian major carp, *Labeo rohita* (Ham). Asian Fisheries Sci. 1996;9:149–52.

Lakra WS, Goswami M. Development and characterization of a continuous cell line PSCF from *Puntius sophore*. J Fish Biol. 2011;78:987–1001.

Lakra WS, Goswami M, Rajaswaminathan T, Rathore G. Development and characterization of two new cell lines from common carp, *Cyprinus carpio* (Linn). Biol Res. 2010;43:385–92.

Lakra WS, Goswami M, Rajaswaminathan T, Joy KP. Development, characterization, conservation and storage of fish cell lines: a review. Fish Physiol Biochem. 2011;37:1–20.

Lannan CN. Fish cell culture: A protocol for quality control. J Tissue Cult Methods. 1994; 16:95–8.

Lanza RP, Cibelli JB, Diaz F, Moraes CT, Farin PW, Farin CE, et al. Cloning of an endangered species (*Bos gaurus*) using interspecies nuclear transfer. Cloning. 2000;2:79–90.

Laskey RA. The use of antibiotics in the preparation of amphibian cell cultures from highly contaminated material. J Cell Sci. 1970;7:653–9.

Le Bail P-Y, Depince A, Chenais N, Mahe S, Maisse G, Labbe C. Optimization of somatic cell injection in the perspective of nuclear transfer in goldfish. BMC Dev Biol. 2010;10:64. doi:10.1186/1471-213X-10-64.

Lee HS, Yu XF, Bang JI, Cho SJ, Deb GK, Kim BW, et al. Enhanced histone acetylation in somatic cells induced by a histone deacetylase inhibitor improved inter-generic cloned leopard cat blastocysts. Theriogenology. 2010;74:1439–49.

Lee E, Bhuiyan MM, Watanabe H, Matsuoka K, Fujise Y, Ishikawa H, et al. Production of cloned sei whale (*Balaenoptera borealis*) embryos by interspecies somatic cell nuclear transfer using enucleated pig oocytes. J Vet Sci. 2009;10:285–92.

Lee B, Wirtu GG, Damiani P, Pope E, Dresser BL, Hwang W, et al. Blastocyst development after intergeneric nuclear transfer of mountain bongo antelope somatic cells into bovine oocytes. Cloning Stem Cells. 2003;5:25–33.

Lee K-Y, Huang H, Ju B, Yang Z, Lin S. Cloned zebrafish by nuclear transfer from long-term cultured cells. Nat Biotechnol. 2002;20:795–9.

Lees CM, Wilcken J. Sustaining the ark: The challenges faced by zoos in maintaining viable populations. Int Zoo Yb. 2009;43:6–18.

León-Quinto T, Simón MA, Sánchez A, Martín F, Soria B. Cryobanking the genetic diversity in the critically endangered Iberian lynx (*Lynx pardinus*) from skin biopsies. Investigating the cryopreservation and culture ability of highly valuable explants and cells. Cryobiology. 2011;62:145–51.

Li Y, Dai Y, Du W, Zhao C, Wang H, Wang L, et al. Cloned endangered species takin (*Budorcas taxicolor*) by inter-species nuclear transfer and comparison of the blastocyst development with yak (*Bos grunniens*) and bovine. Mol Reprod Dev. 2006a;73:189–95.

Li Z, Sun X, Chen J, Liu X, Wisely SM, Zhou Q, et al. Cloned ferrets produced by somatic cell nuclear transfer. Dev Biol. 2006b;293:439–48.

Lister R, Pelizzola M, Kida YS, Hawkins RD, Nery JR, Hon G, et al. Hotspots of aberrant epigenomic reprogramming in human induced pluripotent stem cells. Nature. 2011;471:68–73.

Liu P-C, Wang C-Y, Lin S-L, Lin C-C, Hung S-W, Chang C-H, et al. Establishment of a soft shell turtle, *Pelodiscus sinensis*, embryo primary cell culture for studies of soft shell turtle poxvirus-like virus replication and characteristics. Afr J Microbiol Res. 2012;6:960–7.

Liu S-Z, Zhou Z-M, Chen T, Zhang Y-L, Wen D-C, Kou Z-H, et al. Blastocysts produced by nuclear transfer between chicken blastodermal cells and rabbit oocytes. Mol Reprod Dev. 2004;69:296–302.

Loi P, Matsukawa K, Ptak G, Clinton M, Fulka Jr J, Nathan Y, et al. Freeze-dried somatic cells direct embryonic development after nuclear transfer. PLoS One. 2008;3:e2978. doi:10.1371/journal.pone.0002978.

Loi P, Beaujean N, Khochbin S, Fulka Jr J, Ptak G. Asymmetric nuclear reprogramming in somatic cell nuclear transfer? Bioessays. 2007;30:66–74.

Loi P, Ptak G, Barboni B, Fulka Jr J, Cappai P, Clinton M. Genetic rescue of an endangered mammal by cross-species nuclear transfer using post-mortem somatic cells. Nat Biotechnol. 2001;19:962–4.

Lu Y, West FD, Jordan BJ, Mumaw JL, Jordan ET, Gallegos-Cardenas A, et al. Avian-induced pluripotent stem cells derived using human reprogramming factors. Stem Cells Dev. 2012; 21:394–403.

Luvoni GC, Chigioni S, Beccaglia M. Embryo production in dogs: from in vitro fertilization to cloning. Reprod Domest Anim. 2006;41:286–90.

Mancia A, Spyropoulos DD, McFee WE, Newton DA, Baatz JE. Cryopreservation and in vitro culture of primary cell types from lung tissue of a stranded pygmy sperm whale (*Kogia breviceps*). Comp Biochem Phys C. 2012;155:136–42.

Manoli DS, Subramanyam D, Carey C, Sudin E, Van Westerhuyzen JA, Bales KL, et al. Generation of induced pluripotent stem cells from the prairie vole. PLoS One. 2012;7:e38119. doi:10.1371/journal.pone.0038119.

Mastromonaco GF, Paris MCJ, Krisher RL, Paris DBBP. Consensus discussion on artificial insemination in companion animal, non-domestic and endangered species (CANDES). Embryo Transfer Newsletter. 2011;29(2):10–2.

Mastromonaco GF, King WA. Cloning in companion animal, non-domestic and endangered species: can the technology become a practical reality? Reprod Fertil Dev. 2007;19:748–61.

Mastromonaco GF, Favetta LA, Smith LC, Filion F, King WA. The influence of nuclear content on developmental competence of gaur x cattle hybrid in vitro fertilized and somatic cell nuclear transfer embryos. Biol Reprod. 2007;76:514–23.

Mastromonaco GF, Perrault SD, Betts DH, King WA. Role of chromosome stability and telomere length in the production of viable cell lines for somatic cell nuclear transfer. BMC Dev Biol. 2006;6:41. doi:10.1186/1471-213X-6-41.

Mauger P-E, Le Bail PY, Labbé C. Cryobanking of fish somatic cells: optimizations of fin explants culture and fin cell cryopreservation. Comp Biochem Phys B. 2006;144:29–37.

Maya-Soriano MJ, Holt WV, Lloyd RE. Biobanked amphibian samples confirmed to species level using 16S rRNA DNA barcodes. Biopreserv Biobank. 2012;10:22–8.
Minematsu T, Tajima A, Kanai Y. Attempt to produce nuclear transferred primordial germ cells using electrofusion in domestic chicken. Anim Sci J. 2004;75:271–4.
Moore MK, Work TM, Balazs GH, Docherty DE. Preparation, cryopreservation, and growth of cells prepared from the green turtle (*Chelonia mydas*). Methods Cell Sci. 1997;19:161–8.
Moresco KA, Stallknecht DE, Swayne DE. Evaluation and attempted optimization of avian embryos and cell culture methods for efficient isolation and propagation of low pathogenicity avian influenza virus. Avian Dis. 2010;54:622–6.
Morgan HD, Santos F, Green K, Dean W, Reik W. Epigenetic reprogramming in mammals. Hum Mol Genet. 2005;14:R47–58.
Na RS, Bai CY, Jin DP, Su XH, Feng BG, Guan WJ, et al. Establishment and biological characteristics of Qingyuan partridge chicken fibroblast line. Poult Sci. 2010;89:1207–16.
Nakagawa M, Koyanagi M, Tanabe K, Takahashi K, Ichisaka T. Generation of induced pluripotent stem cells without Myc from mouse and human fibroblasts. Nat Biotechnol. 2007;26:101–6.
Nayernia K, Nolte J, Michelmann HW, Lee JH, Rathsack K, Drusenheimer N, et al. In vitro-differentiated embryonic stem cells give rise to male gametes that can generate offspring mice. Dev Cell. 2006;11:125–32.
Nel-Themaat L, Gomez MC, Pope CE, Lopez M, Wirtu G, Jenkins JA, et al. Cloned embryos from semen. Part 2: Intergeneric nuclear transfer of semen-derived eland (*Taurotragus oryx*) epithelial cells into bovine oocytes. Cloning Stem Cells. 2008;10:161–72.
Nel-Themaat L, Gómez MC, Damiani P, Wirtu G, Bl D, Bondioli KR, et al. Isolation, culture and characterization of somatic cells derived from semen and mild of endangered sheep and eland antelope. Reprod Fertil Dev. 2007;19:576–84.
Ng S-C, Chen N, Yip W-Y, Liow S-L, Tong G-Q, Martelli B, et al. The first cell cycle after transfer of somatic cell nuclei in a non-human primate. Development. 2004;131:2475–84.
Nowak JA, Fuchs E. Isolation and culture of epithelial stem cells. Methods Mol Biol. 2009;482:215–32.
Oh HT, Kim MK, Jang G, Kim HJ, Hong SG, Park JE, et al. Cloning endangered gray wolves (*Canis lupus*) from somatic cells collected postmortem. Theriogenology. 2008;70:638–47.
Oh B-C, Kim J-T, Shin N-S, Kwon S-W, Kang S-K, Lee B-C, et al. Production of blastocysts after intergeneric nuclear transfer of goral (*Naemorhedus goral*) somatic cells into bovine oocytes. J Vet Med Sci. 2006;68:1167–71.
Okita K, Nakagawa M, Hyenjong H, Ichisaka T, Yamanaka S. Generation of mouse induced pluripotent stem cells without viral vectors. Science. 2008;322:949–52.
Okita K, Ichisaka T, Yamanaka S. Generation of gemline-competent induced pluripotent stem cells. Nature. 2007;448:313–8.
Ono T, Mizutani E, Li C, Wakayama T. Nuclear transfer preserves the nuclear genome of freeze-dried mouse cells. J Reprod Dev. 2008;54:486–91.
Peat JR, Reik W. Incomplete methylation reprogramming in SCNT embryos. Nat Genet. 2012;44:965–6.
Pereira RJ, Napolitano A, Garcia-Pereira FL, Baldo CF, Suhr ST, King LE, et al. Conservation of avian germplasm by xenogeneic transplantation of spermatogonia from sexually mature donors. Stem Cells Dev. 2012. doi:10.1089/scd.2012.0497.
Peruffo A, Panin M, Suman M, Mazzariol S, Ballarin C, Giurisato M, et al. Research strategies and development possibilities of a marine mammal tissue bank: conservation biology and biomolecular science. In: Isobe T, Nomiyama K, Subramanian A, Tanabe S, editors. Interdisciplinary studies on environmental chemistry, Environmental Specimen Bank: exploring possibilities of setting up ESB's in developing countries, vol. 4. Tokyo: Terrapub; 2010. p. 87–93.
Petitte JN. Avian germplasm preservation: embryonic stem cells or primordial germ cells? Poultry Sci. 2006;85:237–42.

Portz C, Lopes de Almeida L, Bianco Jr A, Reck H, Franco AC, Canal CW. Comparison of different cell cultures for replication of infectious laryngotracheitis virus from chickens. Acta Sci Vet. 2008;36:101–5.
Prasanna I, Lakra WS, Ogale SN, Bhonde RR. Cell culture from fin explants of endangered golden mahseer, *Tor putitora* (Hamilton). Curr Sci. 2000;79:93–8.
Rathore G, Kumar G, Rajaswaminathan T, Sood N, Singh V, Abidi R, et al. Primary cell culture from fin explants of *Labeo rohita* (Ham.). Indian J Fish. 2007;54:93–7.
Rawson D. Fish biobanking: current activies, lessons, and opportunities. Biopreserv Biobank. 2012;10:68 [abstract].
Rideout III WM, Eggan K, Jaenisch R. Nuclear cloning and epigenetic reprogramming of the genome. Science. 2001;293:1093–8.
Rolf HJ, Kierdorf U, Kierdorf H, Schulz J, Seymour N, Schliephake H, et al. Localization and characterization of STRO-1+ cells in the deer pedicle and regenerating antler. PLoS One. 2008;3:e2064. doi:10.1371/journal.pone.0002064.
Sangalli JR, De Bem TH, Perecin F, Chiaratti MR, Oliveira LJ, de Araujo RR, et al. Treatment of nuclear-donor cells or cloned zygotes with chromatin-modifying agents increases histone acetylation but does not improve full-term development of cloned cattle. Cell Reprogram. 2012;14:235–47.
Sansinena MJ, Hylan D, Hebert K, Denniston RS, Godke RA. Banteng (*Bos javanicus*) embryos and pregnancies produced by interspecies nuclear transfer. Theriogenology. 2005;63:1081–91.
Saragusty J. Genome banking for vertebrates wildlife conservation. In: Katkov I, editor. Current frontiers in cryobiology. Croatia: Intech; 2012. p. 293–368.
Signoret J, David J-C, Lefresne J, Houillon C. Control of DNA ligase molecular forms in nucleocytoplasmic combinations of axolotl and Pleurodeles. Dev Biol. 1983;80:3368–71.
Silva RC, Costa GM, Lacerda SM, Batlouni SR, Soares JM, Avelar GF, et al. Germ cell transplantation in felids: a potential approach to preserving endangered species. J Androl. 2012; 33:264–76.
Sinzelle L, Thuret R, Hwang H-Y, Herszberg B, Paillard E, Bronchain OJ, et al. Characterization of a novel *Xenopus tropicalis* cell line as a model for in vitro studies. Genesis. 2012;50: 316–24.
Slimane Bureau W, Bordignon V, Leveilée C, Smith LC, King WA. Assessment of chromosomal abnormalities in bovine nuclear transfer embryos and in their donor cells. Cloning Stem Cells. 2003;5:123–32.
Srirattana K, Imsoonthornruksa S, Laowtammathron C, Sangmalee A, Tunwattana W, Thongprapai T, et al. Full-term development of gaur-bovine interspecies somatic cell nuclear transfer embryos: effect of trichostatin A treatment. Cell Reprogram. 2012;14:248–57.
Stadtfeld M, Nagaya M, Utikal J, Weir G, Hochedlinger K. Induced pluripotent stem cells generated without viral integration. Science. 2008;322:945–9.
Stephenson NG. Effects of temperature on reptilian and other cells. J Embryol Exp Morph. 1966;16:455–67.
Su X, Jin D, Li C, Yue H, Pu Y, Ma Y, et al. Establishment of Big Bone chicken fibroblast cell bank and study of its biological characteristics. Belg J Zool. 2011;141:14–23.
Sun Y-H, Chen S-P, Wang Y-P, Hu W, Zhu Z-Y. Cytoplasmic impact on cross-genus cloned fish derived from transgenic common carp (*Cyrprinus carpio*) nuclei and goldfish (*Carassius auratus*) enucleated eggs. Biol Reprod. 2005;72:510–5.
Sweat JM, Dunigan DD, Wright SD. Characterization of kidney epithelial cells from the Florida manatee, *Trichechus manatus latirostris*. In Vitro Cell Dev Biol Anim. 2001;37:386–94.
Tanaka D, Takahashi A, Takai A, Ohta H, Ueno K. Attempt at cloning high-quality goldfish breed 'Ranchu' by fin-cultured cell nuclear transplantation. Zygote. 2010;20:79–85.
Takahashi K, Tanabe K, Ohnuke M, Narita M, Ichisaka T, Tomoda K, et al. Induction of pluripotent stem cells from adult human fibroblasts by defined factors. Cell. 2007;131:861–72.
Tao Y, Liu J, Zhang Y, Zhang M, Fang J, Han W, et al. Fibroblast cell line establishment, cryopreservation and interspecies embryos reconstruction in red panda (*Ailurus fulgens*). Zygote. 2009;17:117–24.

Tayfur Tecirlioglu R, Trounson AO. Embryonic stem cells in companion animals (horses, dogs and cats): present status and future prospects. Reprod Fert Dev. 2007;19:740–7.

Thongphakdee A, Siriaroonrat B, Manee-in S, Klincumhom N, Kamolnorranath S, Chatdarong K, et al. Intergeneric somatic cell nucleus transfer in marbled cat and flat-headed cat. Theriogenology. 2010;73:120–8.

Thongphakdee A, Numchaisrika P, Omsongkram S, Chatdarong K, Kamolnorranath S, Dumnui S, et al. In vitro development of marbled cat embryos derived from interspecies somatic cell nuclear transfer. Reprod Dom Anim. 2006;41:219–26.

Till JE, McCulloch EA. A direct measurement of the radiation sensitivity of normal mouse bone marrow cells. Radiat Res. 1961;14:213–22.

Tilly JL, Telfer EE. Purification of germline stem cells from adult mammalian ovaries: a step closer towards control of the female biological clock? Mol Hum Reprod. 2009;15:393–8.

Torvar H, Navarrete F, Rodriguez L, Skewes O, Castro FO. Cold storage of biopsies from wild endangered native Chilean species in field conditions and subsequent isolation of primary culture cell lines. In Vitro Cell Dev Biol. 2008;44:309–20.

Travis AJ, Kim Y, Meyers-Wallen V. Development of new stem cell-based technologies for carnivore reproduction research. Reprod Domest Anim. 2009;44 Suppl 2:22–8.

Tryon RC, Johnson SL. Clonal and lineage analysis of melanocyte stem cells and their progeny in the zebrafish. Methods Mol Biol. 2012;916:181–95.

Ty LV, Hanh NV, Uoc NT, Duc NG, Thanh NT, Bui LC, et al. Preliminary results of cell cryobanking and embryo production of black bear (*Ursus thibetanus*) by interspecies somatic cell nuclear transfer. Theriogenology. 2003;59:290 [abstract].

Venegas F, Guillomot M, Vignon X, Servely J-L, Audouard C, Montiel E, et al. Obtaiment of pudu (*Pudu pudu*) deer embryos by the somatic nuclear transfer technique. Int J Morphol. 2006; 24:285–92.

Veprintsev BN, Rott NN. Conserving genetic resources of animal species. Nature. 1979;280:633–4.

Verma R, Holland MK, Temple-Smith P, Verma PJ. Inducing pluripotency in somatic cells from the snow leopard (*Panthera uncia*), an endangered felid. Theriogenology. 2012;77:220–8.

Wagers AJ, Weissman IL. Plasticity of adult stem cells. Cell. 2004;116:639–48.

Wang K, Beyhan Z, Rodriguez RM, Ross PJ, Iager AE, Kaiser GG, et al. Bovine ooplasm partially remodels primate somatic nuclei following somatic cell nuclear transfer. Cloning Stem Cells. 2009;11:187–202.

Wang L, Peng T, Zhu H, Lv Z, Liu T, Shuai Z, et al. In vitro development of reconstructed ibex (*Capra ibex*) embryos by nuclear transfer using goat (*Capra hircus*) oocytes. Small Ruminant Res. 2007;73:135–41.

Wani NA, Wernery U, Hassan FAH, Wernery R, Skidmore JA. Production of the first cloned camel by somatic cell nuclear transfer. Biol Reprod. 2010;82:373–9.

Wakamatsu Y. Novel method for the nuclear transfer of adult somatic cells in medaka fish (*Oryzias latipes*): use of diploidized eggs as recipients. Develop Growth Differ. 2008;50:427–36.

Wen D-C, Bi C-M, Xu Y, Yang C-X, Zhu Z-Y, Sun Q-Y, et al. Hybrid embryos produced by transferring panda or cat somatic nuclei into rabbit MII oocytes can develop to blastocyst in vitro. J Exp Zool Part A. 2005;303A:689–97.

White KL, Bunch TD, Mitalipov S, Reed WA. Establishment of pregnancy after the transfer of nuclear transfer embryos produced from the fusion of argali (*Ovis ammon*) nuclei into domestic sheep (*Ovis aries*) enucleated oocytes. Cloning. 1999;1:47–54.

Wildt DE. Genetic resource banks for conserving wildlife species. Justification, examples and becoming organized on a global basis. Anim Reprod Sci. 1992;28:247–57.

Williams JB, Shin T, Liu L, Flores-Foxworth G, Romano J, Blue-McClendon A, et al. Cloning of exotic/endangered species: desert bighorn sheep. Methods Mol Biol. 2006;348:169–82.

Wilmut I, Schnieke AE, McWhir J, Kind AJ, Campbell AH. Viable offspring derived from fetal and adult mammalian cells. Nature. 1997;385:810–3.

Wu DM, Schneiderman T, Burgett J, Gokhale P, Barthel L, Raymond PA. Cones regenerate from retinal stem cells sequestered in the inner nuclear layer of adult goldfish retine. Invest Ophthalmol Vis Sci. 2001;42:2115–24.

Xi Y, Nada Y, Soh T, Fujihara N, Hattori M. Establishment of feather follicle stem cells as potential vehicles for delivering exogenous genes in birds. J Reprod Dev. 2003;49:213–9.

Xiong X-R, Wang L-J, Zi X-D, Ma L, Xu W-B, Wang Y-S, et al. Epigenetic reprogramming of yak iSCNT embryos after donor cell pre-treatment with oocyte extracts. Anim Reprod Sci. 2012;133:229–36.

Yan SY, Tu M, Yang HY, Mao ZG, Zhao ZY, Fu LJ, et al. Developmental incompatibility between cell nucleus and cytoplasm as revealed by nuclear transplantation experiments in teleost of different families and orders. Int J Dev Biol. 1990;34:255–66.

Yan SY, Lu DY, Du M, Li GS, Lin LT, Jin GQ, et al. Nuclear transplantation in teleosts. Hybrid fish from the nucleus of crucian and the cytoplasm of carp. Sci Sin B. 1984;27:1029–34.

Yang CX, Han ZM, Wen DC, Sun QY, Zhang KY, Zhang LS, et al. In vitro development and mitochondrial fate of macaca-rabbit cloned embryos. Mol Reprod Dev. 2003;65:396–401.

Yin X, Lee Y, Lee H, Kim N, Kim L, Shin H, et al. In vitro production and initiation of pregnancies in inter-genus nuclear transfer embryos derived from leopard cat (*Prionailurus bengalensis*) nuclei fused with domestic cat (*Felis silverstris catus*) enucleated oocytes. Theriogenology. 2006;66:275–82.

Yoshizaki G, Okutsu T, Morita T, Terasawa M, Yazawa R, Takeuchi Y. Biological characteristics of fish germ cells and their application to developmental biotechnology. Reprod Domest Anim. 2012;47 Suppl 4:187–92.

Zeng CJ, Qing YE, Fang SG. Establishment and characterization of liver, heart and muscel cell lines derived from the Chinese alligator (*Alligator sinensis*). Chinese Sci Bull. 2011;56:2576–9.

Zhao Z-J, Ouyang Y-C, Nan C-L, Lei Z-L, Song X, Sun Q-Y, et al. Rabbit oocyte cytoplasm supports development of nuclear transfer embryos derived from the somatic cells of the camel and Tibetan antelope. J Reprod Dev. 2006;52:449–59.

Zhou Q, Yang SH, Ding CH, He XC, Xie YH, Hildebrandt TB, et al. A comparative approach to somatic cell nuclear transfer in the rhesus monkey. Hum Reprod. 2006;21:2564–71.

Zhu H, Craig JA, Dyce PW, Sunnen N, Li J. Embryos derived from porcine skin-derived stem cells exhibit enhanced preimplantation development. Biol Reprod. 2004;71:1890–7.

Chapter 17
Biosafety in Embryos and Semen Cryopreservation, Storage, Management and Transport

A. Bielanski

Abstract This chapter summarizes pertinent procedures, data and opinions on the potential hazards of disease transmission through liquid nitrogen (LN)-cryopreserved and banked germplasm and tissues for somatic cell nuclear transfer (SCNT) The importance of applying internationally adopted sanitary washing procedures to germplasm as a crucial step towards their successful microbial-free cryopreservation and storage is emphasised. Special attention is given to the survival of pathogens in LN, variety of vitrification methods, sterility of LN, risks associated with the use of straws and cryovials, and LN Dewars including dry shippers. It was experimentally demonstrated that cross-contamination between LN and embryos may occur, when infectious agents are present in LN and if embryos are not protected by use of a sealed container. It is important, therefore, to prevent direct contact of germplasm and reproductive tissues with LN during cryopreservation and their storage as a mandatory measure for reducing the risk of contamination. This includes the usage of hermetically sealed high quality shatter proof freezing containers and/or the application of a secondary enclosure such as "double bagging or straw in straw". A periodic disinfection of cryo-Dewars should be considered as an additional precaution to diminish the potential for inadvertent cross-contamination. It would be advisable to use separate LN Dewars to quarantine embryos derived from infected donors of valuable genotypes or from unknown health status, extinction-threatened species.

Keywords Cryopreservation • Disease transmission • Embryo transfer • Semen • Liquid nitrogen • Contamination • ART

A. Bielanski, D.V.M., Ph.D. (✉)
Animal Diseases Research Institute, 3851 Fallowfield Road, Ottawa, ON, Canada, K2H 8P9
e-mail: Andrzej.Bielanski@inspection.gc.ca

W.V. Holt et al. (eds.), *Reproductive Sciences in Animal Conservation*,
Advances in Experimental Medicine and Biology 753, DOI 10.1007/978-1-4939-0820-2_17,

1 Introduction

This chapter describes the potential hazards of disease transmission through embryo transfer (ET) of cryopreserved and banked embryos and semen based on data obtained through experimentation with germplasm of domestic mammals as a model for wildlife species. As there is insufficient pertinent background on the reproductive biology of most wild species the lessons learnt from domestic mammals can provide valuable guidance. This includes lack of information on the morpho-physiological properties of gametes (e.g. zona pellucida (ZP) of oocytes and embryos and plasma membrane of spermatozoa), which is crucial for the determination of their interaction with pathogens. With regard to microbial agents there is a long list of viral, bacterial and parasitic diseases affecting both wild and farm animals (Williams and Barker 2001). According to the risk of pathogens transmission via *in vivo*-derived embryos, the International Embryo Transfer Society (IETS) classified diseases into four categories (Table 17.1). In category 1 are agents with evidence of the negligible risk of disease transmission (e.g. bluetongue, foot-and-mouth disease virus (FMDV), bovine spongiform encephalopathy (BSE), *Brucella abortus* in cattle). Category 2 group contains diseases for which the risk of transmission is negligible. But further experimentation with additional transfer is required to verify existing data (e.g. scrapie), and category 3 lists diseases for which preliminary evidence indicates negligible risk of transmission (e.g. Rindepest, *Mycobacterium paratuberculosis*). Category 4 comprises agents for which some studies have been done, but no conclusions are yet possible or the risk of transmission via ET might not be negligible even if the embryos are handled between collection and transfer according to the IETS sanitary protocol (e.g. vesicular stomatitis virus, bovine anaplasmosis, African swine fever, porcine circovirus). For details see IETS (Stringfellow and Givens 2010).

For background on basic cryobiology, germplasm and ovarian tissue cryopreservation, readers are referred to the publication by Karrow and Critser (1997) and the more recent articles on new methods of vitrification by Vajta and Nagy (2006), Tucker and Liebermann (2007), Yavin and Arav (2007), Kuwayama (2007), Mazur et al. (2008), Mazur and Seki (2011), Saragusty and Arav (2011), Seli and Agarwal (2011), Zhang et al. (2011), Arav and Natan (2012) and Liu et al. (2012). Comparative cryobiological requirements for germplasm and ovarian tissue of wildlife and threatened species were reviewed by Leibo and Songsasen (2002), Santos et al. (2010) and Comizzoli et al. (2012). Factors to be considered in developing a genome resources bank (GRB) for a taxon/species also were identified by Wild (1997), Woelders et al. (2012), Agca (2012) and Mara et al. (2013). Lastly, managing the potential for cross contamination of germplasm during cryobanking in the liquid (LN) and vapour phases of nitrogen (VPLN) was previously discussed by several authors (Rall 2003; Bielanski and Vajta 2009; Pomeroy et al. 2010; Criado 2012).

Table 17.1 Risk of infectious agents transmission by ET and semen in domestic animals and suggested potential for association of agents with embryos of wildlife counterparts

Disease or infectious agent	Presence in ovary, oviduct or uterus	Association with washed embryos	Transmission by ET	Transmission by semen	Investigated species (source/reference-as superscript)	Potential wildlife counterparts
Viral						
SVSV[3]	–	+	?	+	[a, b]Sus scrofa	Suidae
PRRS[3]	+	–	–	+	[e, n, s]Sus scrofa	Suidae
PrV	+	+	±	±	[a, b]Sus scrofa	Suidae
PCV[4]	+	+	–	+	[t]Sus scrofa	Suidae
HCV[2]	+	+	–	±	[a, b]Sus scrofa	Suidae
ASFV[4]	–	+	?	+	[a, b]Sus scrofa	Suidae
PPV[4]	+	+		±	[a, b]Sus scrofa	Suidae
FMDV[3]	–	+		+	[a, b]Sus scrofa	Suidae
FMDV[1]	+	–	–	+	[a, b]Bos Taurus	Bovidae, Cervidae, Camelide
FMDV[3]	–	–	–	+	[h]Ovis aries	
FMDV[3]	–	–	–	+	[h]Capra hircus	
Maedi-Visna[3]	+	–	–	?	[r]Ovis aries	Unknown
Maedi-Visna	–	+	–	?	[h]Capra hircus	Unknown
BVDV CP[3]	+	–	–	+	[a, b]Bos Taurus	Some ruminants
BVDV NCP[3]	+	+	–	+	[o]Bos Taurus	
RPV[3]	–	+	–	±	[a, b]Bos Taurus	Bovidae; Cervidae
PI-3V[4]	–	+	?	?	[a, b]Bos Taurus	Various spp.
BIV[3]	–	–	–	+	[i]Bos Taurus	Unknown
BHV[1]	+	+	–	+	[a, b]Bos Taurus	Bovidae
AKV[4]	–	–	?	+	[a, b]Bos Taurus	Some ruminants
BLV[1]	–	–	–	–	[a, b]Bos Taurus	Unknown
VSV[4]		+	–	±	[a, b]Bos Taurus	
BTV[1]	+	?	–	+	[a, b]Bos Taurus	Bovidae; Cervidae
BTV-8[4]	–	?	?	+	[h]Capra hircus	
BTV[2]		–	–		[h]Ovis aries	
FIV	?	?	?	+	[c]Felis catus	Felidae
CAEVC[2]	+	–	–	–	[h]Capra hircus	Small ruminants
EAV	+	±	±	+	[p]Equus ferus caballus	Equidae

(continued)

Table 17.1 (continued)

Disease or infectious agent	Presence in ovary, oviduct or uterus	Association with washed embryos	Transmission by ET	Transmission by semen	Investigated species (source/reference-as superscript)	Potential wildlife counterparts
TSEs						
BSE[1]	–	–	–	–	[m]Bos Taurus	Bovidae; Felidae
CVD	?	?	?	–	[m]Odocoileus sp.	Cervidae
Scrapie[2] (typical)[4]	?	?	–	–	[m]Ovis aries	Unknown species
			–	–	[m]Capra hircus	
Bacterial						
M. bovis[4]	–	–	?	+	[f]Bos Taurus	Bovidae
B. abortus[1]	–	–	–	+	[a, b]Bos Taurus	Bovidae; Cervidae;
B. ovis	–	–	?	±	[a, b]Ovis aries	
Leptospirosis[4]	+	+	?	+	[a, d]Bos Taurus	Bovidae; Suidae
			?	+	[a]Sus scrofa	
Yohnes disease[3]	+	+	–	±	[a, l]Bos Taurus	Bovidae; Cervidae
Mycoplasmas[4]		+	?	+	[a, b, g]Bos Taurus	All mammals
Vibriosis		+	?	+	[a]Bos Taurus	Bovidae
Parasitic						
Trichomoniasis[4]	+	+	?	+	[a, i]Bos Taurus	Unknown
Piroplasmosis	–	?	±	?	[a]Bos Taurus	Bovidae
Neosporosis[3]	–	+	–	?	[a, j]Bos Taurus	Bovidae; Cervidae Felidae; Canidae

?=unknown; ±=contradicted data; +=positive; –=negative; agent with superscript numbers=risk of disease transmission (category, IETS; Stringfellow and Givens 2010)

BHV-1 bovine herpesvirus-1; *BVDV CP* bovine viral diarrhea virus, cytopathic strain; *BVDV NCP* bovine viral diarrhea virus, noncytopathic strain; *BIV* bovine immunodeficiency virus; *BLV* bovine leukemia virus; *BTV* bluetongue virus; *FMDV* foot-and-mouth disease virus; *AV* akabane virus; *RPV* Rinderpest virus; *PI-3V* parainfluenza virus; *PPV* porcine parvovirus; *PRRS* porcine reproductive and respiratory syndrome virus; *PrV* porcine pseudorabies virus (Aujesky' virus); *PCV-2* porcine circovirus-2; *HCV* hog cholera virus (classical swine fever virus); *SVSV* swine vesicular stomatitis virus; *VSV* vesicular stomatitis virus; *CAEV* caprine arthritis-encephalitis virus; *EAV* equine arteritis virus; *TSEs* transmissible spongiform encephalopathies; *BSE* bovine spongiform encephalopathy agent; *CVD* chronic wasting disease; *M. paratib. Mycobacterium paratuberculosis*; *M. bovis Mycobacterium bovis*; *B. abortus Brucella abortus*

[a]Hare (1985); [b]Stringfellow and Seidel (1998); [c]Jordan et al. (1998); [d]Bielanski et al. (1998b); [e]Randall et al. (1999); [f]Bielanski et al. (1999b); [g]Bielanski et al. (2000b); [h]Thibier and Guerin (2000); [i]Bielanski et al. (2001); [j]Bielanski et al. (2002); [k]Bielanski et al. (2004); [l]Bielanski et al. (2006); [m]Wrathall et al. (2008); [n]Maes et al. (2008); [o]Bielanski et al. (2009); [p]Broaddus et al. (2011); [r]Cortez-Romero et al. (2011); [s]Gregg et al. (2011); [t]Bielanski et al. (2013)

2 The Origin of Microbial Contamination in Germplasm and Somatic Cells Prior to Cryopreservation

2.1 Semen

It has been recognized that systemic and local infections of the reproductive tract, as well as the inadvertent introduction of microorganisms during processing, may potentially contribute to the contamination of semen. In general terms, microorganisms can already be present in the semen of an infected male when it is ejaculated or they can gain entry during collection, processing or storage. Spermatozoa can become infected by a microorganism in the testes or during their transit through the epididymis, ductus deferens and urethra. Other occasions when microorganisms can be present in semen are when they are associated with blood cells or there is inflammation or trauma of the accessory glands (prostate, seminal vesicle or bulbourethral gland). Furthermore, there are some microorganisms that can contaminate semen due to their high concentration in urine or in the preputial cavity.

In addition, some potential contaminants (e.g. mycoplasmas) may be introduced into semen with animal derived supplements used in diluents and extenders (egg yolk, milk). Environmental microbes may also contribute to semen contamination or a result of poor laboratory hygiene (Schiewe and Hasler 2010; Wrathall et al. 2007).

Frequently, ejaculated semen is not free from bacterial flora. The saprophytic bacteria of the prepuce in a healthy male comprise numerous species that may become associated with the semen. Some of these bacteria may behave as opportunistic pathogens (e.g. *Pseudomonas aeruginosa*) and may be a potential risk to the inseminated female. For example, the most common potentially pathogenic microorganisms isolated from bull semen are *P. aeruginosa. Streptococcus* spp., *Staphylococcus* spp., *Proteus* spp., and *Bacillus* spp. (Wierzbowski 1985). The notion that spermatozoa could function to transport surface-bound bacteria has been reported for *Chlamydia trachomatis, Chlamydia psittaci, Escherichia coli, Neisseria gonorrhoea, Veillonella parvula, Peptostreptococcus spp., Ureaplasma urealyticum, Mycoplasma spp.,* and *Candida albicans* (Toth et al. 1982).

A number of viral pathogens have also been identified in association with the semen of infected animals and humans (Hare 1985; Dejucq-Rainsford and Jegou 2004; Bielanski 2006; Zea-Mazo et al. 2010). Some of the viruses can adhere to the surface of spermatozoa and others are associated with the seminal plasma or non-sperm cells present in the semen. Several reports, some of which are controversial, have suggested an ability of some viruses to penetrate the sperm head and integrate their nuclei acid into the sperm genome. These viruses include, Simian virus 40 (Brackett et al. 1971), bluetongue virus (Foster et al. 1980), bovine herpesvirus-1 (BHV-1) (Elazhary et al. 1980), human hepatitis B virus (Hadchouel et al. 1985), Rous sarcoma virus (Habrova et al. 1996), herpes simplex 1 and 2 (Kotronias and Kapranos 1998), murine cytomegalovirus (Magnano et al. 1998), human immunodeficiency virus-1 (HIV-1) (Piomboni and Baccetti 2000), and human papilloma virus (Rintala et al. 2005). For this reason, the complete elimination of these viral agents from semen and sperm cells may be difficult or even not possible to achieve.

2.2 Embryos

Prior to ovulation oocytes may become infected by contact with an infectious agent present in either the ovarian granulosa cells or the follicular fluid, probably during viremia at the acute stage of a disease (Van Soom et al. 2010a). At this stage viruses may be present in the blood and other body fluids and spread to various tissues and organs (Table 17.1). For example, in cattle, microorganisms have been found in follicular fluid a few days after natural and experimental exposure to bovine viral diarrhea virus (BVDV) and BHV-1 (Bielanski and Dubuc 1994; Bielanski et al. 1998a). This indicates that the collection of oocytes for *in vitro* fertilization (IVF) at this stage of the disease may result in embryos contaminated with the pathogenic agent. This hazard may be substantial when ovaries are harvested from asymptomatic persistently or latently infected donors (e.g. BVDV, BHV-1). Post-ovulation, oocytes may become infected by a spermatozoon during fertilization, or by contact with a pathogen that has been excreted into the oviduct or uterus (Booth et al. 1995). Other sources of contamination include agents introduced with culture supplements of biological origin such as serum, trypsin, supporting co-culture cells or cell lines for nuclear transfers (Van Os et al. 1991; Brock 1998; Shin et al. 2000; Drew et al. 2002; Schiewe and Hasler 2010; Nikfarjam and Farzaneh 2012). Environmental microbes associated with an operator, abattoir origin or the laboratory may pose risks during the production of embryos *in vitro* when pooled materials are used. In this regard, inadvertent inclusion of the follicular fluid from an infected animal into the pool may pose a risk of the cross-contamination of all clean oocytes and consequently lead to batches of contaminated embryos (Bielanski and al. 1993; Bielanski and Stewart 1996; Marquant-Leguienne et al. 2000). Also transvaginally collected oocytes are potential sources of microbial contamination of the in *vitro* fertilization (IVF) and embryo transfer (ET) culture systems. Cottell et al. (1996) reported that various microorganism (*Mycoplasma hominis*, *Ureaplasma urealyticum*, *Staphylococcus epidermis*, *Lactobacilli* sp., *Difteroids*) were isolated from approximately one-third of the needle flushes after oocyte recovery and from more than one-third of the follicular fluids aspirated from the first follicle punctures on each ovary.

In general, there are two periods prior to gamete retrieval that affect success of gamete preservation and their sanitary status. The first is the interval from animal's death to necropsy and the second is from gonad retrieval to gamete recovery in laboratories (Chatdarong 2011). It can be expected that harvesting somatic tissues and germplasm from wildlife animals in their natural habitat may occur hours or days after their death. It is well known that gastro-intestinal bacteria are capable of migrating from the gut to any other region of the body by using the lymphatic system and blood vessels as early as 5 min after death (Melvin et al. 1984; Heimesaat et al. 2012). How rapidly microbial invasion reaches the reproductive tract, and particularly testes and ovaries, have been not established. It is predictable however, that in climatic regions with high environmental temperatures and humidity, the process of carcass decomposition will be accelerated. Nevertheless, it appeared that sperm cells harvested after death may remain viable longer than oocytes (Johnston et al. 1991; Stoops et al. 2011).

2.3 Tissues for Somatic Cells Nuclear Transfer (SCNT)

In recent years SCNT procedures have become an important tool for regeneration of valuable genetically species as part of the assisted reproductive technologies (ART) (Holt et al. 2004; Andrabi and Maxwell 2007; Galli and Lazzari 2008). There are three main potential routes for the introduction of pathogens during cloning embryos prior to cryopreservation. These include contaminated donor somatic cells, oocytes and culture system used for reconstructed embryos. Another important source of contamination can be laboratory environment and personnel, who handle cell cultures. Most frequently mycoplasma and other bacteria from human skin, airborne fungal spores and saprophytic microorganisms are isolated from cell cultures (Cobo and Concha 2007).

A variety of somatic cells such as mammary epithelium, adult and fetal fibroblast, granulosa, cumulus cells and other, have been used to produce animal clones (Rodriguez-Osorio et al. 2012; Men et al. 2012). More recently, induced pluripotent stem (iPS) cells derived from somatic cells opened a potential source of reprogrammed donor cells for nuclear transfer and gene banking of domestic and endangered species (Malaver-Ortega et al. 2012; Verma et al. 2012).

Cells for SCNT are usually obtained either from the existing established cell lines or from live animals with desirable phenotypes. When somatic cells are harvested, health status of donors should be taken into consideration, since microorganisms can be present in blood and many tissues in acutely and persistently infected animals (Fields 2006). For example, persistently BVDV-infected cattle represent a special hazard since they constitute reservoirs of infectious virus in serum, ovarian follicular fluid, gametes, and somatic cells as well as fetal tissues and serum (Brownlie 1990). Stringfellow et al. (2005) reported that 5 of 39 fetal fibroblast cell lines used for cloning research were positive for BVDV as determined by the reverse transcription-polymerase chain reaction (RT-PCR) assay. The risk and consequences of introducing BVDV by infected somatic cells in nuclear transfer was demonstrated by Shin et al. (2000). At days 40–55 all cloned fetuses produced from BVDV-infected fetal fibroblast cell lines were positive. In addition, pregnancy loss was significantly greater in fetuses derived from BVDV-infected cell lines as compared to non-infected cell lines.

It is relevant to mention that there is a theoretical risk of reactivation of endogenous retroviruses (ERV) (dormant viruses) via introduction of a foreign nucleus into an enucleated oocyte. However, to date there have not been any reports of such outcomes occurring during the cloning process (Thibier 2001; Kochhar and Rudenko 2010). Experimentation with bovine clones revealed that ERV sequences were not transcribed and no RNA was detected in the blood of clones, donor animals or controls (Heyman et al. 2007; Anon 2008a).

Health risks associated with the procedure of oocytes recovery from live animals using ultrasonography or from abattoir materials and their handling are of equal importance to those encountered during IVF (Bielanski 2010). Oocytes for SCNT recovered from follicular fluid pools of slaughtered animals with unknown health status may be a potential source of infectious agents for cloned embryos. Some of those agents [e.g. caprine arthritis encephalitis virus (CAEV), BVDV, BHV-1] may adhere to the ZP of oocytes and may also have ability to replicate in oocyte cumulus cells (OCC) (Vanroose 1999; Stringfellow et al. 2000; Ali Al Ahmad et al. 2005). Therefore, it is of paramount importance to remove OCC and wash oocytes prior to puncturing the ZP and transfer a somatic cell into an enucleated oocyte. Nevertheless, the risk of transmission of some viral agents such as lentiviruses [e.g. HIV, bovine immunodeficiency virus (BIV), feline immunodeficiency virus (FIV), and equine infectious anemia virus (EIAV)] by SCNT seems to be unlikely, since oocytes and early embryonic cells are not susceptible to infection with those agents (Gregg and Polejaeva 2009).

To reduce the risk of producing infected cloned embryos, using defined media free of biologically derived supplements [e.g. fetal bovine serum (FBS), porcine trypsin] for culture of somatic cell lines and reconstructed embryos is advisable (Givens et al. 2004; Bielanski 2007; Wang et al. 2012; Schiewe and Hasler 2010). A major concern is the introduction of BVDV to culture system through FBS. Other most common viral contaminants in FBS are BHV-1, and parainfluenza virus-3 (Ericson et al. 1991). Standards for quality testing of FBS have been defined in WHO Technical Report Series No. 673 (Anon, WHO 1982). Use of bovine serum albumin (BSA) may reduce considerably the risk of contamination due to the way that BSA is manufactured (Schiewe and Hasler 2010).

While there are steps in the SCNT technique which differ from the *in vitro* fertilisation procedure, no specific health risks related to oocyte enucleation, the fusion of oocyte with a somatic cell nucleus or the injection of the somatic cell nucleus directly into the cytoplasm of the enucleated oocyte have been reported.

The animal health risk associated with SCNT cloning technology has been addressed in the Terrestrial Animal Health Code (OIE, Chapter 4.11.) and by European Food Safety Authority (Anon 2008a, 2011). Both publications provide a scientific basis for recommendations on health and welfare of animals involved in SCNT cloning with other assisted reproductive technologies. In general, the guidelines for sanitary precautions of collection of oocytes and processing of reconstructed embryos are similar to those recommended by IETS with regard to the production *in vivo* and *in vitro* fertilized embryos (Stringfellow and Givens 2010). In particular, the importance of sanitary precautions in the following contexts is highlighted:

1. During oocyte collections—to minimize blood contamination, to test each pooled batch of follicular fluid for the presence of infectious agents, to use pathogen-free serum and protein components, and proper antibiotics in culture media to control bacterial microbes.
2. During donor cell processing—to harvest cells under proper sanitary conditions, to test master cell lines for the presence of bacteria, fungi, mycoplasmas and viruses.
3. During cloning—if a co-culture system (e.g. oviductal cells) is used to culture reconstructed embryos, proper screening of the cells should be performed. A sample of cells should be tested for bacterial, fungal, mycoplasmal or viral components (e.g. BVDV). Proper washing and cryopreservation of the reconstructed embryos should be followed as recommended by IETS (Stringfellow and Givens 2010).

Although the above listed sanitary recommendations were put together for processing somatic cells and embryos of livestock, it appears they also may find practical application in cloning procedures for wildlife species.

Other parts of publications deal with issues of animal health risk related to surrogate dams, born clones and their offspring.

So far the risk for disease transmission by SCNT has been determined only for a few pathogens of livestock animals. It was concluded that if strict sanitary precautions, as recommended by OIE and IETS, are followed, the hazard of EIAV (and probably other lentiviruses), porcine reproductive and respiratory syndrome virus (PRRSV), and BVDV would be small or absent (Gregg and Polejaeva 2009; Gregg et al. 2010a, b, 2011; Asseged et al. 2012). The risk of transmission of other infectious agents by SCNT, which may affect health status of clones of both livestock and wildlife species, remains to be investigated.

3 Fundamental Methods of Rendering Gametes Free of Pathogens Prior to Cryopreservation

3.1 Washing Procedure for Embryos

Procedures and requirements for washing of ova/embryos are outlined in detail in the Manual of the IETS (Stringfellow 2010). Zona pellucida-intact embryos free of adherent material are transferred through 5 washes of phosphate–buffered saline (PBS), containing antibiotics and 0.4 % bovine serum albumin (BSA), and then through 2 aliquots of 0.25 % sterile porcine-origin trypsin for a total 60–90 s. Next, embryos are transferred through an additional 5 sterile washes of PBS containing antibiotics and 0.4 % BSA or 2 % FBS (pathogens free). Where washing did not remove an agent from the ZP, enzymes have been used to inactivate the agent or loosen its attachment to the ZP (e.g. herpes viruses) (Bielanski 2007).

The embryos should be washed with at least 100-fold dilution between each wash, and a fresh pipette should be used for transferring the embryos through each wash. An example of embryo washing efficiency on reduction of pathogenic agents associated with ZP is illustrated in Fig. 17.1.

3.2 Washing Procedures for Semen

When raw semen is diluted and extended for commercial use, the number of potential micro-organisms per unit volume can be practically decreased to below the minimum infective artificial insemination (AI) dose (Hare 1985). In contrast to AI, for IVF only motile sperm has been used, which requires application of further antimicrobial procedures to eliminate the potential for disease transmission via semen (Bielanski 2007).

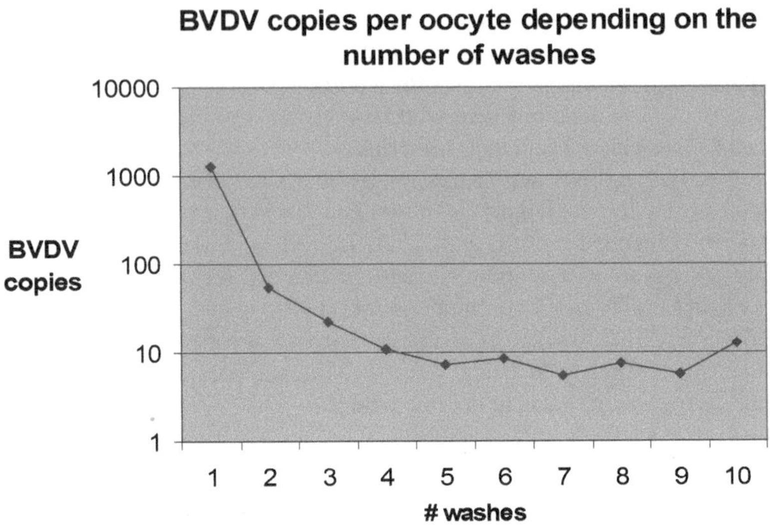

Fig. 17.1 Effect of sequential washing of contaminated oocytes with bovine viral diarrhea virus (BVDV) on a number of viral copies remained in association with a single oocyte (based on data Lalonde and Bielanski 2011)

Over the years different forms of washing have been developed for both human and animal spermatozoa, and, as with embryos, washing has become the most important procedure for the control and elimination of microorganisms in ART clinics.

Although the transmission of infectious agents to embryos by contaminated semen was only proven experimentally (BHV and BVDV), such potential should not be overlooked when cryopreserved semen is used for the generation of embryos using superovulation and AI or IVF (Wrathall et al. 2006; Bielanski et al. 2008, 2009). At present, to reduce such a risk, it is commonly the practice to use a discontinuous gradient centrifugation and/or swim up procedures to remove or reduce the load of various viral and bacterial agents from frozen-thawed semen prior to fertilization. It has been shown that this procedure diminishes the risk of disease transmission to recipients and contamination of *in vitro* fertilization systems, e.g. HIV, HCV, porcine reproductive and respiratory syndrome virus (PRRSV), equine arteritis virus (EAV) (Maertens et al. 2004; Morfeld et al. 2005; Loskutoff et al. 2005; Morrell and Geraghty 2006). It appears that these procedures may be less effective in removing pathogens from animal semen than from human semen because patients can be treated with antiviral agents before ART (Bielanski 2007).

3.3 Disinfection Procedures for Semen and Embryos

Besides trypsin treatment, a variety of other antimicrobial procedures (e.g. antiviral agents, photosensitive dyes, immunological methods) are currently available for disinfecting semen and embryos, but some of them are still at an experimental stage of development. However, none of them seems to fulfil the requirement for a universal disinfectant (for review see Bielanski 2007).

In addition to the IETS sanitary protocol and OIE health codes for collection and processing of semen and embryos, general sanitary recommendations for cryoconservation of animal genetic resources have been summarized in FAO guidelines (Anon 2010).

4 Factors Facilitating Contamination of Gametes During Cryopreservation

There are several critical factors which may influence the contamination of embryos with pathogens during cryopreservation and some of them are also applicable to semen processing. These include the integrity of the embryonic ZP, the cooling method, loading and sealing of the freezing container and the sterility of the LN and the Dewar storage container.

4.1 Integrity of ZP

Unfertilized oocytes, as well as oviductal and uterine-stage embryos, up to approximately day 8 after fertilization, are surrounded by an acellular, glycoprotein shell with a sponge-like surface, the ZP (Herrler and Beier 2000). The specific structural and chemical nature of the ZP is a major factor with regard to interaction with pathogenic microorganisms and its role in disease transmission (Van Soom et al. 2010b).

It is known that the intact ZP of uterine stage and IVF embryos is an effective barrier against penetration by most pathogens even though some may adhere firmly to the surface (Fig. 17.2). The ZP protects the entire surface of the embryo until more advanced blastocyst (expanded) stages. Soon after, the ZP surface is stretched to such a degree that microvilli of the embryonic trophectoderm project through gaps in ZP and thus enable direct contact to occur between the embryonic cells and potential microorganisms (Fig. 17.3). New IVF related ART procedures such as ICSI, embryo sexing, embryo cloning and gene transfer involve cracking and puncturing the ZP or removing the entire ZP from embryos to permit manipulation of the embryonic cells. This is often followed by an extended period of in vitro culture. These procedures increase the risk of exposure of embryonic cells to pathogens and their contamination.

Maintenance of the intact ZP throughout freezing and post-thaw manipulation is crucial for the prevention of embryonic cell contamination and infection. Under certain freezing and thawing conditions, more than 50 % of mammalian embryos may have ZP fractures (Bielanski et al. 1986; Rall and Meyer 1989). These conditions depend on the speed of cooling and warming, the type of freezing container and the cryoprotectant used (Schiewe et al. 1991). In addition to the risk of embryonic cell contamination, damage to the ZP may have a negative influence on further embryonic development. The short exposure of straws in LN vapour

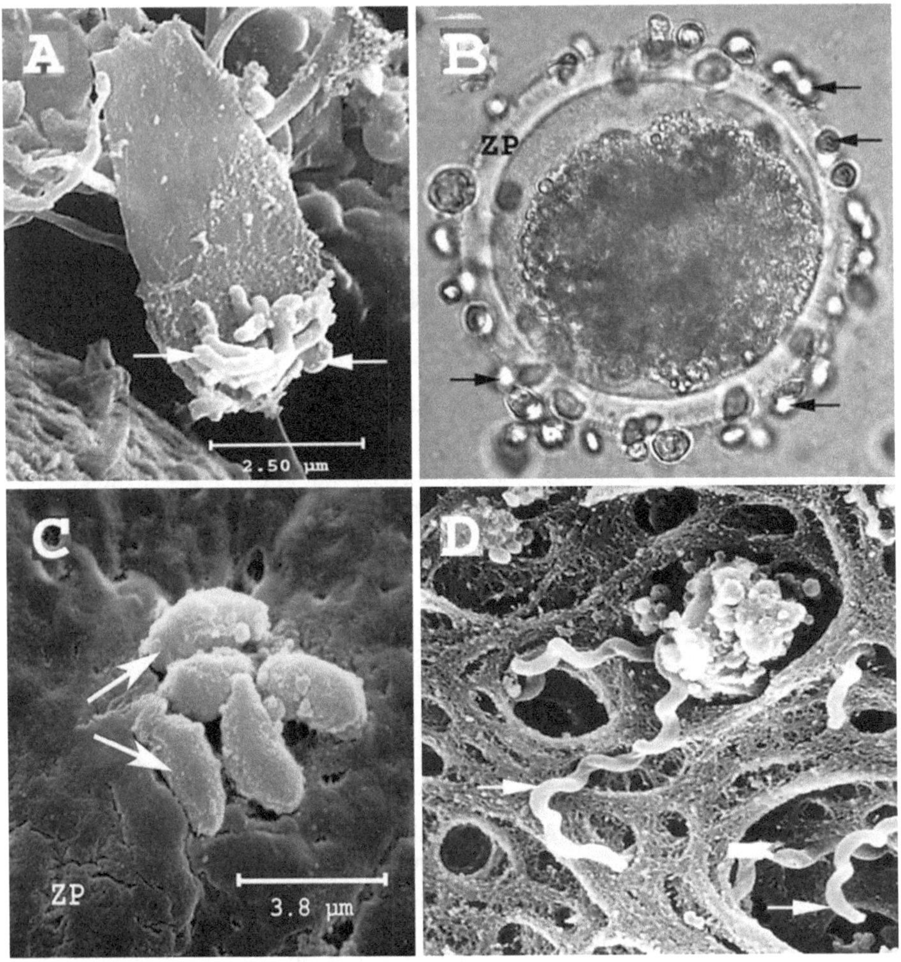

Fig. 17.2 Scanning electron microscopy (SEM) figure of *Campylobacter fetus* (*arrows*) attached to the acrosomal region of a sperm cell on the zona pellucida (ZP) of bovine embryo (**a**); *Tritrichomonas foetus* (light micrograph) (**b**); *Neospora caninum* (**c**); and *Leptospira borgpetersenii* serovar *hardjobovis* (**d**) (*arrows*) on the surface of ZP (from A. Bielanski collection)

(−150 °C) before plunging into LN may reduce fracture of the ZP. Similarly, during warming, the exposure of straws for a few seconds in air before immersion into warm water may have a beneficial effect (Kasai et al. 1996). More damage to the ZP was observed when low cooling rates and glass containers were used than fast cooling by vitrification and 0.25 ml straws (Schiewe et al. 1991). A small sample volume and adjusted post-freeze warming temperature parameters may protect the ZP against fractures. It appears that vitrification procedures cause less damage to the ZP as compared to standard, slow cooling methods (Schiewe et al. 1991; Kasai et al. 1996). Thus, from a practical viewpoint, the choice of a method of freezing-thawing

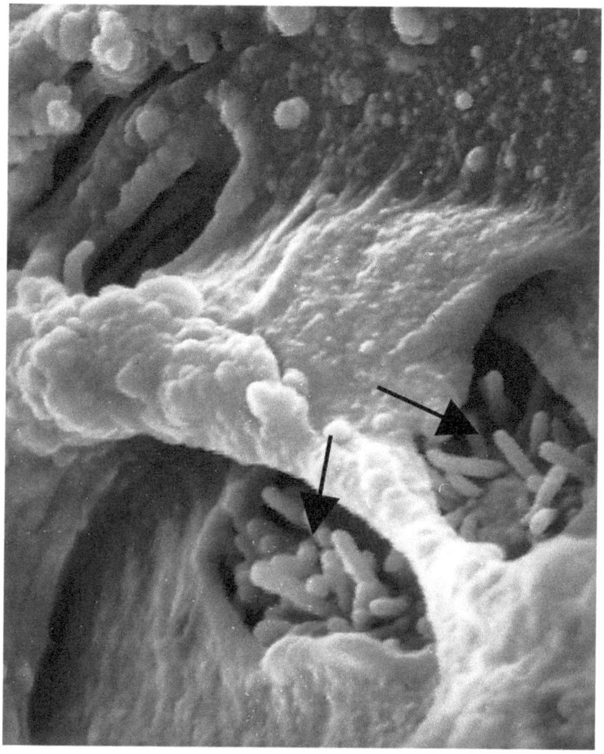

Fig. 17.3 Bovine expanded blastocyst showing gaps in the zona pellucida penetrated by trophectoderm microvilli (*arrows*) (from A. Bielanski collection)

which ensures maximum embryonic survival and minimum ZP damage should be considered, especially when the embryos are produced for international movement.

It has been shown that *in vitro* culture and cryopreservation processes may cause alterations in the structure of the ZP (Moreira da Silva and Metelo 2005). It was observed that the number and the diameter of the pores in the ZP decrease due to slow cooling and it is even more reduced after vitrification. Whether it has any effect on microbial adherence to the ZP or their protection from penetration is unknown. Nevertheless, it is conceivable that before cryopreservation, rapid changes in osmotic pressure due to the high concentration of cryoprotectant may cause the passage of a viral agent through an intact- ZP. However, it was shown that the exposure of embryos to BVDV (60 μm) in a 30 % suspension of either DMSO, ethylene glycol, glycerol, or 2 M sucrose for 10 min did not cause the contamination of embryonic cells (Bielanski et al. 1999a). Whether this observation is valid for other smaller viral agents, e.g. foot-and-mouth disease virus (FMDV, 24 μm) or porcine circo virus (PCV, 27 μm) and various mixtures of vitrification solutions remain to be investigated. Further modification in ZP structure during cryopreservation might be influenced by a high concentration of cryoprotectants used for vitrification of oocytes. This may induce hardening of ZP and prevent its penetration by sperm required for fertilization (Vincent et al. 1990). Whether these alterations in ZP surface influence an interaction with microorganisms during cryopreservation is unknown.

It should also be noted that studies on pathogen interaction with ZP-intact and ZP-free camelid embryos (e.g. Dromedary Camels, Llamas, Alpacas, Guanacos, and Vicunas) have not been carried out. Since camelid embryos have already hatched from the ZP prior to entering the uterus (Picha et al. 2013), it would be unrealistic to stipulate that such embryos can be used for cryopreservation without any risk of contamination. Similarly, little is known about transmission of pathogens by cryopreserved embryos of some species of perissodactyla, marsupialia and lagomorpha, which are enclosed in capsule, shell membrane or mucin coats respectively (Selwood 2000; Herrler and Beier 2000; Stout et al. 2005).

4.2 Cooling Methods

Comparatively, cryopreservation by a modified slow cooling method of farm animals embryos as described by Willadsen (1977) and by vitrification originally described by Rall and Fahy (1985), as well as by Stachecki et al. (2008), offers a relatively lower risk of contamination of samples during cooling due to lack of direct contact with potentially infected LN, than the recently invented, so called "open system vitrification". In both preceding methods, embryos in hermetically sealed vials or straws are exposed to vapours of LN before their plunge in LN for storage. Efficiency of sealing and quality of the freezing containers as well as their resistance to cracking at low temperatures will determine the risk of contamination of embryos during their banking.

During application of vitrification using "open system" methodology, embryos at first can be cooled in a small volume of "clean" LN, followed by placing the original sample into a secondary sealed container before plunging into LN for storage [e.g. open pulled straws (OPS) (Vajta et al. 1997), cut standard 0.25 ml straws (Isachenko et al. 2007), Cryotops® (Kitazato BioPharma Co., Japan) (Kuwayama 2007), Cryoloops (Lane et al. 1999), hemi-straw system (Vanderzwalmen et al. 2003) and a plastic blade (Sugiyama et al. 2010)]. As an example, the risk of contaminating embryos vitrified in OPS using the commercial kit Vit-Set™ (Minitube Canada, Ingersoll, Ontario, Canada) was tested under experimental conditions (Bielanski and Hanniman 2007). The Vit Set™ consists of three stainless steel chambers for cooling the 0.5 ml protective straws, vitrification of OPS straws, and for loading OPS into the protective straws respectively. Bovine embryos were vitrified and protected against contamination as described originally by Vajta et al. (1998). Briefly, when embryos contaminated with cultures of *Escherichia coli (E. coli), Pseudomonas aeruginosa* or non-cytopathic New York (NY) strain of BVDV, were vitrified in OPS straws and then protected by 0.5 ml Cryo Bio System (CBS™) straws, no cross-contamination to clean embryos or LN was detected. It was concluded that the potential for cross contamination of samples by application of the Vit-Set™ for vitrification of embryos using OPS is negligible if: (1) LN in the chambers is frequently replaced and the chambers are disinfected between embryo donors, (2) the protective straws are applied over OPS and are hermetically sealed. The straws should be thermally sealed and used without a cotton plug.

More recently, many researchers, recognizing the strict sanitary requirements for the generation of cryopreserved oocytes/embryos free of infectious agents, have developed entirely "closed systems" of vitrification without the necessity of exposing embryos or semen to LN (Kuleshova and Shaw 2000; Isachenko et al. 2005a; Kuwayama et al. 2005; Hirayama et al. 2007; Schiewe 2010; Larman and Gardner 2011; Criado et al. 2011). Using such a system, embryos are positioned on Cryoloops, Cryotips® (Irvine Sci.) into OPS, CBS™ straws, microcapillary tubes or on a stainless steel needle chip and then into pre-cooled cryovials or 0.25 ml plastic straws ("straw in straw") to accommodate the sample during vitrification by super-cool air inside the container before immersion in LN. Similar techniques have been reported earlier for the freezing of sperm, embryos and ovarian tissue in small drops on super-cooled aluminium foil or steel blocks and cubes (solid surface vitrification); Cryologic vitrification method™ (Cryologic; Mulgrave, Australia) before placement into a protective container and immersion in LN (Dinnyes et al. 2000; Lindemans et al. 2004; Isachenko et al. 2005b; Aerts et al. 2008). Furthermore, Yavin et al. (2009) applied liquid nitrogen slush for vitrification of murine embryos in sealed pulled straws (SPS), to facilitate a high cooling rate and reduced toxicity of cryoprotectants.

Presently, there are cryo-sets allowing avoidance of contact with LN available on the market, which include LN containers, straws, sealers and reagents required for vitrification of embryos [e.g. Vitrolife Rapid-I™ vitrification system, Vitrolife, Goteborg, Sweden; Cryologic CVM Kit™; OPS, RVT, Cairns, Australia; Ultravit, Criado, Spain; CBS™ High Security Straws (IMV, France)]. At this point it is relevant to note that some of the "closed" vitrification systems, which require exposure of embryos or protective containers to LN vapours, may pose some risk of sample contamination when LN is not sterile (Parmegiani and Vajta 2011). Above described methods of cryopreservation, however, involve troublesome LN handling and its danger for contamination of specimens. On the contrary, Faszer et al. (2006) and Morris et al. (2006) described the application of a liquid nitrogen-free Stirling Cycle Cryocooler (EF 600, Asymptote Ltd, Cambridge, UK) for the cryopreservation of mouse and stallion spermatozoa and human embryonic stem cells by slow cooling rates. Survival rates of all cell types frozen in the Stirling Cycle freezer were similar to samples frozen in the LN freezers. The freezer works based on the expansion and compression of helium in a sealed cylinder and can be used for the freezing of large volume samples (e.g. up to 15 ml semen bags) without the risk of contamination. A similar programmable LN-free freezer, the "Pulse Tube" cryocooler with a special low vibration engine, is under development at the "Sapienza" University of Rome (Lopez et al. 2012).

Whether cooling/thawing rates and the cryoprotectant used are relevant to the survival of microorganisms associated with embryos was investigated (Bielanski and Lalonde 2009). It was found that both methods, namely cryopreservation by slow cooling and vitrification, significantly reduced titers of BVDV and BHV-1, but did not however, render embryos free from infectious viruses after thawing.

In general, it can be assumed that procedures which involve the step-wise dilution of the cryoprotectant provide an opportunity to inspect the ZP for fractures and wash the potential pathogens from the embryos. Thus, the one-step dilution of cryoprotectants in straws followed by direct ET should be considered as more hazardous from the sanitary point of view.

4.3 *Freezing Containers*

In general, regardless of the type of containers used, the operator's safety and prevention of cross-contamination are key factors, which must be considered. For such reasons, glass ampoules previously used for germplasm cryopreservation have become obsolete due to the explosion hazard when not sealed properly.

4.3.1 Cryovials

All of the above-mentioned hazards, however, may also, to a lower extent be applied to various plastic cryovials (Cryo Tube™, Nunc A/S, Roskilde, Denmark; Nalgene Nunc International, Naperville, IL, USA) which—despite packages marked "for vapour storage only"—are often stored in LN. The most common cause of contamination is the faulty seal, leak or breakage of these containers in LN. Clarke (1999) reported that 45 % of cryovials without O-ring (Nunc) and 58 % of cryovials with an O-ring (Iwaki, Japan) absorbed LN during 3 h immersion in LN. Although manufacturers strongly recommend the use of a second skin (Cryoflex™ tubing, Nunc; or Nescofilm™, Merk Ltd, Dorset, UK), these measures are rarely applied in everyday practice.

4.3.2 Straws

Regarding plastic straws, three main types are available on the market: those made from polyvinylchloride (PVC) (Minitub GmbH, Germany), polyethylene terephthalate glycol (PETG; IMV, L'Aigle, France) and ionomeric resin (CBS™ High Security Straws; CryoBioSystem, Paris, France). To prevent contamination, CBS™ straws are loaded into a special heat sealer (Syms Sealer™, CryoBioSystem, Paris, France), and sealed thermally at both ends. According to the manufacturers claim, the straws are impermeable to pathogenic agents and have recently been validated for sanitary properties by experimental contamination with HIV-1 and HCV (Benifla et al. 2000; Letur-Konirsch et al. 2003; Loskutoff et al. 2005). CBS™ straws conform to ISO 9002 standards and have been cleared by the U.S. Food and Drug Administration (FDA) for human applications in assisted reproductive technology (Mortimer 2004).

The filling and sealing methods (e.g. polyvinyl alcohol (PVA) powder, plastic spheres, and metal balls) may have even more influence on biosafety than the material of the straw. The potential for contamination of samples by straws containing suspensions of *E. coli* and Newcastle disease virus or ethylene blue and sealed by different methods was investigated by Russell et al. (1997). Samples sealed by a traditional "dip and wipe" method of immersing a tip of the PVC straw into the dry powder or the solution with the polyvinyl alcohol (PVA) powder (in a multi-use container), demonstrated a significantly higher degree of contamination as compared to straws filled aseptically with a syringe. This was probably a consequence of retained residue solution on the inside of the straw tip which subsequently contaminated the sealing plug. It was concluded that, in a real scenario, PVA could accumulate microbes from a number of individuals which can then be introduced to the inside of the straws of another donor. Therefore, it is important to aliquot PVA into tubes which can be used for one donor only and leave an air-gap of at least 1 cm to allow for expansion during freezing. According to a publication of Letur-Könirsch et al. (2003) CBS™ straws prevented HIV-1 contamination, while some of the samples cryopreserved in PVC and PETG straws had become infected. The above authors, however, suggested that the main factor was the different sealing method (thermal vs. ultrasonic sealing for CBS™ vs. PVC and PETG straws, respectively).

Thermal sealing of straws by use of a specially designed device would be most recommended from a sanitary point of view. In addition, to reduce the risk of translocation of contaminants, the sample container should not only be closed hermetically, but its outside surfaces disinfected before freezing and after thawing (e.g. 3 % hypochlorite, 70 % ethanol). When programmed rate alcohol freezers are used this risk is automatically eliminated.

4.3.3 Sterility of LN and Survival of Microorganisms

The majority of cryopreservation techniques utilize LN and involve storage either during the liquid (−196 °C) or vapour (−150 °C) phase. There are no available data provided by manufactures on the microbial load of recently produced LN. Nevertheless, based on limited observations, it can be assumed that the level of contamination of LN is low and limited to ubiquitous microorganisms. For example Morris (1999) referred to less than 100 colony forming units (CFU) of aerobes and to less than 10 CFU anaerobic microorganisms per 10 kg of LN. Likewise, Radnot and Farkas (1966) found one bacterium in every 5–10 ml of LN. It was also speculated that major microbial contamination of LN takes place probably during its distribution to Dewars in clinics (e.g. via pipe nozzles).

As for pathogenic agents, it could be speculated that only severe rodent infestations or staff infected with airborne agents (e.g. Hantavirus, variola, *Myc. tuberculosis*, and anthrax) could possibly contaminate LN in LN production facilities. Although such a risk of contamination seems unlikely it has never been determined experimentally.

It should also be noted that methods exist to produce LN in a completely sterile way and there are commercial companies which offer devices for such production but for pharmaceutical purposes (e.g. Veriseq, Linde AG, Pullach, Germany). To this author's knowledge, at the present time, there is no commercial supplier of sterile LN or of a portable device producing LN suitable for the assisted reproductive technologies (ART).

At the laboratory level, filtration (0.22 μm filter) of LN has been suggested when a small amount is needed for vitrification (Vajta et al. 1998). Larger ceramic filters (Ceralin on Line™, Ceralin, Air Liquide) can be installed at the end of LN delivery lines.

The efficacy of the application of UV radiation as means of sterilization of LN in ART practice has been investigated recently. Parmegiani et al. (2010) reported that when LN (500 ml) was experimentally contaminated with high titres of *Stenotrophomonas maltophilia, Pseudomonas aeruginosa,* and *Escherichia coli* and *Aspergillus Niger* and then exposed to UV radiation (253.7 nm) at 15 cm above the LN surface for 15 min, microbial growth did not result. It appears that it can be an inexpensive and simple method to render small volumes of LN free from at least the ubiquitous microorganisms. It can be applicable for vitrification of oocytes/embryos when a direct contact with LN is required prior to its banking.

Basic research on the effect of subzero temperatures, cooling and thawing rates on survival of microorganisms was pioneered by Mazur (1960) and Doebbler and Rinfret (1963).

Most microorganisms can survive storage at low temperatures, including in LN in the form of "clean" cultures or in association with germplasm (Bielanski et al. 2003; Garcia et al. 1981; Mirabet et al. 2012; Wrathall et al. 2007). Many ingredients of embryo culture media and semen extenders may act as stabilizers for microorganisms at freezing temperatures (e.g. milk, egg yolk, blood serum or serum albumin, sucrose, sorbitol, and other sugars). The most successful cryoprotectants for freezing of viral microorganisms have been DMSO, methanol, ethylene glycol, propylene glycol, while glycerol and polyethylene glycol are less successful (Hubalek 2003; Tedeschi and De Paoli 2011). For example, concentrations of DMSO as low as 5 % effectively protect the enveloped viruses (e.g. vesicular stomatitis virus, herpes viruses) against the trauma of freezing. The non-enveloped viruses (e.g. adenovirus, poliovirus) are not inactivated by freezing and thawing even in the absence of protective agents (Wallis and Melnik 1968).

On the bacterial side, many microorganisms (e.g. *Acinetobacter spp., Corynebacterium spp., Bacillus spp.,* and *Streptomyces spp.*) tolerate very high DMSO concentrations without visible toxic effects and some are even capable of multiplication in a growth medium containing 2–45 % DMSO (Hubalek 2003). Cryopreservation may, however, reduce the concentration of some bacterial contaminants. It has also been shown, for example, that with a concentrated suspension of *Brucella bovis*, 64 % of the organisms failed to survive a simple freezing and thawing cycle in the antibiotic and cryoprotectant-free embryo culture medium (Stringfellow et al. 1986). Loss of viability was reduced 15 % in the presence of 1.4 M glycerol and 18 % in the presence of 1.5 M DMSO.

The bacteria, nevertheless, possess a relatively high tolerance to the freezing procedure and the toxicity of high concentrations of cryoprotectants (e.g. DMSO), in contrast to fungi which proved very sensitive. For example, cryopreservation reduced the concentration of fungi in human semen by more than 90 % (Glander et al. 1983).

It should be noticed that the minimum infectivity titres for most of the pathogenic agents associated with cryopreserved semen and embryos transferred *in utero* to recipients remains unknown. Therefore, the potential for disease transmission by even a residual amount of the agents associated with the zona pellucida (ZP), spermatozoa or culture supplements of biological origin may still exist during cryopreservation and banking. This view is supported by few reports on suspected transmission of BVDV, BHV-1, and HBV to recipients by cryopreserved embryos and semen (Kupferschmied et al. 1986; Van Os et al. 1991; Lindberg et al. 2000; Drew et al. 2002).

So far, in absence of specific methods for cryopreservation of wildlife embryos, the existing cooling technology developed for domestic and laboratory species has been utilized (Saragusty and Arav 2011). Consequently, application of the "closed" system to wildlife embryos during cryopreservation and banking should be considered as a preventive measure of contamination and cross-contamination.

5 Hazard of Contamination and Cross-contamination of Germplasm via LN

Here, cross-contamination refers to the pathogen transmission between cryopreserved germplasm samples during storage in LN.

The clinical risk of transmission of viral agents such as hepatitis B virus (HBV), Herpes simplex, adenovirus and papillomavirus to humans via dermatologic practices of direct exposure of patients' skin to LN has been recognized for three decades (Schaffer et al. 1976; Jones and Darville 1989). Cross-contamination of semen samples during storage in LN within the same container with BHV-1 was reported by Straub (1990). However, the safety of cryopreserved germplasm received a great degree of attention after the discovery of a case of transmission of human hepatitis B *via* bone marrow transplants cryopreserved in LN (Tedder et al. 1995). Experimentally, Piasecka-Serafin (1972) was first to demonstrate the possibility of translocation of bacteria from infected semen pellets to sterile ones in LN. Ninety four percent of sterile samples became infected with *Escherichia coli* and *Staphylococcus aureus* within 2 h after placing them in a container holding contaminated LN. More recently, as a model for human and animal viral pathogens, three bovine viruses, namely bovine viral diarrhea virus (BVDV), bovine herpes virus-1 (BHV) and bovine immunodeficiency virus (BIV), were used to study the potential for their transmission to embryos frozen and stored in open freezing containers (Bielanski et al. 2000a). Bovine embryos in a mixture of 20 % ethylene glycol, 20 % DMSO and 0.6 % sucrose were vitrified in either unsealed standard 0.25 ml, modified open pulled PVC straws or in plastic cryovials and then plunged into contaminated LN. After 3–5 weeks of storage in LN, embryos were thawed, sequentially washed and only those with intact-ZP were pooled together and tested in batches of three for viral contamination. From this pool of 83 batches, 13 out of 61 (21.3 %) batches exposed to BVDV and BHV-1 tested positive for viral association while all 22 batches exposed to BIV in unsealed containers tested negative. All control embryos vitrified in sealed cryovials or straws were free from viral contamination. Also, retrospective studies in which commercial LN cryotanks were examined after 35 continuous years of service revealed various bacterial and fungal contaminations in the LN detritus. Many of the identified bacteria isolated in this study were ubiquitous environmental microorganisms and rare opportunistic pathogens of low significance in producing disease in humans or animals (Bielanski et al. 2003; Mirabet et al. 2012). It should be acknowledged that some of the isolates may have resulted from laboratory contamination during semen and embryo processing for cryopreservation rather than genuinely being present within the sample. It is interesting to observe that although *P. aeruginosa* is a frequently isolated contaminant of bull semen, *Stenotrophomonas maltophilia* was the most prevalent bacterial strain detected in this study in association with cryopreserved germplasm and LN samples (Bielanski et al. 2003). Implications of the introduction of antibiotic resistant strains of *Pseudomonas* spp., *Enterobacter cloacae, Staphylococcus sciuri, Acinetobacter calcoaceticus* and Flavobacterium spp. by contaminated cryopreserved semen into IVF systems has been reported by Stringfellow et al. (1997). Particularly relevant is the demonstration that *Stenotrophomonas maltophilia* can affect sperm motility and severely suppress embryonic development (Bielanski et al. 2003; Stringfellow et al. 1997).

Liquid nitrogen can be a cause for spreading mycoplasmas. It is significant that mycoplasmas can survive in liquid nitrogen even without cryopreservation.

While mycoplasmas do not proliferate in liquid nitrogen, they are able to contaminate cell cultures and germplasm stored in liquid nitrogen (Wrathall et al. 2007; Nikfarjam and Farzaneh 2012).

A quantified risk assessment for cross-contamination between embryos of livestock and wildlife species in LN has not been evaluated as of yet. However, taking into consideration that hundreds of thousands of livestock embryos have been frozen and transferred annually, with some of the donors being of unknown health status, there have been no compelling reports on disease transmission. It can be speculated, therefore, that the risk of cross-contamination between embryos stored in LN may be negligible as compared to experimental embryos where the worst case scenario was created by a very high titer of microbial agents introduced into the cryo-system (Bielanski et al. 2000a).

6 Banking of Somatic Cells and Germplasm

6.1 Storage in the Freezers

Conventional low temperature freezers operating in the range of −20 to −80 °C temperatures have usually been used for short term storage of some perishable reagents and biological products for which preservation cryogenic temperatures are not required. Most of the time, for successful long term storage of cells and tissues sustained temperatures below −130 °C are necessary. On the contrary, viral, bacterial and fungal microorganisms can survive well in a broad range of subzero temperatures without affecting their viability. The effectiveness of a simple method of storage and preservation of viral agents at temperatures below −100 °C was investigated in some detail by Van der Maaten (1987). In this study the aliquots of embryo collection fluid were spiked with various isolates of IBRV, BVDV, bovine parainfluenza virus type 3 (PI3) and bluetongue virus, and then placed directly in storage at 4 °C, −20 °C and −70 °C without using controlling cooling rates. In general, all viruses retained their titres for 6 days at 4 °C. After storage at −20 °C for up to 3 months, some decrease in viral titers was observed, but all of the viruses tested seemed to survive well during storage at −70 °C.

Despite of what was said above, there are a few reports on successful preservation of tissues without any treatment (e.g. cryoprotectants) for the somatic cell nuclear transfer (SCNT) within a similar range of temperatures. This "grab and store it" method resulted in healthy clones derived from tissues harvested post-mortem or from dead animals and then kept at −20 and −80 °C up to 16 years (Wakayama et al. 2008, 2010; Hoshino et al. 2009). However, in these reports there is no description of any sanitary measures or protocols related to tissue processing and storage. It can be assumed that the risk of contamination or cross-contamination of those tissues may depend on subsequent accumulation of microorganisms in ice post collection in a freezer chamber and package integrity over a storage period.

6.2 Storage in Liquid Phase of LN

The readers are directed to the Tomlinson's and Morroll's (2008) publication on risk, hazard and cryogenic practices applicable to the manipulation of cryopreserved cells and germplasm. Working with cryopreserved germplasm requires complying with the best cryogenic practices to minimise the risk of personal injury and also losses of irreplaceable germplasm stored in LN Dewars. Much can be learned from guidelines and codes issued by other disciplines (Benson 2008), intergovernmental organizations (e.g. OIE, WHO, FDA) and scientific associations experienced with cryobanking such as the Society for Low Temperature Biology, the Human Fertilization and Embryology Authority (HFEA) in the United Kingdom; the International Society for Biological and Environmental Repositories (ISBER), the American Association of Tissue Banks Committee and the International Federation of Fertility Societies.

In general, the laboratory staff should be trained to observe safety rules for handling LN according to the appropriate Material Safety Data Sheets (MSDS) and the laboratory Standard Operation Procedure (SOP). Considering that there is no method to sterilize large quantities of LN, all cryotanks used for the storage of biological samples should be considered potentially contaminated with at least environmental microorganisms. Most germplasm are stored in large capacity cryotanks, some of which may accommodate hundreds of thousands of straws or other sample containers. This may create a potential cross-contamination of clean samples in case of breaking or leaking of infected samples into the LN. Also, over the storage time, due to the exposure of cryotanks to the laboratory environment during refilling and handling of specimens, ice crystals will form on the walls of the vessels. Aggregated ice and sediment may entrap viruses, bacteria, fungal spores, and debris posing a risk of microbial transmission to stored samples (Morris 2003, 2005). Nevertheless, a recent study seems to show that the long term banking of germplasm in the LN phase, even over 35 years, is a safe technique for the preservation of genetic materials with a low potential risk of cross-contamination when the specimens are properly sealed (Bielanski et al. 2003; Mirabet et al. 2012). No viral agents, but a number of ubiquitous bacterial microorganisms, were isolated from liquid nitrogen and ice sediments in Dewar vessels. In another study (Mirabet et al. 2012), mostly environmental bacteria and fungi were identified in almost all samples of LN sediment and ice collected from three Dewars used for the storage of human embryos in sealed glass vials. However, it was pointed out that some of them such as *Acinetobacter baumannii,* and *Chryseomonas luteola* are capable of causing nosocomial infections in humans (Morris 2003). The microbial load did not correlate with the period the Dewar had been used (7–15 years).

Usually, for large scale commercial purposes, germplasm and somatic tissues have been stored in accredited cryobanks, professionally managed and appropriately staffed, in compliance with industry or government Standards and/or Regulations (Mortimer 2004). Often germplasm is stored in environmentally controlled facilities and large capacity Dewars furnished with auto-fill systems and alarm/sensor devices to monitor LN level. Furthermore, in some human ART settings there are regulatory polices imposing mandatory requirements for germplasm quarantine for a specific period of time to reduce the risk of transmitting infectious pathogens (e.g. HIV, HBV, HCV) to recipients (e.g. HFEA; the American Society for Reproductive Medicine and Society for Assisted Reproductive Technology, USA). According to data resulting from the survey of human ART units, the majority of embryos were stored in PTEG straws and semen in polypropylene vials (Tomlinson and Morroll 2008). Cryovials were also used despite almost 70 % of respondents witnessing them explode and 72 % observing LN boiling inside. Likewise, 80 % had observed either broken or unsealed straws, which again indicated samples remaining in direct contact with LN (Tomlinson and Morroll 2008).

On the other hand, thousands of animal embryos have been produced by individual ET practitioners and stored at local clinics with a limited inventory of LN Dewars, which might not allow for segregated storage of embryos from different donors. In contrast to human ART, there are no specific cryobanking regulatory policies in animal ET practices and embryos from multiple donors can be stored together until recipients are available. However, prior to storage, practitioners are required to follow mandatory sanitary guidelines for processing and freezing embryos recommended by IETS and OIE (Stringfellow and Givens 2010; Anon 2011). Moreover, embryos collected from donors infected with e.g. FMDV, or bovine tuberculosis, in order to salvage the genes, should be stored under quarantine in a separate LN Dewar. It should be strongly emphasized that in both human and animal embryo cryobanking settings no case of disease transmission has been reported which was attributed to the storage of embryos in LN. Furthermore, Pomeroy et al. (2010) after reviewing data on embryo cryopreservation in IVF/ART, concluded a negligible risk of human embryo cross-contamination during storage in LN.

6.3 Storage in Vapour Phase of Nitrogen

As an alternative to the LN phase, the vapour phase of liquid nitrogen (VPLN) has been proposed as a safe method for the storage of germplasm. However, these kinds of refrigerators are prone to various difficulties during prolonged storage and handling of specimens and it is difficult to maintain steady temperatures at −150 to −190 °C (Cobo et al. 2010). Nevertheless, successful storage of human semen and oocytes in the vapour phase has been practiced in some clinics (Clarke 1999; Cobo et al. 2010).

Frequent opening and high air humidity may also cause ice formation and frosting on lids and walls of storage Dewars and thus attract the environmental microorganisms such as *Aspergillus* spp. (Fountain et al. 1997; Grout and Morris 2009).

In the future, from the sanitary point of view, application of LN-free mechanical cryogenic freezers with sustained temperatures of −150 °C for storage of tissues may be safer while minimizing the level of contamination or cross-contamination of samples (e.g. Thermo Scientific Revco Ultima II™ −150 °C Cryogenic Chest Freezers; Sanyo V.I.P. PLUS™ Cryogenic Series −150 °C).

Until recently, liquid preservation containers

7 Biosafety of Cryopreserved Germplasm and Reproductive Tissues During Transportation

The sanitary regulations for shipping semen and embryos depend on where germplasm is being shipped to and from and may differ between countries. It must be shipped in a cooled container to be viable upon delivery. Most countries require documentation (a health certificate) to ensure that the germplasm does not carry disease. Even when semen can be imported from multiple countries, the procedures are often different. A particular country may allow one type of sperm but disallow sperm from a different species to be imported from another country. Some countries allow sperm to be imported from only specific regions.

In general, Dewars containing LN, even without biological specimens, when moved by air or land transportation, are often categorized as diagnostic specimens and subjected to regulation as a “Hazardous Material” (IATA). Conversely, “dry shippers” or Dewars that do not contain free LN are classified as “non-hazardous” throughout the world (USDOT).

Various Dewar containers (dry shippers), designated for storage of tissues in LN vapours during transportation, are available on the market (e.g. Taylor–Wharton; MVE).

The risk of contamination of germplasm in VPLN with the use of dry shipper-Dewars during short storage was investigated by Bielanski (2005a). Most of the dry shippers are capable of maintaining nitrogen vapours at approximately −150 °C for about 14 days without the risk of spilling LN during transportation. In our experiments, none of the embryos or semen samples exposed to LN vapours in previously contaminated dry shippers tested positive for the presence of selected bovine bacterial and viral agents Bielanski (2005a). There was no transmission of these agents (e.g., BVDV, BHV-1, *Pseudomonas aeruginosa, Escherichia coli* and *Staphylococcus aureus*) between contaminated and non-contaminated germplasm stored in proximity in open containers in the vapour phase of LN. This finding may suggest that, in contrast to the LN phase, the VPLN can be used as a safer means of short term storage and transportation of germplasm even in the proximity of pathogenic agents.

Although the detailed mechanism of the microbial movements within dry shippers is unknown, it could be assumed that there is no, or very limited circulation of frozen particles within the Dewar chamber by the LN vapours in the absence of LN phase. This contrasts to the environment within the conventional LN Dewars where extensive movement of both vapours and boiling LN takes place particularly during temperature changes and tank refilling (Morris 2005). This view could be supported by the report of Fountain et al. (1997) who detected environmental or waterborne bacteria and fungi in the vapour phase above LN in a Dewar. This vapour contamination reflected the findings of bacterial cultures isolated from LN and was likely the result of aerosolization by boiling LN collecting contaminants over a long period of Dewar service without its periodic decontamination. More recently this view was supported by Grout and Morris (2009), who experimentally induced contamination of LN vapour phase by spiking LN with fungal spores of *Sclerotinia minor*. Authors also indicated that these small particles could be suspended for at least 24 h in the LN phase and transmitted via vapours used to cool programmable freezers. It is difficult to assess the risk of embryo contamination by LN vapours, but prior to germplasm storage it should be taken into consideration by practitioners.

It remains to be established whether more advanced LN vapour freezers and mechanical freezers, which can provide air-phase at −140 and −150 °C without the need for LN, can prevent cross-contamination of germplasm over a long period of storage without compromising germplasm post-thaw viability of germplasm.

8 Practical Considerations for Germplasm Cryostorage

8.1 Decontamination/Disinfection LN Dewars

Although it is cumbersome from a practical standpoint, cryotanks require periodic decontamination using an efficient disinfectant to decrease the risk of cross-contamination.

From information obtained from the manufacturers and suppliers, regular cryotanks can be sanitised with any solution that does not react with aluminium or stainless-steel. In most cases, bleach, any household detergent or a mild soap solution is suitable. The generally accepted practice of using 10 % household chlorine bleach with 90 % water solution holds as the best method for decontamination and it is recommended by the cryotank manufacturers (Anon 2008b). Other cleaners and disinfectants that can be safely used include 3–6 % hydrogen peroxide, and 37 % denatured alcohol. It is important that after exposure to disinfectant (15–30 min) the inner vessel is thoroughly rinsed with sterile water and all cleaner residues have been removed. Spraying the solution into the inner vessel is preferred, although agitation of the solution inside the inner vessel will suffice. Other disinfectants, for example Virkon S™ (sodium chloride/potassium peroxymonosulfate; DuPont), are federally approved (US Environmental Protection Agency, EPA, USA) for use against viruses of highly contagious diseases such as FMDV, avian influenza, African swine fever virus (ASFV), vesicular stomatitis virus (VSV), and others. Frequency of decontamination and servicing of LN Dewars would depend on its volume, presence of infectious samples, the number of stored embryos (straws, vials), and frequency of LN refilling and moving germplasm in and out. It should be kept in mind that non-sterile LN, common air pollutants and microorganisms attached to the outside of embryo containers will contribute to the accumulation of contaminants in a Dewar over time (approximately by a factor of 100 within 10 years) (Morris 2003). Since a regulatory policy has not been established (IETS, OIE), it remains up to ET practitioners to take into consideration the above factors and to determine appropriate intervals between periodic decontamination of LN Dewars. Should "clean embryos" be stored in a small volume of LN (up to 10 L) it would be prudent to decontaminate Dewars every 6 months and once a year when in larger volumes When embryos are infected or their health status is uncertain, it is reasonable to decontaminate LN Dewars after these embryos have been used, and before storage of new embryos from other donors.

Special attention should be given to Dewars potentially contaminated by transmissible spongiform encephalopathies (TSE) such as Creutzfeldt-Jacob disease (CJD), bovine spongiform encephalopathy agent (BSE), scrapie and chronic wasting disease (CVD). In general these agents are extremely resistant to inactivation by standard physical and chemical treatments such as dry heat and radiation, and many chemical disinfectants (McDonnell and Burke 2003).

Prions have been shown to bind avidly to steel surfaces. Within specific decontamination procedures for metallic instruments, prion destruction can be achieved by applying corrosive agents such as 2 M NaOH or sodium hypochlorite 20,000 ppm for at least 1 h. Recently, the use of alkaline detergents and enzyme-based disinfectants (Rely On PI™, Du Pont Corp; Priozyme™, Genencor; Klenzyme™, Steris; Septo Clean™, Septo-Clean) has shown to be effective and appears suitable for Dewars decontamination (Edgeworth et al. 2011; Rutala and Weber 2010).

8.2 Decontamination of Dry Shippers

In contrast, decontamination of dry shippers is more difficult due to their inner construction. Vapour shipper units will require filling the inner vessel to its full capacity with the cleaning mixture and then rinsing. Allow the unit to dry thoroughly by inverting it under a laminar flow hood or if unavailable in the dust-free area.

The method of disinfection of dry shippers, with two different types of a LN absorbent was investigated (Bielanski 2005b). In general, it was demonstrated that shippers containing a hydrophobic absorbent (e.g. SC2/V1™, Minnesota Valley Engineering, Inc, MN, USA) were suitable for disinfection using liquid biocide solutions, while those with a non-hydrophobic insert (e.g. CX100™ Taylor Wharton) could only be disinfected by the application of vapour sterilization. An attempt to use liquid solutions on the latter resulted in permanent damage to the LN absorbent. In the above study, the dry shippers were heavily contaminated with high titers of cultures of *Pseudomonas aeruginosa, Escherichia coli, Staphylococcus aureus*, BVDV and BHV-1 to create a worst-case scenario. The concentrations of bacterial and viral cultures to which the vapour shippers were exposed exceeded potential titers which can be expected to occur during the storage and transportation of germplasm or other contaminated biological material inadvertently harbouring infectious pathogens. During our investigation we were able to identify two biocidal products which are not suitable for the decontamination of dry shippers furnished with a hydrophobic membrane. The application of Viralex™ (Alda Pharmaceut. Inc. Canada) or ethanol may result in irreversible physical damage to the LN absorbent and properties of the dry shipper. Other biocides used, such as Virkon S™ (Antec Inter.Inc. UK) and 1-Stroke™ (A.P.A., USA) were not fully effective as disinfectants when introduced into the chamber of the dry shippers. Presumably, their foaming properties caused by the surfactants or other ingredients, in combination with the air contained in the LN insert, and may have reduced contact time with the contaminated surface of the absorbent membrane.

The application of gas sterilization using ethylene oxide to both types of dry shippers was fully effective as a means of disinfection (Bielanski 2005b). The advantages of using ethylene oxide is its broad spectrum of antimicrobial activity as well as the elimination of the introduction of liquid solutions into the dry shipper chamber which may damage the LN absorbent . Ethylene oxide is widely used for items which cannot be sterilized with steam. However, due to its health hazards, it must be used in controlled facilities capable of sterilizing such large items.

It should be pointed out that it remains unknown whether the disinfectants selected in our experiments would be effective against the causal agents of transmissible spongiform encephalopathies (TSEs), bacterial spores or very small viral agents (e.g., porcine parvovirus or FMDV). Based on the results presented in this study, it appears that solutions of sodium hypochlorite and ethylene oxide are equally useful for the disinfection of dry shippers constructed with a hydrophobic LN absorbent. In contrast, for dry shippers without a hydrophobic LN absorbent it is advisable to use only gas sterilization for decontamination in order to avoid their damage by liquid disinfectants. Adequate aeration of dry shippers prior to filling with LN should be allowed to avoid potential toxic effect of ethylene oxide residues (Schiewe and Hasler 2010).

8.3 Segregation of Germplasm

Storage of both semen and embryos in the same Dewar may pose more risk for contamination of LN and cross-contamination between samples because of a high microbial load and a large volume of semen containers, which may be more prone to damage and leaking, as compared to smaller ones required for storage of embryos.

It is advisable that germplasm collected from suspected or infected donors (e.g. for the purpose of salvage of genetics) be stored under quarantine in separate LN Dewars until donors have been tested for seroconversion and/or samples have been tested for the presence of infectious agent(s).

Master cell lines for SCNT should be tested for the presence of pathogenic agents and then banked in separate Dewars.

Table 17.2 Potential hazard of germplasm contamination prior, during and post cryopreservation and its worst-case-scenarios outcome

Stage of germplasm processing	Source of hazard	Outcome
Harvesting	Viremic donors; donors latently infected; contaminated environment facilities/equipment	Contaminated ova/semen; contaminated follicular fluid
Fertilization	Lack or deficient washing procedure of embryos and semen; infected sperm cells	Titre of infectious agent not reduced; replication of agent in culture cells or embryos; contaminated embryos; embryonic death
Cryopreservation and storage	Contaminated media and or containers; unsealed or leaked containers; contaminated LN or Dewars The same Dewar for banking contaminated and non-contaminated germplasm	Risk of cross-contamination of samples during storage
Post-cryopreservation	Lack or deficient decontamination of containers surface retrieved from LN Dewars; deficient washing of post-thawed embryos	Risk of cross-contamination Risk of disease transmission to ET recipients; early embryonic mortality or disease transmission to offspring

9 Conclusion

For the safe and successful cryopreservation of semen and embryos, the freezing method and sanitary procedures must be chosen carefully to ensure not only a high post-thaw survival of gametes, but also to minimise the risk of disease transmission when those gametes are used for AI and ET. These procedures have been applied successfully to livestock germplasm over the last four decades. Since basic data on interaction of pathogenic agents with wildlife germplasm is unavailable, preventive measures of disease transmission developed for livestock embryos and semen should be considered.

Potential hazard of germplasm contamination prior, during and post cryopreservation is shown in Table 17.2.

References

Aerts JM, De Clercq JB, Andries S, Leroy JL, Van Aelst S, Bols PE. Follicle survival and growth to antral stages in short-term murine ovarian cortical transplants after Cryologic solid surface vitrification or slow-rate freezing. Cryobiology. 2008;57:163–9.

Agca Y. Genome resource banking of biomedically important laboratory animals. Theriogenology. 2012;78:1653–65.

Ali Al Ahmad MZ, Fieni F, Martignat L, Chatagnon G, Baril G, Bouvier F, Chebloune Y. Proviral DNA of caprine arthritis encephalitis virus (CAEV) is detected in cumulus oophorus cells but not in oocytes from naturally infected goats. Theriogenology. 2005;64:1656–66.

Andrabi SM, Maxwell WM. A review on reproductive biotechnologies for conservation of endangered mammalian species. Anim Reprod Sci. 2007;99:223–43.

Anon. A review of tests on virus vaccines. Geneva, Switzerland: World Health Organization (WHO); 1982. Technical Report Series No. 673.

Anon. Scientific Opinion of the Scientific Committee Food Safety, Animal Health and Welfare and Environmental Impact of Animals derived from Cloning by Somatic Cell Nucleus Transfer (SCNT) and their Offspring and Products Obtained from those Animals. EFSA J. 2008a;767:1–49.

Anon. Cryogenic. Liquid nitrogen dewar repair and decontamination. 2008b. princetoncryo.blogspot.com/…/liquid-nitrogen-dewar-repair-and.htm

Anon. FAO guidelines for the cryoconservation of animal genetic resources. Rome, Italy: FAO; 2010. pp. 87–90.

Anon. Somatic cell nuclear transfer in production livestock and horses. In: The Terrestrial Animal Health Code, Chapter 4.11., vol.1. Paris: The World Animal Health Organization (OIE); 2011.

Arav A, Natan D. Directional freezing of reproductive cells and organs. Reprod Domest Anim. 2012;47 Suppl 4:193–6.

Asseged BD, Habtemariam T, Tameru B, Nganwa D. The risk of introduction of equine infectious anemia virus into USA via cloned horse embryos imported from Canada. Theriogenology. 2012;77:445–58.

Benifla JL, Letur-Konirsch H, Collin G, Devaux A, Kuttenn F, Madelenat P, Brun-Vezinet F, Feldmann G. Safety of cryopreservation straws for human gametes or embryos: a preliminary study with human immunodeficiency virus-1. Hum Reprod. 2000;15:2186–9.

Benson EE. Cryopreservation of phytodiversity: a critical appraisal of theory and practice. Crit Rev Plant Sci. 2008;27:141–219.

Bielanski A, Schneider U, Pawlyshyn VP, Mapletoft RJ. Factors affecting survival of deep frozen bovine embryos *in vitro*: the effect of freezing container and method of removing cryoprotectant. Theriogenology. 1986;25:429–37.

Bielanski A, Loewen KS, Del Campo MR, Sirard MA, Willadsen S. Isolation of bovine herpesvirus-1 (BHV-1) and bovine viral diarrhea virus (BVDV) in association with the *in vitro* production of bovine embryos. Theriogenology. 1993;40:531–8.

Bielanski A, Dubuc C. *In vitro* fertilization and culture of ova from heifers infected with bovine herpesvirus-1 (BHV-1). Theriogenology. 1994;41:1211–7.

Bielanski A, Stewart B. Ubiquitous microbes isolated from *in vitro* fertilization (IVF) system. Theriogenology. 1996;45:269. Abstr.

Bielanski A, Sapp T, Lutzewallace C. Association of bovine embryos produced by *in vitro* fertilization with a noncytopathic strain of bovine viral diarrhea virus type II. Theriogenology. 1998a;49:1231–8.

Bielanski A, Surujballi O, Thomas EG, Tanaka E. Sanitary status of oocytes and embryos collected from heifers experimentally exposed to *Leptospira borgpetersenii* serovar *hardjobovis*. Anim Reprod Sci. 1998b;54:65–73.

Bielanski A, Sapp T, Lutze-Wallace C, Palasz A. The effect of high concentrations of cryoprotectants on the passage of bovine viral diarrhea virus through the zona pellucida of *in vitro* fertilized embryos. Anim Reprod Sci. 1999a;55:83–90.

Bielanski A, Hutching D, Turcotte C. Status of embryos and ova collected from superovulated heifers experimentally infected with *Mycobacterium bovis*. Theriogenology. 1999b;51:270. Abstr.

Bielanski A, NadinDavis S, Sapp T, Lutze-Wallace C. Viral contamination of embryos cryopreserved in liquid nitrogen. Cryobiology. 2000a;40:110–6.

Bielanski A, Devenish J, Phipps-Todd B. Effect of *Mycoplasma bovis* and *Mycoplasma bovigenitalium* in semen on fertilization and association with in vitro produced morula and blastocyst stage embryos. Theriogenology. 2000b;53:1213–23.

Bielanski A, NadinDavis S, Simard C, Maxwell P, Algire J. Experimental collection and transfer of embryos from bovine immunodeficiency virus (BIV) infected cattle. Theriogenology. 2001;55:641–8.

Bielanski A, Robinson J, Phipps-Todd B. Effect of *Neospora caninum* on *in vitro* development of preimplantation stage bovine embryos and adherence to the zona pellucida. Vet Rec. 2002;150:316–8.

Bielanski A, Bergeron H, Lau PC, Devenish J. Microbial contamination of embryos and semen during long term banking in liquid nitrogen. Cryobiology. 2003;46:46–152.

Bielanski A, Ghazi DF, Phipps-Toodd B. Observations on the fertilization and development of preimplantation bovine embryos *in vitro* in the presence of *Tritrichomonas foetus*. Theriogenology. 2004;61:821–9.

Bielanski A. Non-transmission of bacterial and viral microbes to embryos and semen stored in the vapour phase of liquid nitrogen in dry shippers. Cryobiology. 2005a;50:206–10.

Bielanski A. Experimental microbial contamination and disinfection of dry (vapour) shipper dewars designed for short-term storage and transportation of cryopreserved germplasm and other biological specimens. Theriogenology. 2005b;63:1946–57.

Bielanski A. The potential for animal and human germplasm contamination through assisted reproductive technologies. Trends Reprod Biol. 2006;2:13–36.

Bielanski A, Algire J, Randall GC, Surujballi O. Risk of transmission of *Mycobacterium avium* ssp. paratuberculosis by embryo transfer of in vivo and in vitro fertilized bovine embryos. Theriogenology. 2006;66:260–6.

Bielanski A. Disinfection procedures for controlling microorganisms in the semen and embryos of human and farm animals. Theriogenology. 2007;68:1–22.

Bielanski A, Hanniman A. Non-cross-contamination of bovine embryos with microbes using the OPS vitrification system. Reprod Fert Dev. 2007;19:232. Abstr.

Bielanski A, Lalonde A, Algire J. Risk of transmission of bovine herpesvirus (BHV-1) by infected semen to recipients and embryos. Reprod Fertil Dev. 2008;21:173.

Bielanski A, Lalonde A, Algire J. Transmission of noncytopathic bovine diarrhea virus by infected semen to embryo donors and by embryos to embryo transfer recipients and offspring. Reprod Fert Develop. 2009;22:251.

Bielanski A, Vajta G. Risk of contamination of germplasm during cryopreservation and cryobanking in IVF units. Hum Reprod. 2009;24:2457–68.

Bielanski A, Lalonde A. Effect of cryopreservation by slow cooling and vitrification on viral contamination of IVF embryos experimentally exposed to bovine viral diarrhea virus and bovine herpesvirus-1. Theriogenology. 2009;72:919–25.

Bielanski A. Potential for disease control or transmission by transfer of embryos produced *in vitro*. A review. In: Stringfellow DA, Givens D, editors. Manual of the International Embryo Transfer Society. 4th ed. Champaign, IL: IETS; 2010. p. 41–50.

Bielanski A, Algire J, Lalonde A, Garceac A, Pollard JW, Plante C. Non-transmission of Porcine Circovirus 2 (PCV2) by embryo transfer. Theriogenology. 2013;80(2):77–83.

Booth PJ, Stevens DA, Coolins ME, Brownlie J. Detection of bovine diarrhoea virus (BVDV) in ovarian and oviductal tissue. J Reprod Fertil. 1995;105:17–24.

Brackett BG, Baranska W, Sawicki W, Koprowski H. Uptake of heterologous genome by mammalian spermatozoa and its transfer to ova through fertilization. Proc Natl Acad Sci U S A. 1971;2:353–7.

Brock K. Quality control for materials of animal origin used in embryo production and transfer. In: Stringfellow D, Seidel S, editors. IETS Manual; 1998. p. 135–39.

Broaddus CC, Balasuriya UB, Timoney PJ, White JL, Makloski C, Torrisi K, Payton M, Holyoak GR. Infection of embryos following insemination of donor mares with equine arteritis virus infective semen. Theriogenology. 2011;76:47–60.

Brownlie J. The pathogenesis of bovine virus diarrhoea virus infections. Rev Sci Tech. 1990;9:43–59.

Chatdarong K. Gamete rescues from gonads of wild animal post-mortem. Thai J Vet Med. 2011;41:99–102.

Clarke GN. Sperm cryopreservation: is there a significant risk of cross- contamination? Hum Reprod. 1999;14:2941–3.

Cobo F, Concha A. Environmental microbial contamination in a stem cell bank. Lett Appl Microbiol. 2007;44:379–86.

Cobo A, Romero JL, Perez S, Los Santos MJ, Meseguer M, Remohi J. Storage of human oocytes in the vapor phase of nitrogen. Fertil Steril. 2010;94:1903–7.

Comizzoli P, Songsasen N, Hagedorn M, Wildt DE. Comparative cryobiological traits and requirements for gametes and gonadal tissues collected from wildlife species. Theriogenology. 2012;78:1666–81.

Cortez-Romero C, Fieni F, Russo P, Pepin M, Roux C, Pellerin JL. Presence of Maedi Visna virus (MVV)-proviral DNA in the genital tissues of naturally infected ewes. Reprod Domest Anim. 2011;46:1–6.

Cottell E, McMorrow J, Lennon B, Fawsy M, Cafferkey M, Harrison RF. Microbial contamination in an in vitro fertilization-embryo transfer system. Fertil Steril. 1996;66:776–80.

Criado E, Moalli F, Polentarutti N, Albani E, Morreale G, Menduni F, Levi-Setti PE. Experimental contamination assessment of a novel closed ultravitrification device. Fertil Steril. 2011;95:1777–9.

Criado E. The problem of contamination: open vs. closed vs. Semi-closed vitrification systems. In: Katkov I, editor. Current frontiers in cryopreservation. Rijeka: Intech Publishers; 2012. p. 105–36.

Dejucq-Rainsford N, Jegou B. Viruses in semen and male genital tissues consequences for the reproductive system and therapeutic perspectives. Curr Pharm Des. 2004;10:557–75.

Dinnyes A, Dai YP, Jiang S, Yang XZ. High developmental rates of vitrified bovine oocytes following parthenogenetic activation, in vitro fertilization, and somatic cell nuclear transfer. Biol Reprod. 2000;63:513–8.

Doebbler GF, Rinfret AP. Survival microorganisms after ultrarapid freezing and thawing. J Bacteriol. 1963;85:485.

Drew TW, Sandvik P, Wakeley P, Jones T, Howard P. BVD virus genotype 2 detected in British cattle. Vet Rec. 2002;151:551–2.

Edgeworth JA, Sicilia A, Linehan J, Brandner S, Jackson GS, Collinge J. A standardized comparison of commercially available prion decontamination reagents using the Standard Steel-Binding Assay. J Gen Virol. 2011;92:718–26.

Elazhary MA, Lamothe P, Silim A, Roy RS. Bovine herpesvirus 1 in the sperm of a bull from a herd with fertility problems. Can Vet J. 1980;21:336–9.

Erickson GA, Bolin SR, Landgraf JG. Viral contamination of fetal bovine serum used for tissue culture: risks and concerns. Dev Biol Stand. 1991;75:173–5.

Faszer K, Draper D, Green JE, Morris J, Grout BW. Cryopreservation of horse semen under laboratory and field conditions using a Stirling Cycle freezer. Cryo Letters. 2006;27:179–86.

Fields BN. Fields virology. 5th ed. Philadelphia, PA: Lippincott Williams & Wilkins; 2006.

Foster NM, Alders MA, Luedke AJ, Walton TE. Abnormalities and virus-like particles in spermatozoa from bull latently infected with bluetongue virus. Am J Vet Res. 1980;41:1045–8.

Fountain D, Ralston M, Higgins N, Gorlin JB, Wheeler C, Antin JH, Churchill WH. Liquid nitrogen freezers: a potential source of microbial contamination of hematopoietic stem cell components. Transfusion. 1997;37:585–91.

Galli C, Lazzari G. The manipulation of gametes and embryos in farm animals. Reprod Domest Anim. 2008;43 Suppl 2:1–7.

Garcia A, Sierra MF, Friberg J. Survival of bacteria after freezing of human semen in liquid nitrogen. Fertil Steril. 1981;35:549–51.

Givens MD, Stringfellow DA, Dykstra CC, Riddell KP, Galik PK, Sullivan E, Robl J, Kasinathan P, Kumar A, Boykin DW. Prevention and elimination of bovine viral diarrhea virus infections in fetal fibroblast cells. Antiviral Res. 2004;64:113–8.

Glander HJ, Schonborn C, Rytter M. Microbiological investigations on cryopreserved human semen. Int J Androl. 1983;6:358–66.

Gregg K, Polejaeva I. Risk of equine infectious anemia virus disease transmission through *in vitro* embryo production using somatic cell nuclear transfer. Theriogenology. 2009;72:289–99.

Gregg K, Riddell KP, Chen SH, Galik PK, Xiang T, Guerra T, Marley MS, Polejaeva I, Givens MD. Risk and prevention of bovine viral diarrhea virus (BVDV) transmission through embryo production via somatic cell nuclear transfer (SCNT) using oocytes from persistently infected donors. Theriogenology. 2010a;74:1–10.

Gregg K, Gosch G, Guerra T, Chen SH, Xiang T, Broek D, Bruner B, Polejaeva I. Large scale *in vivo* risk assessment of bovine viral diarrhea virus (BVDV) transmission through transfer of bovine embryos produced via somatic cell nuclear transfer (SCNT). Theriogenology. 2010b;74:1264–70.

Gregg K, Xiang T, Arenivas SS, Hwang E, Arenivas F, Chen SH, Walker S, Picou A, Polejaeva I. Risk assessment of porcine reproductive and respiratory syndrome virus (PRRSV) transmission via somatic cell nuclear transfer (SCNT) embryo production using oocytes from commercial abattoirs. Anim Reprod Sci. 2011;125:148–57.

Grout BW, Morris GJ. Contaminated liquid nitrogen vapour as a risk factor in pathogen transfer. Theriogenology. 2009;71:1079–82.

Habrova V, Takac M, Navratil J, Macha J, Ceskova N, Jonak J. Association of Rous sarcoma virus DNA with *Xenopus laevis* spermatozoa and its transfer to ova through fertilization. Mol Reprod Dev. 1996;44:332–42.

Hadchouel M, Scotto J, Huret JL, Molinie C, Villa E, Degos F, Brechot C. Presence of HBV DNA in spermatozoa: a possible vertical transmission of HBV via the germ line. J Med Virol. 1985;16:61–6.

Hare WCD. Diseases transmissible by semen and embryo transfer techniques. Technical series no. 4. Paris: O. I. E; 1985. p. 1–117.

Heimesaat MM, Boelke S, Fischer A, Haag LM, Loddenkemper C, Kuhl AA, Gobel UB, Bereswill S. Comprehensive postmortem analyses of intestinal microbiota changes and bacterial translocation in human flora associated mice. PLoS One. 2012;7:e40758.

Herrler A, Beier HM. Early embryonic coats: morphology, function, practical applications. An overview. Cells Tissues Organs. 2000;166:233–46.
Heyman Y, Chavatte-Palmer P, Berthelot V, Fromentin G, Hocquette JF, Martignat L, Renard JP. Assessing the quality of products from cloned cattle: an integrative approach. Theriogenology. 2007;67:134–41.
Hirayama Y, Ohkubo M, Misumi K, Imai K. Successful production of piglets derived from vitrified morulae and early blastocysts without direct exposure to liquid nitrogen. Reprod Fertil Dev. 2007;19:177.
Holt WV, Pickard AR, Prather RS. Wildlife conservation and reproductive cloning. Reproduction. 2004;127:317–24.
Hoshino Y, Hayashi N, Taniguchi S, Kobayashi N, Sakai K, Otani T, Iritani A, Saeki K. Resurrection of a bull by cloning from organs frozen without cryoprotectant in a -80 degrees C freezer for a decade. PLoS One. 2009;4:e4142.
Hubalek Z. Protectants used in the cryopreservation of microorganisms. Cryobiology. 2003;46:205–29.
Isachenko V, Montag M, Isachenko E, Zaeva V, Krivokharchenko I, Shafei R, van der Ven H. Aseptic technology of vitrification of human pronuclear oocytes using open-pulled straws. Hum Reprod. 2005a;20:492–6.
Isachenko V, Isachenko E, Montag M, Zaeva V, Krivokharchenko I, Nawroth F, Dessole S, Katkov II, van der Ven H. Clean technique for cryoprotectant-free vitrification of human spermatozoa. Reprod Biomed Online. 2005b;10:350–4.
Isachenko V, Katkov II, Yakovenko S, Lulat AG, Ulug M, Arvas A, Isachenko E. Vitrification of human laser treated blastocysts within cut standard straws (CSS): novel aseptic packaging and reduced concentrations of cryoprotectants. Cryobiology. 2007;54:305–9.
Johnston LA, Donoghue AM, O'Brien SJ, Wildt DE. Rescue and maturation in vitro of follicular oocytes collected from nondomestic felid species. Biol Reprod. 1991;45:898–906.
Jones SK, Darville JM. Transmission of virus particles by cryotherapy and multi-use caustic pencils: a problem to dermatologists? Br J Dermatol. 1989;121:481–6.
Jordan HL, Howard J, Barr MC, Kennedy-Stoskopf S, Levy JK, Tompkins WA. Feline immunodeficiency virus is shed in semen from experimentally and naturally infected cats. AIDS Res Hum Retroviruses. 1998;14:1087–92.
Karrow AM, Critser JK. Reproductive tissue banking. Scientific principles. San Diego: Academic; 1997.
Kasai M, Zhu SE, Pedro PB, Nakamura K, Sakurai T, Edashige K. Fracture damage of embryos and its prevention during vitrification and warming. Cryobiology. 1996;33:459–64.
Kochhar HS, Rudenko L. Safety of food derived from animal clones. In: Stringfellow DA, Givens D, editors. Manual of the International Embryo Transfer Society. 4th ed. Champaign, IL: IETS; 2010. p. 114–24.
Kotronias D, Kapranos N. Detection of herpes simplex virus DNA in human spermatozoa by in situ hybridization technique. In Vivo. 1998;12:391–4.
Kuleshova LL, Shaw JM. A strategy for rapid cooling of mouse embryos within a double straw to eliminate the risk of contamination during storage in liquid nitrogen. Hum Reprod. 2000;15:2604–9.
Kupferschmied HU, Kihm U, Bachman P, Muller KH, Ackerman M. Transmission of IBR/IPV virus in bovine semen: a case report. Theriogenology. 1986;25:439–43.
Kuwayama M. Highly efficient vitrification for cryopreservation of human oocytes and embryos: the Cryotop method. Theriogenology. 2007;67:73–80.
Kuwayama M, Vajta G, Leda S, Kato O. Comparison of open and closed methods for vitrification of human embryos and the elimination of potential contamination. Reprod Biomed Online. 2005;11:608–14.
Lalonde A, Bielanski A. Efficacy of the International Embryo Transfer Society (IETS) washing procedure for rendering oocytes matured in vitro free of bovine viral diarrhea virus (BVDV) 1. Theriogenology. 2011;76:261–6.

Lane M, Bavister BD, Lyons EA, Forest KT. Containerless vitrification of mammalian oocytes and embryos. Nat Biotechnol. 1999;17:1234–6.

Larman MG, Gardner DK. Vitrification of mouse embryos with super-cooled air. Fertil Steril. 2011;95:462–1466.

Leibo SP, Songsasen N. Cryopreservation of gametes and embryos of non-domestic species. Theriogenology. 2002;57:303–26.

Lindberg A, Ortman K, Alenius S. Seroconversion of bovine viral diarrhea virus (BVDV) in dairy heifers after embryo transfer. Proceedings of the 1st International Congress on Animal Reproduction, Stockholm, Sweden; 2000. p. 250.

Lindemans W, Sangalli L, Kick A, Earl CR, Fry RC. Vitrification of bovine embryos using the CLV method. Reprod Fertil Dev. 2004;16:174. Abstr.

Liu J, Phy J, Yeomans E. Theoretic considerations regarding slow cooling and vitrification during cryopreservation. Theriogenology. 2012;78:1641–52.

Lopez E, Cipri K, Naso V. Technologies for cryopreservation: Overview and innovation. In: Katkov I, editor. Current frontiers in cryobiology; 2012. pp. 528–43. ISBN: 978-953-51-0191-8,

Letur-Konirsch H, Collin G, Sifer C, Devaux A, Kuttenn F, Madelenat P, Brun-Vezinet F, Feldmann G, Benifla JL. Safety of cryopreservation straws for human gametes or embryos: a study with human immunodeficiency virus-1 under cryopreservation conditions. Hum Reprod. 2003;18:140–4.

Loskutoff NM, Huyser C, Singh R, Morfeld KA, Walker DL, Thornhill AR, Smith M, Morris L, Webber L. Removing HIV-1 and hepatitis C virus from semen: an improved method for washing sperm using a trypsinized silica particle suspension and a novel tube insert. Fertil Steril. 2005;84:1001–10.

Maertens A, Bourlet T, Plotton N, Pozzetto B, Levy R. Validation of safety procedures for the cryopreservation of semen contaminated with hepatitis C virus in assisted reproductive technology. Hum Reprod. 2004;19:1554–7.

Maes D, Nauwynck H, Rijsselaere T, Mateusen B, Vyt P, de Kruif A, Van Soom A. Diseases in swine transmitted by artificial insemination: an overview. Theriogenology. 2008;70:1337–45.

Magnano AR, Giordano R, Moscufo N, Baccetti B, Spadafora C. Sperm/DNA interaction: integration of foreign DNA sequences in the mouse sperm genome. J Reprod Immunol. 1998;41:187–96.

Malaver-Ortega LF, Sumer H, Liu J, Verma PJ. The state of the art for pluripotent stem cells derivation in domestic ungulates. Theriogenology. 2012;78:1749–62.

Mara L, Casu S, Carta A, Dattena M. Cryobanking of farm animal gametes and embryos as a means of conserving livestock genetics. Anim Reprod Sci. 2013;138:25–38.

Marquant-Leguienne B, Delalleau N, Harlay T, Allietta M, Diemert S, Le Tallec B, Guerin B. Results of a four-year survey on viral and bacterial contamination in a bovine *in vitro* embryo production system when using slaughterhouse material. Theriogenology. 2000;53:321. Abstr.

Mazur P, Leibo SP, Seidel Jr GE. Cryopreservation of the germplasm of animals used in biological and medical research: importance, impact, status, and future directions. Biol Reprod. 2008;78:2–12.

Mazur P, Seki S. Survival of mouse oocytes after being cooled in a vitrification solution to -196 degrees °C at 95 degrees to 70,000 degrees °C/min and warmed at 610 degrees to 118,000 degrees °C/min: a new paradigm for cryopreservation by vitrification. Cryobiology. 2011;62:1–7.

Mazur P. Physical factors implicated in the death of microorganisms at subzero temperatures. Ann N Y Acad Sci. 1960;85:610–29.

McDonnell G, Burke P. The challenge of prion decontamination. Clin Infect Dis. 2003;36:1152–4.

Melvin Jr JR, Cronholm LS, Simson Jr LR, Isaacs AM. Bacterial transmigration as an indicator of time of death. J Forensic Sci. 1984;29:412–7.

Men H, Walters EM, Nagashima H, Prather RS. Emerging applications of sperm, embryo and somatic cell cryopreservation in maintenance, relocation and rederivation of swine genetics. Theriogenology. 2012;78:1720–9.

Mirabet V, Alvarez M, Solves P, Ocete D, Gimeno C. Use of liquid nitrogen during storage in a cell and tissue bank: contamination risk and effect on the detectability of potential viral contaminants. Cryobiology. 2012;64:121–3.

Moreira da Silva F, Metelo R. Relation between physical properties of the zona pellucida and viability of bovine embryos after slow-freezing and vitrification. Reprod Domest Anim. 2005;40:205–9.

Morfeld KA, White B, Mills G, Krisher R, Mellencamp MA, Loskutoff NM. A novel sperm processing method for eliminating porcine reproductive and respiratory syndrome virus (PRRSv) from boar semen and its effects on embryo development *in vitro* and *in vivo*. Reprod Fertil Dev. 2005;17:186. Abstr.

Morrell JM, Geraghty RM. Effective removal of equine arteritis virus from stallion semen. Equine Vet J. 2006;38:224–9.

Mortimer D. Current and future concepts and practices in human sperm cryobanking. Reprod Biomed Online. 2004;9:134–51.

Morris J. Contamination issues. In: McLaughlin E, Pacey A, Elliot T, editors. Safe cryopreservation of gametes and embryos. Morley, WA: Ladybrook Publishing; 1999. p. 26.

Morris J. Cryopreservation and contamination. A.R.T. & Science. 2003;2:1–3.

Morris GJ. The origin, ultrastructure, and microbiology of the sediment accumulating in liquid nitrogen storage vessels. Cryobiology. 2005;50:231–8.

Morris GJ, Acton E, Faszer K, Franklin A, Yin H, Bodine R, et al. Cryopreservation of murine embryos, human spermatozoa and embryonic stem cells using a liquid nitrogen-free, controlled rate freezer. Reprod Biomed Online. 2006;13:421–6.

Nikfarjam L, Farzaneh P. Prevention and detection of Mycoplasma contamination in cell culture. Cell J. 2012;13:203–12.

Parmegiani L, Accorsi A, Cognigni GE, Bernardi S, Troilo E, Filicori M. Sterilization of liquid nitrogen with ultraviolet irradiation for safe vitrification of human oocytes or embryos. Fertil Steril. 2010;94:1525–8.

Parmegiani L. Vajta G. Fertil Steril: Contamination of single-straw carrier for vitrification; 2011. doi:10.1016/jfertnstert.2011.04.036.

Piasecka-Serafin M. The effect of the sediment accumulated in containers under experimental conditions on the infection of semen stored directly in liquid nitrogen (-196 degree °C). Bull Acad Pol Sci Biol. 1972;20:263–7.

Picha Y, Tibary A, Memon M, Kasimanickam R, Sumar J. Chronology of early embryonic development and embryo uterine migration in alpacas. Theriogenology. 2013;79:702–8.

Piomboni P, Baccetti B. Spermatozoon as a vehicle for HIV-1 and other viruses: a review. Mol Reprod Dev. 2000;56:238–42.

Pomeroy KO, Harris S, Conaghan J, Papadakis M, Centola G, Basuray R, Battaglia D. Storage of cryopreserved reproductive tissues: evidence that cross-contamination of infectious agents is a negligible risk. Fertil Steril. 2010;94:1181–8.

Radnot D, Farkas E. Sterility and cryosurgery. Arch Ophthalmol. 1966;75:302–3.

Rall WF, Fahy GM. Ice-free cryopreservation of mouse embryos at -196 degrees °C by vitrification. Nature. 1985;313:573–5.

Rall WF. Avoidance of microbial cross-contamination of cryopreserved gametes, embryos, cells and tissues during storage in liquid nitrogen. Embryologists' Newslett. 2003;6:2–15.

Rall WF, Meyer TK. Zona fracture damage and its avoidance during the cryopreservation of mammalian embryos. Theriogenology. 1989;31:83–692.

Randall AE, Pettitt MJ, Plante C, Buckrell BC, Randall GCB, Henderson JM, Larochelle R, Magar R, Polard JW. Elimination of porcine reproductive and respiratory syndrome virus through embryo transfer. Theriogenology. 1999;51:274. Abstr.

Russell PH, Lyaruu VH, Millar JD, Curry MR, Watson PF. The potential transmission of infectious agents by semen packaging during storage for artificial insemination. Anim Reprod Sci. 1997;4:337–42.

Rintala MA, Grenman SE, Puranen MH, Isolauri E, Ekblad U, Kero PO, Syrjanen SM. Transmission of high-risk human papillomavirus (HPV) between parents and infant: a prospective study of HPV in families in Finland. J Clin Microbiol. 2005;43:376–81.

Rodriguez-Osorio N, Urrego R, Cibelli JB, Eilertsen K, Memili E. Reprogramming mammalian somatic cells. Theriogenology. 2012;78:1869–86.

Rutala WA, Weber DJ. Guideline for disinfection and sterilization of prion-contaminated medical instruments. Infect Control Hosp Epidemiol. 2010;31:107–17.

Santos RR, Amorim C, Cecconi S, Fassbender M, Imhof M, Lornage J, Paris M, Schoenfeldt V, Martinez-Madrid B. Cryopreservation of ovarian tissue: an emerging technology for female germline preservation of endangered species and breeds. Anim Reprod Sci. 2010;122:151–63.

Saragusty J, Arav A. Current progress in oocyte and embryo cryopreservation by slow freezing and vitrification 5. Reproduction. 2011;141:1–19.

Schaffer TW, Everett J, Silver GH, Came PE. Biohazard potential: recovery of infectious virus from the liquid nitrogen of a virus repository. Health Lab Sci. 1976;13:23–4.

Schiewe M, Hasler J. General hygiene and quality control practices in an embryo-production laboratory. In: Stringfellow D, Givens D, editors. Manual of the International Embryo Transfer Society. 4th ed. Champaign, IL: IETS; 2010. p. 73–86.

Schiewe MC. Microsecure vitrification (μS-VTF) procedure: optimum simplicity, security, cost-saving and effectiveness combaning FDA-approved products. J Clin Embryol. 2010;13:33–41.

Schiewe MC, Rall WF, Stuart LD, Wild DE. Analysis of cryoprotectant, cooling rate and in situ dilution using conventional or vitrification for cryopreserving sheep embryos. Theriogenology. 1991;36:270–93.

Seli E, Agarval A. Fertility preservation in females: emerging technologies and clinical applications. In: Seli E, Agarval A Editors, Ashok, Springer; 2011.

Selwood L. Marsupial egg and embryo coats. Cells Tissues Organs. 2000;166:208–19.

Shin T, Sneed L, Hill JR, Westhusin ME. High incidence of developmental failure in bovine fetuses derived by cloning bovine viral diarrhea virus-infected cells. Theriogenology. 2000;53:243. Abstr.

Stachecki JJ, Garrisi J, Sabino S, Caetano JPJ, Wiemer KE, Cohen J. A new safe, simple and successful vitrification method for bovine and human blastocysts. Reprod Biomed Online. 2008;17:360–7.

Stoops MA, O'Brien JK, Roth TL. Gamete rescue in the African black rhinoceros (Diceros bicornis). Theriogenology. 2011;76:1258–65.

Stout TA, Meadows S, Allen WR. Stage-specific formation of the equine blastocyst capsule is instrumental to hatching and to embryonic survival in vivo. Anim Reprod Sci. 2005;87:269–81.

Straub OC. Infectious bovine rhinotracheitis virus. In: Dinter Z, Morein B, editors. Virus infections of ruminants. Amsterdam: Elsevier Science; 1990. p. 71–108.

Stringfellow DA, Wolfe DF, McGuire JA, Lauerman LH, Gray BW, Sparling PH. Effects of embryo-freezing and thawing techniques on the survivability of *Brucella abortus*. Theriogenology. 1986;26:553–9.

Stringfellow JS, Hathcock TL, Riddel KP, Stringfellow DA, Galik PK, Riddel MGJ, Carson R. Introduction of *Stenotrophomonas maltophilia* through semen used for *in vitro* production of bovine embryos. Theriogenology. 1997;47:382. Abstr.

Stringfellow D, Seidel S. Manual of the International Embryo Transfer Society. 3rd ed. Savoy, Illinois: IETS; 1998.

Stringfellow DA, Riddell KP, Galik PK, Damiani P, Bishop MD, Wright JC. Quality controls for bovine viral diarrhea virus-free IVF embryos. Theriogenology. 2000;53:827–39.

Stringfellow DA, Riddell KP, Givens MD, Galik PK, Sullivan E, Dykstra CC, Robl J, Kasinathan P. Bovine viral diarrhea virus (BVDV) in cell lines used for somatic cell cloning. Theriogenology. 2005;63:1004–13.

Stringfellow D. Recommendations for the sanitary handling of *in vivo*-derived embryos. In: Stringfellow DA, Givens D, editors. Manual of the International Embryo Transfer Society. 4th ed. Champaign, IL: IETS; 2010. p. 65–9.

Stringfellow DA, Givens D. Manual of the International Embryo Transfer Society. 4th ed. Champaign, IL: IETS; 2010.
Sugiyama R, Nakagawa K, Shirai A, Nishi Y, Kuribayashi Y, Inoue M. Clinical outcomes resulting from the transfer of vitrified human embryos using a new device for cryopreservation (plastic blade). J Assist Reprod Genet. 2010;27:161–7.
Tedder RS, Zuckerman MA, Goldstone AH, Hawkins AE, Fielding A, Briggs EM, Irvin D, Blair S, Gorman AM, Patterson KG, Linch DC, Heptonstall J, Brink NS. Hepatitis B transmission from contaminated cryopreservation tank. Lancet. 1995;346:137–9.
Tedeschi R, De Paoli P. Collection and preservation of frozen microorganisms. Methods Mol Biol. 2011;675:313–26.
Thibier M, Guerin B. Embryo transfer in small ruminants: the method of choice for health control in germplasm exchanges. Livest Prod Sci. 2000;62:253–70.
Thibier M. Identified and unidentified challenges for reproductive biotechnologies regarding infectious diseases in animal and public health. Theriogenology. 2001;56:1465–81.
Tomlinson M, Morroll D. Risks associated with cryopreservation: a survey of assisted conception units in the UK and Ireland. Hum Fertil (Camb). 2008;11:33–42.
Toth A, O'Leary WM, Ledger W. Evidence for microbial transfer by spermatozoa. Obstet Gynecol. 1982;59:556–9.
Tucker MJ, Liebermann J, editors. Vitrification in assisted reproduction: a user's manual and trouble-shooting guide. London: Informa Healthcare; 2007.
Vajta G, Booth PJ, Holm P, Greve T, Callesen H. Successful vitrification of early stage bovine *in vitro* produced embryos with the open pulled straw (OPS) method. Cryo-Letters. 1997;18: 191–5.
Vajta G, Lewis IM, Kuwayama M, Greve T, Callesen H. Sterile application of the open pulled straw (OPS) vitrification method. Cryo-Letters. 1998;19:389–92.
Vajta G, Nagy ZP. Are programmable freezers still needed in the embryo laboratory? Review on vitrification. Reprod Biomed Online. 2006;12:779–96.
Van der Maaten MJ. Determining the health status of embryos for international movement. In: Hare WCD, Seidel SM, editors. International embryo movement. Montreal: IETS; 1987. p. 147–61.
Vanderzwalmen P, Bertin G, Debauche C, Standaert V, Bollen N, van Roosendaal E, Vandervorst M, Schoysman R, Zech N. Vitrification of human blastocysts with the hemi-straw carrier: application of assisted hatching after thawing. Hum Reprod. 2003;18:1504–11.
Van Os HC, Drogendijk AC, Fetter WP, Heijtink RA, Zeilmaker GH. The influence of contamination of culture medium with hepatitis B virus on the outcome of *in vitro* fertilization pregnancies. Am J Obstet Gynecol. 1991;165:152–9.
Vanroose G. Interactions of bovine herpesvirus-1 and bovine viral diarrhea virus with bovine gametes and *in-vitro*-produced embryos. Dissertation University of Ghent; 1999.
Van Soom A, Nauwynck H, Wrathall A. Scientific foundations for the epidemiologigal safety of embryo transfer. In: Stringfellow DA, Givens D, editors. Manual of the International Embryo Transfer Society. 4th ed. Champaign, IL: IETS; 2010a. p. 13–41.
Van Soom A, Wrathall AE, Herrler A, Nauwynck HJ. Is the zona pellucida an efficient barrier to viral infection? Reprod Fertil Dev. 2010b;22:21–31.
Verma R, Holland MK, Temple-Smith P, Verma PJ. Inducing pluripotency in somatic cells from the snow leopard (*Panthera uncia*), an endangered felid. Theriogenology. 2012;77:220–8.
Vincent C, Pickering SJ, Johnson MH. The hardening effect of dimethylsulphoxide on the mouse zona pellucida requires the presence of an oocyte and is associated with a reduction in the number of cortical granules present. J Reprod Fertil. 1990;89:253–9.
Wakayama S, Ohta H, Hikichi T, Mizutani E, Iwaki T, Kanagawa O, Wakayama T. Production of healthy cloned mice from bodies frozen at -20 degrees C for 16 years. Proc Natl Acad Sci U S A. 2008;105:17318–22.
Wakayama S, Mizutani E, Wakayama T. Production of cloned mice from somatic cells, ES cells, and frozen bodies. Methods Enzymol. 2010;476:151–69.

Wallis C, Melnik JL. Stabilization of enveloped viruses by dimethyl sulfoxide. J Virol. 1968;2:953–4.

Wang LJ, Xiong XR, Zhang H, Li YY, Li Q, Wang YS, Xu WB, Hua S, Zhang Y. Defined media optimization for in vitro culture of bovine somatic cell nuclear transfer (SCNT) embryos. Theriogenology. 2012;78:2110–9.

Wierzbowski S. Bull semen opportunistic pathogen and ubiquitary microflora. In: Disease control in semen and embryos, vol. 23. FAO Animal Production and Health Paper; 1985. pp. 21–28

Wild DE. Genome resource banking. In: Karrow AM, Critser JK, editors. Reproductive tissue banking. Scientific principles. San Diego, CA: Academic; 1997. p. 399–438.

Willadsen SM. Factors affecting the survival of sheep embryos during deep-freezing and thawing. In: Elliot K, Whelan J, editors. The freezing of mammalian embryos. Amsterdam: Elsevier-Exerpta Medica; 1977. p. 175–201.

Williams ES, Barker IK. Infectious diseases of wild mammals. 3rd ed. Ames, IA: Blackwell Publishing; 2001.

Woelders H, Windig J, Hiemstra SJ. How developments in cryobiology, reproductive technologies and conservation genomics could shape gene banking strategies for (farm) animals. Reprod Domest Anim. 2012;47 Suppl 4:264–73.

Wrathall AE, Simmons HA, Van Soom A. Evaluation of risks of viral transmission to recipients of bovine embryos arising from fertilisation with virus-infected semen. Theriogenology. 2006;65:247–74.

Wrathall AE, Ayling RD, Simmons HA. Risks of transmitting mycoplasmas by semen and embryo transfer techniques in cattle, sheep, goats and pigs. CAB Rev Perspect Agric Vet Sci Nutr Nat Resour. 2007;2:036, 1–31.

Wrathall AE, Holyoak GR, Parsonson IM, Simmons HA. Risks of transmitting ruminant spongiform encephalopathies (prion diseases) by semen and embryo transfer techniques. Theriogenology. 2008;70:725–45.

Yavin S, Arav A. Measurement of essential physical properties of vitrification solutions. Theriogenology. 2007;67:81–9.

Yavin S, Aroyo A, Roth Z, Arav A. Embryo cryopreservation in the presence of low concentration of vitrification solution with sealed pulled straws in liquid nitrogen slush. Hum Reprod. 2009;24:797–804.

Zhang X, Catalano PN, Gurkan UA, Khimji I, Demirci U. Emerging technologies in medical applications of minimum volume vitrification. Nanomedicine (Lond). 2011;6:1115–29.

Zea-Mazo JW, Negrette-Mejia YA, Cardona-Maya W. Virus of sexual transmission: semen and virus relationship. Actas Urol Esp. 2010;34:845–53.

Chapter 18
Fertility Control in Wildlife: Review of Current Status, Including Novel and Future Technologies

Deborah Garside, Ayman Gebril, Manal Alsaadi, and Valerie A. Ferro

Abstract Overpopulation of selected groups of animals is widely recognised as an issue that can have adverse effects on several current global problems, such as animal and human health, conservation and environmental changes. This review will, therefore, focus on recent novel contraception together with future technologies that may provide additional contraceptive methods.

Keywords Immunocontraception • Sperm antibodies • Sperm antigens • Nanoparticles • Liposomes • Bacteriophage • Pest control • Conservation

Increased populations cause a problem for both companion and wild animals. For example, in the United States it is estimated that over 20 million unwanted dogs and cats are annually euthanized, while several million additional animals starve to death each year. Worldwide, there are significant problems with excessive numbers of a species, including wild horses, deer, grey squirrel, mink, geese, feral swine and elephants, many of which can cause significant damage to agricultural land, impact on rare or indigenous species, or compete with humans for limited resources, particularly in developing countries. Additionally, some captive animals, such as large cats, can become overpopulated in specific zoological collections and, for management purposes, their breeding success has to be controlled.

D. Garside, B.Sc., Ph.D.
Department of Medicine, Imperial College London, South Kensington, London, UK

A. Gebril, B.V.Sc., Ph.D. • V.A. Ferro, B.Sc., Ph.D. (✉)
Strathclyde Institute of Pharmacy and Biomedical Sciences,
University of Strathclyde, Glasgow, UK
e-mail: v.a.ferro@strath.ac.uk

M. Alsaadi, Ph.D.
Department of Industrial Pharmacy, Faculty of Pharmacy, University of Tripoli, Tripoli, Libya

W.V. Holt et al. (eds.), *Reproductive Sciences in Animal Conservation*, Advances in Experimental Medicine and Biology 753, DOI 10.1007/978-1-4939-0820-2_18,

Hence, a long-standing goal has been the search for safe and effective means to temporarily or permanently eliminate fertility in wild or feral animals. Various methodologies have been used over the years, including surgical sterilisation. However, surgical methods are very expensive, invasive, require individual capture and have been found lacking in terms of effectiveness. More recently, steroids, hormonal implants and contraceptive vaccines (immunocontraceptives) have been successfully employed to manage population growth of several species in relation to issues such as conservation and pest control. As such, this review focuses on the current state of anti-fertility methods for wildlife population management. It will cover the main areas of wildlife population growth that have used contraception to facilitate a reduction in numbers, the types of contraception used, the issues that have been encountered and future technologies.

1 Key Areas and Reasons for Wildlife Fertility Control

There are several areas of wildlife management that require fertility control of various species. While contraceptive developments have made progress for human and veterinary applications, wildlife population control has unique hurdles. The key areas where contraception has been used to control fertility are conservation, pest control, companion animals and zoological institutions, each of which provides specific management problems. In wildlife situations where the animals are free-roaming the use of contraception presents difficult challenges, as most animals in this situation can only be captured and treated once in their lives. This makes ensuring that the fertility of sufficient animals is controlled over a significant time period a difficult task. The ideal wildlife contraceptive, therefore, is one that can be given once and is long-acting, rather than one that requires frequent administration. Importantly, it must also be safe and not impact on the animal's welfare, either as an individual or as a group. Welfare is considered less important in relation to pest control. For zoos, animals can be captured and treated repeatedly, which makes the use of short-acting contraceptives feasible.

From a population biology perspective, controlling reproduction in females is the most important factor in regulating population size. An equivalent level of infertility in males often has little effect (Budke and Slater 2009). However, mortality is often also considered to be an equally important factor. This is mainly in relation to the control of populations, such as rats by the use of poisons like rodenticides. The toxicity of these products can affect non-target species, if accidentally ingested (Rattner et al. 2012). In addition to controlling reproduction, a secondary goal of fertility control is to reduce adverse health impacts and undesirable behaviours associated with sex hormones in both sexes.

Reasons that fertility control are desired for many wild species are summarised below.

1.1 Conservation

An increase in certain wildlife and free-ranging animals poses and causes serious conservation problems, not dissimilar to those associated with the increase in human populations. For example, in Sri Lanka and Thailand, wild elephants often destroy local crops in their search for food. This obviously impacts on local communities, who try and scare the elephants away from their fields by chasing or firing guns into the air. This can sometimes lead to accidental injury and the elephants may later die from infections related to such injuries. In some areas, translocation of 'problem elephants' has been employed in an effort to keep such animals away from local communities (Fernando 2011). This is often not a successful solution, as some elephants (known as 'homers') frequently return to the capture site (Baskran and Desai 1996; Fernando et al. 2008). Interestingly, translocation generally causes wider propagation and intensification of human–elephant-conflict, increasing elephant mortality. Translocation would seem to defeat both reducing human–elephant-conflict and elephant conservation goals in these situations (Fernando et al. 2012). In such cases, programmes using immunocontraception (anti-zona pellucida [ZP] protein vaccines) have been employed successfully (Delsink et al. 2002).

Traditional lethal control programmes (indirect or direct intervention, for example by culling, poisoning, translocation) are not always safe, legal or publicly acceptable and alternative approaches are therefore required. Given the range of species for which fertility control is desired, the development of a single contraceptive method is not straightforward. Additionally, the many differences in species reproductive strategies, required outcome of contraceptive programmes (e.g. reversible or irreversible; male or female contraception) and the lack of reproductive biology information for many species, means that detailed species-specific investigations are required for different conservation programmes.

For many conservation population control programmes it is desirable to use a long-acting contraceptive, as frequent access to the animals may be difficult, dangerous or impact on the animal's welfare. For this reason, long-acting methods such as steroid and non-hormonal implants have been used to control koala (Hynes et al. 2010) and kangaroo populations (Bertschinger et al. 2002), while immunocontraceptives, such as anti-zona-pellucida vaccines (Spayvac™) have been used for controlling deer (Locke et al. 2007a), wild horse (Kirkpatrick and Turner 2008) and elephant (Delsink et al. 2007) populations, with reasonable success (Kirkpatrick et al. 1997, 2011).

1.2 Companion Animals

Worldwide, there are increasing numbers of cats and dogs without owners, abandoned, strays and those given to animal charities for re-homing. This is often caused by the lack of responsible fertility control by owners, particularly with regard to cats. Apart from the animal welfare and cost issues, there are also many consequences to society of pet overpopulation: stray dogs are annually responsible for serious bites of the public and feral cats and dogs contribute significantly to sanitation problems in large cities (Dubna et al. 2007). In developing countries, stray dogs are an important vector of the rabies virus, resulting in the deaths of several thousand people annually (Knobel et al. 2005). Environmentally, the overpopulation of both owned and feral cats is thought to have an impact on the reduction of native species (e.g. birds) via competition, predation and infectious diseases (Jessup 2004). This is particularly a problem on islands, where cats have been responsible for major reductions in the local bird populations (Blackburn et al. 2004). As with dogs, there is often resistance from the public to culling large populations of feral cats in areas where they impact on local bird numbers. The use of contraception to reduce cat numbers in such instances, therefore, has welfare and ethical advantages (Levy and Crawford 2004; Levy 2011).

The surgical sterilisation of both unwanted cats and dogs is an effective non-lethal population control method, particularly for those in animal charity rescue centres. However, it has limitations for use with feral and stray animals, due to expense and logistical impediments. Alternative practical, cost effective and long-acting methods are therefore currently being researched, particularly for feral cats and dogs. An ideal feral cat or dog immunocontraceptive would induce long-term or permanent contraception following a single treatment. For instance, one model suggests that a contraceptive with 3-year duration of effect may be successful in controlling cat populations (Budke and Slater 2009). Those therefore being investigated for cat and dog use include immunocontraception (e.g. anti-zona pellucida and anti-GnRH vaccines) (Munks 2012), steroids and non-hormonal implants (Kutzler and Wood 2006).

1.3 Pest Control

Several species can be considered a 'pest' in relation to their habitat, due to excessive numbers. For example the grey squirrel, rabbit, monkey, wild boar, deer, mink, badger and rodent are considered pests in various countries, due to the damage they do to agriculture, indigenous species or impact on human or animal health (Rao et al. 2002; Olivera et al. 2010; Mayle and Broome 2013). This includes non-native invasive species, which often have negative effects on biodiversity and ecosystem function of native wildlife (Shackelford et al. 2013). In several situations, the traditional methods of controlling population numbers are by poison, trapping or shooting. As with the control of feral cats and dogs, these methods are often considered unsuitable due to welfare issues, practicalities and cost. Currently, contraceptive methods being investigated are non-steroidal (e.g. dopamine agonist in foxes and stoats; Marks 2001)) immunocontraception (Hardy et al. 2006) and chemical compounds that disrupt spermatogenesis (Dell'Omo and Palmery 2002). In many cases of pest fertility control, the contraceptive agent is administered in 'bait', which has consequences for ensuring that the contraceptive reaches the targeted species and in the relevant effective dose. This will be discussed in the sections on individual contraceptive methods.

1.3.1 Zoo Applications

Many zoo and species survival programmes rely on managed reproduction to maintain both the genetic integrity and physiological health of individuals. The key genetic objectives are to maximise population heterozygosity and equalise genetic founder representation. Both require either the prevention of breeding by some individuals and/or the enhancement of breeding by others. Often, breeding can be simply prevented by separation of the sexes. However, in some cases contraception is the preferred option due to reasons such as lack of space, animal welfare and the maintenance of family groups and established population hierarchies. Contraception is therefore an important contributor to zoo breeding programmes and the key reproductive issue in zoos. Again the main contraceptives that have been used are steroid and non-steroidal injections and implants (Wheaton et al. 2011), and immunocontraception (Powers et al. 2011). A comprehensive list of contraceptive products recommended for different species can be found at http://www.stlzoo.org/animals/scienceresearch/contraceptioncenter/contraceptionrecommendatio/contraceptionmethods/.

2 Review of Wildlife Fertility Control Methods

As indicated, several methods have been investigated for the control of wildlife fertility, some of which are still being researched and others which are currently being used. The methods are aimed at disrupting the reproductive process by inhibiting the key hormones involved or specific reproductive processes. The common targets investigated for the prevention of reproduction in males and females are shown in Fig. 18.1.

The main methods used for wildlife contraception will be reviewed in the following section, together with their advantages and disadvantages, particularly in relation to different species.

3 Hormonal Methods

3.1 Steroid Implants

The use of steroid hormones to inhibit female reproduction has generally been found to be successful in several species. Contraceptives, such as the synthetic progestin melangestrol acetate (MGA), have been used in zoos to control the fertility of captive large cats, while levonorgestrel (developed for human female contraception), has been found to be a suitable long-term contraceptive in tammar wallabies (Herbert et al. 2007) and koalas (Hynes et al. 2010). Levonorgestrel would seem to

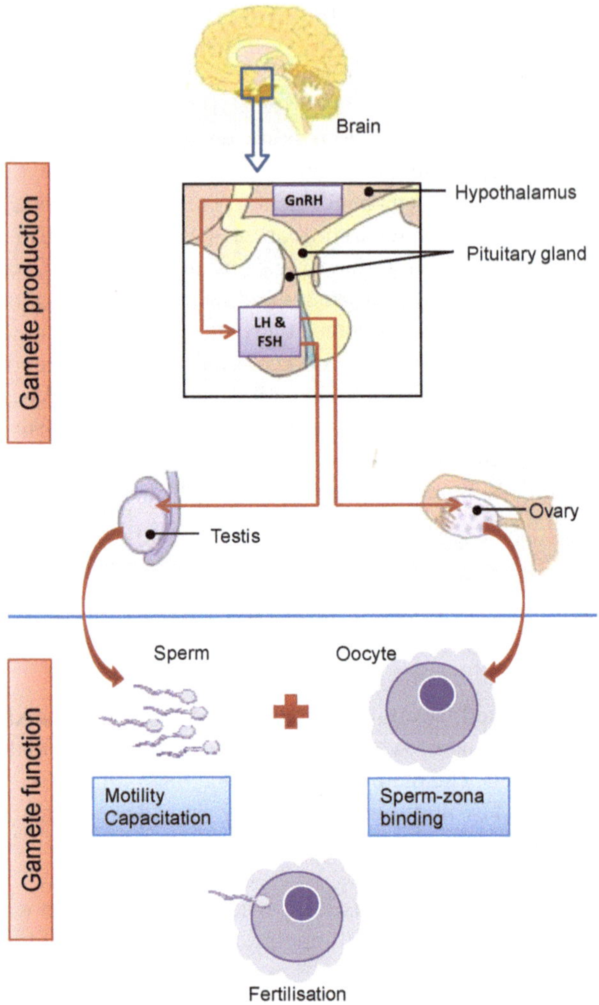

Fig. 18.1 Targets for reproductive inhibition

act in koalas by inhibiting ovulation, not by preventing follicular development and cycling. Interestingly, etonogestrel was not found to be an effective contraceptive in this species. In captive and wild populations of eastern grey kangaroos and tammar wallabies, levonorgestrel successfully prevents reproduction by inhibiting oestrus (Nave et al. 2002). It is not completely understood how levonorgestrel acts in both species, as it does not appear to inhibit follicular development. However, ovulation may be affected, as there is a suggestion that the pre-ovulatory surge of luteinising hormone is inhibited (Hynes et al. 2007). Levonorgestrel does not affect the reactivation and subsequent development of blastocysts in diapause that are conceived prior to treatment. Lactation is also unimpaired, as young were reared to weaning in both species. It would appear that levonorgestrel implants provide a safe, effective and long-term method of fertility control for macropodid marsupials and can be used for the management of overabundant captive and selected wild populations of these animals (Nave et al. 2000). However, in general it is recognised that steroid use is limited for free-ranging wildlife for a variety of reasons including: environmental toxicity, passage through the food chain, adverse effects on social behaviours, high cost of application (particularly for delivering the steroids remotely), health risks in pregnant animals and regulatory issues (Kirkpatrick et al. 2011).

3.2 Non-steroid Methods

Non-steroidal contraceptives, such as deslorelin acetate, an anti GnRH agonist, have also been used to control fertility in various wild carnivores and domestic dogs and cats. Deslorelin (Ovuplant®), was initially developed as an ovulation-inducing agent in mares (Farquhar et al. 2002), but it has more recently been used as a subcutaneous long-acting implant for the suppression of reproduction in both male and female wild and captive animals (Bertschinger et al. 2002; Johnson et al. 2003; Junaidi et al. 2003; Patton et al. 2006). It acts by down-regulating follicle stimulating hormone (FSH) and luteinizing hormone (LH), thereby suppressing ovulation and sperm production. For example, it has been found to be effective in female tammar wallabies in Australia (Herbert et al. 2005) and semi-captive male cheetahs in South Africa (Bertschinger et al. 2006). In the latter, deslorelin implants (6 mg) reliably suppressed fertility for a year, without side effects.

3.3 Unwanted Side Effects of Hormonal Methods

Although, steroids have proved an effective contraceptive in many species, some have also induced unwanted side effects. levonorgestrel, for example, has been implicated in side effects associated with the reproductive tract and also metabolic changes (Monier 1988; Zook et al. 2001). For example, in rabbits, those with levonorgestrel implants developed endometrial decidualisation or deciduosarcomas, especially with increased doses (+233 μg/day) of levonorgestrel, or if oestrogens were also included at doses over 60 μg/day (Janne et al. 2001). In primates, levonorgestrel combined with ethinyl oestradiol has some effect on female basal metabolic rate (Edelman et al. 2011). It also seems likely that metabolic changes seen in women who take oral contraceptives, such as increased blood pressure, insulin insensitivity and raised insulin and lipid levels, could be induced in animals.

Deslorelin implants appear to be safe for administration to males, although, in some species there is evidence of adverse reactions in females. In bitches there have been cases of prolonged oestrus, ovarian cysts and pyometra after deslorelin implants (Arlt et al. 2011). The risk of induction of oestrus can be reduced when such implants are administered at concentrations of progesterone in plasma of $\geq$16.0 pmol/L. More detailed studies are required in female dogs to confirm safety and currently it is suggested that a complete gynaecological examination be performed before implanting deslorelin. Additionally, it is recommended that the implant should be positioned subcutaneously, close to the umbilicus, in order to allow relocation and excision if necessary.

In the female tammar wallaby, the duration of contraception induced by deslorelin is highly variable amongst individuals and has been associated with a significant reduction in basal LH concentrations and a cessation of oestrous cycles (Herbert et al. 2005). Also, there is some evidence to suggest that aspects of blastocyst survival, luteal reactivation, pregnancy or birth may be affected by deslorelin treatment in some animals. Interestingly, in the male, deslorelin appears to have no contraceptive effect (Herbert et al. 2004). In this study, there was no evidence of a treatment-induced decline in plasma testosterone concentration or basal LH concentrations. These studies highlight the problems related to choosing a suitable long-acting contraceptive for a particular application, given the variation in contraceptive effects on individuals, species and gender.

Regarding the use of long-acting steroid and hormonal contraceptives in conservation scenarios, there are specific matters that need to be considered, particularly for those animals that live in groups. Altering hormone levels, both male and female, has the potential to alter social behaviour, including dominance hierarchies, which can influence access to food. If used for a lengthy period, therefore, contraception can result in a breakdown of the group structure, safety, health and ultimately negate the conservation goals (Pukazhenthi et al. 2006; Druce et al. 2011).

Practically, the use of contraceptive implants such as steroids and non-steroids like deslorelin are generally limited to small populations and captive animals rather than large scale population control. This is due mainly to the contraceptives being expensive and also, in wildlife situations (such as for the wallaby and kangaroo), often cannot be applied without first anaesthetising the animal. It is hoped such contraceptives will become cheaper and more easily applied to wildlife situations, if methods can be developed for reliably darting wild free-ranging animals from a distance without the need for anaesthesia.

A list of the types of contraception used for the control of various wildlife species is indicated in Table 18.1.

4 Immunocontraception: Review of Current Vaccines and Uses

Immunocontraception is the use of vaccines to prevent the process of fertilisation. This involves harnessing the immune system to disrupt the reproductive process by targeting key components of the reproductive system, such as the reproductive hormones and gametes (sperm and oocyte). Research has mainly focussed on three areas: the reproductive hormones GnRH (Herbert and Trigg 2005; Schneider et al. 2006) and FSH (Moudgal et al. 1992; Delves and Roitt 2005; Yang et al. 2011); the disruption of fertilisation by preventing sperm–egg binding and thirdly, the

Table 18.1 Contraceptives used to control both wild and captive wildlife populations

Species	Contraceptive	Reference
Cheetah, *Acinonyx jubatus*	Deslorelin	Bertschinger et al. (2002)
Domestic cat, *Felis catus*	Anti-GnRH Melengestrol	Asa et al. (1996), Robbins et al. (2004)
Domestic dog, *Canis lupus familiaris*	Anti-GnRH vaccine Testosterone	Simmons and Hamner (1973), Ladd et al. (1994)
Elk, *Cervus canadensis*	Anti-ZP vaccine Diethyl stillboestrol (DES)	Asa et al. (1996), Shideler et al. (2002)
Elephant (African), *Loxodonta africana*	Anti ZP Progesterone	Delsink et al. (2002), Lincoln et al. (2004)
Leopard (African), *Panthera pardus pardus*	Deslorelin	Bertschinger et al. (2002), Bertschinger et al. (2007)
Baboon, *Papio cynocephalus*	Levonorgestrel Anti-GnRH Anti-LDH	O'Hern et al. (1995), Asa et al. (1996), Nie et al. (1997)
Rabbit, *Oryctolagus cuniculus*	Meloxicam (Cox 2 inhibitor) Anti-PH20 Anti-LDH	Holland et al. (1997), Salhab et al. (2001) Naz et al. (1984)
Deer, *Odocoileus virginianus*	DES Anti-ZP vaccine	Asa et al. (1996), Curtis et al. (2002)
Grey squirrel, *Sciurus carolinensis*	Anti-sperm	Moore et al. (1997)
Mink, *Neovison vison*	DES	Asa et al. (1996)
Rhesus monkey, *Macaca mulatta*	Anti-FSH Norethindrone Anti-FSH GnRH analogue	Wickings and Nieschlag (1980), Srinath et al. (1983), Asa et al. (1996)

inhibition of sperm function and motility (Suri 2004; Naz 2011). The latter two areas have involved the identification of oocyte and sperm proteins involved in sperm function (e.g. sperm–egg binding and sperm motility), which have then been used as targets for vaccine development. The three individual immunocontraceptive research areas referenced above have been extensively reviewed and will not be covered in detail in this chapter.

Although the research to develop safe and effective immunocontraceptives for animal use is ongoing, only two are currently marketed for animal use: GonaCon™ and SpayVac™. As such, the use of immunocontraceptives for animal population control has centred on both of these: GonaCon™, an anti-GnRH vaccine and SpayVac™, which targets a specific oocyte zonapellucida (ZP) protein, thereby preventing fertilisation. They have proved successful in a large variety of species, both at the population level and for individual animals and have been used in both captive and free-ranging species. In particular, they have been used for the population management of African elephants, wild horses, bison and deer. However, although both immunocontraceptives have been used to control feral cat populations, only anti-GnRH vaccines have been successful in this species (Levy 2011). Immunisation against GnRH has resulted in long-term contraception in both male and female cats

Table 18.2 Sperm proteins researched for potential immunocontraception

Sperm target	Reference	Sperm function affected
PH20	Primakoff et al. (1988)	Sperm–egg interaction
SP17	Lea et al. (1998)	Sperm motility
LDH C4	Goldberg and Herr (2000)	Sperm motility
SP10	Herr et al. (1990)	Sperm–egg binding
SAMP14	Shetty et al. (2003)	Sperm–egg binding
SPAG9	Shankar et al. (1998)	Sperm–egg fusion, fertilisation
TSA-1	Santhanam and Naz (2001)	Sperm motility, capacitation
hCRISP1	Ellerman et al. (2010)	Sperm–egg binding
Izumo	Wang et al. (2008)	Sperm–egg fusion, fertilisation
CatSper1	Li et al. (2012)	Intracellular Ca (2+) concentration, motility

following a single dose. GnRH is an ideal contraceptive target for feral cats, as it regulates both pituitary and gonadal hormone responses in males and females. This, therefore, has the added benefit of suppressing nuisance behaviours associated with sex hormones (spraying, fighting, mating calls) in addition to preventing pregnancy. In contrast, the use of anti-ZP vaccines such as SpayVac™ has shown little success in controlling feral cat fertility.

Although not applicable to all areas of population control, GonaCon™ and SpayVac™ have shown that immunocontraception is a feasible and effective tool for controlling fertility in animals. However, research continues to develop additional immunocontraceptives that may have more ideal profiles for fertility control. New targets are being found, such as the discovery of vesicle-associated protein 1 (a novel target isolated from the vesicle-rich hemisphere of the brushtail possum oocyte) (Nation et al. 2008), or a uterine-secreted protein CP4 that can affect conceptus development (Menkhorst et al. 2008). Preventing fertilisation is currently the most popular research area, as it offers a method that specifically prevents fertilisation without effects on the endocrine system and other physiological processes. This has advantages in that it does not affect reproductive hormone levels, negating effects on social behaviour, and the resultant breakdown of the group structure, safety and health. Anti-sperm vaccines have the additional benefit of potentially being effective in both males and females, although the majority of current research has been in the female. The anti-sperm vaccine targets are also sperm specific, thereby limiting side effects, and highly immunogenic which enhances their efficacy and likelihood of being more robust, long-acting and with less response variability between individuals. They are also more suitable for being administered in bait to a population, as they are active in both males and females, potentially reducing populations more quickly.

Numerous sperm proteins have been investigated for contraceptive purposes, many of which are listed in Table 18.2. More recent novel sperm targets include CatSper, Eppin (Chen et al. 2011), Izumo (Wang et al. 2009) and also the epididymal target, SFP2 (Khan et al. 2011). Although still at the research stage, the sperm protein targets potentially provide exciting possibilities for future wildlife contraception.

Current immunocontraceptive research is also investigating novel vaccine approaches, for example, the use of more than one target in an immunocontraceptive. This has the potential to make it more effective with fewer non-responders within a population. In addition, there is the opportunity to develop immunocontraceptives that combine contraception with disease control. Modern vaccine technology enables the development of combined vaccines and scientists are currently investigating the possible development of a GnRH vaccine, that also provides rabies protection for use in feral dog and fox populations (Wu et al. 2009).

The goal for the ideal immunocontraceptive includes a wide margin of safety for target animals and the environment, fast acting with a long duration of activity following a single treatment in males and females of all ages. The development, therefore, of improved anti-GnRH and anti-ZP immunocontraceptives, together with anti-sperm and combined vaccines, offers exciting new possibilities for effective, safe, long-acting practical wildlife fertility control. However, the barrier to success is a current key limitation for the funding of well-designed research programmes to develop immunocontraceptive fertility control products.

4.1 Immunocontraceptive Challenges

The challenges in the application of the current marketed vaccine-based wildlife contraceptives are similar to those of the steroid and hormonal contraceptives: that is differences in efficacy across species, safety and the need for practical and cost effective delivery systems for wild and free-ranging animals. Promisingly, to date, both GonaCon™ and SpayVac™ have been shown to be successful in controlling particular species, such as deer (Locke et al. 2007b; Miller et al. 2008) and elk (Killian et al. 2009), with few issues. Indeed, for wild and feral applications, immunocontraception has the potential to be a more practical and cost-effective method of fertility control than that of steroids and hormones. However, there are problems concerning immunity. For example, can sufficiently strong long-acting immune responses be provoked against the vaccine targets (immunogens) of gametes or reproductive hormones to cause contraception in a sufficient number of animals to produce effective population control? One problem relates to long-term use of immunocontraception in particular populations, whereby a resistant population of low or non-responders arises. This has led to research into understanding the immune responses in species such as brushtail possums, which has shown that there are different major histocompatibility complex (MHC) haplotypes that correspond to non-responsiveness (Holland et al. 2009).

New adjuvants that enhance the mucosal immune response are now available and more are being developed (Zaman et al. 2013). In particular, adjuvants that specifically target the reproductive tract immune response to enable good immune responses against sperm and egg target proteins have made preventing fertilisation using vaccines more effective (Zhang et al. 2007; Zaman et al. 2013). In some cases, combination of drug delivery systems with inherent adjuvant properties, also have proven advantages.

The development of improved anti-GnRH and anti-ZP immunocontraceptives, together with anti-sperm and combined vaccines, offers exciting new possibilities for effective, safe, long-acting, practical wildlife fertility control. However, the current lack of funding for well designed research programmes is a major barrier to success and a current key limitation for the development of advanced immunocontraceptive products.

Apart from the scientific questions, immunocontraception will need to meet the regulatory requirements for use in the environment and on the biological and economical feasibility of their use. In addition, widespread use will also depend on the health and safety requirements of individual countries and scenarios and on public acceptance of the techniques. To date, the success of GonaCon™ and SpayVac™ has gone someway to paving the way for further immunocontraceptives to be used for wildlife population control.

5 The Future of Immunocontraception: New Technologies

The drive in developments incorporating wildlife specifications has come from a pressing need to control disease such as rabies and tuberculosis. For these diseases as with immunocontraception of free-roaming populations, oral delivery using bait, remains the most convenient and cost-effective means of delivering vaccines to large or widely spread populations of animals. It is non-invasive and so concerns about injection site adverse effects are negated. Recently oral vaccination field trials of badgers to prevent tuberculosis have been reported (Gormley and Corner 2013). Nasal or inhalation delivery may emerge in the future, but currently there are only a limited number of reports of technologies with aerosolised potential (Corner et al. 2001; Corner et al. 2008).

5.1 Bait Delivery Related to Different Species

Based on a highly successful vaccine that has been established for many years and used extensively in the prevention of rabies, RABORAL V-RG®, demonstrates the success of vectors in vaccines, utilizing a vaccine-strain of Vaccinia virus as the live vector. This vector system allows delivery of a thermostable rabies vaccine contained in edible baits. Prior to being licensed by the USDA, the vaccine was approved for environmental release and a range of safety tests were carried out in non-target animals.

To be effective, oral formulations need to be stable under different field conditions (temperature, moisture, pH effects), but can be contained in weather-proof packaging (Cross et al. 2009). Flavouring or chewing stimulants to encourage species-specific and effective sampling may also be an option (Bergman et al. 2008;

Ballesteros et al. 2009c), while baits can be placed underground or in trees to limit the species that have access to them (Buddle et al. 2011). Selective feeders have also been designed to reduce non-target consumption of bait (Ballesteros et al. 2011; Telford et al. 2011). Changing the consistency of an oral vaccine, such as using a viscous material that also acts as an adjuvant can improve vaccination rate in some species. Again, the latter has been demonstrated with inclusion of chitosan and methylated derivatives, with RABORAL V-RG® (Fry et al. 2012). Examples of various baits designed for use in different species include synthetic grit consisting of plasticizer and cross-linking agents targeted at Mallards (Hurley and Johnston 2002), lipid-based baits for a variety of potential TB reservoir species (Nol et al. 2008; Corner et al. 2010; Buddle et al. 2011) and cereal-based products to attract pigs and wild boar (Ballesteros et al. 2009a). Over-dosing should not endanger target or non-target species and it is also important to be able to evaluate vaccination coverage. Therefore, chemical markers such as the use of iophenoxic acid incorporated in the formulation can be detected easily in serum (Massei et al. 2009). Where baits are distributed aerially (using helicopters for example), computerised and geographical information systems are invaluable for tracking bait location (Mulatti et al. 2011), although the cost to benefit ratio must be evaluated carefully.

Thus it can be seen that creative designs are an essential part of wildlife vaccine formulation. However, effectiveness is key. To further enhance immunogenicity, different types of delivery system may be required and these include living and non-living technologies.

5.2 Living Vectors

Many reported successful studies vaccinating wild animals against infectious diseases use viral or bacterial vectors (Rocke et al. 2008; Ballesteros et al. 2009b). Therefore, it stands to reason that similar vector application could also provide an important strategy for contraceptive antigen delivery. This approach was initially investigated in Australia for the control of population in rabbits. As such, the strategy was to select a target molecule capable of inducing an immunocontraceptive response and insert the gene encoding the target into the myxoma virus for distribution into the rabbit population. This raised several concerns regarding safety and the possible transmission to other non-targeted species (Tyndale-Biscoe 1994). In the current research, species-specific viruses are now being genetically engineered to produce contraceptive vaccines for pest animals such as mice, rabbits and foxes. Laboratory experiments have shown that high levels of infertility can be induced in mice infected with recombinant murine cytomegalovirus and ectromelia virus expressing reproductive antigens as well as in rabbits using myxoma virus vectors (Gu et al. 2004; Hardy 2007; Nikolovski et al. 2009). However, safety and social concerns will need to be allayed prior to the use of these vaccines to control wildlife pests.

5.3 Non-living Particulates

In general cost of production and delivery is a prime consideration in the development of wildlife vaccines, while minor/short-term side-effects might be less of an issue. Another factor is genetics and genetic diversity of the target species, particularly as there is limited knowledge of this and impact on immune response. For example porcine ZP (pZP) vaccine tested in wild horses and deer was 85 % effective, whereas in African elephants efficacy fell by 10 % (Kirkpatrick et al. 2011). However, other studies in African elephants have indicated that the efficacy of pZP is equivalent to that in wild horses (Fayrer-Hosken et al. 1999). Although there are reported differences in vaccine efficacy between species, many particulate vaccine delivery systems are seen as a particularly attractive method. The main reasons are that they can be produced cheaply and generally target the innate immune system so do not have to involve sophisticated or knowledge-based inventions. Particulate systems include: organic (e.g. lipid-based) and inorganic (e.g. inert beads) technologies.

5.3.1 Lipid-Based Technologies

Several methodologies have been investigated in the quest to induce antibodies for the effective immunisation against diseases. One such are liposomes and non-ionic surfactant vesicles which have been shown to possess the ability to induce antibodies in human studies and are also safe (Alving et al. 1995). These usually consist of a lipid bilayer with an aqueous core. Antigens (protein, peptide, nucleic acid and indeed drugs such as contraceptive agents) can be entrapped within the bilayer or core and the vesicles can be co-administered with immune enhancers such as saponin derivatives, monophosphoryl lipid A and chitosan (Ferro 2011). These are able to potentiate strong immune responses after mucosal vaccination (Figueiredo et al. 2012). Intranasal immunisation of mice using such formulations successfully elicits mucosal, humoral and cellular immune responses, including Th1 and Th2 mediated immune responses and high IgG2a antibody titres. Viral envelope proteins or bacterial membrane lipids can be integrated into the vesicles to form virosomes or archaeosomes (Felnerova et al. 2004). An emerging field in lipid-based platforms is the development of synthetic exosome-like particles (Seow and Wood 2009). This ability to provide a delivery system that induces strong mucosal immune responses, gives liposome-based vaccines the exciting possibility of inducing effective immune responses in the reproductive tract (the male and female reproductive tract is part of the mucosal immune system). This could facilitate the development of novel oral and nasal delivered contraceptive vaccines that target sperm function and fertilisation. To date, however, the current liposome technology has not been used for immunocontraception (Webster et al. 2013). Earlier liposomal delivery systems have been applied to cat contraception; these used ZP antigens isolated from five mammalian species to investigate immunocontraceptive activity in the cat (*Feliscatus*). Vaccines were constructed using ZP from the five species encapsulated in liposomes suspended in saline and emulsified with Freund's complete adjuvant (SpayVac™). This method showed variation in immunogenicity amongst the cats vaccinated. Also, although an antibody response was initiated in the cats, fertility was not suppressed (Levy et al. 2005). It is possible that the more recent lipid-based vaccine delivery systems may be more effective and robust for immunocontraceptive applications, or more effective in different species.

5.3.2 Micro-organism Based Particles

A non-living bacterial ghost technology is being trialled in New Zealand that targets both systemic and mucosal immune systems (Walcher et al. 2008). Bacterial ghosts are cell envelopes that contain a bacteriophage ϕφX174 lysis gene *E* that enables cytoplasmic contents to be expelled. The resulting ghosts are still antigenic but contain minimal DNA. Foreign proteins, such as ZP proteins can also be expressed on or within the cell envelope of the ghost. Similarly, virus-like particles (VLP) are made up of the structural proteins of a virus, without containing nucleic acids. GnRH or ZP antigens can be incorporated into or attached to the VLP (Choudhury et al. 2009). However, there are loading capacity limitations based on the size of the particles (Seow and Wood 2009).

5.3.3 Inorganic Particles

One way around the challenge of a single-dose vaccine is to use slow release systems based on inorganic substances, such as metals (gold or silver), metal oxides (e.g. iron oxide) carbon, silica, dendrimers, and organic–inorganic hybrids (Sekhon and Kamboj 2010). In this case, antigen can be conjugated or entrapped into structures formed from these materials, enabling controlled release of antigen over a period of time. These are different from the hormonal drug contraceptive implants (currently available) that consist of a plastic rod that needs to be removed after a certain period and are more often in the nanoparticle size range. The inorganic particles can be manufactured in different sizes in order to target specific immune cells i.e. small particles can be trafficked via dendritic cells to induce antibody as well as cell mediated immunity (Scheerlinck and Greenwood 2006; Mann et al. 2009). The most likely administration route for these will be via an implant (for example using a dart) although delivery via bait would be a viable option if excretion of the particles could be slowed down sufficiently for uptake across the gastrointestinal tract to occur. One prohibitive factor for mainstream development of inorganic particle technology would be cost, particularly if using silver and gold nanoparticles.

6 Summary

The need for control of animal populations in various scenarios is an on-going problem and is likely to increase as the human population expands. Current lethal methods for population control such as poisoning, trapping, shooting and the deliberate introduction of disease (e.g. myxomatosis) are often unacceptable and animal welfare is now a key consideration for control programmes. In addition, the complete removal of a species from its native environment is also not a solution. The control of population size is the key for a balanced and healthy ecosystem. As such, many new methods being investigated for population control are now including animal welfare in the success criteria. However, often the animal welfare implications of new population control technologies must be balanced against the existing inhumane lethal methods used, particularly for cost and practicability. This has helped drive research into new methods towards a more objective selection of the most effective and humane approaches, that are also practical and cost effective and also take into account animal social structures, especially in relation to conservation.

References

Alving CR, Koulchin V, Glenn GM, Rao M. Liposomes as carriers of peptide antigens – induction of antibodies and cytotoxic T-lymphocytes to conjugated and unconjugated peptides. Immunol Rev. 1995;145:5–31.

Arlt SP, Spankowsky S, Heuwieser W. Follicular cysts and prolonged oestrus in a female dog after administration of a deslorelin implant. N Z Vet J. 2011;59:87–91.

Asa C, Porton I, Baker A, Plotka E. Contraception as a management tool for controlling surplus animals. In: Kleiman M, Allen M, Thompson K, Lumpkin S, editors. Wild mammals in captivity. Chicago, IL: University of Chicago Press; 1996.

Ballesteros C, Carrasco-Garcia R, Vicente J, Carrasco J, Lasagna A, de la Fuente J, et al. Selective piglet feeders improve age-related bait specificity and uptake rate in overabundant Eurasian wild boar populations. Wildl Res. 2009a;36:203–12.

Ballesteros C, Garrido JM, Vicente J, Romero B, Galindo RC, Minguijon E, et al. First data on Eurasian wild boar response to oral immunization with BCG and challenge with a Mycobacterium bovis field strain. Vaccine. 2009b;27:6662–8.

Ballesteros C, Gortazar C, Canales M, Vicente J, Lasagna A, Gamarra JA, et al. Evaluation of baits for oral vaccination of European wild boar piglets. Res Vet Sci. 2009c;86:388–93.

Ballesteros C, Vicente J, Morriss G, Jockney I, Rodriguez O, Gortazar C, et al. Acceptance and palatability for domestic and wildlife hosts of baits designed to deliver a tuberculosis vaccine to wild boar piglets. Prev Vet Med. 2011;98:198–203.

Baskran N, Desai AA. Ranging behavior of the Asian elephant (*Elephas maximus*) in the Nilgiri biosphere reserve, South India. Gajah. 1996;15:41–57.

Bergman D, Bender S, Wenning K, Slate D, Rupprecht C, Heuser C, et al. Bait acceptability for delivery of oral rabies vaccine to free-ranging dogs on the Navajo and Hopi Nations. Dev Biol. 2008;131:145–50.

Bertschinger HJ, Trigg TE, Jochle W, Human A. Induction of contraception in some African wild carnivores by downregulation of LH and FSH secretion using the GnRH analogue deslorelin. Reproduction. 2002;41–52.

Bertschinger HJ, Jago M, Nothling JO, Human A. Repeated use of the GnRH analogue deslorelin to down-regulate reproduction in male cheetahs (Acinonyx jubatus). Theriogenology. 2006;66:1762–7.

Bertschinger, H. J., A. Delsink, J.J. van Altena, D. Grobler, J.F. Kirkpatrick, M. Bates & Burke, T. Effects of porcine zona pellucida immunocontraception on annual calving percentages of six discrete African Elephant populations. 6th International Conference on Fertility Control for Wildlife. York, UK. 2007.

Blackburn TM, Cassey P, Duncan RP, Evans KL, Gaston KJ. Avian extinction and mammalian introductions on oceanic islands. Science. 2004;305:1955–8.

Buddle BM, Wedlock DN, Denis M, Vordermeier HM, Hewinson RG. Update on vaccination of cattle and wildlife populations against tuberculosis. Vet Microbiol. 2011;151:14–22.

Budke CM, Slater MR. Utilization of matrix population models to assess a 3-year single treatment nonsurgical contraception program versus surgical sterilization in feral cat populations. J Appl Anim Welf Sci. 2009;12:277–92.

Chen Z, He W, Wu Y, Yan P, He H, Zhang J, et al. A highly specific antibody response after protein prime-peptide boost immunization with Eppin/B-cell epitope in mice. Hum Vaccines. 2011;7:849–55.

Choudhury S, Kakkar V, Suman P, Chakrabarti K, Vrati S, Gupta SK. Immunogenicity of zona pellucida glycoprotein-3 and spermatozoa YLP12 peptides presented on Johnson grass mosaic virus-like particles. Vaccine. 2009;27:2948–53.

Corner LA, Buddle BM, Pfeiffer DU, Morris RS. Aerosol vaccination of the brushtail possum (Trichosurus vulpecula) with bacille Calmette-Guerin: the duration of protection. Vet Microbiol. 2001;81:181–91.

Corner LA, Costello E, Lesellier S, O'Meara D, Gormley E. Vaccination of European badgers (Meles meles) with BCG by the subcutaneous and mucosal routes induces protective immunity against endobronchial challenge with Mycobacterium bovis. Tuberculosis (Edinb). 2008;88:601–9.

Corner LAL, Costello E, O'Meara D, Lesellier S, Aldwell FE, Singh M, et al. Oral vaccination of badgers (Meles meles) with BCG and protective immunity against endobronchial challenge with Mycobacterium bovis. Vaccine. 2010;28:6265–72.

Cross ML, Henderson RJ, Lambeth MR, Buddle BM, Aldwell FE. Lipid-formulated bcg as an oral-bait vaccine for tuberculosis: vaccine stability, efficacy, and palatability to brushtail possums (Trichosurus vulpecula) in New Zealand. J Wildl Dis. 2009;45:754–65.

Curtis PD, Pooler RL, Richmond ME, Miller LA, Mattfeld GF, Quimby FW. Comparative effects of GnRH and porcine zona pellucida (PZP) immunocontraceptive vaccines for controlling reproduction in white-tailed deer (Odocoileus virginianus). Reproduction. 2002;131–141.

Dell'Omo G, Palmery M. Fertility control in vertebrate pest species. Contraception. 2002;65:273–5.

Delsink AK, van Altena JJ, Kirkpatrick J, Grobler D, Fayrer-Hosken RA. Field applications of immunocontraception in African elephants (Loxodonta africana). Reproduction. 2002; 117–124.

Delsink AK, van Altena JJ, Grobler D, Bertschinger HJ, Kirkpatrick JF, Slotow R. Implementing immunocontraception in free-ranging African elephants at makalali conservancy. J S Afr Vet Assoc. 2007;78:25–30.

Delves PJ, Roitt IM. Vaccines for the control of reproduction—status in mammals, and aspects of comparative interest. Dev Biol. 2005;121:265–73.

Druce HC, Mackey RL, Slotow R. How Immunocontraception Can Contribute to Elephant Management in Small. Enclosed Reserves: Munyawana Population as a Case Study. Plos One; 2011. p. 6.

Dubna S, Langrova I, Jankovska I, Vadlejch J, Pekar S, Napravnik J, et al. Contamination of soil with Toxocara eggs in urban (Prague) and rural areas in the Czech Republic. Vet Parasitol. 2007;144:81–6.

Edelman A, Jensen JT, Bulechowsky M, Cameron J. Combined oral contraceptives and body weight: do oral contraceptives cause weight gain? A primate model. Hum Reprod. 2011;26: 330–6.

Ellerman DA, Cohen DJ, Weigel Munoz M, Da Ros VG, Ernesto JI, Tollner TL, et al. Immunologic behavior of human cysteine-rich secretory protein 1 (hCRISP1) in primates: prospects for immunocontraception. Fertil Steril. 2010;93:2551–6.

Farquhar VJ, McCue PM, Carnevale EM, Nett TM, Squires EL. Deslorelin acetate (Ovuplant) therapy in cycling mares: effect of implant removal on FSH secretion and ovarian function. Equine Vet J. 2002;34:417–20.

Fayrer-Hosken RA, Bertschinger HJ, Kirkpatrick JF, Grobler D, Lamberski N, Honneyman G, et al. Contraceptive potential of the porcine zona pellucida vaccine in the African elephant (Loxodonta africana). Theriogenology. 1999;52:835–46.

Felnerova D, Viret JF, Gluck R, Moser C. Liposomes and virosomes as delivery systems for antigens, nucleic acids and drugs. Curr Opin Biotechnol. 2004;15:518–29.

Fernando P. Managing 'problem elephants'. Loris [Online]. 2011;25. http://www.ccrsl.org/Downloads/Fernando-10-ProblemEles.pdf.

Fernando P, Wikramanayake ED, Janaka HK, Jayasinghe LKA, Gunawardena M, Kotagama SW, et al. Ranging behavior of the Asian elephant in Sri Lanka. MAMBIO Mamm Biol. 2008;73:2–13.

Fernando P, Pastorini J, Leimgruber P, Prasad T. Problem-elephant translocation: translocating the problem and the elephant? PLoS One. 2012;7:e50917.

Ferro VA. Adjuvants and vaccine delivery systems. *In:Novel approaches to vaccine research.* Hefferon, K. L. (ed.)^(eds). Kerala, India: Research Signpost; 2011.

Figueiredo L, Cadete A, Goncalves LMD, Corvo ML, Almeida AJ. Intranasal immunisation of mice against Streptococcus equi using positively charged nanoparticulate carrier systems. Vaccine. 2012;30:6551–8.

Fry T, Van Dalen K, Hurley J, Nash P. Mucosal adjuvants to improve wildlife rabies vaccination. J Wildl Dis. 2012;48:1042–6.

Goldberg E, Herr J. LDH-C4 as a contraceptive vaccine. In: Gupta S, editor. *Reproductive immunology*. Amsterdam: Springer; 2000.

Gormley E, Corner LAL. Control Strategies for wildlife tuberculosis in Ireland. Transbound Emerg Dis. 2013;60:128–35.

Gu W, Holland M, Janssens P, Seamark R, Kerr P. Immune response in rabbit ovaries following infection of a recombinant myxoma virus expressing rabbit zona pellucida protein B. Virology. 2004;318:516–23.

Hardy CM. Current status of virally vectored immunocontraception for biological control of mice. Soc Reprod Fertil Suppl. 2007;63:495–506.

Hardy CM, Hinds LA, Kerr PJ, Lloyd ML, Redwood AJ, Shellam GR, et al. Biological control of vertebrate pests using virally vectored immunocontraception. J Reprod Immunol. 2006;71:102–11.

Herbert CA, Trigg TE. Applications of GnRH in the control and management of fertility in female animals. Anim Reprod Sci. 2005;88:141–53.

Herbert CA, Trigg TE, Renfree MB, Shaw G, Eckery DC, Cooper DW. Effects of a gonadotropin-releasing hormone agonist implant on reproduction in a male marsupial, Macropus eugenii. Biol Reprod. 2004;70:1836–42.

Herbert CA, Trigg TE, Renfree MB, Shaw G, Eckery DC, Cooper DW. Long-term effects of deslorelin implants on reproduction in the female tammar wallaby (Macropus eugenii). Reproduction. 2005;129:361–9.

Herbert CA, Eckery DC, Trigg TE, Cooper DW. Chronic treatment of male tammar wallabies with deslorelin implants during pouch life: effects on development, puberty, and reproduction in adulthood. Biol Reprod. 2007;76:1054–61.

Herr JC, Flickinger CJ, Homyk M, Klotz K, John E. Biochemical and morphological characterization of the intraacrosomal antigen Sp-10 from human sperm. Biol Reprod. 1990;42:181–93.

Holland MK, Andrews J, Clarke H, Walton C, Hinds LA. Selection of antigens for use in a virus-vectored immunocontraceptive vaccine: PH-20 as a case study. Reprod Fertil Dev. 1997;9:117–24.

Holland OJ, Cowan PE, Gleeson DM, Duckworth JA, Chamley LW. MHC haplotypes and response to immunocontraceptive vaccines in the brushtail possum. J Reprod Immunol. 2009;82:57–65.

Hurley JC, Johnston JJ. Poly(methyl methacrylate) synthetic grit formulations sustain the delivery of nicarbazin, a contraceptive agent, in pest waterfowl. J Contr Release. 2002;85:135–43.

Hynes EF, Nave CD, Shaw G, Renfree MB. Effects of levonorgestrel on ovulation and oestrous behaviour in the female tammar wallaby. Reprod Fertil Dev. 2007;19:335–40.

Hynes EF, Handasyde KA, Shaw G, Renfree MB. Levonorgestrel, not etonogestrel, provides contraception in free-ranging koalas. Reprod Fertil Dev. 2010;22:913–9.
Janne OA, Zook BC, Didolkar AK, Sundaram K, Nash HA. The roles of estrogen and progestin in producing deciduosarcoma and other lesions in the rabbit. Toxicol Pathol. 2001;29:417–21.
Jessup DA. The welfare of feral cats and wildlife. JAVMA J Am Vet Med Assoc. 2004;225:1377–83.
Johnson CA, Thompson DL, Cartmill JA. Effects of deslorelin acetate implants in horses: single implants in stallions and steroid-treated geldings and multiple implants in mares. J Anim Sci. 2003;81:1300–7.
Junaidi A, Williamson PE, Cummins JM, Martin GB, Blackberry MA, Trigg TE. Use of a new drug delivery formulation of the gonadotrophin-releasing hormone analogue Deslorelin for reversible long-term contraception in male dogs. Reprod Fertil Dev. 2003;15:317–22.
Khan SA, Jadhav SV, Suryawanshi AR, Bhonde GS, Gajbhiye RK, Khole VV. Evaluation of contraceptive potential of a novel epididymal sperm protein SFP2 in a mouse model. Am J Reprod Immunol. 2011;66:185–98.
Killian G, Kreeger TJ, Rhyan J, Fagerstone K, Miller L. Observations on the use of GonaCon in captive female elk (Cervus elaphus). J Wildl Dis. 2009;45:184–8.
Kirkpatrick JF, Turner A. Achieving population goals in a long-lived wildlife species (Equus caballus) with contraception. Wildl Res. 2008;35:513–9.
Kirkpatrick JF, Turner JW, Liu IKM, FayrerHosken R, Rutberg AT. Case studies in wildlife immunocontraception: wild and feral equids and white-tailed deer. Reprod Fertil Dev. 1997;9:105–10.
Kirkpatrick JF, Lyda RO, Frank KM. Contraceptive vaccines for wildlife: a review. Am J Reprod Immunol. 2011;66:40–50.
Knobel DL, Cleaveland S, Coleman PG, Fevre EM, Meltzer MI, Miranda MEG, et al. Re-evaluating the burden of rabies in Africa and Asia. Bull World Health Organ. 2005;83:360–8.
Kutzler M, Wood A. Non-surgical methods of contraception and sterilization. Theriogenology. 2006;66:514–25.
Ladd A, Tsong YY, Walfield AM, Thau R. Development of an antifertility vaccine for pets based on active immunization against luteinizing-hormone-releasing hormone. Biol Reprod. 1994;51:1076–83.
Lea IA, van Lierop MJC, Widgren EE, Grootenhuis A, Wen Y, van Duin M, et al. A chimeric sperm peptide induces antibodies and strain-specific reversible infertility in mice. Biol Reprod. 1998;59:527–36.
Levy JK. Contraceptive vaccines for the humane control of community cat populations. Am J Reprod Immunol. 2011;66:63–70.
Levy JK, Crawford PC. Humane strategies for controlling feral cat populations. JAVMA J Am Vet Med Assoc. 2004;225:1354–60.
Levy JK, Mansour M, Crawford PC, Pohajdak B, Brown RG. Survey of zona pellucida antigens for immunocontraception of cats. Theriogenology. 2005;63:1334–41.
Li H, Ding X, Guo C, Guan H, Xiong C. Immunization of male mice with B-cell epitopes in transmembrane domains of CatSper1 inhibits fertility. Fertil Steril. 2012;97:445–52.
Lincoln G, Anderson R, Anderson H. Progestagen depot with or without testosterone: a strategy for regulating fertility in male elephants. Expert consultation on the control of wild elephant populations. Utrecht: Office for International Co-operation, Faculty of Veterinary Medicine, Utrecht University, The Netherlands; 2004.
Locke SL, Cook MW, Harveson LA, Davis DS, Lopez RR, Silvy NJ, et al. Effectiveness of Spayvac (R) for reducing white-tailed deer fertility. J Wildl Dis. 2007a;43:726–30.
Locke SL, Cook MW, Harveson LA, Davis DS, Lopez RR, Silvy NJ, et al. Effectiveness of Spayvac for reducing white-tailed deer fertility. J Wildl Dis. 2007b;43:726–30.
Mann JF, Shakir E, Carter KC, Mullen AB, Alexander J, Ferro VA. Lipid vesicle size of an oral influenza vaccine delivery vehicle influences the Th1/Th2 bias in the immune response and protection against infection. Vaccine. 2009;27:3643–9.

Marks CA. Bait-delivered cabergoline for the reproductive control of the red fox (Vulpes vulpes): estimating mammalian non-target risk in south-eastern Australia. Reprod Fertil Dev. 2001;13:499–510.

Massei G, Jones A, Platt T, Cowan DP. Iophenoxic acid as a long-term marker for wild boar. J Wildl Manage. 2009;73:458–61.

Mayle BA, Broome AC. Changes in the impact and control of an invasive alien: the grey squirrel (Sciurus carolinensis) in Great Britain, as determined from regional surveys. Pest Manag Sci. 2013;69:323–33.

Menkhorst EM, Cui S, Selwood L. Novel immunocontraceptive targets in mammals: uterine secretions and the conceptus; a marsupial approach. Reproduction. 2008;136:471–80.

Miller LA, Gionfriddo JP, Fagerstone KA, Rhyan JC, Killian GJ. The single-shot GnRH immunocontraceptive vaccine (GonaCon) in white-tailed deer: comparison of several GnRH preparations. Am J Reprod Immunol. 2008;60:214–23.

Monier L. Metabolic risks of oral contraception. Rev Int Pediatr. 1988;178:64–5.

Moore HDM, Jenkins NM, Wong C. Immunocontraception in rodents: a review of the development of a sperm-based immunocontraceptive vaccine for the grey squirrel (Sciurus carolinensis). Reprod Fertil Dev. 1997;9:125–9.

Moudgal NR, Ravindranath N, Murthy GS, Dighe RR, Aravindan GR, Martin F. Long-term contraceptive efficacy of vaccine of ovine follicle-stimulating hormone in male bonnet monkeys (Macaca radiata). J Reprod Fertil. 1992;96:91–102.

Mulatti P, Ferre N, Patregnani T, Bonfanti L, Marangon S. Geographical information systems in the management of the 2009-2010 emergency oral anti-rabies vaccination of foxes in north-eastern Italy. Geospat Health. 2011;5:217–26.

Munks MW. Progress in development of immunocontraceptive vaccines for permanent non-surgical sterilization of cats and dogs. Reprod Domest Anim. 2012;47:223–7.

Nation A, Cui S, Selwood L. Vesicle-associated protein 1: a novel ovarian immunocontraceptive target in the common brushtail possum, Trichosurus vulpecula. Reproduction. 2008;136:657–65.

Nave CD, Shaw G, Short RV, Renfree MB. Contraceptive effects of levonorgestrel implants in a marsupial. Reprod Fertil Dev. 2000;12:81–6.

Nave CD, Coulson G, Short RV, Poiani A, Shaw G, Renfree MB. Long-term fertility control in the kangaroo and the wallaby using levonorgestrel implants. Reproduction. 2002;71–80.

Naz RK. Antisperm contraceptive vaccines: where we are and where we are going? Am J Reprod Immunol. 2011;66:5–12.

Naz RK, Alexander NJ, Isahakia M, Hamilton MS. Monoclonal-antibody to a human germ-cell membrane glycoprotein that inhibits fertilization. Science. 1984;225:342–4.

Nie GY, Butt AR, Salamonsen LA, Findlay JK. Hormonal and non-hormonal agents at implantation as targets for contraception. Reprod Fertil Dev. 1997;9:65–76.

Nikolovski S, Lloyd ML, Harvey N, Hardy CM, Shellam GR, Redwood AJ. Overcoming innate host resistance to vaccination: employing a genetically distinct strain of murine cytomegalovirus avoids vector-mediated resistance to virally vectored immunocontraception. Vaccine. 2009;27:5226–32.

Nol P, Palmer MV, Waters WR, Aldwell FE, Buddle BM, Triantis JM, et al. Efficacy of oral and parenteral routes of Mycobacterium bovis bacille Calmette-Guerin vaccination against experimental bovine tuberculosis in white-tailed deer (Odocoileus virginianus): a feasibility study. J Wildl Dis. 2008;44:247–59.

O'Hern PA, Bambra CS, Isahakia M, Goldberg E. Reversible contraception in female baboons immunized with a synthetic epitope of sperm-specific lactate dehydrogenase. Biol Reprod. 1995;52:331–9.

Olivera P, Menezes D, Trout R, Buckle A, Geraldes P, Jesus J. Successful eradication of the European rabbit (Oryctolagus cuniculus) and house mouse (Mus musculus) from the island of Selvagem Grande (Macaronesian archipelago), in the Eastern Atlantic. Integr Zool. 2010;5:70–83.

Patton ML, Bashaw MJ, del Castillo SM, Jochle W, Lamberski N, Rieches R, et al. Long-term suppression of fertility in female giraffe using the GnRH agonist deslorelin as a long-acting implant. Theriogenology. 2006;66:431–8.

Powers JG, Baker DL, Davis TL, Conner MM, Lothridge AH, Nett TM. Effects of gonadotropin-releasing hormone immunization on reproductive function and behavior in captive female rocky mountain Elk (Cervus elaphus nelsoni). Biol Reprod. 2011;85:1152–60.
Primakoff P, Lathrop W, Woolman L, Cowan A, Myles D. Fully effective contraception in male and female guinea-pigs immunized with the sperm protein Ph-20. Nature. 1988;335:543–6.
Pukazhenthi B, Comizzoli P, Travis AJ, Wildt DE. Applications of emerging technologies to the study and conservation of threatened and endangered species. Reprod Fertil Dev. 2006;18:77–90.
Rao KS, Maikhuri RK, Nautiyal S, Saxena KG. Crop damage and livestock depredation by wildlife: a case study from Nanda Devi Biosphere Reserve, India. J Environ Manage. 2002;66:317–27.
Rattner BA, Horak KE, Lazarus RS, Eisenreich KM, Meteyer CU, Volker SF, et al. Assessment of toxicity and potential risk of the anticoagulant rodenticide diphacinone using Eastern screech-owls (Megascops asio). Ecotoxicology. 2012;21:832–46.
Robbins SC, Jelinski MD, Stotish RL. Assessment of the immunological and biological efficacy of two different doses of a recombinant GnRH vaccine in domestic male and female cats (Felis catus). J Reprod Immunol. 2004;64:107–19.
Rocke TE, Smith SR, Stinchcomb DT, Osorio JE. Immunization of black-tailed prairie dog against plague through consumption of vaccine-laden baits. J Wildl Dis. 2008;44:930–7.
Salhab AS, Gharaibeh MN, Shomaf MS, Amro BI. Meloxicam inhibits rabbit ovulation. Contraception. 2001;63:329–33.
Santhanam R, Naz RK. Novel human testis-specific cDNA: molecular cloning, expression and immunobiological effects of the recombinant protein. Mol Reprod Dev. 2001;60:1–12.
Scheerlinck JP, Greenwood DL. Particulate delivery systems for animal vaccines. Methods. 2006;40:118–24.
Schneider F, Tomek W, Grundker C. Gonadotropin-releasing hormone (GnRH) and its natural analogues: a review. Theriogenology. 2006;66:691–709.
Sekhon BS, Kamboj SR. Inorganic nanomedicine-Part 2. Nanomed Nanotechnol Biol Med. 2010;6:612–8.
Seow Y, Wood MJ. Biological gene delivery vehicles: beyond viral vectors. Mol Ther. 2009;17:767–77.
Shackelford N, Renton M, Perring MP, Hobbs RJ. Modeling disturbance-based native invasive species control and its implications for management. Ecol Appl. 2013;23:1331–44.
Shankar S, Mohapatra B, Suri A. Cloning of a novel human testis mRNA specifically expressed in testicular haploid germ cells, having unique palindromic sequences and encoding a leucine zipper dimerization motif. Biochem Biophys Res Commun. 1998;243:561–5.
Shetty J, Wolkowicz MJ, Digilio LC, Klotz KL, Jayes FL, Diekman AB, et al. SAMP14, a novel, acrosomal membrane-associated, glycosylphosphatidylinositol-anchored member of the Ly-6/urokinase-type plasminogen activator receptor superfamily with a role in sperm-egg interaction. J Biol Chem. 2003;278:30506–15.
Shideler SE, Stoops MA, Gee NA, Howell JA, Lasley BL. Use of porcine zona pellucida (PZP) vaccine as a contraceptive agent in free-ranging Tule Elk (Cervus elaphus nannodes). Reproduction. 2002;60:169–76.
Simmons JG, Hamner CE. Inhibition of estrus in the dog with testosterone implants. Am J Vet Res. 1973;34:1409–19.
Srinath BR, Wickings EJ, Witting C, Nieschlag E. Active immunization with follicle-stimulating-hormone for fertility-control – a 4 1/2-year study in male rhesus-monkeys. Fertil Steril. 1983;40:110–7.
Suri A. Sperm specific proteins-potential candidate molecules for fertility control. Reprod Biol Endocrinol. 2004;2:10.
Telford 3rd SR, Cunningham JA, Waltari E, Hu L. Nest box-deployed bait for delivering oral vaccines to white-footed mice. Ticks Tick Borne Dis. 2011;2:151–5.
Tyndale-Biscoe CH. Virus-vectored immunocontraception of feral mammals. Reprod Fertil Dev. 1994;6:281–7.

Walcher P, Cui XL, Arrow JA, Scobie S, Molinia FC, Cowan PE, et al. Bacterial ghosts as a delivery system for zona pellucida-2 fertility control vaccines for brushtail possums (Trichosurus vulpecula). Vaccine. 2008;26:6832–8.

Wang DG, Huang TH, Xie QD, An G. Investigation of recombinant mouse sperm protein izumo as a potential immunocontraceptive antigen. Am J Reprod Immunol. 2008;59:225–34.

Wang M, Lv Z, Shi J, Hu Y, Xu C. Immunocontraceptive potential of the Ig-like domain of Izumo. Mol Reprod Dev. 2009;76:794–801.

Webster DM, Sundaram P, Byrne ME. Injectable nanomaterials for drug delivery: carriers, targeting moieties, and therapeutics. Eur J Pharm Biopharm [Online]. 2013. http://www.ncbi.nlm.nih.gov/pubmed/23313176. Accessed Jan 8 2013

Wheaton CJ, Savage A, Shukla A, Neiffer D, Qu W, Sun Y, et al. The use of long acting subcutaneous levonorgestrel (LNG) gel depot as an effective contraceptive option for cotton-top Tamarins (Saguinus oedipus). Zoo Biol. 2011;30:498–522.

Wickings EJ, Nieschlag E. Suppression of spermatogenesis over two years in rhesus monkeys actively immunized with follicle-stimulating hormone. Fertil Steril. 1980;34:269–74.

Wu X, Hu R, Zhang Y, Dong G, Rupprecht CE. Reemerging rabies and lack of systemic surveillance in People's Republic of China. Emerg Infect Dis. 2009;15:1159–64.

Yang LH, Li JT, Yan P, Liu HL, Zeng SY, Wu YZ, et al. Follicle-stimulating hormone receptor (FSHR)-derived peptide vaccine induced infertility in mice without pathological effect on reproductive organs. Reprod Fertil Dev. 2011;23:544–50.

Zaman M, Chandrudu S, Toth I. Strategies for intranasal delivery of vaccines. Drug Deliv Transl Res. 2013;3:100–9.

Zhang LH, Tian XS, Zhou FZ. Intranasal administration of CpG oligonucleotides induces mucosal and systemic Type 1 immune responses and adjuvant activity to porcine reproductive and respiratory syndrome killed virus vaccine in piglets in vivo. Int Immunopharmacol. 2007;7:1732–40.

Zook BC, Janne OA, Abraham AA, Nash HA. The development and regression of deciduosarcomas and other lesions caused by estrogens and progestins in rabbits. Toxicol Pathol. 2001;29:411–6.

Chapter 19
Cloning the Mammoth: A Complicated Task or Just a Dream?

Pasqualino Loi, Joseph Saragusty, and Grazyna Ptak

Abstract Recently there has been growing interest in applying the most advanced embryological tools, particularly cloning, to bring extinct species back to life, with a particular focus on the woolly mammoth (*Mammuthus primigenius*). Mammoth's bodies found in the permafrost are relatively well preserved, with identifiable nuclei in their tissues. The purpose of this chapter is to review the literature published on the topic, and to present the strategies potentially suitable for a mammoth cloning project, with a frank assessment of their feasibility and the ethical issues involved.

Keywords Somatic cell nuclear transfer (SCNT) • Cloning • Mammoth • Elephant

1 Introduction

Surprisingly, writers and moviemakers have anticipated some of the most extraordinary scientific breakthroughs. In the movie *Sleeper*, directed by Woody Allen in 1973, the protagonists were asked to clone a dictator, killed by a bomb, using a fragment of his nose. Twenty-four years later the transfer of a somatic cell into an enucleated oocyte cloned the first mammal, a sheep named Dolly (Wilmut et al. 1997). In *Jurassic Park*, a 1993 movie by Steven Spielberg based on the 1990 book written by Michael Crichton, DNA extracted from fossil amber was used to generate

P. Loi, Ph.D. (✉) • G. Ptak, Ph.D.
Department of Comparative Biomedical Sciences, University of Teramo,
Piazza Aldo Moro 45, Teramo 64100, Italy
e-mail: ploi@unite.it

J. Saragusty, D.V.M., Ph.D. (✉)
Department of Reproduction Management, Leibniz Institute for Zoo and Wildlife Research,
Alfred-Kowalke-Straße 17, Berlin 10315, Germany
e-mail: saragusty@izw-berlin.de

W.V. Holt et al. (eds.), *Reproductive Sciences in Animal Conservation*, Advances in Experimental Medicine and Biology 753, DOI 10.1007/978-1-4939-0820-2_19,

a large proportion of the fauna of that era. This movie was made over 19 years ago, and it still remains science fiction, but for how long?

Bringing back to life extinct species appears to be a common wish among humans. When we step in front of a well-preserved fossil or stuffed specimen, instinctively we start imagining what the living animal looked like.

Cloning by somatic cells nuclear transfer (SCNT) indeed offers the possibility to materialize such a dream. An essential requirement for cloning is the availability of soft, or otherwise well-preserved tissue with identifiable nuclei. Hence, fossil skulls or skeletons are not the ideal material to start with, but frozen mammoth bodies found in the permafrost fulfil the minimum requirements for SCNT. In fact, every time a mammoth body is found, a timely exercise from the mass media is to speculate about cloning it. At the beginning, these forums were confined to the everyday people, but recently even developmental biologists have started considering bringing mammoth back to life through cloning. Whether projects to clone a mammoth have genuine scientific basis, or reflect commercial enterprises we do not know.

In this chapter our effort will be to critically analyse the few published reports on cloning the mammoth, then we propose what might be, in our view at least, a realistic approach to clone a mammoth, using the scientific knowledge currently available. We conclude with a few thoughts on ethical issues involved in such a project.

2 Cloning the Mammoth: What Has Been Done?

Leaving aside abstracts or poster communications in international meetings, only one ISI publication is available on mammoth cloning (Kato et al. 2009); hence, review of the state-of-the-art is a very easy task. So far only one group, led by Akira Iritani, a Japanese scientist, is officially engaged in a mammoth cloning project, but according to some press releases, a second group of South Korean and Russian scientists is also competing for the task (The Telegraph, UK; 13 March 2012).

In the published data available, Kato et al. (2009) used somatic cells from a 15,000 years old mammoth calf. The source of cells was a leg, from which epithelial and muscle cells were collected. The first surprise was that the tissues had actually maintained their structure over the years, as shown in haematoxylin-eosin stained histological specimen (Fig. 19.1). Nuclei were isolated from these two tissues (Fig. 19.2). The details were not revealed in the paper, but we presume that nuclei were mechanically dissected by micromanipulation from histological sections not mounted on resin and slides. No indications were given on how many nuclei were harvested, however, in our opinion, that must have been a very time consuming and cumbersome effort. The nuclei isolated from muscle and skin tissues were injected into enucleated mouse oocytes, which were activated and monitored throughout the first cell cycle. The two mammoth somatic nuclei sources, muscle and epithelial cells, were not modified by the oocytes, whereas the control oocytes injected with nuclei from frozen mouse bone marrow cells formed well-shaped pronuclei (Fig. 19.3). The message that Kato et al. (2009) paper conveys is very important because single nuclei are identifiable and retrievable from 15,000 years old mammoth sample.

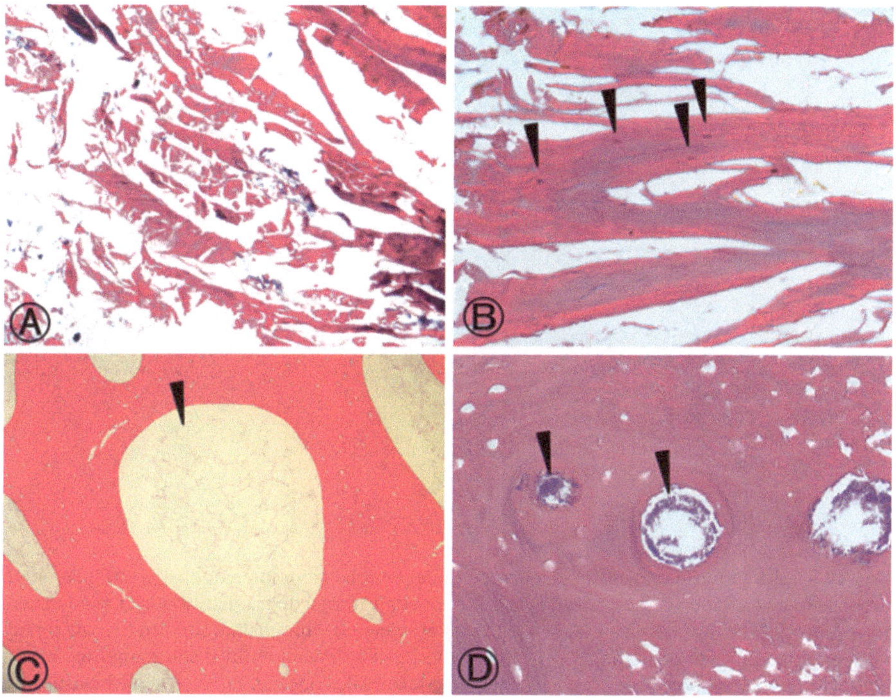

Fig. 19.1 Thin sections of mammoth tissues: (**a**) skin (×100), (**b**) muscle (×400), (**c**, **d**) bone and bone marrow (**c** ×100 and **d** ×400) samples stained by hematoxylin-eosin double staining method. There were many cell nuclei in the muscle (Fig. 19.1b). In the medullary cavity of the bone marrow there were many foam shaped structures (Fig. 19.1c) and blood cells or epithelial cells in the bone (Fig. 19.1d) (Reproduced with permission from Kato et al. 2009)

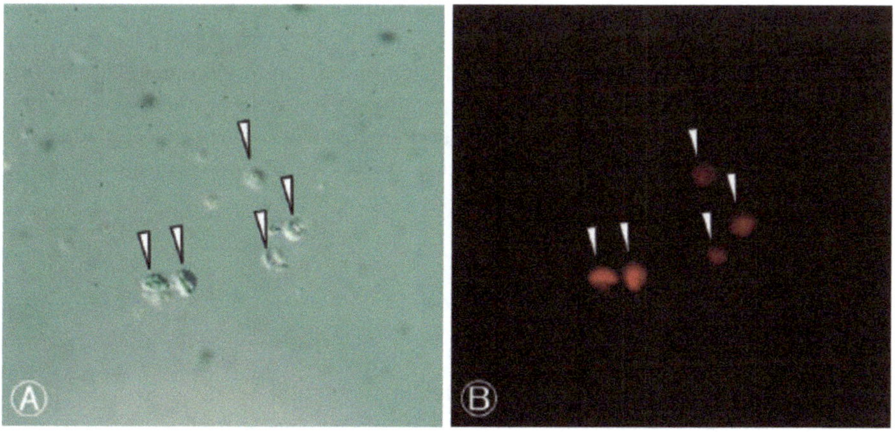

Fig. 19.2 Nuclei isolated from mammoth tissue. A - bright field; B, P.I. fluorescence. From Kato et al., 2009, with permission

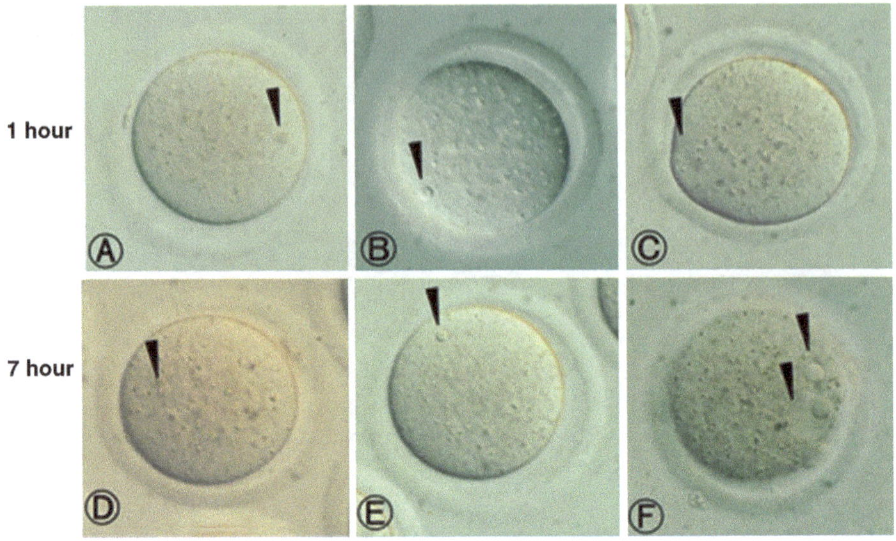

Fig. 19.3 (**a–c**) Mouse oocytes injected with nuclei derived from mammoth skin (**a**), mammoth muscle (**b**) and mouse bone marrow (**c**) at 1 h after nuclear injection. Injected nuclei were visible (*arrows*). (**d–f**) Mouse oocytes injected with nucleus derived from mammoth skin (**d**), mammoth muscle (**e**) and mouse bone marrow (**f**) at 7 h after nuclear injection. In **d** and **e**, injected nuclei without any change were still visible (*arrows*). Meanwhile, oocytes injected with mouse bone marrow derived nucleus transformed into 2 pronuclear-like structure (*arrows*) (Reproduced with permission from Kato et al. 2009)

This is a remarkable finding indeed. The lack of nuclear remodelling detected could be attributable to the inaccessibility of oocyte DNA remodelling factors to the mammoth's nuclei caused by structural modification during storage.

This is the present state-of-the-art of mammoth cloning, impressive but not really sound. The Japanese group dealing with the mammoth cloning project is not an off-the-cuff team. This group has a robust reputation in the field of embryo manipulation and is one of the leading Japanese laboratories working on SCNT; hence the project must have some chances for success, as will be discussed in the following pages.

2.1 Cloning the Mammoth

There are two possible strategies to resurrect a mammoth:

1. *"Synthetic" genome assembly AND nuclear transfer.*
2. *"Canonical" Interspecies Somatic Cell Nuclear Transfer (ISCNT).*

2.1.1 "Synthetic" Genome Assembly AND Nuclear Transfer

The "*Synthetic*" approach relies on the recent annotation of the mammoth genomic DNA (Miller et al. 2008). Since the application of next generation DNA sequencing technologies (Metzker 2010) to ancient DNA (aDNA), the entire sequences of the mammoth genomic (Miller et al. 2008) and mitochondrial DNA (Gilbert et al. 2008) DNA have become available. Hence, we have the DNA recipe on which a new mammoth genome can be *ex novo* synthesized. The entire procedure has been brilliantly described by Henry Nicholls in a special Nature issue celebrating Darwin's centenary (Nicholls 2008).

The first step is to synthesize the mammoth genome; secondly, the genome should be allocated into individual chromosomes, using the elephant genome and karyotype as a guide. This in itself is a formidable task given that elephants have 56 chromosomes. Of course the chromosomes must be canonically organized, with telomeres, centrosomes, and all vital sequences required for DNA replication and accurate segregation in mitosis; by all means an overwhelming task.

Let's imagine we have accomplished the task of arranging the DNA into the respective chromosomes. There is a crucial organelle that is still missing: the centrosome. The centrosome is essential in cell division, therefore a centrosome, presumably obtained from elephant tissues, has to be somehow associated with the chromosome set. Obtaining purified centrosomes with subcellular fractioning through gradient centrifugation is an established procedure (Moritz and Alberts 1998), but the trouble would be to package and hold together the chromosomes and the centrosome in the mammoth-unique structure.

Given that the structure should also allow the transfer of the chromosomes/centriole into an egg, a synthetic lipid monolayer appears to be the most appropriated solution. Artificial membranes have been a reality for many years, and the state-of-the-art is particularly advanced thanks to recent development in the production of artificial cells (Zagnoni 2012). The electro-mediated fusion of the artificial chromosomes along with the centriole into an enucleated elephant egg will finalize the procedure. If all goes well, we will have mammoth/elephant hybrid embryos, with elephant mtDNA, which offers a realistic possibility for development.

In theory, this is an interesting approach. Technical problems such copying errors during DNA synthesis, which could jeopardize the reading of the artificial genome by the oocyte's transcriptional machinery can occur and will need to be dealt with. An additional problem could be lack of communication between mammoth DNA and elephant mtDNA. The only remaining strategy left would be the generation of synthetic mammoth mtDNA, the sequence of which is available (Gilbert et al. 2008), and packaging it into artificial mitochondria. In the era of synthetic biology and artificial cells (Hammer and Kamat 2012), this might be technically feasible, but will certainly complicate the task. From a nuclear reprogramming point of view, paradoxically this might be an extraordinary possibility to improve the mammoth genome. We can actually confer upon the new DNA a structure compatibility using the reprogramming machinery of the oocyte, thus enabling complete reprogramming and hence normal development.

2.1.2 "Canonical" Interspecific Somatic Cells Nuclear Transfer (ISCNT)

In Interspecific Somatic Cell Nuclear Transfer (ISCNT) a nucleus taken from a target species is transplanted by electro-fusion or direct injection into an enucleated oocyte from a different species. ISCNT is a general term, often misused because in some reports nuclear transfer is accomplished between cell and oocyte donors belonging to different families, orders, or even classes (Loi et al. 2011).

SCNT and its variant ISCNT are potentially powerful tools for the production of unlimited numbers of offspring from a dead animal, in practice a real "asexual" reproduction, which has a tremendous appeal for multiplying endangered animals. The trouble though is that the outcome of the procedure in terms of offspring born is unpredictable and commonly low, about 1–5 %. The reason for the limited efficiency is an incomplete "*nuclear reprogramming*" of the differentiated nucleus. Put simply, the oocyte is unable in most cases to erase completely the differentiation memory accumulated in the form of epigenetic changes on the genome during development. Many excellent reviews are available on the topics, to which the reader is recommended (Gurdon and Wilmut 2011; Maruotti et al. 2010; Loi et al. 2008); in this chapter we would like to address exclusively the technical and biological aspects concerned with the cloning of a mammoth through a "canonical" ISCNT.

Kato's paper (Kato et al. 2009) has demonstrated that nuclei can be isolated from mammoth tissue. The first step in our opinion would be to verify the state of preservation of mammoth DNA. It is likely that DNA will be mostly degraded after 15,000 years of permanence in the permafrost, probably worsened by cycles of thawing and freezing. The dynamics of DNA degradation over time is a constitutive branch of ancient DNA (aDNA), a relatively new scientific discipline (Hofreiter et al. 2001) launched by Svante Pääbo through his pioneer study on DNA extracted from Egyptian mummies (Pääbo 1985). The temperature and general conditions in permafrost are tolerated relatively well by the DNA, although its structural decay cannot be avoided. DNA degradation starts with deamination and depurination, followed by single and double strand breaks and deletion of large portion of the genome (Briggs et al. 2007). It is plausible that mammoth DNA, although stored in permafrost, has undergone the same kind of damage. Not all tissues however are equally exposed to damaging condition.

The best laboratory practice in aDNA for obtaining good quality DNA for sequencing studies consists of grinding compact bones and extracting DNA from the powder (Briggs et al. 2007). Bone mechanically protects cells and the DNA within from adverse conditions. Kato's paper (Kato et al. 2009) shows that nuclei might be identified and dissected from mammoth's tissues; hence we have a starting point for our cloning project. It is possible in fact that some, but not all, cells have a well-preserved nucleus so the first decision to be taken is how to assess genome preservation in isolated nuclei. In our opinion there are no better choices than an empirical approach using a biological assay: the transfer of the mammoth nuclei into enucleated oocytes.

The second decision is the source of oocytes. The long gone mammoth has living relatives, the elephants, particularly the Asian species; hence, our choice should be the use of Asian elephant oocytes as recipients of mammoth nuclei. This is indeed the best option, though not a problem-free route to success. To date there are no reports of oocyte collection in elephants. Attempts at superstimulation and oocyte recovery in other megaherbivores such as the rhinoceros have been reported (Hermes et al. 2009) so there is good reason to believe that the procedure, at least the oocyte retrieval part, in elephants is not insurmountable. The size of the elephant and the location of the ovaries make such a procedure technically challenging but not impossible. In elephants only one, and rarely two oocytes mature and ovulate at the end of each estrus cycle. With only three to four cycles per year (in the absence of pregnancy), the number of oocytes that each female elephant can "donate" for the mammoth cloning project is very limited (Hildebrandt et al. 2011). If thousands of oocytes will be needed, hundreds of elephants will have to be subjected to repeated oocyte collection procedures. Finding such a large number of female elephants for the procedures is probably going to be an impossible task. The entire world captive elephant population (Asian and African combined) in zoos and circuses is in the range of 1,500–2,000 animals of both sexes and of all age groups (Saragusty 2012). The alternative would be to seek captive elephants in range countries. Although the number of these throughout Southeast Asia is in the range of 15,000 animals, many are inaccessible or unsuitable (including males, immature females, and elephants in temples). Furthermore, by collecting oocytes from an elephant, we will prevent it from the possibility of becoming pregnant, thus putting the female at high risk of developing reproductive pathologies in the future (Hermes et al. 2004) and, by preventing animals from reproducing, putting the entire captive Asian elephant population at higher risk of extinction (Saragusty 2012). Once a procedure for ovarian superstimulation in elephants has been developed, this could be used in an attempt to somewhat reduce the number of elephants needed. The alternative approach would be to search for a different source of elephant oocytes. We think there are two leading options available for consideration.

In some regions of Africa, elephant populations have grown beyond the carrying capacity of the habitat in which they live. One of the measures employed by population managers is culling, often of whole family groups. When culling takes place, the ovaries can be retrieved and oocytes can be harvested. The drawbacks of this option are that, at any point in time, the vast majority of wild female elephants are not cycling either because they are pregnant or because of lactational anoestrus. The ovarian follicular reserve in African elephants is constituted almost entirely of early- and late-primary follicles (Stansfield et al. 2010) so *in vitro* culture and maturation protocols will need to be developed to bring the oocytes to a sufficiently mature stage to be used for the SCNT procedure. These *in vitro* techniques are not yet available. The other drawback of this option is that culling takes place only in Africa, so only African elephant oocytes may become available this way.

The alternative is to collect ovarian tissue slices from deceased Asian elephant cows and either maintain them *in vitro* or transplant them into host animals so that their circulation and hormones will support follicular growth *in vivo*. This has been attempted once when cryopreserved African elephant ovarian tissue slices were transplanted into nude mice (Gunasena et al. 1998). These mice supported the development of antral follicles but oocytes were generally of poor quality. As good quality oocytes will be needed for the SCNT procedures, improvements of the technique, or finding an alternative host that will give better support to follicular development, will be needed. For the number of oocytes required, a large number of immune-deficient animals will have to be maintained at very high costs. Once harvested, the good quality oocytes will still need to be matured so an *in vitro* maturation protocol will need to be developed. Despite all foreseen and unforeseen difficulties, we think that this approach stands a better chance of success.

Regardless of the approach taken, we think that it would be best to restrict the number of elephant oocytes needed to the very minimum, given the technical difficulties and the high costs involved in collecting and maintaining them. The probability of finding nuclei bearing intact DNA in a mammoth tissue is not very high, so a large number of oocytes, most probably in the order of thousands, will have to be injected with isolated nuclei to find the very few that might be reactivated and start development. As an alternative approach, we propose the use of easily available oocytes for the first round of cloning. A potential candidate for oocyte donation might be the bovine. The state-of-the-art of *in vitro* maturation and culture in cattle is the most advanced amongst all farm animals (Lonergan 2007), and a large number of oocytes can be conveniently collected from ovaries taken from slaughtered cows. Objections, however, can be raised against this option as it clashes with the established concept of ISCNT, namely genomic/mtDNA compatibility and the high probability of Zygotic Genome Activation (ZGA) failure in the "extreme" mammoth/bovine hybrid embryos (Loi et al. 2011).

3 Genomic/mtDNA Incompatibility in ISCNT

Mitochondrial DNA codes for proteins responsible for the production of cellular energy. Given the low fidelity of the mtDNA replication machinery, some of the mitochondrial crucial genes are secured in the nucleus where they are safely duplicated and expressed (Amarnath et al. 2011). Therefore, a coordinated mt/genomic DNA cross talk is necessary for normal embryo development. Mammoth/bovine hybrid cloned embryos have no chance of developing normally, but early cleavages might occur since energy production and mtDNA replication do not occur before the blastocyst stage (Thundathil et al. 2005).

Following our suggested approach, successfully cleaving cloned embryos will be used at the morula or even 4–8-cell stage, before the unavoidable ZGA failure, for a second round of cloning, but this time using elephant oocytes. The role of the first round of nuclear transfer is to probe a large number of mammoth nuclei, and select those with a genome capable of directing embryonic development. The second round of nuclear transfer will transfer the mammoth nuclei into elephant cytoplasm, where an appropriated mt/genomic DNA cross talk, as well as a successful activation of the mammoth "embryonic genome" will probably take place. Technically it is very easy. The blastomeres of the cloned embryo will be disaggregated and electro-fused into enucleated elephant oocytes, essentially, the standard nuclear transfer procedure for embryonic cells.

In our opinion, this is the best approach to the mammoth cloning project, assuming we, the scientists, are in charge. Experienced developmental biologists familiar with nuclear transfer would surely spot further advantages offered by the strategy we pro-

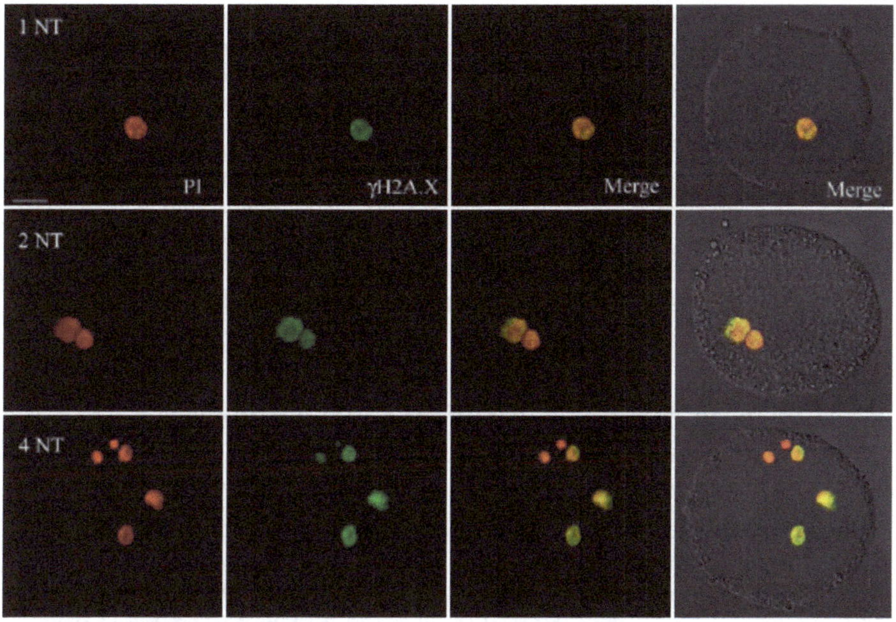

Fig. 19.4 Immuno-localization of histone variant gamma H2AX, which is recruited at sites of DNA repair, in lyophilized cells injected into enucleated sheep oocytes. The DNA repairing activity of the oocyte is not diluted even in case four somatic nuclei are injected (4 NT; the nucleus in the upper part of the oocyte has divided, an event which often occurs in SCNT), suggesting DNA repairing activity of the oocyte is highly redundant; hence it might turn out to be a mighty allied in a mammoth cloning project (Iuso et al. 2013)

pose. First, it has been known since the dawn of cloning that a serial nuclear transfer improves the efficiency of nuclear reprogramming. This is quite logical, if we consider that the nucleus is exposed twice to the oocyte's reprogramming machinery. In case the epigenetic memory of the differentiated cell is partially maintained, the second nuclear transfer will remove it, thus improving the nuclear reprogramming efficiency.

The second advantage comes from the DNA repairing capacity of the oocyte. It is indeed unrealistic to believe that mammoth nuclei will have intact DNA. Single and double strand breaks will probably be scattered throughout the genome. DNA repair is the latest of the oocyte properties revealed, and researchers are still debating its real implications (Ménézo et al. 2010). The redundancy of DNA repairing machinery is by all means unexpected and surprising. In our recent study (Iuso et al. 2013) on nuclear transfer of lyophilized cells, we addressed the issue of DNA lesions caused by the freeze-drying process and whether they are repaired after nuclear transfer. DNA damage was indeed observed in dried lymphocytes, but after nuclear transfer the resulting pronucleus was stuffed with foci of active DNA repair (identified with an antibody raised against modified histone recruited at sites of DNA repair (Podhorecka et al. 2010)). To our surprise, the signal was undiluted even when five somatic nuclei were injected into a single oocyte, indicating redundancy of DNA repairing enzymes in oocytes (Fig. 19.4). Therefore, a second round of nuclear transfer would have the additional advantage of more complete DNA repairing.

Hence, hybrid mammoth/elephant embryos will be allowed to develop to blastocyst stage using the culture conditions most appropriate for elephant embryo. What these culture conditions might be is still a big question mark since any of the *in vitro* techniques related to elephant oocytes and embryos have never been attempted, or at least was not reported on. In the absence of elephant embryo *in vitro* culture protocols, model animals will have to be relied upon to first develop the entire procedure in elephants. Once elephant *in vitro* embryo production and culture protocols have been developed, one can consider attempting to culture mammoth/elephant hybrid embryos. The production of blastocyst stage mammoth/elephant embryos would be already an incredible achievement, but still there is a long way to making the baby mammoth.

4 Transfer of Mammoth/Elephant Cloned Embryos into Elephant Foster Mothers

This might be another leap in the dark. We know that Asian (*Elephas maximus*) and African (*Loxodonta* sp.) elephants can interbreed. These two genera separated about 4.2–9.0 million years ago (mya). About 2.6–5.6 mya the African elephants split into the African savanna elephant (*L. africana*) and the African forest elephant (*L. cyclotis*). During this same time range the mammoth and the Asian elephant split into two genera—*Elephas* and *Mammuthus* (Rohland et al. 2010). When genomic DNA is compared, the Asian elephant is closer to the mammoth than it is to the African elephants. Some still consider the African savannah and forest elephants as being the same species whereas the mammoth and Asian elephant were assigned into different genera, and yet the ratio of genetic divergence of the two African elephants to the Asian elephant-mammoth is close to unity (Rohland et al. 2010). This genetic analysis suggest that it is highly plausible to assume that Asian elephants and mammoths could interbreed, thus increasing the probability of succeeding in transferring hybrid embryos into elephant foster mothers.

Like all other techniques relevant to the handling of oocytes and embryos in elephants, embryo transfer has never been attempted in this species. Assuming all other hurdles have been overcome—oocytes harvested, matured *in vitro*, and injected with mammoth nuclei to produce embryos that have been cultured successfully to the blastocyst stage, performing embryo transfer in elephants would pose primarily a technical problem. Artificial insemination in elephants has been in practice for about 16 years now (Hildebrandt et al. 1998; Hildebrandt et al. 1999). For this procedure, a flexible 3.0-m long customized video chip endoscope is used to place the insemination dose in the vagina, close to the cervical opening. A much longer endoscope will be required to attempt to go through the 15-cm long folded cervix, which in itself is going to be a formidable task, and into the uterus to transfer the embryo.

The alternative would be to conduct embryo transfer by laparoscopy. Elephants, however, do not have a pleural cavity (Brown et al. 1997), so inflation of the abdominal cavity will most likely lead to the collapse of the animal's lungs, resulting in its death. Laparoscopy will thus have to be done without inflating the abdominal cavity, a technique that is already in practice in human medicine (Paolucci et al. 1995). Due to the enormous weight and size of the elephant's abdominal wall, some of the tools will need to be modified but otherwise, it can be assumed that the technique can be applied to elephants as well.

Twins in captive elephants occur at a rate of about 1 % and experience indicate that when twins do occur, the mother and both fetuses are at high risk of perishing during parturition (Hermes et al. 2008). So, in order not to put the elephants at elevated risk of mortality, only a single embryo will most likely be transferred to each foster elephant cow, thus considerably decreasing the probability of success and increasing the number of animals needed for the project.

Another issue to consider with respect to embryo transfer is the sheer number of elephants that will be needed to conduct such a study. Animals will be needed to first develop oocyte collection and embryo transfer in elephants and then attempt to transfer the product of ISCNT—mammoth/elephant hybrid embryos. To achieve success, a large number of transfers to a large number of elephants will probably be needed before the first offspring will be produced, going through many failures along the way. Elephants are not laboratory mice and no place around the world maintains a large enough number of elephants for this purpose. This means that any such study will either have to rely on a small number of animals, conducting many repeated procedures on each, or recruiting participants from zoos and elephant camps around the world. Either way, the number of facilities that will let their elephants participate will probably not be large. To get the embryos to all these different participating locations, the embryos will need to be transported around the world. This will drastically increase the costs of the project and may compromise the embryos, thus decreasing the rate of success. But, with large enough number of attempts, and after surmounting all the hurdles on the way, pregnancies might be achieved and some might even be carried to term.

5 Some Ethical Questions Associated with Such a Prospective Mammoth Cloning Project

Resurrecting the mammoth is a very attractive and catchy topic but it also brings forth some ethical questions. Suppose this whole endeavour proves successful. With the currently low efficiency of SCNT, and even more so of ISCNT, we would be very fortunate if any research group were successful in producing a single specimen, dedicating much time and huge budgets for the task. Is it really justified to spend all this time and money on resurrecting a single specimen of an extinct species? Given the anticipated success rate, there would probably be many futile attempts to transfer embryos to surrogate mothers, some of which will become pregnant and carry them for different duration of time. The minority of these might carry the pregnancy through its full 22-months length only to deliver a stillborn, or a calf that will survive only hours or days. Is it really fair to all these surrogate mothers who will certainly bond with the developing foetus in their uterus?

Assuming the offspring survives, should we leave it with the surrogate mother to raise it? Should we let the family unit (the herd) interact with it? Will we, by doing so, be fair to the mother and herd? Do we really know the needs of mammoth when kept in captivity? After all, it has never been done before. Or will the calf be separated from its surrogate mother and herd and become an isolated research subject, undoubtedly a stressful (and unfair) handling of all animals involved? What will we do with this animal once it matures? Should we put it in a zoo as a tourist attraction? Will it become a reproduction machine—semen "donor" if it is a male or oocyte "donor" if it is a female? Or should it travel around the world to generate more funds? Or maybe we should release it in a place somewhat more suitable for it, such as Siberia, Greenland or Alaska? These are just some of the ethical questions that come to mind when considering the resurrection of the woolly mammoth. Considering all stakes involved, is this scientific endeavour justified?

6 Conclusions

From the two approaches described for cloning a mammoth, the "synthetic" one is, in our opinion, the most advanced approach likely to succeed, but impossible to be implemented at the moment. A newly synthesized genome, bright new and virtually devoid of damage/mutations, would be ideal, far better than the damaged nuclei found in mammoth tissues. Hence, we could transfer the artificial membrane-containing mammoth's DNA and the centriole directly into elephant oocytes. A further advantage is offered by the possibility to confer upon the naked, "naïve" DNA, an organization that is easily "readable" by the oocyte, thus resulting in an improved nuclear reprogramming, and in turn, development to term. Of course it must be granted that no further complications will arise, such as incompatibility between mammoth and elephant mt/genomic DNA. Likewise, we have to trust that mammoth and elephant are genetically close enough to allow ZGA activation.

This "ideal" approach, however, is still far from our grasp, so the only way to tackle the mammoth cloning project would be a "canonical" Interspecies Somatic Cells Nuclear Transfer approach, as we have described. This approach, we believe, has realistic chances of success. Whether such a project should get under way, knowing the enormous costs and animal welfare issues involved, is still under debate and will probably remain so for many years to come while work in other, more conventional species, aim to improve SCNT and ISCNT efficiency and find solutions to the many pertaining problems involved.

Acknowledgments The research leading to these results received funding from the European Research Council under the European Community Seventh Framework Programme (FP7/2007-2013)/ERC grant agreement no 210103 to GP, and PRIN 2007, no 2007MY2M92 to GP. PL acknowledges the support of the EU FP7-KBBE -2009-3 Programme, project no 244356 (NextGene) and PRIN MIUR founding (project no 2009JE3CHM). Funds from the Bank Foundation Tercas (Teramo, Italy) and funds from project "Genhome", Miur/Firb are acknowledged. The authors are grateful to the Japanese Academy for the permit to reproduce figures published by Kato et al., on Proc. Jpn. Acad. Ser B., Phys Biol Sci. 2009; 85(7):240–7.

References

Amarnath D, Choi I, Moawad AR, Wakayama T, Campbell KHS. Nuclear-cytoplasmic incompatibility and inefficient development of pig-mouse cytoplasmic hybrid embryos. Reproduction. 2011;142(2):295–307. doi:10.1530/rep-11-0044.

Briggs AW, Stenzel U, Johnson PLF, Green RE, Kelso J, Prüfer K, Meyer M, Krause J, Ronan MT, Lachmann M, Pääbo S. Patterns of damage in genomic DNA sequences from a Neandertal. Proc Nat Acad Sci U S A. 2007;104(37):14616–21. doi:10.1073/pnas.0704665104.

Brown RE, Butler JP, Godleski JJ, Loring SH. The elephant's respiratory system: adaptations to gravitational stress. Resp Physiol. 1997;109(2):177–94. doi:10.1016/S0034-5687(97)00038-8.

Gilbert MTP, Drautz DI, Lesk AM, Ho SYW, Qi J, Ratan A, Hsu C-H, Sher A, Dalén L, Götherström A, Tomsho LP, Rendulic S, Packard M, Campos PF, Kuznetsova TV, Shidlovskiy F, Tikhonov A, Willerslev E, Iacumin P, Buigues B, Ericson PGP, Germonpré M, Kosintsev P, Nikolaev V, Nowak-Kemp M, Knight JR, Irzyk GP, Perbost CS, Fredrikson KM, Harkins TT, Sheridan S, Miller W, Schuster SC. Intraspecific phylogenetic analysis of Siberian woolly mammoths using complete mitochondrial genomes. Proc Nat Acad Sci U S A. 2008;105(24): 8327–32. doi:10.1073/pnas.0802315105.

Gunasena KT, Lakey JRT, Villines PM, Bush M, Raath C, Critser ES, McGann LE, Critser JK. Antral follicles develop in xenografted cryopreserved African elephant (*Loxodonta africana*) ovarian tissue. Anim Reprod Sci. 1998;53(1–4):265–75. doi:10.1016/S0378-4320(98)00132-8.

Gurdon JB, Wilmut I. Nuclear transfer to eggs and oocytes. Cold Spring Harb Perspect Biol. 2011;3(6):a002659. doi:10.1101/cshperspect.a002659.

Hammer DA, Kamat NP. Towards an artificial cell. FEBS Lett. 2012;586(18):2882–90. doi:10.1016/j.febslet.2012.07.044.

Hermes R, Göritz F, Portas TJ, Bryant BR, Kelly JM, Maclellan LJ, Keeley T, Schwarzenberger F, Walzer C, Schnorrenberg A, Spindler RE, Saragusty J, Kaandorp S, Hildebrandt TB. Ovarian superstimulation, transrectal ultrasound-guided oocyte recovery, and IVF in rhinoceros. Theriogenology. 2009;72(7):959–68. doi:10.1016/j.theriogenology.2009.06.014.

Hermes R, Hildebrandt TB, Göritz F. Reproductive problems directly attributable to long-term captivity-asymmetric reproductive aging. Anim Reprod Sci. 2004;82–83:49–60. doi:10.1016/j.anireprosci.2004.05.015.

Hermes R, Saragusty J, Schaftenaar W, Goritz F, Schmitt DL, Hildebrandt TB. Obstetrics in elephants. Theriogenology. 2008;70(2):131–44. doi:10.1016/j.theriogenology.2008.04.003.

Hildebrandt TB, Göritz F, Pratt NC, Lehnhardt JL, Brown JL, Howard JG, Schmitt DL, Jewgenow K, Cambre RC, Montali RJ Artificial insemination (AI) of an Asian elephant (*Elephas maximus*). In: Lauria A, Gandolfi F, Enne G, Gianaroli L, editors. 50th anniversary of ICAR, gametes: development and function, Milan, Italy, 14–16 Sept 1998. p. 546 (abstract).

Hildebrandt TB, Göritz F, Schnorrenberg A, Hermes R, Schmitt DL, Hagan D, Peterson J, Brown JL, Loskutoff NM, Pratt NC, Lehnhardt JL, Miller G, Montali RJ, Olson D. Successful artificial insemination of African nulliparous elephants at the Indianapolis Zoo. In: Verhandlungsbericht des 39 Internationalen Symposiums über die Erkrankungen der Zoo- und Wildtiere, Vienna, Austria, 12–16 May 1999. Institut für Zoo- und Wildtierforschung, Berlin. pp. 41–46.

Hildebrandt TB, Lueders I, Hermes R, Goeritz F, Saragusty J. Reproductive cycle of the elephant. Anim Reprod Sci. 2011;124(3–4):176–83. doi:10.1016/j.anireprosci.2010.08.027.

Hofreiter M, Serre D, Poinar HN, Kuch M, Paabo S. Ancient DNA. Nat Rev Genet. 2001;2(5): 353–9. doi:10.1038/35072071.

Kato H, Anzai M, Mitani T, Morita M, Nishiyama Y, Nakao A, Kondo K, Lazarev PA, Ohtani T, Shibata Y, Iritani A. Recovery of cell nuclei from 15,000 years old mammoth tissues and its injection into mouse enucleated matured oocytes. Proc Jpn Acad Ser B Phys Biol Sci. 2009; 85(7):240–7. doi:10.2183/pjab.85.240.

Loi P, Matsukawa K, Takahashi S, Akagi S, Ptak G. Nuclear reprogramming: what has been done and potential avenues for improvements. Cent Eur J Biol. 2008;3(3):211–23. doi:10.2478/s11535-008-0025-6.

Loi P, Modlinski JA, Ptak G. Interspecies somatic cell nuclear transfer: a salvage tool seeking first aid. Theriogenology. 2011;76(2):217–28. doi:10.1016/j.theriogenology.2011.01.016.

Lonergan P. State-of-the-art embryo technologies in cattle. Soc Reprod Fertil Suppl. 2007;64: 315–25.

Maruotti J, Jouneau A, Renard J-P. Faithful reprogramming to pluripotency in mammals—what does nuclear transfer teach us? Int J Dev Biol. 2010;54(11–12):1609–21. doi:10.1387/ijdb.103195jm.

Ménézo Y, Dale B, Cohen M. DNA damage and repair in human oocytes and embryos: a review. Zygote. 2010;18(04):357–65. doi:10.1017/S0967199410000286.

Iuso D, Czernick M, Di Egidio F, Sampino S, Zacchini F, Bochenek M, Smorag Z, Modlinski JA, Ptak G, Loi P. Maintenance of genomic integrity in freeze-dried somatic cells. PLoS One; 2013;8(1):e51317. doi: 10.1371/journal.pone.0051317.

Metzker ML. Sequencing technologies—the next generation. Nat Rev Genet. 2010;11(1):31–46. doi:10.1038/nrg2626.

Miller W, Drautz DI, Ratan A, Pusey B, Qi J, Lesk AM, Tomsho LP, Packard MD, Zhao F, Sher A, Tikhonov A, Raney B, Patterson N, Lindblad-Toh K, Lander ES, Knight JR, Irzyk GP, Fredrikson KM, Harkins TT, Sheridan S, Pringle T, Schuster SC. Sequencing the nuclear genome of the extinct woolly mammoth. Nature. 2008;456(7220):387–90. doi:10.1038/nature07446.

Moritz M, Alberts BM. Isolation of centrosomes from drosophila embryos. In: Rieder CL, editor. Methods in cell biology, vol. 61. London: Academic; 1998. p. 1–12. doi:10.1016/S0091-679X(08)61972-X.

Nicholls H. Darwin 200: Let's make a mammoth. Nature. 2008;456(7720):310–4. doi:10.1038/456310a.

Pääbo S. Molecular cloning of ancient Egyptian mummy DNA. Nature. 1985;314(6012):644–5. doi:10.1038/314644a0.

Paolucci V, Gutt CN, Schaeff B, Encke A. Gasless laparoscopy in abdominal surgery. Surg Endosc. 1995;9(5):497–500. doi:10.1007/bf00206834.

Podhorecka M, Skladanowski A, Bozko P. H2AX phosphorylation: its role in DNA damage response and cancer therapy. J Nucl Acids 2010. ID: 920161. doi: 10.4061/2010/920161

Rohland N, Reich D, Mallick S, Meyer M, Green RE, Georgiadis NJ, Roca AL, Hofreiter M. Genomic DNA sequences from mastodon and woolly mammoth reveal deep speciation of forest and savanna elephants. PLoS Biol. 2010;8(12):e1000564. doi:10.1371/journal.pbio.1000564.

Saragusty J. Elephants under human care: fertility, birth rate, infant mortality and survivorship. In: Aranovich M, Dufresne O, editors. Elephants: ecology, behavior and conservation. New York: Nova Publishers; 2012. p. 1–48.

Stansfield FJ, Picton HM, Nöthling JO. Early primary-rather than primordial follicles constitute the main follicular reserve in the African elephant (*Loxodonta africana*). Anim Reprod Sci. 2010;123(1–2):112–8. doi:10.1016/j.anireprosci.2010.11.003.

Thundathil J, Filion F, Smith LC. Molecular control of mitochondrial function in preimplantation mouse embryos. Mol Reprod Dev. 2005;71(4):405–13. doi:10.1002/mrd.20260.

Wilmut I, Schnieke AE, McWhir J, Kind AJ, Campbell KH. Viable offspring derived from fetal and adult mammalian cells. Nature. 1997;385(6619):810–3. doi:10.1038/385810a0.

Zagnoni M. Miniaturised technologies for the development of artificial lipid bilayer systems. Lab Chip. 2012;12(6):1026–39. doi:10.1039/C2LC20991H.

Chapter 20
Conclusions: Environmental Change, Wildlife Conservation and Reproduction

William V. Holt, Janine L. Brown, and Pierre Comizzoli

Abstract Our intention when planning this book was to explore the diverse ways that reproductive science is inextricably tied to many aspects of biodiversity conservation, using the opportunity to present a vast amount of specialised information in a way that forms a coherent and important body of work. Some of the chapters were therefore concerned with understanding how taxonomic groups and species are being affected by globally important environmental changes, mostly caused through anthropogenic influences. Others were more focused on monitoring and understanding the physiology of wild species, with the aim of better understanding mechanisms underlying responses to captive conditions and environmental change, in both wild and captive animals. We also wanted to review advances in technological measures that are being actively developed to support the breeding and management of wildlife. In a few cases we have presented specific case studies that highlight the amount of effort required for the successful development of assisted reproductive technologies for wild species. Viewed overall, the outcome is spectacular; the last decade has seen enormous progress in many aspects of the sciences and technologies relevant to the topic. It is also clear that the boundaries between different scientific disciplines are becoming ever more blurred, and it is no longer easy or even possible to remain focused on a highly specialized topic in reproduction or conservation, without having at least some understanding of allied subjects. Here we

W.V. Holt, Ph.D. (✉)
Academic Department of Reproductive and Developmental Medicine,
University of Sheffield, Jessop Wing, Tree Root Walk, Sheffield S10 2SF, UK
e-mail: Bill.holt@sheffield.ac.uk

J.L. Brown
Center for Species Survival, Smithsonian Conservation Biology Institute,
1500 Remount Road, Front Royal, VA, USA

P. Comizzoli
Smithsonian Conservation Biology Institute, National Zoological Park,
Washington, DC, USA

W.V. Holt et al. (eds.), *Reproductive Sciences in Animal Conservation*,
Advances in Experimental Medicine and Biology 753, DOI 10.1007/978-1-4939-0820-2_20,

present a few concluding comments about what we have learnt, and how the various topics interact with each other. We also emphasize that, as far as we know, no similarly comprehensive consideration of the contribution of reproductive science to wildlife conservation has been published within the last decade.

Keywords Assisted colonization • Climate change • Cryobiology • Epigenetics • Extinction • Genomics

1 Is Reproductive Science Helpful for Conservation?

The global environment is under increasing pressure from expanding human activities and, despite the efforts of climate change deniers and sceptics, the outcomes are widely accepted by the scientific community (Rosenau 2012). Species of all types, especially those that cannot migrate, have to face increasing environmental temperatures, atmospheric and marine pollution, ocean acidification, desertification and other threats. Ecosystems are therefore changing in response to these problems, and despite a few notable achievements in terms of protection of rare species and habitats, traditional conservation approaches have been largely powerless to reverse or stabilize the underlying rhythms of the global economy. Documenting these problems has largely been the preserve of ecologists, whose work often identifies long-term biological trends associated with environmental change, and whose advice is then fed into the formulation of policies for habitat protection (Bino et al. 2013). However, the formulation of those policies tends to be in the hands of politicians who, under pressure from interested lobby groups, are free to ignore the well intentioned advice while species go extinct. The precarious state of the endangered Leadbeater's possum in Victoria, Australia, is a topical example of a mammal likely to undergo extinction because of timber logging (see http://www.depi.vic.gov.au/environment-and-wildlife/wildlife/leadbeaters-possum).

One might therefore ask whether reproductive scientists can really play a useful part in solving these problems. The causes of imminent extinctions of wild species are not, after all, always caused by simple failure to breed. This was, however, the problem with the Florida panther, an endangered subspecies of puma (*Felis concolor coryi*), where some years ago the population was known to be shrinking as individual animals showed one or more of a variety of physiological, reproductive, endocrine, and immune system defects (Facemire et al. 1995). Initial hypotheses for the decline of this species blamed the presence of environmental endocrine disrupters, but eventually it was determined that the reproductive abnormalities were caused largely by inbreeding depression. The introduction of new genetics into the population, by the importation of unrelated males of another subspecies (*F. c. stanleyana*), successfully reversed the situation and resulted in a new generation of healthy animals (Mansfield and Land 2002; Hostetler et al. 2012). Without input from reproductive biologists, the problems associated with abnormal sexual differentiation and defective sperm function would not have been recognised.

The recognition that abnormal sexual differentiation (i.e., feminization) in fish within UK rivers was caused by the presence of endocrine disrupting chemicals (EDCs) originating from sewage was also a landmark contribution to conservation made by reproductive biologists (Jobling et al. 2002; van Aerle et al. 2001; Rodgers-Gray et al. 2001), and, as explored in this book, the negative consequences of endocrine disruption exposure by chemicals in the environment are widely recognised in many different taxonomic groups and monitored all over the world.

It is also important to note that humans are not exempt from the deleterious consequences of EDC exposure, and there is a widespread consensus that sperm output in men has been declining globally over the last several decades. This was first noted, and widely dismissed, in a controversial paper by Carlsen et al. (1992), but now has been confirmed by reanalysis and updated with new sperm concentration data from many countries (Swan et al. 2000). Detailed studies on laboratory species have revealed potential mechanisms for the decline in sperm production, the most likely of which is exposure to industrial chemicals at important stages of sexual differentiation. These findings have certainly had high level impacts and have stimulated global regulatory initiatives aimed at safety assessments for newly manufactured chemicals, and the enforcement of safe disposal methods. This particular example also emphasises the important interdisciplinary benefits that emerge when related fields in wildlife science and human clinical medicine collide or collaborate. It is increasingly recognised that mechanisms exist whereby exposure to an EDC such as Bisphenyl A in one generation produces reproductive and other effects that persist down the generations (Sofiane et al. 2013).

Similarly, the reported rapid changes to the breeding seasons of wild species, coupled with loss of their normal synchronization with the seasonal production of their appropriate foodstuffs, parallels the linkages between inadequate nutrition and early embryonic development in mammals. It is now well established that the occurrence of many adult diseases such as diabetes and hypertension, can be traced back to mismatches between the poor availability of nutrition at the embryonic stage and richer diets later in life (Barker et al. 2010; Barker 1999). Such problems have been replicated in laboratory species, and it seems logical to expect that the relatively rapid environmental changes observed today would induce similar effects in wild species.

These relationships have received very little attention in terms of wildlife, but we suspect they will soon be recognised as being highly influential modulators of species survival. An improved understanding of these interactions is, however, not easy to achieve. It requires a great deal of scientific effort and is problematic in unusual species, especially those whose genomes have not yet been sequenced, microarrays for gene expression studies have not yet been developed, and epigenetic chromatin patterns are not known. Nevertheless, over the last decade, biologically-based technologies of all descriptions have leapt forward in ways that were previously impossible to foresee, most notably in the field of genomics. Whole genomes can now be sequenced relatively cheaply, specific DNA and RNA sequences can be easily detected on a massive scale using array technologies and when these are combined with proteomic technologies and computational power, the ability to generate data in basic cell and molecular biology is staggering. Projects such as "Genome 10k" (http://genome10k.soe.ucsc.edu/), which aims to evaluate the DNA sequences for 10,000 vertebrate species, will contribute towards these advances and their progress will hopefully accelerate as time passes.

Knowing the DNA sequence is, of course, only a first step in understanding the ways that cells and whole animals work. How species with similar DNA sequences, such as humans, chimpanzees and bonobos that share 98 % of their alignable genomic sequence, develop into phenotypically distinct animals is still poorly understood. A recent study into this question (Marchetto et al. 2013) showed that the expression and "within-genome" mobility of a retrotransposon, known as "long-interspersed element-1", may have exerted a profound influence in primate evolution. It is of interest in the context of this book that this study, whose aim was mainly to study processes in evolution, was only possible because the authors were able to generate pluripotent iPS cells from an archive of frozen chimpanzee and bonobo fibroblasts. Within this book we have stressed the importance of cryopreserving cells and tissues from wild species, partly for their potential use in assisted reproductive technologies, but also because of their value for scholarly and applied research. This represents just one example of an important insight in basic biology that would not have been possible without access to frozen cells and highlights the unforeseeable value of maintaining cryopreserved cell and tissue collections from wild species. While the immediate benefit of establishing such collections has not always been clear, especially to funding bodies with focused conservation or biomedical goals in mind, the consistent pace of advances in genomics emphasises the benefits of being able to study the entire cell, with its elaborate compartmentation, complex system of regulatory RNA molecules and its interplay with mitochondrial DNA, rather than only having a genomic DNA sequence available for examination.

Over the past 60 years the practice of cell and tissue preservation has relied heavily on freezing technology, with the result that repositories of cells and tissues around the world are reliant on electrical or liquid nitrogen-cooled freezers of various sorts. These are not only expensive to purchase and maintain, but their use is completely dependent on reliable availability of power or regular supplies of liquid nitrogen. Even short-term sample storage in remote regions of the world is therefore difficult, while even well-established repositories are vulnerable to power failures and equipment breakdown. In an effort to bypass these problems a number of research groups are investigating the feasibility of drying the samples and then keeping them in sealed containers at room temperature. Loi et al. (2008) showed that if freeze-dried nuclei were injected into enucleated sheep oocytes they would undergo the early stages of embryonic development, and several groups have shown that freeze-dried mouse spermatozoa can generate embryos after intracytoplasmic sperm injection (ICSI) of oocytes (Li et al. 2007a, b; Kusakabe et al. 2008). More recently Graves-Herring et al. (2013) have shown that isolated germinal vesicles can be dried and stored without losing the ability to resume meiosis after rehydration. Developments in this field, recently reviewed by Loi et al. (2013) would revolutionize cell preservation in its many forms. Human infertility clinics, animal breeding facilities, museums, and cell repositories of all types could dispense with their collections of freezers and liquid nitrogen containers in favour of sealed tubes stored at room temperature. No doubt this is an optimistic prediction when considered in the ever-present light of species differences, but the potential benefits of this approach are so important that they cannot be ignored. It is interesting to see that the pioneers in this field are largely drawn from the community of conservation-focused reproductive scientists.

Scientific insights and new technologies develop a synergy of their own, and in this book we have highlighted several examples directly relevant to conservation biology. After a good deal of concentrated effort, Janine L. Brown and her colleagues eventually managed to develop a successful method for artificial insemination in elephants (Brown et al. 2004; Hermes et al. 2007); however, the practical implementation of this method was preceded by focused research into the ovarian dynamics of the elephant, and in turn this has led to further basic research into relationships between body condition and reproduction. These investigations are fundamentally important in elephant management, but it is of interest that they are also important in understanding the likely implications of habitat modification and dietary changes. The elephant research is also relevant to human medicine, where the negative consequences of human obesity on the chances of conception are matters of considerable interest to infertility clinics.

The last decade has also seen advances in technologies that can be used to the benefit of dwindling and threatened wild populations, and it is clear that there has been a great deal of international effort invested in gaining more understanding of wild species and developing technologies to help prevent their extinction. Motivations vary; some major projects have underlying economic aims, such as the "Millennium Seed Bank" established by Kew in the UK (Madeley 1999), which aims to conserve seeds from most of the world's plants and the National Animal Genetic resource program in the USA (Blackburn 2004, 2012). Other projects are often founded for nationalistic or cultural reasons; conservation of the giant panda (*Ailuropoda melanoleuca*) in China, the Mauritius pink pigeon (*Nesoenas mayeri*), Asian elephants in Thailand and nearby countries, the Global Tiger Initiative, and seahorse preservation in many different locations, exemplify projects taking place around the world. Many of these projects, whether established for economic, cultural or other reasons, can benefit from reproductive technologies and knowledge of species biology gained over recent years. Technologies such as artificial insemination and semen freezing still have a long way to go, mainly because success is species-dependent in many respects, even with closely related species within a taxon (e.g., Felidae). Long-term and well-focused studies are beginning to pay off where they are supported by detailed background research. Extensive work on amphibian-related technologies over the last decade is one example of a field that has progressed remarkably since the first reports of sperm motility recovery in cane toads, used in that instance as a model species (Browne et al. 1998) to the industrial scale production and reintroduction of Wyoming and Boreal toads in Colorado, reported in this book by Clulow and colleagues. The latter is an excellent demonstration that a laboratory based technique, mainly used for solving fertility problems in individuals, can be transformed into a tool for the conservation of whole populations. The preservation of coral cells and gametes is another example of eventual success being achieved through long-term efforts. Rather like the amphibian work, the aims of coral cryopreservation are directly focused on the preservation of coral reefs, but are focused on a population (the reef as an assemblage) rather than individuals. If corals can be maintained *ex situ*, with cryopreservation as a genetically supportive tool, it may be possible to repopulate extant coral reefs if and when the environment recovers. Incidentally, prospects for coral recovery are being assisted by studies of their related microflora (Hunt et al. 2012; Kvennefors et al. 2012) as there is some hope for coral restoration through rebalancing abnormal bacterial populations with methods such as antibiotic treatment (Sheridan et al. 2013). This demonstrates the complexity and unexpectedness of conservation measures in different fields.

The potential importance to reproductive success of maintaining well-adjusted microflora in corals parallels the situation in mammals where reproductive tract function is disturbed by infections. Studies in female dogs (bitches) have revealed that uterine infections interfere with sperm transport by reducing the sperm-binding ability of the uterine epithelia (England et al. 2012). More generally, reproductive tract infections are known to interfere with implantation and, much like periconception pollutant exposure, exert negative impacts on embryonic development with long-term epigenetic effects lasting into adulthood (Kwak-Kim et al. 2010; Silva et al. 2012; Yang et al. 2011). Recent genomic studies in humans have shown that the reproductive tract, especially the vagina, contains a highly complex but relatively constant microflora, but that shifts in the balance between bacterial species and communities can result in reproductive problems. After conducting a review of the literature, Reid et al. (2013) proposed that supplementing the altered microbiome with probiotic lactobacilli, which are antagonistic to pathogens, would improve reproductive health and reduce the number of pregnancy-associated complications. Speculatively, these insights probably should have practical relevance for assisted breeding technologies applied to non-domestic species, especially if they become stressed through handling and captivity. They may, however, explain some instances of poor reproductive success in the face of environmental stress.

Technologies for monitoring wild populations through faecal, urinary and blood analyses have remained rather laborious and need improvements. This is not entirely a technical problem as assays already exist for laboratory species and humans. The difficulty relates more to the scale of investment needed to develop and validate suitable antibodies, when the potential market is small and uneconomical. Perhaps advances in synthetic biology will enable the development of assays that are effective across multiple species. In fact, as noted by several authors in this book, hormone assay techniques are crucially important in understanding reproductive dynamics in many species. However, apart from the gradual transition from radioimmunoassay techniques to enzyme-linked immunoassays, there has not been a revolution in assay technology that matches some of the genomic techniques available today. It is now feasible to take a small, inexpensive and integrated PCR machine into the field to monitor multiple DNA sequences in river or pond water, thus obtaining an instant snapshot of the range of species present. By comparison, collecting multiple faecal samples in the field, drying and transporting them to a laboratory for analysis is a lengthy and labour-intensive process that in the wrong hands often fails for no apparent reason. Technological advances in this area would be very welcome. Future developments in this field will involve biosensors that can be implanted for long periods in free ranging species so that data can be collected remotely and uploaded via satellite links. Data such as body temperature, blood pressure, pH, blood glucose, etc. should be relatively easy to develop as the technologies already exist; biosensors capable of steroid hormone measurement already exist to some extent (Posthuma-Trumpie et al. 2009); for example, progesterone biosensors in milking lines can measure the hormone within 5 min (Friggens et al. 2008; Kappel et al. 2007), but have yet to be developed for long term remote use. Biosensor technologies would be especially useful for studies with wild marine species whose ranges enclose thousands of square miles; in fact, simpler devices that detect and report on swimming depth and location are already available and are being used with species such as turtles, sharks (Hays et al. 2001; Wang et al. 2012; Wearmouth et al. 2013) and eels (Breukelaar et al. 2009). Such an ability to remotely monitor reproductive or other health biomarkers would revolutionize our ability to assess physiological function in both captive and free-living species.

2 Novel Ideas in Conservation: Crazy or Worth a Try?

Most of the chapters in this book are aimed at gaining a deeper understanding of biological principles that can eventually be transformed into applications for species conservation. Assisted breeding technologies fall into this category, but inevitably the level of complexity associated with such methodology covers a huge range, from simple artificial insemination with fresh semen at one extreme to cloning technologies at the more complex end. It is important to note, however, that attitudes among conservation managers towards the acceptability of these techniques vary enormously among individuals, and especially between countries. The apparently simple collection and mixing of amphibian eggs and sperm as a strategy for breeding threatened frogs and toads in the UK is regarded as unacceptable by zoo managers and species coordinators because of the need to handle animals (Holt 2008). It appears that they would prefer to see species go extinct than try any kind of technological fix! In contrast to this irrational level of conservatism, some technologists go too far the other way, embracing extremely hi-tech, expensive and unpredictable methods and proposing them as ways to prevent extinction (Ryder and Benirschke 1997; Kim et al. 2007; Li et al. 2007c; Smits et al. 2012). As shown in this book by Mastromonaco and Comizzoli, and discussed previously by Wells (2005), these techniques tend to result in offspring with poor survival prospects. Indeed, it is then somewhat surprising that there is a current fashion among some biologists for taking the nuclear transfer technology several stages further and proposing that extinct species could potentially be resurrected (Church 2013; Zimmer 2013) and used to repopulate whole ecosystems. Given the ever decreasing availability of habitat for extant, but threatened, species this approach seems to be illogical in the context of conservation biology. If a handful of extinct animals were to be cloned successfully, how could they survive in a modern world? They would immediately be subject to the deleterious effects of inbreeding, they would be poorly adapted to their "new" environment, and their only hope of survival would be in a zoo that could provide significant veterinary support.

The concept behind the repopulation of whole ecosystems with technologically resurrected species is reminiscent of a different conservation technique known as "Assisted Colonization" (AC), another rather extreme strategy that is certainly at odds with the extreme conservatism of many conservation biologists. In essence, AC is relatively straightforward: it proposes that if an endangered species is threatened with extinction because of global warming, it follows that the species could be saved by moving all or part of the population to a more suitable habitat not necessarily in the same country (Minteer and Collins 2010). A surprisingly extreme example discussed in a thoughtful review by Albrecht et al. (2013) is the proposal to relocate polar bears from the Arctic, where the sea ice is eventually expected to melt completely, to the Antarctic where they could live in relative safety. A second case concerns the Australian mountain pygmy-possum (*Burramys parvus*), which is likely to be among the first wave of species to be rendered extinct by anthropogenic climate change. Three isolated populations of the possum live in boulder fields at high elevations in the highest parts of south-eastern Australia and appear to rely on the migratory Bogong moth (*Agrostis infusa*) for food. As these possums cannot move to higher and cooler habitats in Australia they could, in theory, be moved to similar mountain habitats in New Zealand. These proposals raise complex ethical issues, not least about the safety of penguin and seal populations in the Antarctic if they are suddenly confronted with unfamiliar predators such as polar bears, but also about the health and wellbeing of the translocated species themselves in their new environment.

A different but related use of modern genomics to support species conservation was recently proposed by Thomas et al. (2013), who suggested that genetic engineering could be harnessed to insert or modify the genomes of threatened species, thus helping them to adapt to changing environments. This was met with dismay and disapproval from some well-known conservation geneticists (Hedrick et al. 2013), who argued that such modifications would almost certainly cause deleterious genetic effects. The emerging field of epigenetics, which is currently showing that genetic fitness depends on more than the bare DNA sequence, would certainly suggest that the outcomes of such well-intentioned engineering will be hard to predict.

In a review aimed at assessing the potential role of "synthetic biology" in wildlife conservation, Redford et al. (2013) recently wrote that

> … The limited and timid engagement of conservation science and policy with the development of synthetic biology is unfortunate, because the technology is likely to transform the operating space within which conservation functions, and therefore the prospects for maintaining biodiversity into the future.

These sentiments show an unusual degree of open-mindedness about modern technologies. At the same time, however, it is surprising that while these authors were impressed by the exciting hi-tech possibilities on offer, they, like many other conservation scientists and managers, have consistently failed to recognise that numerous, simpler, technological opportunities to support conservation programmes already exist.

3 Does the Control of Unwanted Species Count as Conservation? Arguably Important, but Widely Ignored

As human population growth is a global problem responsible for many environmental problems including loss of wildlife habitats and species, the development of contraception methods that can be applied widely and cheaply can be as a powerful tool for wildlife conservation, although it is not always seen in that context. Steroid-based contraceptives for human females are well established, whereas male-based steroid contraception treatments for humans remain unreliable (Grimes et al. 2012; Nieschlag 2011). While human contraceptive techniques have to be reversible, the same does not apply to many wildlife scenarios where the aim is often to find a humane method of controlling animal populations. In zoos, where individuals can be handled, steroid based methods are suitable, but they are not easily transferred to free-ranging animals. Suitably effective techniques are still being sought, despite many different approaches having been studied. Potential targets for blocking reproduction in wild species have included inhibiting epididymal function with antibodies and chemicals, preventing sperm transport in the female reproductive tract by attempting to induce anti-sperm antibodies, and down-regulating reproductive function by the use of antibodies against GnRH or zona pellucida proteins. These are not always successful though, so novel thinking is needed to find better ways of controlling pest species, such as the cane toad in Australia, which can produce 30,000 oocytes in a single ovulation event.

4 Concluding Remarks

The merits of germplasm biobanking have been discussed many times before, and it is interesting to realise that some of the first and most detailed proposals for a variety of biobanks emerged from the former Soviet Union in the late 1970s (Veprintsev and Rott 1980, 1979). Apart from proposing the biobanking of germplasm, these authors suggested conserving an enormous variety of biomaterials, even including human gut contents. At the time it was difficult to see the relevance of some of these suggestions, but now, more than three decades later we can see treatments in human health that involve transplanting gut bacteria between individuals. This demonstrates the difficulty of seeing into the future and dismissing ideas because they are difficult to comprehend. The same principle applies today, and it is interesting to see the growth of the biobanking movement into unexpected directions. In a similar vein, back in 1999 one of the editors (WVH) suggested to a group of ecologists that they should invest in state of the art genetic technologies in order to study "environmental genomics". This idea was dismissed because it was very clear to those present that the only DNA sequences relevant to ecologists were microsatellites. It is thus satisfying to note that *Nature* recently published a call for ecology to embrace environmental genomics (Mace 2013). Similar experiences could probably be found in other scientific fields, and historically the role of prejudice in science has been far worse (cf. Semmelweiss and the resistance to ideas of sepsis in maternity wards (Funk et al. 2009)).

Throughout this book, we have consistently argued that conservation biology requires inputs from as many disciplines as possible, and we hope that readers are now convinced. As suitable habitats for wildlife continue to shrink and a variety of ecological threats loom, long-term species survival may well rely, at least in part, on the herculean efforts of reproductive biologists and the tools they develop.

References

Albrecht GA, Brooke C, Bennett DH, Garnett ST. The ethics of assisted colonization in the age of anthropogenic climate change. J Agric Environ Ethics. 2013;26(4):827–45. doi:10.1007/s10806-012-9411-1.

Barker DJ. Early growth and cardiovascular disease. Arch Dis Child. 1999;80(4):305–7.

Barker DJ, Gelow J, Thornburg K, Osmond C, Kajantie E, Eriksson JG. The early origins of chronic heart failure: impaired placental growth and initiation of insulin resistance in childhood. Eur J Heart Fail. 2010;12(8):819–25. doi:10.1093/eurjhf/hfq069. hfq069 [pii].

Bino G, Ramp D, Kingsford RT. Improving bioregional frameworks for conservation by including mammal distributions. Austral Ecol. 2013;38(4):393–404. doi:10.1111/j.1442-9993.2012.02423.x.

Blackburn HD. Development of national animal genetic resource programs. Reprod Fertil Dev. 2004;16(1–2):27–32. doi:10.10371/RD03075.

Blackburn HD. Genetic selection and conservation of genetic diversity*. Reprod Domest Anim. 2012;47 Suppl 4:249–54. doi:10.1111/j.1439-0531.2012.02083.x.

Breukelaar AW, Ingendahl D, Vriese FT, de Laak G, Staas S, Klein Breteler JG. Route choices, migration speeds and daily migration activity of European silver eels *Anguilla anguilla* in the River Rhine, north-west Europe. J Fish Biol. 2009;74(9):2139–57. doi:10.1111/j.1095-8649.2009.02293.x.

Brown JL, Goritz F, Pratt-Hawkes N, Hermes R, Galloway M, Graham LH, Gray C, Walker SL, Gomez A, Moreland R, Murray S, Schmitt DL, Howard J, Lehnhardt J, Beck B, Bellem A, Montali R, Hildebrandt TB. Successful artificial insemination of an Asian elephant at the National Zoological Park. Zoo Biol. 2004;23(1):45–63. doi:10.1002/zoo.10116.

Browne RK, Clulow J, Mahony M, Clark A. Successful recovery of motility and fertility of cryopreserved cane toad (*Bufo marinus*) sperm. Cryobiology. 1998;37(4):339–45. doi:10.1006/cryo.1998.2129.

Carlsen E, Giwercman A, Keiding N, Skakkebaek NE. Evidence for decreasing quality of semen during past 50 years. BMJ. 1992;305(6854):609–13.

Church G. Please reanimate. Reviving mammoths and other extinct creatures is a good idea. Sci Am. 2013;309(3):12.

England GCW, Burgess CM, Freeman SL. Perturbed sperm–epithelial interaction in bitches with mating-induced endometritis. Vet J. 2012;194(3):314–8. http://dx.doi.org/10.1016/j.tvjl.2012.04.031.

Facemire CF, Gross TS, Guillette Jr LJ. Reproductive impairment in the Florida panther: nature or nurture? Environ Health Perspect. 1995;103 Suppl 4:79–86.

Friggens NC, Bjerring M, Ridder C, Hojsgaard S, Larsen T. Improved detection of reproductive status in dairy cows using milk progesterone measurements. Reprod Domest Anim. 2008;43 Suppl 2:113–21. doi:10.1111/j.1439-0531.2008.01150.x.

Funk DJ, Parrillo JE, Kumar A. Sepsis and septic shock: a history. Crit Care Clin. 2009;25(1):83–101. doi:10.1016/j.ccc.2008.12.003. viii.

Graves-Herring JE, Wildt DE, Comizzoli P. Retention of structure and function of the cat germinal vesicle after air-drying and storage at suprazero temperature. Biol Reprod. 2013;88(6):139. doi:10.1095/biolreprod.113.108472.

Grimes DA, Lopez LM, Gallo MF, Halpern V, Nanda K, Schulz KF. Steroid hormones for contraception in men. Cochrane Database Syst Rev. 2012;3. doi:10.1002/14651858.CD004316.pub4. CD004316.

Hays GC, Akesson S, Broderick AC, Glen F, Godley BJ, Luschi P, Martin C, Metcalfe JD, Papi F. The diving behaviour of green turtles undertaking oceanic migration to and from Ascension Island: dive durations, dive profiles and depth distribution. J Exp Biol. 2001;204(Pt 23):4093–8.

Hedrick PW, Allendorf FW, Waples RS. Ecology: Genetic engineering in conservation. Nature. 2013;502(7471):303. doi:10.1038/502303b.

Hermes R, Goritz F, Streich W, Hildebrandt T. Assisted reproduction in female rhinoceros and elephants—current status and future perspective. Reprod Domest Anim. 2007;42 Suppl 2:33–44. doi:10.1111/j.1439-0531.2007.00924.x. RDA924 [pii].

Holt WV. Cryobiology, wildlife conservation and reality. Cryo Letters. 2008;29:43–52.

Hostetler JA, Onorato DP, Jansen D, Oli MK. A cat's tale: the impact of genetic restoration on Florida panther population dynamics and persistence. J Anim Ecol. 2012;82(3):608–20. doi:10.1111/1365-2656.12033.

Hunt LR, Smith SM, Downum KR, Mydlarz LD. Microbial regulation in gorgonian corals. Mar Drugs. 2012;10(6):1225–43. doi:10.3390/md10061225.

Jobling S, Beresford N, Nolan M, Rodgers-Gray T, Brighty GC, Sumpter JP, Tyler CR. Altered sexual maturation and gamete production in wild roach (*Rutilus rutilus*) living in rivers that receive treated sewage effluents. Biol Reprod. 2002;66(2):272–81.

Kappel ND, Proll F, Gauglitz G. Development of a TIRF-based biosensor for sensitive detection of progesterone in bovine milk. Biosens Bioelectron. 2007;22(9–10):2295–300. doi:10.1016/j.bios.2006.11.030.

Kim MK, Jang G, Oh HJ, Yuda F, Kim HJ, Hwang WS, Hossein MS, Kim JJ, Shin NS, Kang SK, Lee BC. Endangered wolves cloned from adult somatic cells. Cloning Stem Cells. 2007;9(1):130–7. doi:10.1089/clo.2006.0034.

Kusakabe H, Yanagimachi R, Kamiguchi Y. Mouse and human spermatozoa can be freeze-dried without damaging their chromosomes. Hum Reprod. 2008;23(2):233–9.

Kvennefors EC, Sampayo E, Kerr C, Vieira G, Roff G, Barnes AC. Regulation of bacterial communities through antimicrobial activity by the coral holobiont. Microb Ecol. 2012;63(3):605–18. doi:10.1007/s00248-011-9946-0.

Kwak-Kim J, Park JC, Ahn HK, Kim JW, Gilman-Sachs A. Immunological modes of pregnancy loss. Am J Reprod Immunol. 2010;63(6):611–23. doi:10.1111/j.1600-0897.2010.00847.x.

Li MW, Biggers JD, Elmoazzen HY, Toner M, McGinnis L, Lloyd KC. Long-term storage of mouse spermatozoa after evaporative drying. Reproduction. 2007a;133(5):919–29.

Li MW, Biggers JD, Toner M, Griffey SM, Lloyd KC. Phenotypic analysis of C57BL/6J and FVB/NJ mice generated using evaporatively dried spermatozoa. Comp Med. 2007b;57(5):469–75.

Li Y, Dai Y, Du W, Zhao C, Wang L, Wang H, Liu Y, Li R, Li N. In vitro development of yak (*Bos grunniens*) embryos generated by interspecies nuclear transfer. Anim Reprod Sci. 2007c;101(1–2):45–59. doi:10.1016/j.anireprosci.2006.09.018.

Loi P, Iuso D, Czernik M, Zacchini F, Ptak G. Towards storage of cells and gametes in dry form. Trends Biotechnol. 2013;31(12):688–95. doi:10.1016/j.tibtech.2013.09.004.

Loi P, Matzukawa K, Ptak G, Natan Y, Fulka Jr J, Arav A. Nuclear transfer of freeze-dried somatic cells into enucleated sheep oocytes. Reprod Domest Anim. 2008;43 Suppl 2:417–22. doi:10.1111/j.1439-0531.2008.01193.x.

Mace G. Global change: ecology must evolve. Nature. 2013;503(7475):191–2.

Madeley J. Seeds for the future. Millennium trailblazers 2: Geoffrey Hawtin and Jules Pretty. People Planet. 1999;8(4):18–9.

Mansfield KG, Land ED. Cryptorchidism in Florida panthers: prevalence, features, and influence of genetic restoration. J Wildl Dis. 2002;38(4):693–8.

Marchetto MC, Narvaiza I, Denli AM, Benner C, Lazzarini TA, Nathanson JL, Paquola AC, Desai KN, Herai RH, Weitzman MD, Yeo GW, Muotri AR, Gage FH. Differential L1 regulation in

pluripotent stem cells of humans and apes. Nature. 2013;503(7477):525–9. doi:10.1038/nature12686.
Minteer BA, Collins JP. Move it or lose it? The ecological ethics of relocating species under climate change. Ecol Appl. 2010;20(7):1801–4.
Nieschlag E. The struggle for male hormonal contraception. Best Pract Res Clin Endocrinol Metab. 2011;25(2):369–75. doi:10.1016/j.beem.2010.08.008.
Posthuma-Trumpie GA, van Amerongen A, Korf J, van Berkel WJH. Perspectives for on-site monitoring of progesterone. Trends Biotechnol. 2009;27(11):652–60. http://dx.doi.org/10.1016/j.tibtech.2009.07.008.
Redford KH, Adams W, Mace GM. Synthetic biology and conservation of nature: wicked problems and wicked solutions. PLoS Biol. 2013;11(4):e1001530. doi:10.1371/journal.pbio.1001530.
Reid JNS, Bisanz JE, Monachese M, Burton JP, Reid G. The rationale for probiotics improving reproductive health and pregnancy outcome. Am J Reprod Immunol. 2013;69(6):558–66. doi:10.1111/aji.12086.
Rodgers-Gray TP, Jobling S, Kelly C, Morris S, Brighty G, Waldock MJ, Sumpter JP, Tyler CR. Exposure of juvenile roach (*Rutilus rutilus*) to treated sewage effluent induces dose-dependent and persistent disruption in gonadal duct development. Environ Sci Technol. 2001;35(3):462–70.
Rosenau J. Science denial: a guide for scientists. Trends Microbiol. 2012;20(12):567–9. doi:10.1016/j.tim.2012.10.002.
Ryder OA, Benirschke K. The potential use of "cloning" in the conservation effort. Zoo Biol. 1997;16:295–300.
Sheridan C, Kramarsky-Winter E, Sweet M, Kushmaro A, Leal MC. Diseases in coral aquaculture: causes, implications and preventions. Aquaculture. 2013;396–399:124–35. http://dx.doi.org/10.1016/j.aquaculture.2013.02.037.
Silva E, Henriques S, Brito S, Ferreira-Dias G, Lopes-da-Costa L, Mateus L. Oestrous cycle-related changes in production of Toll-like receptors and prostaglandins in the canine endometrium. J Reprod Immunol. 2012;96(1–2):45–57. doi:10.1016/j.jri.2012.07.003.
Smits K, Hoogewijs M, Woelders H, Daels P, Van Soom A. Breeding or assisted reproduction? Relevance of the horse model applied to the conservation of endangered equids. Reprod Domest Anim. 2012;47 Suppl 4:239–48. doi:10.1111/j.1439-0531.2012.02082.x.
Sofiane B, Raymond B, Claire C, Mireille F, Laurence D, Bruno P, Latifa AN, Marie-Chantal CL. A multi-generational study on low-dose BPA exposure in Wistar rats: effects on maternal behavior, flavor intake and development. Neurotoxicol Teratol. 2013. doi:10.1016/j.ntt.2013.11.002.
Swan SH, Elkin EP, Fenster L. The question of declining sperm density revisited: an analysis of 101 studies published 1934–1996. Environ Health Perspect. 2000;108:961–6.
Thomas MA, Roemer GW, Donlan CJ, Dickson BG, Matocq M, Malaney J. Ecology: gene tweaking for conservation. Nature. 2013;501(7468):485–6.
van Aerle R, Nolan TM, Jobling S, Christiansen LB, Sumpter JP, Tyler CR. Sexual disruption in a second species of wild cyprinid fish (the gudgeon, *Gobio gobio*) in United Kingdom freshwaters. Environ Toxicol Chem. 2001;20(12):2841–7.
Veprintsev BN, Rott NN. Conserving genetic resources. Nature. 1979;280:633–4.
Veprintsev BN, Rott NN. Genome conservation. Puschino: Academy of Sciences of the USSR; 1980.
Wang Y, Li W, Zeng X, Cui Y. A short note on the horizontal and vertical movements of a whale shark, *Rhincodon typus*, tracked by satellite telemetry in the South China Sea. Integr Zool. 2012;7(1):94–8. doi:10.1111/j.1749-4877.2011.00278.x.
Wearmouth VJ, Southall EJ, Morritt D, Sims DW. Identifying reproductive events using archival tags: egg-laying behaviour of the small spotted catshark *Scyliorhinus canicula*. J Fish Biol. 2013;82(1):96–110. doi:10.1111/j.1095-8649.2012.03473.x.
Wells DN. Animal cloning: problems and prospects. Rev Sci Tech. 2005;24(1):251–64.
Yang Z, Kong B, Mosser DM, Zhang X. TLRs, macrophages, and NK cells: our understandings of their functions in uterus and ovary. Int Immunopharmacol. 2011;11(10):1442–50. doi:10.1016/j.intimp.2011.04.024.
Zimmer C. Bringing them back to life. National Geographic. 2013;223(4):28.

Index

W.V. Holt et al. (eds.), *Reproductive Sciences in Animal Conservation*,
Advances in Experimental Medicine and Biology 753, DOI 10.1007/978-1-4939-0820-2,

B

J

K

FSC
www.fsc.org
MIX
Papier | Fördert gute Waldnutzung
FSC® C083411

Kolibri 360 GmbH
Ferdinand-Jühlke-Straße 7
99095 Erfurt, Deutschland
produktsicherheit@kolibri360.de